Integration and Cubature Methods

A Geomathematically Oriented Course

MONOGRAPHS AND RESEARCH NOTES IN MATHEMATICS

Published Titles

Actions and Invariants of Algebraic Groups, Second Edition, Walter Ferrer Santos and Alvaro Rittatore

Analytical Methods for Kolmogorov Equations, Second Edition, Luca Lorenzi

Application of Fuzzy Logic to Social Choice Theory, John N. Mordeson, Davender S. Malik and Terry D. Clark

Blow-up Patterns for Higher-Order: Nonlinear Parabolic, Hyperbolic Dispersion and Schrödinger Equations, Victor A. Galaktionov, Enzo L. Mitidieri, and Stanislav Pohozaev

Bounds for Determinants of Linear Operators and Their Applications, Michael Gil′

Complex Analysis: Conformal Inequalities and the Bieberbach Conjecture, Prem K. Kythe

Computation with Linear Algebraic Groups, Willem Adriaan de Graaf

Computational Aspects of Polynomial Identities: Volume I, Kemer's Theorems, 2nd Edition Alexei Kanel-Belov, Yakov Karasik, and Louis Halle Rowen

A Concise Introduction to Geometric Numerical Integration, Fernando Casas and Sergio Blanes

Cremona Groups and Icosahedron, Ivan Cheltsov and Constantin Shramov

Delay Differential Evolutions Subjected to Nonlocal Initial Conditions Monica-Dana Burlică, Mihai Necula, Daniela Roşu, and Ioan I. Vrabie

Diagram Genus, Generators, and Applications, Alexander Stoimenow

Difference Equations: Theory, Applications and Advanced Topics, Third Edition Ronald E. Mickens

Dictionary of Inequalities, Second Edition, Peter Bullen

Elements of Quasigroup Theory and Applications, Victor Shcherbacov

Finite Element Methods for Eigenvalue Problems, Jiguang Sun and Aihui Zhou

Integration and Cubature Methods: A Geomathematically Oriented Course, Willi Freeden and Martin Gutting

Introduction to Abelian Model Structures and Gorenstein Homological Dimensions Marco A. Pérez

Iterative Methods without Inversion, Anatoly Galperin

Iterative Optimization in Inverse Problems, Charles L. Byrne

Line Integral Methods for Conservative Problems, Luigi Brugnano and Felice Iavernaro

Lineability: The Search for Linearity in Mathematics, Richard M. Aron, Luis Bernal González, Daniel M. Pellegrino, and Juan B. Seoane Sepúlveda

Published Titles Continued

Modeling and Inverse Problems in the Presence of Uncertainty, H. T. Banks, Shuhua Hu, and W. Clayton Thompson

Monomial Algebras, Second Edition, Rafael H. Villarreal

Noncommutative Deformation Theory, Eivind Eriksen, Olav Arnfinn Laudal, and Arvid Siqveland

Nonlinear Functional Analysis in Banach Spaces and Banach Algebras: Fixed Point Theory Under Weak Topology for Nonlinear Operators and Block Operator Matrices with Applications, Aref Jeribi and Bilel Krichen

Optimization and Differentiation, Simon Serovajsky

Partial Differential Equations with Variable Exponents: Variational Methods and Qualitative Analysis, Vicenţiu D. Rădulescu and Dušan D. Repovš

A Practical Guide to Geometric Regulation for Distributed Parameter Systems Eugenio Aulisa and David Gilliam

Reconstruction from Integral Data, Victor Palamodov

Signal Processing: A Mathematical Approach, Second Edition, Charles L. Byrne

Sinusoids: Theory and Technological Applications, Prem K. Kythe

Special Integrals of Gradshteyn and Ryzhik: the Proofs – Volume I, Victor H. Moll

Special Integrals of Gradshteyn and Ryzhik: the Proofs – Volume II, Victor H. Moll

Spectral and Scattering Theory for Second-Order Partial Differential Operators, Kiyoshi Mochizuki

Stochastic Cauchy Problems in Infinite Dimensions: Generalized and Regularized Solutions, Irina V. Melnikova

Submanifolds and Holonomy, Second Edition, Jürgen Berndt, Sergio Console, and Carlos Enrique Olmos

Symmetry and Quantum Mechanics, Scott Corry

The Truth Value Algebra of Type-2 Fuzzy Sets: Order Convolutions of Functions on the Unit Interval, John Harding, Carol Walker, and Elbert Walker

Variational-Hemivariational Inequalities with Applications, Mircea Sofonea and Stanislaw Migórski

Willmore Energy and Willmore Conjecture, Magdalena D. Toda

Forthcoming Titles

Groups, Designs, and Linear Algebra, Donald L. Kreher

Handbook of the Tutte Polynomial, Joanna Anthony Ellis-Monaghan and Iain Moffat

Microlocal Analysis on R^n and on NonCompact Manifolds, Sandro Coriasco

Practical Guide to Geometric Regulation for Distributed Parameter Systems, Eugenio Aulisa and David S. Gilliam

MONOGRAPHS AND RESEARCH NOTES IN MATHEMATICS

Integration and Cubature Methods

A Geomathematically Oriented Course

Willi Freeden
Martin Gutting

CRC Press
Taylor & Francis Group
Boca Raton London New York

CRC Press is an imprint of the
Taylor & Francis Group, an **informa** business
A CHAPMAN & HALL BOOK

CRC Press
Taylor & Francis Group
6000 Broken Sound Parkway NW, Suite 300
Boca Raton, FL 33487-2742

Printed on acid-free paper

International Standard Book Number-13: 978-1-138-71882-1 (Hardback)

Library of Congress Cataloging-in-Publication Data

Names: Freeden, W., author. | Gutting, Martin, 1978- author.
Title: Integration and cubature methods : a geomathematically oriented course / Willi Freeden and Martin Gutting.
Description: Boca Raton : CRC Press, [2018] | Series: Chapman & Hall/CRC monographs and research notes in mathematics | Includes bibliographical references and index.
Identifiers: LCCN 2017026232 | ISBN 9781138718821 (hardcover : alk. paper) | ISBN 9781315195674 (ebook) | ISBN 9781351764766 (ebook) | ISBN 9781351764759 (ebook) | ISBN 9781351764742 (ebook)
Subjects: LCSH: Cubature formulas. | Numerical integration. | Differential equations, Partial. | Earth sciences–Mathematics.
Classification: LCC QA299.4.C83 F74 2018 | DDC 515/.43–dc23
LC record available at https://lccn.loc.gov/2017026232

Visit the Taylor & Francis Web site at
http://www.taylorandfrancis.com

and the CRC Press Web site at
http://www.crcpress.com

Printed at CPI on sustainably sourced paper

Contents

Preface

An essential goal of geomathematics is the investigation of qualitative and quantitative structures of the Earth's system. In this respect, we are confronted with the challenge to solve differential equations involving certain laws of physics (e.g., for gravitational, magnetic, electric, and elastic fields, seismic waves, fluid flow, heat flow) under realistic assumptions. As a consequence, integral expressions involving a "potato-like" Earth or geophysically relevant manifolds (such as sphere, ellipsoid, certain equipotential surfaces, actual Earth's surface) must be attacked under the knowledge of a finite set of discrete data. At the present stage, the complexity and the typical character of such integrals and their appropriate replacement by adaptive sums are striking obstacles for many geoscientific purposes.

Looking at today's *integration and cubature procedures* in geosciences we see that an approximate spherical shape of the Earth is used in almost all global applications. In fact, the maximum deviation of the actual Earth's surface from the average Earth's radius is known to be less than 0.4%. Moreover, almost all modern satellite missions collect their data on nearly spherical orbits. This is the reason why a "spheriodization", i.e., an approximate sum approach simply in a spherical context is at least acceptable for a large number of geoscientific problems. The good news in the area of integration is that surface integrals over a sphere or volume integrals over a ball can be replaced by iterated one-dimensional integrals, e.g., in terms of polar coordinates for which powerful quadrature rules are available. The bad news is that there is no representation free of singularities when using a polar coordinate system for the whole sphere. Coordinate-free integration rules, however, are indispensable for many global purposes. Even more, if the attempt is made to split a multi-dimensional integral into iterated one-dimensional settings, physical structures inherently occurring in the data are often ignored. For example, vectorial or tensorial rotation-invariant patterns such as satellite-to-satellite tracking (SST) and satellite gravity gradiometry (SGG) data systems lead over to anisotropic one-dimensional conglomerates usually without any physical interpretability. This is the reason why undesirable effects caused by iterated one-dimensional integration should be kept in mind in geoscientifically relevant spheriodization.

Another form of integration that allows a canonical reformulation by one-dimensional quadrature techniques relates to polyhedral volumes, characterizing local parts of the Earth, e.g., in gravimetry, seismic exploration, earthquake modeling and prediction, ocean circulation, etc. Very often, when multivariate lattice point based procedures are used in volume integration, the resulting cubature expressions intrinsically show periodic structures ("periodization"). The connecting links between integrability and periodicity are Euler and Poisson summation formulas. In fact, "multivariate counterparts of Bernoulli polynomials", i.e., Green's functions corresponding to "periodic boundary conditions" and elliptic differential operators enable us to express key volume integrals in geosciences (such as Newton, Helmholtz, Mie, elastic potentials) by lattice point conglomerates that discretely fill out the integration domain in an equidistributed way. A serious problem in numerical integration by lattice point summation is the appropriate treatment of a "potato-shaped" boundary. Clearly, the overall objective of *cubature* is to derive finite summation formulas as proper substitutes for integrals. The integral formulas are usually generated by the conversion of differential equations into integral equations using Green's functions as bridging tools. By aid of Green's functions, the Euler summation formulas allow comparison of the weighted sum of functional values over a certain nodal point system with its corresponding integral, plus a remainder term normally occurring in integral form thereby containing derivatives in the integrand. In parallel, the Poisson summation formulas provide remainder terms free of derivatives. Error terms are useful for many purposes, for example, sampling formulas, polynomial and spline exact integration rules, low discrepancy formulas, and best approximate integration formulas. In addition, singular kernels obtained from Green's functions by differentiation canonically lead to multiscale integration rules providing data reduction, noise cancellation, signature decorrelation, etc.

All in all, the present textbook does not represent a survey compendium over all facets of approximate integration in the vast literature. Instead, it allows the reader to concentrate on special topics on georeflected integration and cubature. The schedule basically is characterized by a subdivision into a pair of topics, namely periodization and spheriodization. As preparatory material, classical one-dimensional quadrature is recognized from the viewpoint of Green's function theory and resulting integral formulas possessing remainder term expressions. Our multivariate approach is intended to be selective, although a broad area of applications can be specified. The content is restricted to topics which are important in (geo-)mathematics such as spherical integration, integration over regular surfaces (such as spheres, equipotential surfaces, geological layers), lattice point cubature over regular regions (such as ball, ellipsoid, geoid, Earth's "potato-like" body). Most of the material under explicit consideration for the sphere is based on our own research, in particular in geophysics and (satellite) geodesy, where the observables are of isotropic (scalar, vectorial, or tensorial) nature and integration rules have to reflect scattered preassigned measurement locations. Moreover, Romberg integration over regular regions is an innovative component in numerical integration with significant applications. Furthermore, Shannon-type sampling and Paley–Wiener integration over volumes breaks a new promising area of research in Earth's sciences.

The cover illustration includes a model of the geoid (i.e., the equipotential surface of the Earth, at sea level). The "geoidal potato" constitutes a typical geophysically relevant regular region as discussed in this work. We are obliged to R. Haagmans, Head, Earth Surfaces and Interior Section, Mission Science Division, ESA-European Space Agency, ESTEC, Noordwijk, the Netherlands, for providing us with the image (ESA ID number SEMLXEOA90).

The idea of writing this book first occurred to us while teaching graduate lectures given in the geomathematics course during the last years at the Universities of Kaiserslautern and Siegen. The preparation of the final version of this work was supported by important remarks and suggestions of many colleagues. We are deeply obliged to M.Z. Nashed, Orlando, FL for friendly collaboration and continuous support over the last years. It is a great pleasure to express our particular appreciation to our colleagues C. Gerhards, Vienna, Austria; V. Michel, Siegen, Germany; and M. Schreiner, Buchs, Switzerland for establishing and continuing geomathematical concepts and study programmes.

Kaiserslautern, Siegen — Willi Freeden, Martin Gutting

About the Authors

Willi Freeden: Studies in mathematics, geography, and philosophy at the RWTH Aachen, 1971 diploma in mathematics, 1972 Staatsexamen in mathematics and geography, 1975 PhD in mathematics, 1979 Habilitation in mathematics, 1981/1982 visiting research professor at The Ohio State University, Columbus (Department of Geodetic Science and Surveying), 1984 professor of mathematics at the RWTH Aachen (Institute of Pure and Applied Mathematics), 1989 professor of technomathematics (industrial mathematics), 1994 head of the Geomathematics Group, 1999 Fellow of the International Association of Geodesy (IAG), 2002–2006 vice-president for Research and Technology at the University of Kaiserslautern, 2009 editor-in-chief of the *International Journal on Geomathematics (GEM)*, 2010 editor of the *Handbook of Geomathematics*, 2014 editor-in-chief of the book series *Geosystems Mathematics*, 2014 editor of the *Handbuch Tiefe Geothermie*, 2015 editor of the second edition of the *Handbook of Geomathematics*, 2016 editor of the *Handbook of Geodesy*, 2017 editor of the *Handbuch Oberflächennahe Geothermie*, member of the editorial board of several international journals, since 2005 member of the German Geodetic Commission of the Bavarian Academy of Sciences, Munich.

Martin Gutting: Studies in mathematics at the University of Kaiserslautern, 2003 diploma in mathematics, focus on geomathematics, 2007 PhD in mathematics, postdoc researcher at the University of Kaiserslautern, lecturer in the course of geomathematics (in particular for constructive approximation, special functions, and inverse problems), 2011 lecturer for engineering mathematics at the University of Kaiserslautern and DHBW Mannheim, 2013 substitute professor for analysis, numerics, and geomathematics at the University of Siegen.

About the Authors

List of Symbols

x, y, z .... elements of $\mathbb{R}^3$
$x \cdot y$ .... scalar product of vectors in $\mathbb{R}^3$
$x \otimes y$ .... tensor product in $\mathbb{R}^3$
$x \wedge y$ .... vector product in $\mathbb{R}^3$
$|x|$ .... Euclidean norm of $x \in \mathbb{R}^3$
ε^i, $i = 1, 2, 3$ .... canonical orthonormal basis vectors in $\mathbb{R}^3$
$\mathbf{A}$, $\mathbf{B}$ .... matrices (e.g. in $\mathbb{R}^{N\times N}$)
$\delta_{i,j}$ .... Kronecker symbol
$\oplus$ .... orthogonal sum
$\lfloor x \rfloor$ .... the largest integer $\leq x$
$\lceil x \rceil$ .... the smallest integer $\geq x$
o, O .... Landau symbols
$\asymp$ .... asymptotically equal
$\approx$ .... approximately equal
$\Gamma(s)$ .... Gamma function at s
$\zeta(s)$ .... Riemann zeta function at s
F, G .... scalar-valued functions
f, g .... vector-valued functions
$F|M$ .... restriction of the function F to M
$\mathrm{C}^{(k)}, \mathrm{L}^p$ .... classes of scalar functions
$\mathrm{c}^{(k)}, \mathrm{l}^p$ .... classes of vector functions
$\mathrm{Pol}_{0,\ldots,m}$ .... polynomials of degree $\leq m$
$\tau\mathbb{Z}$ .... 1D τ-dilated lattice ($\tau > 0$)
$\mathrm{C}^{(k)}_{\tau\mathbb{Z}}(\mathbb{R}), \mathrm{L}^p_{\tau\mathbb{Z}}(\mathbb{R})$ .... classes of scalar, $\tau\mathbb{Z}$-periodic functions
$\mathrm{Pol}^{\mathbb{Z}}_{0,\ldots,m}$ .... periodic polynomials of degree $\leq m$
Λ .... lattice in $\mathbb{R}^q$
Λ^{-1} .... inverse lattice to Λ
$\mathcal{F}_\Lambda$ .... fundamental cell of the lattice Λ
$\|\mathcal{F}_\Lambda\|$ .... volume of the fundamental cell $\mathcal{F}_\Lambda$
$\mathrm{C}^{(k)}_\Lambda, \mathrm{L}^p_\Lambda$ .... classes of scalar, Λ-periodic functions
$\mathrm{Pol}^{\Lambda}_{\Xi}$ .... set of Λ-periodic polynomials (with index set $\Xi \subset \mathbb{R}^q$)
$\mathbb{S}^2$ .... unit sphere in $\mathbb{R}^3$ around the origin
$\mathbb{S}^2_R(y)$ .... sphere in $\mathbb{R}^3$ with radius R around y

r, t, φ	spherical coordinates in $\mathbb{R}^3$
ξ, η, ζ	elements of $\mathbb{S}^2$
$\mathbf{t} \in O(3)$	orthogonal transformation
Γ	regular region on $\mathbb{S}^2$
$\partial\Gamma$	boundary curve of Γ on $\mathbb{S}^2$
$\overline{\Gamma} = \Gamma \cup \partial\Gamma$	closure of Γ on $\mathbb{S}^2$
Γ^c	open complement of Γ in $\mathbb{S}^2$
$\|\Gamma\|$	"volume" of a regular region Γ on $\mathbb{S}^2$
$\alpha_\Gamma(\xi)$	solid angle at ξ subtended by $\partial\Gamma$
$\Gamma_\rho(\xi)$	spherical cap with radius ρ and center $\xi \in \mathbb{S}^2$
∇	gradient
L	curl gradient
$\nabla\cdot$	divergence
$\mathrm{L}\cdot$	curl divergence
Δ	Laplace operator
$\mathrm{Spect}_\Delta(\tau\mathbb{Z})$	eigenspectrum of Δ (with respect to $\tau\mathbb{Z}$)
$\mathrm{Spect}_\Delta(\Lambda)$	eigenspectrum of Δ (with respect to Λ)
∇^*	surface gradient
L^*	surface curl gradient
$\nabla^*\cdot$	surface divergence
$\mathrm{L}^*\cdot$	surface curl divergence
Δ^*	Beltrami operator on $\mathbb{S}^2$
dx	volume element
dS	surface element
$d\sigma$	line element
$\mathcal{G}$	regular region in $\mathbb{R}^3$
$\Sigma = \partial\mathcal{G}$	boundary of $\mathcal{G}$
$\|\Sigma\| = \|\partial\mathcal{G}\|$	surface of $\partial\mathcal{G}$
$\overline{\mathcal{G}} = \mathcal{G} \cup \partial\mathcal{G}$	closure of $\mathcal{G}$
$\alpha_\mathcal{G}(\xi)$	solid angle at ξ subtended by $\partial\mathcal{G}$
$\mathcal{G}^c$	open complement of $\mathcal{G}$ in $\mathbb{R}^3$
$\|\mathcal{G}\|$	volume of the regular region $\mathcal{G}$ in $\mathbb{R}^3$
$\mathbb{B}_R(y)$	ball in $\mathbb{R}^3$ with radius R around y
$\mathbb{B}_{\rho,R}(y)$	spherical shell with radii ρ and R around y
$X_N^\mathcal{U}$	ensemble of $\mathcal{U} \subset \mathbb{S}^2$
$\mathcal{P}(X_N^\mathcal{U})$	associated partition of $\mathcal{U} \subset \mathbb{S}^2$
$\Pi(X_N^\mathcal{U})$	set of all equidistributed partitions $\mathcal{P}(X_N^\mathcal{U})$ of $\mathcal{U} \subset \mathbb{S}^2$
$\sigma(X_N^\mathcal{U})$	partition size of the ensemble $X_N^\mathcal{U}$ of $\mathcal{U}$
$\#M$	number of elements in the set M

Introduction

This Introduction is concerned with the necessity of quadrature, some historical stages of approximate integration, and the concept as well as layout of the book.

0.1 Necessity of Quadrature

Approximate integration is the study of how the numerical value of an integral, i.e., a "continuous sum" can be found by a "discrete sum". To be more concrete, an integral over a one-dimensional interval $[a, b]$ has to be approximated by a linear combination of values of the integrand in the form

$$\mathcal{I}F = \int_a^b F(x)\,dx \approx \mathcal{L}_n F = \sum_{k=0}^{n} w_k F(x_k). \tag{0.1}$$

In the integration formula (0.1), $\{x_0, \ldots, x_n\}$ is the point set of *knots* (or *nodes*) usually chosen so as to lie in the interval $[a, b]$ of integration, and the numbers $w_k, k = 0, \ldots, n$, are *weights* accompanying these points.

Remark 0.1. *Occasionally, values of the derivatives of the integrand appear on the right hand side of (0.1).*

A *quadrature rule*, i.e., a formula of the type

$$\mathcal{I}F = \int_a^b F(x)\,dx \approx \mathcal{L}_n F = \sum_{k=0}^{n} w_k F(x_k), \quad x_0 < x_1 < \ldots < x_n, \tag{0.2}$$

is said to be of *closed type* if the function F is evaluated at the end points of the interval, that is, if $x_0 = a$ and $x_n = b$. A quadrature formula is said to be of *open type* if both of the end points are omitted from the evaluation. A formula of open type, so to speak, performs an extrapolation of the function to the whole interval and then integrates forward and backward beyond the last known values. Open formulas are not necessarily restricted to equidistant points. For example, the Gaussian integration formulas to be discussed later on in Chapter 8 are of open type.

Remark 0.2. *Because of the well-known transformation*

$$\int_a^b F(x)\,dx = \frac{b-a}{2}\int_{-1}^1 F\left(\frac{a+b}{2} + t\frac{b-a}{2}\right)\,dt \tag{0.3}$$

one-dimensional quadrature may be restricted to the interval $[-1,1]$.

The reasons for (one-dimensional) numerical integrations can be described as follows:

(i) F possesses an indefinite integral, i.e., (an antiderivative) I_F

$$\frac{dI_F(x)}{dx} = I_F'(x) = F(x), \quad x \in [a,b], \tag{0.4}$$

that cannot be calculated in explicit form, for example,

$$F(x) = \frac{\sin x}{x}, \tag{0.5}$$

$$F(x) = e^{-x^2}, \tag{0.6}$$

$$F(x) = \sqrt{1-k^2\sin^2 x}, \quad k \in (0,1). \tag{0.7}$$

(ii) The number of computations that must be carried out with the "exact" formula is substantial. A standard example to be mentioned (see, e.g., H. Engels [1980]) is

$$\int_0^b \frac{dt}{1+t^4} = \frac{1}{8}\sqrt{2}\ln\left(\frac{b^2+\sqrt{2}b+1}{b^2-\sqrt{2}b+1}\right) + \frac{1}{4}\sqrt{2}\arctan\left(\frac{\sqrt{2}b}{1-b^2}\right), \tag{0.8}$$

where b is a real number of the interval $(0,1)$.

(iii) In practice, we are usually confronted with the problem of integrating experimental or observational data, i.e., F is known only in discrete points $x_k \in [a,b]$, $k = 0,\ldots,n$.

The most important fields of approximate integration arise within the numerical treatment of certain problem areas, where analytical methods are usually out of question, for example, *integral transforms* such as Fourier, Bessel, and Laplace transform (see, e.g., R. Cools [1997], P.J. Davis, P. Rabinowitz [1984], P. Deuflhard, A. Hohmann [1991], M. Gräf et al. [2009], H.N. Mhaskar [2004], V. Michel [2011], M. R. Schäferkotter, P.K. Kythe [2005], H.J. Stetter [1966], J. Stoer [1989], J. Werner [1991]), *solutions of initial/boundary-value problems* from discrete data (see, e.g., P. Deuflhard, F. Bornemann [1994, 2002], W. Freeden, V. Michel [2004b], J. Stoer, R. Bulirsch [1980], J. Werner [1991]), and integral expressions corresponding to certain measurements resulting from *partial differential equations* (see, e.g., N. Bellomo, R. Gatignol [2003], R.E. Caflisch [1998], W. Dahmen [1997], M.J. Fengler, W. Freeden

[2005], C. Felippa [2004], M. Ganesh, V. Dominguez [2013], M. Ganesh et al. [2011], R.J. Hinde [2000], A. Palczewski et al. [1997], R.F. Tichy [1988], A. Townsend et al. [2016]), *image analysis and computer aided geometric design (CAGD)* (see, e.g., P. Alfeld et al. [1996], G.E. Fasshauer [1998], M. Gräf et al. [2012], M.Z. Nashed, O. Scherzer [2002]), and *ill-posed and inverse problems* in geosciences, geoengineering and satellite technology (see, e.g., K. Atkinson, A. Sommariva [2005], K. Atkinson, W. Han [2012], W. Freeden et al. [1998], W. Freeden [1999], W. Freeden, K. Hesse [2002], W. Freeden, M.Z. Nashed [2017], W. Freeden, H. Nutz [2011a], W. Freeden, F. Schneider [1998], W. Freeden, M. Schreiner [2009], C. Gerhards [2015], K. Hesse et al. [2010], V. Michel [2013], K.A. Whaler [1994]) and the references therein (note that the list is rather incomplete and subjective because of the restriction to geoscientifically relevant framework; for more details the reader is referred to references in alternatively targeted textbooks and handbook publications about approximate integration).

0.2 Historical Roots of the Book

Historical remarks on numerical integration can be found in many textbooks, for example, C. Runge, F.A. Willers [1915], A. Kowalewski [1917], J. Albrecht, L. Collatz [1958], M. Kline [1972], H.H. Goldstine [1977], H. Engels [1980], W. Gautschi [1981], G. Evans [1993], V. Katz [1998], T. Sonar [2011]: In principle, all work starts with the late neolithic period. The earliest attempts at measuring, e.g., the area of a circle have been undertaken by the Babylonians and the Egyptians. Further progress came during the Greek period. Archimedes calculated the first strict bounds for the number π. The treatment of isolated and completely determined curves, such as parabolas or circles, is typical of the ancient way of thinking. A large amount of work was done in order to find more and more complicated plane regions whose areas are rational numbers.

G. Snellius (1470–1552) and C. Huygens (1629–1695) were among the researchers who refined the idea of doubling the edges of an inscribed polygon in order to compute better approximations for π. The technique was that of extrapolation to the limit zero for the length of the polygonal side. The approach was also applied more generally to raise the order of accuracy of methods inherently showing periodically reflected structures. J. Kepler (1571–1630) became engaged with the correct determination of the capacity of wine barrels. His proposal in J. Kepler [1615] was to compute the volume of the cask by discretizing the volume into a sum of cylinders. B. Cavalieri (1635) introduced the approach of splitting areas into parallel line segments and volumes into parallel plane areas. T. Simpson (1743) dealt with simple geometric means to bound an area by straight segments.

Quadrature. The ancient period was finally overcome by the invention of functions in the modern sense of mappings. From the late Middle Ages it made sense to study mathematical procedures such as interpolation or approximation as general problems rather than those connected to one special topic or geometric figure. The concept of integration was born. In the early modern time, it was realized that not every function could be integrated analytically and, hence, the necessity for developing classes of approximate methods arose. I. Newton (1642–1726) handled functions at equidistant arguments and integrated the interpolators, thus becoming the father of the famous Newton–Cotes quadrature. In principle, I. Newton (1679) started, like Archimedes, on approximating the arcs of the functions piecewise by secants in the most simple case. However, Newton generalized the method by replacing the secants by non-linear substitutes, e.g., parabolas of increasing degree. Quadratures with equally spaced sampling points such as trapezoidal rule or Simpsons's rule suggest the formal Riemannian definition of an integral in the year 1854.

In the modern era, a large number of quadratures was developed by abstraction and subsequent concretization. These formulas were improved by various corrections and by judicious combinations of existing formulas. A milestone (see L. Euler [1736a,b]) in the development of one-dimensional lattice point summation and error estimates is the work by L. Euler (1707–1783). Independently, C. McLaurin (1698–1746) found his famous integration formula about 10 years later (cf. C. McLaurin [1742]). In the year 1955, this principle worked out in the context of Bernoulli polynomials was rediscovered and improved for numerical quadrature by W. Romberg [1955] and many others. This work presents a multi-dimensional generalization of Romberg integration for geoscientifically relevant volumes such as the geoid (cf. our cover illustration).

While for a long time it was regarded that the evaluation points should be preassigned equidistantly or with some other lattice distribution, C.F. Gauss (1777–1855) was the first to notice that a suitable variation of these points led to a better accuracy, at least seen from polynomial precision. The fundamental result of Gauss stimulated many variations and generalizations. An outline of the developments in numerical quadrature that took place in a period of approximately 200 years from Newton via Gauss to early developments in the twentieth century can be found in C. Runge, F.A. Willers [1915], and a German edition of the four principal memoirs (of Newton, Cotes, Gauss, and Jacobi) is due to A. Kowalewski [1917]. A more detailed survey on developments up to the second half of the twentieth century is given by W. Gautschi [1981]. Toward the end of the century, it inspired K. Heun [1900] to generalize Gauss's idea to ordinary differential equations, which in turn led to significant developments in the numerical solution of differential equations, notably the discovery of the Runge–Kutta method (cf. W. Kutta [1901]). Whether the Gauss method had actually been widely used in practice up to the seventies of the last century is a matter of some doubt, since the method requires the evaluation of functions at irrational arguments, hence, tedious interpolation. All this changed when powerful digital computers entered the scene, which generated a phase of renewed interest in Gaussian quadrature. The formulas began to be routinely applied, and increased usage, in turn, led to important new theoretical developments. W. Freeden [1980a] gave a new approach to Gaussian integration that is based on the concept of Green's functions with respect to (iterated) Legendre operators. Unfortunately, the theory of spherical designs shows that the Gaussian idea to approximate integration does not apply to the spheres in Euclidean space $\mathbb{R}^3$.

Spline quadratures were proposed by W. Quade, L. Collatz [1938], and I.J. Schoenberg [1964a] in the last century. By virtue of Hilbert's theory of linear integral equations (see D. Hilbert [1912]), Peano's kernel (cf. G. Peano [1913]) could be understood and generalized as an expression in the context of spline functions. The relationship between best approximate integration in the sense of A. Sard [1949, 1963] and spline exact integration became obvious (see, e.g., T.N.E. Greville [1967, 1969], I.J. Schoenberg [1964a,b]). Numerous new spline quadrature formulas were found, but it should be pointed out from algorithmic point of view that optimality with respect to spline accuracy always is at the cost of speed. In addition, high order optimality comes at the cost of smoothness, i.e., oscillation phenomena come into play in spite of the imposed curvature norm. In turn, mathematicians of the last century became aware of the fact that their abstraction processes to build up numerical integration rules will never lead to an overall optimal result. Instead only solutions optimal with respect to specifically predetermined criteria may be guaranteed.

Integration over Spheres. In spherical theory, based on the work of P.S. de Laplace [1785] and A.M. Legendre [1785], a great incentive came from the celebrated studies by C.F. Gauss [1826] to handle globally available geomagnetic data in terms of spherical harmonic expansions. Nowadays, global reference models for geopotentials are widely known by tables of Fourier coefficients of its spherical harmonic series. However, it should be pointed out that spherical harmonic developments form polynomial manifestations with all advantages, but also stronger calamities than known from one-dimensional trigonometric theory. Although truncated Fourier expansions of moderately large degree turn out to be of tremendous importance to resolve the "trend" of a signature, their application to model high(er) resolution phenomena is critical. This difficulty is also well-known in theoretical physics, e.g., when describing monochromatic waves: Plane waves with fixed frequencies (ideal frequency (momentum) localization, no space localization) do certainly not reflect the physical reality. As a remedy, plane waves of different frequencies are superposed to so-called wave-packages which show a certain amount of space localization, while losing their spectral localization. Correspondingly, a suitable superposition of spherical harmonics leads to so-called kernel functions, i.e., functions with a reduced frequency, but increased space localization. As a consequence, in view of the uncertainty principle (see, e.g., W. Freeden, M. Schreiner [2009] and the references therein), by using kernels of different scales reflecting different stages of frequency (momentum)/space localization the modeling process can be adapted to the specific band-structure in the signatures of a signal.

P. Alfeld et al. [1996], T. Lyche and L.L. Schumaker [2000], G.E. Fasshauer [1998], M.-J. Lai, and L. Schumaker [2007] deal with tensor spherical splines and use efficient one-dimensional tools. Isotropic spherical splines, as introduced independently by W. Freeden [1981a] and G. Wahba [1981], show a fixed stage of frequency/space localization. Thus, they offer a cap-structured interface between global and local modeling. Since most of the observables in geophysics and geodesy may be understood as rotation-invariant pseudodifferential operators, these spline types are particularly suited in geophysical/geodetic constructive approximation and modeling of inverse problems (such as "downward continuation" of satellite data).

Spherical wavelets show various methodologies and have many sources (see, e.g., J.-P. Antoine et al. [2002], J.-P. Antoine, P. Vandergheynst [1999], S. Dahlke et al. [1995], S. Dahlke, P. Maass [1996], I. Daubechies, W. Sweldens [1999], W. Freeden, U. Windheuser [1996, 1997], W. Freeden et al. [1998], W. Freeden, K. Hesse [2002], W. Freeden, M. Gutting [2013], W. Freeden, M. Schreiner [2007], M. Holschneider [1996], I. Pesenson [2015], D. Potts, M. Tasche [1995], D. Potts et al. [1998], P. Schröder and W. Sweldens [1995]). The power of spherical wavelets lies in the fact that kernel functions with variable frequency/space localization come into use. Wavelet transform can be understood as a space reflected replacement of Fourier transform, providing a space-varying frequency-package distribution. This is the reason why the urgent demand, for example, in geopotential modeling today, may be realized by concepts simultaneously satisfying the following two requirements, namely, first to be consistent with global trend approaches in terms of spherical harmonics, and second to allow a natural transition from the globally available framework to more and more localized settings within an efficient and economical "zooming-in" process. In this respect the combination of globally available satellite data, well-suited for global purposes, with regional/local airborne and/or terrestrial data, reflecting local features, seems to be the decisive step forward to be realized in the near future.

Applications where data are given inherently on the sphere are found in *planetary science* (see, e.g., D.L. Turcotte et al. [1981], M.A. Wieczorek [2006] and the references therein), *gravitational and geomagnetic field determination* (see, e.g., W. Freeden et al. [2008], W. Freeden, C. Gerhards [2013], W. Freeden, M. Gutting [2013], W. Freeden, T. Maier [2002], W. Freeden, V. Michel [1999], W. Freeden, M. Schreiner [2009], C. Gerhards [2011, 2012], E.W. Grafarend et al. [1999], Ilk et al. [2005], V. Michel [1999, 2015], R. Rummel [1992, 1997], F.J. Simons, A. Plattner [2015] and the references therein), *material behavior and texture* (see, e.g., R. Hielscher et al. [2010] and the references therein) *elastic (surface) deformation* (see, e.g., M.K. Abeyratne et al. [2003], W. Freeden et al. [1990], E.W. Grafarend [1986]), *ocean (surface) flow* (see, e.g., T. Fehlinger et al. [2007], R.S. Nerem et al. [1990]), *meteorology* (see, e.g., M.J. Fengler, W. Freeden [2005], M. Ganesh et al. [2011], J. Pedlowsky

[1979] and the references therein), *geothermal research* (see, e.g., M. Augustin et al. [2012], B.L. Biondi [2006], W. Freeden, H. Nutz [2011b], I. Ostermann [2011]), *satellite technology* (see, e.g., W. Freeden et al. [2002], W. Freeden, H. Nutz [2011a], C. Gerhards [2011], R. Rummel et al. [1993]), *astrophysics* (cf. A.J. Banday et al. [2001], C.L. Bennett et al. [1996], and the references therein), to quote only a few for purposes of entry. Algorithms based on spherical harmonic transforms have received considerable attention. Many realizations correspond to exact sampling theorems for a bandlimited signal. Clearly, the fundamental property of any spherical sampling is the total number of samples required to represent a bandlimited function exactly.

Integration over "Potato Skins". No doubt, spherical splines and wavelets represent very promising structures to be studied in their conceptual characteristics and applicabilities. Nonetheless, it should be observed that all these tools are not always directly transferable to a geoscientifically relevant manifold such as an equipotential surface or the actual Earth's surface. As a consequence, because of the lack of present concepts, the adequate treatment of integration methods for manifolds of geoscientific shape is a strong challenge for future work (see, e.g., I. Pesenson [2004, 2015] and the references therein). Nowadays, all settings for more complicated surfaces such as geoid, (real) Earth's surface are still in infancy, far-reaching improvements will be indispensable. So, the field of numerical integration of surface integrals is far from being completely explored.

Two different variants of geoscientifically significant surface integration are conceivable: First, since most of the relevant surfaces (e.g., ellipsoidal, telluroidal, geoidal surfaces, etc.) are star shaped we are able to deal with transformation procedures back to the sphere so that all spherical concepts may be taken into account. In a second step, Green's concept of integral formulas with respect to the Beltrami operator may be applied to general surfaces. Its study, however, demands the knowledge of differential geometric tools for purposes of approximate integration such as best approximate rules or low discrepancy methods.

Integration over "Potato-like" Regions. While the quadrature problem has always profited greatly from one-dimensional Fourier theory, approximation theory, and the theory of ordinary differential equations which have been highly developed and, in particular, from the fact that the structure of the underlying one-dimensional application is usually simple, the cubature problem is far from having sufficient progress, especially for geoscientifically relevant "potato-like" geometries. The difficulties are numerous (cf. H. Engels [1980], A. Stroud [1971]): the results and formulas are not only dependent on the domain of integration, but also on the specific physical constraints, data sources, and number as well as character of occurring parameters, etc. Iterations of one-dimensional results do not reflect specific multi-dimensional features. The multi-dimensional nature has to be taken into account. This is the reason why the interest in volume integration by use of problem-adapted rules such as Euler and Poisson summation formulas related to partial differential operators and "potato-like" bodies continues to grow rapidly, and the efforts and successes are challenging.

As in the one-dimensional theory, "multivariate Bernoulli polynomials" may be understood as Green's functions for the Laplace operator Δ and the "boundary condition of Λ-lattice periodicity". Unfortunately, the Green's function for the Laplace operator and the "boundary condition of Λ-periodicity", briefly called Λ-lattice function $G(\Delta;\cdot)$, is not available in closed form for dimensions $q \geq 2$. Instead, we have to base Euler summation with respect to elliptic partial differential operators on the constituting properties of the associated Green's functions, i.e., differential equation, characteristic singularity, boundary condition, and normalization. From Hilbert's theory of linear (weakly) singular integral equations (cf. D. Hilbert [1912]) we are able to deduce that each iteration of the Laplace operator, i.e., each convolution over the fundamental cell of the lattice, reduces the order of the singularity of the resulting Green's function by two, such that $G(\Delta^m;\cdot)$, $m > \frac{q}{2}$, turns out to be a continuous function in $\mathbb{R}^q$ in accordance with the Sobolev embedding theorem. So, based on the properties of the Λ-lattice function for iterated Laplace operators, the "multivariate Bernoulli polynomials" occurring in extended Euler summation formulas allow uniformly and absolutely convergent Fourier expansions under the condition $m > \frac{q}{2}$. As a result, in the framework of periodization, spline interpolation and smoothing are appropriately described in reproducing kernel Sobolev spaces. The associated spline spaces may be used as reference spaces for reproducing kernel reflected sampling.

The problem of reconstructing a signal from measurements or samples taken at discrete points is of considerable practical importance. There are a number of ways to realize the manner in which data can be extracted from a signal. Concerning one-dimensional theory, the famous Shannon sampling theorem (cf. C. E. Shannon [1949]) gives an answer to the question how a time signal bandlimited to a subinterval $\mathcal{G}$ of the fundamental cell $\mathcal{F}_{\mathbb{Z}}$ of the lattice $\mathbb{Z}$ can be reconstructed from discrete values in the lattice points of $\mathbb{Z}$. Based on periodization, i.e., Euler summation formulas, the problem is attackable how a space signal bandlimited to a region $\mathcal{G} \subset \mathbb{R}^q$, $q \geq 2$, allows a reconstruction from discrete values in the lattice points of a (general) lattice Λ. Weighted Hardy–Landau lattice point formulas are created to allow explicit characterizations of over- and undersampling procedures, thereby specifying not only the occurrence, but also the type of aliasing in thorough mathematical description. An essential tool for the proof of the weighted Hardy–Landau identities in lattice point theory is the extension of the Euler summation formula to multi-dimensional Helmholtz-type operators involving associated Green's functions with respect to the "boundary condition of periodicity". In order to circumvent convergence difficulties and slow convergence in multi-dimensional lattice point summation, some summability methods are unavoidable, for example, lattice ball and Gauss–Weierstrass averaging. In consequence, multi-dimensional lattice sampling becomes available in the proposed summability context to accelerate the summation of the cardinal series. Finally, some aspects of approximate integration resulting from Shannon-type sampling may be deduced from the associated Paley–Wiener framework.

0.3 Own Roots and Concept of the Book

Our own roots of the book are fourfold:

(i) The basic results due to L.J. Mordell [1928a,b, 1929] on one-dimensional Euler and Poisson summation formulas yielding certain one-dimensional error expressions in terms of the derivatives of the integrand in the Euler summation case and being free of derivatives for the error terms in the Poisson summation case,

(ii) The work by C. Müller [1954a,b,c] on two-dimensional periodic Green's functions with respect to Helmholtz operators and on spherical harmonics (see C. Müller [1966]) bearing the opportunity of adaptive integration rules (cf. W. Freeden [1980a, 1981c], W. Freeden, J. Fleck [1987]),

(iii) Our own geomathematical work on spherical Green's functions (in our approach briefly called sphere functions) within the context of spherical harmonics (see W. Freeden [1978c, 1979, 1980b, 1981a, 1982a, 1987, 1988], W. Freeden, R. Reuter [1982, 1983], W. Freeden et al. [1998]) and their generalizations to the vectorial and tensorial context (cf. W. Freeden, M. Schreiner [2009]), pseudodifferential equations and radial basis function theory (see, e.g., W. Freeden et al. [1996, 1998], W. Freeden, V. Michel [2004b], W. Freeden, C. Gerhards [2013], W. Freeden, M. Gutting [2013], V. Michel [2013]).

(iv) Our own geomathematical work on Green's functions corresponding to elliptic partial differential operators and satisfying the "boundary conditions of lattice periodicity" (in our approach briefly called lattice functions), leading to cubature over potato-like bodies (see W. Freeden [1982a, 1988], W. Freeden, P. Hermann [1985], W. Freeden, J. Fleck [1987], W. Freeden, M. Gutting [2013], W. Freeden, I. Ostermann [2012]) and its manifestations in analytic theory of numbers (cf. W. Freeden [1975, 1978a,b, 2011]) and sampling theory (cf. W. Freeden, M.Z. Nashed [2015]).

The work of the book follows an increase in dimension:

(i) Part I is concerned with classical quadrature and some preparatory 1D-material to get more insight into our multivariate integration techniques.

(ii) Part II discusses 2D-spherical integration procedures. It contains conventional latitude-longitude integration as well best-approximate integration for scattered data point systems and rules based on the idea of equidistribution (in the sense of H. Weyl [1916]). Of particular interest are spherical spline and wavelet methods.

(iii) Part III makes an attempt to transfer some of the spherical concepts to regular 2D-surfaces. Auxiliary tools are based on differential geometric means.

(iv) Part IV deals with cubature rules over 3D-volumes within the framework of elliptic partial differential equations. A central role is played by the Euler and Poisson summation formulas in their adaptation to the specific properties of the integrands.

The concept of this book is intended not only to be a textbook for students at graduate and (at least for the quadrature part) undergraduate levels, but also to indicate the "status quo" in geomathematical research areas of today's numerical integration. With the geoscientific background in mind, the person who is researching this area should find much of the material ready for appropriate use and helpful for numerous practical applications of quadrature and cubature.

Since much numerical analysis and constructive approximation today are accomplished within the framework of partial differential equations and functional analysis, these disciplines are also involved in our approach here. In particular, the eigenvalue theory of differential operators, integral formulas, reproducing kernel space expressions as well as remainder term specifications in Sobolev spaces play an eminent role. However, for reasons of readability and applicability, it is not assumed here that a deeper knowledge of the aforementioned topics is required in an *a priori* manner. As a matter of fact, they are subsequently explained as they occur in course of content.

Part I

Preparatory 1D-Integration

Chapter 1

Algebraic Polynomial Integration

1.1 Interpolatory Integration Rules

The integration formulas are obtained by replacing the integral

$$\mathcal{I}F = \int_a^b F(x)\,dx, \quad F \in \mathrm{C}^{(0)}[a,b], \tag{1.1}$$

by an expression

$$\int_a^b P_n^F(x)\,dx, \tag{1.2}$$

where P_n^F is the uniquely determined interpolating polynomial of degree n at the nodal system $\{x_k\}_{k=0,\ldots,n} \subset [a,b]$. Well-known from classical textbooks (see, e.g., P.J. Davis, P. Rabinowitz [1967], V.I. Krylov [1962], J. Stoer [1989], J. Werner [1991]) is the following theorem.

Theorem 1.1. *Let $x_k \in [a,b]$, $k = 0,\ldots,n$, with*

$$a \le x_0 < x_1 < \ldots < x_{n-1} < x_n \le b \tag{1.3}$$

be given knots. Suppose that $F(x_k)$, $k = 0,\ldots,n$, are known from a function of class $\mathrm{C}^{(0)}[a,b]$. Then there exists a uniquely determined interpolatory quadrature formula

$$\mathcal{L}_n F = \sum_{k=0}^{n} w_k F(x_k), \tag{1.4}$$

i.e.,

$$\int_a^b P(x)\,dx = \sum_{j=0}^{n} w_j P(x_j) \tag{1.5}$$

for all $P \in \mathrm{Pol}_{0,\ldots,n}$, i.e., for all polynomials of degree $\le n$. The coefficients (weights) w_j of the quadrature formula are given in the form

$$w_j = \int_a^b L_j(x)\,dx, \tag{1.6}$$

where L_j *are the* Lagrange polynomials. L_j *given by*

$$L_j(x) = \prod_{\substack{k=0 \\ k \neq j}}^{n} \frac{x - x_k}{x_j - x_k} \tag{1.7}$$

satisfy $L_j(x_k) = \delta_{j,k}$.

Theorem 1.1 leads to the following consequence:

$$\mathcal{L}_n F = \sum_{j=0}^{n} \underbrace{\left(\int_a^b L_j(x)\, dx \right)}_{=w_j} F(x_j) = \int_a^b \underbrace{\left(\sum_{j=0}^{n} L_j(x) F(x_j) \right)}_{=P_n^F(x)} dx. \tag{1.8}$$

In other words, $\mathcal{L}_n F$ is obtained by integrating the interpolating polynomial P_n^F over $[a, b]$.

Newton–Cotes Formulas. The *Newton–Cotes formulas* (cf. R. Cotes [1722]) are interpolatory quadratures to *equidistant* knots

$$x_j^{(n)} = a + j\tau, \quad \tau = \frac{b-a}{n}, \quad j = 0, \ldots, n. \tag{1.9}$$

The associated weights are calculated by

$$\begin{aligned} w_j^{(n)} &= \int_a^b \prod_{\substack{k=0 \\ k \neq j}}^{n} \frac{x - x_k^{(n)}}{x_j^{(n)} - x_k^{(n)}}\, dx = \int_a^b \prod_{\substack{k=0 \\ k \neq j}}^{n} \frac{x - (a + k\tau)}{\tau(j-k)}\, dx \qquad (1.10) \\ &= \tau \int_0^n \prod_{\substack{k=0 \\ k \neq j}}^{n} \frac{t-k}{j-k}\, dt = \tau\, \alpha_j^{(n)} \end{aligned}$$

with

$$\alpha_j^{(n)} = \frac{(-1)^{n-j}}{j!(n-j)!} \int_0^n \prod_{\substack{k=0 \\ k \neq j}}^{n} (t-k)\, dt. \tag{1.11}$$

See Examples 1.2 and 1.3 for the derivation of the cases $n = 1$ and $n = 2$. Further cases are summarized in Table 1.1.

Example 1.2. *For* $n = 1$ *we obtain the* (Simple) Trapezoidal rule, *i.e.*,

$$\alpha_0^{(1)} = -\int_0^1 (t-1)\, dt = \frac{1}{2}, \tag{1.12}$$

$$\alpha_1^{(1)} = \int_0^1 t\, dt = \frac{1}{2}. \tag{1.13}$$

The integrand is approximated by the linear function passing through the points $(a, F(a))$ *and* $(b, F(b))$*. By integrating the linear function we obtain*

$$\mathcal{L}_1 F = \frac{b-a}{2}\left(F(a) + F(b)\right). \tag{1.14}$$

Example 1.3. *For* $n = 2$, Simpson's rule *yields*

$$\alpha_0^{(2)} = \frac{1}{2}\int_0^2 (t-1)(t-2)\,dt = \frac{1}{3}, \tag{1.15}$$

$$\alpha_1^{(2)} = -\int_0^2 t(t-2)\,dt = \frac{4}{3}, \tag{1.16}$$

$$\alpha_2^{(2)} = \frac{1}{2}\int_0^2 t(t-1)\,dt = \frac{1}{3}. \tag{1.17}$$

It is derived by approximating the integrand F *by the quadratic polynomial passing through* $(a, F(a))$, $(\frac{a+b}{2}, F(\frac{a+b}{2}))$, *and* $(b, F(b))$*. J. Kepler [1615] used this method over 100 years prior. Hence, in German nomenclature, the method is sometimes called "Keplersche Fassregel".*

$$\mathcal{L}_2 F = \frac{b-a}{6}\left(F(a) + 4F\left(\frac{a+b}{2}\right) + F(b)\right). \tag{1.18}$$

TABLE 1.1: Newton–Cotes formulas for $n = 1, \ldots, 4$.

n			$\alpha_j^{(n)}$			Name
1	$\frac{1}{2}$	$\frac{1}{2}$				(Simple) Trapezoidal rule
2	$\frac{1}{6}$	$\frac{4}{6}$	$\frac{1}{6}$			Simpson's rule
3	$\frac{1}{8}$	$\frac{3}{8}$	$\frac{3}{8}$	$\frac{1}{8}$		Newton's 3/8-rule
4	$\frac{7}{90}$	$\frac{32}{90}$	$\frac{12}{90}$	$\frac{32}{90}$	$\frac{7}{90}$	Milne's rule

Remark 1.4. *For* n *large, the weights are large. Moreover, they are of mixed sign for* $n = 8$ *and* $n > 10$ *(which is usually not useful). Therefore, Newton–Cotes formulas of higher orders are not often in use. They must be seen with caution.*

Example 1.5. *In Table 1.2 the value of the integral*

$$2\int_0^1 \frac{1}{1+x^2}\,dx = \pi \tag{1.19}$$

is calculated by a sequence of Newton–Cotes rules (following P.J. Davis, P. Rabinowitz [1967]) for n*-point coefficients from* $n = 2$ *up to* $n = 21$*. The exact value is listed in the last line.*

TABLE 1.2: Newton–Cotes rule applied to the integral (1.19).

$n = 2$	3.0000000	$n = 12$	3.1415925
$n = 3$	3.1333333	$n = 13$	3.1415926
$n = 4$	3.1384615	$n = 14$	3.1415920
$n = 5$	3.1421176	$n = 15$	3.1415932
$n = 6$	3.1418781	$n = 16$	3.1415925
$n = 7$	3.1415708	$n = 17$	3.1415962
$n = 8$	3.1415789	$n = 18$	3.1415935
$n = 9$	3.1415925	$n = 19$	3.1415896
$n = 10$	3.1415926	$n = 20$	3.1415920
$n = 11$	3.1415925	$n = 21$	3.1415775
Exact value			3.1415926

1.2 Peano's Theorem

In discussing the truncation error of approximate integration, it is useful to regard the error (cf. P.J. Davis, P. Rabinowitz [1967]) as a *linear functional* defined over a certain class of functions. In the considerations that follow, we consider integrands F of class $\mathrm{C}^{(m+1)}[a,b]$ and linear functionals of integral type (1.1). Note that a generalization to linear functionals of the type

$$\mathcal{I}F = \int_a^b \sum_{j=0}^{m} a_j(x)F^{(j)}(x)\,dx + \sum_{j=0}^{m}\sum_{i=1}^{n_j} b_{ij}F^{(j)}(x_{ij}) \tag{1.20}$$

is obvious. Here, it is assumed that the functions a_j are piecewise continuous over $[a,b]$, $b_{ij} \in \mathbb{R}$, and the points x_{ij} lie in the interval $[a,b]$.

Remark 1.6. *Evidently included as particular cases of (1.20) are the integral over $[a,b]$ or any subinterval, the r-th derivative of F, $r \leq m$, evaluated at a point out of $[a,b]$, and any linear combination of ordinates of F with abscissas in $[a,b]$.*

We consider an approximation of $\mathcal{I}$ of the form (1.20) by a (quadrature) functional $\mathcal{L}$ of the form

$$\mathcal{L}F = \sum_{k=0}^{n} w_k F(x_k), \quad F \in \mathrm{C}^{(m+1)}[a,b]. \tag{1.21}$$

We shall say that $\mathcal{L}$ is *exact for the degree* m if $\mathcal{L}F = \mathcal{I}F$ for all polynomials of degree $\leq m$, i.e., $F \in \mathrm{Pol}_{0,\ldots,m}$. The *error*, or *remainder*, when $\mathcal{L}$ is used to approximate $\mathcal{I}$ is a linear functional $\mathcal{E}$ defined by

$$\mathcal{E} = \mathcal{I} - \mathcal{L}. \tag{1.22}$$

We shall say that $\mathcal{E}$ *annihilates* F if

$$\mathcal{E}F = 0. \tag{1.23}$$

Theorem 1.7 (Peano's Theorem). *Let* $\mathcal{E}P = 0$ *whenever* $P \in \mathrm{Pol}_{0,\ldots,m}$. *Then, for all* $F \in \mathrm{C}^{(m+1)}[a,b]$,

$$\mathcal{E}F = \int_a^b F^{(m+1)}(t)K(t)\ dt, \tag{1.24}$$

where

$$K(t) = \frac{1}{m!}\mathcal{E}_x(x-t)_+^m \tag{1.25}$$

and

$$(x-t)_+^m = \begin{cases} (x-t)^m, & x \geq t, \\ 0, & x < t. \end{cases} \tag{1.26}$$

The result is due to G. Peano [1913, 1914]. The function K is called the *Peano kernel* for the linear functional $\mathcal{E}$.

Remark 1.8. *Note that the notation* $\mathcal{E}_x$ *means the linear functional* $\mathcal{E}$ *is applied to the* x*-variable in the expression* $(x-t)_+^m$.

Proof. From the one-dimensional Taylor formula with explicit representation of the remainder term we obtain

$$F(x) = \sum_{k=0}^{m} \frac{F^{(k)}(a)}{k!}(x-a)^k + \frac{1}{m!}\int_a^x F^{(m+1)}(t)(x-t)^m\ dt. \tag{1.27}$$

By virtue of the setting (1.26), the integral remainder may be rewritten as

$$\frac{1}{m!}\int_a^x F^{(m+1)}(t)(x-t)^m\ dt = \frac{1}{m!}\int_a^b F^{(m+1)}(t)(x-t)_+^m\ dt. \tag{1.28}$$

We apply $\mathcal{E}$ to both sides of (1.27) and use the fact that $\mathcal{E}$ annihilates all members of the class $\mathrm{Pol}_{0,\ldots,m}$. This yields

$$\mathcal{E}F(x) = \frac{1}{m!}\mathcal{E}_x \int_a^b F^{(m+1)}(t)(x-t)_+^m\ dt. \tag{1.29}$$

Now, the type of functional we are working with allows the interchange of $\mathcal{E}$ and the integral. Thus it follows that

$$\mathcal{E}F(x) = \int_a^b F^{(m+1)}(t)\frac{1}{m!}\mathcal{E}_x(x-t)_+^m\ dt. \tag{1.30}$$

This is the required result. □

1.3 Error Truncation

From Peano's Theorem (Theorem 1.7) we immediately obtain

Corollary 1.9. *For* $F \in \mathrm{C}^{(m+1)}[a,b]$,

$$|\mathcal{E}F| \leq \max_{x\in[a,b]} |F^{(m+1)}(x)| \int_a^b |K(t)|\, dt. \tag{1.31}$$

If the Peano kernel does not change its sign over the interval $[a,b]$, then $\mathcal{E}$ may be expressed essentially as $F^{(m+1)}$ evaluated at an intermediate point. More explicitly, we have the following result.

Corollary 1.10. *If, in addition, K does not change its sign on $[a,b]$, then*

$$\mathcal{E}F = \frac{F^{(m+1)}(\xi)}{(m+1)!}\, \mathcal{E}(x^{m+1}), \quad a < \xi < b. \tag{1.32}$$

Proof. Under the additional hypotheses we may apply the mean-value theorem for integrals. This leads to

$$\mathcal{E}F = F^{(m+1)}(\xi) \int_a^b K(t)\, dt. \tag{1.33}$$

Now, we insert $x \mapsto F(x) = x^{m+1}$, $x \in [a,b]$, into (1.33) and obtain

$$\mathcal{E}(x^{m+1}) = (m+1)! \int_a^b K(t)\, dt. \tag{1.34}$$

This yields the assertion (1.32). □

For an arbitrary rule of approximate integration (0.1) there is no reason for the Peano kernel to have one sign. However, for integration rules of Newton–Cotes type, it can be shown that the kernel does have one sign.

Example 1.11 (Remainder in the (Simple) Trapezoidal Rule)**.** *If we set*

$$\mathcal{E}F = \int_a^b F(x)\, dx - \frac{b-a}{2}\left(F(a) + F(b)\right), \tag{1.35}$$

then $\mathcal{E}F$ is the truncation error in the (simple) trapezoidal rule. We may select $m = 1$ in Peano's Theorem and obtain

$$\begin{aligned} K(t) &= \int_a^b (x-t)_+\, dx - \frac{b-a}{2}\left((a-t)_+ + (b-t)_+\right) \\ &= \int_t^b (x-t)\, dx - \frac{b-a}{2}(b-t) = \frac{1}{2}(a-t)(b-t). \end{aligned} \tag{1.36}$$

Therefore,

$$\int_a^b F(x)\,dx - \frac{b-a}{2}(F(a)+F(b)) = \frac{1}{2}\int_a^b F''(t)(a-t)(b-t)\,dt. \tag{1.37}$$

The kernel K, as determined by (1.36), is non-positive throughout $[a,b]$; hence we may apply Corollary 1.10. Now it is easy to see that

$$\mathcal{E}(x^2) = \int_a^b x^2\,dx - \frac{b-a}{2}(a^2+b^2) = -\frac{1}{6}(b-a)^3. \tag{1.38}$$

Consequently,

$$\int_a^b F(x)\,dx = \frac{b-a}{2}(F(a)+F(b)) - \frac{1}{12}(b-a)^3 F''(\xi), \quad a<\xi<b. \tag{1.39}$$

Example 1.12 (Remainder in Simpson's rule). *We consider an integral on the interval $[-1,1]$. Let*

$$\mathcal{E}F = \int_{-1}^1 F(x)\,dx - \frac{1}{3}F(-1) - \frac{4}{3}F(0) - \frac{1}{3}F(1). \tag{1.40}$$

Now, $\mathcal{E}P = 0$ whenever $P \in \mathrm{Pol}_{0,\dots,3}$. Hence, we may take $m=3$ in Peano's Theorem. We find that $K(t) = \frac{1}{6}\mathcal{E}_x(x-t)_+^3$. Explicitly written out we have

$$K(t) = \begin{cases} -\frac{1}{72}(1-t)^3(3t+1), & 0 \le t \le 1, \\ -\frac{1}{72}(1+t)^3(-3t+1), & -1 \le t \le 0. \end{cases} \tag{1.41}$$

Obviously, $K(t) \le 0$ in the interval $[-1,1]$ so that Corollary 1.10 is applicable. Since $\mathcal{E}(x^4) = -\frac{4}{15}$, this leads to the error in Simpson's rule:

$$\int_{-1}^1 F(x)\,dx = \frac{1}{3}F(-1) + \frac{4}{3}F(0) + \frac{1}{3}F(1) - \frac{F^{(4)}(\xi)}{90}, \quad -1<\xi<1. \tag{1.42}$$

Chapter 2

Algebraic Spline Integration

Polynomials have long been the functions most widely used to approximate other functions, mainly because of their simple mathematical structures. However, it is a well-known observation called Runge-phenomenon (cf. C. Runge [1901], C. Runge, F.A. Willers [1915]) that a polynomial of moderately high degree fitted to a fairly large number of given data points tends to exhibit numerous and severe undulations (cf. Example 1.5 and Table 1.2). There is now considerable evidence that in many circumstances a spline function is a more adaptable approximating function than a polynomial. This conclusion is based in part on actual numerical experience, and in part on mathematical demonstrations that the solutions of a variety of problems of "best approximation" actually turn out to be spline functions.

2.1 Spline Integration Formulas

A particularly simple type of a spline function (cf. (1.26)) is the *truncated power function* x_+^m defined by

$$x_+^m = \begin{cases} x^m, & x > 0 \\ 0, & x \leq 0. \end{cases} \tag{2.1}$$

For $m = 0$ this is the well-known *Heaviside function.*

Roughly spoken, a *spline* can be understood as a function of the class $C^{(m-1)}(\mathbb{R})$, $m \geq 1$, whose $(m-1)$-th derivative is piecewise differentiable with constant derivative in each subinterval, i.e., its m-th derivative is a step function. Concisely, a spline of degree m is any m-th order indefinite integral of a step function. Thus, a spline S of degree m with the knots $x_1, \ldots, x_n \in X_n \subset \mathbb{R}$, $x_j < x_{j+1}$, $j = 1, \ldots, n-1$, is a function defined on $\mathbb{R}$ having the following properties:

(i) In each interval (x_j, x_{j+1}), $j = 0, \ldots, n$ $(x_0 = -\infty, x_{n+1} = \infty)$, S is a polynomial of degree m or less,

(ii) S and its derivatives of order $1, \ldots, m-1$ are continuous everywhere.

In other words, any *spline function* $S \in \mathrm{Spline}_m^{\mathrm{Pol}}(X_n)$ with the knots $x_1, \ldots, x_n \in X_n$ has a unique representation of the form

$$S(x) = P(x) + \sum_{j=1}^{n} c_j (x - x_j)_+^m, \quad x \in \mathbb{R}, \tag{2.2}$$

where P is a member of the class $\mathrm{Pol}_{0,\ldots,m}$. Obviously, the set $\mathrm{Spline}_m^{\mathrm{Pol}}(X_n)$ of spline functions of degree m having the nodal system X_n includes the space $\mathrm{Pol}_{0,\ldots,m}$ of all polynomials of degree m or less.

A spline S of odd degree $2k-1$ with the knots $x_1, x_2, \ldots, x_n \in X_n$ is called a *natural spline* (see, e.g., T.N.E. Greville [1967, 1969]) if it is given in each of the two intervals $(-\infty, x_1)$, (x_n, ∞) by some polynomial of degree $k-1$ (rather than $2k-1$) or less (in general, not the same polynomial in the two intervals). It is a matter of simple algebra to show that S is a natural spline of degree $2k-1$ with the indicated knots if and only if the representation (2.2) assumes the form

$$S(x) = P(x) + \sum_{j=1}^{n} c_j (x - x_j)_+^{2k-1}, \quad P \in \mathrm{Pol}_{0,\ldots,k-1}, \tag{2.3}$$

and the coefficients $c_j \in \mathbb{R}$ satisfy the relations

$$\sum_{j=1}^{n} c_j x_j^r = 0, \quad r = 0, 1, \ldots, k-1. \tag{2.4}$$

We denote by $\mathrm{Spline}_{2k-1}(X_n)$ the class of natural splines of degree $2k-1$ having the knots $x_1, x_2, \ldots, x_n \in X_n$. Evidently, this class contains $\mathrm{Pol}_{0,\ldots,k-1}$.

Hereafter we shall be concerned with integrals in which a spline function appears as a kernel function. The following lemma facilitates the evaluation of such integrals. In this connection, the reader should observe that the derivative of a spline function of degree greater than zero is a spline function of the next lower degree with the same knots. Consequently, k-fold differentiation of a spline of degree k or more reduces the degree by k.

Lemma 2.1. *Let S be given by (2.2) with $m = 2k-1$, where*

$$a \le x_1 < x_2 < \ldots < x_n \le b, \tag{2.5}$$

and let F be a function having the following properties:

(i) $F \in C^{(k-1)}[a,b]$ and $F^{(k)}$ is continuous in each open interval (x_j, x_{j+1}), $j = 0, \ldots, n$, with $x_0 = a$ and $x_{n+1} = b$, when appropriate.

(ii)

$$F^{(k-r-1)}(x) S^{(k+r)}(x) = 0, \quad r = 0, \ldots, k-2, \quad x = a, b, \tag{2.6}$$

(iii)

$$F(a) S^{(2k-1)}(a-0) = F(b) S^{(2k-1)}(b+0) = 0. \tag{2.7}$$

Then the spline integration formula

$$\int_a^b F^{(k)}(x)S^{(k)}(x)\,dx = (-1)^k(2k-1)!\sum_{j=1}^{n} c_j F(x_j). \tag{2.8}$$

holds true.

Proof. By successive integration by parts, we have

$$\begin{aligned}\int_a^b F^{(k)}(x)S^{(k)}(x)\,dx &= (-1)^{k-1}\int_a^b F'(x)S^{(2k-1)}(x)\,dx \\ &+ \sum_{r=0}^{k-2}(-1)^r\left(F^{(k-r-1)}(b)S^{(k+r)}(b) - F^{(k-r-1)}(a)S^{(k+r)}(a)\right).\end{aligned} \tag{2.9}$$

The sum on the right member of (2.9) vanishes because of property (ii). Since $S^{(2k-1)}$ is a step function with the same knots as S, the integral in the right member of (2.9) is a sum of integrals of the form

$$\eta_j \int_{x_j}^{x_{j+1}} F'(x)\,dx = \eta_j\left(F(x_{j+1}) - F(x_j)\right), \tag{2.10}$$

where η_j is the constant value of $S^{(2k-1)}$ in (x_j, x_{j+1}). Summing up the right member of (2.10) with respect to j and rearranging the occurring terms gives

$$\begin{aligned}&\sum_{j=1}^{n} F(x_j)\left(S^{(2k-1)}(x_j-0) - S^{(2k-1)}(x_j+0)\right) \\ &\qquad + F(b)S^{(2k-1)}(b+0) - F(a)S^{(2k-1)}(a-0).\end{aligned} \tag{2.11}$$

Now, the last two terms of (2.11) vanish by property (iii), while successive differentiation of (2.2) gives

$$S^{(2k-1)}(x_j+0) - S^{(2k-1)}(x_j-0) = (2k-1)!c_j, \quad j = 1, 2, \ldots, n. \tag{2.12}$$

In view of (2.11) and (2.12), Equation (2.9) reduces to (2.8), as required. □

From Lemma 2.1 the following corollary can be easily deduced.

Corollary 2.2. *Let (2.5) hold, let S be a natural spline with $k \geq 1$, and let $F \in \mathrm{C}^{(k-1)}[a,b]$ be such that $F^{(k)}$ is continuous in each interval (x_j, x_{j+1}), $j = 0, 1, \ldots, n$ with $x_0 = a$ and $x_{n+1} = b$, when appropriate. Then*

$$\int_a^b F^{(k)}(x)S^{(k)}(x)\,dx = (-1)^k(2k-1)!\sum_{j=1}^{n} c_j F(x_j). \tag{2.13}$$

If, in addition, F vanishes at every knot of S,

$$\int_a^b F^{(k)}(x)S^{(k)}(x)\,dx = 0. \tag{2.14}$$

Proof. Since S is a member of $\text{Spline}_{2k-1}(X_n)$, it follows that $S^{(k)}$ is a spline of degree $k-1$ that vanishes identically for $x \le x_1$ and for $x \ge x_n$. Consequently, conditions (ii) and (iii) of Lemma 2.1 are fulfilled, and the first conclusion follows. The last statement is obvious. □

2.2 Spline Interpolation

Now we show that for a given set of n data points

$$(x_1, y_1), \dots, (x_n, y_n) \tag{2.15}$$

with distinct abscissas there is, if $1 \le k \le n$, a unique function $S \in \text{Spline}_{2k-1}(X_n)$ that interpolates the given data points. Following T.N.E. Greville [1969] we start with an auxiliary lemma.

Lemma 2.3. *If a function $G \in \mathrm{C}^{(k)}[a, b]$ vanishes for $n \ge k$ distinct arguments in $[a, b]$ and $G^{(k)}$ is identically zero, then G is identically zero.*

Proof. Obviously, the remainder term of the Taylor expansion of G vanishes. Therefore, $G \in \text{Pol}_{0,\dots,k-1}$ and the conclusion follows. □

Theorem 2.4. *If $1 \le k \le n$ and the abscissas $x_1, x_2, \dots, x_n$ are distinct, there is a unique $S \in \text{Spline}_{2k-1}(X_n)$ that interpolates the data points (2.15).*

Proof. Clearly, there is no loss of generality in assuming that (2.5) holds.

The statement that S interpolates the points (2.15) is tantamount to the assertion that the relations

$$S(x_j) = y_j, \quad j = 1, 2, \dots, n \tag{2.16}$$

are satisfied. Expression (2.3) contains a total of $n + k$ coefficients, and substitution of this expression, with $x = x_j$, in the left member of (2.16) gives n linear equations in these coefficients, while (2.4) provides k further linear equations. The theorem will be proved if it can be shown that the overall linear system is non-singular. This will be established if it can be shown that the corresponding homogeneous system has only the trivial solution in which all $n + k$ coefficients vanish. In other words, we need only show that the only $S_0 \in \text{Spline}_{2k-1}(x_1, x_2, \dots, x_n)$ that interpolates the data points

$$(x_1, 0), (x_2, 0), \dots, (x_n, 0) \tag{2.17}$$

is the trivial one that is identically zero everywhere. To verify this, let S_0 be a natural spline with the required properties, and consider the integral given by

$$\|S_0^{(k)}\|^2_{\mathrm{L}^2[a,b]} = \int_a^b \left(S_0^{(k)}(x)\right)^2 \, dx. \tag{2.18}$$

By Corollary 2.2 identifying both F and S with S_0, we conclude that $\|S_0^{(k)}\|^2_{\mathrm{L}^2[a,b]} = 0$. Since S_0 is a natural spline, this clearly implies that $S_0^{(k)}$ vanishes identically. Consequently, by Lemma 2.3, S_0 is identically zero, as required. □

Next we consider the problem of finding the *smoothest interpolating function* G for the n data points (2.16), where by "smoothest" we mean that the integral (i.e., the semi-norm)

$$\|G^{(k)}\|_{\mathrm{L}^2[a,b]} = \left(\int_a^b \left(G^{(k)}(x)\right)^2 \, dx \right)^{1/2} \tag{2.19}$$

is minimized, with $k \geq 1$. We shall consider only interpolating functions $G \in \mathrm{C}^{(k-1)}[a,b]$ such that $G^{(k)}$ is piecewise continuous.

Remark 2.5. *Our spline approach can be weakened somewhat by use of the Sobolev space* $\mathrm{H}^{(k)}[a,b] = \overline{\mathrm{C}^{(k)}[a,b]}^{\|\cdot\|_{\mathrm{H}^{(k)}[a,b]}}$, *however, this aspect will not be discussed in the preparatory part.*

Remark 2.6. *If $k = n$, it is well-known that there is a unique polynomial L of degree $k-1$ that interpolates the points (2.15). This is given by Lagrange's formula using the Lagrange basis polynomials L_j of* (1.7).

$$L = \sum_{j=1}^{n} y_j L_j. \tag{2.20}$$

For the Lagrange interpolant, (2.19) gives $\|L^{(k)}\|_{\mathrm{L}^2[a,b]} = 0$, which is evidently its smallest possible value. Conversely, $\|L^{(k)}\|_{\mathrm{L}^2[a,b]} = 0$ implies that G is in $\mathrm{Pol}_{0,\ldots,k-1}$, *and L is known to be the unique interpolating function of this class.*

Remark 2.7. *The case $k > n$ is not interesting, as there is then an infinite set of interpolating polynomials in* $\mathrm{Pol}_{0,\ldots,k-1}$.

We shall show that for $k < n$ there is a unique smoothest interpolating function, which is, in fact, precisely that interpolating natural spline of degree $2k-1$ having the abscissas of the given data points as knots, whose existence is guaranteed by Theorem 2.4.

Lemma 2.8. *Let F_1, F_2 be piecewise continuous and such that*

$$\int_a^b F_1(x) F_2(x) \, dx = 0, \tag{2.21}$$

and let $F = F_1 + F_2$. Then

$$\|F_1\|^2_{\mathrm{L}^2[a,b]} \leq \|F\|^2_{\mathrm{L}^2[a,b]} \tag{2.22}$$

with equality only if $F = F_1$ at all points of continuity of F_2.

Proof. In view of (2.21) and (2.22),

$$\|F\|^2_{L^2[a,b]} = \|F_1 + F_2\|^2_{L^2[a,b]} = \|F_1\|^2_{L^2[a,b]} + \|F_2\|^2_{L^2[a,b]}, \tag{2.23}$$

hence, (2.22) is valid. Equality holds in (2.22) only if $\|F_2\|^2_{L^2[a,b]} = 0$. But this implies $F_2 = 0$ at all points of continuity, and consequently $F = F_1$ at such points. □

Corollary 2.9. *Let $F_1, F_2 \in C^{(k-1)}[a,b]$ have piecewise continuous k-th derivatives such that*

$$\int_a^b F_1^{(k)}(x) F_2^{(k)}(x)\, dx = 0, \tag{2.24}$$

and let F_2 vanish for $n \geq k$ distinct arguments in $[a,b]$. Let F be defined by $F = F_1 + F_2$. Then

$$\|F_1^{(k)}\|^2_{L^2[a,b]} \leq \|F^{(k)}\|^2_{L^2[a,b]}, \tag{2.25}$$

with equality only if $F = F_1$.

Proof. Inequality (2.25) follows immediately from Lemma 2.8 with $F_1^{(k)}$, $F_2^{(k)}$, $F^{(k)}$ assuming the roles of F_1, F_2, F, respectively. If equality holds in (2.25), then, by the continuity of $F^{(k-1)}$, $F^{(k)}(x) = 0$. By Lemma 2.3 we find that $F_2 = 0$, and so $F = F_1$. □

Theorem 2.10. *Let (2.5) hold, let k satisfy $1 \leq k \leq n$, and let S be the unique natural spline interpolating function for the data points (2.15) given by Theorem (2.4). Let F be any interpolating function for the data points (2.15) such that $F \in C^{(k-1)}[a,b]$, and $F^{(k)}$ is piecewise continuous. Then*

$$\|S^{(k)}\|^2_{L^2[a,b]} \leq \|F^{(k)}\|^2_{L^2[a,b]} \tag{2.26}$$

with equality only if $F = S$.

Proof. Applying Corollary 2.9 and replacing F by $F - S$, we are able to conclude that

$$\int_a^b S^{(k)}(x) \left(F^{(k)}(x) - S^{(k)}(x) \right) dx = 0. \tag{2.27}$$

We next apply Corollary 2.9 replacing F_1 by S and F_2 by $F - S$. In view of (2.27), the requirements of the corollary are guaranteed. The conclusion of the theorem follows at once. □

Remark 2.11. *The natural spline functions S of the class* $\mathrm{Spline}_3(x_1, \ldots x_n)$, *i.e., $k = 2$, are called* natural cubic splines. *In this case, the integral*

$$\int_a^b |F''(x)|^2\, dx = \int_a^b |\Delta_x F(x)|^2\, dx \tag{2.28}$$

admits a physical interpretation. It is the potential energy of a statically deflected thin beam which indeed is proportional to the integral taken over the square of the linearized curvature of the beam. Moreover, note that the term "spline" is intended to suggest that the graph of such a function is similar to a curve drawn by a mechanical device used by a draftsman so that it can be constrained to pass through certain points.

2.3 Best Approximation and Spline Exact Formulas

We now come to the role of Peano's Theorem in the theory of spline functions. For that purpose, consider again the approximation of a functional $\mathcal{I}$ of the form (1.20) by a functional $\mathcal{L}_{X_n}$ of the form

$$\mathcal{L}_{X_n} F = \sum_{j=1}^{n} b_j F(x_j), \tag{2.29}$$

where $a \le x_1 < x_2 < \ldots < x_{n-1} < x_n \le b$ and b_j, $j = 1, \ldots, n$ are real values. If the approximation of $\mathcal{I}$ by $\mathcal{L}_{X_n}$ is exact for the (polynomial) degree m and F is a member of class $\mathrm{C}^{(m+1)}[a, b]$, the error functional $\mathcal{E} = \mathcal{I} - \mathcal{L}_{X_n}$ satisfies the hypothesis of Peano's Theorem, and, therefore, we have

$$\mathcal{E}F = (\mathcal{I} - \mathcal{L}_{X_n})F = \int_a^b F^{(m+1)}(t) K(t)\, dt, \tag{2.30}$$

where

$$K(t) = \frac{1}{m!} \mathcal{E}_x (x-t)_+^m = \frac{1}{m!} (\mathcal{I} - \mathcal{L}_{X_n})_x (x-t)_+^m. \tag{2.31}$$

According to A. Sard [1948, 1949, 1963], $\mathcal{L}_{X_n}$ is called a *best approximation* to $\mathcal{I}$ of the form (2.29) exact for the degree m, if within the class of such approximations, it is one for which the quantity

$$J = \int_a^b (K(t))^2\, dt = \frac{1}{m!} \int_a^b \left((\mathcal{I} - \mathcal{L}_{X_n})(\,\cdot\, - t)_+^m \right)^2\, dt \tag{2.32}$$

assumes its minimum value.

For a given $\mathcal{I}$ of the form (1.20) and given abscissas $x_1, x_2, \ldots, x_n \in X_n$, an approximation $\mathcal{L}_{X_n}$ of the form (2.29) is not uniquely determined by the requirement that it be exact for the (polynomial) degree r, provided $r < n-1$. We shall see, however, that for $1 \le k \le n$, $\mathcal{L}_{X_n}$ is, in fact, uniquely determined by the stronger requirement that $\mathcal{L}_{X_n} S = \mathcal{I} S$ whenever S is a (natural) spline of degree $2k-1$, i.e., S is a member of the class $\mathrm{Spline}_{2k-1}(X_n)$.

Theorem 2.12. *Let $\mathcal{I}$ be of the form (1.20) with $m = k-1$, let k satisfy $1 \le k \le n$, and let $a \le x_1 < x_2 < \ldots < x_{n-1} < x_n \le b$. Then there is a unique $\mathcal{L}_{X_n}$ of the form (2.29) having the property that*

$$\mathcal{L}_{X_n} F = \mathcal{I} F, \quad F \in \mathrm{Spline}_{2k-1}(X_n). \tag{2.33}$$

For a given $F \in \mathrm{C}^{(k)}[a,b]$, *let* S_F *denote the unique element of the class* $\mathrm{Spline}_{2k-1}(X_n)$ *that interpolates the data points*

$$(x_1, F(x_1)), (x_2, F(x_2)), \ldots, (x_n, F(x_n)). \tag{2.34}$$

Then $\mathcal{L}_{X_n}$ *is also uniquely determined by the property that* $\mathcal{L}_{X_n}F = \mathcal{I}S_F$ *for every* $F \in \mathrm{C}^{(k)}[a,b]$.

Finally, $\mathcal{L}_{X_n}$ *is uniquely determined by the two conditions that* $\mathcal{L}_{X_n}F = \mathcal{I}F$ *whenever* $F \in \mathrm{Pol}_{0,\ldots,k-1}$ *and that there exists a unique polynomial* $T \in \mathrm{Pol}_{0,\ldots,k-1}$ *such that*

$$\mathcal{E}_x\,(x - x_i)_+^{2k-1} = T(x_i), \quad i = 1, \ldots, n, \tag{2.35}$$

where, as usual, $\mathcal{E} = \mathcal{I} - \mathcal{L}_{X_n}$.

Proof. For convenience, let

$$\begin{aligned} \mathcal{I}(x^r) &= \alpha_r, \quad r = 0, 1, \ldots, k-1, \\ \mathcal{I}_x\,(x - x_i)_+^{2k-1} &= \beta_i, \quad i = 1, 2, \ldots, n. \end{aligned} \tag{2.36}$$

Then, if $\mathcal{L}_{X_n}$ is given by (2.29), we are led to the $(n+k) \times (n+k)$ linear system

$$\sum_{j=1}^{n} b_j x_j^r = \alpha_r, \quad r = 0, 1, \ldots, k-1, \tag{2.37}$$

$$\sum_{j=1}^{n} b_j (x_j - x_i)_+^{2k-1} + T(x_i) = \beta_i, \quad i = 1, 2, \ldots, n, \tag{2.38}$$

for determining the polynomial T and the coefficients b_j. Inspection of this linear system reveals that $\mathcal{L}_{X_n}$ and T are uniquely determined by (2.37) and (2.38) (note that the corresponding homogeneous system admits only the trivial solution, since the coefficient matrix is the transpose of that considered in the proof of Theorem 2.4). We shall now show that, for the $\mathcal{L}_{X_n}$ so obtained, (2.33) implies $\mathcal{L}_{X_n}F = \mathcal{I}F$.

In fact, (2.33) implies that F is a member of $\mathrm{Spline}_{2k-1}(X_n)$ with coefficients c_j satisfying $\sum_{j=1}^{n} c_j x_j^r = 0$, $r = 0, \ldots, k-1$. Thus, (2.36) and (2.38) give

$$\begin{aligned} \mathcal{I}F &= \mathcal{I}T + \sum_{i=1}^{n} c_i \beta_i \\ &= \mathcal{L}_{X_n}T + \sum_{i=1}^{n} c_i \Big(\sum_{j=1}^{n} b_j (x_j - x_i)_+^{2k-1} + T(x_i) \Big). \end{aligned} \tag{2.39}$$

Noting that

$$\sum_{i=1}^{n} c_i T(x_i) = 0 \tag{2.40}$$

and reversing the order of summation with respect to i and j yields $\mathcal{I}F = \mathcal{L}_{X_n}F$.

Conversely, let $\mathcal{L}'_{X_n}$ be any functional of the form (2.29) having the property that $\mathcal{L}'_{X_n}F = \mathcal{I}F$ whenever F satisfies (2.33). Then, for any $F \in \mathrm{C}^{(k)}[a,b]$, $\mathcal{L}'_{X_n}S_F = \mathcal{L}'_{X_n}F$. But, by (2.29) and the definition of S_F, $\mathcal{L}'_{X_n}S_F = \mathcal{L}'_{X_n}F$, and so $\mathcal{L}'_{X_n}F = \mathcal{I}S_F$.

In particular, for $\mathcal{L}'_{X_n} = \mathcal{L}_{X_n}$, we have $\mathcal{L}_{X_n}F = \mathcal{I}S_F$, and we have $\mathcal{L}'_{X_n}S_F = \mathcal{I}S_F$ such that

$$\mathcal{L}'_{X_n}F = \mathcal{I}S_F = \mathcal{L}_{X_n}F \tag{2.41}$$

whenever $F \in \mathrm{C}^{(k)}[a,b]$. But this implies $\mathcal{L}'_{X_n} = \mathcal{L}_{X_n}$.

Thus, $\mathcal{L}_{X_n}$ is uniquely determined either by the property that $\mathcal{L}_{X_n}F = \mathcal{I}F$ whenever F is a function of class $\mathrm{Spline}_{2k-1}(X_n)$ or by the property that $\mathcal{L}_{X_n}F = \mathcal{I}S_F$ whenever $F \in \mathrm{C}^{(k)}[a,b]$. □

Remark 2.13. *It should be remarked that the preceding proof is a rearrangement of portions of the work due to T.N.E. Greville [1967, 1969], which itself is based on the approach by I.J. Schoenberg [1964a,b].*

For a given functional $\mathcal{I}$ of the form (1.20) with $m = k-1$, if we desire to find explicitly the unique approximation $\mathcal{L}_{X_n}$ specified by Theorem 2.12 that is exact for all $F \in \mathrm{Spline}_{2k-1}(X_n)$, we see that we can do so by solving the linear system consisting of (2.37) and (2.38). However, if we merely want $\mathcal{L}_{X_n}F$ for some particular F, an alternative procedure is to operate with $\mathcal{L}_{X_n}$ on the spline S_F that interpolates the data points (2.34). I.J. Schoenberg [1964a] has made the remarkable observation that the best approximation (in the sense of Sard) exact for the degree $m = k-1$, to a functional $\mathcal{I}$ is precisely the unique approximation given by Theorem 2.12 that is exact for functions $F \in \mathrm{Spline}_{2k-1}(X_n)$.

Theorem 2.14 (Schoenberg's Theorem)**.** *Let $\mathcal{I}$ be any functional of the form (1.20) (with $m = k-1 < n$), and let $a \le x_1 < x_2 < \ldots < x_{n-1} < x_n \le b$. Let $\mathcal{L}'_{X_n}$ be any approximation to $\mathcal{I}$ of the form (2.29) exact for the degree $k-1$, and let $\mathcal{L}_{X_n}$ denote the approximation given by Theorem 2.12; let J and J' be the corresponding quantities given by (2.32). Then $J \le J'$, with equality only if $\mathcal{L}'_{X_n} = \mathcal{L}_{X_n}$.*

Proof. We introduce a functional $\mathcal{L}''_{X_n}$ by

$$\mathcal{L}''_{X_n} = \mathcal{L}_{X_n} - \mathcal{L}'_{X_n}. \tag{2.42}$$

In view of Theorem 2.12 and the hypotheses on $\mathcal{L}'_{X_n}$,

$$\mathcal{L}_{X_n}F = \mathcal{I}F = \mathcal{L}'_{X_n}F \tag{2.43}$$

whenever $F \in \mathrm{Pol}_{0,\ldots,k-1}$, and thus $\mathcal{L}''_{X_n}$ satisfies the hypothesis of Peano's Theorem, with $m = k-1$. Hence, for $F \in \mathrm{C}^{(k)}[a,b]$,

$$\mathcal{L}''_{X_n} F = \int_a^b K_2(t) F^{(k)}(t)\, dt, \tag{2.44}$$

where

$$K_2(t) = \frac{1}{(k-1)!} \mathcal{L}''_{X_n}(\,\cdot\, - t)_+^{k-1}. \tag{2.45}$$

As $\mathcal{L}_{X_n}$ and $\mathcal{L}'_{X_n}$ are both of the form (2.29). $\mathcal{L}''_{X_n}$ is also a linear combination of ordinates, and it follows from (2.45) that $K_2 \in \mathrm{C}^{(k-2)}[a,b]$, while its $(k-1)$-th derivative is a step function with discontinuities at most at the points $x_1, \ldots, x_n$. Thus, K_2 is a function of the class $\mathrm{Spline}_{k-1}(X_n)$. Moreover, by (2.45), K_2 vanishes identically for $t \geq x_n$, while, for $t \leq x_1$,

$$K_2(t) = \frac{1}{(k-1)!} \mathcal{L}''_{X_n}(\,\cdot\, - t)^{k-1} = 0, \tag{2.46}$$

since $\mathcal{L}''_{X_n}$ annihilates polynomials in $\mathrm{Pol}_{0,\ldots,k-1}$.

Now, let G be any function such that $G^{(k)} = K_2$. Then, evidently,

$$G \in \mathrm{Spline}_{2k-1}(X_n). \tag{2.47}$$

Let $\mathcal{E}$ be given by $\mathcal{E} = \mathcal{I} - \mathcal{L}_{X_n}$ and let $\mathcal{E}'$ denote the corresponding functional associated with $\mathcal{L}'_{X_n}$. Also, let K and K_1 be the Peano kernel functions given by (2.31) corresponding to $\mathcal{L}_{X_n}$ and $\mathcal{L}'_{X_n}$, respectively. From (2.45) we get

$$\mathcal{L}''_{X_n} = (\mathcal{I} - \mathcal{E}) - (\mathcal{I} - \mathcal{E}') = \mathcal{E}' - \mathcal{E}, \tag{2.48}$$

and, therefore, (2.31) gives $K_2 = K_1 - K$, or

$$K_1 = K + K_2. \tag{2.49}$$

Altogether, we have

$$\int_a^b K(t) K_2(t)\, dt = \int_a^b K(t) G^{(k)}(t)\, dt = \mathcal{E} G = 0, \tag{2.50}$$

by (2.30), (2.47), and Theorem 2.12. The conclusion of the theorem now follows immediately from (2.49) and (2.50). □

Finally we deal with error bounds in the approximation of the linear functionals. Let $F \in \mathrm{C}^{(k-1)}[a,b]$ and let $F^{(k)}$ be piecewise continuous on $[a,b]$. A. Sard [1963] has pointed out that if $F^{(k)}$ satisfies a bound of the form

$$\int_a^b \left(F^{(k)}(x)\right)^2\, dx \leq M, \tag{2.51}$$

then the application of the Cauchy–Schwarz inequality to (2.30) and (2.32) gives

$$|\mathcal{E}F| \leq \sqrt{J}\sqrt{M}. \tag{2.52}$$

We note that K, and therefore J, depends on the knots x_i and on the operator $\mathcal{I}$, but not on the function F. In view of (2.52), the numerical determination of J is of some interest. The following approach was described in T.N.E. Greville [1969]: Let a spline function S be given by

$$S(x) = \mathcal{L}_{X_n}(\,\cdot\, - x)_+^{2k-1} + T(x), \tag{2.53}$$

where $\mathcal{L}_{X_n}$ and T are determined by (2.37) and (2.38). Equations (2.35) can now be written in the form

$$S(x_i) = \mathcal{I}_t(t - x_i)_+^{2k-1}, \quad i = 1, 2, \ldots, n. \tag{2.54}$$

Now, let a function H be defined by

$$H(x) = \frac{(-1)^k}{(2k-1)!}\left(\mathcal{I}_t(t-x)_+^{2k-1} - S(x)\right). \tag{2.55}$$

It follows from (2.54) and (2.55) that

$$H(x_i) = 0, \quad i = 1, 2, \ldots, n. \tag{2.56}$$

Moreover, k-fold differentiation of (2.55) and (2.53) gives

$$H^{(k)}(x) = \frac{1}{(k-1)!}\mathcal{E}_t(t-x)_+^{k-1} = K(x), \tag{2.57}$$

by (2.31). By (2.56) and (2.29), $\mathcal{L}_{X_n}H = 0$, and therefore $\mathcal{E}H = \mathcal{I}H$. But, by (2.30) and (2.57),

$$\mathcal{E}H = \int_a^b K(t)H^{(k)}(t)\,dt = \int_a^b (K(t))^2\,dt = J. \tag{2.58}$$

Thus,

$$J = \mathcal{E}H. \tag{2.59}$$

Remark 2.15. *It is usually easier to obtain J by (2.59) than by (2.32). A detailed discussion for the case $k = 2$ (leading to interpolation by cubic splines) is due to T.N.E. Greville [1967].*

The linear systems arising in spline problems tend to be very ill conditioned if the basis is chosen as in (2.3). This may cause difficulty if the attempt is made to solve these systems directly in order to obtain the required parameters. The numerical instabilities encountered increase with the dimensions of the linear system involved and are related to the mathematical properties of the truncated power functions. The difficulties can be overcome by adopting a different basis for the classes of splines dealt with. The most desirable basis would consist of splines whose support is finite, consisting of the smallest possible number of intervals between knots. Schoenberg's investigations (see, e.g., T.N.E. Greville [1967, 1969] for more details) lead to a construction of B-spline bases for the class $\text{Spline}_{2k-1}(X_n)$, which are, in a sense, the splines of minimal support for a given degree. The details will be omitted here. They can be found, e.g., in T.N.E. Greville [1967, 1969] and L. Schumaker [2007].

Chapter 3

Periodic Polynomial Integration

After the well-known material involving algebraic polynomials and splines, presented here only for recapitulation and motivation, we are interested in explaining different types of integration involving periodic polynomials in one-dimensional Euclidean space $\mathbb{R}$. We start with the classical Euler summation formula for the operator of the second derivative, i.e., the Laplace operator in one dimension and "periodic boundary conditions". Later on, we extend the Euler summation formula to Helmholtz operators. Both variants will be used to develop sufficient conditions for the validity of the Poisson summation formula in one dimension. As a matter of fact, Euler and Poisson summation formulas turn out to be equivalent in one dimension (see the introduction of W. Freeden [2011]). However, in the context of approximate integration, they take a different role. Remainder terms in Euler summation formulas include derivatives of the integrand under consideration, whereas error terms in Poisson summation formulas are free of derivatives.

3.1 Integer Lattice and Periodic Polynomials

Let $\mathbb{Z}$ denote the one-dimensional lattice (cf. Figure 3.1), i.e., the additive group of points in $\mathbb{R}$ having integral coordinates (the addition being, of course, the one derived from the vector structure of $\mathbb{R}$). The *fundamental cell $\mathcal{F}$ (more accurately, $\mathcal{F}_{\mathbb{Z}}$) of the integer lattice* $\mathbb{Z}$ is given by

$$\mathcal{F}_{\mathbb{Z}} = \left\{ x \in \mathbb{R} : -\frac{1}{2} \leq x < \frac{1}{2} \right\} = \left[-\frac{1}{2}, \frac{1}{2} \right). \tag{3.1}$$

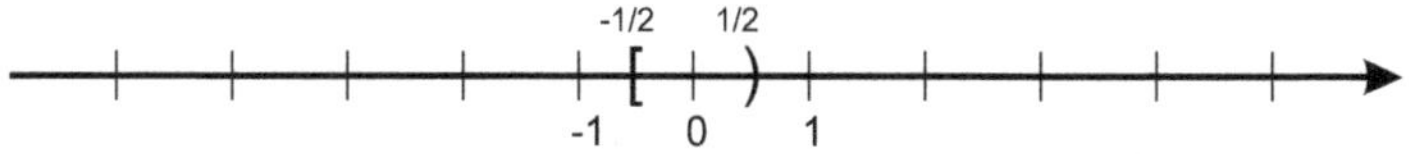

FIGURE 3.1: The fundamental cell $\mathcal{F}_{\mathbb{Z}}$ of the integer lattice $\mathbb{Z}$.

Definition 3.1. *A function $F : \mathbb{R} \to \mathbb{C}$ is called* $\mathbb{Z}$-periodic *if $F(x+g) = F(x)$ holds for all $x \in \mathcal{F}_{\mathbb{Z}}$ and $g \in \mathbb{Z}$.*

Example 3.2. *The function $\Phi_h : \mathbb{R} \to \mathbb{C}$, $h \in \mathbb{Z}$, given by $x \mapsto \Phi_h(x) = e(hx) = e^{2\pi ihx}$ is $\mathbb{Z}$-periodic: $\Phi_h(x+g) = e^{2\pi ih(x+g)} = e^{2\pi ihx}e^{2\pi ihg} = e^{2\pi ihx} = \Phi_h(x)$ for all $x \in \mathcal{F}_\mathbb{Z}$ and all $g \in \mathbb{Z}$.*

The space of all $F \in \mathrm{C}^{(m)}(\mathbb{R})$ that are $\mathbb{Z}$-periodic is denoted by $\mathrm{C}^{(m)}_\mathbb{Z}(\mathbb{R})$, $0 \leq m \leq \infty$. $\mathrm{L}^2_\mathbb{Z}(\mathbb{R})$ is the space of all $F : \mathbb{R} \to \mathbb{C}$ that are $\mathbb{Z}$-periodic and are Lebesgue-measurable on $\mathcal{F}_\mathbb{Z}$ with

$$\|F\|_{\mathrm{L}^2_\mathbb{Z}(\mathbb{R})} = \left(\int_{\mathcal{F}_\mathbb{Z}} |F(x)|^2 \, dx \right)^{1/2} < \infty. \tag{3.2}$$

Clearly, the space $\mathrm{L}^2_\mathbb{Z}(\mathbb{R})$ is the completion of $\mathrm{C}^{(0)}_\mathbb{Z}(\mathbb{R})$ with respect to the norm $\|\cdot\|_{\mathrm{L}^2_\mathbb{Z}(\mathbb{R})}$:

$$\mathrm{L}^2_\mathbb{Z}(\mathbb{R}) = \overline{\mathrm{C}^{(0)}_\mathbb{Z}(\mathbb{R})}^{\|\cdot\|_{\mathrm{L}^2_\mathbb{Z}(\mathbb{R})}}. \tag{3.3}$$

An easy calculation shows that the system $\{\Phi_h\}_{h\in\mathbb{Z}}$ is orthonormal with respect to the $\mathrm{L}^2_\mathbb{Z}(\mathbb{R})$-inner product

$$\langle \Phi_h, \Phi_{h'} \rangle_{\mathrm{L}^2_\mathbb{Z}(\mathbb{R})} = \int_{\mathcal{F}_\mathbb{Z}} \Phi_h(x)\overline{\Phi_{h'}(x)} \, dx = \delta_{h,h'} = \begin{cases} 1, & h = h', \\ 0, & h \neq h'. \end{cases} \tag{3.4}$$

An elementary calculation yields (with $\nabla_x = \frac{d}{dx}$ as the one-dimensional gradient and $\Delta_x = \left(\frac{d}{dx}\right)^2$ as the one-dimensional Laplacian)

$$\nabla_x \Phi_h(x) = \frac{d}{dx}\Phi_h(x) = \frac{d}{dx}e^{2\pi ihx} = 2\pi ih\Phi_h(x) \tag{3.5}$$

such that for $h \in \mathbb{Z}$, $x \in \mathbb{R}$

$$\Delta_x \Phi_h(x) = \left(\frac{d}{dx}\right)^2 \Phi_h(x) = (2\pi ih)^2\Phi_h(x) = \underbrace{-4\pi^2h^2}_{=-\Delta^\wedge(h)} \Phi_h(x). \tag{3.6}$$

By convention we say that λ is an *eigenvalue of the lattice $\mathbb{Z}$ with respect to the operator* Δ of the second order derivative (i.e., the one-dimensional Laplace operator), if there is a non-trivial solution U of $(\Delta + \lambda)U = 0$ satisfying the "boundary condition of periodicity" $U(x+g) = U(x)$ for all $x \in \mathcal{F}_\mathbb{Z}$ and $g \in \mathbb{Z}$. From classical Fourier analysis (see, e.g., P.L. Butzer, R. Nessel [1971], R. Courant, D. Hilbert [1924], M.A. Pinsky [2002]) we know that the operator Δ has a half-bounded and discrete eigenspectrum $\{\Delta^\wedge(h)\}_{h\in\mathbb{Z}} \subset \mathbb{R}$ such that $(\Delta_x + \Delta^\wedge(h))\,\Phi_h(x) = 0, x \in \mathcal{F}_\mathbb{Z}$, with eigenvalues $\Delta^\wedge(h)$ given by $\Delta^\wedge(h) = 4\pi^2h^2, h \in \mathbb{Z}$, and eigenfunctions $\Phi_h(x) = e^{2\pi ihx}, h \in \mathbb{Z}, x \in \mathcal{F}_\mathbb{Z}$. The *eigenspectrum of the operator* Δ (with respect to $\mathbb{Z}$) is given by

$$\mathrm{Spect}_\Delta(\mathbb{Z}) = \left\{\Delta^\wedge(h) : \Delta^\wedge(h) = 4\pi^2h^2, h \in \mathbb{Z}\right\}. \tag{3.7}$$

Remark 3.3. *We consistently write $\Delta^\wedge(h)$ instead of $\Delta^\wedge_\mathbb{Z}(h)$, $h \in \mathbb{Z}$, if no confusion is likely to arise.*

The system $\{\Phi_h\}_{|h|\leq m}$ of (eigen)functions $\Phi_h : x \mapsto \Phi_h(x) = e^{2\pi i h x}$, $x \in \mathbb{R}$ is orthonormal in the space $\mathrm{Pol}^{\mathbb{Z}}_{0,\ldots,m}$ of periodic polynomials of degree $\leq m$, such that any $F \in \mathrm{Pol}^{\mathbb{Z}}_{0,\ldots,m}$ can be represented by its Fourier series

$$F(x) = \sum_{\substack{|h|\leq m \\ g\in\mathbb{Z}}} \int_{\mathcal{F}_{\mathbb{Z}}} F(y)\overline{\Phi_h(y)}\, dy\ \Phi_h(x), \quad x \in \mathbb{R}. \tag{3.8}$$

Let $X_{2m+1} = \{x_{-m}, \ldots, x_0, \ldots, x_m\}$ be a set of distinct points of the fundamental cell $\mathcal{F}_{\mathbb{Z}}$. Then we are able to formulate a counterpart to the *Shannon sampling theorem.*

Theorem 3.4 (Shannon Sampling Theorem in $\mathrm{Pol}^{\mathbb{Z}}_{0,\ldots,m}$). *Any member $F \in \mathrm{Pol}^{\mathbb{Z}}_{0,\ldots,m}$ can be reconstructed from its samples at points in X_{2m+1} by the following interpolation formula*

$$F(x) = \sum_{\substack{|g|\leq m \\ g\in\mathbb{Z}}} F(x_g) L_g(x), \quad x \in \mathbb{R}, \tag{3.9}$$

where $L_g \in \mathrm{Pol}^{\mathbb{Z}}_{0,\ldots,m}$ is given by

$$L_g(x) = \sum_{\substack{|h|\leq m \\ h\in\mathbb{Z}}} a_h^g \Phi_h(x), \quad x \in \mathbb{R}, g \in \mathbb{Z}, |g| \leq m, \tag{3.10}$$

and the coefficients $a_h^g \in \mathbb{C}$ are determined by

$$\sum_{\substack{|h|\leq m \\ h\in\mathbb{Z}}} a_h^g \Phi_h(x_l) = \delta_{g,l}, \quad g, l \in \mathbb{Z},\ |g| \leq m,\ |l| \leq m. \tag{3.11}$$

Proof. Interpreting (3.11) as matrix product rule we equivalently obtain its transposed counterpart

$$\sum_{\substack{|k|\leq m \\ k\in\mathbb{Z}}} a_h^k \Phi_l(x_k) = \delta_{h,l}, \quad h, l \in \mathbb{Z},\ |h| \leq m,\ |l| \leq m. \tag{3.12}$$

Thus, it is not hard to see that

$$\sum_{\substack{|k|\leq m \\ k\in\mathbb{Z}}} F(x_k) L_k(x) = \sum_{\substack{|h|\leq m \\ h\in\mathbb{Z}}} \sum_{\substack{|k|\leq m \\ k\in\mathbb{Z}}} F(x_k) a_h^k \Phi_h(x) \tag{3.13}$$

$$= \sum_{\substack{|h|\leq m \\ h\in\mathbb{Z}}} \sum_{\substack{|l|\leq m \\ l\in\mathbb{Z}}} \int_{\mathcal{F}_{\mathbb{Z}}} F(z)\overline{\Phi_l(z)}\, dz \underbrace{\sum_{\substack{|k|\leq m \\ k\in\mathbb{Z}}} a_h^k \Phi_l(x_k)}_{=\delta_{h,l}} \Phi_h(x)$$

$$= \sum_{\substack{|h|\leq m \\ h\in\mathbb{Z}}} \int_{\mathcal{F}_{\mathbb{Z}}} F(z)\overline{\Phi_h(z)}\, dz\ \Phi_h(x) = F(x). \tag{3.14}$$

This guarantees Theorem 3.4. □

The orthonormal system $\{\Phi_h\}_{h\in\mathbb{Z}}$ of (eigen)functions is closed in the space $\mathrm{C}^{(0)}_{\mathbb{Z}}(\mathbb{R})$; i.e., for every $\varepsilon > 0$ and every $F \in \mathrm{C}^{(0)}_{\mathbb{Z}}(\mathbb{R})$ there exist an integer $N(= N(\varepsilon))$ and a linear combination in $\mathrm{Pol}^{\mathbb{Z}}_{0,\ldots,N}$ with coefficients $a_h \in \mathbb{C}$, $h \in \mathbb{Z}$, $|h| \leq N$, such that

$$\sup_{x\in\mathcal{F}_\mathbb{Z}} \left| F(x) - \sum_{\substack{|h|\leq N\\ h\in\mathbb{Z}}} a_h\Phi_h(x)\right| \leq \varepsilon. \tag{3.15}$$

By virtue of the norm estimate we obtain for $F \in \mathrm{C}^{(0)}_{\mathbb{Z}}(\mathbb{R})$ that

$$\|F\|_{\mathrm{L}^2_\mathbb{Z}(\mathbb{R})} = \left(\int_{\mathcal{F}_\mathbb{Z}} |F(x)|^2\,dx\right)^{1/2} \leq \sup_{x\in\mathcal{F}_\mathbb{Z}} |F(x)| = \|F\|_{\mathrm{C}^{(0)}_\mathbb{Z}(\mathbb{R})}, \tag{3.16}$$

i.e., the closure of the system $\{\Phi_h\}_{h\in\mathbb{Z}}$ in $\left(\mathrm{C}^{(0)}_{\mathbb{Z}}(\mathbb{R}), \|\cdot\|_{\mathrm{C}^{(0)}_\mathbb{Z}(\mathbb{R})}\right)$ implies the closure in $\left(\mathrm{C}^{(0)}_{\mathbb{Z}}(\mathbb{R}), \|\cdot\|_{\mathrm{L}^2_\mathbb{Z}(\mathbb{R})}\right)$. Since $\mathrm{C}^{(0)}_{\mathbb{Z}}(\mathbb{R})$ is dense in $\mathrm{L}^2_\mathbb{Z}(\mathbb{R})$ with respect to the norm $\|\cdot\|_{\mathrm{L}^2_\mathbb{Z}(\mathbb{R})}$, the validity of the following equivalences is well-known.

Theorem 3.5. *The following statements are equivalent:*

(i) The system $\{\Phi_h\}_{h\in\mathbb{Z}}$ is closed in $\mathrm{L}^2_\mathbb{Z}(\mathbb{R})$; i.e., for every $\varepsilon > 0$ and every $F \in \mathrm{L}^2_\mathbb{Z}(\mathbb{R})$ there exist an index $N(= N(\varepsilon))$ and coefficients $a_h \in \mathbb{C}$ such that

$$\left(\int_{\mathcal{F}_\mathbb{Z}} \left|F(x) - \sum_{\substack{|h|\leq N\\ h\in\mathbb{Z}}} a_h\Phi_h(x)\right|^2 dx\right)^{1/2} \leq \varepsilon. \tag{3.17}$$

(ii) The Fourier series of $F \in \mathrm{L}^2_\mathbb{Z}(\mathbb{R})$

$$\sum_{h\in\mathbb{Z}} F^\wedge_{\mathcal{F}_\mathbb{Z}}(h)\Phi_h(x) \tag{3.18}$$

with the "Fourier coefficients"

$$F^\wedge_{\mathcal{F}_\mathbb{Z}}(h) = \langle F, \Phi_h\rangle_{\mathrm{L}^2_\mathbb{Z}(\mathbb{R})} = \int_{\mathcal{F}_\mathbb{Z}} F(x)\overline{\Phi_h(x)}\,dx, \quad h \in \mathbb{Z}, \tag{3.19}$$

converges in the $\|\cdot\|_{\mathrm{L}^2_\mathbb{Z}(\mathbb{R})}$-norm:

$$\lim_{N\to\infty} \left(\int_{\mathcal{F}_\mathbb{Z}} \left|F(x) - \sum_{\substack{|h|\leq N\\ h\in\mathbb{Z}}} F^\wedge_{\mathcal{F}_\mathbb{Z}}(h)\Phi_h(x)\right|^2 dx\right)^{1/2} = 0. \tag{3.20}$$

(iii) Parseval's identity holds. That is, for any $F \in \mathrm{L}^2_\mathbb{Z}(\mathbb{R})$,

$$\|F\|^2_{\mathrm{L}^2_\mathbb{Z}(\mathbb{R})} = \sum_{h\in\mathbb{Z}} \left|F^\wedge_{\mathcal{F}_\mathbb{Z}}(h)\right|^2. \tag{3.21}$$

(iv) The extended Parseval identity holds. That is, for any $F, H \in \mathrm{L}^2_{\mathbb{Z}}(\mathbb{R})$,

$$\langle F, H\rangle_{\mathrm{L}^2_{\mathbb{Z}}(\mathbb{R})} = \sum_{h\in\mathbb{Z}} F^{\wedge}_{\mathcal{F}_{\mathbb{Z}}}(h)\overline{H^{\wedge}_{\mathcal{F}_{\mathbb{Z}}}(h)}. \tag{3.22}$$

(v) There is no strictly larger orthonormal system containing the orthonormal system $\{\Phi_h\}_{h\in\mathbb{Z}}$.

(vi) The system $\{\Phi_h\}_{h\in\mathbb{Z}}$ *has the completeness property. That is,* $F \in \mathrm{L}^2_{\mathbb{Z}}(\mathbb{R})$ *and* $F^{\wedge}_{\mathcal{F}_{\mathbb{Z}}}(h) = 0$ *for all* $h \in \mathbb{Z}$ *implies* $F = 0$.

(vii) An element $F \in \mathrm{L}^2_{\mathbb{Z}}(\mathbb{R})$ *is uniquely determined by its orthogonal coefficients. That is, if* $F^{\wedge}_{\mathcal{F}_{\mathbb{Z}}}(h) = H^{\wedge}_{\mathcal{F}_{\mathbb{Z}}}(h)$ *for all* $h \in \mathbb{Z}$, *then* $F = H$.

For more details on the system $\{\Phi_h\}_{h\in\mathbb{Z}}$ the reader is referred to monographs on Fourier theory in Euclidean spaces, for example, P.L. Butzer, R. Nessel [1971], E.M. Stein, G. Weiss [1971], M.A. Pinsky [2002]. The proof of Theorem 3.5 can be found in every textbook on constructive approximation (for example, P.J. Davis, P. Rabinowitz [1975]).

3.2 Lattice Functions

The following comments only serve as heuristic preparation for the introduction of the so-called $\mathbb{Z}$-lattice function, that plays a particular role in Euler as well as Poisson summation formulas. Both summation formulas constitute appropriate means for comparing sums and integrals over finite as well as infinite intervals. The rigorous mathematical treatment of the $\mathbb{Z}$-lattice function and its consequences to establish extensions of Euler and Poisson summation formulas starts from Definition 3.8. As an immediate consequence, trapezoidal rules form equidistant numerical realizations of the Euler and Poisson summation formulas in the area of approximate integration.

Motivating Heuristic Remarks. In the early 1730s, L. Euler (1707–1783) discovered his powerful "summation formula"

$$\sum_{\substack{0\leq g\leq n\\ g\in\mathbb{Z}}} F(g) - \frac{1}{2}\left(F(0) + F(n)\right) \tag{3.23}$$

$$= \int_0^n F(x)\, dx + \frac{1}{12}\left(F'(n) - F'(0)\right) + \int_0^n \underbrace{\left(-\frac{1}{2}B_2(x)\right)}_{=G(\Delta;x)} F''(x)\, dx,$$

where B_2 given by

$$B_2(x) = (x - \lfloor x\rfloor)^2 - (x - \lfloor x\rfloor) + \frac{1}{6} = -2G(\Delta; x) \tag{3.24}$$

is the “Bernoulli function” of degree 2 (note that, for $x \in \mathbb{R}$, the symbol $\lfloor x \rfloor$ means that integer in $\mathbb{Z}$ for which $\lfloor x \rfloor \leq x < \lfloor x \rfloor + 1$) and the integral on the right hand involving B_2 acts as the remainder term in approximate integration. There is a huge literature concerned with the Euler summation formula (see, e.g., G.H. Hardy [1915], K. Knopp [1990], T.M. Apostol [1999], W. Dunham [1999] for a historic explanation of elementary constituents, and P.J. Davis, P. Rabinowitz [1984], P.L. Butzer, R.L. Stens [1983], J. Stoer [1989], J. Werner [1991] and the references therein for the one-dimensional relationship to numerical integration). The essential idea of the Euler summation formula in our work is based on aspects of mathematical physics (cf. L.J. Mordell [1928a,b], C. Müller [1954a], W. Freeden [2011]), namely the interpretation of the Bernoulli function B_2 occurring in (3.23) by means of mathematical physics (apart from a multiplicative constant) as the Green's function $G(\Delta;\cdot)$ for the (one-dimensional) Laplace operator Δ corresponding to the “boundary condition of $\mathbb{Z}$-periodicity” (once more, as usual in one dimension, $\Delta = (\frac{d}{dx})^2$ is the operator of the second order derivative). More concretely, the periodic Green's function $G(\Delta;\cdot)$ for the Laplace operator Δ is constructed so as to have the bilinear expansion

$$G(\Delta;x) = \sum_{\substack{\Delta^\wedge(h)\neq 0\\ h\in\mathbb{Z}}} \frac{\Phi_h(x)}{-\Delta^\wedge(h)}, \quad x \in \mathbb{R}. \tag{3.25}$$

Remark 3.6. *The Fourier expansion (3.25) enables us to relate $G(\Delta;\cdot)$ heuristically to the Dirac function(al). Indeed, we formally have*

$$\Delta_x G(\Delta;x) = \Delta_x \sum_{\substack{h\neq 0\\ h\in\mathbb{Z}}} \frac{1}{-\Delta^\wedge(h)} \Phi_h(x) = \sum_{\substack{h\neq 0\\ h\in\mathbb{Z}}} \Phi_h(x) = \sum_{h\in\mathbb{Z}} \Phi_h(x) - 1 = \delta(x) - 1, \tag{3.26}$$

where the δ-expansion $\delta(x) = \sum_{h\in\mathbb{Z}} \Phi_h(x)$ has to be understood in distributional sense (see, e.g., E. Berz [1967], M.J. Lighthill [1959], L. Schwarz [1950, 1951]).

In other words (see the 1D-part of W. Freeden [2011]), the Bernoulli function – in the jargon of mathematical physics, the Green's function of the operator of the second derivative with respect to the “boundary condition of periodicity” – acts as a connecting tool to convert a differential equation involving the Laplace operator corresponding to “periodic boundary conditions” into an associated integral equation, i.e., the Euler summation formula (3.23). Observing the special values

$$G(\Delta;0) = G(\Delta;n) = \sum_{\substack{\Delta^\wedge(h)\neq 0\\ h\in\mathbb{Z}}} \frac{1}{-\Delta^\wedge(h)} = -\frac{1}{12}, \quad n \in \mathbb{Z}, \tag{3.27}$$

and the explicit representation of the Fourier series expansion (3.25) we are able to reformulate the Euler summation formula (3.23). Integration by parts yields (by use of $F'(x) = \nabla F(x)$, $F''(x) = \Delta F(x)$)

$$\begin{aligned} &\sum_{\substack{0\le g\le n\\ g\in\mathbb{Z}}} F(g) - \frac{1}{2}\left(F(0)+F(n)\right) \\ &= \int_0^n F(x)\,dx + \sum_{\substack{h\neq 0\\ h\in\mathbb{Z}}} \left(\frac{\nabla F(n) - \nabla F(0)}{4\pi^2 h^2} - \int_0^n \frac{\Phi_h(x)}{4\pi^2 h^2}\Delta F(x)\,dx\right) \\ &= \int_0^n F(x)\,dx - \lim_{N\to\infty} \sum_{\substack{|h|\le N\\ h\neq 0}} \frac{1}{2\pi i h}\int_0^n \nabla F(x)\,\Phi_h(x)\,dx, \end{aligned} \tag{3.28}$$

such that the Poisson summation formula comes into play

$$\sum_{\substack{0\le g\le n\\ g\in\mathbb{Z}}} F(g) - \frac{1}{2}\left(F(0)+F(n)\right) = \lim_{N\to\infty} \sum_{\substack{|h|\le N\\ h\neq 0, h\in\mathbb{Z}}} \int_0^n F(x)\,\Phi_h(x)\,dx. \tag{3.29}$$

A simple manipulation shows that the Poisson summation formula can be rewritten in the form

$$\begin{aligned} \sum_{\substack{0\le g\le n\\ g\in\mathbb{Z}}} F(g) - \frac{1}{2}\left(F(0)+F(n)\right) &= \int_0^n F(x)\,dx \\ &\quad + \lim_{N\to\infty} \sum_{\substack{0<|h|\le N\\ h\in\mathbb{Z}}} \int_0^n F(x)\,\Phi_h(x)\,dx, \end{aligned} \tag{3.30}$$

where the sum on the right-hand side of (3.30) is a remainder term free of derivatives. All in all, in spite of their apparent dissimilarity, the Euler summation formula (3.23) and the Poisson summation formula (3.29) are equivalent for twice continuously differentiable functions on the interval $[0, n]$. Moreover, the Green's function for the Laplace operator and the "boundary condition of $\mathbb{Z}$-periodicity" acts as the canonical bridge between both identities. Clearly, both summation formulas play an important role in numerical integration by opening the perspective to compare the integral $\int_0^n F(x)\,dx$ with the sum $\sum_{\substack{0\le g\le n\\ g\in\mathbb{Z}}} F(g)$.

Remark 3.7. *The "building blocks" of the bridge between the two equivalent formulas (3.23) and (3.29) are the* defining constituents of the Green's function $G(\Delta;\cdot)$ *as given in Definition 3.8.*

Rigorous Formulation. After these heuristic remarks concerning the introduction of $\mathbb{Z}$-periodic functions (i.e., functions with period 1) and Euler as well as Poisson summation formulas we come to the rigorous mathematical formulation. We begin with the definition of the $\mathbb{Z}$-lattice function with respect to the one-dimensional Laplacian $\Delta = \nabla^2$, i.e., Green's function with respect to Δ corresponding to "$\mathbb{Z}$-periodic boundary conditions" as a prototype of forthcoming considerations. Based on the defining properties of the $\mathbb{Z}$-lattice function the Euler summation formula can be developed using integration by parts.

Definition 3.8. *A function $G(\Delta; \cdot) : \mathbb{R} \to \mathbb{R}$ is called the* Green's function with respect to the operator Δ corresponding to the $\mathbb{Z}$-periodicity (in brief, $\mathbb{Z}$-lattice function for Δ) *if it satisfies the following properties:*

*(i) (*Periodicity*) $G(\Delta; \cdot)$ is continuous in $\mathbb{R}$, and*

$$G(\Delta; x + g) = G(\Delta; x) \tag{3.31}$$

for all $x \in \mathbb{R}$ and $g \in \mathbb{Z}$.

*(ii) (*Differential Equation*) G is twice continuously differentiable with*

$$\Delta G(\Delta; x) = -1 \tag{3.32}$$

for all $x \notin \mathbb{Z}$.

*(iii) (*Characteristic Singularity*)*

$$G(\Delta; x) - \frac{1}{2} x \ \mathrm{sign}(x) \tag{3.33}$$

is continuously differentiable for all $x \in \mathcal{F}_{\mathbb{Z}}$.

*(iv) (*Normalization*)*

$$\int_{\mathcal{F}_{\mathbb{Z}}} G(\Delta; x) \, dx = 0. \tag{3.34}$$

Indeed, we are able to show that the $\mathbb{Z}$-*lattice function for the operator of the second derivative*, i.e., for the one-dimensional Laplacian, is uniquely determined by its constituting properties.

Lemma 3.9. *$G(\Delta; \cdot)$ is uniquely determined by the properties (i)–(iv).*

Proof. Denote by $D(\Delta; \cdot)$ the difference between two $\mathbb{Z}$-lattice functions for Δ. Then we have the following properties:

(i) D is continuous in $\mathbb{R}$, and for all $x \in \mathbb{R}$ and $g \in \mathbb{Z}$

$$D(\Delta; x + g) = D(\Delta; x). \tag{3.35}$$

(ii) $D(\Delta; \cdot)$ is twice continuously differentiable for all $x \notin \mathbb{Z}$ with

$$\Delta D(\Delta; \cdot) = 0. \tag{3.36}$$

(iii) $D(\Delta;\cdot)$ is continuously differentiable in $\mathbb{R}$.

(iv)

$$\int_{\mathcal{F}_{\mathbb{Z}}} D(\Delta;x)\,dx = 0. \tag{3.37}$$

The properties (i)–(iii) show that $D(\Delta;\cdot)$ is a constant function. The last condition (iv) shows us that the constant must be zero. Thus, G is uniquely determined. □

Theorem 3.10. *$G(\Delta;\cdot)$ given by (3.24) satisfies all defining properties (i)–(iv) of the $\mathbb{Z}$-lattice function for the one-dimensional Laplace operator Δ.*

In addition, observing the $\mathbb{Z}$-periodicity *(i)* and the characteristic singularity (iii) of the $\mathbb{Z}$-lattice function for Δ we readily obtain by applying integration by parts

$$\langle G(\Delta;\cdot),\Phi_h\rangle_{\mathrm{L}^2_{\mathbb{Z}}(\mathbb{R})} = \int_{\mathcal{F}_{\mathbb{Z}}} G(\Delta;x)\overline{\Phi_h(x)}\,dx = -\frac{1}{4\pi^2h^2} = -\frac{1}{\Delta^\wedge(h)} \tag{3.38}$$

provided that $h \neq 0$. Thus, the classical representation theorem of one-dimensional Fourier theory yields

Lemma 3.11. *For all $x \in \mathbb{R}$*

$$G(\Delta;x) = \sum_{\substack{h\neq 0\\ h\in\mathbb{Z}}} \frac{1}{-\Delta^\wedge(h)}\Phi_h(x) = \sum_{\substack{h\neq 0\\ h\in\mathbb{Z}}} \frac{1}{-\Delta^\wedge(h)}\overline{\Phi_h(x)} = \overline{G(\Delta;x)}, \tag{3.39}$$

and the series on the right side is absolutely and uniformly convergent on each compact interval $I \subset \mathbb{R}$.

More concretely, for all $x \in \mathbb{R}$, we have (cf. Figure 3.2)

$$\begin{aligned} G(\Delta;x) &= \sum_{n=1}^{\infty} \frac{1}{-4\pi^2n^2}\left(e^{2\pi inx}+e^{-2\pi inx}\right) = -\frac{1}{2\pi^2}\sum_{n=1}^{\infty}\frac{1}{n^2}\cos(2\pi nx) \\ &= -\frac{(x-\lfloor x\rfloor)^2}{2} + \frac{x-\lfloor x\rfloor}{2} - \frac{1}{12}. \end{aligned} \tag{3.40}$$

Figure 3.2 gives an illustration of the $\mathbb{Z}$-lattice function for Δ. It shows that the graph of $G(\Delta;\cdot)$ consists of parabolic arcs that are continuously put together at integer points.

Remark 3.12. *Obviously, $G(\Delta;x-y) = \overline{G(\Delta;x-y)}$ and $G(\Delta;x-y) = G(\Delta;y-x)$.*

Since the Fourier series of $G(\Delta;\cdot)$ converges absolutely and uniformly for all $x \in \mathbb{R}$, an elementary differentiation yields

$$\nabla_x G(\Delta;x) = G(\nabla;x) = -(x-\lfloor x\rfloor) + \frac{1}{2}, \quad x \in \mathbb{R}\setminus\mathbb{Z}. \tag{3.41}$$

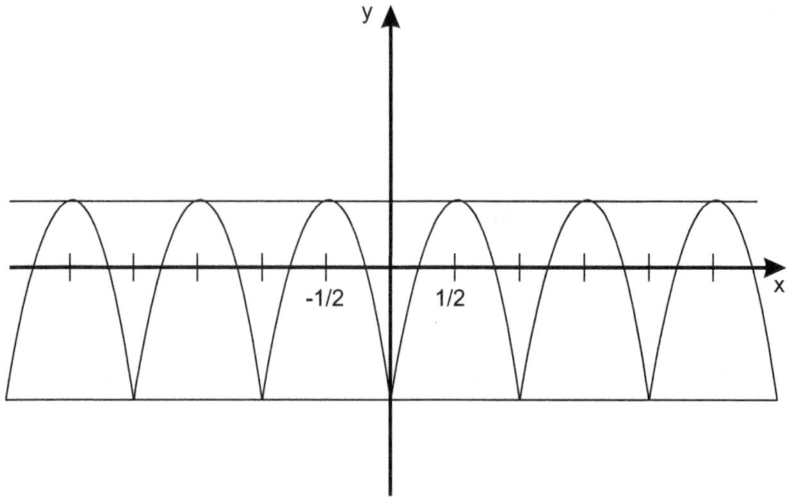

FIGURE 3.2: The illustration of the $\mathbb{Z}$-lattice function $G(\Delta;\cdot)$.

Remark 3.13. *The function* $-G(\nabla;\cdot)$ *is called the* Bernoulli function B_1 of degree *1 given by* $x \mapsto B_1(x) = x - \lfloor x \rfloor - \frac{1}{2}, \quad x \in \mathbb{R}$.

Using the well-known representation of the sin-function we are led to the expansion

$$\sum_{\substack{h\neq 0\\ h\in\mathbb{Z}}} \frac{1}{2\pi i h}\Phi_h(x) = \sum_{n=1}^{\infty} \frac{1}{2i}\left(\frac{e^{2\pi i n x}}{\pi n} - \frac{e^{-2\pi i n x}}{\pi n}\right) = \sum_{n=1}^{\infty} \frac{\sin(2\pi n x)}{\pi n}. \tag{3.42}$$

We mention some known results (see, e.g., L.J. Mordell [1928a,b]) concerning the convergence of the series (3.42) and its explicit representation (see Figure 3.3 for its illustration).

Lemma 3.14. *The series (3.42) converges uniformly in each compact interval* $I \subset (g, g+1)$, $g \in \mathbb{Z}$. *Moreover, for* $x \in I$ *with* I *a compact subset of* $(g, g+1)$, $g \in \mathbb{Z}$, *we have*

$$\underbrace{-(x - \lfloor x \rfloor - \frac{1}{2})}_{=B_1(x)} = \frac{d}{dx}\left(-\sum_{n=1}^{\infty} \frac{\cos(2\pi n x)}{2\pi^2 n^2}\right) = \sum_{n=1}^{\infty} \frac{\sin(2\pi n x)}{\pi n}. \tag{3.43}$$

3.3 Euler Summation Formulas

Next we want to verify the *Euler summation formula* in its classical form related to the one-dimensional operator of the first order derivative, i.e., the

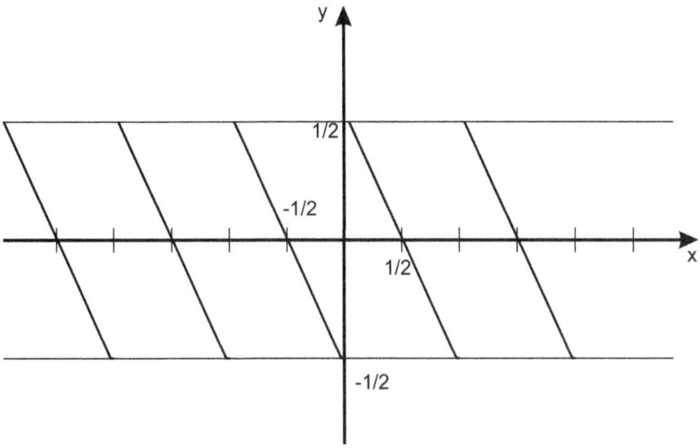

FIGURE 3.3: The derivative $G(\nabla;\cdot) = \nabla G(\Delta;\cdot)$ of the $\mathbb{Z}$-lattice function.

gradient ∇, and the one-dimensional operator of the second order derivative, i.e., the Laplace operator Δ, respectively.

Theorem 3.15 (Euler Summation Formula). *If* $F \in \mathrm{C}^{(1)}[0,n]$, $n \in \mathbb{N}$, *then*

$$\sum_{\substack{g\in[0,n]\\ g\in\mathbb{Z}}} F(g) = \int_0^n F(x)\, dx + \frac{1}{2}\Big(F(n)+F(0)\Big) - \int_0^n G(\nabla;x)\nabla F(x)\, dx. \tag{3.44}$$

If $F \in \mathrm{C}^{(2)}[0,n]$, $n \in \mathbb{N}$, *then*

$$\sum_{\substack{g\in[0,n]\\ g\in\mathbb{Z}}} F(g) = \int_0^n F(x)\, dx + \frac{1}{2}\Big(F(n)+F(0)\Big) \tag{3.45}$$

$$+ \frac{1}{12}\Big((\nabla F)(n) - (\nabla F)(0)\Big) + \int_0^n G(\Delta;x)\Delta F(x)\, dx.$$

Let $F : [a,b] \to \mathbb{C}$, $a < b$, *be a twice continuously differentiable function, i.e.,* $F \in \mathrm{C}^{(2)}[a,b]$. *Then*

$$\sum_{\substack{g\in[a,b]\\ g\in\mathbb{Z}}}{}' F(g) = \int_a^b F(x)\, dx + \int_a^b G(\Delta;x)\Delta F(x)\, dx \tag{3.46}$$

$$+ \big(F(x)\nabla G(\Delta;x) - G(\Delta;x)\nabla F(x)\big)\Big|_a^b,$$

where

$$\sum_{\substack{g\in[a,b]\\ g\in\mathbb{Z}}}{}' F(g) = \sum_{\substack{g\in(a,b)\\ g\in\mathbb{Z}}} F(g) + \frac{1}{2} \sum_{\substack{g=a,g=b\\ g\in\mathbb{Z}}} F(g), \tag{3.47}$$

and the last sum in (3.47) occurs only if a and/or b are lattice points. Otherwise it is assumed to be zero.

Proof. We restrict ourselves to the proof of the formula (3.46). Moreover, we are concerned with the case that both endpoints a, b are no integers. Integrating by parts we find, for every (sufficiently small) $\varepsilon > 0$, by cutting out one-dimensional ε-balls $\mathbb{B}^1_\varepsilon(g) = (g - \varepsilon, g + \varepsilon)$ that

$$\int_{\substack{x\in[a,b] \\ x\notin\bigcup_{g\in\mathbb{Z}}\mathbb{B}^1_\varepsilon(g)}} (G(\Delta;x)\Delta F(x) - F(x)\Delta G(\Delta;x))\ dx \tag{3.48}$$

$$= \big(G(\Delta;x)\nabla F(x) - F(x)\nabla G(\Delta;x)\big)\Big|_a^b$$

$$+ \sum_{\substack{g\in(a,b) \\ g\in\mathbb{Z}}} \big(G(\Delta;x)\nabla F(x) - F(x)\nabla G(\Delta;x)\big)\Big|_{g+\varepsilon}^{g-\varepsilon}.$$

By application of the differential equation $\Delta G(\Delta;x) = -1$, $x \in \mathbb{R}\setminus\mathbb{Z}$, it follows that

$$\int_{\substack{x\in[a,b] \\ x\notin\bigcup_{g\in\mathbb{Z}}\mathbb{B}^1_\varepsilon(g)}} F(x)\Delta G(\Delta;x)\ dx = -\int_{\substack{x\in[a,b] \\ x\notin\bigcup_{g\in\mathbb{Z}}\mathbb{B}^1_\varepsilon(g)}} F(x)\ dx. \tag{3.49}$$

Letting $\varepsilon \to 0$ and observing the (limit) values of the $\mathbb{Z}$-lattice function for Δ and its derivatives in the lattice points we obtain

$$\sum_{\substack{g\in(a,b) \\ g\in\mathbb{Z}}} F(g) = \int_a^b F(x)\ dx + \int_a^b G(\Delta;x)\Delta F(x)\ dx \tag{3.50}$$

$$+ \big(F(x)\nabla G(\Delta;x) - G(\Delta;x)\nabla F(x)\big)\Big|_a^b.$$

This is the desired result. The cases where a and/or b are integers, i.e., members of the lattice $\mathbb{Z}$, follow by obvious modifications. □

Note that in Theorem 3.15, the term $G(\nabla;x)F(x)|_a^b$ is understood in the sense $G(\nabla;b-0)F(b) - G(\nabla;a+0)F(a)$. As a special case we obtain the following lemma.

Lemma 3.16. *For $a, b \in \mathbb{R}$, $a < b$, we have the so-called one-dimensional Hardy–Landau identity*

$$\sum_{\substack{g\in[a,b] \\ g\in\mathbb{Z}}}{}' 1 = (b-a) + G(\nabla;x)\Big|_a^b. \tag{3.51}$$

In other words, the total number of integer points in the interval $[a, b]$ can be compared with the length $b - a$ of the interval under explicit knowledge of the remainder term $G(\nabla; x)|_a^b = \nabla G(\Delta; x)|_a^b$.

Next we want to present an extension of the Euler summation formula. This can be achieved by replacing the lattice $\mathbb{Z}$ by a "translated lattice" $\mathbb{Z}+\{x\}$ based at $x \in \mathcal{F}_{\mathbb{Z}}$.

Corollary 3.17. *Let x be a point of the fundamental cell $\mathcal{F}_{\mathbb{Z}}$ of $\mathbb{Z}$. Suppose that F is of class $\mathrm{C}^{(2)}[a, b]$. Then*

$$\sum_{\substack{g+x\in[a,b]\\ g\in\mathbb{Z}}}{}' F(g+x) = \int_a^b F(y)\, dy + \int_a^b G(\Delta; x-y)\Delta_y F(y)\, dy \tag{3.52}$$
$$+ \left(F(y)\nabla_y G(\Delta; x-y) - G(\Delta; x-y)\nabla_y F(y)\right)\Big|_a^b .$$

Note that the operators $\nabla_y = \frac{d}{dy}$ and $\Delta_y = \nabla_y^2 = (\frac{d}{dy})^2$ apply to the y-variable.

3.4 Euler Summation Formulas for Periodic Functions

From Corollary 3.17 we obtain by specializing the interval $[a, b]$ to the fundamental cell $\mathcal{F}_{\mathbb{Z}} = [-\frac{1}{2}, \frac{1}{2})$ of $\mathbb{Z}$.

Corollary 3.18. *Let x be a point of the fundamental cell $\mathcal{F}_{\mathbb{Z}}$ of the lattice $\mathbb{Z}$. Suppose that F is of class $\mathrm{C}^{(2)}(\overline{\mathcal{F}_{\mathbb{Z}}})$. Then*

$$F(x) = \int_{\mathcal{F}_{\mathbb{Z}}} F(y)\, dy + \int_{\mathcal{F}_{\mathbb{Z}}} G(\Delta; x-y)\Delta_y F(y)\, dy \tag{3.53}$$
$$+ \left(F(y)\nabla_y G(\Delta; x-y) - G(\Delta; x-y)\nabla_y F(y)\right)\Big|_{-\frac{1}{2}}^{\frac{1}{2}} .$$

This immediately leads to the following result involving $\mathbb{Z}$-periodic functions.

Corollary 3.19. *Assume that $x \in \mathcal{F}_{\mathbb{Z}}$ and $F \in \mathrm{C}_{\mathbb{Z}}^{(2)}(\mathbb{R})$. Then*

$$F(x) = \int_{\mathcal{F}_{\mathbb{Z}}} F(y)\, dy + \int_{\mathcal{F}_{\mathbb{Z}}} G(\Delta; x-y)\Delta_y F(y)\, dy. \tag{3.54}$$

The formula (3.54) is of exactness degree 0, i.e., (3.54) is exact for all constant functions.

Next, following Hilbert's approach, we are interested in iterated Green's functions which turn out to be Green's function to iterated differential operators. We begin with the definition of iterations by convolution.

Definition 3.20. *The iterated kernel* $G^{(2)}(\Delta;\ \cdot\ )$ *is defined by the convolution as follows:*

$$G^{(2)}\left(\Delta; x-y\right) = \int_{\mathcal{F}_{\mathbb{Z}}} G(\Delta; x-z)G(\Delta; z-y)\, dz, \quad x, y \in \mathbb{R}. \tag{3.55}$$

Obviously, the bilinear series admits the following manipulations

$$\begin{aligned} G^{(2)}(\Delta; x-y) &= \sum_{\substack{k\neq 0\\ k\in\mathbb{Z}}}\sum_{\substack{h\neq 0\\ h\in\mathbb{Z}}} \frac{\Phi_h(x)}{-\Delta^\wedge(h)} \frac{\overline{\Phi_k(y)}}{-\Delta^\wedge(k)} \int_{\mathcal{F}_{\mathbb{Z}}} \overline{\Phi_h(z)}\Phi_k(z)\, dz \qquad (3.56)\\ &= \sum_{\substack{k\neq 0\\ k\in\mathbb{Z}}}\sum_{\substack{h\neq 0\\ h\in\mathbb{Z}}} \frac{\Phi_h(x)}{-\Delta^\wedge(h)} \frac{\overline{\Phi_k(y)}}{-\Delta^\wedge(k)}\, \delta_{h,k}\\ &= \sum_{\substack{h\neq 0\\ h\in\mathbb{Z}}} \frac{1}{(-\Delta^\wedge(h))^2}\Phi_h(x-y), \end{aligned}$$

so that

$$G^{(2)}(\Delta; x-y) = G(\Delta^2; x-y). \tag{3.57}$$

Moreover, it follows that

$$\Delta_y G\left(\Delta^2; x-y\right) = \nabla_y^2 G\left(\Delta^2; x-y\right) = G(\Delta; x-y). \tag{3.58}$$

As already announced, $G(\Delta^2;\cdot)$ is the lattice (Green's) function to the iterated Laplace operator $\Delta^2 = \Delta\Delta$, i.e., the operator of the fourth derivative.

Lemma 3.21. *The function* $G\left(\Delta^2;\cdot\right)$ *with respect to the operator* $\Delta^2 = \nabla^4$ *given by*

$$G\left(\Delta^2; x\right) = \frac{1}{8\pi^4}\sum_{h=1}^{\infty}\frac{1}{h^4}\cos(2\pi h x) \tag{3.59}$$

is $\mathbb{Z}$*-periodic, twice continuously differentiable in* $\mathbb{R}$ *such that*

$$\Delta G\left(\Delta^2; x\right) = G(\Delta; x), \quad x \in \mathbb{R}, \tag{3.60}$$

and we have the representation

$$G\left(\Delta^2; x\right) = -\frac{1}{24}(x-\lfloor x\rfloor)^4 + \frac{1}{12}(x-\lfloor x\rfloor)^3 - \frac{1}{24}(x-\lfloor x\rfloor)^2 + \frac{1}{720}. \tag{3.61}$$

Corollary 3.19 may be rewritten in the form

$$F(x) = \int_{\mathcal{F}_{\mathbb{Z}}} F(y)\, dy + \int_{\mathcal{F}_{\mathbb{Z}}} \Delta_y G(\Delta^2; x-y)\, \Delta_y F(y)\, dy. \tag{3.62}$$

provided that $F \in \mathrm{C}^{(2)}_{\mathbb{Z}}(\mathbb{R})$. Indeed, the identity (3.62) can be used for both best approximate integration of $\mathbb{Z}$-periodic functions and periodic spline interpolation (see W. Freeden, R. Reuter [1981], W. Freeden, J. Fleck [1987], W. Freeden [1988], and the references therein). In addition, Corollary 3.19 leads back to the (well-known) solutions of the following differential equations corresponding to "periodic boundary conditions".

Corollary 3.22. *Assume that F is of class $\mathrm{C}^{(2)}_{\mathbb{Z}}(\mathbb{R})$ satisfying $\Delta F(y) = 0$, $y \in \mathcal{F}_{\mathbb{Z}}$. Then F is a constant function. More explicitly,*

$$F(x) = \int_{\mathcal{F}_{\mathbb{Z}}} F(y)\, dy, \quad x \in \mathcal{F}_{\mathbb{Z}}. \tag{3.63}$$

Corollary 3.23. *Assume that H is of class $\mathrm{C}^{(0)}_{\mathbb{Z}}(\mathbb{R})$ such that*

$$\int_{\mathcal{F}_{\mathbb{Z}}} H(y)\, dy = 0. \tag{3.64}$$

Let $F \in \mathrm{C}^{(2)}_{\mathbb{Z}}(\mathbb{R})$ satisfy $\Delta F(y) = H(y)$, $y \in \mathcal{F}_{\mathbb{Z}}$, such that

$$\int_{\mathcal{F}_{\mathbb{Z}}} F(y)\, dy = 0. \tag{3.65}$$

Then

$$F(x) = \int_{\mathcal{F}_{\mathbb{Z}}} G(\Delta; x-y) H(y)\, dy, \quad x \in \mathcal{F}_{\mathbb{Z}}. \tag{3.66}$$

Remark 3.24. *By virtue of the Cauchy–Schwarz inequality we obtain from Corollary 3.19 the* error estimate:

$$\left| F(x) - \int_{\mathcal{F}_{\mathbb{Z}}} F(y)\, dy \right| \leq \left(G\left(\Delta^2; 0\right)\right)^{\frac{1}{2}} \left(\int_{\mathcal{F}_{\mathbb{Z}}} |\Delta F(y)|^2\, dy \right)^{1/2}, \tag{3.67}$$

where

$$G\left(\Delta^2; 0\right) = \frac{1}{720}. \tag{3.68}$$

Chapter 4

Periodic Spline Integration

Analogously to the case of algebraic polynomials we are able to go over to periodic spline functions. For simplicity, we restrict ourselves to splines with respect to Δ-derivatives in this preparatory part. As reference space for spline approximation we use a Sobolev-like space, in which spline interpolation may be understood as minimum norm procedure under consistency to the given data points. Best approximate integration is seen in the context of integrating periodic spline interpolants.

4.1 Best Approximate Integration in Sobolev Spaces

Our purpose is to introduce a Sobolev space corresponding to $\mathrm{C}^{(2)}_{\mathbb{Z}}(\mathbb{R})$. To this end wee start from the sequence $\{a_h\}_{h\in\mathbb{Z}}$ defined by

$$a_h = \begin{cases} 1, & h = 0, \\ -\Delta^\wedge(h), & h \neq 0. \end{cases} \tag{4.1}$$

Remember

$$\Delta_x \Phi_h(x) = -\Delta^\wedge(h)\Phi_h(x) = -4\pi^2 h^2 \Phi_h(x), \quad h \in \mathbb{Z},\ x \in \mathbb{R}. \tag{4.2}$$

Consider the set $\mathrm{E}(\{a_h\};\mathbb{R})$ of all F of class $\mathrm{C}^{(\infty)}_{\mathbb{Z}}(\mathbb{R})$ satisfying the condition

$$\sum_{h\in\mathbb{Z}} |a_h|^2 \left|F^\wedge_{\mathcal{F}_\mathbb{Z}}(h)\right|^2 < \infty, \tag{4.3}$$

where the Fourier coefficients $F^\wedge_{\mathcal{F}_\mathbb{Z}}(h)$ are defined (cf. (3.19)) as

$$F^\wedge_{\mathcal{F}_\mathbb{Z}}(h) = \int_{\mathcal{F}_\mathbb{Z}} F(x)\overline{\Phi_h(x)}\, dx. \tag{4.4}$$

From the Cauchy–Schwarz inequality it immediately follows that

$$\left|\sum_{h\in\mathbb{Z}} |a_h|^2 F^\wedge_{\mathcal{F}_\mathbb{Z}}(h)\overline{G^\wedge_{\mathcal{F}_\mathbb{Z}}(h)}\right|^2 \leq \left(\sum_{h\in\mathbb{Z}} |a_h|^2 |F^\wedge_{\mathcal{F}_\mathbb{Z}}(h)|^2\right)\left(\sum_{h\in\mathbb{Z}} |a_h|^2 |G^\wedge_{\mathcal{F}_\mathbb{Z}}(h)|^2\right) \tag{4.5}$$

for all functions F, G of class $\mathrm{E}(\{a_h\};\mathbb{R})$. As both sums on the right-hand side are finite for elements of class $\mathrm{E}(\{a_h\};\mathbb{R})$, the left-hand side is also finite. Hence, we are able to introduce an inner product $\langle\cdot,\cdot\rangle_{\mathrm{H}_{\mathbb{Z}}^{(2)}(\mathbb{R})}$ on $\mathrm{E}(\{a_h\};\mathbb{R})$

$$\langle F, G\rangle_{\mathrm{H}_{\mathbb{Z}}^{(2)}(\mathbb{R})} = \sum_{h\in\mathbb{Z}} |a_h|^2 F^{\wedge}_{\mathcal{F}_{\mathbb{Z}}}(h)\overline{G^{\wedge}_{\mathcal{F}_{\mathbb{Z}}}(h)}. \tag{4.6}$$

The corresponding norm is given by

$$\|F\|^2_{\mathrm{H}_{\mathbb{Z}}^{(2)}(\mathbb{R})} = \sum_{h\in\mathbb{Z}} |a_h|^2 |F^{\wedge}_{\mathcal{F}_{\mathbb{Z}}}(h)|^2. \tag{4.7}$$

This enables us to introduce the Sobolev space $\mathrm{H}_{\mathbb{Z}}^{(2)}(\mathbb{R})$ as follows.

Definition 4.1. *The* Sobolev space $\mathrm{H}_{\mathbb{Z}}^{(2)}(\mathbb{R})$ *is the completion of* $\mathrm{E}(\{a_h\};\mathbb{R})$ *under the* $\mathrm{H}_{\mathbb{Z}}^{(2)}(\mathbb{R})$*-norm, i.e.,*

$$\mathrm{H}_{\mathbb{Z}}^{(2)}(\mathbb{R}) = \overline{\mathrm{E}(\{a_h\};\mathbb{R})}^{\|\cdot\|_{\mathrm{H}_{\mathbb{Z}}^{(2)}(\mathbb{R})}}. \tag{4.8}$$

The space $\mathrm{H}_{\mathbb{Z}}^{(2)}(\mathbb{R})$ equipped with the inner product (4.6) is a Hilbert space. Considering the series $\{a_h\}_{h\in\mathbb{Z}}$ of (4.1) and remembering Parseval's identity in Hilbert spaces we can rewrite the inner product in integral form

$$\sum_{h\in\mathbb{Z}} |a_h|^2 F^{\wedge}_{\mathbb{Z}}(h)\overline{G^{\wedge}_{\mathbb{Z}}(h)} = \int_{\mathcal{F}_{\mathbb{Z}}} F(x)\,dx \overline{\int_{\mathcal{F}_{\mathbb{Z}}} G(x)\,dx} + \int_{\mathcal{F}_{\mathbb{Z}}} \Delta F(x)\,\overline{\Delta G(x)}\,dx. \tag{4.9}$$

The corresponding norm is then given by

$$\|F\|^2_{\mathrm{H}_{\mathbb{Z}}^{(2)}(\mathbb{R})} = \left|\int_{\mathcal{F}_{\mathbb{Z}}} F(x)\,dx\right|^2 + \int_{\mathcal{F}_{\mathbb{Z}}} |\Delta F(x)|^2\,dx. \tag{4.10}$$

Indeed, every $F \in \mathrm{H}_{\mathbb{Z}}^{(2)}(\mathbb{R})$ can be represented by its Fourier series

$$F(x) = \sum_{h\in\Lambda} \int_{\mathcal{F}_{\mathbb{Z}}} F(z)\overline{\Phi_h(z)}\,dz\ \Phi_h(x). \tag{4.11}$$

Splitting the series and observing the fact, that $\Delta\Phi_h = -\Delta^{\wedge}(h)\Phi_h$ we obtain

$$F(x) = \int_{\mathcal{F}_{\mathbb{Z}}} F(z)\,dz + \sum_{\substack{h\neq 0\\ h\in\mathbb{Z}}} \frac{1}{-\Delta^{\wedge}(h)} \int_{\mathcal{F}_{\mathbb{Z}}} F(z)\Delta\overline{\Phi_h(z)}\,dz\ \Phi_h(x). \tag{4.12}$$

Because of the $\mathbb{Z}$-periodicity of F we obtain

$$F(x) = \int_{\mathcal{F}_{\mathbb{Z}}} F(z)\,dz + \sum_{\substack{h\neq 0\\ h\in\mathbb{Z}}} \frac{1}{-\Delta^{\wedge}(h)} \int_{\mathcal{F}_{\mathbb{Z}}} \Delta_z F(z)\overline{\Phi_h(z)}\,dz\ \Phi_h(x). \tag{4.13}$$

Observing the absolute and uniform convergence of the bilinear expansion of $G(\Delta; x-z)$ we finally arrive at the *Euler summation formula within the periodic framework of the Sobolev space* $\mathrm{H}_{\mathbb{Z}}^{(2)}(\mathbb{R})$.

Theorem 4.2 (Integral Formula). *For all F of class $\mathrm{H}_{\mathbb{Z}}^{(2)}(\mathbb{R})$ the integral formula*

$$F(x) = \int_{\mathcal{F}_{\mathbb{Z}}} F(z)\, dz + \int_{\mathcal{F}_{\mathbb{Z}}} \Delta_z G(\Delta^2; x - z)\, \Delta_z F(z)\, dz \tag{4.14}$$

holds true. Moreover, the identity (4.14) is of exactness degree 0, i.e., (4.14) is exact for all constant functions.

A consequence of this periodic variant of the Euler summation formula is the following result.

Corollary 4.3. *The kernel of the operator Δ in $\mathrm{H}_{\mathbb{Z}}^{(2)}(\mathbb{R})$ is the set of constants, i.e., if $F \in \mathrm{H}_{\mathbb{Z}}^{(2)}(\mathbb{R})$ fulfills the Laplace equation $\Delta F = 0$, then*

$$F = \int_{\mathcal{F}_{\mathbb{Z}}} F(x)\, dx = const. \tag{4.15}$$

Theorem 4.2 can be used *for approximate integration of functions of class* $\mathrm{H}_{\mathbb{Z}}^{(2)}(\mathbb{R})$. We merely present some introductory framework about best approximate integration. If $a_1, \ldots, a_N \in \mathbb{R}$ satisfy $\sum_{k=1}^{N} a_k = 1$, we have for points $x_1, \ldots, x_N \in X_N \subset \mathcal{F}_{\mathbb{Z}}$

$$\sum_{k=1}^{N} a_k F(x_k) - \int_{\mathcal{F}_{\mathbb{Z}}} F(y)\, dy = \int_{\mathcal{F}_{\mathbb{Z}}} \sum_{k=1}^{N} a_k G(\Delta; x_k - y)\, \Delta F(y)\, dy \tag{4.16}$$

provided that F is of class $\mathrm{H}_{\mathbb{Z}}^{(2)}(\mathbb{R})$. By use of the Cauchy–Schwarz inequality, we obtain from (4.16)

$$\begin{aligned} &\left| \int_{\mathcal{F}_{\mathbb{Z}}} F(y)\, dy - \sum_{k=1}^{N} a_k F(x_k) \right| \\ &\quad \le \left(\sum_{p=1}^{N} \sum_{q=1}^{N} a_p a_q G(\Delta^2; x_p - x_q) \right)^{1/2} \left(\int_{\mathcal{F}_{\mathbb{Z}}} |\Delta_x F(x)|^2\, dx \right)^{1/2}. \end{aligned} \tag{4.17}$$

Thus, the *"best approximation" formula* (in the sense of A. Sard [1949, 1963]) corresponding to the given knots $X_N = \{x_1, \ldots, x_N\} \subset \mathcal{F}_{\mathbb{Z}}$

$$\mathcal{L}_{X_N} F = \sum_{k=1}^{N} \mathring{a}_k F(x_k) \tag{4.18}$$

to the integral

$$\mathcal{I}(F) = \int_{\mathcal{F}_{\mathbb{Z}}} F(x)\, dx \tag{4.19}$$

is obtained by the (unique) solution of the quadratic optimization problem

$$\sum_{p=1}^{N}\sum_{q=1}^{N} a_p a_q G(\Delta^2; x_p - x_q) \to \min \tag{4.20}$$

under the constraints

$$\sum_{k=1}^{N} a_k = 1. \tag{4.21}$$

In accordance with Lagrange's method of multipliers, the solution $\mathring{a} = (\mathring{a}_1, \ldots, \mathring{a}_N)^T$ of the best approximation formula solves the linear system

$$\begin{array}{rcl} G(\Delta^2; x_1 - x_1)\mathring{a}_1 + \ldots + G(\Delta^2; x_N - x_1)\mathring{a}_N - \mathring{\lambda} &=& 0, \\ \vdots \quad \vdots \qquad\qquad \vdots \quad \vdots && \\ G(\Delta^2; x_1 - x_N)\mathring{a}_1 + \ldots + G(\Delta^2; x_N - x_N)\mathring{a}_N - \mathring{\lambda} &=& 0, \\ \mathring{a}_1 + \ldots + \qquad\qquad \mathring{a}_N \qquad &=& 1 \end{array} \tag{4.22}$$

such that

$$\mathring{\lambda} = \sum_{p=1}^{N}\sum_{q=1}^{N} \mathring{a}_p \mathring{a}_q G(\Delta^2; x_p - x_q). \tag{4.23}$$

In consequence, the linear system does not only provide the coefficients $\mathring{a}_1, \ldots, \mathring{a}_N$ of the best approximation, but also the *a posteriori* accuracy of the integration formula via the Lagrange's multiplier (4.23):

$$\left| \int_{\mathcal{F}_{\mathbb{Z}}} F(y)\, dy - \sum_{k=1}^{N} \mathring{a}_k F(x_k) \right| \le \mathring{\lambda}^{1/2} \left(\int_{\mathcal{F}_{\mathbb{Z}}} |\Delta_x F(x)|^2 \, dx \right)^{1/2}. \tag{4.24}$$

4.2 Spline Lagrange Basis

Next we are interested in the introduction of $\mathrm{H}_{\mathbb{Z}}^{(2)}(\mathbb{R})$-splines. Although one dimensional $\mathbb{Z}$-periodic splines are well-understood, our purpose is to work out the essential concepts of the spline theory again as preparation for the generalization to the multi-dimensional context, thereby observing the structure of the Sobolev space

$$\mathrm{H}_{\mathbb{Z}}^{(2)}(\mathbb{R}) = \overline{\mathrm{C}_{\mathbb{Z}}^{(2)}(\mathbb{R})}^{\|\cdot\|_{\mathrm{H}_{\mathbb{Z}}^{(2)}(\mathbb{R})}}, \tag{4.25}$$

where $\|\cdot\|_{\mathrm{H}_{\mathbb{Z}}^{(2)}(\mathbb{R})}$ is given by (4.10).

Remark 4.4. *Once again, the semi-norm in* $\mathrm{H}_{\mathbb{Z}}^{(2)}(\mathbb{R})$ *given by*

$$\|\Delta F\|_{\mathrm{L}_{\mathbb{Z}}^2(\mathbb{R})} = \left(\int_{\mathcal{F}_{\mathbb{Z}}} |\Delta_x F(x)|^2 \, dx \right)^{1/2}, \quad F \in \mathrm{H}_{\mathbb{Z}}^{(2)}(\mathbb{R}), \tag{4.26}$$

may be physically interpreted (at least in linearized sense) as the bending energy of a (thin) beam, F denotes the deflection normal to the rest position supposed, of course, to be horizontal. This physical model is suggested by the classical interpretation of the potential energy of a statically deflected thin beam which indeed is proportional to the integral taken over the square of the linearized curvature of the beam.

Definition 4.5. *Suppose that $X_N = \{x_1, \ldots, x_N\} \subset \mathcal{F}_\mathbb{Z}$ is a set of mutually distinct points. Then, any function S of the form*

$$S(x) = C_0 \underbrace{\Phi_0(x)}_{=1} + \sum_{i=1}^{N} a_i G(\Delta^2; x - x_i), \quad x \in \mathbb{R}, \quad C_0 \in \mathbb{R}, \tag{4.27}$$

with

$$\sum_{i=1}^{N} a_i = 0 \tag{4.28}$$

is denoted a $\mathrm{H}_\mathbb{Z}^{(2)}(\mathbb{R})$-spline function *relative to the point set X_N. The space of all $\mathrm{H}_\mathbb{Z}^{(2)}(\mathbb{R})$-splines is designated by* $\mathrm{Spline}_{\mathrm{H}_\mathbb{Z}^{(2)}(\mathbb{R})}(X_N)$.

By virtue of Theorem 4.2 and the particular definition of a spline S we are led to

Corollary 4.6 (Spline Integration Formula). *Suppose that H is of class $\mathrm{H}_\mathbb{Z}^{(2)}(\mathbb{R})$. Let $x_1, \ldots, x_N \in \mathcal{F}_\mathbb{Z}$ be any (prescribed) set of points. Then the formula*

$$\sum_{i=1}^{N} a_i H(x_i) = \int_{\mathcal{F}_\mathbb{Z}} \Delta_z \left(C_0 \Phi_0(z) + \sum_{i=1}^{N} a_i G(\Delta^2; x_i - z) \right) \Delta_z H(z)\, dz \tag{4.29}$$

holds for any constant C_0, and any selection of weights $a_1, \ldots, a_N \in \mathbb{R}$ satisfying

$$\sum_{i=1}^{N} a_i = 0. \tag{4.30}$$

Combining the properties of Definition 4.5 and Equation (4.29) we obtain the so-called $\mathrm{H}_\mathbb{Z}^{(2)}(\mathbb{R})$*-spline integration formula.*

Lemma 4.7. *Suppose that S is an element of class* $\mathrm{Spline}_{\mathrm{H}_\mathbb{Z}^{(2)}(\mathbb{R})}(X_N)$, *then*

$$\sum_{i=1}^{N} a_i H(x_i) = \int_{\mathcal{F}_\mathbb{Z}} \Delta S(x)\, \Delta H(x)\, dx \tag{4.31}$$

holds true for all $H \in \mathrm{H}_\mathbb{Z}^{(2)}(\mathbb{R})$.

Now we want to show, that there exists a unique spline function of class $\mathrm{Spline}_{\mathrm{H}_{\mathbb{Z}}^{(2)}(\mathbb{R})}(X_N)$ which interpolates prescribed data $(x_i, y_i) \in X_N \times \mathbb{R}$, $i = 1, \ldots, N$ with $F(x_i) = y_i$, $i = 1, \ldots, N$ for some $F \in \mathrm{H}_{\mathbb{Z}}^{(2)}(\mathbb{R})$.

Theorem 4.8 (Uniqueness of Interpolation). *Given $(x_i, y_i) \in X_N \times \mathbb{R}$ with $F(x_i) = y_i, i = 1, \ldots, N$ for some $F \in \mathrm{H}_{\mathbb{Z}}^{(2)}(\mathbb{R})$. Then there exists a unique element $S \in \mathrm{Spline}_{\mathrm{H}_{\mathbb{Z}}^{(2)}(\mathbb{R})}(X_N)$ which fulfills the interpolation property*

$$S(x_i) = F(x_i) = y_i, \quad i = 1, \ldots, N. \tag{4.32}$$

Proof. Any spline S of class $\mathrm{Spline}_{\mathrm{H}_{\mathbb{Z}}^{(2)}(\mathbb{R})}(X_N)$ of the form

$$S(x) = C_0\Phi_0(x) + \sum_{i=1}^{N} a_i G(\Delta^2; x - x_i) \tag{4.33}$$

with

$$\sum_{i=1}^{N} a_i = 0, \tag{4.34}$$

contains $N+1$ unknowns. The requirements $S(x_i) = y_i$, $i = 1, \ldots, N$, provide N linear equations. Equation (4.34) gives another condition.

For the proof of the uniqueness of the interpolation property it suffices to show that the corresponding homogeneous system has just the trivial solution, i.e., $S_0(x) = 0$, $x \in \mathbb{R}$. Indeed, Lemma 4.7 tells us that for $H \in \mathrm{H}_{\mathbb{Z}}^{(2)}(\mathbb{R})$ satisfying $H(x_i) = 0, i = 1, \ldots, N$,

$$\int_{\mathcal{F}_{\mathbb{Z}}} \Delta S_0(x)\, \Delta H(x)\, dx = 0. \tag{4.35}$$

If we substitute F by S_0, we get the following identity $\int_{\mathcal{F}_{\mathbb{Z}}} |\Delta S_0(x)|^2\, dx = 0$. In other words, ΔS_0 is equal to zero for all $x \in \mathbb{R}$. Because of the periodicity and the homogeneous data, i.e., $S_0(x_i) = 0$, $i = 1, \ldots, N$, it follows that $S_0(x) = 0$, $x \in \mathbb{R}$. This proves the theorem. □

In matrix notation, the linear system occurring in spline interpolation can be reformulated as

$$\begin{pmatrix} & & & \Phi_0(x_1) \\ & \mathbf{A} & & \vdots \\ & & & \Phi_0(x_N) \\ \Phi_0(x_1) & \ldots & \Phi_0(x_N) & 0 \end{pmatrix} \begin{pmatrix} a^F \\ C_0^F \end{pmatrix} = \begin{pmatrix} y \\ 0 \end{pmatrix}, \tag{4.36}$$

where $y = (y_1, \ldots, y_N)^T$ is the given data vector, $a^F = (a_1^F, \ldots, a_N^F)^T$ is the solution vector of the linear system (4.36). Furthermore,

$$\mathbf{A} = \left(G(\Delta^2; x_i - x_j)\right)_{i,j=1,\ldots,N} \tag{4.37}$$

is a symmetric matrix of Gram type, since we have

$$G(\Delta^2; x_i - x_j) = G^{(2)}(\Delta; x_i - x_j) = \int_{\mathcal{F}_{\mathbb{Z}}} G(\Delta; x - x_i) G(\Delta; x - x_j)\, dx. \quad (4.38)$$

Note that $\Phi_0(x) = 1$ for all x. Thus, the matrix $\mathbf{A}$ in (4.37) is positive definite because of the linear independence of the functions $G(\Delta; \cdot - x_i)$, $i = 1, \ldots, N$, (see, e.g., P.J. Davis [1963]).

Now, let $(x_i, F(x_i)) \in X_N \times \mathbb{R}$ with $F(x_i) = y_i$, $i = 1, \ldots, N$, be a set of known data for some $F \in \mathrm{H}_{\mathbb{Z}}^{(2)}(\mathbb{R})$. We denote the space of all functions H of class $\mathrm{H}_{\mathbb{Z}}^{(2)}(\mathbb{R})$ interpolating the data as $\mathcal{I}_N$, i.e.,

$$\mathcal{I}_N = \{H \in \mathrm{H}_{\mathbb{Z}}^{(2)}(\mathbb{R}) : H(x_i) = F(x_i) = y_i,\ i = 1, \ldots, N, \}. \quad (4.39)$$

We are interested in the statement that the spline of class $\mathrm{Spline}_{\mathrm{H}_{\mathbb{Z}}^{(2)}(\mathbb{R})}(X_N)$ interpolating the data in the points $X_N = (x_1, \ldots, x_N)$ constitutes the "smoothest" interpolation function within the Sobolev space $\mathrm{H}_{\mathbb{Z}}^{(2)}(\mathbb{R})$.

Theorem 4.9 (First Minimum Property). *Let S_N^F denote the unique member in $\mathrm{Spline}_{\mathrm{H}_{\mathbb{Z}}^{(2)}(\mathbb{R})}(X_N)$ which fulfills the interpolation property $S(x_i) = F(x_i) = y_i$, $i = 1, \ldots, N$, i.e., S_N^F is of class $\mathcal{I}_N \cap \mathrm{Spline}_{\mathrm{H}_{\mathbb{Z}}^{(2)}(\mathbb{R})}(X_N)$. Then the interpolation problem*

$$\int_{\mathcal{F}_{\mathbb{Z}}} |\Delta S_N^F(z)|^2\, dz = \min_{H \in \mathcal{I}_N} \int_{\mathcal{F}_{\mathbb{Z}}} |\Delta H(z)|^2\, dz \quad (4.40)$$

is well-posed in the sense that its solution exists, is unique, and depends continuously on the data.

Proof. By virtue of Lemma 4.7 it follows for all $H \in \mathcal{I}_N$ that

$$\int_{\mathcal{F}_{\mathbb{Z}}} \Delta S_N^F(x) \Delta(H(x) - S_N^F(x))\, dx = \sum_{i=1}^{N} a_i^F (H(x_i) - S_N^F(x_i)) = 0, \quad (4.41)$$

where S_N^F is given by

$$S_N^F = C_0^F \Phi_0 + \sum_{i=1}^{N} a_i^F G(\Delta^2; \cdot - x_i), \quad C_0^F \in \mathbb{R}, \quad (4.42)$$

with $\sum_{i=1}^{N} a_i^F = 0$. Observing $\Delta H(x) = \Delta S_N^F(x) + \Delta(H(x) - S_N^F(x))$ we get by elementary calculations

$$\begin{aligned} \int_{\mathcal{F}_{\mathbb{Z}}} |\Delta H(x)|^2\, dx = & \int_{\mathcal{F}_{\mathbb{Z}}} |\Delta S_N^F(x)|^2\, dx + \int_{\mathcal{F}_{\mathbb{Z}}} |\Delta \left(H(x) - S_N^F(x)\right)|^2\, dx \\ & + 2 \underbrace{\int_{\mathcal{F}_{\mathbb{Z}}} \Delta S_N^F(x) \Delta(H(x) - S_N^F(x))\, dx}_{=0}. \end{aligned} \quad (4.43)$$

This provides the desired result. □

Next we come to another minimum property of spline theory.

Theorem 4.10 (Second Minimum Property). *Let F be a function of class $\mathcal{I}_N$ and S_N^F the unique spline of class $\mathcal{I}_N \cap \mathrm{Spline}_{\mathrm{H}_{\mathbb{Z}}^{(2)}(\mathbb{R})}(X_N)$ represented by (4.42). Then*

$$\int_{\mathcal{F}_{\mathbb{Z}}} |\Delta(S_N^F(z) - F(z))|^2 \, dz \tag{4.44}$$

$$= \min \left\{ \int_{\mathcal{F}_{\mathbb{Z}}} |\Delta(S(z) - F(z))|^2 \, dz : S \in \mathrm{Spline}_{\mathrm{H}_{\mathbb{Z}}^{(2)}(\mathbb{R})}(X_N) \right\}.$$

Proof. First we see that

$$\begin{aligned}
&\int_{\mathcal{F}_{\mathbb{Z}}} |\Delta(S(z) - F(z))|^2 \, dz \\
&= \int_{\mathcal{F}_{\mathbb{Z}}} \Delta(S(z) - F(z)) \, \Delta(S(z) - F(z) + S_N^F(z) - S_N^F(z)) \, dz \\
&= \int_{\mathcal{F}_{\mathbb{Z}}} \Delta(S(z) - F(z)) \, \Delta(S(z) - S_N^F(z)) \, dz \\
&\quad + \int_{\mathcal{F}_{\mathbb{Z}}} \Delta(S(z) - F(z)) \, \Delta(S_N^F(z) - F(z)) \, dz.
\end{aligned} \tag{4.45}$$

Obviously, we find by Lemma 4.7 that

$$\int_{\mathcal{F}_{\mathbb{Z}}} \Delta(S(z) - F(z)) \, \Delta(S(z) - S_N^F(z)) \, dz = \sum_{i=1}^{N} (a_i - a_i^F)(S(x_i) - F(x_i)). \tag{4.46}$$

Moreover, by Lemma 4.7, we obtain

$$\sum_{i=1}^{N} (a_i - a_i^F)(S(x_i) - S_N^F(x_i)) = \int_{\mathcal{F}_{\mathbb{Z}}} \Delta(S(z) - S_N^F(z)) \, \Delta(S(z) - S_N^F(z)) \, dz. \tag{4.47}$$

Note that $F(x_i) = S_N^F(x_i)$. Furthermore, the identity

$$\int_{\mathcal{F}_{\mathbb{Z}}} \Delta(S(z) - F(z)) \, \Delta(S_N^F(z) - F(z)) \, dz = \int_{\mathcal{F}_{\mathbb{Z}}} |\Delta(S_N^F(z) - F(z))|^2 \, dz \tag{4.48}$$

holds true because of

$$\int_{\mathcal{F}_{\mathbb{Z}}} \Delta S(z) \, \Delta(S_N^F(z) - F(z)) \, dz = \int_{\mathcal{F}_{\mathbb{Z}}} \Delta S_N^F(z) \, \Delta(S_N^F(z) - F(z)) \, dz = 0. \tag{4.49}$$

Combining the results we finally get

$$\begin{aligned}
\int_{\mathcal{F}_{\mathbb{Z}}} |\Delta(S(z) - F(z))|^2 \, dz &= \int_{\mathcal{F}_{\mathbb{Z}}} |\Delta(S(z) - S_N^F(z))|^2 \, dz \\
&\quad + \int_{\mathcal{F}_{\mathbb{Z}}} |\Delta(S_N^F(z) - F(z))|^2 \, dz.
\end{aligned} \tag{4.50}$$

This guarantees the result of Theorem 4.10. □

Next we are interested in a result that can be regarded as a spline variant of the *Shannon sampling theorem.*

Theorem 4.11 (Sampling Theorem in $\mathrm{Spline}_{\mathrm{H}_{\mathbb{Z}}^{(2)}(\mathbb{R})}(X_N)$). *Any member* $F \in \mathrm{Spline}_{\mathrm{H}_{\mathbb{Z}}^{(2)}(\mathbb{R})}(X_N)$ *can be represented from its samples at points of* $X_N = \{x_1, \ldots, x_N\} \subset \mathcal{F}_{\mathbb{Z}}$ *by the spline interpolation formula*

$$F(x) = \sum_{k=1}^{N} F(x_k) L_k(x), \quad x \in \mathbb{R}, \tag{4.51}$$

where the Lagrange basis functions $L_k \in \mathrm{Spline}_{\mathrm{H}_{\mathbb{Z}}^{(2)}(\mathbb{R})}(X_N)$, $k = 1, \ldots, N$, *are given by*

$$L_k(x) = C_0^k + \sum_{i=1}^{N} a_i^k G(\Delta^2; x_i - x) \tag{4.52}$$

and the coefficients C_0^k, a_i^k *satisfy the linear equations for* $i, k = 1, \ldots, N$:

$$\sum_{l=1}^{N} a_l^k G(\Delta^2, x_i - x_l) + C_0^k = \delta_{i,k}, \tag{4.53}$$

$$\sum_{l=1}^{N} a_l^k = 0. \tag{4.54}$$

Proof. As an element of $\mathrm{Spline}_{\mathrm{H}_{\mathbb{Z}}^{(2)}(\mathbb{R})}(X_N)$, the function F is uniquely determined by

$$F(x) = C_0 + \sum_{l=1}^{N} a_l G(\Delta^2; x - x_l), \quad x \in \mathbb{R}, \tag{4.55}$$

such that

$$\sum_{l=1}^{N} a_l G(\Delta^2, x_i - x_l) + C_0 = F(x_i), \qquad i = 1, \ldots, N, \tag{4.56}$$

$$\sum_{l=1}^{N} a_l = 0. \tag{4.57}$$

Now we let

$$\sum_{k=1}^{N} F(x_k) L_k(x) = C_0 + \sum_{l=1}^{N} a_l G(\Delta^2; x - x_l), \tag{4.58}$$

where

$$C_0 = \sum_{k=1}^{N} C_0^k F(x_k) \tag{4.59}$$

and

$$a_l = \sum_{k=1}^{N} a_l^k F(x_k). \tag{4.60}$$

Then, in connection with (4.53), it follows that

$$\begin{aligned} C_0 + \sum_{l=1}^{N} a_l G(\Delta^2; x_i - x_l) &= \sum_{k=1}^{N} F(x_k) \left(C_0^k + \sum_{l=1}^{N} a_l^k G(\Delta^2; x_i - x_l) \right) \\ &= F(x_i) \end{aligned} \tag{4.61}$$

and

$$\sum_{l=1}^{N} a_l C_0 = \sum_{k=1}^{N} F(x_k) \sum_{l=1}^{N} a_l^k C_0 = 0. \tag{4.62}$$

This proves Theorem 4.11. □

Remark 4.12. *All in all, the key point of departure for both methods, namely, spline interpolation and best approximate integration, is the same integral formula (derived from Theorem 4.14)*

$$\sum_{k=1}^{N} a_k F(x_k) = \sum_{k=1}^{N} a_k \int_{\mathcal{F}_{\mathbb{Z}}} F(z)\, dz + \int_{\mathcal{F}_{\mathbb{Z}}} \Delta_z \sum_{k=1}^{N} a_k G(\Delta^2; x_k - z)\, \Delta_z F(z)\, dz. \tag{4.63}$$

Whereas best approximate integration is characterized by a selection of coefficients $a_1, \ldots, a_N$ under the constraint $\sum_{k=1}^{N} a_k = 1$, i.e.,

$$\sum_{k=1}^{N} a_k F(x_k) = \int_{\mathcal{F}_{\mathbb{Z}}} F(z)\, dz + \int_{\mathcal{F}_{\mathbb{Z}}} \Delta_z \sum_{k=1}^{N} a_k G(\Delta^2; x_k - z)\, \Delta_z F(z)\, dz \tag{4.64}$$

spline interpolation is based on selections $a_1, \ldots, a_N$ with $\sum_{k=1}^{N} a_k = 0$, i.e.,

$$\sum_{k=1}^{N} a_k F(x_k) = \int_{\mathcal{F}_{\mathbb{Z}}} \Delta_z \sum_{k=1}^{N} a_k G(\Delta^2; x_k - z)\, \Delta_z F(z)\, dz \tag{4.65}$$

Already these formal observation explains why both methods show an intrinsic relationship, that allows a more detailed concretization in the next section.

4.3 Peano's Theorem

Our purpose is to give a more detailed investigation of the relationship between best approximation and spline interpolation. In fact, it is shown that the calculation of interpolating splines is strongly related to the best approximation of bounded linear functionals $\mathcal{I}$ on $\mathrm{H}_{\mathbb{Z}}^{(2)}(\mathbb{R})$, i.e., linear functionals $\mathcal{I}$ such that there exists a constant M with

$$|\mathcal{I}F| \leq M \; \|F\|_{\mathrm{H}_{\mathbb{Z}}^{(2)}(\mathbb{R})} \tag{4.66}$$

for all $F \in \mathrm{H}_{\mathbb{Z}}^{(2)}(\mathbb{R})$.

As particular examples of *bounded linear functionals on* $\mathrm{H}_{\mathbb{Z}}^{(2)}(\mathbb{R})$, we mention

(i) $\mathcal{I}F = \int_{\mathcal{F}_{\mathbb{Z}}} F(t)\; dt$,

(ii) $\mathcal{I}F = \int_{\mathcal{G}} F(t)\; dt, \quad \mathcal{G} \subset \mathcal{F}_{\mathbb{Z}}$,

(iii) $\mathcal{I}F = F(x), \quad x \in \mathcal{F}_{\mathbb{Z}}$,

where F is of class $\mathrm{H}_{\mathbb{Z}}^{(2)}(\mathbb{R})$.

Theorem 4.13 (Peano's Theorem). *Let $\mathcal{I}$ be a bounded linear functional on $\mathrm{H}_{\mathbb{Z}}^{(2)}(\mathbb{R})$ such that $\mathcal{I}\Phi_0 = 0$. Then, for all $F \in \mathrm{H}_{\mathbb{Z}}^{(2)}(\mathbb{R})$,*

$$\mathcal{I}F = \int_{\mathcal{F}_{\mathbb{Z}}} K(x)\; \Delta F(x)\; dx, \tag{4.67}$$

where

$$K(x) = \mathcal{I}_x \Delta_y G(\Delta^2; x - y) \tag{4.68}$$

(note that $\mathcal{I}_x$ means that the linear functional is applied to the x variable in $G(\Delta^2; x-y)$. The function $K(x)$ is called Peano kernel *for the linear functional $\mathcal{I}$).*

Proof. We start form the integral formula

$$F(x) = \int_{\mathcal{F}_{\mathbb{Z}}} F(y)\; dy + \int_{\mathcal{F}_{\mathbb{Z}}} G(\Delta; x-y)\; \Delta F(y)\; dy. \tag{4.69}$$

Applying the linear functional to both sides of the equation yields

$$\mathcal{I}_x F(x) = \mathcal{I}_x \int_{\mathcal{F}_{\mathbb{Z}}} \Delta_y G(\Delta^2; x-y)\; \Delta_y F(y)\; dy. \tag{4.70}$$

Now, we are allowed to interchange the linear functional and the integration. This leads us to

$$\mathcal{I}_x F(x) = \int_{\mathcal{F}_{\mathbb{Z}}} \mathcal{I}_x \Delta_y G(\Delta^2; x-y)\; \Delta_y F(y)\; dy, \tag{4.71}$$

which is the wanted result. □

Next we approximate a bounded linear functional $\mathcal{I}$ by a linear combination $\mathcal{L}$ given by

$$\mathcal{L}F = \sum_{i=1}^{N} a_i F(x_i) \tag{4.72}$$

with F of class $\mathrm{H}_{\mathbb{Z}}^{(2)}(\mathbb{R})$, $x_i \in X_N$ and $a_i \in \mathbb{R}$ for $i = 1, \ldots, N$.

The functional is called *exact for the degree* 0, if $\mathcal{I}\Phi_0 = \mathcal{L}\Phi_0$. Using $\mathcal{L}$ to approximate $\mathcal{I}$ the remainder term $\mathcal{E}$ is defined by

$$\mathcal{E} = \mathcal{I} - \mathcal{L}. \tag{4.73}$$

If the approximation is exact for all polynomials of degree 0 and $F \in \mathrm{H}_{\mathbb{Z}}^{(2)}(\mathbb{R})$, then $\mathcal{E}$ satisfies the assumptions of Peano's Theorem. Thus we obtain

$$\mathcal{E}F = \int_{\mathcal{F}_{\mathbb{Z}}} K(y)\Delta F(y)\, dy \tag{4.74}$$

where

$$K(y) = \mathcal{E}_x \Delta_y G(\Delta^2; x - y). \tag{4.75}$$

Applying the Schwarz inequality to Equation (4.74) we are led to

$$|\mathcal{E}F| \leq \left(\int_{\mathcal{F}_{\mathbb{Z}}} |K(y)|^2\, dy \right)^{1/2} \left(\int_{\mathcal{F}_{\mathbb{Z}}} |\Delta F(y)|^2\, dy \right)^{1/2}, \tag{4.76}$$

where the equality holds true if the functions K and F are linearly dependent. Now, we specify the quantity

$$J = \int_{\mathcal{F}_{\mathbb{Z}}} |K(z)|^2\, dz = \mathcal{E}_x \mathcal{E}_y \int_{\mathcal{F}_{\mathbb{Z}}} \Delta_z G(\Delta^2; x - z) \Delta_z G(\Delta^2; y - z)\, dz. \tag{4.77}$$

From the properties of the lattice function we obtain

$$J = \mathcal{E}_x \mathcal{E}_y G(\Delta^2; x - y). \tag{4.78}$$

Using the definition of the linear functional $\mathcal{E}$ we end up with

Lemma 4.14.

$$J = \mathcal{I}_x \mathcal{I}_y G(\Delta^2; x - y) - 2\mathcal{I}_x \mathcal{L}_y G(\Delta^2; x - y) + \mathcal{L}_x \mathcal{L}_y G(\Delta^2; x - y). \tag{4.79}$$

Collecting our results we obtain the *a priori estimate*

Theorem 4.15. *Let $\mathcal{I}$ be a bounded linear function on $\mathrm{H}_{\mathbb{Z}}^{(2)}(\mathbb{R})$, and $\mathcal{L}$ an approximation to $\mathcal{I}$ of the form (4.72), with $\mathcal{L}$ being exact for the degree zero. If F is of class $\mathrm{H}_{\mathbb{Z}}^{(2)}(\mathbb{R})$, then it follows that*

$$|\mathcal{E}F| \leq \left(\left(\mathcal{I}_x \mathcal{I}_y - 2\mathcal{I}_x \mathcal{L}_y + \mathcal{L}_x \mathcal{L}_y \right) G(\Delta^2; x - y) \right)^{1/2} \left(\int_{\mathcal{F}_{\mathbb{Z}}} |\Delta F(y)|^2\, dy \right)^{1/2}. \tag{4.80}$$

Remembering the definition of the linear functional $\mathcal{L}$ (4.72) we obtain from Theorem 4.15

Theorem 4.16. *Let F of class $\mathrm{H}_{\mathbb{Z}}^{(2)}(\mathbb{R})$ and X_N be a point set of mutually distinct points in $\mathcal{F}_{\mathbb{Z}}$. Then the estimate*

$$\left|\mathcal{I}F - \sum_{i=1}^{N} a_i F(x_i)\right| \tag{4.81}$$
$$\leq \left(\mathcal{I}_x\mathcal{I}_y G(\Delta^2; x-y) - 2\sum_{i=1}^{N} a_i \mathcal{I}_x G(\Delta^2; x-x_i)\right.$$
$$\left. + \sum_{i=1}^{N}\sum_{k=1}^{N} a_i a_k G(\Delta^2; x_i - x_k)\right)^{1/2} \left(\int_{\mathcal{F}_{\mathbb{Z}}} |\Delta F(y)|^2 \, dy\right)^{1/2}$$

holds for all choices $a_1, \ldots, a_N \in \mathbb{R}$ such that

$$\sum_{i=1}^{N} a_i \Phi_0(x_i) = \mathcal{I}\Phi_0. \tag{4.82}$$

Theorem 4.15 and Theorem 4.16 provide an estimate for the remainder term of the approximation $\mathcal{L}$ to the linear functional $\mathcal{I}$. To establish the best approximation to $\mathcal{I}$, we have to minimize the error of the approximation, and therefore the remainder. As the term $\int_{\mathcal{F}_{\mathbb{Z}}} |F(y)|^2 \, dy$ is not dependent on the linear functionals $\mathcal{L}$ and $\mathcal{I}$, we are led to the following quadratic optimization problem

$$J = \mathcal{I}_x\mathcal{I}_y G(\Delta^2; x-y) - 2\sum_{i=1}^{N} a_i \mathcal{I}_x G(\Delta^2; x-x_i) \tag{4.83}$$
$$+ \sum_{i=1}^{N}\sum_{k=1}^{N} a_i a_k G(\Delta^2; x_i - x_k) \qquad \longrightarrow \min$$

under the constraint that

$$\sum_{i=1}^{N} a_i \Phi_0(x_i) = \sum_{i=1}^{N} a_i = \mathcal{I}\Phi_0. \tag{4.84}$$

According to Lagrange method of multipliers applied to (4.83) we find that

$$\frac{\partial}{\partial a_l}\left(\frac{1}{2}\mathcal{I}_x\mathcal{I}_y G(\Delta^2; x-y) - \sum_{i=1}^{N} a_i \mathcal{I}_x G(\Delta^2; x-x_i)\right. \tag{4.85}$$
$$\left. + \frac{1}{2}\sum_{i=1}^{N}\sum_{k=1}^{N} a_i a_k G(\Delta^2; x_i - x_k) - \lambda \sum_{i=1}^{N} (a_i \Phi_0(x_i) - \mathcal{I}\Phi_0)\right) = 0$$

for all $l = 1, \ldots, N$ and

$$\underbrace{\frac{\partial}{\partial\lambda}\left(\frac{1}{2}\mathcal{I}_x\mathcal{I}_y G(\Delta^2; x-y) - \ldots\right)}_{=0} - \frac{\partial}{\partial\lambda}\left(\lambda\sum_{i=1}^{N}(a_i\Phi_0(x_i) - \mathcal{I}\Phi_0)\right) = 0. \quad (4.86)$$

This leads us to the $(N+1)\times(N+1)$ equation system

$$\mathbf{A}a - \begin{pmatrix}\lambda\\ \vdots\\ \lambda\end{pmatrix} = \begin{pmatrix}\mathcal{I}_x G(\Delta^2; x-x_1)\\ \vdots\\ \mathcal{I}_x G(\Delta^2; x-x_N)\end{pmatrix}, \quad (4.87)$$

$$\sum_{i=1}^{N} a_i\Phi_0(x_i) = \mathcal{I}\Phi_0.$$

As already known, the matrix $\mathbf{A}$ of (4.37) is non-singular. Therefore, we get the solution of the linear system as

$$a = \mathbf{A}^{-1}\left(\begin{pmatrix}\mathcal{I}_x G(\Delta^2; x-x_1)\\ \vdots\\ \mathcal{I}_x G(\Delta^2; x-x_N)\end{pmatrix} + \lambda d\right) \quad (4.88)$$

with $d = (\Phi_0(x_1), \ldots, \Phi_0(x_N))^T = (1, \ldots, 1)^T \in \mathbb{R}^N$ the vector with all coefficients equal to one and the parameter λ given by

$$\lambda = \left(d^T \cdot \mathbf{A}^{-1}d\right)^{-1}\left(\mathcal{I}\Phi_0 - d^T\cdot\mathbf{A}^{-1}\begin{pmatrix}\mathcal{I}_x G(\Delta^2; x-x_1)\\ \vdots\\ \mathcal{I}_x G(\Delta^2; x-x_N)\end{pmatrix}\right). \quad (4.89)$$

The vector $a = (a_1, \ldots, a_N)^T$ in Equation (4.88) is the only solution, which minimizes the problem.

Theorem 4.17. *Let $\mathcal{I}$ be a bounded linear functional on $\mathrm{H}_{\mathbb{Z}}^{(2)}(\mathbb{R})$. Suppose that $a \in \mathbb{R}^N$, $\lambda \in \mathbb{R}$, be the solution of the linear system (4.87). Then, for every function $F \in \mathrm{H}_{\mathbb{Z}}^{(2)}(\mathbb{R})$, the linear functional $\mathcal{L}$ given by*

$$\mathcal{L}F = \sum_{i=1}^{N} a_i F(x_i) \quad (4.90)$$

represents the best approximation to $\mathcal{I}$ (in the $\mathrm{H}_{\mathbb{Z}}^{(2)}(\mathbb{R})$-sense). For the approximate formula

$$\mathcal{I}F = \sum_{i=1}^{N} a_i F(x_i) + \mathcal{E}F \quad (4.91)$$

the error estimate for $\mathcal{E}F$ is given by

$$|\mathcal{E}F| = \left| \mathcal{I}F - \sum_{i=1}^{N} a_i F(x_i) \right| \tag{4.92}$$
$$\leq \left(\mathcal{I}_x \mathcal{I}_y G(\Delta^2; x-y) - \sum_{i=1}^{N} a_i \mathcal{I}_x G(\Delta^2; x-x_i) + \lambda \mathcal{I}\Phi_0 \right)^{1/2}$$
$$\times \left(\int_{\mathcal{F}_{\mathbb{Z}}} (\Delta F(y))^2 \, dy \right)^{1/2} .$$

4.4 Best Approximation and Spline Exact Formulas

For a given linear bounded functional $\mathcal{I}$ on $\mathrm{H}_{\mathbb{Z}}^{(2)}(\mathbb{R})$ and X_N being a system of mutually distinct points, the approximation $\mathcal{L}$ of the form (4.72) is not uniquely determined with the condition, that $\mathcal{L}$ is exact for constant functions. However, it can be shown that the linear functional $\mathcal{L}$ is uniquely determined by the stronger requirement $\mathcal{I}S = \mathcal{L}S$ whenever $S \in \mathrm{Spline}_{\mathrm{H}_{\mathbb{Z}}^{(2)}(\mathbb{R})}(X_N)$.

Theorem 4.18. *Let $X_N \subset \mathcal{F}_{\mathbb{Z}}$ be a system mutually distinct points and let $\mathcal{I}$ be a bounded linear functional on $\mathrm{H}_{\mathbb{Z}}^{(2)}(\mathbb{R})$. Then there exists a unique linear functional $\mathcal{L}$ of the form (4.72) satisfying the properties*

(i) $\mathcal{I}\Phi_0 = \mathcal{L}\Phi_0$,

(ii) there exists a unique constant $\tilde{C}$ such that

$$\mathcal{I}_x G(\Delta^2; x-x_i) - \mathcal{L}_x G(\Delta^2; x-x_i) = \tilde{C} \tag{4.93}$$

for $x_i \in X_N$, $i = 1, \ldots, N$.

For a given function $F \in \mathrm{H}_{\mathbb{Z}}^{(2)}(\mathbb{R})$, let S_N^F be the unique spline function in $\mathrm{H}_{\mathbb{Z}}^{(2)}(\mathbb{R})$, which interpolates the data points $(x_i, F(x_i))_{i=1,\ldots,N}$. Then $\mathcal{L}$ is also uniquely determined by the requirement that $\mathcal{L}F = \mathcal{I}S_N^F$ for every $F \in \mathrm{H}_{\mathbb{Z}}^{(2)}(\mathbb{R})$.

Finally, $\mathcal{I}$ is uniquely determined by the property that $\mathcal{I}S = \mathcal{L}S$ for all $S \in \mathrm{Spline}_{\mathrm{H}_{\mathbb{Z}}^{(2)}(\mathbb{R})}(X_N)$.

Proof. If $\mathcal{L}$ is a linear functional on $\mathrm{H}_{\mathbb{Z}}^{(2)}(\mathbb{R})$, the conditions (i),(ii) as stated in the first part of Theorem 4.18 are equivalent to the following linear system

$$\sum_{i=1}^{N} a_i G(\Delta^2; x_j - x_i) + C_0 = \mathcal{I}_x G(\Delta^2; x - x_j) \tag{4.94}$$

and

$$\sum_{i=1}^{N} a_i \Phi_0(x_i) = \mathcal{I}\Phi_0 \tag{4.95}$$

for determining the constant C_0 and the coefficients $a_1, \ldots, a_N$. But this equation system uses the same matrix as in Equation (4.36), which is known to be non-singular. Thus, C_0 and $a = (a_1, \ldots, a_N)^T$ are uniquely determined by (4.94) and (4.95).

Next we prove that the solution of this equation system implies

$$\mathcal{I}S = \mathcal{L}S \tag{4.96}$$

for all $S \in \mathrm{Spline}_{\mathrm{H}_{\mathbb{Z}}^{(2)}(\mathbb{R})}(X_N)$. As we know, S fulfills

$$S(x) = \sum_{i=1}^{N} b_i G(\Delta^2; x - x_i) + C_1 \Phi_0(x) \tag{4.97}$$

and

$$\sum_{i=1}^{N} b_i = 0 \tag{4.98}$$

with $x_i \in X_N$. Therefore,

$$\mathcal{I}S = \underbrace{C_1 \mathcal{I}\Phi_0}_{=C_1 \mathcal{L}\Phi_0} + \sum_{i=1}^{N} b_i \mathcal{I}_x G(\Delta^2; x - x_i) \tag{4.99}$$

holds. Using (4.94) we get

$$\mathcal{I}S = C_1 \mathcal{L}\Phi_0(x) + \sum_{i=1}^{N} b_i \left(\sum_{j=1}^{N} a_j G(\Delta^2; x_i - x_j) + C \right). \tag{4.100}$$

Exchanging the summation over i and j we find that

$$\begin{aligned} \mathcal{I}S &= C_1 \mathcal{L}\Phi_0(x) + \sum_{j=1}^{N} a_j \sum_{i=1}^{N} b_i G(\Delta^2; x_i - x_j) + \underbrace{\sum_{i=1}^{N} b_i C \Phi_0(x_i)}_{=0} \\ &= \mathcal{L}\left(C_1 + \sum_{i=1}^{N} b_i G(\Delta^2; x - x_i) \right) = \mathcal{L}S. \end{aligned} \tag{4.101}$$

Now, we take a linear functional $\mathcal{L}'$ of the form (4.72), having the property

$$\mathcal{L}'S = \mathcal{I}S, \qquad S \in \mathrm{Spline}_{\mathrm{H}_{\mathbb{Z}}^{(2)}(\mathbb{R})}(X_N). \tag{4.102}$$

Then it follows that

$$\mathcal{L}'S_N^F = \mathcal{I}S_N^F \tag{4.103}$$

with $S_N^F, F \in \mathrm{H}_{\mathbb{Z}}^{(2)}(\mathbb{R})$ satisfying $S_N^F(x_i) = F(x_i), i = 1, \ldots, N$. Therefore, S_N^F is the unique interpolating spline of F. This implies

$$\mathcal{L}' S_N^F = \mathcal{L}' F. \tag{4.104}$$

Combining the solutions we are led to

$$\mathcal{I} S_N^F = \mathcal{L}' F. \tag{4.105}$$

As we know, $\mathcal{I} S_N = \mathcal{L} F$ for all $F \in \mathrm{H}_{\mathbb{Z}}^{(2)}(\mathbb{R})$ interpolating the data. Therefore, we obtain

$$\mathcal{L} F = \mathcal{I} S_N^F = \mathcal{L}' F \tag{4.106}$$

whenever $F \in \mathrm{H}_{\mathbb{Z}}^{(2)}(\mathbb{R})$, i.e.,

$$\mathcal{L} = \mathcal{L}' \tag{4.107}$$

on $\mathrm{H}_{\mathbb{Z}}^{(2)}(\mathbb{R})$. This proves that $\mathcal{L}$ is uniquely determined either by the assumption, that $\mathcal{I} S = \mathcal{L} S$ for all $S \in \mathrm{Spline}_{\mathrm{H}_{\mathbb{Z}}^{(2)}(\mathbb{R})}(X_N)$ or by the property that the identity

$$\mathcal{L} F = \mathcal{I} S_N^F \tag{4.108}$$

is valid for all F of class $\mathrm{H}_{\mathbb{Z}}^{(2)}(\mathbb{R})$. □

Next we are interested in proving that the best approximation $\mathcal{L}$ as discussed in Theorem 4.18, exact for constants, to the linear functional $\mathcal{I}$ is the unique approximation that is exact for functions $S \in \mathrm{Spline}_{\mathrm{H}_{\mathbb{Z}}^{(2)}(\mathbb{R})}(X_N)$.

Theorem 4.19 (Schoenberg's Theorem). *Suppose that $\mathcal{I}$ is a bounded linear functional on $\mathrm{H}_{\mathbb{Z}}^{(2)}(\mathbb{R})$. Let $\mathcal{L}'$ be an approximation to $\mathcal{I}$ of the form (4.72) exact for constants and let $\mathcal{L}$ be the linear functional fulfilling the conditions of Theorem 4.18. Let J (respectively J') be the quantities given by (4.77) corresponding to $\mathcal{E}$ or $\mathcal{E}'$. Then the estimate*

$$J \leq J' \tag{4.109}$$

holds with equality if and only if $\mathcal{L} = \mathcal{L}'$.

Proof. Let $\mathcal{L}''$ be the functional given by

$$\mathcal{L}'' = \mathcal{L} - \mathcal{L}'. \tag{4.110}$$

From Theorem 4.18 we get

$$\mathcal{L} C_0 = \mathcal{I} C_0 = \mathcal{L}' C_0 = 0 \tag{4.111}$$

with C_0 being an arbitrary constant. This means, that the linear functional $\mathcal{L}''$ fulfills the conditions of Peano's theorem. From Peano's Theorem it follows

$$\mathcal{L}'' F = \int_{\mathcal{F}_{\mathbb{Z}}} K_2(x)\, \Delta F(x)\, dx, \tag{4.112}$$

for every F of class $\mathrm{H}_{\mathbb{Z}}^{(2)}(\mathbb{R})$. The kernel K_2 is then defined by

$$K_2(x) = \mathcal{L}''_x \Delta_x G(\Delta^2; x - y). \tag{4.113}$$

As $\mathcal{L}''$ is the difference between $\mathcal{L}$ and $\mathcal{L}'$ it is also of the form (4.72). Therefore, we obtain the representation

$$K_2(x) = \sum_{i=1}^{N} a_i \Delta_x G(\Delta^2; x - x_i) \tag{4.114}$$

with $a_1, \ldots, a_N \in \mathbb{R}$ fulfilling $\sum_{i=1}^{N} a_i = 0$. Considering the functionals $\mathcal{E} = \mathcal{I} - \mathcal{L}$ and $\mathcal{E}' = \mathcal{I} - \mathcal{L}'$, we get

$$\mathcal{L}'' = (\mathcal{I} - \mathcal{E}) - (\mathcal{I} - \mathcal{E}') = \mathcal{E}' - \mathcal{E}. \tag{4.115}$$

Let now K and K_1 be the Peano Kernel with respect to the linear functionals $\mathcal{E}$ respectively $\mathcal{E}'$. This yields following representation

$$K_2 = K_1 - K. \tag{4.116}$$

Combining Equation (4.77) and Equation (4.114) we are able to conclude that

$$\int_{\mathcal{F}_{\mathbb{Z}}} K(x) K_2(x)\, dx = \mathcal{E}_x \sum_{i=1}^{N} a_i G(\Delta^2; x - x_i). \tag{4.117}$$

Thus, by Theorem 4.18 we see that

$$\int_{\mathcal{F}_{\mathbb{Z}}} K(x) K_2(x)\, dx = 0. \tag{4.118}$$

From Equation (4.116) and (4.118) we know that

$$\int_{\mathcal{F}_{\mathbb{Z}}} |K_1(x)|^2\, dx = \int_{\mathcal{F}_{\mathbb{Z}}} |K_2(x)|^2\, dx + \int_{\mathcal{F}_{\mathbb{Z}}} |K(x)|^2\, dx. \tag{4.119}$$

But this means that $J \leq J'$ is true, with equality only if

$$\int_{\mathcal{F}_{\mathbb{Z}}} |K_2(x)|^2\, dx = \sum_{i=1}^{N} \sum_{k=1}^{N} a_i a_k G(\Delta^2; x_i - x_k) = 0. \tag{4.120}$$

As the matrix $\mathbf{A} = \left(G(\Delta^2; x_i - x_k)\right)_{i,k=1,\ldots,N}$ is positive definite (see also (4.37)) it follows that $a_1 = \ldots = a_N = 0$. This proves the theorem. □

4.5 Smoothing Splines for Erroneous Data Points

Up to now we have constructed spline functions which interpolate data points. In practical applications the measurements will be affected by errors or irregularities. Therefore, the interpolation of these data points will not lead to good results. What we need are splines, which smooth the data (cf. C. Reinsch [1967]). For example, we seek a function F of class $\mathrm{H}_{\mathbb{Z}}^{(2)}(\mathbb{R})$ which minimizes the quantity

$$\sigma_{\beta,\delta}(F) = \sum_{i=1}^{N} \left(\frac{F(x_i) - y_i}{\beta_i} \right)^2 + \delta \int_{\mathcal{F}_{\mathbb{Z}}} |\Delta F(z)|^2 \, dz \tag{4.121}$$

in $\mathrm{H}_{\mathbb{Z}}^{(2)}(\mathbb{R})$ with $(x_i, y_i)_{i=1,\ldots,N} \in \mathcal{F}_{\mathbb{Z}} \times \mathbb{R}$ being the given set of measurements. The set of parameters $\beta_1, \ldots, \beta_N \in \mathbb{R}$ are understood to be positively chosen weights, which should be fitted to the standard deviation of the measurements and δ is a positive constant chosen by the user. The constant δ can be used to influence the smoothness of F. Using a small value of δ emphasizes the measured data points at the expense of smoothness, while a large value does the opposite. Setting δ to zero leads back to the interpolation problem.

Theorem 4.20. *Let δ be a positive constant and $\beta = (\beta_1, \ldots, \beta_N)^T$ some positive coefficients. Then there exists a spline function $S \in \mathrm{Spline}_{\mathrm{H}_{\mathbb{Z}}^{(2)}(\mathbb{R})}(X_N)$, which is unique and fulfills the identity*

$$\sigma_{\beta,\delta}(S) \leq \sigma_{\beta,\delta}(F) \tag{4.122}$$

for all $F \in \mathrm{H}_{\mathbb{Z}}^{(2)}(\mathbb{R})$. If S has the form of Equation (4.27), S fulfills the linear system

$$S(x_i) + \delta \beta_i^2 a_i^S = y_i, \quad i = 1, \ldots, N, \tag{4.123}$$

with $(x_i, y_i)_{i=1,\ldots,N}$ constituting the point set which represents the data under consideration.

Proof. Let S be an $\mathbb{Z}$-periodic spline fulfilling the conditions

$$y_i - S(x_i) = \delta \beta_i^2 a_i^S \tag{4.124}$$

for $i = 1, \ldots, N$. Suppose that F is of class $\mathrm{H}_{\mathbb{Z}}^{(2)}(\mathbb{R})$. Multiplying this with $F(x_i)$, $i = 1, \ldots, N$ and summing over i we obtain

$$\sum_{i=1}^{N} F(x_i) \frac{y_i - S(x_i)}{\beta_i^2} = \delta \sum_{i=1}^{N} a_i^S F(x_i). \tag{4.125}$$

With Lemma 4.7 we therefore find that

$$\sum_{i=1}^{N} F(x_i) \frac{y_i - S(x_i)}{\beta_i^2} = \delta \int_{\mathcal{F}_{\mathbb{Z}}} \Delta S(x) \Delta F(x) \, dx. \tag{4.126}$$

To prove that the $(N+1) \times (N+1)$ linear system is uniquely solvable, we have to verify that the homogeneous system

$$\sum_{i=1}^{N} a_i G(\Delta^2; x_i - x_k) + \delta \beta_k^2 a_k + C = 0, \quad k = 1, \ldots, N \tag{4.127}$$

and

$$\sum_{i=1}^{N} a_i = 0 \tag{4.128}$$

has just the trivial solution. To be able to write this equation system in matrix-vector multiplication we additionally introduce the diagonal matrix

$$\mathbf{B} = \begin{pmatrix} \beta_1^2 & 0 & \cdots & 0 \\ 0 & \beta_2^2 & \cdots & 0 \\ \vdots & & \ddots & \vdots \\ 0 & 0 & \cdots & \beta_N^2 \end{pmatrix}. \tag{4.129}$$

Then we can write the linear system by aid of $\mathbf{A}$ as introduced in (4.37) in the form

$$\begin{pmatrix} 0 \\ \vdots \\ \vdots \\ 0 \end{pmatrix} = \begin{pmatrix} & \mathbf{A} & & \Phi_0(x_1) \\ & & & \vdots \\ & & & \Phi_0(x_N) \\ \Phi_0(x_1) & \cdots & \Phi_0(x_N) & 0 \end{pmatrix} \begin{pmatrix} a_1 \\ \vdots \\ a_N \\ C \end{pmatrix} + \delta \begin{pmatrix} \mathbf{B} & 0 \\ 0 & 0 \end{pmatrix} \begin{pmatrix} a_1 \\ \vdots \\ a_N \\ C \end{pmatrix}. \tag{4.130}$$

Let $S_0 \in \mathrm{Spline}_{\mathrm{H}_{\mathbb{Z}}^{(2)}(\mathbb{R})}(X_N)$ be the spline function, which fulfills (4.130), then from $y_i = 0$ for all $i = 1, \ldots, N$ it follows by virtue of (4.126) that

$$\sum_{i=1}^{N} \frac{|S_0(x_i)|^2}{\beta_i} + \delta \int_{\mathcal{F}_{\mathbb{Z}}} |\Delta S_0(x)|^2 \, dx = 0. \tag{4.131}$$

Since S_0, δ and all summands are positive, this equation is just fulfilled, if

$$\Delta S_0(x) = 0, \quad x \in \mathbb{R} \tag{4.132}$$

and

$$S_0(x_i) = 0, \quad i = 1, \ldots, N. \tag{4.133}$$

But this means, that S_0 is identically zero in $\mathbb{R}$.

Observing (4.121) we find that

$$\sigma_{\beta,\delta}(F) = \sum_{i=1}^{N} \frac{|F(x_i) - S(x_i) + S(x_i) - y_i|^2}{\beta_i^2} \tag{4.134}$$

$$+ \delta \int_{\mathcal{F}_{\mathbb{Z}}} |\Delta(F(x) - S(x) + S(x))|^2 \, dx.$$

Using the binomial formula we therefore obtain

$$\sigma_{\beta,\delta}(F) = \underbrace{\sum_{i=1}^{N} \left| \frac{S(x_i) - y_i}{\beta_i} \right|^2 + \delta \int_{\mathcal{F}_{\mathbb{Z}}} |\Delta S(x)|^2 \, dx}_{\sigma_{\beta,\delta}(S)} \tag{4.135}$$

$$+ \sum_{i=1}^{N} \left| \frac{F(x_i) - S(x_i)}{\beta_i} \right|^2 + \delta \int_{\mathcal{F}_{\mathbb{Z}}} |\Delta(F(x) - S(x))|^2 \, dx.$$

Substituting $F(x_i)$ by $(F(x_i) - S(x_i))$ in (4.126) we have

$$-\sum_{i=1}^{N} \frac{(F(x_i) - S(x_i))(S(x_i) - y_i)}{\beta_i^2} = \delta \int_{\mathcal{F}_{\mathbb{Z}}} (\Delta(F(x) - S(x)))\Delta S(x) \, dx. \tag{4.136}$$

Combining all results we are led to

$$\sigma_{\beta,\delta}(F) = \sigma_{\beta,\delta}(S) + \sum_{i=1}^{N} \left| \frac{F(x_i) - S(x_i)}{\beta_i} \right|^2 + \delta \int_{\mathcal{F}_{\mathbb{Z}}} |\Delta(F(x) - S(x))|^2 \, dx. \tag{4.137}$$

Therefore it follows that $\sigma_{\beta,\delta}(F) \geq \sigma_{\beta,\delta}(S)$. Since we know that the spline function fulfilling the smoothest property is unique, we have the desired result. □

Exactness of Degree m**.** Until now, for simplicity, we were concerned with the *exactness degree* 0 for periodic polynomials. However, it should be mentioned that our considerations also open the perspective of guaranteeing periodic polynomials exact integration rules of degree $m \in \mathbb{N}_0$. In order to demonstrate this assertion we go over to the *$\mathbb{Z}$-lattice function for Δ of the order $m \in \mathbb{N}_0$* given by

$$G^{\perp}_{0,\ldots,m}(\Delta; x - y) = G(\Delta; x - y) - \sum_{\substack{0<|h|\leq m \\ h\in\mathbb{Z}}} \frac{1}{-\Delta^\wedge(h)} \Phi_h(x - y), \quad m \in \mathbb{N}, \tag{4.138}$$

$$G^{\perp}_{0,\ldots,m}(\Delta; x - y) = G(\Delta; x - y), \quad m = 0. \tag{4.139}$$

Remark 4.21. *In* (4.138) *we can also choose any finite set* $\Xi \subset \mathbb{Z} \setminus \{0\}$ *of integers not containing zero instead of the set* $\Xi = \{h \in \mathbb{Z} : 0 < |h| < m\}$ *(see also the multi-dimensional case in Section 20.4).*

Assume that F is of class $\mathrm{H}^{(2)}_{\mathbb{Z}}(\mathbb{R})$. By integration by parts we then obtain

$$\int_{\mathcal{F}_{\mathbb{Z}}} \sum_{\substack{0<|h|\leq m \\ h\in\mathbb{Z}}} \frac{1}{-\Delta^\wedge(h)} \Phi_h(x - y) \Delta_y F(y) \, dy = \sum_{\substack{0<|h|\leq m \\ h\in\mathbb{Z}}} \int_{\mathcal{F}_{\mathbb{Z}}} \Phi_h(x - y) F(y) \, dy. \tag{4.140}$$

Combining (4.140) with Corollary 3.19 we therefore obtain

$$F(x) = \sum_{\substack{|h|\le m\\ h\in\mathbb{Z}}} \underbrace{\int_{\mathcal{F}_\mathbb{Z}} F(y)\overline{\Phi_h(y)}\, dy}_{=F^\wedge_{\mathcal{F}_\mathbb{Z}}(h)}\ \Phi_h(x) + \int_{\mathcal{F}_\mathbb{Z}} G^\perp_{0,\ldots,m}(\Delta; x-y)\Delta_y F(y)\, dy. \tag{4.141}$$

for all $x \in \mathcal{F}_\mathbb{Z}$. Thus we are able to verify the following integration rule.

Corollary 4.22. *Let $X_N = \{x_1, \ldots, x_N\}$ be a given subset of $\mathcal{F}_\mathbb{Z}$. Suppose that F is of class $\mathrm{H}^{(2)}_\mathbb{Z}(\mathbb{R})$. Then,*

$$\sum_{i=1}^N a_i F(x_i) = \int_{\mathcal{F}_\mathbb{Z}} F(y)\, dy + \int_{\mathcal{F}_\mathbb{Z}} \sum_{i=1}^N a_i G^\perp_{0,\ldots,m}(\Delta; x_i - y)\Delta_y F(y)\, dy \tag{4.142}$$

holds true for all $a \in \mathbb{R}^N$, $a = (a_1, \ldots, a_N)^T$, satisfying

$$\sum_{i=1}^N a_i = 1, \tag{4.143}$$

$$\sum_{i=1}^N a_i \Phi_h(x_i) = 0, \quad 0 < |h| \le m. \tag{4.144}$$

The approximate integration formula (4.142) is exact for all periodic polynomials Φ_h with $|h| \le m$. Accordingly, any spline function S of degree m relative to the point set $X_N = \{x_1, \ldots, x_N\}$ is given by

$$S(x) = \sum_{\substack{|h|\le m\\ h\in\mathbb{Z}}} C_h \Phi_h(x) + \sum_{i=1}^N a_i G^\perp_{0,\ldots,m}(\Delta^2; x - x_i), \quad x \in \mathbb{R}, \tag{4.145}$$

with

$$G^\perp_{0,\ldots,m}(\Delta^2; x-y) = G(\Delta^2; x-y), \quad m = 0, \tag{4.146}$$

$$G^\perp_{0,\ldots,m}(\Delta^2; x-y) = G(\Delta^2; x-y) - \sum_{\substack{0<|h|\le m\\ h\in\mathbb{Z}}} \frac{1}{(-\Delta^\wedge(h))^2}\Phi_h(x-y), \quad m \in \mathbb{N}, \tag{4.147}$$

where the coefficients $a_1, \ldots, a_N$ satisfy the constraints

$$\sum_{i=1}^N a_i = 1, \tag{4.148}$$

$$\sum_{i=1}^N a_i \Phi_h(x_i) = 0, \quad 0 < |h| \le m. \tag{4.149}$$

Best approximate integration, spline interpolation, and spline smoothing of exactness degree m can be handled analogously to the aforementioned approach of exactness 0. The details are left to the reader.

Remark 4.23. *W. Freeden, B. Witte [1982] propose an algorithm that simultaneously provides interpolation of error-free data and smoothing of error-affected data.*

Chapter 5

Trapezoidal Rules

First we deal with lattice functions corresponding to higher order differential operators. We present some preparatory material on the Riemann zeta function in its relationship to lattice functions (for early references see, e.g., L.J. Mordell [1928a,b, 1929]). It should be mentioned that the one-dimensional Riemann zeta function is introduced here in such a way that its extension to the multivariate case becomes apparent. The trapezoidal rule is seen in its relationship to Euler and Poisson summation formulas. The classical Romberg integration rule is recapitulated. Finally, remainder terms for the trapezoidal rule are estimated in terms of special values of the Riemann zeta function.

5.1 Riemann Zeta Function and Lattice Function

The study of the role of the Zeta function in the context of Euler summation has a long tradition (see R. Ayoub [1974], G.H. Hardy [1915], H. Rademacher [1973], E.C. Titchmarsh [1966] and many others). Our interest is in the relationship to the lattice function (cf. W. Freeden [2011]).

In analogy to the setting of (3.55) we define the *$\mathbb{Z}$-lattice functions* $G(\Delta^l; \cdot)$ *for the operator* Δ^l, $l \in \mathbb{N}$, by letting

$$G\left(\Delta^{l+1}; x-y\right) = \int_{\mathcal{F}_{\mathbb{Z}}} G\left(\Delta^l; x-z\right) G(\Delta; z-y)\, dz, \quad x, y \in \mathbb{R}. \tag{5.1}$$

Clearly, we have $\Delta G\left(\Delta^{l+1}; \cdot\right) = G(\Delta^l; \cdot)$, so that the l-iterated $\mathbb{Z}$-lattice function is nothing more than the $\mathbb{Z}$-lattice function to the iterated operator Δ^l

$$G\left(\Delta^l; x\right) = \sum_{\substack{h \neq 0 \\ h \in \mathbb{Z}}} \frac{1}{(-\Delta^\wedge(h))^l}\, \Phi_h(x), \quad x \in \mathbb{R} \tag{5.2}$$

(note that, by convention, $\nabla\Delta^l = \nabla^{2l+1}$). Furthermore, it is clear that $\nabla G\left(\Delta^l; x\right) = G\left(\nabla^{2l-1}; x\right)$, $x \in \mathbb{R}$, for every $l \in \mathbb{N}$. An easy calculation shows that

$$G\left(\Delta^l; x\right) = 2(-1)^l \sum_{\substack{h > 0 \\ h \in \mathbb{Z}}} \frac{\cos(2\pi h x)}{(2\pi h)^{2l}} \tag{5.3}$$

and

$$\nabla G\left(\Delta^l; x\right) = 2(-1)^{l-1} \sum_{\substack{h>0 \\ h\in\mathbb{Z}}} \frac{\sin(2\pi hx)}{(2\pi h)^{2l-1}}, \tag{5.4}$$

such that

$$G\left(\nabla^k; x\right) = -2\frac{1}{(2\pi)^k} \sum_{m=1}^{\infty} \frac{1}{m^k} \cos\left(2\pi mx - k\frac{\pi}{2}\right). \tag{5.5}$$

Remark 5.1. *The functions B_k defined by $B_k(x) = -k!\,G\left(\nabla^k; x\right)$, $x \in \mathbb{R}$, are known as* Bernoulli functions of degree k. *B_k is a piecewise polynomial of degree k, and it is $\mathbb{Z}$-periodic. Moreover, we have (see, e.g., W. Magnus et al. [1966])*

$$B'_{k+1}(x) = (k+1)B_k(x), \quad k = 0, 1, \dots, \tag{5.6}$$

$$B_k(x+1) - B_k(x) = kx^{k-1}, \quad k = 0, 1, \dots, \tag{5.7}$$

$$B_{2l+1}(0) = B_{2l+1}(1) = 0, \quad l = 1, 2, \dots. \tag{5.8}$$

Furthermore, the values B_k, $k = 2, 3, \dots$, given by $B_k = B_k(0) = B_k(1)$ are called *Bernoulli numbers*. Obviously,

$$G\left(\nabla^k; 0\right) = -\frac{1}{k!} B_k = G\left(\nabla^k; 1\right), \tag{5.9}$$

and

$$B_{2l+1} = 0, \quad l = 1, 2, \dots, \tag{5.10}$$

$$B_{2l} = 2(-1)^{l+1}(2l)! \sum_{k=1}^{\infty} \left(\frac{1}{2\pi k}\right)^{2l}, \quad l = 1, 2, \dots. \tag{5.11}$$

Note that the integer values have to be exempted for the Bernoulli function of degree 1 because of the discontinuity, i.e., the characteristic singularity of $G(\nabla; \cdot)$ for $x \in \mathbb{Z}$.

Now, for $s \in \mathbb{C}$ with $\mathrm{Re}(s) > 1$, we consider the *(Riemann) zeta function* ζ given by the series

$$s \mapsto \zeta(s) = \sum_{n=1}^{\infty} \frac{1}{n^s}. \tag{5.12}$$

The series is absolutely convergent since it can be majorized by $\sum_{n=1}^{\infty} n^{-\mathrm{Re}(s)}$, which shows that the convergence of (5.12) is uniform in every half-plane $\{s \in \mathbb{C} : \mathrm{Re}(s) \geq 1 + \delta,\ \delta > 0\}$. Thus, ζ is holomorphic in these half-planes (note that we define c^s, where $s \in \mathbb{C}$ with $\mathrm{Re}(s) > 0$, as $e^{s\ln(c)}$ with $-\frac{\pi}{2} < \mathrm{Im}(\ln(s)) < \frac{\pi}{2}$). For the function $x \mapsto \frac{1}{x^s}$, $x > 0$, we have

$$\Delta_x \frac{1}{x^s} = s(s+1)x^{-s-2}, \quad x > 0. \tag{5.13}$$

The Euler summation formula (Theorem 3.15) yields

$$\sum_{\substack{\rho\le n\le N\\ n\in\mathbb{Z}}}{}' \frac{1}{n^s} = \left(\frac{1}{1-s}x^{1-s}\right)\Big|_\rho^N + s(s+1)\int_\rho^N G(\Delta;x)\frac{1}{x^{s+2}}\,dx \tag{5.14}$$
$$- \left(\frac{1}{x^s}\left(\nabla G(\Delta;x)\right) + s\frac{1}{x^{s+1}}G(\Delta;x)\right)\Big|_\rho^N$$

for all $\rho \in \mathbb{R}$ with $0 < \rho \le 1$. The explicit representation of the lattice function gives

$$\frac{1}{x^s}\nabla G(\Delta;x)\Big|_\rho^N = \frac{1}{x^s}\left(-(x-\lfloor x\rfloor)+\frac{1}{2}\right)\Big|_\rho^N \tag{5.15}$$

and

$$s\frac{1}{x^{s+1}}G(\Delta;x)\Big|_\rho^N = s\frac{1}{x^{s+1}}\left(-\frac{(x-\lfloor x\rfloor)^2}{2}+\frac{x-\lfloor x\rfloor}{2}-\frac{1}{12}\right)\Big|_\rho^N. \tag{5.16}$$

$|G(\nabla;\cdot)|$ and $|G(\Delta;\cdot)|$ are bounded such that the terms (5.15) and (5.16) do not contribute as $N\to\infty$. For $s\in\mathbb{C}$ with $\mathrm{Re}(s)>1$ we therefore obtain

$$\zeta(s) = \frac{1}{s-1}\rho^{1-s} + s(s+1)\int_\rho^\infty G(\Delta;x)\frac{1}{x^{s+2}}\,dx \tag{5.17}$$
$$+\frac{1}{\rho^s}\left(\rho-\frac{1}{2}\right)+s\frac{1}{\rho^{s+1}}\left(-\frac{\rho^2}{2}+\frac{\rho}{2}-\frac{1}{12}\right).$$

With the help of the Fourier expansion of $G(\Delta;\cdot)$ (cf. Lemma 3.11) the integral

$$\int_\rho^N G(\Delta;x)\frac{1}{x^{s+2}}\,dx \tag{5.18}$$

permits the representation

$$-\sum_{\substack{h\neq 0\\ h\in\mathbb{Z}}}\frac{1}{4\pi^2h^2}\int_\rho^N \frac{1}{x^{s+2}}e^{2\pi ihx}\,dx. \tag{5.19}$$

Therefore, the integral

$$\int_\rho^\infty G(\Delta;x)\frac{1}{x^{s+2}}\,dx \tag{5.20}$$

converges not only in the half-plane of all $s\in\mathbb{C}$ with $\mathrm{Re}(s)>1$, but also for all $s\in\mathbb{C}$ with $\mathrm{Re}(s)>-1$. Thus (5.17) furnishes an analytic continuation of ζ into the half-plane $\{s\in\mathbb{C}:\mathrm{Re}(s)>-1\}$ showing as the only singularity the pole at $s=1$. In addition, the expression (5.18) is convergent as $\rho\to 0$ provided that $\mathrm{Re}(s)<0$; hence, ζ can be continued by (5.17) to any point in the s-plane, and ζ emerges a meromorphic function with the simple pole $s=1$. Even more, we are able to formulate the following lemma (cf. W. Freeden [2011]).

Lemma 5.2. *For $s \in \mathbb{C}$ with $-1 < \mathrm{Re}(s) < 0$ we have*

$$\zeta(s) = s(s+1)\int_0^\infty G(\Delta;x)\frac{1}{x^{s+2}}\,dx. \tag{5.21}$$

Now, if $s \in \mathbb{C}$ with $-1 < \mathrm{Re}(s) < 0$, we get by integration by parts

$$\zeta(s) = \sum_{\substack{h\neq 0\\ h\in\mathbb{Z}}} \int_0^\infty e^{2\pi i h x}\frac{1}{x^s}\,dx. \tag{5.22}$$

For $h \neq 0$ we have (see, e.g., L.J. Mordell [1928b])

$$\int_0^\infty e^{2\pi i h x}\frac{1}{x^s}\,dx = \begin{cases} \left(\frac{1}{2h\pi}\right)^{1-s}\Gamma(1-s)e^{\frac{\pi i}{2}(1-s)}, & h>0,\\ \left(\frac{1}{2(-h)\pi}\right)^{1-s}\Gamma(1-s)e^{-\frac{\pi i}{2}(1-s)}, & h<0, \end{cases} \tag{5.23}$$

where Γ denotes the Gamma function (see, e.g., M. Abramowitz, I.A. Stegun [1972], W. Freeden, M. Gutting [2013]). For all $s \in \mathbb{C}$ with $-1 < \mathrm{Re}(s) < 0$ it therefore follows that

$$\zeta(s) = 2^s\pi^{s-1}\Gamma(1-s)\sin\left(\frac{\pi s}{2}\right)\sum_{h=1}^\infty \frac{1}{h^{1-s}}. \tag{5.24}$$

Now, the left side of (5.24) is a meromorphic function (of the variable s) with the only pole at $s=1$, so that this equation provides an analytic continuation of the right hand member as a meromorphic function over the whole s-plane, and ζ appears as a meromorphic function only with the simple pole at $s=1$.

Theorem 5.3 (Functional Equation of the Riemann zeta Function). *The zeta function ζ given by*

$$\zeta(s) = \sum_{n=1}^\infty \frac{1}{n^s}, \quad s\in\mathbb{C}, \quad \mathrm{Re}(s) > 1, \tag{5.25}$$

can be extended analytically to a meromorphic function with the pole $\frac{1}{s-1}$ to the whole complex plane $\mathbb{C}$. Moreover, ζ satisfies the functional equation

$$\zeta(s) = 2^s\pi^{s-1}\Gamma(1-s)\sin\left(\frac{\pi s}{2}\right)\zeta(1-s). \tag{5.26}$$

The functional equation can be put into a more illuminating form if we make use of the duplicator formula of the Gamma function (see, e.g., W. Freeden, M. Gutting [2013])

$$\Gamma(s)\Gamma\left(s+\frac{1}{2}\right) = 2\pi^{\frac{1}{2}}2^{-2s}\Gamma(2s). \tag{5.27}$$

Replacing $2s$ by $1-s$ we obtain

$$\Gamma(1-s)\pi^{\frac{1}{2}}2^s \sin\left(\frac{\pi s}{2}\right) = \Gamma\left(\frac{1}{2}-\frac{s}{2}\right)\Gamma\left(1-\frac{s}{2}\right)\sin\left(\frac{\pi s}{2}\right) = \pi\frac{\Gamma(\frac{1-s}{2})}{\Gamma(\frac{s}{2})}. \tag{5.28}$$

Thus, Theorem 5.3 can be reformulated in the form of

Corollary 5.4. *Under the assumptions of Theorem 5.3*

$$\Gamma\left(\frac{s}{2}\right)\zeta(s) = \pi^{s-\frac{1}{2}}\Gamma\left(\frac{1-s}{2}\right)\zeta(1-s). \tag{5.29}$$

Remark 5.5. *Note that the function ξ given by $\xi(s) = \pi^{-\frac{s}{2}}\Gamma\left(\frac{s}{2}\right)\zeta(s)$ fulfills the functional equation $\xi(s) = \xi(1-s)$.*

The identity (5.17) is also valid for $s \in \mathbb{C}$ with $\mathrm{Re}(s) > -1$, and we get from (5.17) the result

$$\zeta(0) = -\frac{1}{2}. \tag{5.30}$$

Even more, our considerations enable us to deduce that

$$\zeta(-n) = -\frac{B_{n+1}}{n+1}, \quad n = 1, 2, \ldots\ , \tag{5.31}$$

$$\zeta(-2n) = 0, \quad n = 1, 2, \ldots\ , \tag{5.32}$$

$$\zeta(-2n+1) = -\frac{B_{2n}}{2n}, \quad n = 1, 2, \ldots\ . \tag{5.33}$$

In addition, an easy calculation yields

$$\zeta'(0) = -\frac{1}{2}\ln(2\pi). \tag{5.34}$$

Clearly, $\zeta(2n)$ is proportional to the Bernoulli numbers $B_{2n} = -(2n)!G(\Delta^n;0)$ (cf. (5.11))

$$\zeta(2n) = (-1)^{n+1}\frac{(2\pi)^{2n}}{2(2n)!}B_{2n} = \frac{1}{2}(-4\pi^2)^n G(\Delta^n;0). \tag{5.35}$$

Furthermore, we have

$$\zeta(-2n+1) = -\frac{1}{2n}B_{2n} = (2n-1)!G(\Delta^n;0). \tag{5.36}$$

The zeros of ζ at $-2l$, $l = 1, 2, \ldots$ are often called the "trivial zeros" as they are easily found. The role of the "trivial zeros" of ζ is evident in (5.29). For $\mathrm{Re}(s) < 0$ we have $\mathrm{Re}(1-s) > 1$ so that the right-hand side of (5.29) is regular. However, $\Gamma(\frac{s}{2})$ has poles for $s = -2n$, $n \in \mathbb{N}$, which are just neutralized by the zeros of ζ. Only for $s = 0$ we have a pole of first order on both sides of (5.29) since $\zeta(0) = -\frac{1}{2}$.

Setting $s = 1 + t$ we obtain from Corollary 5.4

$$\zeta(1+t) = \frac{\pi^{t-\frac{1}{2}}\Gamma(-\frac{t}{2})}{t\Gamma(\frac{t-1}{2})}\zeta(-t). \tag{5.37}$$

In the neighborhood of $t = 0$ we have with (5.30) and (5.34)

$$\zeta(t) = -\frac{1}{2} - \frac{1}{2}\ln(2\pi)t + \ldots. \tag{5.38}$$

Furthermore,

$$\frac{\pi^{t-\frac{1}{2}}\Gamma\left(-\frac{t}{2}\right)}{\Gamma\left(\frac{t-1}{2}\right)} = -2 + (\ln(2\pi) - \Gamma'(1))\,t + \ldots \tag{5.39}$$

such that

$$\zeta(1+t) = \frac{1}{t} - \Gamma'(1) + \ldots. \tag{5.40}$$

With $-\Gamma'(1) = C$ (Euler constant) we therefore obtain the so-called *Kronecker limit formula.*

Lemma 5.6.

$$\lim_{s\to 1}\left(\zeta(s) - \frac{1}{s-1}\right) = C, \tag{5.41}$$

where C is the Euler constant ($C = 0.57721\ldots$).

Combining Lemma 5.6 and the identity (5.17) we find with $\rho = 1$

$$C = \lim_{s\to 1}\left(\frac{1-1}{s-1} + s(s+1)\int_1^\infty G(\Delta;x)\frac{1}{x^{s+2}}\,dx + \frac{1}{2} - \frac{s}{12}\right) \tag{5.42}$$

such that

$$\lim_{s\to 1}\left(\zeta(s) - \frac{1}{s-1}\right) = C = 2\int_1^\infty G(\Delta;x)\frac{1}{x^3}\,dx + \frac{5}{12}. \tag{5.43}$$

Hence, from the Euler summation formula (Theorem 3.15), we end up with the integration formula

$$\lim_{N\to\infty}\left(\sum_{\substack{n\in\mathbb{Z}\\1\le n\le N}}{}' \frac{1}{n} - \int_1^N \frac{1}{x}\,dx\right) = C = 2\int_1^\infty G(\Delta;x)\frac{1}{x^3}\,dx + \frac{5}{12}. \tag{5.44}$$

5.2 Classical Trapezoidal Sums for Finite Intervals

Next our purpose is to discuss trapezoidal rules in the classical framework of Euler summation, thereby involving the Riemann zeta function for an estimate of the remainder term (see, e.g., P.J. Davis, P. Rabinowitz [1984], R.W. Freund, R.W. Hoppe [2007], J. Stoer [1989], J. Werner [1991]). In doing so we start from conventional trapezoidal rules of closed type where endpoints are included. Later on we are also concerned with non-standard numerical rules of not-necessarily closed type, where the starting point is up to the user. These formulas play an important role in multivariate generalizations.

By the *classical trapezoidal rule* we mean a sum of the form

$$\begin{aligned} \mathcal{L}_\tau F &= \tau\left(\frac{F(a)}{2} + \sum_{j=1}^{n-1} F(a+j\tau) + \frac{F(b)}{2}\right) \\ &= \tau \sum_{j=0}^{n}{}' F(a+\tau j) = \tau \sum_{j=0}^{n} F(a+\tau j) - \frac{\tau}{2}\left(F(a)+F(b)\right) \end{aligned} \tag{5.45}$$

where $F : [a,b] \to \mathbb{R}$ is a Riemann integrable function and the nodal width τ is given by

$$\tau = \frac{b-a}{n}. \tag{5.46}$$

From the Euler summation formula (Theorem 3.15) we easily obtain by integration by parts, for any function $H \in \mathrm{C}^{(3)}[0,n]$, $n \in \mathbb{N}$, the identity

$$\begin{aligned} &\frac{1}{2}H(0) + \sum_{j=1}^{n-1} H(j) + \frac{1}{2}H(n) \\ &= \int_0^n H(x)\, dx + \frac{1}{12}(\nabla H(n) - \nabla H(0)) - \int_0^n G(\nabla^3; x)\nabla^3 H(x)\, dx, \end{aligned} \tag{5.47}$$

where

$$G(\nabla^3; x) = 2\sum_{h=1}^{\infty} \frac{\sin(2\pi h x)}{(2\pi h)^3}. \tag{5.48}$$

This fact together with (5.45) lead to the following statement.

Theorem 5.7. *Let $F \in \mathrm{C}^{(3)}[a,b]$. Set $\tau = (b-a)/n$. Then*

$$\begin{aligned} \mathcal{L}_\tau F = &\int_a^b F(x)\, dx + \frac{1}{12}\tau^2(\nabla F(b) - \nabla F(a)) \\ &- \tau^4 \int_a^b G\left(\nabla^3; n\frac{x-a}{b-a}\right) \nabla^3 F(x)\, dx. \end{aligned} \tag{5.49}$$

Proof. Apply the Euler summation formula to $H : x \mapsto H(x) = F(a+\tau x)$, $x \in [0,n]$. □

Theorem 5.7 provides a relation of the trapezoidal sum to the zeta function.

Corollary 5.8. *Assume for $F \in C^{(3)}[a,b]$ there exists a constant $M_{(3)} > 0$ such that the following estimate holds for the third derivative:*

$$\sup_{x\in[a,b]} \left|\nabla^3 F(x)\right| \leq M_{(3)}. \tag{5.50}$$

If $\mathcal{L}_\tau F$ designates the trapezoidal sum as defined in (5.45), then

$$\left| \int_a^b F(x)\, dx - \mathcal{L}_\tau F + \frac{1}{12}\tau^2 \left(\nabla F(b) - \nabla F(a)\right) \right| \leq \frac{2(b-a)M_{(3)}}{(2\pi)^3}\tau^4 \zeta(3). \tag{5.51}$$

Proof. In case of the boundary conditions $\nabla F(a) = \nabla F(b) = 0$ we obtain from Theorem 5.7

$$\left| \int_a^b F(x)\, dx - \mathcal{L}_\tau F \right| \leq \tau^4 \int_a^b \left| G\left(\nabla^3; n\frac{x-a}{b-a}\right)\right| \left|\nabla^3 F(x)\right|\, dx. \tag{5.52}$$

Together with (5.50) this provides the desired statement. □

For $H \in C^{(2m+1)}[0,n]$, further integration by parts in connection with the differential relation $\nabla_x G(\Delta^k; x) = \nabla_x G(\nabla^{2k}; x) = G(\nabla^{2k-1}; x)$ and the characterization of the Bernoulli numbers (cf. (5.11)) yields

$$\begin{aligned} &\frac{1}{2}H(0) + \sum_{j=1}^{n-1} H(j) + \frac{1}{2}H(n) \\ &\quad = \int_0^n H(x)\, dx + \sum_{l=1}^{m} \frac{B_{2l}}{(2l)!}\left(\nabla^{2l-1}H(n) - \nabla^{2l-1}H(0)\right) \\ &\quad\quad - \int_0^n G(\nabla^{2m+1}; x)\, \nabla^{2m+1} H(x)\, dx. \end{aligned} \tag{5.53}$$

The identity (5.53) immediately leads to the following result.

Theorem 5.9. *Suppose that $F : [a,b] \to \mathbb{R}$ is of class $C^{(2m+1)}[a,b]$. Set $\tau = (b-a)/n$. Then*

$$\begin{aligned} \mathcal{L}_\tau F = &\int_a^b F(x)\, dx + \sum_{l=1}^{m} \frac{B_{2l}}{(2l)!}\left(\nabla^{2l-1}F(b) - \nabla^{2l-1}F(a)\right)\tau^{2l} \\ &- \tau^{2m+2}\int_a^b G\left(\nabla^{2m+1}; \frac{x-a}{\tau}\right) \nabla^{2m+1} F(x)\, dx. \end{aligned} \tag{5.54}$$

Proof. Theorem 5.9 follows by application of the summation formula (5.53) to $F(a+\tau x)$, $\tau = (b-a)/n$. □

Theorem 5.9 enables us to formulate the following estimates.

Corollary 5.10. *Assume that for $F \in C^{(2m+1)}[a,b]$ with*

$$\nabla^{2k+1}F(b) = \nabla^{2k+1}F(a), \quad k = 0, \cdots, m-1, \tag{5.55}$$

there exists a constant $M_{(2m+1)} > 0$ such that for its $(2m+1)$-th derivative holds

$$\sup_{x \in [a,b]} \left| \left(\nabla^{2m+1} F \right)(x) \right| \leq M_{(2m+1)}. \tag{5.56}$$

If $\mathcal{L}_\tau F$ designates the trapezoidal sum as defined in (5.45), then

$$\left| \int_a^b F(x)\, dx - \mathcal{L}_\tau F \right| \leq \frac{2(b-a)M_{(2m+1)}}{(2\pi)^{2m+1}} \tau^{2m+2} \zeta(2m+1). \tag{5.57}$$

5.3 Romberg Integration

The trapezoidal sums discussed until now form rules of closed type, i.e., the τ-lattice is chosen in such a way that start- and endpoint are part of the lattice. The same property is a basic assumption in classical Romberg integration (cf. W. Romberg [1955]) as explained in this section. However, it should be pointed out that this assumption cannot be realized for geoscientifically relevant volumes such as the geoid (see our cover illustration), or the actual Earth's body. In this case, lattice and integration area are not in consistency.

Theorem 5.9, in combination with Corollary 5.10, shows that we are confronted with an asymptotic expansion dependent on τ of the form

$$\int_a^b F(x)\, dx - \mathcal{L}_\tau F = \sum_{l=1}^{m} a_{2l} \tau^{2l} + O\left(\tau^{2m+2}\right), \tag{5.58}$$

where the trapezoidal sum $\mathcal{L}_\tau F$ is given by (5.45) and the constituting coefficients are given as follows:

$$a_{2l} = \frac{B_{2l}}{(2l)!} \left(\nabla^{2l-1}F(a) - \nabla^{2l-1}F(b) \right), \quad l = 1, \ldots, m. \tag{5.59}$$

Now, for values $q \in (0,1)$ we have

$$\int_a^b F(x)\, dx - \mathcal{L}_{q\tau} F = \sum_{l=1}^{m} a_{2l} (q\tau)^{2l} + O\left(\tau^{2m+2}\right), \tag{5.60}$$

such that

$$\int_a^b F(x)\, dx - \frac{\mathcal{L}_{q\tau}F - q^2 \mathcal{L}_\tau F}{1-q^2} = \sum_{l=1}^{m-1} \tilde{a}_{2l+2} \tau^{2l+2} + O\left(\tau^{2m+2}\right) \tag{5.61}$$

with

$$\tilde{a}_{2l+2} = -q^2 \frac{1-q^{2l}}{1-q^2} a_{2l+2}, \quad l = 1, \ldots, m-1. \tag{5.62}$$

The linear combination

$$\mathcal{L}_\tau^{(1)} F = \frac{\mathcal{L}_{q\tau} F - q^2 \mathcal{L}_\tau F}{1-q^2} \tag{5.63}$$

represents a quadrature formula that shows a better asymptotic behavior in terms of τ. As indicated by the Euler summation formula, the process of taking a smaller nodal width $q\tau$ and subsequently forming a linear combination with improved asymptotic error behavior with respect to τ can be continued.

The *classical Romberg scheme* proceeds by successively halving the values, i.e., $q = \frac{1}{2}$ (see W. Romberg [1955]). The advantage of this scheme is that all functional values are used at each stage. The setting $\tau_i = q^i \tau$ finally allows us to construct the following triangular array, which is called the *table of Romberg integration.*

$$\begin{array}{c|llll}
 & k=0 & k=1 & k=2 & k=3 \\
\hline
\tau_0^2 & \mathcal{L}_{\tau_0} F = \mathcal{L}_0^{(0)} F & & & \\
 & & \mathcal{L}_1^{(1)} F & & \\
\tau_1^2 & \mathcal{L}_{\tau_1} F = \mathcal{L}_1^{(0)} F & & \mathcal{L}_2^{(2)} F & \\
 & & \mathcal{L}_2^{(1)} F & & \mathcal{L}_3^{(3)} F \\
\tau_2^2 & \mathcal{L}_{\tau_2} F = \mathcal{L}_2^{(0)} F & & \mathcal{L}_3^{(2)} F & \\
 & & \mathcal{L}_3^{(1)} F & & \\
\tau_3^2 & \mathcal{L}_{\tau_3} F = \mathcal{L}_3^{(0)} & & &
\end{array} \tag{5.64}$$

where, for $i = 0, \ldots, m$,

$$\mathcal{L}_i^{(0)} F = \mathcal{L}_{\tau_i} F \tag{5.65}$$

and, for $k = 1, \ldots, m$, $i = k, \ldots, m$

$$\mathcal{L}_i^{(k)} F = \mathcal{L}_i^{(k-1)} F + \frac{\mathcal{L}_i^{(k-1)} F - \mathcal{L}_{i-1}^{(k-1)} F}{\left(\frac{\tau_{i-k}}{\tau_i}\right)^2 - 1}, \quad \tau_i = q^i \tau. \tag{5.66}$$

A result due to R. Burlisch [1964] guarantees that the quadrature rules $\mathcal{L}_m^{(m)} F$ obtained by the Romberg scheme tends to the integral, i.e.,

$$\int_a^b F(x)\, dx = \lim_{m \to \infty} \mathcal{L}_m^{(m)} F, \tag{5.67}$$

provided that F is Riemann-integrable on the interval $[a, b]$.

Remark 5.11. *In the classical Romberg scheme (based on* $q = \frac{1}{2}$*), the number of values computed increases exponentially. This is the reason* *why variations of the Romberg method have been proposed (for practical purposes, the Rutishauser-sequence* $\{\tau_i\}_{i\in\mathbb{N}_0}$ *with* $\tau_{2i} = 3^{-\frac{i}{\tau}}$, $\tau_{2i+1} = \frac{1}{2}3^{-i}\tau$ *or the Bulirsch-sequence* $\{\tau_i\}_{i\in\mathbb{N}_0}$ *with* $\tau_0 = \tau$, $\tau_{2i-1} = 2^{-i}\tau$, $\tau_{2i} = \frac{1}{3}2^{-i+1}$ *are in use). In fact, there is a large literature about nodal width extrapolation procedures (see, for example, R.W. Freund, R.W. Hoppe [2007], J. Stoer, R. Bulirsch [1980], J. Werner [1991] and the references therein).*

5.4 Poisson Summation Based Integration

First we remember some variants of the one-dimensional *Poisson summation formula* that is related to work by L.J. Mordell [1928a,b, 1929].

Poisson Summation Based Integration over Finite Intervals. By going over to dilated lattices $\tau\mathbb{Z}$ we are subsequently led to not-necessarily endpoint variants of trapezoidal rules, i.e., integration formulas of open type.

Theorem 5.12 (Poisson Summation Formula – Variant 1). *Let F be twice continuously differentiable, i.e., $F \in \mathrm{C}^{(2)}[a,b]$. Then*

$$\sum_{\substack{g\in[a,b]\\ g\in\mathbb{Z}}}{}' F(g) = \sum_{h\in\mathbb{Z}} \int_a^b F(x)\overline{\Phi_h(x)}\, dx. \tag{5.68}$$

i.e., the integral over F can be expressed by the $\mathbb{Z}$-lattice sum over $[a,b]$ with a remainder term free of derivatives of F in the form

$$\sum_{\substack{g\in[a,b]\\ g\in\mathbb{Z}}}{}' F(g) = \int_a^b F(x)\, dx + \sum_{\substack{h\neq 0\\ h\in\mathbb{Z}}} \int_a^b F(x)\overline{\Phi_h(x)}\, dx \tag{5.69}$$

Proof. Using the Fourier expansion of the $\mathbb{Z}$-lattice function $G(\Delta;\cdot)$ (see Lemma 3.11) we get

$$\int_a^b G(\Delta;x)\Delta F(x)\, dx = \sum_{\substack{h\neq 0\\ h\in\mathbb{Z}}} \int_a^b (\Delta F(x))\, \frac{\overline{\Phi_h(x)}}{(2\pi i h)^2}\, dx \tag{5.70}$$

(note that the Fourier expansion is absolutely and uniformly convergent in $\mathbb{R}$; hence, summation and integration can be interchanged). As already pointed out earlier, integration by parts yields with $(\Delta_x + \Delta^\wedge(h))\Phi_h(x) = 0$, $x \in \mathbb{R}$, and $\Delta^\wedge(h) = 4\pi^2 h^2$

$$\begin{aligned}\int_a^b (\Delta F(x))\,\overline{\Phi_h(x)}\,dx &= (\nabla F(x))\overline{\Phi_h(x)}\Big|_a^b - (2\pi i h)\int_a^b (\nabla F(x))\overline{\Phi_h(x)}\,dx \\ &= \left((\nabla F(x))\overline{\Phi_h(x)} - (2\pi i h)F(x)\overline{\Phi_h(x)}\right)\Big|_a^b \\ &\quad + (2\pi i h)^2 \int_a^b F(x)\overline{\Phi_h(x)}\,dx. \end{aligned} \tag{5.71}$$

Consequently, we obtain from (5.70)

$$\begin{aligned}\int_a^b G(\Delta;x)\Delta F(x)\,dx &= \left(G(\Delta;x)\nabla F(x) - F(x)\nabla G(\Delta,x)\right)\Big|_a^b \\ &\quad + \sum_{\substack{h\neq 0\\ h\in\mathbb{Z}}}\int_a^b F(x)\overline{\Phi_h(x)}\,dx. \end{aligned} \tag{5.72}$$

Combining all results we arrive at the Poisson summation formula

$$\sum_{\substack{g\in[a,b]\\ g\in\mathbb{Z}}}{}' F(g) = \int_a^b F(x)\,dx + \sum_{\substack{h\neq 0\\ h\in\mathbb{Z}}}\int_a^b F(x)\overline{\Phi_h(x)}\,dx, \tag{5.73}$$

as announced. □

Remark 5.13. *Unfortunately, Theorem 5.12 cannot be generalized to the multivariate case in a straightforward way, since the step* (5.70) *is not transferable because of convergence deficits of higher-dimensional lattice functions.*

Example 5.14. *We know from Variant 1 of the Poisson summation formula (Theorem 5.12) that*

$$\sum_{\substack{g\in[a,b]\\ g\in\mathbb{Z}}}{}' 1 = \sum_{h\in\mathbb{Z}}\int_a^b \overline{\Phi_h(x)}\,dx. \tag{5.74}$$

Equivalently,

$$\sum_{\substack{g\in[a,b]\\ g\in\mathbb{Z}}}{}' 1 = \int_a^b dx + \sum_{\substack{h\neq 0\\ h\in\mathbb{Z}}}\int_a^b \overline{\Phi_h(x)}\,dx. \tag{5.75}$$

In particular, for $[a,b] = [-R,R]$, $R > 0$, *we obtain as a* one-dimensional counterpart of the so-called Hardy–Landau identity *(cf. G.H. Hardy [1915], G.H. Hardy, E. Landau [1924])*

$$\sum_{\substack{g\in[-R,R]\\ g\in\mathbb{Z}}}{}' 1 = 2R + 2R\sum_{\substack{h\neq 0\\ h\in\mathbb{Z}}}\frac{\sin(2\pi h R)}{2\pi h R}. \tag{5.76}$$

In other words, we have

$$\sum_{\substack{g\in[-R,R]\\ g\in\mathbb{Z}}}{}' \; 1 = 2R\sum_{h\in\mathbb{Z}} \operatorname{sinc}(2\pi hR), \tag{5.77}$$

where

$$\operatorname{sinc}(2\pi hR) = \frac{\sin(2\pi hR)}{2\pi hR}. \tag{5.78}$$

It should be noted that the *one-dimensional Hardy–Landau series* on the right hand side of (5.76) is alternating. Nevertheless, the (pointwise) convergence of the alternating series (5.76) can be readily seen from the derived Poisson summation formula for the interval $[-R, R]$. Moreover, it is uniformly convergent on each compact interval $I \subset (g, g+1)$, $g \in \mathbb{Z}$.

Poisson Summation Based Integration over Infinite Intervals. We start with a discussion of a summation formula over the non-negative real line.

Theorem 5.15 (Poisson Summation Formula – Variant 2). *Let F be twice continuously differentiable in the interval $[0,\infty)$. Moreover, suppose that*

$$F(x) \to 0, \quad \nabla F(x) \to 0, \qquad x \to \infty. \tag{5.79}$$

Furthermore, assume that the limits

$$\int_0^\infty F(x)\; dx = \lim_{N\to\infty} \int_0^N F(x)\; dx, \tag{5.80}$$

$$\int_0^\infty |\Delta F(x)|\; dx = \lim_{N\to\infty} \int_0^N |\Delta F(x)|\; dx \tag{5.81}$$

exist. Then

$$\sum_{\substack{0\le g\\ g\in\mathbb{Z}}}{}' F(g) = \int_0^\infty F(x)\; dx + \sum_{\substack{h\neq 0\\ h\in\mathbb{Z}}} \int_0^\infty F(x)\overline{\Phi_h(x)}\; dx. \tag{5.82}$$

Proof. For every $\varepsilon > 0$ there exists an $N(= N(\varepsilon))$ such that

$$\left| G(\Delta; x) - \sum_{\substack{0<|h|\le N\\ h\in\mathbb{Z}}} \frac{1}{-\Delta^\wedge(h)}\overline{\Phi_h(x)} \right| \le \varepsilon \tag{5.83}$$

for all $x \in \mathbb{R}$. From the Euler summation formula (Theorem 3.15) it is not difficult to show that

$$\sum_{\substack{0\le g\\ g\in\mathbb{Z}}} F(g) = \int_0^\infty F(x)\; dx - \frac{1}{12}\left(\nabla F\right)(0) + \frac{1}{2}F(0) + \int_0^\infty G(\Delta; x)\Delta F(x)\; dx. \tag{5.84}$$

It is clear that

$$\int_0^\infty \left| G(\Delta;x) - \sum_{\substack{0<|h|\leq N \\ h\in\mathbb{Z}}} \frac{1}{-\Delta^\wedge(h)} \overline{\Phi_h(x)} \right| |\Delta F(x)| \ dx \leq \varepsilon \int_0^\infty |\Delta F(x)| \ dx, \tag{5.85}$$

where the integral on the right side of (5.85) is assumed to be convergent in accordance with (5.81). Thus we find

$$\lim_{N\to\infty} \int_0^\infty \left(G(\Delta;x) - \sum_{\substack{0<|h|\leq N \\ h\in\mathbb{Z}}} \frac{1}{-\Delta^\wedge(h)} \overline{\Phi_h(x)} \right) \Delta F(x) \ dx = 0. \tag{5.86}$$

Finally we see that

$$\begin{aligned} \int_0^\infty G(\Delta;x) \ \Delta F(x) \ dx &= \sum_{\substack{0<|h| \\ h\in\mathbb{Z}}} \int_0^\infty \frac{1}{-\Delta^\wedge(h)} \left(\Delta F(x)\right) \overline{\Phi_h(x)} \ dx \qquad (5.87) \\ &= \frac{1}{12} \left(\nabla F\right)(0) - \frac{1}{2} F(0) + \sum_{\substack{h\neq 0 \\ h\in\mathbb{Z}}} \int_0^\infty F(x) \overline{\Phi_h(x)} \ dx. \end{aligned}$$

From (5.84) we therefore obtain the desired result. □

Next, sufficient conditions for the validity of the Poisson summation formula in the Euclidean space $\mathbb{R}$ can be formulated such that the absolute convergence is not automatically implied for the series $\sum_{g\in\mathbb{Z}} F(g)$, hence, the summation process under consideration becomes important for purposes of series convergence.

Theorem 5.16 (Poisson Summation Formula – Variant 3a). *Let F be twice continuously differentiable in $\mathbb{R}$. Moreover, suppose that*

$$F(x) \to 0, \quad \nabla F(x) \to 0, \qquad |x| \to \infty. \tag{5.88}$$

Furthermore, assume that the limit

$$\int_{\mathbb{R}} |\Delta F(x)| \ dx = \lim_{N\to\infty} \int_{-N}^{N} |\Delta F(x)| \ dx \tag{5.89}$$

exists. Then

$$\lim_{N\to\infty} \left(\sum_{\substack{|g|\leq N \\ g\in\mathbb{Z}}} F(g) - \int_N^N F(x) \ dx \right) = \int_{\mathbb{R}} G(\Delta;x) \Delta F(x) \ dx. \tag{5.90}$$

Proof. First, from Theorem 3.15, we obtain for all $N > 0$

$$\sum_{\substack{|g|\leq N \\ g\in\mathbb{Z}}}{}' F(g) = \int_{-N}^{N} F(x)\ dx + \int_{-N}^{N} G(\Delta;x)\Delta F(x)\ dx \tag{5.91}$$

$$+ \left(F(x)\nabla G(\Delta;x) - G(\Delta;x)\nabla F(x)\right)\Big|_{-N}^{N}.$$

Suppose that the function F satisfies the conditions (5.88). Then it follows that

$$\sum_{\substack{|g|=N \\ g\in\mathbb{Z}}} F(g) = o(1), \quad N \to \infty. \tag{5.92}$$

The $\mathbb{Z}$-lattice function $G(\Delta;\cdot)$ and its first order derivative are bounded in $\mathbb{R}$; i.e., there exists a constant C such that

$$|G(\Delta;x)| \leq C, \qquad |\nabla G(\Delta;x)| \leq C \tag{5.93}$$

for all $x \in \mathbb{R}$. By virtue of (5.88) we therefore are able to verify, for $N \to \infty$,

$$\left(F(x)\nabla G(\Delta;x) - G(\Delta;x)\nabla F(x)\right)\Big|_{-N}^{N} = o(1). \tag{5.94}$$

For $\varepsilon > 0$ and for all $N > 0$, the asymptotic estimate (5.89) leads to

$$\left|\int_{-N}^{N} G(\Delta;x)\Delta F(x)\ dx\right| \leq C\int_{-N}^{N} |\Delta F(x)|\ dx \leq C\int_{\mathbb{R}} |\Delta F(x)|\ dx < \infty. \tag{5.95}$$

This shows that, for a function $F \in \mathrm{C}^{(2)}(\mathbb{R})$ satisfying (5.88) and (5.89), we have

$$\lim_{N\to\infty}\left(\sum_{\substack{|g|\leq N \\ g\in\mathbb{Z}}} F(g) - \int_{-N}^{N} F(x)\ dx\right) = \int_{\mathbb{R}} G(\Delta;x)\Delta F(x)\ dx. \tag{5.96}$$

This is the desired result. □

Theorem 5.17 (Poisson Summation Formula – Variant 3b). *Let F be twice continuously differentiable in $\mathbb{R}$. Moreover, suppose that*

$$F(x) \to 0, \quad \nabla F(x) \to 0, \qquad |x| \to \infty. \tag{5.97}$$

Furthermore, assume that the limits

$$\int_{\mathbb{R}} F(x)\ dx = \lim_{N\to\infty}\int_{-N}^{N} F(x)\ dx, \tag{5.98}$$

$$\int_{\mathbb{R}} |\Delta F(x)|\ dx = \lim_{N\to\infty}\int_{-N}^{N} |\Delta F(x)|\ dx \tag{5.99}$$

exist. Then we have

$$\sum_{g\in\mathbb{Z}} F(g) = \int_{\mathbb{R}} F(x)\ dx + \sum_{\substack{h\neq 0\\ h\in\mathbb{Z}}} \int_{\mathbb{R}} F(x)\overline{\Phi_h(x)}\ dx. \tag{5.100}$$

Proof. We start from Theorem 5.16. Observing the absolute and uniform convergence of the Fourier expansion of $G(\Delta; x)$ we find by Lebesgue's theorem

$$\int_{\mathbb{R}} G(\Delta; x)\Delta F(x)\ dx = \sum_{\substack{h\neq 0\\ h\in\mathbb{Z}}} \frac{1}{-\Delta^\wedge(h)} \int_{\mathbb{R}} \overline{\Phi_h(x)}\Delta F(x)\ dx. \tag{5.101}$$

For all $h \in \mathbb{Z}$ with $h \neq 0$ integration by parts, in connection with the asymptotic relations (5.98), yields

$$\int_{\mathbb{R}} \overline{\Phi_h(x)}\Delta F(x)\ dx = -\Delta^\wedge(h) \int_{\mathbb{R}} F(x)\overline{\Phi_h(x)}\ dx. \tag{5.102}$$

This gives the limit relation

$$\lim_{N\to\infty} \left(\sum_{\substack{|g|\leq N\\ g\in\mathbb{Z}}} F(g) - \int_{-N}^{N} F(x)\ dx \right) = \sum_{\substack{h\neq 0\\ h\in\mathbb{Z}}} \int_{\mathbb{R}} F(x)\overline{\Phi_h(x)}\ dx. \tag{5.103}$$

By virtue of (5.98) we finally obtain

$$\sum_{g\in\mathbb{Z}} F(g) = \sum_{h\in\mathbb{Z}} \int_{\mathbb{R}} F(x)\overline{\Phi_h(x)}\ dx. \tag{5.104}$$

□

Modifications of Variant 3b are formulated in the next corollary.

Corollary 5.18. *Let $y \mapsto F(x+y)$, $x \in \mathbb{R}$, satisfy the assumptions of Theorem 5.17. Then*

$$\sum_{g\in\mathbb{Z}} F(g+x) = \sum_{h\in\mathbb{Z}} \int_{\mathbb{R}} F(y)\overline{\Phi_h(y)}\ dy\ \Phi_h(x). \tag{5.105}$$

Proof. From Theorem 5.17 it follows that

$$\sum_{g\in\mathbb{Z}} F(g+x) = \sum_{h\in\mathbb{Z}} \int_{\mathbb{R}} F(x+y)\overline{\Phi_h(y)}\ dy. \tag{5.106}$$

Clearly, under the assumptions of Theorem 5.17,

$$\int_{\mathbb{R}} F(x+y)\overline{\Phi_h(y)}\ dy = \Phi_h(x) \int_{\mathbb{R}} F(y)\overline{\Phi_h(y)}\ dy. \tag{5.107}$$

This is the desired result of Corollary 5.18. □

5.5 Trapezoidal Sums over Dilated Lattices

Next we deal with trapezoidal sums of not-necessarily closed type which is canonically transferable to the multivariate case: To this end we first list some obvious results on "τ-dilated lattices"

$$\tau\mathbb{Z} = \{\tau g : g \in \mathbb{Z}, \tau > 0\} \tag{5.108}$$

with fundamental cell $\mathcal{F}_{\tau\mathbb{Z}}$ of the lattice $\tau\mathbb{Z}$ given by

$$\mathcal{F}_{\tau\mathbb{Z}} = \left\{x \in \mathbb{R} : -\frac{1}{2}\tau \le x < \frac{1}{2}\tau\right\} = \left[-\frac{1}{2}\tau, \frac{1}{2}\tau\right]. \tag{5.109}$$

Integration over Finite Intervals. A function $F : \mathbb{R} \to \mathbb{C}$ is called *$\tau\mathbb{Z}$-periodic* if $F(x+g) = F(x)$ holds for all $x \in \mathcal{F}_{\tau\mathbb{Z}}$ and $g \in \tau\mathbb{Z}$. The function $\Phi_h : \mathbb{R} \to \mathbb{C}$, $h \in \frac{1}{\tau}\mathbb{Z}$, given by

$$x \mapsto \Phi_h(x) = \tau^{-1/2} e^{2\pi i h x} \tag{5.110}$$

is $\tau\mathbb{Z}$-periodic.

The space of all $F \in \mathrm{C}^{(m)}(\mathbb{R})$ that are $\tau\mathbb{Z}$-periodic is denoted by $\mathrm{C}^{(m)}_{\tau\mathbb{Z}}(\mathbb{R})$, $0 \le m \le \infty$. Clearly, the space $\mathrm{L}^2_{\tau\mathbb{Z}}(\mathbb{R})$ is the completion of $\mathrm{C}^{(0)}_{\tau\mathbb{Z}}(\mathbb{R})$ with respect to the norm $\|\cdot\|_{\mathrm{L}^2_{\tau\mathbb{Z}}(\mathbb{R})}$. An easy calculation shows that the system $\{\Phi_h\}_{h\in\frac{1}{\tau}\mathbb{Z}}$ is orthonormal with respect to the $\mathrm{L}^2_{\tau\mathbb{Z}}(\mathbb{R})$-inner product.

By convention we say that λ is an *eigenvalue of the lattice $\tau\mathbb{Z}$ with respect to the operator* Δ of the second order derivative, if there is a non-trivial solution U of the differential equation $(\Delta+\lambda)U = 0$ satisfying the "boundary condition of $\tau\mathbb{Z}$-periodicity" $U(x+g) = U(x)$ for all $x \in \mathcal{F}_{\tau\mathbb{Z}}$ and $g \in \tau\mathbb{Z}$. From classical Fourier analysis we know that the operator Δ has a half-bounded and discrete eigenspectrum $\{\Delta^\wedge(h)\}_{h\in\frac{1}{\tau}\mathbb{Z}} \subset \mathbb{R}$ such that $(\Delta_x + \Delta^\wedge(h))\,\Phi_h(x) = 0$, $x \in \mathcal{F}_{\tau\mathbb{Z}}$, with eigenvalues $\Delta^\wedge(h)$ given by $\Delta^\wedge(h) = 4\pi^2 h^2$, $h \in \frac{1}{\tau}\mathbb{Z}$, and eigenfunctions $\Phi_h(x) = \tau^{-1/2}e^{2\pi i h x}$, $h \in \frac{1}{\tau}\mathbb{Z}$, $x \in \mathcal{F}_{\tau\mathbb{Z}}$. Consequently, the *eigenspectrum of the operator* Δ (with respect to $\tau\mathbb{Z}$) is given by

$$\mathrm{Spect}_\Delta(\tau\mathbb{Z}) = \left\{\Delta^\wedge(h) : \Delta^\wedge(h) = 4\pi^2 h^2,\ h \in \frac{1}{\tau}\mathbb{Z}\right\}. \tag{5.111}$$

Definition 5.19. *A function $G(\tau\mathbb{Z};\Delta;\cdot) : \mathbb{R} \to \mathbb{R}$ is called* Green's function with respect to the operator Δ corresponding to the $\tau\mathbb{Z}$-periodicity *(in brief,* $\tau\mathbb{Z}$-lattice function for Δ*) if it satisfies the following properties:*

(i) (Periodicity) *$G(\tau\mathbb{Z};\Delta;\cdot)$ is continuous in $\mathbb{R}$, and*

$$G(\tau\mathbb{Z};\Delta;x+g) = G(\tau\mathbb{Z};\Delta;x) \tag{5.112}$$

for all $x \in \mathbb{R}$ and $g \in \tau\mathbb{Z}$.

(ii) (Differential Equation) *G is twice continuously differentiable with*

$$\Delta G(\tau\mathbb{Z};\Delta;x) = -\frac{1}{\tau} \tag{5.113}$$

for all $x \notin \tau\mathbb{Z}$.

(iii) (Characteristic Singularity)

$$G(\tau\mathbb{Z};\Delta;x) - \frac{1}{2}x \ \mathrm{sign}(x) \tag{5.114}$$

is continuously differentiable for all $x \in \mathcal{F}_{\tau\mathbb{Z}}$.

(iv) (Normalization)

$$\int_{\mathcal{F}_{\tau\mathbb{Z}}} G(\tau\mathbb{Z};\Delta;x)\ dx = 0. \tag{5.115}$$

The $\tau\mathbb{Z}$*-lattice function for the operator of the second derivative*, i.e., for the one-dimensional Laplacian, is uniquely determined by its constituting properties.

Lemma 5.20. *For all $x, y \in \mathbb{R}$, $G(\tau\mathbb{Z};\Delta;x-y)$ has the bilinear expansion*

$$G(\tau\mathbb{Z};\Delta;x-y) = \sum_{\substack{h\neq 0\\ h\in\frac{1}{\tau}\mathbb{Z}}} \frac{1}{-\Delta^\wedge(h)}\Phi_h(x)\overline{\Phi_h(y)} = G(\tau\mathbb{Z};\Delta;y-x). \tag{5.116}$$

The *Euler summation formula* in $\tau\mathbb{Z}$-nomenclature allows the following formulation.

Theorem 5.21. *Let $F : [a,b] \to \mathbb{C}$, $a < b$, be a twice continuously differentiable function, i.e., $F \in \mathrm{C}^{(2)}[a,b]$. Then*

$$\begin{aligned}\sum_{\substack{g\in[a,b]\\ g\in\tau\mathbb{Z}}}{}' F(g) = \frac{1}{\tau}\int_a^b F(x)\ dx + \int_a^b G(\tau\mathbb{Z};\Delta;x)\Delta F(x)\ dx \\ + \left(F(x)\nabla G(\tau\mathbb{Z};\Delta;x) - G(\tau\mathbb{Z};\Delta;x)\nabla F(x)\right)\Big|_a^b,\end{aligned} \tag{5.117}$$

where

$$\sum_{\substack{g\in[a,b]\\ g\in\tau\mathbb{Z}}}{}' F(g) = \sum_{\substack{g\in(a,b)\\ g\in\tau\mathbb{Z}}} F(g) + \frac{1}{2}\sum_{\substack{g=a,g=b\\ g\in\tau\mathbb{Z}}} F(g), \tag{5.118}$$

and the last sum in (5.118) occurs only if a and/or b are lattice points of $\tau\mathbb{Z}$. Otherwise it is assumed to be zero.

For purposes of numerical integration we are allowed to deduce that

$$\int_a^b F(x)\,dx = \tau \sum_{\substack{g\in[a,b]\\ g\in\tau\mathbb{Z}}}{}' F(g) \tag{5.119}$$
$$- \tau \left(F(x)\nabla G(\tau\mathbb{Z};\Delta;x) - G(\tau\mathbb{Z};\Delta;x)\nabla F(x)\right)\Big|_a^b + \mathcal{E}F,$$

where the remainder term $\mathcal{E}F$ can be estimated as follows:

$$|\mathcal{E}F| = \left| -\tau \int_a^b G(\tau\mathbb{Z};\Delta;x)\Delta F(x)\,dx \right| \le \frac{2\tau^2}{(4\pi^2)^2}\zeta(2) \int_a^b |\Delta F(x)|\,dx. \tag{5.120}$$

After these considerations based on the Euler summation formula for the Laplace operator (i.e., the second derivative) we are prepared to formulate the *Poisson relation formula relative to the lattice* $\tau\mathbb{Z}$.

Theorem 5.22 (Poisson Summation Formula). *Let F be twice continuously differentiable, i.e., $F \in \mathrm{C}^{(2)}[a,b]$. Then, for $\tau > 0$,*

$$\sum_{\substack{g\in[a,b]\\ g\in\tau\mathbb{Z}}}{}' F(g) = \frac{1}{\tau}\int_a^b F(x)\,dx + \frac{1}{\sqrt{\tau}} \sum_{\substack{h\neq 0\\ h\in\frac{1}{\tau}\mathbb{Z}}} \int_a^b F(x)\overline{\Phi_h(x)}\,dx. \tag{5.121}$$

Proof. In Theorem 5.12, the Fourier expansion of the Green's function with respect to the operator Δ corresponding to the $\mathbb{Z}$-periodicity has to be replaced by the Fourier expansion of the Green's function with respect to the operator Δ corresponding to the $\tau\mathbb{Z}$-periodicity. The result of Theorem 5.22 then follows by the same arguments as presented in Theorem 5.12. □

Numerical Rules based on Dilation over Finite Intervals. In numerical integration we are therefore led to the following situation: Under twice continuous differentiability of F on $[a,b]$ and for a dilation parameter $\tau > 0$ we therefore have

$$\int_a^b F(x)\,dx = \tau \sum_{\substack{g\in[a,b]\\ g\in\tau\mathbb{Z}}}{}' F(g) + \mathcal{E}F, \tag{5.122}$$

where the remainder term is explicitly available in the form

$$\mathcal{E}F = \sqrt{\tau} \sum_{\substack{h\neq 0\\ h\in\frac{1}{\tau}\mathbb{Z}}} \int_a^b F(x)\overline{\Phi_h(x)}\,dx = \sum_{\substack{h\neq 0\\ h\in\frac{1}{\tau}\mathbb{Z}}} \int_a^b F(x)e^{-2\pi i h x}\,dx. \tag{5.123}$$

This is a remarkable result, which provides the remainder term in the form a series free of a certain differential operator. However, we see a striking difference between the rules (5.119) and (5.122), since the second term at the right side of (5.119) is included in the remainder term (5.123).

Numerical Rules based on Dilation over Infinite Intervals. We are able to formulate Theorem 5.16 and Theorem 5.17 within the framework of the dilated lattice $\tau\mathbb{Z}$, $\tau > 0$.

Corollary 5.23. *Let F satisfy the assumptions of Theorem 5.16. Then*

$$\lim_{N\to\infty}\left(\tau\sum_{\substack{|g|\le N\\ g\in\tau\mathbb{Z}}} F(g) - \int_{-N}^{N} F(y)\,dy\right) = \mathcal{E}F = \tau\int_{\mathbb{R}} G(\tau\mathbb{Z};\Delta;x)\Delta F(x)\,dx. \tag{5.124}$$

The remainder term admits the error estimate

$$\begin{aligned}|\mathcal{E}F| = \left|\tau\int_{\mathbb{R}} G(\tau\mathbb{Z};\Delta;x)\Delta F(x)\,dx\right| &\le \left(\sum_{\substack{h\neq 0\\ h\in\frac{1}{\tau}\mathbb{Z}}}\frac{1}{4\pi^2h^2}\right)\int_{\mathbb{R}}|\Delta F(x)|\,dx\\ &\le \frac{\tau^2}{2\pi^2}\zeta(2)\int_{\mathbb{R}}|\Delta F(x)|\,dx. \end{aligned}\tag{5.125}$$

A formulation free of derivatives follows from Theorem 5.17.

Corollary 5.24. *Let F satisfy the assumptions of Theorem 5.17. Then*

$$\int_{\mathbb{R}} F(x)\,dx = \tau\sum_{g\in\tau\mathbb{Z}} F(g) + \sum_{h\in\frac{1}{\tau}\mathbb{Z}}\int_{\mathbb{R}} F(x)e^{-2\pi ihx}\,dx. \tag{5.126}$$

Chapter 6

Adaptive Trapezoidal Rules

The numerical integration of oscillating functions shows serious difficulties if we automatically apply one of the usual approximate integration procedures. The problems essentially arise from the following facts:

(1) During the calculation of alternating functions many positive and negative values have to be added together. In a large proportion of instances, the sum of the positive values is nearly equal to the absolute value of the sum of the negative values. The resulting cancellation is attended by instabilities and a significant loss of accuracy.

(2) For oscillating functions, the absolute values of certain derivatives often increase considerably as the degree of differentiation increases. Thus, the growth of the derivatives in remainder terms causes slow convergence.

For example, $\Delta^m e(h \cdot x) = (-4\pi^2 h^2)^m e(h \cdot x)$, $e(h \cdot x) = e^{2\pi i (h \cdot x)}$, ($\Delta$: one-dimensional Laplace-operator, $h \in \mathbb{Z}$). Hence, apart from a multiplicative constant, we have the absolute bound $|h|^{2m}$ for the m-th operator Δ^m, therefore, this estimate can be regarded as non-adaptive to problems of estimating the error for large numbers $|h|$. Choosing $(\Delta + \lambda)^m$ instead of Δ^m, however, we obtain $(\Delta + \lambda)^m e(h \cdot x) = (\lambda - 4\pi^2 h^2)^m e(h \cdot x)$. The operator can be adapted to the specific properties (i.e., the frequency) of the oscillating function. Especially, for $\lambda = 4\pi^2 h^2$ the bound is zero.

As a consequence, we are concerned with the idea of circumventing instability and slow convergence in integration problems by appropriate adaptation of differential operators. Simply structured tools are Euler summation formulas with respect to Helmholtz operators. The numerical method is based on the observation that the remainder terms in Euler summation formulas can be formulated explicitly in integral form for the Helmholtz operator $\Delta + \lambda$, $\lambda \in \mathbb{R}$, thus opening new perspectives in estimating the error terms by adaptive procedures. Altogether, the philosophy of the procedure can be explained briefly as follows:

(i) Investigate the characteristic oscillating or alternating properties of the integrand,

(ii) Choose an appropriate operator, e.g., in adaptation to the frequency of an oscillating process,

(iii) Use the corresponding Euler summation formula thereby obtaining an appropriate integration formula and an estimate of the error at the same time.

The method is applicable for finite and infinite intervals in $\mathbb{R}$. Moreover, it can be extended to the multivariate case in cubature problems. Of particular interest are computations of Fourier transforms. As special examples, discontinuous integrals resulting in Fourier transforms over alternating integrands can be investigated in more detail.

6.1 Lattice Functions for Helmholtz Operators

It is helpful to begin with a heuristic argument to demonstrate that the $\tau\mathbb{Z}$-lattice function $G(\Delta + \lambda; \cdot)$ for the Helmholtz operator $\Delta + \lambda$ and the "boundary condition of $\tau\mathbb{Z}$-periodicity" exists in two variants (cf. W. Freeden [2011]) dependent on the choice of the parameter $\lambda \in \mathbb{R}$ under consideration:

(i) For $\lambda \notin \mathrm{Spect}_\Delta(\tau\mathbb{Z})$, i.e., $\lambda \neq \Delta^\wedge(h) = 4\pi^2 h^2$ for all $h \in \frac{1}{\tau}\mathbb{Z}$, the $\tau\mathbb{Z}$-lattice function shows the essential ingredients of a classical Green's function, namely, homogeneous Helmholtz differential equation, characteristic singularity (i.e., a finite jump of its derivative in lattice points), and ($\tau\mathbb{Z}$-periodic) boundary condition. Following mathematical physics we have to describe $G(\tau\mathbb{Z}; \Delta + \lambda; \cdot)$ in formal consistency with the following identities (of course, to be understood more precisely in a distributional sense)

$$(\Delta_x + \lambda)G(\tau\mathbb{Z}; \Delta + \lambda; x) = (\Delta_x + \lambda)\frac{1}{\sqrt{\tau}} \sum_{h \in \frac{1}{\tau}\mathbb{Z}} \frac{1}{-(\Delta + \lambda)^\wedge(h)} \Phi_h(x)$$

$$= \frac{1}{\sqrt{\tau}} \sum_{h \in \frac{1}{\tau}\mathbb{Z}} \Phi_h(x) = \delta_{\tau\mathbb{Z}}(x), \tag{6.1}$$

where $\delta_{\tau\mathbb{Z}}$ is the *Dirac function(al)*, $(\Delta + \lambda)^\wedge(h) = 4\pi^2 h^2 - \lambda$, $h \in \frac{1}{\tau}\mathbb{Z}$, and $\Phi_h(x) = \tau^{-1/2} e(h \cdot x) = \tau^{-1/2} e^{2\pi i h x}$.

(ii) For $\lambda \in \mathrm{Spect}_\Delta(\tau\mathbb{Z})$, i.e., $\lambda = \Delta^\wedge(h) = 4\pi^2 h^2$ for some $h \in \frac{1}{\tau}\mathbb{Z}$, in the terminology of D. Hilbert [1912], the $\mathbb{Z}$-lattice function is a Green's function in an enlarged sense; i.e., the right side of the differential equation (6.1) must be modified which implies an additional normalization condition to ensure the uniqueness. Again, in accordance with mathematical physics, we have to specify $G(\Delta+\lambda;\cdot)$ in such a way that

$$(\Delta_x+\lambda)G(\tau\mathbb{Z};\Delta+\lambda;x) = (\Delta_x+\lambda)\frac{1}{\sqrt{\tau}}\sum_{\substack{(\Delta+\lambda)^\wedge(h)\neq 0\\ h\in\frac{1}{\tau}\mathbb{Z}}}\frac{1}{-(\Delta+\lambda)^\wedge(h)}\Phi_h(x)$$

$$= \frac{1}{\sqrt{\tau}}\sum_{\substack{(\Delta+\lambda)^\wedge(h)\neq 0\\ h\in\frac{1}{\tau}\mathbb{Z}}}\Phi_h(x) \tag{6.2}$$

holds true (in a distributional sense) such that

$$(\Delta_x+\lambda)G(\tau\mathbb{Z};\Delta+\lambda;x) = \delta_{\tau\mathbb{Z}}(x) - \frac{1}{\sqrt{\tau}}\sum_{\substack{(\Delta+\lambda)^\wedge(h)=0\\ h\in\frac{1}{\tau}\mathbb{Z}}}\Phi_h(x). \tag{6.3}$$

In other words, for $\lambda\in\operatorname{Spect}_\Delta(\tau\mathbb{Z})$, the differential equation becomes inhomogeneous with the right side indicated by (6.3).

Remark 6.1. *It should be noted that the classical Bernoulli function B_2 of degree 2 is (apart from a factor) the one-dimensional $\mathbb{Z}$-lattice function corresponding to the* eigenvalue $0\in\operatorname{Spect}_\Delta(\mathbb{Z})$. *In consequence, in the jargon of our approach, the standard theory of numerical integration usually deals with a Green's function corresponding to an eigenvalue, namely $\lambda=0$.*

The *definition of the $\tau\mathbb{Z}$-lattice function with respect to the operator $\Delta+\lambda$,* $\lambda\in\mathbb{R}$, is straightforward after our preparations.

Definition 6.2. *A function $G(\tau\mathbb{Z};\Delta+\lambda;\cdot):\mathbb{R}\to\mathbb{R}$ is called the* Green's function for the Helmholtz operator $\Delta+\lambda$, $\lambda\in\mathbb{R}$ with respect to the lattice $\tau\mathbb{Z}$ (in brief, $\tau\mathbb{Z}$-lattice function for $\Delta+\lambda$), *if it fulfills the following properties:*

*(i) (*Periodicity*) $G(\tau\mathbb{Z};\Delta+\lambda;\cdot)$ is continuous in $\mathbb{R}$, and*

$$G(\tau\mathbb{Z};\Delta+\lambda;x+g) = G(\tau\mathbb{Z};\Delta+\lambda;x) \tag{6.4}$$

for all $x\in\mathbb{R}$ and $g\in\tau\mathbb{Z}$.

*(ii) (*Differential Equation*) $G(\tau\mathbb{Z};\Delta+\lambda;\cdot)$ is twice continuously differentiable for all $x\notin\tau\mathbb{Z}$ with*

$$(\Delta+\lambda)G(\tau\mathbb{Z};\Delta+\lambda;x) = 0 \tag{6.5}$$

provided that $\lambda\notin\operatorname{Spect}_\Delta(\tau\mathbb{Z})$, $G(\tau\mathbb{Z};\Delta+\lambda;\cdot)$ is twice continuously differentiable for all $x\notin\tau\mathbb{Z}$ with

$$(\Delta+\lambda)G(\tau\mathbb{Z};\Delta+\lambda;x) = -\frac{1}{\sqrt{\tau}}\sum_{\substack{(\Delta+\lambda)^\wedge(h)=0\\ h\in\frac{1}{\tau}\mathbb{Z}}}\Phi_h(x) \tag{6.6}$$

provided that $\lambda\in\operatorname{Spect}_\Delta(\tau\mathbb{Z})$ (note that the summation on the right side of (6.6) is to be taken over all lattice points $h\in\frac{1}{\tau}\mathbb{Z}$ satisfying $(\Delta+\lambda)^\wedge(h)=0$, i.e., $4\pi^2h^2-\lambda=0$).

(iii) (Characteristic Singularity)

$$x \mapsto G(\tau\mathbb{Z}; \Delta + \lambda; \cdot) - \frac{1}{2} x \ \mathrm{sign}(x) \tag{6.7}$$

is continuously differentiable for all $x \in \mathcal{F}_{\tau\mathbb{Z}}$.

(iv) (Normalization) *For all* $h \in \frac{1}{\tau}\mathbb{Z}$ *with* $(\Delta + \lambda)^\wedge(h) = 0$,

$$\int_{\mathcal{F}_{\tau\mathbb{Z}}} G(\tau\mathbb{Z}; \Delta + \lambda; x)\overline{\Phi_h(x)} \ dx = 0. \tag{6.8}$$

The Fourier expansion is given by

$$G(\tau\mathbb{Z}; \Delta + \lambda; x) = \frac{1}{\sqrt{\tau}} \sum_{\substack{(\Delta+\lambda)^\wedge(h)\neq 0 \\ h\in\frac{1}{\tau}\mathbb{Z}}} \frac{1}{-(\Delta + \lambda)^\wedge(h)} \Phi_h(x), \quad x \in \mathbb{R}. \tag{6.9}$$

Existence and uniqueness follow from the classical Hilbert theory of linear integral equations (cf. D. Hilbert [1912]).

6.2 Adaptive Trapezoidal Sums over Finite Intervals

By the standard procedure of integration by parts we obtain in usual manner

Theorem 6.3 ($\tau\mathbb{Z}$-Euler Summation Formula for the Helmholtz Operator $\Delta + \lambda$, $\lambda \in \mathbb{R}$). *Let F be of class $\mathrm{C}^{(2)}[a, b]$, $a < b$. Suppose that x is a point of $\mathcal{F}_{\tau\mathbb{Z}}$. Then, the following identity*

$$\begin{aligned}\sum_{\substack{g+x\in[a,b] \\ g\in\tau\mathbb{Z}}}{}' F(g + x) = & \sum_{\substack{(\Delta+\lambda)^\wedge(h)=0 \\ h\in\frac{1}{\tau}\mathbb{Z}}} \int_a^b F(y)\overline{\Phi_h(y)} \ dy \ \Phi_h(x) \\ & + \int_a^b G(\tau\mathbb{Z}; \Delta + \lambda; x - y)(\Delta_y + \lambda)F(y) \ dy \\ & + \left(F(y)\nabla_y G(\tau\mathbb{Z}; \Delta + \lambda; x - y) - G(\tau\mathbb{Z}; \Delta + \lambda; x - y)\nabla_y F(y)\right)\Big|_a^b,\end{aligned} \tag{6.10}$$

holds true, where the sum on the right side is to be taken over all points $h \in \frac{1}{\tau}\mathbb{Z}$ for which $(\Delta+\lambda)^\wedge(h) = 4\pi^2h^2 - \lambda = 0$. In case of $(\Delta+\lambda)^\wedge(h) = 4\pi^2h^2 - \lambda \neq 0$ for all $h \in \frac{1}{\tau}\mathbb{Z}$, this sum is understood to be zero.

As already mentioned, the case $\lambda = 0$ leads back to the already known classical concept for the Laplace operator (cf. Theorem 3.15). This is the reason why we restrict ourselves to $\lambda \neq 0$ throughout the remaining part of this section. Under this assumption $\lambda \neq 0$ we introduce

$$G_0^{\perp}(\tau\mathbb{Z};\Delta+\lambda;x) = G(\tau\mathbb{Z};\Delta+\lambda;x) + \frac{1}{\lambda\tau}. \tag{6.11}$$

The Fourier series of $G_0^{\perp}(\tau\mathbb{Z};\Delta+\lambda;x)$ is given by

$$G_0^{\perp}(\tau\mathbb{Z};\Delta+\lambda;x) = \frac{1}{\sqrt{\tau}} \sum_{\substack{(\Delta+\lambda)^\wedge(h)\neq 0 \\ h\neq 0, h\in\frac{1}{\tau}\mathbb{Z}}} \frac{1}{-(\Delta+\lambda)^\wedge(h)} \Phi_h(x), \quad x\in\mathbb{R}. \tag{6.12}$$

We therefore obtain from Theorem 6.3 an integral formula that is an appropriate preparation for adapted quadrature.

Corollary 6.4. *Let x be an arbitrary point of $\mathcal{F}_{\tau\mathbb{Z}}$. Suppose that F is of class $C^{(2)}[a,b]$. Then, for $\lambda \neq 0$,*

$$\begin{aligned}\sum_{\substack{g+x\in[a,b]\\ g\in\tau\mathbb{Z}}}{}' F(g+x) &= \frac{1}{\tau}\int_a^b F(y)\,dy + \sum_{\substack{(\Delta+\lambda)^\wedge(h)=0\\ h\in\frac{1}{\tau}\mathbb{Z}}} \int_a^b F(y)\overline{\Phi_h(y)}\,dy\ \Phi_h(x)\\ &\quad + \left(F(y)\nabla_y G_0^{\perp}(\tau\mathbb{Z};\Delta+\lambda;x-y) - G_0^{\perp}(\tau\mathbb{Z};\Delta+\lambda;x-y)\nabla_y F(y)\right)\Big|_a^b\\ &\quad + \int_a^b G_0^{\perp}(\tau\mathbb{Z};\Delta+\lambda;x-y)(\Delta_y+\lambda)F(y)\,dy. \end{aligned} \tag{6.13}$$

Classification of Quadrature Rules. We are now able to classify the following quadrature rules: For $\lambda \notin \mathrm{Spect}_{\Delta}(\tau\mathbb{Z})$ and every $x\in\mathcal{F}_{\tau\mathbb{Z}}$, an adaptive trapezoidal rule of the following kind

$$\begin{aligned}\int_a^b F(y)\,dy &= \tau \sum_{\substack{g+x\in[a,b]\\ g\in\tau\mathbb{Z}}}{}' F(g+x) + \mathcal{E}F \\ &\quad - \tau\left(F(y)(\nabla_y G_0^{\perp}(\tau\mathbb{Z};\Delta+\lambda;x-y)) - G_0^{\perp}(\tau\mathbb{Z};\Delta+\lambda;x-y)\nabla_y F(y)\right)\Big|_a^b. \end{aligned} \tag{6.14}$$

holds true, where the remainder term $\mathcal{E}F$ given by

$$\mathcal{E}F = \tau\int_a^b G_0^{\perp}(\tau\mathbb{Z};\Delta+\lambda;x-y)(\Delta_y+\lambda)F(y)\,dy \tag{6.15}$$

is of the order $O(\tau^2)$. Note that the formula requires the knowledge of $F(y)$ and $\nabla_y F(y)$ at the endpoints a and b of the interval $[a,b]$. Especially, for $x=0$, we have

$$\begin{aligned}\int_a^b F(y)\,dy &= \tau \sum_{\substack{g\in[a,b]\\ g\in\tau\mathbb{Z}}}{}' F(g) + \mathcal{E}F \\ &\quad - \tau\left(F(y)\nabla_y G_0^{\perp}(\tau\mathbb{Z};\Delta+\lambda;y) - G_0^{\perp}(\tau\mathbb{Z};\Delta+\lambda;y)\nabla_y F(y)\right)\Big|_a^b. \end{aligned} \tag{6.16}$$

For $\lambda \in \operatorname{Spect}_\Delta(\tau\mathbb{Z}) \setminus \{0\}$ and any prescribed nodal system $x_1, \ldots, x_N \subset \mathcal{F}_{\tau\mathbb{Z}}$ of $N \geq 2$ mutually distinct points and any selection of weights $a_1, \ldots, a_N$ satisfying $\sum_{i=1}^N a_i = 1$ and $\sum_{i=1}^N a_i \Phi_h(x_i) = 0$ for all $h \in \frac{1}{\tau}\mathbb{Z}$ such that $(\Delta + \lambda)^\wedge(h) = 0$ the following adaptive trapezoidal rule

$$\tau \sum_{i=1}^N a_i \sum_{\substack{g+x_i \in [a,b] \\ g \in \tau\mathbb{Z}}}{}' F(g + x_i) - \tau \sum_{i=1}^N a_i \left(F(y) \nabla_y G_0^\perp(\tau\mathbb{Z}; \Delta + \lambda; x_i - y)\right)\Big|_a^b$$

$$+ \tau \sum_{i=1}^N a_i \left(G_0^\perp(\tau\mathbb{Z}; \Delta + \lambda; x_i - y) \nabla_y F(y)\right)\Big|_a^b + \mathcal{E}F = \int_a^b F(y)\, dy \qquad (6.17)$$

is valid, where the remainder term $\mathcal{E}F$ given by

$$\mathcal{E}F = \tau \sum_{i=1}^N a_i \int_a^b G_0^\perp(\tau\mathbb{Z}; \Delta + \lambda; x_i - y)(\Delta_y + \lambda)F(y)\, dy \qquad (6.18)$$

is of the order $O(\tau^2)$. Note that the formula requires the knowledge of $F(y)$ and $\nabla_y F(y)$ at the endpoints a and b of the interval $[a, b]$, too.

Remark 6.5. *Clearly, $\tau\mathbb{Z}$-Euler summation formulas and associated trapezoidal rules showing improved asymptotic behavior can be formulated for iterated Helmholtz operators $(\Delta + \lambda)^m$, similarly to iterated Laplace operators Δ^m. The details are left to the reader.*

6.3 Adaptive Poisson Summation Formula over Infinite Intervals

Next we are interested in a variant of the Poisson summation formula involving the (one-dimensional) Helmholtz differential operator $\Delta + \lambda$, $\lambda \in \mathbb{R}$, in order to specify adaptive validity. Our approach as presented in Theorem 6.6 is formulated in such a way that its generalization to higher dimensional Euclidean spaces $\mathbb{R}^q$, $q \geq 2$, becomes immediately obvious (see W. Freeden [2011] for number theoretical obligations involving Hardy–Landau identities).

Theorem 6.6 (Poisson Summation Formula – Modified Variant 3a)**.** *For given values $\varepsilon > 0$, $\lambda \in \mathbb{R}$, let F be a member of class $\mathrm{C}^{(2)}(\mathbb{R})$ satisfying the following properties:*

(i) F obeys the asymptotic relations

$$F(x) \to 0, \quad \nabla F(x) \to 0, \qquad |x| \to \infty, \qquad (6.19)$$

(ii) $(\Delta + \lambda)F$ *fulfills the asymptotic relation*

$$(\Delta + \lambda)F(x) = O\left(|x|^{-(1+\varepsilon)}\right), \quad |x| \to \infty. \tag{6.20}$$

Then

$$\lim_{N\to\infty} \left(\sum_{\substack{|g|\le N \\ g\in\tau\mathbb{Z}}} F(g) - \sum_{\substack{(\Delta+\lambda)^\wedge(h)=0 \\ h\in\frac{1}{\tau}\mathbb{Z}}} \int_{-N}^{N} F(x)\overline{\Phi_h(x)}\, dx \right) \tag{6.21}$$

$$= \int_{\mathbb{R}} G(\tau\mathbb{Z}; \Delta + \lambda; x)(\Delta + \lambda)F(x)\, dx.$$

Proof. First, from Theorem 6.3, we obtain for every $N > 0$

$$\sum_{\substack{|g|\le N \\ g\in\tau\mathbb{Z}}}{}' F(g) = \sum_{\substack{(\Delta+\lambda)^\wedge(h)=0 \\ h\in\frac{1}{\tau}\mathbb{Z}}} \int_{-N}^{N} F(x)\overline{\Phi_h(x)}\, dx$$

$$+ \int_{-N}^{N} G(\tau\mathbb{Z}; \Delta + \lambda; x)(\Delta + \lambda)F(x)\, dx \tag{6.22}$$

$$+ \left(F(x)\nabla G(\tau\mathbb{Z}; \Delta + \lambda; x) - G(\tau\mathbb{Z}; \Delta + \lambda; x)\nabla F(x)\right)\Big|_{-N}^{N}.$$

Assume that F satisfies (6.19). Then it follows that

$$\sum_{\substack{|g|=N \\ g\in\mathbb{Z}}} F(g) = o(1), \quad N \to \infty. \tag{6.23}$$

For each $\lambda \in \mathbb{R}$, the $\mathbb{Z}$-lattice function $G(\tau\mathbb{Z}; \Delta + \lambda; \cdot)$ and its first order derivative are bounded in $\mathbb{R}$; i.e., there exists a constant C such that

$$|G(\tau\mathbb{Z}; \Delta + \lambda; x)| \le C, \tag{6.24}$$

$$|\nabla G(\tau\mathbb{Z}; \Delta + \lambda; x)| \le C \tag{6.25}$$

for all $x \in \mathbb{R}$. By virtue of (6.19) we therefore are able to verify, for $N \to \infty$,

$$\left(F(x)\nabla G(\tau\mathbb{Z}; \Delta + \lambda; x) - G(\tau\mathbb{Z}; \Delta + \lambda; x)\nabla F(x)\right)\Big|_{-N}^{N} = o(1). \tag{6.26}$$

For the specified parameters $\varepsilon > 0$, $\lambda \in \mathbb{R}$, and for all $N > 0$, the asymptotic estimate (6.20) leads to

$$\left| \int_{-N}^{N} G(\tau\mathbb{Z}; \Delta + \lambda; x)(\Delta + \lambda)F(x)\, dx \right| \le C \int_{-N}^{N} |(\Delta + \lambda)F(x)|\, dx \tag{6.27}$$

$$\le C \int_{\mathbb{R}} |(\Delta + \lambda)F(x)|\, dx < \infty\,.$$

This shows that, for a function $F \in \mathrm{C}^{(2)}(\mathbb{R})$ satisfying (6.19) and (6.20), we have

$$\lim_{N\to\infty}\left(\sum_{\substack{|g|\le N\\ g\in\tau\mathbb{Z}}} F(g) - \sum_{\substack{(\Delta+\lambda)^\wedge(h)=0\\ h\in\frac{1}{\tau}\mathbb{Z}}} \int_{-N}^{N} F(x)\overline{\Phi_h(x)}\,dx\right) \tag{6.28}$$

$$= \int_{\mathbb{R}} G(\tau\mathbb{Z};\Delta+\lambda;x)(\Delta+\lambda)F(x)\,dx.$$

All in all, this is the desired result. □

Theorem 6.7 (Poisson Summation Formula – Modified Variant 3b). *For given values $\varepsilon > 0$, $\lambda \in \mathbb{R}$, let F be a member of class $\mathrm{C}^{(2)}(\mathbb{R})$ satisfying the following properties:*

(i) F obeys the asymptotic relations (cf. (6.19))

$$F(x) \to 0, \quad \nabla F(x) \to 0, \qquad |x| \to \infty, \tag{6.29}$$

(ii) $(\Delta+\lambda)F$ fulfills the asymptotic relation (cf. (6.20))

$$(\Delta+\lambda)F(x) = O\left(|x|^{-(1+\varepsilon)}\right), \quad |x| \to \infty. \tag{6.30}$$

(iii) For all $h \in \frac{1}{\tau}\mathbb{Z}$ with $(\Delta+\lambda)^\wedge(h) = 0$, i.e., $4\pi^2h^2 = \lambda$, the integrals

$$\int_{\mathbb{R}} F(x)\overline{\Phi_h(x)}\,dx < \infty \tag{6.31}$$

exist in the sense

$$\int_{\mathbb{R}} \cdots \quad = \quad \lim_{N\to\infty}\int_{-N}^{N} \cdots . \tag{6.32}$$

Then the Poisson summation formula

$$\lim_{N\to\infty} \sum_{\substack{|g|\le N\\ g\in\tau\mathbb{Z}}} F(g) = \sum_{h\in\frac{1}{\tau}\mathbb{Z}} \int_{\mathbb{R}} F(x)\overline{\Phi_h(x)}\,dx. \tag{6.33}$$

holds true.

Proof. We start from the identity (6.21). Observing the absolute and uniform convergence of the Fourier expansion of $G(\Delta+\lambda;x)$ we obtain by Lebesgue's theorem

$$\int_{\mathbb{R}} G(\tau\mathbb{Z};\Delta+\lambda;x)(\Delta+\lambda)F(x)\,dx \tag{6.34}$$

$$= \sum_{\substack{(\Delta+\lambda)^\wedge(h)\neq 0\\ h\in\frac{1}{\tau}\mathbb{Z}}} \frac{1}{-(\Delta+\lambda)^\wedge(h)} \int_{\mathbb{R}} \overline{\Phi_h(x)}(\Delta+\lambda)F(x)\,dx.$$

For all $h \in \frac{1}{\tau}\mathbb{Z}$ with $(\Delta+\lambda)^\wedge(h) \neq 0$ integration by parts, in connection with the asymptotic relations (6.29), gives

$$\int_{\mathbb{R}} \overline{\Phi_h(x)}(\Delta+\lambda)F(x)\,dx = -(\Delta+\lambda)^\wedge(h)\int_{\mathbb{R}} F(x)\overline{\Phi_h(x)}\,dx. \tag{6.35}$$

This yields

$$\begin{aligned}\lim_{N\to\infty}\Bigg(\sum_{\substack{|g|\leq N\\ g\in\tau\mathbb{Z}}} F(g) - \sum_{\substack{(\Delta+\lambda)^\wedge(h)=0\\ h\in\frac{1}{\tau}\mathbb{Z}}}\int_{-N}^{N} F(x)\overline{\Phi_h(x)}\,dx\Bigg) \\ = \sum_{\substack{(\Delta+\lambda)^\wedge(h)\neq 0\\ h\in\frac{1}{\tau}\mathbb{Z}}}\int_{\mathbb{R}} F(x)\overline{\Phi_h(x)}\,dx.\end{aligned} \tag{6.36}$$

We treat two cases:

(i) If $\lambda \notin \mathrm{Spect}_\Delta(\tau\mathbb{Z})$ and $F \in \mathrm{C}^{(2)}(\mathbb{R})$ satisfies (6.29) and (6.30), then we have

$$\lim_{N\to\infty}\sum_{\substack{|g|\leq N\\ g\in\tau\mathbb{Z}}} F(g) = \sum_{h\in\frac{1}{\tau}\mathbb{Z}}\int_{\mathbb{R}} F(x)\overline{\Phi_h(x)}\,dx. \tag{6.37}$$

(ii) If $\lambda \in \mathrm{Spect}_\Delta(\tau\mathbb{Z})$ and $F \in \mathrm{C}^{(2)}(\mathbb{R})$ satisfies (6.29), (6.30), and (6.31), then we find

$$\begin{aligned}\lim_{N\to\infty}\sum_{\substack{|g|\leq N\\ g\in\tau\mathbb{Z}}} F(g) = & \sum_{\substack{(\Delta+\lambda)^\wedge(h)=0\\ h\in\frac{1}{\tau}\mathbb{Z}}}\int_{\mathbb{R}} F(x)\overline{\Phi_h(x)}\,dx \\ & + \sum_{\substack{(\Delta+\lambda)^\wedge(h)\neq 0\\ h\in\frac{1}{\tau}\mathbb{Z}}}\int_{\mathbb{R}} F(x)\overline{\Phi_h(x)}\,dx.\end{aligned} \tag{6.38}$$

Summing up the expressions on the right side of (6.38) we obtain the desired result. □

Going over to shifted lattices $\{x\}+\tau\mathbb{Z}$ we obtain the following corollaries.

Corollary 6.8. *Let x be a point in $\mathbb{R}$. Then, under the assumptions of Theorem 6.6,*

$$\begin{aligned}\lim_{N\to\infty}\Bigg(\sum_{\substack{|x+g|\leq N\\ g\in\tau\mathbb{Z}}} F(x+g) - \sum_{\substack{(\Delta+\lambda)^\wedge(h)=0\\ h\in\frac{1}{\tau}\mathbb{Z}}}\int_{-N}^{N} F(y)\overline{\Phi_h(y)}\,dy\;\Phi_h(x)\Bigg) \\ = \int_{\mathbb{R}} G(\tau\mathbb{Z};\Delta+\lambda;x-y)(\Delta+\lambda)F(y)\,dy.\end{aligned} \tag{6.39}$$

Corollary 6.9. *Let x be a point in $\mathbb{R}$. Then, under the assumptions of Theorem 6.7,*

$$\lim_{N\to\infty} \sum_{\substack{|x+g|\le N \\ g\in\tau\mathbb{Z}}} F(x+g) = \sum_{h\in\frac{1}{\tau}\mathbb{Z}} \int_{\mathbb{R}} F(y)\overline{\Phi_h(y)}\, dy\ \Phi_h(x). \tag{6.40}$$

Clearly, the Poisson summation formula (Theorem 6.7) can be extended by use of iterated operators $(\Delta+\lambda)^m$, $\lambda \in \mathbb{R}$, $m \in \mathbb{N}$, in a straightforward way.

6.4 Adaptive Trapezoidal Sums over Infinite Intervals

The criteria leading to the Poisson summation formula for Helmholtz operators turn out to be extremely useful to formulate adaptive trapezoidal rules over infinite intervals. Again we distinguish two cases which are derivable from Corollary 6.8:

(i) $\lambda \notin \mathrm{Spect}_\Delta(\tau\mathbb{Z})$: Then, for every $x \in \mathbb{R}$, an adaptive trapezoidal rule of the following kind

$$\tau \sum_{g\in\tau\mathbb{Z}} F(x+g) = \int_{\mathbb{R}} F(y)\, dy + \tau \int_{\mathbb{R}} G_0^{\perp}(\tau\mathbb{Z};\Delta+\lambda; x-y)(\Delta+\lambda)F(y)\, dy. \tag{6.41}$$

is available, where the remainder term is of the order $O(\tau^2)$.

(ii) $\lambda \in \mathrm{Spect}_\Delta(\tau\mathbb{Z})\setminus\{0\}$: Then, for any prescribed nodal system $x_1,\ldots,x_N$ of $N\ge 2$ mutually distinct points and any selection of weights $a_1,\ldots,a_N$ satisfying $\sum_{i=1}^N a_i = 1$ and $\sum_{i=1}^N a_i\Phi_h(x_i) = 0$ for all $h \in \frac{1}{\tau}\mathbb{Z}$ with $(\Delta+\lambda)^\wedge(h) = 0$ the following adaptive trapezoidal rule

$$\tau \sum_{i=1}^N a_i \sum_{g\in\tau\mathbb{Z}} F(x_i+g) = \int_{\mathbb{R}} F(y)\, dy + \tau \sum_{i=1}^N a_i \int_{\mathbb{R}} G_0^{\perp}(\tau\mathbb{Z};\Delta+\lambda; x_i-y)(\Delta+\lambda)F(y)\, dy. \tag{6.42}$$

is valid, where the remainder term is of the order $O(\tau^2)$.

All in all, trapezoidal rules are simply structured and easy to handle. The convergence, however, is not rapid (exceptions must be made, e.g., in the case of certain periodic functions).

6.5 Discontinuous Integrals of Hardy–Landau Type

In what follows we make the attempt to apply the assumptions (6.19) and (6.20) of Theorem 6.6 to the function

$$F(x) = \operatorname{sinc}(2\pi Rx) = \frac{\sin(2\pi Rx)}{2\pi Rx}, \quad x \in \mathbb{R},\ R > 0. \tag{6.43}$$

An elementary calculation yields the identities

$$\nabla F(x) = \frac{\cos(2\pi Rx)}{x} - \frac{\sin(2\pi Rx)}{2\pi Rx^2} \tag{6.44}$$

and

$$\Delta F(x) = -4\pi^2 R^2 \frac{\sin(2\pi Rx)}{2\pi Rx} - 2\frac{\cos(2\pi Rx)}{x^2} + 2\frac{\sin(2\pi Rx)}{2\pi Rx^3}. \tag{6.45}$$

In other words, we have

$$\left(\Delta + 4\pi^2 R^2\right) \frac{\sin(2\pi Rx)}{2\pi Rx} = O\left(\frac{1}{x^2}\right), \quad x \to \pm\infty. \tag{6.46}$$

Therefore, F is an arbitrarily often differentiable function in $\mathbb{R}$ satisfying (6.20). Furthermore, the integral

$$\int_{\mathbb{R}} F(x)\, dx = \lim_{R\to\infty} \int_{-R}^{R} F(x)\, dx \tag{6.47}$$

exists. Unfortunately, *the Laplace derivative, i.e., the second order derivative (6.45) is not absolutely integrable over* $\mathbb{R}$. From (6.45), however, we see that the Helmholtz derivative $(\Delta + 4\pi R^2)F$ is absolutely integrable over $\mathbb{R}$, i.e.,

$$\int_{\mathbb{R}} \left| \left(\Delta + 4\pi^2 R^2\right) F(x) \right| \, dx < \infty. \tag{6.48}$$

As a consequence, choosing

$$\varepsilon = 1, \quad \lambda = 4\pi^2 R^2, \tag{6.49}$$

we see that the sinc-function F, given by (6.43), satisfies all properties (6.29), (6.30), and (6.31) listed in Theorem 6.7. Therefore we are allowed to deduce that

$$\begin{aligned}\lim_{N\to\infty} \Bigg(\sum_{\substack{|x+g|\le N \\ g\in\tau\mathbb{Z}}} F(x+g) - \sum_{\substack{(\Delta+4\pi^2R^2)^\wedge(h)=0 \\ h\in\frac{1}{\tau}\mathbb{Z}}} \int_{-N}^{N} F(y)\overline{\Phi_h(y)}\, dy\ \Phi_h(x) \Bigg) \\ = \int_{\mathbb{R}} G(\tau\mathbb{Z}; \Delta + 4\pi^2R^2; x-y)(\Delta + 4\pi^2 R^2)F(y)\, dy. \end{aligned} \tag{6.50}$$

We distinguish two cases derivable from (6.50):

Example 6.10. *Let $4\pi^2R^2 \notin \mathrm{Spect}_\Delta(\tau\mathbb{Z})$. For every $x \in \mathbb{R}$ an adaptive trapezoidal rule reads*

$$\tau \sum_{g\in\tau\mathbb{Z}} \frac{\sin(2\pi R(x+g))}{2\pi R(x+g)} + \mathcal{E}F = \int_{\mathbb{R}} \frac{\sin(2\pi Ry)}{2\pi Ry}\, dy \tag{6.51}$$

is available, where the remainder term

$$\mathcal{E}F = \sum_{\substack{h\neq 0\\ h\in\frac{1}{\tau}\mathbb{Z}}} \frac{1}{2\pi^2(R^2-h^2)} \int_{\mathbb{R}} e^{2\pi i h(x-y)} \left(\frac{\sin(2\pi Ry)}{2\pi Ry^3} - \frac{\cos(2\pi Ry)}{y^2}\right) dy \tag{6.52}$$

is of the order $O(\tau^2)$.

Example 6.11. *Let now $4\pi^2R^2 \in \mathrm{Spect}_\Delta(\tau\mathbb{Z})$. For any prescribed nodal system $x_1,\ldots,x_N$ of $N \geq 2$ mutually distinct points and any selection of weights $a_1,\ldots,a_N$ satisfying $\sum_{i=1}^N a_i = 1$ and $\sum_{i=1}^N a_i\Phi_h(x_i) = 0$ for all $h \in \frac{1}{\tau}\mathbb{Z}$ with $|h| = R$ an adaptive trapezoidal rule reads as follows*

$$\tau\sum_{i=1}^N a_i \sum_{g\in\tau\mathbb{Z}} \frac{\sin(2\pi Rx_i + g)}{2\pi Rx_i + g} + \mathcal{E}F = \int_{\mathbb{R}} \frac{\sin(2\pi Ry)}{2\pi Ry}\, dy \tag{6.53}$$

is valid, where the remainder term

$$\mathcal{E}F = \sum_{i=1}^N a_i \sum_{\substack{h\in\frac{1}{\tau}\mathbb{Z}\\ h\neq 0, |h|\neq R}} \frac{1}{2\pi^2(R^2-h^2)} \int_{\mathbb{R}} e^{2\pi i h(x_i-y)} \left(\frac{\sin(2\pi Ry)}{2\pi Ry^3} - \frac{\cos(2\pi Ry)}{y^2}\right) dy \tag{6.54}$$

is of the order $O(\tau^2)$.

Finally it is worth mentioning that, in accordance with the criteria stated in Theorem 6.7, we are able to deduce that

$$\sum_{g\in\tau\mathbb{Z}} \frac{\sin(2\pi Rg)}{2\pi Rg} = \int_{\mathbb{R}} \frac{\sin(2\pi Rx)}{2\pi Rx}\, dx + \sum_{\substack{h\neq 0\\ h\in\frac{1}{\tau}\mathbb{Z}}} \int_{\mathbb{R}} \frac{\sin(2\pi Rx)}{2\pi Rx} e^{-2\pi i h x}\, dx. \tag{6.55}$$

Moreover, it should be noted that the (discontinuous) integral on the right side of (6.55) can be calculated explicitly by elementary manipulations. Indeed, we have (see, e.g., M. Abramowitz, I.A. Stegun [1972])

$$\lim_{N\to\infty} \int_{-N}^{N} \frac{\sin(2\pi Rx)}{\pi x} e^{-2\pi i h x}\, dx = \begin{cases} 1, & |h| < R, \quad h \in \frac{1}{\tau}\mathbb{Z}, \\ \frac{1}{2}, & |h| = R, \quad h \in \frac{1}{\tau}\mathbb{Z}, \\ 0, & |h| > R, \quad h \in \frac{1}{\tau}\mathbb{Z}. \end{cases} \tag{6.56}$$

Thus, we end up with the *one-dimensional Hardy–Landau identity* (cf. W. Freeden [2011])

$$\sum_{\substack{|h|\leq R \\ h\in\tau\mathbb{Z}}}{}' 1 = 2R + 2R \sum_{\substack{h\neq 0 \\ h\in\frac{1}{\tau}\mathbb{Z}}} \frac{\sin(2\pi Rh)}{2\pi Rh}. \tag{6.57}$$

6.6 Periodic Polynomial Accuracy

We start from the adaptive Euler summation formula in Theorem 6.3. Under the assumption $\lambda \in \mathbb{R} \setminus \mathbb{L}_m$ with $\mathbb{L}_m = \{\Delta^\wedge(k) : k \geq m, k \in \mathbb{N}\}$ we introduce

$$G^{\perp}_{0,\ldots,m-1}(\tau\mathbb{Z};\Delta+\lambda;x) = G(\tau\mathbb{Z};\Delta+\lambda;x) - \frac{1}{\sqrt{\tau}} \sum_{\substack{\lambda\neq\Delta^\wedge(h) \\ h\leq m-1, h\in\frac{1}{\tau}\mathbb{Z}}} \frac{1}{\lambda - \Delta^\wedge(h)}\Phi_h(x). \tag{6.58}$$

The Fourier series of $G^{\perp}_{0,\ldots,m-1}(\tau\mathbb{Z};\Delta+\lambda;x)$, $x \in \mathbb{R}$, is given by

$$G^{\perp}_{0,\ldots,m-1}(\tau\mathbb{Z};\Delta+\lambda;x) = \frac{1}{\sqrt{\tau}} \sum_{\substack{\lambda\neq\Delta^\wedge(h) \\ h\geq m, h\in\frac{1}{\tau}\mathbb{Z}}} \frac{1}{\lambda - \Delta^\wedge(h)}\Phi_h(x). \tag{6.59}$$

Integrating by parts we therefore obtain from Theorem 6.3 and Corollary 6.4.

Corollary 6.12 (Extended Trapezoidal Rule)**.** *Let x be an arbitrary point of $\mathbb{R}$. Suppose that F is of class $C^{(2)}[a,b]$. Then, for $\lambda \in \mathbb{R}\setminus\mathbb{L}_m$,*

$$\begin{aligned}\sum_{\substack{g+x\in[a,b] \\ g\in\tau\mathbb{Z}}}{}' F(g+x) &= \frac{1}{\tau}\int_a^b F(y)\,dy + \sum_{\substack{0\leq h\leq m-1 \\ \Delta^\wedge(h)=\lambda, h\in\frac{1}{\tau}\mathbb{Z}}} \int_a^b F(y)\overline{\Phi_h(y)}\,dy\;\Phi_h(x) \\ &+ \left(F(y)\nabla_y G^{\perp}_{0,\ldots,m-1}(\tau\mathbb{Z};\Delta+\lambda;x-y)\right)\Big|_a^b \\ &- \left(G^{\perp}_{0,\ldots,m-1}(\tau\mathbb{Z};\Delta+\lambda;x-y)\nabla_y F(y)\right)\Big|_a^b \\ &+ \int_a^b G^{\perp}_{0,\ldots,m-1}(\tau\mathbb{Z};\Delta+\lambda;x-y)(\Delta_y+\lambda)F(y)\,dy. \end{aligned} \tag{6.60}$$

Integration of 2π-Periodic Functions. We choose $a = 0$ and $b = 2\pi$ (cf. V.I. Krylov [1962], R. von Mises [1933], J. Radon [1935]). Moreover, we let F be of class $C^{(2)}_{2\pi\mathbb{Z}}(\mathbb{R})$. Then we are able to deduce from Corollary 6.12

that, for $\tau = \frac{2\pi}{m}$, $m \in \mathbb{N}$,

$$\frac{2\pi}{m} \sum_{\substack{g\in[0,2\pi] \\ g\in\frac{2\pi}{m}\mathbb{Z}}}{}' F(g) = \int_0^{2\pi} F(y)\, dy + \frac{2\pi}{m} \sum_{\substack{0\le h\le m-1 \\ 4\pi^2h^2\neq\lambda, h\in\frac{m}{2\pi}\mathbb{Z}}} \int_0^{2\pi} F(y)\overline{\Phi_h(y)}\, dy\ \Phi_h(x)$$

$$+ \frac{2\pi}{m} \int_0^{2\pi} G^{\perp}_{0,\ldots,m-1}\left(\frac{2\pi}{m}\mathbb{Z}; \Delta+\lambda; y\right)(\Delta_y+\lambda)F(y)\, dy \tag{6.61}$$

holds for all $\lambda \in \mathbb{R} \setminus \mathbb{L}_m$. Because of the 2π-periodicity of F we are able to write

$$\frac{2\pi}{m} \sum_{\substack{g\in[0,2\pi] \\ g\in\frac{2\pi}{m}\mathbb{Z}}}{}' F(g) = \frac{2\pi}{m} \sum_{k=1}^{m} F\left((k-1)\frac{2\pi}{m}\right). \tag{6.62}$$

Furthermore, we notice that the integration rule

$$\int_0^{2\pi} F(y)\, dy = \frac{2\pi}{m}\sum_{k=1}^{m} F\left((k-1)\frac{2\pi}{m}\right) = \frac{2\pi}{m}\sum_{k=0}^{m-1} F\left(k\frac{2\pi}{m}\right) \tag{6.63}$$

holds true for all periodic polynomials F up to degree $m-1$ in the interval $[0, 2\pi]$ (see, e.g., R. von Mises [1933], P. Henrici [1964], J. Stoer [1989]). In order to guarantee this it is sufficient to verify that (6.63) will be exact for the functions $x \mapsto e^{ilx}$, $x \in \mathbb{R}$, $l = 1, \ldots, m-1$. For $m = 0$, the assertion is evidently true. Choosing $l \in \{1, \ldots m-1\} \subset \mathbb{N}$ we obtain

$$\int_0^{2\pi} e^{ily} dy = \frac{1}{il}\left(e^{2\pi il} - 1\right) = 0 \tag{6.64}$$

and

$$\sum_{k=0}^{m-1} e^{il(k\frac{2\pi}{m})} = \frac{e^{ilm\frac{2\pi}{m}} - 1}{e^{il\frac{2\pi}{m}} - 1} = \frac{e^{2\pi il} - 1}{e^{il\frac{2\pi}{m}} - 1} = 0. \tag{6.65}$$

As an immediate consequence of (6.61) we therefore find in combination with our integral formulas that

$$\int_0^{2\pi} F(y)\, dy = \sum_{k=0}^{m-1} \frac{2\pi}{m} F\left(\frac{2\pi k}{m}\right) \tag{6.66}$$

$$- \frac{2\pi}{m} \int_0^{2\pi} G^{\perp}_{0,\ldots,m-1}\left(\frac{2\pi}{m}\mathbb{Z}; \Delta+\lambda; y\right)(\Delta_y+\lambda)F(y)\, dy$$

is valid for all $F \in C^{(2)}_{2\pi\mathbb{Z}}(\mathbb{R})$ and $\lambda \in \mathbb{R} \setminus \mathbb{L}_m$. Note that in this case $\mathbb{L}_m = \{m^2l^2 : l \in \mathbb{N}_0\}$ since $\Delta^\wedge(h) = 4\pi^2h^2$, $h \in \frac{1}{\tau}\mathbb{Z}$, and $\tau = \frac{2\pi}{m}$, i.e. $\Delta^\wedge(l) = m^2l^2$ with $l \in \mathbb{Z}$.

The remainder term

$$\mathcal{E}_{m-1}F = -\int_0^{2\pi} G^{\perp}_{0,\ldots,m}\left(\frac{2\pi}{m}\mathbb{Z}; \Delta+\lambda; y\right)(\Delta_y+\lambda)F(y)\, dy \tag{6.67}$$

of the quadrature depends on the properties of the integrand F. It adaptively provides an appropriate representation if the parameter λ can be chosen such that $(\Delta + \lambda)F$ is small (with respect to a certain topology).

An adaptive estimate in the sense of A. Sard [1949, 1963] is obtainable within the $L^2_{2\pi\mathbb{Z}}(0, 2\pi)$-topology as follows:

$$|\mathcal{E}_{m-1}F| \leq \sqrt{(G^{\perp}_{0,\ldots,m})^{(2)}\left(\frac{2\pi}{m}\mathbb{Z}; \Delta + \lambda, 0\right)} \sqrt{\int_0^{2\pi} |(\Delta_y + \lambda)F(y)|^2 \, dy}, \tag{6.68}$$

where

$$(G^{\perp}_{0,\ldots,m-1})^{(2)}\left(\frac{2\pi}{m}\mathbb{Z}; \Delta + \lambda; 0\right) = \frac{m}{2\pi} \sum_{\substack{m^2l^2 \neq \lambda \\ l \geq m}} \left(\frac{1}{\lambda - m^2 l^2}\right)^2 . \tag{6.69}$$

Chapter 7

Legendre Polynomial Reflected Integration

In order to discuss integral formulas for Gaussian quadrature it is helpful to have some basic knowledge about classical Legendre polynomials. We start with its explicit representation. Then we list some of the properties which turn out to be useful in Gaussian quadrature. More details can be found, e.g., in P.J. Davis, P. Rabinowitz [1984], W. Freeden, M. Gutting [2013], W. Freeden, M. Schreiner [2009], C. Müller [1969, 1998], N.N. Lebedev [1973], G. Szegö [1975].

7.1 Legendre Polynomials

We begin with the explicit definition of the Legendre polynomial.

Definition 7.1. *The function* $P_n : [-1,1] \to \mathbb{R}$, $n = 0, 1, \ldots,$ *defined by*

$$P_n(t) = \sum_{s=0}^{\lfloor \frac{n}{2} \rfloor} (-1)^s \frac{(2n-2s)!}{2^n (n-2s)!(n-s)!s!} t^{n-2s}, \quad t \in [-1,1] \tag{7.1}$$

is called the Legendre polynomial.

$P_n : [-1,1] \to \mathbb{R}$ is uniquely determined by the properties:

(i) P_n is a polynomial of degree n,

(ii) $\int_{-1}^{1} P_n(t) P_m(t) \, dt = 0$ for $n \neq m$,

(iii) $P_n(1) = 1$.

This is easily seen from the usual process of orthogonalization. In particular, we have for $n = 0, \ldots, 4$

$$P_0(t) = 1, \qquad P_1(t) = t, \qquad P_2(t) = \frac{3}{2}t^2 - \frac{1}{2}, \tag{7.2}$$

$$P_3(t) = \frac{5}{2}t^3 - \frac{3}{2}t, \quad P_4(t) = \frac{35}{8}t^4 - \frac{15}{4}t^2 + \frac{3}{8}.$$

A graphical impression of some Legendre polynomials can be found in Figure 7.1.

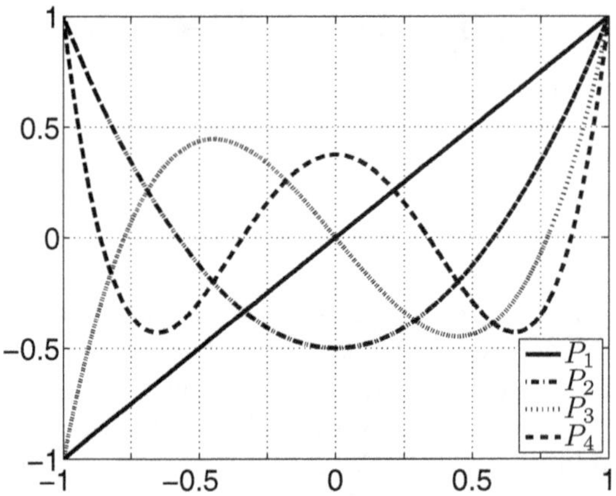

FIGURE 7.1: Legendre polynomials $t \mapsto P_n(t)$, $t \in [-1, 1]$, $n = 1, \ldots, 4$.

Furthermore,

$$\int_{-1}^{1} P_n(t)P_m(t)\, dt = \frac{2}{2n+1}\delta_{n,m}. \tag{7.3}$$

The Legendre polynomial satisfies the estimate

$$|P_n(t)| \leq P_n(1) = 1, \quad t \in [-1, 1]. \tag{7.4}$$

Moreover, we have

$$|P_n^{(k)}(t)| \leq |P_n^{(k)}(1)|, \tag{7.5}$$

where $P_n^{(k)}(1) = O(n^{2k})$. In particular, we have

$$P_n'(1) = \frac{n(n+1)}{2}. \tag{7.6}$$

Furthermore, for $k = 2, 3, \ldots, n$,

$$P_n^{(k)}(1) = \left(\frac{1}{2}\right)^k \frac{1}{k!} \prod_{j=1}^{k} \left(n(n+1) - j(j-1)\right). \tag{7.7}$$

A straightforward calculation yields

$$\left(\mathrm{L}_t + n(n+1)\right) P_n(t) = 0, \quad t \in [-1, 1], \tag{7.8}$$

where

$$\mathrm{L}_t = \left((1-t^2)\left(\frac{d}{dt}\right)^2 - 2t\frac{d}{dt}\right) \tag{7.9}$$

is called the *Legendre operator*. We therefore obtain the following lemma.

Lemma 7.2. *The Legendre polynomial P_n is the only twice differentiable eigenfunction of the "Legendre operator" L as defined in (7.9) on the interval $[-1,1]$, corresponding to the eigenvalues $\lambda_n = -\mathrm{L}^\wedge(n) = -n(n+1)$, $n = 0, 1, \ldots$, that is bounded on $[-1,1]$ with $P_n(1) = 1$.*

The differential equation (7.8) shows that P_n and P_n' cannot vanish simultaneously such that P_n has no multiple zeros. The orthogonality relation for Legendre polynomials implies that P_n has at most k different zeros, $z_1, \ldots z_k$, $k \leq n$, in the interval $(-1,1)$. Letting

$$I_k(t) = \prod_{j=1}^{k} (t - z_j) \tag{7.10}$$

we get $I_k(1) > 0$ and $P_n = J_{n-k} I_k$. The polynomial J_{n-k} is positive in $[-1,1]$, and we have

$$\int_{-1}^{1} P_n(t) I_k(t)\, dt = \int_{-1}^{1} J_{n-k}(t) I_k^2(t)\, dt > 0. \tag{7.11}$$

As P_n is orthogonal to all polynomials of degrees $< n$, this is possible only for the case $k = n$. Thus, we are able to conclude that P_n has n different zeros in the interval $(-1,1)$.

Lemma 7.3. *The Legendre polynomial P_n has n different zeros in the interval $(-1,1)$.*

From the binomial theorem it follows that

$$(t^2 - 1)^n = \sum_{s=0}^{n} (-1)^s \frac{n!}{(n-s)!s!} t^{2n-2s}, \quad n = 0, 1, \ldots. \tag{7.12}$$

For all $s \leq \lfloor n/2 \rfloor$ we find

$$\left(\frac{d}{dt}\right)^n t^{2n-2s} = \frac{(2n-2s)!}{(n-2s)!} t^{n-2s}, \tag{7.13}$$

while for $\lfloor n/2 \rfloor < s \leq n$ we get

$$\left(\frac{d}{dt}\right)^n t^{2n-2s} = 0. \tag{7.14}$$

Therefore we see that

$$\left(\frac{d}{dt}\right)^n (t^2 - 1)^n = \sum_{s=0}^{\lfloor \frac{n}{2} \rfloor} (-1)^s \frac{n!}{(n-s)!s!} \frac{(2n-2s)!}{(n-2s)!} t^{n-2s}. \tag{7.15}$$

By comparison with the definition of the Legendre polynomial we therefore obtain the *Rodriguez formula.*

Lemma 7.4. *For $n = 0, 1, \ldots,$*

$$P_n(t) = \frac{1}{2^n n!}\left(\frac{d}{dt}\right)^n (t^2-1)^n, \quad t \in [-1,1]. \tag{7.16}$$

Using integration by parts we are led to the *Rodriguez rule*

$$\int_{-1}^{1} F(t)P_n(t)\, dt = \frac{1}{2^n n!}\int_{-1}^{1} F^{(n)}(t)(1-t^2)^n\, dt \tag{7.17}$$

for every $F \in \mathrm{C}^{(n)}[-1,1]$. Now we discuss the integral relation that follows by integration by parts from

$$\int_{-1}^{1} P_k(t)(P_n'(t) - P_{n-2}'(t))\, dt = -\int_{-1}^{1} (P_n(t) - P_{n-2}(t))P_k'(t)\, dt. \tag{7.18}$$

Note that $P_n' - P_{n-2}'$ is a polynomial of degree $n-1$ on $[-1,1]$. Consequently the integral (7.18) vanishes for $k \geq n$. On the other hand, P_k' is a polynomial of degree $k-1$. This means that the integral vanishes also for $k < n-1$ so that (7.18) differs from zero only for $k = n-1$. Thus, there is $c_n \in \mathbb{R}$ such that

$$P_n'(t) - P_{n-2}'(t) = c_n P_{n-1}(t), \quad t \in [-1,1]. \tag{7.19}$$

In connection with (7.6) we find

$$P_n'(1) - P_{n-2}'(1) = \frac{n(n+1)}{2} - \frac{(n-2)(n-1)}{2} = 2n-1. \tag{7.20}$$

Therefore, it follows that

$$P_n'(t) - P_{n-2}'(t) = (2n-1)P_{n-1}(t), \quad t \in [-1,1]. \tag{7.21}$$

By similar arguments we are able to show that

$$P_{n+1}'(t) - tP_n'(t) = (n+1)P_n(t), \tag{7.22}$$

$$(t^2-1)P_n'(t) = ntP_n(t) - nP_{n-1}(t), \tag{7.23}$$

$$(n+1)P_{n+1}(t) + nP_{n-1}(t) = (2n+1)tP_n(t). \tag{7.24}$$

The formulas (7.22)–(7.24) are known as the *recurrence formulas for the Legendre polynomials.* Moreover we have the following result.

Lemma 7.5. *For $n = 1, 2, \ldots,$ $t \in [-1,1]$,*

$$(t^2-1)P_n'(t) = \frac{n(n+1)}{2n+1}(P_{n+1}(t) - P_{n-1}(t)). \tag{7.25}$$

Proof. Inserting (7.24) into (7.23) we find

$$\begin{aligned}(t^2-1)P_n'(t) &= n(tP_n(t)-P_{n-1}(t)) \\ &= n\left(\frac{n+1}{2n+1}P_{n+1}(t)+\frac{n}{2n+1}P_{n-1}(t)-P_{n-1}(t)\right) \\ &= n\left(\frac{n+1}{2n+1}P_{n+1}(t)+\frac{n-(2n+1)}{2n+1}P_{n-1}(t)\right) \\ &= \frac{n(n+1)}{2n+1}\left(P_{n+1}(t)-P_{n-1}(t)\right).\end{aligned} \tag{7.26}$$

This is the desired result. □

The power series

$$\phi(h)=\sum_{n=0}^{\infty}P_n(t)h^n, \quad t\in[-1,1], \tag{7.27}$$

is absolutely and uniformly convergent for all h with $|h|\le h_0, h_0\in[0,1)$. By differentiation with respect to h and comparing coefficients according to (7.27) we find

$$(1+h^2-2ht)\phi'(h)=(t-h)\phi(h). \tag{7.28}$$

This differential equation is uniquely solvable under the initial condition $\phi(0)=1$. Since it is not hard to show that

$$h\mapsto(1+h^2-2ht)^{-1/2}, \quad h\in(-1,1), \tag{7.29}$$

solves this initial value problem, we have the following *generating series expansion of the Legendre polynomials.*

Lemma 7.6. *For $t\in[-1,1]$ and all $h\in(-1,1)$*

$$\sum_{n=0}^{\infty}P_n(t)h^n=\frac{1}{\sqrt{1+h^2-2ht}}. \tag{7.30}$$

The power series on the left-hand side in Lemma 7.6 can be differentiated for all $h\in(-1,1)$. Thus it follows that

$$-\frac{h-t}{(1+h^2-2ht)^{3/2}}=\sum_{n=1}^{\infty}nP_n(t)h^{n-1}. \tag{7.31}$$

Now it is easy to see that

$$\frac{1}{\sqrt{1+h^2-2ht}}-\frac{2h^2-2ht}{(1+h^2-2ht)^{3/2}}=\frac{1-h^2}{(1+h^2-2ht)^{3/2}}. \tag{7.32}$$

This gives us the following result.

Lemma 7.7. *For all $t \in [-1,1]$ and $h \in (-1,1)$*

$$\frac{1-h^2}{(1+h^2-2ht)^{3/2}} = \sum_{n=0}^{\infty}(2n+1)h^n P_n(t). \tag{7.33}$$

Lemma 7.6 can be used to prove an integral representation for the Legendre polynomial. To this end we start from the well-known elementary integral

$$\int_0^{\pi} \frac{d\varphi}{1+\gamma\cos\varphi} = \frac{\pi}{\sqrt{1-\gamma^2}}, \quad |\gamma| < 1. \tag{7.34}$$

We set

$$\gamma = -\frac{h\sqrt{t^2-1}}{1-ht}, \qquad |h| < 1. \tag{7.35}$$

On the one hand it follows that

$$\begin{aligned}\int_0^{\pi} \frac{1}{1+\gamma\cos\varphi}\, d\varphi &= (1-ht)\int_0^{\pi} \left[1-h(t+\sqrt{t^2-1}\cos\varphi)\right]^{-1} d\varphi \\ &= (1-ht)\int_0^{\pi} \sum_{n=0}^{\infty}(t+\sqrt{t^2-1}\cos\varphi)^n h^n \, d\varphi. \end{aligned} \tag{7.36}$$

On the other hand we obtain

$$\frac{\pi}{\sqrt{1-\gamma^2}} = \frac{\pi}{\sqrt{1-h^2(t^2-1)/(1-ht)^2}} = \frac{\pi(1-ht)}{\sqrt{1+h^2-2ht}}. \tag{7.37}$$

In connection with Lemma 7.6 this yields

$$\frac{\pi}{\sqrt{1-\gamma^2}} = (1-ht)\pi\sum_{n=0}^{\infty} P_n(t)h^n. \tag{7.38}$$

By comparison we therefore obtain

$$\sum_{n=0}^{\infty}\int_0^{\pi}(t+\sqrt{t^2-1}\cos\varphi)^n \, d\varphi \; h^n = \pi\sum_{n=0}^{\infty} P_n(t)h^n. \tag{7.39}$$

This gives us the *Laplace representation of Legendre polynomials* (cf. C. Müller [1969]).

Lemma 7.8. *For $t \in [-1,1]$ and $n = 0,1,\ldots$*

$$P_n(t) = \frac{1}{\pi}\int_0^{\pi}(t+\sqrt{t^2-1}\cos\varphi)^n \, d\varphi. \tag{7.40}$$

By the representation ($i^2 = -1$)

$$P_n(t) = \frac{1}{2\pi}\int_0^{2\pi}\left(t+i\sqrt{1-t^2}\cos\varphi\right)^n d\varphi \tag{7.41}$$

we obtain an estimate valid for arbitrary $t \in [-1,1]$ (cf. C. Müller [1969]):

$$|P_n(t)| \leq \frac{1}{2\pi} \int_0^{2\pi} |t + i\sqrt{1-t^2}\cos\varphi|^n \, d\varphi \tag{7.42}$$
$$\leq (|t| + |\sqrt{1-t^2}|)^n \leq (|t| + \sqrt{1+|t|^2})^n \leq 2^n(1+|t|^2)^{n/2}.$$

Moreover, it follows by the substitution $s = \cos\varphi$ that (7.41) is equivalent to

$$P_n(t) = \frac{1}{\pi} \int_{-1}^{1} (t + is\sqrt{1-t^2})^n (1-s^2)^{-1/2} \, ds, \tag{7.43}$$

so that we get from $|t + is\sqrt{1-t^2}| = (1-(1-s^2)(1-t^2))^{1/2}$ the estimate

$$|P_n(t)| \leq \frac{1}{\pi} \int_{-1}^{1} e^{\frac{n}{2}\ln(1-(1-t^2)(1-s^2))} (1-s^2)^{-1/2} \, ds. \tag{7.44}$$

Since

$$\ln(1-(1-s^2)(1-t^2)) \leq -(1-s^2)(1-t^2) \tag{7.45}$$

we obtain

$$|P_n(t)| \leq \frac{1}{\pi} \int_{-1}^{1} e^{-\frac{n}{2}(1-s^2)(1-t^2)} (1-s^2)^{-1/2} \, ds. \tag{7.46}$$

hence,

$$|P_n(t)| \leq \frac{2}{\pi} \int_{0}^{1} e^{-\frac{n}{2}(1-s^2)(1-t^2)} (1-s^2)^{-1/2} \, ds. \tag{7.47}$$

By the substitution $s = 1-u$ we find in connection with $u \leq 1-s^2 \leq 2u$

$$|P_n(t)| \leq \frac{\sqrt{2}}{\pi} \int_0^1 e^{-\frac{n}{2}u(1-t^2)} u^{-1/2} \, du$$
$$\leq \frac{\sqrt{2}}{\pi} \int_0^\infty e^{-\frac{n}{2}u(1-t^2)} u^{-1/2} \, du$$
$$= \frac{2}{\pi} \frac{\sqrt{\pi}}{(n(1-t^2))^{1/2}} = \frac{1}{\sqrt{\pi}} \left(\frac{4}{n(1-t^2)} \right)^{1/2}. \tag{7.48}$$

Lemma 7.9 (Estimate of the Legendre polynomial). *For $n = 1, 2, \ldots$ and $t \in (-1,1)$,*

$$|P_n(t)| \leq \frac{1}{\sqrt{\pi}} \left(\frac{4}{n(1-t^2)} \right)^{1/2}. \tag{7.49}$$

7.2 Legendre (Green's) Functions

As already known from Lemma 7.2, the Legendre operator L has a half-bounded and discrete eigenspectrum $\{L^\wedge(n)\}_{n=0,1,\ldots}$ such that

$$(L_t + L^\wedge(n)) P_n^*(t) = 0, \quad t \in [-1,1], \tag{7.50}$$

where P_n^* is the normalized Legendre polynomial of degree n, i.e., due to (7.3)

$$P_n^*(t) = \sqrt{\frac{2n+1}{2}} P_n(t), \tag{7.51}$$

and

$$\mathrm{L}^\wedge(n) = n(n+1), \quad n = 0, 1, \ldots. \tag{7.52}$$

Thus,

$$\mathrm{Spect}_{\mathrm{L}} = \{\mathrm{L}^\wedge(n) \in \mathbb{R} : \mathrm{L}^\wedge(n) = n(n+1), \quad n = 0, 1, \ldots\} \tag{7.53}$$

is the *eigenspectrum of the Legendre operator.*

The Hilbert theory of Green's functions leads to the following definition (cf. W. Freeden [1980a]).

Definition 7.10. *A function* $G(\mathrm{L}+\lambda;\cdot,\cdot)\colon [-1,1]\times[-1,1] \to \mathbb{R}, (x,t) \mapsto G(\mathrm{L}+\lambda;x,t), \lambda \in \mathbb{R}$ *is called* Legendre (Green's) function with respect to the operator $\mathrm{L}+\lambda$, $\lambda \in \mathbb{R}$, *if it satisfies the following properties:*

*(i) (*Boundedness*) For each fixed* $x \in [-1,1]$, $G(L+\lambda;x,\cdot)$ *is a continuous function on* $(-1,1)$ *satisfying the conditions*

$$|G(\mathrm{L}+\lambda;x,1)| < \infty, \tag{7.54}$$

$$|G(\mathrm{L}+\lambda;x,-1)| < \infty. \tag{7.55}$$

*(ii) (*Differential equations*) For each fixed* $x \in [-1,1]$, $G(\mathrm{L}+\lambda;x,\cdot)$ *is twice continuously differentiable in* $[-1,1]\setminus\{x\}$.

For $\lambda \notin \mathrm{Spect}_{\mathrm{L}}$ *and* $t \in [-1,1]\setminus\{x\}$ *we have*

$$(\mathrm{L}_t+\lambda)G(\mathrm{L}+\lambda;x,t) = 0, \quad t \in [-1,1]\setminus\{x\}, \tag{7.56}$$

while, for $\lambda = \mathrm{L}^\wedge(n) \in \mathrm{Spect}_{\mathrm{L}}$, *we have*

$$(\mathrm{L}_t+\lambda)G(\mathrm{L}+\lambda;x,t) = -P_n^*(x)P_n^*(t), \quad t \in [-1,1]\setminus\{x\}. \tag{7.57}$$

*(iii) (*Characteristic singularity*)*

$$(1-x^2)\frac{d}{dt}G(\mathrm{L}+\lambda;x,t)\Big|_{t=x-0}^{t=x+0} = 1. \tag{7.58}$$

*(iv) (*Normalization*) For each* $x \in [-1,1]$ *and* $\lambda = \mathrm{L}^\wedge(n) \in \mathrm{Spect}_{\mathrm{L}}$,

$$\int_{-1}^{1} G(\mathrm{L}+\mathrm{L}^\wedge(n);x,t)P_n^*(t)\,dt = 0. \tag{7.59}$$

The function P_n^* is an eigenfunction of the Green's function with respect to the operator L in the sense of the linear integral equation

$$(1-\delta_{\lambda,\mathrm{L}^\wedge(n)})P_n^*(t) = (\lambda - \mathrm{L}^\wedge(n)) \int_{-1}^{1} G(\mathrm{L}+\lambda; t, u) P_n^*(u)\, du. \tag{7.60}$$

The bilinear expansion of $G(\mathrm{L}+\lambda;\cdot,\cdot)$ therefore reads

$$G(\mathrm{L}+\lambda; x, t) = \sum_{\substack{n=0 \\ \mathrm{L}^\wedge(n)\neq\lambda}}^{\infty} \frac{1}{\lambda - \mathrm{L}^\wedge(n)} P_n^*(x) P_n^*(t), \tag{7.61}$$

where $\Sigma_{\mathrm{L}^\wedge(n)\neq\lambda}$ means that the summation is to be extended over all $n \in \mathbb{N}_0$ with $\mathrm{L}^\wedge(n) \neq \lambda$. Obviously, because of Lemma 7.9 the bilinear expansion (7.61) is absolutely and uniformly convergent both in x and t for every compact subinterval of $(-1,1)$. From the completeness of the system $\{P_n^*\}_{n=0,1,\ldots}$ we easily obtain

Lemma 7.11. *$G(\mathrm{L}+\lambda;\cdot,\cdot)$ is uniquely determined by its defining properties (i)–(iv).*

A particular role is played by the Legendre (Green's) function with respect to the operator L (i.e., $\lambda = 0$). It is explicitly available as elementary function (cf. W. Magnus et al. [1966]):

$$G(\mathrm{L}; x, t) = \begin{cases} \frac{1}{2}\ln((1+t)(1-x)) + \ln 2 - \frac{1}{2}, & x \le t \\ \frac{1}{2}\ln((1-t)(1+x)) + \ln 2 - \frac{1}{2}, & t \le x. \end{cases} \tag{7.62}$$

In the sense of the Fredholm-Hilbert theory of linear integral equations we may interpret Green's function $G(\mathrm{L}+\lambda;\cdot,\cdot)$ for $\lambda \neq 0$ as resolvent of the kernel

$$\begin{aligned} G(\mathrm{L}+\lambda; x, t) &= G(\mathrm{L}; x, t) + \lambda \int_{-1}^{1} G(\mathrm{L}+\lambda; x, u) G(\mathrm{L}; t, u)\, du \\ &\quad - \frac{1}{2\lambda} - \frac{1}{\lambda} P_n^*(x) P_n^*(t) \delta_{\lambda,\mathrm{L}^\wedge(n)}. \end{aligned} \tag{7.63}$$

Inserting the bilinear expansions we therefore obtain the following lemma.

Lemma 7.12. *For each $\lambda \neq 0$*

$$\begin{aligned} G(\mathrm{L}+\lambda; x, t) &= G(\mathrm{L}; x, t) - \lambda {\sum_{n=0}^{\infty}}^{*} \frac{1}{(\lambda - \mathrm{L}^\wedge(n))\mathrm{L}^\wedge(n)} P_n^*(x) P_n^*(t) \\ &\quad - \frac{1}{2\lambda} - \frac{1}{\lambda} P_n^*(x) P_n^*(t) \delta_{\lambda,\mathrm{L}^\wedge(n)}, \end{aligned} \tag{7.64}$$

where the series on the right-hand side converges uniformly and absolutely both in x and t on the interval $[-1,1]$, and the symbol Σ^* means, that the summation is extended over all $n \in \mathbb{N}_0$ satisfying $(\lambda - \mathrm{L}^\wedge(n))\mathrm{L}^\wedge(n) \neq 0$.

Next we introduce Legendre (Green's) functions with respect to the iterated operator $(\mathrm{L}+\lambda)^p$, $p \in \mathbb{N}$.

Definition 7.13. *For* $\lambda \in \mathbb{R}$, *the function* $G((\mathrm{L}+\lambda)^p;\cdot,\cdot)$, $p = 2,3,\ldots$, *defined recursively by*

$$G((\mathrm{L}+\lambda)^p;x,t) = \int_{-1}^{1} G((\mathrm{L}+\lambda)^{p-1};x,u)G(\mathrm{L}+\lambda;t,u)\,du, \qquad (7.65)$$

is called Legendre (Green's) function with respect to the operator $(\mathrm{L}+\lambda)^p$.

$G((\mathrm{L}+\lambda)^p;\cdot,\cdot)$ is symmetric in its arguments, i.e., for all $x,t \in [-1,1]$

$$G((\mathrm{L}+\lambda)^p;x,t) = G((\mathrm{L}+\lambda)^p;t,x). \qquad (7.66)$$

The bilinear expansion of $G((\mathrm{L}+\lambda)^p;\cdot,\cdot)$ reads as follows

$$G((\mathrm{L}+\lambda)^p;x,t) = \sum_{\substack{n=0\\ \lambda\neq \mathrm{L}^\wedge(n)}}^{\infty} \frac{1}{(\lambda-\mathrm{L}^\wedge(n))^p} P_n^*(x)P_n^*(t). \qquad (7.67)$$

The series on the right-hand side is absolutely and uniformly convergent both in $x \in [-1,1]$ and $t \in [-1,1]$. If F is continuously differentiable at the point $t \in (-1,1)$, then

$$(\mathrm{L}_t+\lambda)\int_{-1}^{1} G(\mathrm{L}+\lambda;t,u)F(u)\,du = F(t) - \delta_{\lambda,\mathrm{L}^\wedge(n)}P_n^*(t)\int_{-1}^{1} P_n^*(u)F(u)\,du. \qquad (7.68)$$

In particular, we have

Lemma 7.14. *For* $p = 2,3,\ldots$ *and* $t \neq x$

$$(\mathrm{L}_t+\lambda)^{p-1}G((\mathrm{L}+\lambda)^p;x,t) = G(\mathrm{L}+\lambda;x,t). \qquad (7.69)$$

7.3 Integral Formulas

Suppose that $F : [-1,1] \to \mathbb{R}$ is of the class $\mathrm{C}^{(2)}[-1,1]$. Assume that $\lambda \in \mathbb{R}$, $x \in (-1,1)$, and $\varepsilon > 0$ (sufficiently small). Then integration by parts, i.e., the so-called Green–Lagrange formula gives

$$\int_{-1}^{x-\varepsilon} F(t)(\mathrm{L}_t+\lambda)G(\mathrm{L}+\lambda;x,t) - G(\mathrm{L}+\lambda;x,t)(\mathrm{L}_t+\lambda)F(t)\,dt \qquad (7.70)$$

$$+\int_{x+\varepsilon}^{1} F(x)(\mathrm{L}_t+\lambda)G(\mathrm{L}+\lambda;x,t) - G(\mathrm{L}+\lambda;x,t)(\mathrm{L}_t+\lambda)F(t)\,dt$$

$$= -(1-t^2)\left(F(t)\frac{d}{dt}G(\mathrm{L}+\lambda;x,t) - G(\mathrm{L}+\lambda;x,t)\frac{d}{dt}F(t)\right)\Bigg|_{t=x-\varepsilon}^{t=x+\varepsilon}.$$

Observing the differential equation and the characteristic singularity of the Legendre (Green's) function with respect to $\mathrm{L}+\lambda$ we obtain by letting $\varepsilon \to 0$ the following integral formulas.

Theorem 7.15 (Integral Formula for $\mathrm{L}+\lambda$). *Let x be a point in $(-1,1)$. Suppose that $F \in \mathrm{C}^{(2)}[-1,1]$. Then, for $\lambda \notin \mathrm{Spect}_{\mathrm{L}}$,*

$$F(x) = \int_{-1}^{1} G(\mathrm{L}+\lambda; x, t)(\mathrm{L}_t+\lambda)F(t)\, dt. \tag{7.71}$$

Furthermore, for $\lambda \in \mathrm{Spect}_{\mathrm{L}}$, i.e., $\lambda = \mathrm{L}^\wedge(n) = n(n+1)$,

$$F(x) = P_n^*(x)\int_{-1}^{1} F(t)P_n^*(t)\, dt + \int_{-1}^{1} G(\mathrm{L}+\lambda; x, t)(\mathrm{L}_t+\lambda)F(t)\, dt. \tag{7.72}$$

Repeated application of the Green–Lagrange formula yields

$$\int_{-1}^{1} G(\mathrm{L}+\lambda; x, t)(\mathrm{L}_t+\lambda)F(t)\, dt = \int_{-1}^{1} G((\mathrm{L}+\lambda)^p; x, t)(\mathrm{L}_t+\lambda)^p F(t)\, dt \tag{7.73}$$

$p = 2, 3, \ldots$, provided that F is sufficiently often differentiable. This leads to the following extension of Theorem 7.15.

Theorem 7.16 (Integral Formula for $(\mathrm{L}+\lambda)^p$). *Suppose that $p \in \mathbb{N}$, $\lambda \in \mathbb{R}$, $x \in (-1,1)$, and $F \in \mathrm{C}^{(2p)}[-1,1]$. Then, for $\lambda \notin \mathrm{Spect}_{\mathrm{L}}$*

$$F(x) = \int_{-1}^{1} G((\mathrm{L}+\lambda)^p; x, t)(\mathrm{L}_t+\lambda)^p F(t)\, dt. \tag{7.74}$$

Moreover, for $\lambda \in \mathrm{Spect}_{\mathrm{L}}$, i.e., $\lambda = \mathrm{L}^\wedge(n) = n(n+1)$,

$$F(x) = P_n^*(x)\int_{-1}^{1} F(t)P_n^*(t)\, dt + \int_{-1}^{1} G((\mathrm{L}+\lambda)^p; x, t)(\mathrm{L}_t+\lambda)^p F(t)\, dt. \tag{7.75}$$

Finally, for $m \in \mathbb{N}_0$ and $\lambda \in \mathbb{R}$ with $\lambda \neq \mathrm{L}^\wedge(m+1), \mathrm{L}^\wedge(m+2), \ldots$, we let

$$G^{\perp}_{0,\ldots,m}((\mathrm{L}+\lambda)^p; x, t) = G((\mathrm{L}+\lambda)^p; x, t) - G_{0,\ldots,m}((\mathrm{L}+\lambda)^p; x, t), \tag{7.76}$$

where

$$G_{0,\ldots,m}((\mathrm{L}+\lambda)^p; x, t) = \sum_{\substack{n=0 \\ \lambda \neq \mathrm{L}^\wedge(n)}}^{m} \frac{1}{(\lambda - \mathrm{L}^\wedge(n))^p} P_n^*(x)P_n^*(t). \tag{7.77}$$

$G^{\perp}_{0,\ldots,m}((\mathrm{L}+\lambda)^p; \cdot, \cdot)$ is called *m-th-truncated Legendre (Green's) function with respect to the operator* $(\mathrm{L}+\lambda)^p$. These functions admit an easy reformulation of Theorem 7.16 after application of integration by parts.

Theorem 7.17. *Suppose that* $p \in \mathbb{N}$, $m \in \mathbb{N}_0$, $\lambda \in \mathbb{R}$ *with* $\lambda \neq \mathrm{L}^\wedge(m+1)$, $\mathrm{L}^\wedge(m+2), \ldots,$ *and* $x \in (-1,1)$. *Then, for* $F \in \mathrm{C}^{(2p)}[-1,1]$,

$$F(x) = \sum_{n=0}^{m} P_n^*(x) \int_{-1}^{1} F(t) P_n^*(t)\, dt \tag{7.78}$$
$$+ \int_{-1}^{1} G_{0,\ldots,m}^{\perp}((\mathrm{L}+\lambda)^p; x, t)(\mathrm{L}_t + \lambda)^p F_{0,\ldots,m}^{\perp}(t)\, dt,$$

where

$$F_{0,\ldots,m}^{\perp} = F - \sum_{n=0}^{m} \int_{-1}^{1} F(t) P_n^*(t)\, dt P_n^*. \tag{7.79}$$

Theorem 7.17 enables a comparison between the functional value $F(x)$ at the point x and the m-th truncated orthogonal expansion of F in terms of Legendre polynomials at x with explicit representation of the remainder term in integral form. This formula can be used to formulate estimates of the remainder terms in Gaussian quadrature.

Chapter 8

Gaussian Integration

If we let the knots in the Newton–Cotes formulas of Section 1.1 vary freely and compute the weights accordingly, the question raised by C.F. Gauss [1814] was the following: What is the maximum degree of polynomial exactness that can be achieved and how are the knots to be selected in order to realize this optimum? In fact, Gauss answered this question by means of the theory of continued fractions associated to hypergeometric series (see, e.g., W. Gautschi [1981] for a proof in today's nomenclature).

In this chapter we base our answer on the determination of the weights in the Gaussian quadrature rule by the uniquely solvable linear system (8.5). Our considerations concerning the classical Gaussian quadrature method essentially follow H.J. Stetter [1966], R.W. Freund, R.W. Hoppe [2007], J. Stoer, R. Bulirsch [1980], J. Werner [1991], while the adaptive remainder term representation and the convergence theorem lead back to W. Freeden [1980a].

8.1 Gaussian Quadrature Formulas

Let $x_1^{(n)}, \ldots, x_n^{(n)}$ with $x_1^{(n)} < \ldots < x_n^{(n)}$ be the zeros of the Legendre polynomial P_n of degree n. Then it is well-known that there is precisely one zero of the polynomial P_{n+1} in each interval $(-1, x_1^{(n)}), (x_1^{(n)}, x_2^{(n)}), \ldots, (x_n^{(n)}, 1)$. For later use we want to prove the following result.

Lemma 8.1. *The matrix* $\mathbf{A}$ *given by*

$$\mathbf{A} = \left(P_k^*(x_j^{(n)})\right)_{\substack{k=0,\ldots,n-1 \\ j=1,\ldots,n}}, \tag{8.1}$$

is non-singular.

Proof. Assume the matrix is singular. Then there exist coefficients $C_0, \ldots, C_{n-1}$ with $c = (C_0, \ldots, C_{n-1})^T \neq (0, \ldots, 0)^T$ such that $\mathbf{A}c = 0$, i.e., the following polynomial of degree $\leq n - 1$

$$P(x) = \sum_{l=0}^{n-1} C_l P_l^*(x) \tag{8.2}$$

had as zeros $x_1^{(n)}, \dots, x_n^{(n)}$. Hence, $P = 0$. Now, for $k = 0, \dots, n-1$,

$$0 = \int_{-1}^{1} P(x) P_k^*(x) \, dx = C_k. \tag{8.3}$$

But this is a contradiction. Thus, Lemma 8.1 must be true. □

We are now interested in the *Gaussian integration formula* $\mathcal{G}_n$ (see, e.g., P.J. Davis, P. Rabinowitz [1984], J. Stoer, R. Bulirsch [1980])

$$\mathcal{I}F = \int_{-1}^{1} F(t) \, dt \approx \mathcal{G}_n F = \sum_{i=1}^{n} w_i^{(n)} F(x_i^{(n)}), \tag{8.4}$$

where the knots $x_i^{(n)}$, $i = 1, \dots, n$, are the zeros of the Legendre polynomial P_n and the weights $w_i^{(n)}, i = 1, \dots, n$, are determined by the linear system

$$\sum_{i=1}^{n} w_i^{(n)} P_k^*(x_i^{(n)}) = \int_{-1}^{1} P_k^*(t) P_0(t) \, dt = \sqrt{2}\, \delta_{0,k}, \quad k = 0, \dots, n-1. \tag{8.5}$$

From Lemma 8.1 it is clear that (8.5) is uniquely solvable in the unknowns weights $w_1^{(n)}, \dots, w_n^{(n)}$.

Central for our considerations is the following well-known theorem.

Theorem 8.2. *Let $x_1^{(n)}, \dots x_n^{(n)}$ be the zeros of the Legendre polynomial P_n. Furthermore, let $w_1^{(n)}, \dots, w_n^{(n)}$ be the (unique) solutions of the linear system (8.5).*

(i) Then, the weights are positive, i.e., $w_i^{(n)} > 0$ for $i = 1, \dots, n$, and we have

$$\int_{-1}^{1} P(t) \, dt = \sum_{i=1}^{n} w_i^{(n)} P(x_i^{(n)}) \tag{8.6}$$

for all $P \in \mathrm{Pol}_{0,\dots,2n-1}$.

(ii) Conversely, if (8.6) is valid for real numbers $w_i^{(n)}, x_i^{(n)}$, $i = 1, \dots, n$, then the knots $x_i^{(n)}$, $i = 1, \dots, n$, are the zeros of the Legendre polynomial P_n and the weights $w_i^{(n)}, i = 1, \dots, n$, satisfy the linear equations (8.5).

(iii) Finally, there exist no real numbers $x_i^{(n)}$, $w_i^{(n)}$, $i = 1, \dots, n$, such that (8.6) holds true for all $P \in \mathrm{Pol}_{0,\dots,2n}$.

Proof. Consider a polynomial $P \in \mathrm{Pol}_{0,\dots,2n-1}$. There exist $Q, R \in \mathrm{Pol}_{0,\dots,n-1}$ with

$$Q = \sum_{r=0}^{n-1} A_r P_r^* \tag{8.7}$$

and

$$R = \sum_{r=0}^{n-1} B_r P_r^* \tag{8.8}$$

such that $P = P_n^* Q + R$. Because of the orthogonality of the Legendre polynomials it is clear that

$$\begin{aligned}\int_{-1}^{1} P(t)\,dt &= \int_{-1}^{1} Q(t)P_n^*(t)\,dt + \int_{-1}^{1} R(t)\underbrace{P_0(t)}_{=1}\,dt\\ &= \int_{-1}^{1} R(t)P_0(t)\,dt.\end{aligned}$$

In connection with (8.8) this implies

$$\int_{-1}^{1} P(t)\,dt = \sqrt{2}\,B_0. \tag{8.9}$$

Moreover, because of $P_n^*(x_i^{(n)}) = 0$, $i = 1,\ldots,n$, we obtain using (8.5)

$$\begin{aligned}\sum_{i=1}^{n} w_i^{(n)} P(x_i^{(n)}) &= \sum_{i=1}^{n} w_i^{(n)} \underbrace{P_n^*(x_i^{(n)})}_{=0} Q(x_i^{(n)}) + \sum_{i=1}^{n} w_i^{(n)} R(x_i^{(n)}) \qquad (8.10)\\ &= \sum_{i=1}^{n} w_i^{(n)} \sum_{k=0}^{n-1} B_r P_r^*(x_i^{(n)}) = \sqrt{2}\,B_0.\end{aligned}$$

This shows that (8.6) holds true. Inserting the polynomials $\tilde{T}_i \in \mathrm{Pol}_{0,\ldots,2n-2}$, $i = 1,\ldots,n$, given by

$$\tilde{T}_i(x) = \prod_{\substack{j=1\\ j\neq i}}^{n} \left(x - x_j^{(n)}\right)^2, \quad x \in [-1,1], \tag{8.11}$$

into (8.6) we obtain for $k = 1,\ldots,n$

$$0 < \int_{-1}^{1} \tilde{T}_i(t)\,dt = \sum_{i=1}^{n} w_i^{(n)} T_k(x_i^{(n)}) = \sum_{i=1}^{n} w_i^{(n)} \prod_{\substack{j=1\\ j\neq k}}^{n} \left(x_i^{(n)} - x_j^{(n)}\right)^2, \tag{8.12}$$

hence,

$$w_k^{(n)} > 0, \quad k = 1,\ldots,n. \tag{8.13}$$

This proves (i).

Next we verify (iii). Suppose that $w_i^{(n)}$, $x_i^{(n)}$, $i = 1,\ldots,n$, are given in such a way that (8.6) is valid for all $P \in \mathrm{Pol}_{0,\ldots,2n}$. Choose $H \in \mathrm{Pol}_{0,\ldots,2n}$ of the form

$$H(x) = \prod_{j=1}^{n} \left(x - x_j^{(n)}\right)^2. \tag{8.14}$$

Then

$$0 < \int_{-1}^{1} H(x)\, dx = \sum_{i=1}^{n} w_i^{(n)} H(x_i^{(n)}) = 0. \tag{8.15}$$

This is a contradiction.

In order to guarantee (ii) we apply (8.6) especially to the Legendre polynomials P_k^*, $k = 0, \ldots, n-1$ such that

$$\sum_{i=1}^{n} w_i^{(n)} P_k^*(x_i^{(n)}) = \int_{-1}^{1} P_k^*(x)\, dx = \int_{-1}^{1} P_k^*(x) \underbrace{P_0(x)}_{=1}\, dx. \tag{8.16}$$

In other words, the weights have to satisfy (8.5), i.e.,

$$\sum_{i=1}^{n} w_i^{(n)} P_k^*(x_i^{(n)}) = \sqrt{2}\, \delta_{0,k}. \tag{8.17}$$

We now consider $P = P_k^* P_n^*$ for $k = 0, \ldots, n-1$, and apply (8.6). This gives

$$0 = \int_{-1}^{1} P_k^*(t) P_n^*(t)\, dt = \sum_{i=1}^{n} w_i^{(n)} P_k^*(x_i^{(n)}) P_n^*(x_i^{(n)}) \tag{8.18}$$

for $k = 0, \ldots, n-1$. Consequently, the vector $c = (c_1, \ldots, c_n)^T$ given by

$$(c_1, \ldots, c_n)^T = (w_1^{(n)} P_n^*(x_1^{(n)}), \ldots, w_n^{(n)} P_n^*(x_n^{(n)}))^T \tag{8.19}$$

satisfies the linear system

$$0 = \sum_{i=1}^{n} c_i P_k^*(x_i^{(n)}), \quad k = 0, \ldots, n-1, \tag{8.20}$$

where the coefficient matrix is non-singular. But this means that $c_i = 0$, $i = 1, \ldots, n$, hence, $P_n^*(x_i^{(n)}) = 0$, i.e. $P_n(x_i^{(n)}) = 0$, $i = 1, \ldots, n$, as required. □

8.2 Adaptive Remainder Terms Involving Green's Function

Now, the integral formula (Theorem 7.17) comes into play (cf. W. Freeden [1980a]). For $\lambda \in \mathbb{R} \setminus \mathbb{L}_{2n}$ with $\mathbb{L}_{2n} = \{\mathrm{L}^\wedge(2n), \mathrm{L}^\wedge(2n+1), \ldots\}$ and for $F \in \mathrm{C}^{(2p)}[-1, 1]$, $p \in \mathbb{N}$, we have

$$\sum_{i=1}^{n} w_i^{(n)} F(x_i^{(n)}) = \int_{-1}^{1} F(t)\, dt \tag{8.21}$$

$$+ \sum_{i=1}^{n} w_i^{(n)} \int_{-1}^{1} G^{\perp}_{0,\ldots,2n}((\mathrm{L}+\lambda)^p; x_i^{(n)}, t)(\mathrm{L}_t + \lambda)^p F(t)\, dt.$$

In other words, the remainder $\mathcal{E}_n F = \mathcal{I}F - \mathcal{G}_n F$, when the integral

$$\mathcal{I}F = \int_{-1}^{1} F(t)\, dt, \quad F \in \mathrm{C}^{(2p)}[-1,1], \tag{8.22}$$

is replaced by the *Gaussian n-point formula*

$$\mathcal{G}_n F = \sum_{i=1}^{n} w_i^{(n)} F(x_i^{(n)}), \tag{8.23}$$

is expressible by means of the $2n$-th truncated Legendre (Green's) function in integral form

$$\mathcal{E}_n F = -\sum_{i=1}^{n} w_i^{(n)} \int_{-1}^{1} G^{\perp}_{0,\dots,2n}((\mathrm{L}+\lambda)^p; x_i^{(n)}, t)(\mathrm{L}_t+\lambda)^p F(t)\, dt \tag{8.24}$$

for all $\lambda \in \mathbb{R}\backslash\mathbb{L}_{2n}$ with $\mathbb{L}_{2n} = \{\mathrm{L}^\wedge(2n), \mathrm{L}^\wedge(2n+1), \dots\}$. Applying the Cauchy–Schwarz inequality to (8.24) we obtain for $\lambda \in \mathbb{R} \setminus \mathbb{L}_{2n}$

$$|\mathcal{E}_n F| \le \sqrt{A_\lambda^{(p)}(n)} \sqrt{\int_{-1}^{1} |(\mathrm{L}_t+\lambda)^p F(t)|^2\, dt}, \tag{8.25}$$

where we have used the abbreviation

$$A_\lambda^{(p)}(n) = \sum_{i=1}^{n}\sum_{j=1}^{n} w_i^{(n)} w_j^{(n)} G^{\perp}_{0,\dots,2n}((\mathrm{L}+\lambda)^{2p}; x_i^{(n)}, x_j^{(n)}) \tag{8.26}$$

with

$$\begin{aligned} &G^{\perp}_{0,\dots,2n}((\mathrm{L}+\lambda)^{2p}; x_i^{(n)}, x_j^{(n)}) \\ &= \int_{-1}^{1} G^{\perp}_{0,\dots,2n}((\mathrm{L}+\lambda)^p; x_i^{(n)}, t) G^{\perp}_{0,\dots,2n}((\mathrm{L}+\lambda)^p; t, x_j^{(n)})\, dt. \end{aligned} \tag{8.27}$$

Note that some values $A_\lambda^{(p)}(n)$ are given in Table 8.1.

Summarizing our results we obtain the following theorem.

Theorem 8.3. *Let $x_i^{(n)}$, $i = 1, \dots, n$, be the zeros of the Legendre polynomial $P_n, n \ge 1$, and let $w_i^{(n)}$, $i = 1, \dots, n$, be the solution of the linear system (8.5). Then, for $\lambda \in \mathbb{R} \setminus \mathbb{L}_{2n}$ and for functions $F \in \mathrm{C}^{(2p)}[-1,1]$, $p \in \mathbb{N}$, we have*

$$\left| \int_{-1}^{1} F(t)\, dt - \sum_{i=1}^{n} w_i^{(n)} F(x_i^{(n)}) \right| \le \sqrt{A_\lambda^{(p)}(n)} \sqrt{\int_{-1}^{1} \left|(\mathrm{L}_t+\lambda)^p F(t)\right|^2\, dt}, \tag{8.28}$$

where

$$A_\lambda^{(p)}(n) = \sum_{i=1}^{n}\sum_{j=1}^{n} w_i^{(n)} w_j^{(n)} \sum_{k=2n}^{\infty} \frac{1}{(\lambda - \mathrm{L}^\wedge(k))^{2p}} P_k^*(x_i^{(n)}) P_k^*(x_j^{(n)}). \tag{8.29}$$

TABLE 8.1: $A_\lambda^{(p)}(n)$ from (8.29).

$n=2$	$p=1$	$p=2$	$p=4$	$p=8$
$\lambda=0$	8.05976e-03	1.747841e-05	1.064706e-10	4.153795e-21
$\lambda=2$	9.75068e-03	2.649045e-05	2.472239e-10	2.241631e-20
$\lambda=4.5$	1.2817e-02	4.787625e-05	8.174205e-10	2.452543e-19
$\lambda=8.5$	2.234935e-02	1.567427e-04	8.899805e-09	2.909097e-17
$n=4$	$p=1$	$p=2$	$p=4$	$p=8$
$\lambda=0$	6.955807e-04	1.151926e-07	4.056782e-15	5.554051e-30
$\lambda=4$	7.709229e-04	1.440087e-07	6.397575e-15	1.385326e-29
$\lambda=8.5$	8.724733e-04	1.882346e-07	1.104359e-14	4.144405e-29
$\lambda=16.5$	1.114777e-03	3.191215e-07	3.234008e-14	3.573608e-28
$n=8$	$p=1$	$p=2$	$p=4$	$p=8$
$\lambda=0$	5.270212e-05	6.196683e-10	1.052274e-19	3.375950e-39
$\lambda=8$	5.563827e-05	6.961041e-10	1.333124e-19	5.439077e-39
$\lambda=16.5$	5.905193e-05	7.907972e-10	1.727947e-19	9.175279e-39
$\lambda=32.5$	6.644785e-05	1.017562e-09	2.885629e-19	2.578932e-38

Even more, from (8.24), we get the estimate

$$|\mathcal{E}_n F| \leq \sqrt{A_\lambda^{(p)}(n)} \sqrt{\int_{-1}^{1} \left|(\mathrm{L}_t+\lambda) F_{2n}^{\perp}(t)\right|^2 \, dt} \tag{8.30}$$

where

$$F_{2n}^{\perp}(t) = F(t) - \sum_{n=0}^{2n-1} \underbrace{\int_{-1}^{1} F(u) P_n^*(u)\, du}_{=F^\wedge(n)} \, P_n^*(t), \quad t \in [-1,1]. \tag{8.31}$$

Observing the estimate $|P_n(t)| \leq 1$ for all $n \in \mathbb{N}_0$ and $t \in [-1,1]$ and the properties of the Gaussian weights we get

$$|A_\lambda^p(n)| \leq \frac{1}{2} \sum_{i=1}^{n} \sum_{j=1}^{n} w_i^{(n)} w_j^{(n)} \sum_{k=2n}^{\infty} \frac{2k+1}{(\lambda - \mathrm{L}^\wedge(k))^{2p}} = 2 \sum_{k=2n}^{\infty} \frac{2k+1}{(\lambda - \mathrm{L}^\wedge(k))^{2p}} \tag{8.32}$$

for all $\lambda \in \mathbb{R} \setminus \mathbb{L}_{2n}$. The m-th Peano kernel with $0 \leq m \leq 2n$ (see, e.g., J. Werner [1991])

$$K_m(t) = \frac{1}{m!}\left(\int_{-1}^{1} (x-t)_+^m \, dx - \sum_{j=0}^{n} w_j^{(n)} (x_j^{(n)} - t)_+^m \right) \tag{8.33}$$

changes $2n-m$ times its sign in the interval $[-1,1]$. In particular, K_{2n} admits only one sign on $[-1,1]$. Therefore, Peano's Theorem (see Corollary 1.10) tells us that there exists a point $\xi \in (1,1)$ such that

$$\mathcal{E}_n F = \frac{1}{(2n)!} F^{(2n)}(\xi) \mathcal{E}_n(t^{2n}). \tag{8.34}$$

Now we remember

$$P_n^*(t) = \sqrt{\frac{2n+1}{2}} \left(\frac{1}{2^n} \binom{2n}{n} t^n + \ldots \right), \tag{8.35}$$

i.e.,

$$\alpha_n^* = \sqrt{\frac{2n+1}{2}} \frac{1}{2^n} \binom{2n}{n} \tag{8.36}$$

is the leading coefficient of P_n^*. Consequently, $Q_n \in \mathrm{Pol}_{0,\ldots,n}$ defined by $Q_n(t) = \frac{1}{\alpha_n^*} P_n^*(t)$, $t \in [-1,1]$, admits the decomposition

$$Q_n(t) = \prod_{i=1}^{n} (t - x_i^{(n)}), \quad t \in [-1,1]. \tag{8.37}$$

Thus, we are able to conclude that

$$\mathcal{E}_n(t^{2n}) = \mathcal{E}_n(Q_n^2) = \int_{-1}^{1} Q_n^2(t) \, dt - \underbrace{\mathcal{G}_n(Q_n^2)}_{=0} \tag{8.38}$$

such that, for $F \in \mathrm{C}^{(2n)}[-1,1]$,

$$\mathcal{E}_n F = \frac{F^{(2n)}(\xi)}{2n!} \frac{1}{(\alpha_n^*)^2} \underbrace{\int_{-1}^{1} P_n^*(t) P_n^*(t) \, dt}_{=1} = \frac{1}{(2n)!(\alpha_n^*)^2} F^{(2n)}(\xi). \tag{8.39}$$

This leads to the following result.

Theorem 8.4. *For $F \in \mathrm{C}^{(2n)}[-1,1]$ and $\lambda \in \mathbb{R} \setminus \mathbb{L}_{2n}$,*

$$\int_{-1}^{1} F(t) \, dt - \sum_{i=1}^{n} w_i^{(n)} F(x_i^{(n)})$$

$$= \sum_{i=1}^{n} w_i^{(n)} \int_{-1}^{1} G_{2n}^{\perp}((L+\lambda)^n; x_i^{(n)}, t)(L_t + \lambda)^n F_{2n,\ldots,\infty}(t) \, dt \tag{8.40}$$

$$= \frac{2^{2n+1}(n!)^4}{((2n)!)^3(2n+1)} F^{(2n)}(\xi) \tag{8.41}$$

with some $\xi \in (-1,1)$.

Remark 8.5. *Using (8.37) we find for $j = 1, \ldots, n$ that*

$$w_j^{(n)} = \frac{1}{Q_n'(x_j^{(n)})Q_{n-1}(x_j^{(n)})} \int_{-1}^{1} Q_{n-1}(t)Q_{n-1}(t)\, dt. \tag{8.42}$$

Example 8.6. *We present some examples of Gaussian quadratures in Table 8.2.*

TABLE 8.2: Examples for Gaussian quadrature.

Rule	$\int_{-1}^{1} \frac{1}{1+t^4}\, dt$	$\int_{-1}^{1} \frac{1}{1+e^t}\, dt$	$\int_{-1}^{1} \frac{t}{e^t-1}\, dt$	$\int_{-1}^{1} e^{t^2}\, dt$
$\mathcal{G}_2$	1.8000000000	1	2.0552493429	2.7912248502
$\mathcal{G}_4$	1.7359667360	1	2.0550092773	2.9245409391
$\mathcal{G}_8$	1.7339583832	1	2.0550092682	2.9253034900
$\mathcal{G}_{16}$	1.7339459746	1	2.0550092682	2.9253034918

8.3 Convergence of Gaussian Quadrature

From Theorem 8.3 we obtain with $\lambda = 0$ that

$$\left| \int_{-1}^{1} F(t)\, dt - \sum_{i=1}^{n} w_i^{(n)} F(x_i^{(n)}) \right| \leq \sqrt{A_0^{(p)}(n)} \sqrt{\int_{-1}^{1} (\mathrm{L}_t F(t))^2\, dt}, \tag{8.43}$$

where

$$A_0^{(p)}(n) = \sum_{i=1}^{n} \sum_{j=1}^{n} w_i^{(n)} w_j^{(n)} \sum_{k=2n}^{\infty} \frac{1}{(k(k+1))^{2p}} P_k^*(x_i^{(n)}) P_k^*(x_j^{(n)}). \tag{8.44}$$

The sequence

$$\left\{ \sum_{k=2n}^{\infty} \frac{2k+1}{(k(k+1))^{2p}} \right\}_{n \in \mathbb{N}} \tag{8.45}$$

is (strictly) monotonically decreasing and positive. Furthermore, we have

$$0 \leq A_0^{(p)}(n) \leq 2 \sum_{k=2n}^{\infty} \frac{2k+1}{(k(k+1))^{2p}}. \tag{8.46}$$

Therefore, we come to the following conclusion.

Theorem 8.7. *The sequence* $\{A_0^{(p)}(n)\}_{n\in\mathbb{N}}$ *with* $p \in \mathbb{N}$ *has the limit*

$$\lim_{n\to\infty} A_0^{(p)}(n) = 0 \tag{8.47}$$

such that, for all $F \in \mathrm{C}^{(2p)}[-1,1]$,

$$\lim_{n\to\infty} \mathcal{E}_n F = 0, \tag{8.48}$$

i.e., the Gaussian quadrature rule is convergent

$$\lim_{n\to\infty} \sum_{i=1}^{n} w_i^{(n)} F(x_i^{(n)}) = \int_{-1}^{1} F(t)\, dt. \tag{8.49}$$

Altogether, comparing different numerical integration methods (i.e., polynomial interpolation formulas, spline integration formulas as discussed until now) with the Gaussian method we see that the latter method assures the highest degree of polynomial precision. The critical point in our approach, however, is to find an appropriate estimate for $(\mathrm{L}+\lambda)^{2p}F(t)$ with λ being adapted to the specific properties of the function. Moreover, the point sets $\{x_1^{(n)},\ldots,x_n^{(n)}\}_{n\in\mathbb{N}}$ are not hierarchically distributed.

Note that the theory of Gaussian integration can be generalized to other orthogonal polynomials (cf. W. Freeden, M. Gutting [2013], W. Gautschi [2004]). It is also possible to use one or both endpoints of the integration interval in which case we speak of Gauss–Radau integration or Gauss–Lobatto integration, respectively (see K. Atkinson [1998], P.J. Davis, P. Rabinowitz [1984], W. Gautschi [2004]).

Part II

Integration on 2D-Spheres

Chapter 9

Remainder Terms Involving Beltrami Operators

Today, in the opinion of the authors, there essentially are four approaches for calculating integrals of functions on the sphere (see also the classification presented by A. Townsend et al. [2016]):

(i) Latitude-longitude techniques, matrix-free and so-called double Fourier sphere (DFS) methods,

(ii) Mesh-free (radial basis) methods including polynomial and spline exact methods,

(iii) Equidistributions involving low discrepancy methods,

(iv) Multiscale techniques using spherical wavelet variants.

Unfortunately, there is no universal procedure for all applications. A fast method usually lacks high accuracy, an accurate method usually is at the cost of efficiency and economy, so combinations of the aforementioned integration procedures may be sometimes welcome and helpful. In fact, each method has its own specific advantages and disadvantages, seen from a practical problem to be solved. Numerical analysts interested in solving partial differential equations look for simply structured and fast integration rules. Practitioners making use of scattered data measurements require adequate mesh-free integration rules. Particular criteria for the quality of an approximate integration rule are remainder terms for estimating the integration error as well as algorithms for economically handling given data sets. Altogether, any approximate integration method turns out to be a compromise between the theoretical options of mathematical structures and practical obligations of data type, distribution, and total amount.

The aim of the following chapters on spherical integration is to extend Euler's idea of bridging "continuous sums", i.e., integrals, and discrete sums to the unit sphere $\mathbb{S}^2$. First, some preparatory material about spherical polynomials, i.e., spherical harmonics is presented briefly without proofs (for more specific details the reader is referred to, e.g., W. Freeden, M. Schreiner [2009], W. Freeden, M. Gutting [2013]). Based on the concept of the Green's function with respect to the Beltrami operator, here briefly called $\mathbb{S}^2$-sphere function, a class of integral formulas is presented. The integral formulas offer the perspective of estimating remainder terms in terms of Beltrami derivatives. "Mollified versions" of the integral formulas are developed by use of regularized $\mathbb{S}^2$-sphere functions. Later on, remainder term expressions with respect to rotation-invariant pseudodifferential operators enable us to deduce concepts like best approximate integration, spline and wavelet exact integration formulas as well as discrepancy estimates in equidistributions on the sphere.

9.1 Spherical Framework

As usual, the *unit sphere* in $\mathbb{R}^3$ is denoted by $\mathbb{S}^2 = \{\xi \in \mathbb{R}^3 : |\xi| = 1\}$. We may represent the points $x \in \mathbb{R}^3 \setminus \{0\}$, $x = r\xi$, $\xi \in \mathbb{S}^2$, in *polar coordinates* as follows

$$x = r\xi, \quad r = |x|, \quad \xi = t\varepsilon^3 + \sqrt{1-t^2}\,(\cos\varphi\varepsilon^1 + \sin\varphi\varepsilon^2) \tag{9.1}$$

with $-1 \leq t \leq 1$, $0 \leq \varphi < 2\pi$ ($\vartheta \in [0,\pi]$: (co-)latitude, φ: longitude, $t = \cos\vartheta$: polar distance), so that

$$\xi = (\sin\vartheta\ \cos\varphi,\ \sin\vartheta\ \sin\varphi,\ \cos\vartheta)^T. \tag{9.2}$$

Note that we use lower case Greek letters (like ξ, η, etc.) for elements of $\mathbb{S}^2$.

Spherical Differential Operators. Gradient fields ∇F in $\mathbb{R}^3$ can be decomposed into a radial and a tangential component. More explicitly, the *surface gradient* ∇^* contains the tangential derivatives of the gradient ∇ in the following way:

$$\nabla = \xi\frac{\partial}{\partial r} + \frac{1}{r}\nabla^*. \tag{9.3}$$

Letting $x = r\xi$, $r = |x|$, $\xi \in \mathbb{S}^2$, we find with $\eta \in \mathbb{S}^2$

$$\nabla_x(x\cdot\eta) = \eta = (\xi\cdot\eta)\xi + \nabla^*_\xi(\xi\cdot\eta), \tag{9.4}$$

such that

$$\nabla^*_\xi(\xi\cdot\eta) = \eta - (\xi\cdot\eta)\xi. \tag{9.5}$$

The *surface curl gradient* L^* for $F \in \mathrm{C}^{(1)}(\mathbb{S}^2)$ is defined by

$$\mathrm{L}^*_\xi F(\xi) = \xi \wedge \nabla^*_\xi F(\xi), \qquad \xi \in \mathbb{S}^2. \tag{9.6}$$

According to the definition (9.6), L^*F is a tangential vector field perpendicular to ∇^*F, i.e.,

$$\nabla^*_\xi F(\xi)\cdot \mathrm{L}^*_\xi F(\xi) = 0, \quad \xi \in \mathbb{S}^2. \tag{9.7}$$

$\nabla^* \cdot = \mathrm{div}^*$ and $\mathrm{L}^* \cdot = \mathrm{curl}^*$, respectively, denote the *surface divergence* and the *surface curl* for a vector field $f = (F_1, F_2, F_3)^T \in \mathrm{c}^{(1)}(\mathbb{S}^2)$ given by

$$\nabla^*_\xi \cdot f(\xi) = \sum_{i=1}^{3} \nabla^*_\xi F_i(\xi) \cdot \varepsilon^i \tag{9.8}$$

and

$$\mathrm{L}^*_\xi \cdot f(\xi) = \sum_{i=1}^{3} \mathrm{L}^*_\xi F_i(\xi) \cdot \varepsilon^i. \tag{9.9}$$

Note that the surface curl as defined by (9.9), i.e.,

$$\xi \mapsto \mathrm{L}^*_\xi \cdot f(\xi) = \mathrm{curl}^*_\xi f(\xi) = \mathrm{div}^*_\xi (f(\xi) \wedge \xi) = \nabla^*_\xi \cdot (f(\xi) \wedge \xi), \quad \xi \in \mathbb{S}^2, \tag{9.10}$$

represents a scalar-valued function on the unit sphere $\mathbb{S}^2$ in $\mathbb{R}^3$. The *Beltrami operator* Δ^* is the angle dependent part of the *Laplace operator* Δ

$$\Delta_x = \left(\frac{\partial}{\partial r}\right)^2 + \frac{2}{r}\frac{\partial}{\partial r} + \frac{1}{r^2}\Delta^*_\xi. \tag{9.11}$$

In polar coordinates, the Beltrami operator reads as follows:

$$\Delta^*_\xi = \frac{\partial}{\partial t}\left(1 - t^2\right)\frac{\partial}{\partial t} + \frac{1}{1-t^2}\left(\frac{\partial}{\partial \varphi}\right)^2. \tag{9.12}$$

Moreover, we have in terms of (9.1) the following representations in polar coordinates:

$$\begin{aligned} \nabla^*_\xi &= \frac{1}{\sqrt{1-t^2}}\left(-\sin\varphi\varepsilon^1 + \cos\varphi\varepsilon^2\right)\frac{\partial}{\partial\varphi} && (9.13)\\ &\quad + \sqrt{1-t^2}\left(-t\cos\varphi\varepsilon^1 - t\sin\varphi\varepsilon^2 + \sqrt{1-t^2}\varepsilon^3\right)\frac{\partial}{\partial t}, \\ \mathrm{L}^*_\xi &= \sqrt{1-t^2}\left(\sin\varphi\varepsilon^1 - \cos\varphi\varepsilon^2\right)\frac{\partial}{\partial t} && (9.14)\\ &\quad + \frac{1}{\sqrt{1-t^2}}\left(-t\cos\varphi\varepsilon^1 - t\sin\varphi\varepsilon^2 + \sqrt{1-t^2}\varepsilon^3\right)\frac{\partial}{\partial\varphi}. \end{aligned}$$

It should be mentioned that the operators $\nabla^*, \mathrm{L}^*, \Delta^*$ will always be used here in coordinate-free representation, thereby avoiding any singularity at the poles. Since the operators ∇^*, L^* and $\nabla^*\cdot, \mathrm{L}^*\cdot$ are of particular interest we list some of their properties for F, f sufficiently smooth scalar or vector fields, respectively, and $\xi \in \mathbb{S}^2$:

$$\nabla^*_\xi \cdot \nabla^*_\xi F(\xi) = \Delta^*_\xi F(\xi), \tag{9.15}$$

$$\mathrm{L}^*_\xi \cdot \mathrm{L}^*_\xi F(\xi) = \Delta^*_\xi F(\xi), \tag{9.16}$$

$$\nabla_\xi^* \cdot \mathrm{L}_\xi^* F(\xi) = 0, \tag{9.17}$$
$$\mathrm{L}_\xi^* \cdot \nabla_\xi^* F(\xi) = 0, \tag{9.18}$$
$$\nabla_\xi^* F(\xi) \cdot \mathrm{L}_\xi^* F(\xi) = 0, \tag{9.19}$$
$$\nabla_\xi^* \cdot (F(\xi) f(\xi)) = (\nabla_\xi^* F(\xi)) \cdot f(\xi) + F(\xi)(\nabla_\xi^* \cdot f(\xi)), \tag{9.20}$$
$$\nabla_\xi^* \cdot \xi = 2. \tag{9.21}$$

Moreover, we have

$$\nabla_\xi^* \wedge (F(\xi) f(\xi)) = \nabla_\xi^* F(\xi) \wedge f(\xi) + F(\xi) \nabla_\xi^* \wedge f(\xi), \quad \xi \in \mathbb{S}^2. \tag{9.22}$$

For a given function $F \in \mathrm{C}^{(1)}(\mathbb{S}^2)$, the triple $F(\xi)\xi$, $\nabla_\xi^* F(\xi)$, $\mathrm{L}_\xi^* F(\xi)$, $\xi \in \mathbb{S}^2$, supplies us with a system of three orthogonal vectors at each point $\xi \in \mathbb{S}^2$, provided that $F(\xi) \neq 0$ and $\nabla_\xi^* F(\xi) \neq 0$. If F is of class $\mathrm{C}^{(1)}[-1,1]$ and $F' \in \mathrm{C}^{(0)}[-1,1]$ is its (one-dimensional) derivative, then

$$\nabla_\xi^* F(\xi \cdot \eta) = F'(\xi \cdot \eta)(\eta - (\xi \cdot \eta)\xi), \tag{9.23}$$
$$\mathrm{L}_\xi^* F(\xi \cdot \eta) = F'(\xi \cdot \eta)(\xi \wedge \eta), \tag{9.24}$$

whereas, for $F \in \mathrm{C}^{(2)}[-1,1]$,

$$\Delta_\xi^* F(\xi \cdot \eta) = -2(\xi \cdot \eta) F'(\xi \cdot \eta) + (1 - (\xi \cdot \eta)^2) F''(\xi \cdot \eta). \tag{9.25}$$

Spherical Green's Theorems. A region $\Gamma \subset \mathbb{S}^2$ is called *regular*, if the *surface theorem of Gauss*

$$\int_\Gamma \nabla_\xi^* \cdot f(\xi)\, dS(\xi) = \int_{\partial\Gamma} \nu_\xi \cdot f(\xi)\, d\sigma(\xi), \tag{9.26}$$

is valid, where σ is the arc length along $\partial\Gamma$, ν and τ denote unit surface vector fields (cf. Figure 9.1) normal (outward of Γ) and tangential to $\partial\Gamma$, respectively, and $f : \overline{\Gamma} \to \mathbb{R}$, $\overline{\Gamma} = \Gamma \cup \partial\Gamma$, is continuously differentiable, i.e., $f \in \mathrm{c}^{(1)}(\overline{\Gamma})$, satisfying $f(\xi) \cdot \xi = 0$ for all $\xi \in \overline{\Gamma}$. For a regular region $\Gamma \subset \mathbb{S}^2$ the *surface theorem of Stokes* takes the form

$$\int_\Gamma \mathrm{L}_\xi^* \cdot f(\xi)\, dS(\xi) = \int_{\partial\Gamma} \tau_\xi \cdot f(\xi)\, d\sigma(\xi) \tag{9.27}$$

provided that f is a continuously differentiable vector field on $\overline{\Gamma} = \Gamma \cup \partial\Gamma$ such that $f(\xi) \cdot \xi = 0$, $\xi \in \overline{\Gamma}$.

Applying the Gauss formula (9.26) to $f = F\, \nabla^* G$ with suitable F, G we obtain the *first Green's surface theorem*

$$\int_\Gamma \nabla_\xi^* G(\xi) \cdot \nabla_\xi^* F(\xi)\, dS(\xi) + \int_\Gamma F(\xi) \Delta_\xi^* G(\xi)\, dS(\xi)$$
$$= \int_{\partial\Gamma} F(\xi) \frac{\partial}{\partial \nu_\xi} G(\xi)\, d\sigma(\xi). \tag{9.28}$$

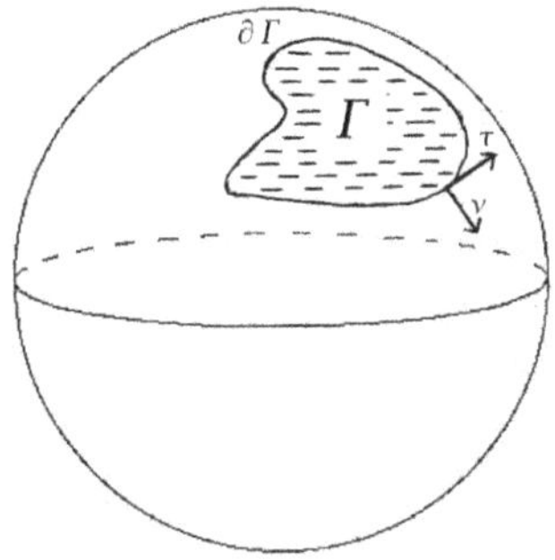

FIGURE 9.1: Γ as regular region of the unit sphere $\mathbb{S}^2$.

In a similar way, applying the Stokes formula (9.27) to $f = F\,\mathrm{L}^*G$ we get

$$\int_\Gamma \mathrm{L}^*_\xi G(\xi)\cdot \mathrm{L}^*_\xi F(\xi)\, dS(\xi) + \int_\Gamma F(\xi)\Delta^*_\xi G(\xi)\, dS(\xi) \\ = \int_{\partial\Gamma} F(\xi)\frac{\partial}{\partial \tau_\xi} G(\xi)\, d\sigma(\xi). \tag{9.29}$$

Observe that $\partial/\partial\nu_\xi = \nu_\xi\cdot\nabla^*_\xi$ and $\partial/\partial\tau_\xi = \tau_\xi\cdot \mathrm{L}^*_\xi$. Interchanging F, G in (9.29) or (9.28) and subtracting this from the original formula yields the *second Green's surface theorem*

$$\int_\Gamma F(\xi)\Delta^*_\xi G(\xi) - G(\xi)\Delta^*_\xi F(\xi)\, dS(\xi) \\ = \int_{\partial\Gamma} F(\xi)\frac{\partial}{\partial\nu_\xi} G(\xi) - G(\xi)\frac{\partial}{\partial\nu_\xi} F(\xi)\, d\sigma(\xi) \\ = \int_{\partial\Gamma} F(\xi)\frac{\partial}{\partial\tau_\xi} G(\xi) - G(\xi)\frac{\partial}{\partial\tau_\xi} F(\xi)\, d\sigma(\xi). \tag{9.30}$$

If $\Gamma \subset \mathbb{S}^2$ is regular, then $\mathbb{S}^2 \setminus \overline{\Gamma}$ is also regular. Therefore, all the integral identities above also hold true on $\mathbb{S}^2 \setminus \overline{\Gamma}$ (under suitable assumptions on the integrands). By addition of the associated integral formulas the boundary integrals vanish (because of their opposite orientation) and for the whole sphere $\mathbb{S}^2$ this leads to

$$\int_{\mathbb{S}^2} f(\xi)\cdot\nabla^*_\xi F(\xi)\, dS(\xi) = -\int_{\mathbb{S}^2} F(\xi)\nabla^*_\xi\cdot f(\xi)\, dS(\xi), \tag{9.31}$$

$$\int_{\mathbb{S}^2} f(\xi)\cdot \mathrm{L}^*_\xi F(\xi)\, dS(\xi) = -\int_{\mathbb{S}^2} F(\xi)\mathrm{L}^*_\xi\cdot f(\xi)\, dS(\xi), \tag{9.32}$$

$$\int_{\mathbb{S}^2} \nabla^*_\xi F(\xi)\cdot\nabla^*_\xi G(\xi)\, dS(\xi) = -\int_{\mathbb{S}^2} F(\xi)\Delta^*_\xi G(\xi)\, dS(\xi) \\ = -\int_{\mathbb{S}^2} G(\xi)\Delta^*_\xi F(\xi)\, dS(\xi). \tag{9.33}$$

Furthermore,

$$\int_{\mathbb{S}^2} \nabla_\xi^* \cdot f(\xi)\, dS(\xi) = 0, \tag{9.34}$$

$$\int_{\mathbb{S}^2} \mathrm{L}_\xi^* F(\xi) \cdot \mathrm{L}_\xi^* G(\xi)\, dS(\xi) = -\int_{\mathbb{S}^2} F(\xi)\Delta_\xi^* G(\xi)\, dS(\xi), \tag{9.35}$$

$$\int_{\mathbb{S}^2} \nabla_\xi^* \cdot (f(\xi) \wedge \xi)\, dS(\xi) = 0, \tag{9.36}$$

provided that $F, G : \mathbb{S}^2 \to \mathbb{R}$ (resp. $f : \mathbb{S}^2 \to \mathbb{R}^3$) are sufficiently often continuously differentiable.

Basic Theory of Spherical Harmonics. Spherical harmonics arise on the unit sphere $\mathbb{S}^2$ in analogous way as periodic polynomials arise on the unit circle (which is also denoted by $\mathbb{S}^1$). Spherical harmonics show many of the same properties as classical Fourier expansions. Many of their properties as well as references to the earlier history of spherical harmonics may be found in E.W. Hobson [1955], J. Lense [1954], A. Wangerin [1921] and E.M. Stein, G. Weiss [1971]. Our approach is based on C. Müller [1966] and W. Freeden [1979], W. Freeden, M. Schreiner [2009] (for more details and q-dimensional generalizations see also W. Freeden, M. Gutting [2013]).

The point of departure here is the theory of homogeneous harmonic polynomials. Let H_n be a homogeneous harmonic polynomial of degree n in $\mathbb{R}^3$. The restriction $Y_n = H_n|\mathbb{S}^2$ is called a *spherical harmonic* of degree n. Using the standard method of separation, we have $H_n(x) = r^n Y_n(\xi), x = r\xi, r = |x|, \xi \in \mathbb{S}^2$. Observing the identity

$$\frac{1}{r^2}\frac{d}{dr}\left(r^2 \frac{d}{dr}\right) r^n = n(n+1)r^{n-2} \tag{9.37}$$

we obtain

$$0 = \Delta_x H_n(x) = r^{n-2} n(n+1) Y_n(\xi) + r^{n-2}\Delta_\xi^* Y_n(\xi). \tag{9.38}$$

Hence, spherical harmonics $Y_n : \xi \mapsto Y_n(\xi)$ of degree $n = 0, 1, \ldots$ are everywhere on the unit sphere $\mathbb{S}^2$ infinitely often differentiable eigenfunctions of the Beltrami operator Δ^* corresponding to the eigenvalues $(\Delta^*)^\wedge(n) = -n(n+1)$, $n = 0, 1, \ldots$. The space of all spherical harmonics of degree n is written as Harm_n. As is well-known (see, e.g., C. Müller [1966], W. Freeden [1979], W. Freeden, M. Gutting [2013] and the references therein), the linear space Harm_n is of dimension $2n+1$. We always assume that $\{Y_{n,j}\}_{j=1,\ldots,2n+1}$ be an orthonormal basis of Harm_n with respect to $\langle \cdot, \cdot \rangle_{\mathrm{L}^2(\mathbb{S}^2)}$. Integration by parts, i.e., Green's theorem, tells us that for $n \neq m$

$$\begin{aligned} \langle \Delta^* Y_{n,j}, Y_{m,l} \rangle_{\mathrm{L}^2(\mathbb{S}^2)} &= \int_{\mathbb{S}^2} \Delta_\xi^* Y_{n,j}(\xi) Y_{m,l}(\xi)\, dS(\xi) \\ &= \int_{\mathbb{S}^2} Y_{n,j}(\xi) \Delta_\xi^* Y_{m,l}(\xi)\, dS(\xi) = \langle Y_{n,j}, \Delta^* Y_{m,l} \rangle_{\mathrm{L}^2(\mathbb{S}^2)}, \end{aligned} \tag{9.39}$$

so that

$$(m(m+1) - n(n+1)) \int_{\mathbb{S}^2} Y_{n,j}(\xi) Y_{m,l}(\xi)\, dS(\xi) = 0. \tag{9.40}$$

In other words, spherical harmonics of different degrees are orthogonal with respect to $\langle \cdot, \cdot \rangle_{\mathrm{L}^2(\mathbb{S}^2)}$. Altogether, we have

$$\langle Y_{n,j}, Y_{m,l} \rangle_{\mathrm{L}^2(\mathbb{S}^2)} = \int_{\mathbb{S}^2} Y_{n,j}(\eta) Y_{m,l}(\eta)\, dS(\eta) = \delta_{n,m}\delta_{j,l}. \tag{9.41}$$

The *addition theorem of scalar spherical harmonics in* Harm_n (see, e.g., W. Freeden [1979], C. Müller [1966]) reads as follows

$$\frac{2n+1}{4\pi} P_n(\xi \cdot \eta) = \sum_{j=1}^{2n+1} Y_{n,j}(\xi) Y_{n,j}(\eta), \quad \xi, \eta \in \mathbb{S}^2, \tag{9.42}$$

where P_n is the Legendre polynomial of degree n.

Let H be a function of class $\mathrm{L}^1[-1,1]$ and Y_n be a spherical harmonic of degree n. Then the *Funk-Hecke formula* states (see, e.g., C. Müller [1966])

$$\int_{\mathbb{S}^2} H(\xi \cdot \eta) Y_n(\eta)\, dS(\eta) = H^\wedge(n) Y_n(\xi), \tag{9.43}$$

where the "*Legendre transform*" of $H \in \mathrm{L}^1[-1,1]$ is given by

$$H^\wedge(n) = 2\pi \int_{-1}^{1} H(t) P_n(t) dt. \tag{9.44}$$

Note that this formula establishes the close connection between the orthogonal invariance of the sphere and the addition theorem.

The *spherical Fourier transform* $H \mapsto (FT)(H)$, $H \in \mathrm{L}^2(\mathbb{S}^2)$, is given by

$$((FT)(H))(n,j) = H^\wedge(n,j) = \langle H, Y_{n,j} \rangle_{\mathrm{L}^2(\mathbb{S}^2)}. \tag{9.45}$$

We use the abbreviation

$$\mathcal{M} = \{(n,j) : n = 0, 1, \ldots,\ j = 1, \ldots, 2n+1\}. \tag{9.46}$$

FT forms a mapping from $\mathrm{L}^2(\mathbb{S}^2)$ into the space $\ell^2(\mathcal{M})$ of all sequences $\{H^\wedge(n,j)\}_{(n,j)\in\mathcal{M}}$ satisfying

$$\sum_{(n,j)\in\mathcal{M}} (H^\wedge(n,j))^2 = \sum_{n=0}^{\infty} \sum_{j=1}^{2n+1} (H^\wedge(n,j))^2 < \infty. \tag{9.47}$$

For $\{H^\wedge(n,j)\}_{(n,j)\in\mathcal{M}} \in \ell^2(\mathcal{M})$ we define the mapping $(FT)^{-1} : \ell^2(\mathcal{M}) \to \mathrm{L}^2(\mathbb{S}^2)$ by

$$(FT)^{-1}\left(\{H^\wedge(n,j)\}_{(n,j)\in\mathcal{M}}\right) = \sum_{n=0}^{\infty} \sum_{j=1}^{2n+1} H^\wedge(n,j) Y_{n,j}. \tag{9.48}$$

Then $(FT)^{-1}(FT) = I_{\mathrm{L}^2(\mathbb{S}^2)}$ and $(FT)(FT)^{-1} = I_{\ell^2(\mathcal{M})}$ (I denotes the identity operator). Moreover, it should be noted that for $G, H \in \mathrm{L}^2(\mathbb{S}^2)$ the relation

$$\lim_{N\to\infty} \left\| G - \sum_{n=0}^{N} \sum_{j=1}^{2n+1} H^\wedge(n,j) Y_{n,j} \right\|_{\mathrm{L}^2(\mathbb{S}^2)} = 0 \tag{9.49}$$

implies $G = H$ almost everywhere on $\mathbb{S}^2$. An approximation of $H \in \mathrm{L}^2(\mathbb{S}^2)$ by its Fourier expansion (in terms of $\mathrm{L}^2(\mathbb{S}^2)$-orthonormal spherical harmonics) of the form

$$\lim_{N\to\infty} \left\| H - \sum_{n=0}^{N} \sum_{j=1}^{2n+1} H^\wedge(n,j) Y_{n,j} \right\|_{\mathrm{L}^2(\mathbb{S}^2)} = 0 \tag{9.50}$$

is certainly true. In addition, T. Gronwall [1914] has shown that H can be recovered by its Fourier expansion in uniform sense

$$\lim_{N\to\infty} \left\| H - \sum_{n=0}^{N} \sum_{j=1}^{2n+1} H^\wedge(n,j) Y_{n,j} \right\|_{\mathrm{C}^{(0)}(\mathbb{S}^2)} = 0 \tag{9.51}$$

if H is assumed to be Lipschitz-continuous. The relation between an element and its Fourier series has been the object of many investigations. Of particular importance for practical purposes are results in the framework of the Hilbert space $(\mathrm{L}^2(\mathbb{S}^2), \langle\cdot,\cdot\rangle_{\mathrm{L}^2(\mathbb{S}^2)})$.

Theorem 9.1. *The closure of the system* $\{Y_{n,j}\}_{(n,j)\in\mathcal{M}}$ *in* $\mathrm{L}^2(\mathbb{S}^2)$ *is equivalent to each of the following statements:*

(a) The orthogonal expansion of any element $H \in \mathrm{L}^2(\mathbb{S}^2)$ *converges in the* $\mathrm{L}^2(\mathbb{S}^2)$*-norm to* H*, i.e.,*

$$\lim_{m\to\infty} \left\| H - \sum_{n=0}^{m} \sum_{j=1}^{2n+1} \langle H, Y_{n,j}\rangle_{\mathrm{L}^2(\mathbb{S}^2)} Y_{n,j} \right\|_{\mathrm{L}^2(\mathbb{S}^2)} = 0. \tag{9.52}$$

(b) Parseval's identity *holds. That is, for any* $H \in \mathrm{L}^2(\mathbb{S}^2)$,

$$\|H\|^2_{\mathrm{L}^2(\mathbb{S}^2)} = \langle H, H\rangle_{\mathrm{L}^2(\mathbb{S}^2)} = \sum_{n=0}^{\infty} \sum_{j=1}^{2n+1} |\langle H, Y_{n,j}\rangle_{\mathrm{L}^2(\mathbb{S}^2)}|^2. \tag{9.53}$$

(c) The extended Parseval identity *holds. That is, for any* $H, K \in \mathrm{L}^2(\mathbb{S}^2)$,

$$\langle H, K\rangle_{\mathrm{L}^2(\mathbb{S}^2)} = \sum_{n=0}^{\infty} \sum_{j=1}^{2n+1} \langle H, Y_{n,j}\rangle_{\mathrm{L}^2(\mathbb{S}^2)} \langle K, Y_{n,j}\rangle_{\mathrm{L}^2(\mathbb{S}^2)}. \tag{9.54}$$

(d) There is no strictly larger orthonormal system containing the orthonormal system $\{Y_{n,j}\}_{(n,j)\in\mathcal{M}}$.

(e) *The system* $\{Y_{n,j}\}_{(n,j)\in\mathcal{M}}$ *has the* spherical harmonics!completeness. *That is,* $H \in \mathrm{L}^2(\mathbb{S}^2)$ *and* $\langle H, Y_{n,j}\rangle_{\mathrm{L}^2(\mathbb{S}^2)} = 0$ *for all* $(n,j) \in \mathcal{M}$, *implies* $H = 0$.

(f) *An element* H *of* $\mathrm{L}^2(\mathbb{S}^2)$ *is determined uniquely by its orthogonal coefficients. That is, if* $\langle H, Y_{n,j}\rangle_{\mathrm{L}^2(\mathbb{S}^2)} = \langle K, Y_{n,j}\rangle_{\mathrm{L}^2(\mathbb{S}^2)}$ *for all* $(n,j) \in \mathcal{M}$, *then* $H = K$.

The proof of Theorem 9.1 is omitted (see, e.g., P.J. Davis, P. Rabinowitz [1975] for the functional analytic background). The property (a) is of great importance for practical purposes. In particular, it tells us that any continuous function may be approximated (in the $\mathrm{L}^2(\mathbb{S}^2)$-sense) by finite truncations of its Fourier (orthogonal) expansion in terms of any $\mathrm{L}^2(\mathbb{S}^2)$-orthonormal system of spherical harmonics $\{Y_{n,j}\}_{(n,j)\in\mathcal{M}}$.

Zonal Kernel Functions. A well-known fact is that the Euclidean distance of two unit vectors is expressible in terms of their inner product:

$$|\xi - \eta|^2 = |\xi|^2 + |\eta|^2 - 2\xi \cdot \eta = 2(1 - \xi \cdot \eta), \quad \xi, \eta \in \mathbb{S}^2. \tag{9.55}$$

Consequently, any function dependent on $|\xi - \eta|$, $\xi, \eta \in \mathbb{S}^2$, i.e., *radial basis function* $K(|\xi - \eta|)$ is equivalently characterized by the property of being dependent only on the inner product $\xi \cdot \eta$ of the unit vectors $\xi, \eta \in \mathbb{S}^2$, i.e.,

$$K(\xi, \eta) = K(|\xi - \eta|) = \hat{K}(\xi \cdot \eta), \quad \xi, \eta \in \mathbb{S}^2, \tag{9.56}$$

and vice versa.

In the theory of special functions of mathematical physics, however, a kernel $\hat{K} : \mathbb{S}^2 \times \mathbb{S}^2 \to \mathbb{R}$ satisfying $\hat{K}(\xi \cdot \eta) = \hat{K}(\mathbf{t}\xi \cdot \mathbf{t}\eta)$, $\xi, \eta \in \mathbb{S}^2$, for all orthogonal transformations $\mathbf{t} \in O(3)$ is known as a *zonal kernel function*. In order to point out the reducibility of $\hat{K}$ to a function defined on the interval $[-1, 1]$, the notation $(\xi, \eta) \mapsto \hat{K}(\xi \cdot \eta)$, $(\xi, \eta) \in \mathbb{S}^2 \times \mathbb{S}^2$, is used throughout this work. The classical addition theorem of spherical harmonics enables us to characterize zonal kernel functions as orthogonal (Fourier) sum expansions in terms of Legendre polynomials. In fact, from the Funk–Hecke formula (9.43) we obtain for all $\xi, \eta \in \mathbb{S}^2$ and $K \in \mathrm{L}^2[-1, 1]$

$$\int_{\mathbb{S}^2} K(\xi \cdot \zeta) P_n(\zeta \cdot \eta)\, dS(\zeta) = K^\wedge(n) P_n(\xi \cdot \eta) \tag{9.57}$$

with Legendre coefficients $K^\wedge(n), n \in \mathbb{N}_0$, given by

$$K^\wedge(n) = 2\pi \int_{-1}^{+1} K(t) P_n(t)\, dt. \tag{9.58}$$

Using both the addition theorem and the Funk–Hecke formula we get a representation of K in terms of a Legendre series (in $\|\cdot\|_{\mathrm{L}^2(\mathbb{S}^2)}$-sense). Explicitly written out we have

$$K(\xi \cdot\ ) = \sum_{n=0}^{\infty} \frac{2n+1}{4\pi} K^\wedge(n) P_n(\xi \cdot\ ), \tag{9.59}$$

where the sequence $\{K^\wedge(n)\}_{n\in\mathbb{N}_0}$ is called the *Legendre symbol* of the zonal kernel $K(\xi\cdot\ )$.

Convolutions. If $H \in \mathrm{L}^1[-1,1]$ and $F \in \mathrm{L}^2(\mathbb{S}^2)$, then the *convolution* of H and F is given (see, e.g., H. Berens et al. [1968]) by

$$(H * F)(\xi) = \int_{\mathbb{S}^2} H(\xi\cdot\eta)F(\eta)\, dS(\eta), \quad \xi \in \mathbb{S}^2. \tag{9.60}$$

Young's inequality (see, e.g., H. Berens et al. [1968], S. Bochner [1954], E.M. Stein, G. Weiss [1971]) yields

$$\|H * F\|_{\mathrm{L}^r(\mathbb{S}^2)} \leq \|F\|_{\mathrm{L}^p(\mathbb{S}^2)}\|H\|_{\mathrm{L}^q[-1,1]} \tag{9.61}$$

provided that $F \in \mathrm{L}^p(\mathbb{S}^2), H \in \mathrm{L}^q[-1,1], 1 \leq p, q < \infty$ satisfying $\frac{1}{r} = \frac{1}{q} + \frac{1}{p} - 1 \geq 0$. In particular, if $F \in \mathrm{C}^{(0)}(\mathbb{S}^2)$ and $H \in \mathrm{L}^1[-1,1]$, then the convolution $H * F$ is a continuous function on $\mathbb{S}^2$, and it follows that

$$\|H * F\|_{\mathrm{C}^{(0)}(\mathbb{S}^2)} \leq \|F\|_{\mathrm{C}^{(0)}(\mathbb{S}^2)}\|H\|_{\mathrm{L}^1[-1,1]}. \tag{9.62}$$

Via the Funk–Hecke formula we are led to compositions of zonal kernels generated by convolution. If $H_1, H_2 \in \mathrm{L}^1[-1,1]$ and $\eta \in \mathbb{S}^2$ fixed, then the function $\xi \mapsto H_2(\xi\cdot\eta)$, $\xi \in \mathbb{S}^2$, is out of class $\mathrm{L}^1(\mathbb{S}^2)$. We let

$$(H_1 * H_2)(\xi\cdot\eta) = (H_1 * H_2(\ \cdot\eta))(\xi) = \int_{\mathbb{S}^2} H_1(\xi\cdot\zeta)H_2(\zeta\cdot\eta)\, dS(\zeta). \tag{9.63}$$

Obviously, $H_1 * H_2$ depends only on the inner product of ξ and η, and we have

$$(H_1 * H_2)^\wedge(n) = H_1^\wedge(n)H_2^\wedge(n), \quad n = 0, 1, \ldots. \tag{9.64}$$

Note that we use the same symbol '$*$' for different specifications of convolutions. Moreover the commutativity in (9.64) should be pointed out, such that the convolution of H against K is equal to the convolution of K against H, i.e.,

$$H * K = K * H. \tag{9.65}$$

Suppose that H, K are members of $\mathrm{L}^2[-1,1]$. Furthermore, assume that F is of class $\mathrm{L}^2(\mathbb{S}^2)$. By virtue of the Cauchy–Schwarz inequality it is not hard to see that $K * F$ is in $\mathrm{L}^2(\mathbb{S}^2)$, whereas $H * K$ is a member of class $\mathrm{C}^{(0)}[-1,1]$. In spectral formulation we have

$$K * F = \sum_{n=0}^{\infty} K^\wedge(n) \sum_{m=1}^{2n+1} F^\wedge(n,m)Y_{n,m}, \tag{9.66}$$

and

$$H * K = \sum_{n=0}^{\infty} \frac{2n+1}{4\pi} H^\wedge(n)K^\wedge(n)P_n. \tag{9.67}$$

Finally, it should be mentioned that $F \in \mathrm{L}^2(\mathbb{S}^2)$ allows the orthogonal decomposition into the projections $P_{\mathrm{Harm}_n} F = \frac{2n+1}{4\pi} P_n * F$.

$$F = \sum_{n=0}^{\infty} P_{\mathrm{Harm}_n} F = \sum_{n=0}^{\infty} \frac{2n+1}{4\pi} P_n * F \tag{9.68}$$

in the topology of $\|\cdot\|_{\mathrm{L}^2(\mathbb{S}^2)}$. An important feature is that the convolution of zonal kernel functions does not affect the property of being a zonal kernel.

Iterated Convolution. Assume that $K \in \mathrm{L}^2[-1,1]$ and $F \in \mathrm{L}^2(\mathbb{S}^2)$. For $(\xi, \zeta) \in \mathbb{S}^2 \times \mathbb{S}^2$ we let

$$K^{(1)}(\xi \cdot \zeta) = K(\xi \cdot \zeta), \tag{9.69}$$

$$K^{(k)}(\xi \cdot \zeta) = \int_{\mathbb{S}^2} K^{(k-1)}(\xi \cdot \eta) K(\zeta \cdot \eta)\, dS(\eta), \quad k = 2, 3 \ldots . \tag{9.70}$$

Then $K^{(k)} * F$ is called the k-th *iterated convolution* of K *against* F. Obviously, the k-th iterated kernel K is a scalar zonal kernel function, and it follows immediately that

$$(K^{(k)})^\wedge(n) = (K^\wedge(n))^k, \quad n = 0, 1, \ldots, \quad k = 1, 2, \ldots . \tag{9.71}$$

9.2 Sphere Functions Involving Beltrami Operators

Our point of departure for discussing the Green's function with respect to the Beltrami operator Δ^* on $\mathbb{S}^2$ is the definition of its constituting properties (cf. W. Freeden [1978c, 1979, 1980b]), which can be given in parallel to one-dimensional settings.

Definition 9.2. *$G(\Delta^*; \cdot, \cdot) : (\xi, \eta) \mapsto G(\Delta^*; \xi, \eta), -1 \leq \xi \cdot \eta < 1$, is called* Green's function on $\mathbb{S}^2$ with respect to the operator Δ^* *(briefly, $\mathbb{S}^2$-sphere function for Δ^*), if it satisfies the following properties:*

(i) (Differential equation) *for every point $\xi \in \mathbb{S}^2$, $\eta \mapsto G(\Delta^*; \xi, \eta)$ is twice continuously differentiable on the set $\{\eta \in \mathbb{S}^2 : -1 \leq \xi \cdot \eta < 1\}$, such that*

$$\Delta^*_\eta G(\Delta^*; \xi, \eta) = -\frac{1}{4\pi}, \quad -1 \leq \xi \cdot \eta < 1. \tag{9.72}$$

(ii) (Characteristic singularity) *for every $\xi \in \mathbb{S}^2$, the function*

$$\eta \mapsto G(\Delta^*; \xi, \eta) - \frac{1}{4\pi} \ln(1 - \xi \cdot \eta) \tag{9.73}$$

is continuously differentiable on $\mathbb{S}^2$. Note that for $-1 \leq \xi \cdot \eta < 1$

$$\ln|\xi - \eta| = \frac{1}{2} \ln(2 - 2\xi \cdot \eta) = \frac{1}{2} \ln(1 - \xi \cdot \eta) + \frac{1}{2} \ln 2. \tag{9.74}$$

*(iii) (*Rotational symmetry*) for every* $\xi \in \mathbb{S}^2$ *and all orthogonal transformations* $\mathbf{t} \in O(3)$

$$G(\Delta^*; \mathbf{t}\xi, \mathbf{t}\eta) = G(\Delta^*; \xi, \eta). \tag{9.75}$$

*(iv) (*Normalization*) for every* $\xi \in \mathbb{S}^2$,

$$\frac{1}{4\pi} \int_{\mathbb{S}^2} G(\Delta^*; \xi, \eta)\, dS(\eta) = 0. \tag{9.76}$$

We first prove the *uniqueness of* $\mathbb{S}^2$*-sphere function with respect to the Beltrami operator* Δ^*.

Lemma 9.3. $G(\Delta^*; \cdot, \cdot)$ *is uniquely determined by its defining properties (i)–(iv).*

Proof. Denote by $D(\Delta^*; \cdot, \cdot)$ the difference between two $\mathbb{S}^2$-sphere functions for Δ^* satisfying (i)–(iv). Then we have the following properties:

(i) For every $\xi \in \mathbb{S}^2$, $D(\Delta^*; \xi, \cdot)$ is twice continuously differentiable for all $\eta \in \mathbb{S}^2$ satisfying $-1 \leq \xi \cdot \eta < 1$, and we have

$$(\Delta^*)_\eta D(\Delta^*; \xi, \eta) = 0, \tag{9.77}$$

(ii) For every $\xi \in \mathbb{S}^2$, $D(\Delta^*; \xi, \cdot)$ is continuously differentiable on $\mathbb{S}^2$,

(iii) for every $\xi \in \mathbb{S}^2$ and for all orthogonal transformations $\mathbf{t} \in O(3)$,

$$D(\Delta^*; \mathbf{t}\xi, \mathbf{t}\eta) = D(\Delta^*; \xi, \eta), \tag{9.78}$$

(iv) For all $\xi \in \mathbb{S}^2$,

$$\int_{\mathbb{S}^2} D(\Delta^*; \xi, \eta)\, dS(\eta) = 0. \tag{9.79}$$

The properties (i)–(iii) show that $D(\Delta^*; \xi, \cdot)$ is an everywhere on the unit sphere $\mathbb{S}^2$ twice continuously differentiable function satisfying the differential equation (i). Therefore, $D(\Delta^*; \xi, \cdot)$ must be a spherical harmonic of degree 0. $D(\Delta^*; \xi, \eta)$ depends only on the scalar product of ξ and η, i.e.,

$$D(\Delta^*; \xi, \eta) = \alpha_0 P_0(\xi \cdot \eta) = \alpha_0. \tag{9.80}$$

From (iv) we obtain

$$\int_{\mathbb{S}^2} D(\Delta^*; \xi, \eta)\, dS(\eta) = 4\pi\alpha_0 = 0. \tag{9.81}$$

Hence, $\alpha_0 = 0$. But this means that the $\mathbb{S}^2$-sphere function $G(\Delta^*; \cdot, \cdot)$ is uniquely determined by the defining properties (i)–(iv). □

An easy calculation shows that

$$(\xi,\eta) \mapsto \frac{1}{4\pi}\ln(1-\xi\cdot\eta) + \frac{1}{4\pi} - \frac{1}{4\pi}\ln 2, \quad -1 \le \xi\cdot\eta < 1 \tag{9.82}$$

satisfies all the defining properties (i)–(iv) of the $\mathbb{S}^2$-sphere function for Δ^*. Therefore, we have the following result.

Lemma 9.4. *For $\xi,\eta \in \mathbb{S}^2$ with $-1 \le \xi\cdot\eta < 1$*

$$G(\Delta^*;\xi,\eta) = \frac{1}{4\pi}\ln(1-\xi\cdot\eta) + \frac{1}{4\pi} - \frac{1}{4\pi}\ln 2. \tag{9.83}$$

Remark 9.5. *Usually, we write $G(\Delta^*;\xi\cdot\eta)$ instead of $G(\Delta^*;\xi,\eta)$, $(\xi,\eta) \in \mathbb{S}^2\times\mathbb{S}^2$. This indicates that $G(\Delta^*;\xi\cdot\eta)$ depends only on the scalar product of ξ and η, i.e., $G(\Delta^*;\cdot)$ is a zonal function, hence, it may be understood as a function defined on the (one-dimensional) interval $[-1,1)$.*

Observing the logarithmic singularity of the $\mathbb{S}^2$-sphere function, we see by applying integration by parts, i.e., Green's surface theorem (9.33) that the spherical harmonics of degree n, i.e., the eigenfunctions of the Beltrami operator Δ^*, are eigenfunctions of the (kernel) function $G(\Delta^*;\cdot)$ in the sense of the integral equation

$$-k(k+1)\int_{\mathbb{S}^2} G(\Delta^*;\xi\cdot\eta)Y_k(\eta)\,dS(\eta) = (1-\delta_{0,k})Y_k(\xi). \tag{9.84}$$

In terms of a (maximal) $\mathrm{L}^2(\mathbb{S}^2)$-orthonormal system $\{Y_{n,m}\}$ of spherical harmonics of degree n and order m, we thus obtain as spectral representation the bilinear expansion (see W. Freeden [1979]) for $-1 \le \xi\cdot\eta < 1$:

$$G(\Delta^*;\xi\cdot\eta) = \sum_{n=1}^{\infty}\sum_{m=1}^{2n+1}\frac{1}{-n(n+1)}Y_{n,m}(\xi)Y_{n,m}(\eta). \tag{9.85}$$

Observing the addition theorem of the spherical harmonics, we find the following series representation in terms of Legendre polynomials

$$G(\Delta^*;\xi\cdot\eta) = \sum_{n=1}^{\infty}\frac{2n+1}{4\pi}\frac{1}{-n(n+1)}P_n(\xi\cdot\eta). \tag{9.86}$$

Space Regularized Sphere Function. Next we are concerned with some "mollification" of the concept of Green's function (cf. W. Freeden, M. Schreiner [2006, 2009]). More concretely, we discuss the *(space) ρ-regularized $\mathbb{S}^2$-sphere function with respect to Δ^** (see Figure 9.2):

$$G^\rho(\Delta^*;\xi\cdot\eta) = \begin{cases} \frac{1}{4\pi}\ln(1-\xi\cdot\eta) + \frac{1}{4\pi}(1-\ln 2), & \xi\cdot\eta < 1-\rho, \\ \frac{1-\xi\cdot\eta}{4\pi\rho} + \frac{1}{4\pi}(\ln\rho - \ln 2), & \xi\cdot\eta \ge 1-\rho. \end{cases} \tag{9.87}$$

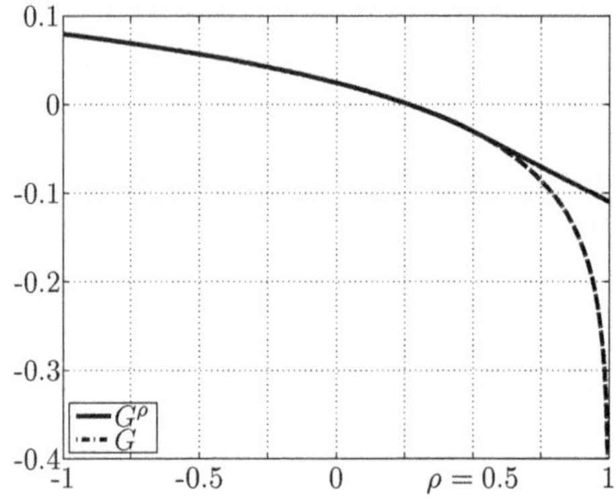

FIGURE 9.2: The regularization of G by G^ρ ($\rho = 0.5$).

Obviously, the kernel function $(\xi, \eta) \mapsto G^\rho(\Delta^*; \xi \cdot \eta)$ only depends on the inner product of ξ and η, hence $G^\rho(\Delta^*; \xi \cdot \eta)$ is a zonal function. According to its construction, $G^\rho(\Delta^*; \ \cdot\eta)$ is a continuously differentiable function on $\mathbb{S}^2$ for every (fixed) $\eta \in \mathbb{S}^2$, $G^\rho(\Delta^*; \xi\cdot \ )$ is a continuously differentiable function on $\mathbb{S}^2$ for every (fixed) $\xi \in \mathbb{S}^2$, and we have

$$t \mapsto G^\rho(\Delta^*; t) = \begin{cases} \frac{1}{4\pi} \ln(1-t) + \frac{1}{4\pi}(1 - \ln 2), & t < 1-\rho, \\ \frac{1-t}{4\pi\rho} + \frac{1}{4\pi}(\ln \rho - \ln 2), & t \geq 1-\rho, \end{cases} \tag{9.88}$$

is a (one-dimensional) continuously differentiable function on the interval $[-1, 1]$. The surface gradient of the ρ-regularized $\mathbb{S}^2$-sphere function kernel is given by

$$\nabla^*_\xi G^\rho(\Delta^*; \xi \cdot \eta) = \begin{cases} -\frac{1}{4\pi} \frac{1}{1-\xi\cdot\eta}(\eta - (\xi \cdot \eta)\xi), & \xi \cdot \eta < 1-\rho, \\ -\frac{1}{4\pi\rho}(\eta - (\xi \cdot \eta)\xi), & \xi \cdot \eta \geq 1-\rho, \end{cases} \tag{9.89}$$

while the surface curl gradient of the ρ-regularized $\mathbb{S}^2$-sphere function kernel reads as follows:

$$\mathrm{L}^*_\xi G^\rho(\Delta^*; \xi \cdot \eta) = \begin{cases} -\frac{1}{4\pi} \frac{1}{1-\xi\cdot\eta}(\xi \wedge \eta), & \xi \cdot \eta < 1-\rho, \\ -\frac{1}{4\pi\rho}(\xi \wedge \eta), & \xi \cdot \eta \geq 1-\rho. \end{cases} \tag{9.90}$$

For graphical illustration see Figures 9.3 and 9.4. In addition, we mention the Beltrami derivative

$$\Delta^*_\xi G^\rho(\Delta^*; \xi \cdot \eta) = \begin{cases} -\frac{1}{4\pi}, & \xi \cdot \eta < 1-\rho, \\ \frac{1}{2\pi\rho}(\xi \cdot \eta), & \xi \cdot \eta \geq 1-\rho. \end{cases} \tag{9.91}$$

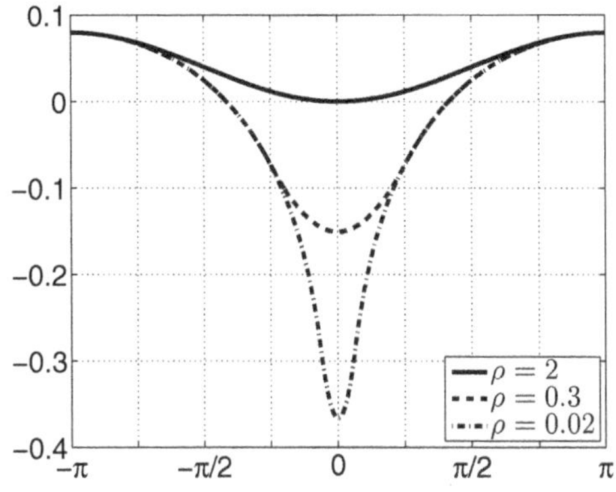

FIGURE 9.3: ρ-regularized $\mathbb{S}^2$-sphere function $\vartheta \mapsto G^\rho(\Delta^*; \cos\vartheta)$ for various values ρ.

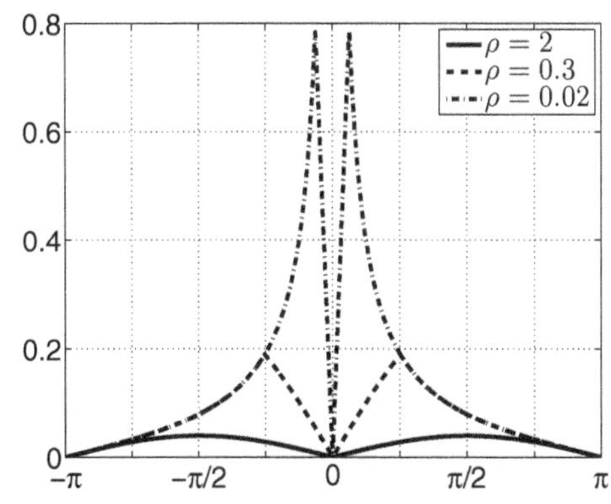

FIGURE 9.4: Absolute value of the surface gradient or surface curl gradient of the ρ-regularized $\mathbb{S}^2$-sphere function $\vartheta \mapsto G^\rho(\Delta^*; \cos\vartheta)$ for various values ρ.

For $F \in \mathrm{C}^{(0)}(\mathbb{S}^2)$, we consider the potential $P^\rho(F)$ given by

$$P^\rho(F)(\xi) = \int_{\mathbb{S}^2} G^\rho(\Delta^*; \xi \cdot \eta) F(\eta)\, dS(\eta), \quad \xi \in \mathbb{S}^2, \tag{9.92}$$

as regularized counterpart to the potential $P(F)$ given by

$$P(F)(\xi) = \int_{\mathbb{S}^2} G(\Delta^*; \xi \cdot \eta) F(\eta)\, dS(\eta), \quad \xi \in \mathbb{S}^2. \tag{9.93}$$

Now we prove the following theorem.

Theorem 9.6. *For (sufficiently small) values $\rho > 0$ and $F \in C^{(0)}(\mathbb{S}^2)$, the potential $P^\rho(F)$ is of class $C^{(1)}(\mathbb{S}^2)$, and we have*

$$\lim_{\rho\to 0} \sup_{\xi\in\mathbb{S}^2} \left| \int_{\mathbb{S}^2} G(\Delta^*;\xi\cdot\eta)F(\eta)\, dS(\eta) - \int_{\mathbb{S}^2} G^\rho(\Delta^*;\xi\cdot\eta)F(\eta)\, dS(\eta) \right| = 0 \tag{9.94}$$

and

$$\lim_{\rho\to 0} \sup_{\xi\in\mathbb{S}^2} \left| \int_{\mathbb{S}^2} \nabla^*_\xi G(\Delta^*;\xi\cdot\eta)F(\eta) dS(\eta) - \nabla^*_\xi \int_{\mathbb{S}^2} G^\rho(\Delta^*;\xi\cdot\eta)F(\eta) dS(\eta) \right| = 0. \tag{9.95}$$

Furthermore,

$$\sup_{\xi\in\mathbb{S}^2} \left| \int_{\mathbb{S}^2} \nabla^*_\xi G(\Delta^*;\xi\cdot\eta)F(\eta)\, dS(\eta) - \nabla^*_\xi \int_{\mathbb{S}^2} G(\Delta^*;\xi\cdot\eta)F(\eta)\, dS(\eta) \right| = 0. \tag{9.96}$$

Proof. First we are concerned with the existence of the occurring integrals. Since F is of class $C^{(0)}(\mathbb{S}^2)$, we easily see that

$$\left| \int_{\mathbb{S}^2} \ln(1-\xi\cdot\eta)F(\eta)\, dS(\eta) \right| \leq 2\pi \|F\|_{C^{(0)}(\mathbb{S}^2)} \int_{-1}^{1} \ln(1-t)\, dt < \infty \tag{9.97}$$

and

$$\left| \int_{\mathbb{S}^2} \nabla^*_\xi \ln(1-\xi\cdot\eta)F(\eta)\, dS(\eta) \right| \leq \|F\|_{C^{(0)}(\mathbb{S}^2)} \int_{\mathbb{S}^2} \frac{\sqrt{1-(\xi\cdot\eta)^2}}{1-\xi\cdot\eta}\, dS(\eta) < \infty. \tag{9.98}$$

For $F \in C^{(0)}(\mathbb{S}^2)$, $P^\rho(F)$ is of class $C^{(1)}(\mathbb{S}^2)$. Thus it is clear that

$$\begin{aligned} \nabla^*_\xi P^\rho(F)(\xi) &= \nabla^*_\xi \int_{\mathbb{S}^2} G^\rho(\Delta^*;\xi\cdot\eta)F(\eta)\, dS(\eta) \\ &= \int_{\mathbb{S}^2} \nabla^*_\xi G^\rho(\Delta^*;\xi\cdot\eta)F(\eta)\, dS(\eta) \end{aligned} \tag{9.99}$$

with $G^\rho(\Delta^*;\xi\cdot\eta)$ and $G(\Delta^*;\xi\cdot\eta)$ differing only on the cap with center ξ and radius ρ, more precisely, on the set $\Gamma_\rho(\xi) = \{\eta \in \mathbb{S}^2 | 1-\xi\cdot\eta < \rho\}$. Thus, we obtain (for sufficiently small values $\rho > 0$)

$$\begin{aligned} &|P^\rho(F)(\xi) - P(F)(\xi)| \\ &\leq \frac{1}{4\pi}\|F\|_{C^{(0)}(\mathbb{S}^2)} \int_{\xi\cdot\eta\geq 1-\rho} \left(|\ln(1-\xi\cdot\eta)| + \ln\rho + 1 + \left|\frac{1-\xi\cdot\eta}{\rho}\right| \right) dS(\eta) \\ &\leq \frac{1}{2}\|F\|_{C^{(0)}(\mathbb{S}^2)} \int_{1-\rho}^{1} (|\ln(1-t)| + 2 + |\ln\rho|)\, dt. \end{aligned} \tag{9.100}$$

Consequently,

$$\sup_{\xi\in\mathbb{S}^2} |P^\rho(F)(\xi) - P(F)(\xi)| = O(\rho\ln\rho), \quad \rho\to 0. \tag{9.101}$$

In other words, for all $\xi \in \mathbb{S}^2$ and $F \in \mathrm{C}^{(0)}(\mathbb{S}^2)$,

$$|P^\rho(F)(\xi) - P(F)(\xi)| = O(\rho \ln \rho), \quad \rho \to 0. \tag{9.102}$$

This proves the first assertion (9.94) of our theorem. In addition, we are able to verify that

$$\begin{aligned} \left|\nabla^*_\xi P^\rho(F)(\xi) - \nabla^*_\xi P(F)(\xi)\right| &\le 2\|F\|_{\mathrm{C}^{(0)}(\mathbb{S}^2)} \int_{\substack{\xi\cdot\eta \ge 1-\rho \\ \eta\in\mathbb{S}^2}} \left|\frac{\eta - (\xi\cdot\eta)\xi}{1-\xi\cdot\eta}\right| \, dS(\eta), \\ &= 2\|F\|_{\mathrm{C}^{(0)}(\mathbb{S}^2)} \int_{\substack{\xi\cdot\eta \ge 1-\rho \\ \eta\in\mathbb{S}^2}} \frac{\sqrt{1-(\xi\cdot\eta)^2}}{1-\xi\cdot\eta} \, dS(\eta), \end{aligned} \tag{9.103}$$

i.e.,

$$\left|\nabla^*_\xi P^\rho(F)(\xi) - \nabla^*_\xi P(F)(\xi)\right| = O(\rho^{1/2}), \quad \rho \to 0. \tag{9.104}$$

Altogether, this yields the desired results stated in Theorem 9.6. □

Analogously to Theorem 9.6, we are able to formulate the following statement.

Theorem 9.7. *For (sufficiently small) values $\rho > 0$ and $F \in \mathrm{C}^{(0)}(\mathbb{S}^2)$, the potential $P^\rho(F)$ is of class $\mathrm{C}^{(1)}(\mathbb{S}^2)$, and we have*

$$\lim_{\rho\to 0} \sup_{\xi\in\mathbb{S}^2} \left| \int_{\mathbb{S}^2} G(\Delta^*;\xi\cdot\eta)F(\eta)\, dS(\eta) - \int_{\mathbb{S}^2} G^\rho(\Delta^*;\xi\cdot\eta)F(\eta)\, dS(\eta)\right| = 0 \tag{9.105}$$

and

$$\lim_{\rho\to 0} \sup_{\xi\in\mathbb{S}^2} \left| \int_{\mathbb{S}^2} \mathrm{L}^*_\xi G(\Delta^*;\xi\cdot\eta)F(\eta)\, dS(\eta) - \mathrm{L}^*_\xi \int_{\mathbb{S}^2} G^\rho(\Delta^*;\xi\cdot\eta)F(\eta)\, dS(\eta)\right| = 0. \tag{9.106}$$

Furthermore,

$$\sup_{\xi\in\mathbb{S}^2} \left| \int_{\mathbb{S}^2} \mathrm{L}^*_\xi G(\Delta^*;\xi\cdot\eta)F(\eta)\, dS(\eta) - \mathrm{L}^*_\xi \int_{\mathbb{S}^2} G(\Delta^*;\xi\cdot\eta)F(\eta)\, dS(\eta)\right| = 0. \tag{9.107}$$

Next, we determine the Legendre (Fourier) coefficients of the regularized $\mathbb{S}^2$-sphere function for the Beltrami operator Δ^* (see W. Freeden, M. Schreiner [2009]). The bilinear expansion of $G^\rho(\Delta^*;\xi\cdot\eta)$ reads as follows

$$G^\rho(\Delta^*;\xi\cdot\eta) = \sum_{n=0}^{\infty} \frac{2n+1}{4\pi} (G^\rho)^\wedge(n) P_n(\xi\cdot\eta), \quad \xi,\eta\in\mathbb{S}^2, \tag{9.108}$$

where

$$(G^\rho)^\wedge(n) = 2\pi \int_{-1}^{1} G^\rho(\Delta^*;t) P_n(t)\, dt. \tag{9.109}$$

In accordance with the definition of the regularized $\mathbb{S}^2$-sphere function for Δ^* we split the integral into two parts:

$$(G^\rho)^\wedge(n) = 2\pi \int_{-1}^{1-\rho} \left(\frac{1}{4\pi} \ln(1-t) + \frac{1}{4\pi} - \frac{1}{4\pi} \ln 2 \right) P_n(t)\, dt \qquad (9.110)$$
$$+ 2\pi \int_{1-\rho}^{1} \left(\frac{1-t}{4\pi\rho} + \frac{1}{4\pi} \ln \rho - \frac{1}{4\pi} \ln 2 \right) P_n(t)\, dt$$

For $n = 0, 1$ we immediately find

$$(G^\rho)^\wedge(0) = \frac{1}{4}\rho \qquad (9.111)$$

and

$$(G^\rho)^\wedge(1) = -\frac{1}{2} + \frac{1}{4}\rho - \frac{1}{24}\rho^2. \qquad (9.112)$$

For $n \geq 2$ we make some auxiliary considerations. For all $s \in (-1,1)$ we know that

$$\int_{-1}^{s} P_n(t)\, dt = \frac{P_{n+1}(s) - P_{n-1}(s)}{2n+1}, \quad n \geq 1. \qquad (9.113)$$

For the first derivative of the Legendre polynomials, we get the identity

$$\int_{-1}^{s} P_n'(t)\, dt = P_n(s) - (-1)^n, \quad n \geq 1, \qquad (9.114)$$

since we know that $P_n(-1) = (-1)^n$. Furthermore, it follows that for the upper limit of the integral $s \in (-1,1)$

$$\int_{-1}^{s} tP_n'(t)\, dt = sP_n(s) + (-1)^n + \frac{P_{n-1}(s) - P_{n+1}(s)}{2n+1}, \quad n \geq 1. \qquad (9.115)$$

Moreover, we have

$$\int_{-1}^{s} (t+1)P_n'(t)\, dt = (1+s)P_n(s) + \frac{P_{n-1}(s) - P_{n+1}(s)}{2n+1}, \quad n \geq 1. \qquad (9.116)$$

Observing these identities, the first integral can be calculated in connection with (7.25) for values $s \in (-1,1)$ as follows:

$$\int_{-1}^{s} \ln(1-t) P_n(t)\, dt \qquad (9.117)$$
$$= \int_{-1}^{s} \ln(1-t) \frac{1}{2n+1} \left(\frac{d}{dt} \right) (P_{n+1}(t) - P_{n-1}(t))\, dt$$
$$= \frac{1}{2n+1} [\ln(1-t)(P_{n+1}(t) - P_{n-1}(t))] \Big|_{-1}^{s}$$
$$+ \int_{-1}^{s} \frac{1}{1-t} \frac{1}{2n+1} (P_{n+1}(t) - P_{n-1}(t))\, dt$$
$$= \frac{1}{2n+1} \ln(1-s) (P_{n+1}(s) - P_{n-1}(s)) - \frac{1}{n(n+1)} \int_{-1}^{s} (t+1)P_n'(t)\, dt.$$

Inserting (9.116), we finally obtain for $s \in (-1,1)$ and $n \geq 1$

$$\int_{-1}^{s} \ln(1-t)P_n(t)\,dt \tag{9.118}$$
$$= \left(\ln(1-s) + \frac{1}{n(n+1)}\right)\frac{P_{n+1}(s) - P_{n-1}(s)}{2n+1} - \frac{(1+s)P_n(s)}{n(n+1)}.$$

For the second integral with $s \in (-1,1)$, we are led to

$$\int_{s}^{1} \left(\ln\rho - 1 + \frac{1-t}{\rho}\right) P_n(t)\,dt \tag{9.119}$$
$$= (\ln\rho - 1)\int_{s}^{1} P_n(t)\,dt + \frac{1}{\rho}\int_{s}^{1}(1-t)P_n(t)\,dt.$$

Moreover,

$$\int_{s}^{1} tP_n'(t)\,dt = 1 - sP_n(s) - \frac{P_{n-1}(s) - P_{n+1}(s)}{2n+1}, \quad n \geq 1, \tag{9.120}$$

and

$$\int_{s}^{1} tP_n(t)\,dt \tag{9.121}$$
$$= \frac{1}{2n+1}\int_{s}^{1} t\left(\frac{d}{dt}\right)(P_{n+1}(t) - P_{n-1}(t))\,dt$$
$$= s\frac{P_{n-1}(s) - P_{n+1}(s)}{2n+1} - \frac{P_n(s) - P_{n+2}(s)}{(2n+1)(2n+3)} + \frac{P_{n-2}(s) - P_n(s)}{(2n+1)(2n-1)},$$

which shows us that

$$\int_{s}^{1} (1-t)P_n(t)\,dt \tag{9.122}$$
$$= (1-s)\frac{P_{n-1}(s) - P_{n+1}(s)}{2n+1} + \frac{P_n(s) - P_{n+2}(s)}{(2n+1)(2n+3)} - \frac{P_{n-2}(s) - P_n(s)}{(2n+1)(2n-1)}.$$

Summarizing our results, we obtain the Legendre coefficients of the regularized $\mathbb{S}^2$-sphere function with respect to Δ^* as follows:

Lemma 9.8. *For $\rho \in (0,2)$, we have*

$$G^\rho(\Delta^*;\xi\cdot\eta) = \sum_{k=0}^{\infty} \frac{2n+1}{4\pi}(G^\rho)^\wedge(n)P_n(\xi\cdot\eta), \quad -1 \leq \xi\cdot\eta \leq 1, \tag{9.123}$$

with

$$(G^\rho)^\wedge(0) = \frac{1}{4}\rho, \tag{9.124}$$

$$(G^\rho)^\wedge(1) = -\frac{1}{2} + \frac{1}{4}\rho - \frac{1}{24}\rho^2, \tag{9.125}$$

$$\begin{aligned}(G^\rho)^\wedge(n) = &\frac{P_{n+1}(1-\rho) - P_{n-1}(1-\rho)}{2n(n+1)(2n+1)} - \frac{(2-\rho)}{2n(n+1)}P_n(1-\rho)\\ &+ \frac{1}{2\rho}\frac{P_n(1-\rho) - P_{n+2}(1-\rho)}{(2n+1)(2n+3)} - \frac{1}{2\rho}\frac{P_{n-2}(1-\rho) - P_n(1-\rho)}{(2n+1)(2n-1)},\end{aligned} \tag{9.126}$$

for $n = 2, 3, \ldots$.

It should be noted that the Legendre coefficients of the regularized $\mathbb{S}^2$-sphere function for Δ^* tend to the Legendre coefficients of the $\mathbb{S}^2$-sphere function for Δ^* as $\rho \to 0$. In more detail,

$$\lim_{\rho \to 0} (G^\rho)^\wedge(0) = 0, \tag{9.127}$$

and

$$\lim_{\rho \to 0} (G^\rho)^\wedge(1) = -\frac{1}{2}. \tag{9.128}$$

Observing the identity (7.6) we find, for $n \geq 2$

$$\lim_{\rho \to 0} \frac{P_{n+1}(1-\rho) - P_{n-1}(1-\rho)}{2n(n+1)(2n+1)} = 0, \tag{9.129}$$

$$\lim_{\rho \to 0} \frac{2-\rho}{2n(n+1)} P_n(1-\rho) = \frac{1}{n(n+1)}, \tag{9.130}$$

$$\lim_{\rho \to 0} \frac{1}{2\rho} \frac{P_n(1-\rho) - P_{n+2}(1-\rho)}{(2n+1)(2n+3)} = 0, \tag{9.131}$$

and

$$\lim_{\rho \to 0} \frac{1}{2\rho} \frac{P_{n-2}(1-\rho) - P_n(1-\rho)}{(2n+1)(2n-1)} = 0, \tag{9.132}$$

which gives us the limit relation

$$\lim_{\rho \to 0} (G^\rho)^\wedge(n) = -\frac{1}{n(n+1)} \tag{9.133}$$

for all integers $n \geq 1$.

Frequency Regularized Sphere Function. Next we come to the *frequency-regularized* $\mathbb{S}^2$*-sphere function with respect to* Δ^*. From (7.25) we obtain that for $t \in [-1, 1]$

$$(t^2 - 1)\frac{2n+1}{n(n+1)} P_n'(t) = P_{n+1}(t) - P_{n-1}(t). \tag{9.134}$$

Hence, for $t \neq 1$, a truncated series of the $\mathbb{S}^2$-sphere function, i.e., a frequency regularization of $G(\Delta^*; t)$, can be expressed as follows:

$$\begin{aligned}
\sum_{n=1}^{N} \frac{2n+1}{4\pi} \frac{1}{-n(n+1)} P_n'(t) &= \frac{1}{4\pi} \frac{1}{1-t^2} \sum_{n=1}^{N} (P_{n+1}(t) - P_{n-1}(t)) \qquad (9.135)\\
&= \frac{1}{4\pi} \frac{1}{1-t^2} \left(\sum_{n=1}^{N-2} P_{n+1}(t) + P_N(t) + P_{N+1}(t) \right)\\
&\quad - \frac{1}{4\pi} \frac{1}{1-t^2} \left(P_0(t) + P_1(t) + \sum_{n=3}^{N} P_{n-1}(t) \right)\\
&= -\frac{1}{4\pi} \frac{1}{1-t^2} (P_0(t) + P_1(t) - P_N(t) - P_{N+1}(t))\\
&\quad - \frac{1}{4\pi} \frac{1}{1-t^2} \left(\sum_{n=2}^{N-1} P_n(t) - \sum_{n=2}^{N-1} P_n(t) \right).
\end{aligned}$$

Since $P_0(t) = 1$ and $P_1(t) = t$ we get

$$\begin{aligned}
\sum_{n=1}^{N} \frac{2n+1}{4\pi} \frac{1}{-n(n+1)} P_n'(t) &= -\frac{1}{4\pi} \frac{1}{(1-t)(1+t)} (1 + t - P_N(t) - P_{N+1}(t))\\
&= -\frac{1}{4\pi} \frac{1}{1-t} + \frac{1}{4\pi} \frac{P_N(t) + P_{N+1}(t)}{1-t^2}. \qquad (9.136)
\end{aligned}$$

Integrating with respect to t, we find, for $-1 < t_0 \leq t < 1$, t_0 fixed,

$$\begin{aligned}
&\sum_{n=1}^{N} \frac{2n+1}{4\pi} \frac{1}{-n(n+1)} P_n(t) - \sum_{n=1}^{N} \frac{2n+1}{4\pi} \frac{1}{-n(n+1)} P_n(t_0) \qquad (9.137)\\
&= \frac{1}{4\pi} \ln(1-t) - \frac{1}{4\pi} \ln(1-t_0) + \frac{1}{4\pi} \int_{t_0}^{t} \frac{P_N(s) + P_{N+1}(s)}{1-s^2}\, ds + C_N(t_0).
\end{aligned}$$

It is not hard to see that

$$\lim_{N\to\infty} \sum_{n=1}^{N} \frac{2n+1}{4\pi} \frac{1}{-n(n+1)} P_n(0) = \frac{1}{4\pi}(1 - \ln 2). \qquad (9.138)$$

Note that $P_{2n+1}(0) = 0$ and $P_{2n}(0) = \frac{(-1)^n}{4^n}\binom{2n}{n}$. As a consequence, we obtain the following lemma.

Lemma 9.9. *For all $\xi, \eta \in \mathbb{S}^2$ with $-1 \leq \xi \cdot \eta < 1$, the N-th frequency regularized $\mathbb{S}^2$-sphere function for Δ^**

$$G^{(N)}(\Delta^*; \xi \cdot \eta) = -\sum_{n=1}^{N} \frac{2n+1}{4\pi} \frac{1}{n(n+1)} P_n(\xi \cdot \eta) \qquad (9.139)$$

satisfies the equation

$$G(\Delta^*;\xi\cdot\eta) - G^{(N)}(\Delta^*;\xi\cdot\eta) = -\frac{1}{4\pi}\int_0^{\xi\cdot\eta}\frac{P_N(s)+P_{N+1}(s)}{1-s^2}\,ds + \frac{1-\ln 2}{4\pi} + \sum_{n=1}^{N}\frac{2n+1}{4\pi}\frac{1}{-n(n+1)}P_n(0). \tag{9.140}$$

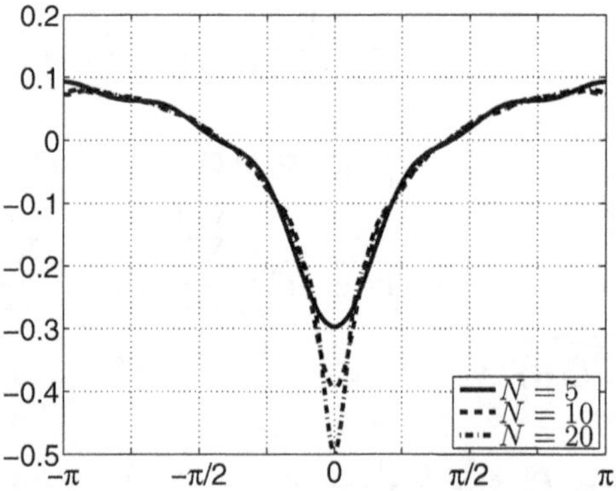

FIGURE 9.5: Frequency regularized $\mathbb{S}^2$-sphere function $\vartheta \mapsto G^{(N)}(\Delta^*;\cos\vartheta)$ for various values N.

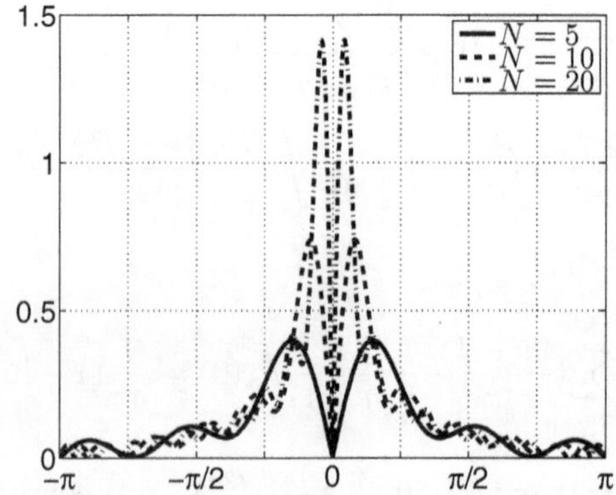

FIGURE 9.6: Absolute value of the surface gradient or surface curl gradient of the frequency regularized $\mathbb{S}^2$-sphere function$\vartheta \mapsto G^{(N)}(\Delta^*;\cos\vartheta)$ for various values N.

Figures 9.5 and 9.6 give graphical impressions of the frequency regularized $\mathbb{S}^2$-sphere function. From Lemma 7.9, we know that for all $s \in [0,t)$, $t < 1$,

$|P_N(s)| = O(N^{-1/2})$, hence, it follows that

$$\lim_{N\to\infty} \int_0^{\xi\cdot\eta} \frac{P_N(s) + P_{N+1}(s)}{1-s^2}\, ds = 0. \tag{9.141}$$

In particular, we are led to the following identities

$$\nabla^*_\xi G^{(N)}(\Delta^*;\xi\cdot\eta) = -\frac{1}{4\pi}\frac{1}{1-\xi\cdot\eta}(\eta - (\xi\cdot\eta)\xi) + \frac{1}{4\pi}\frac{P_N(\xi\cdot\eta)+P_{N+1}(\xi\cdot\eta)}{1-(\xi\cdot\eta)^2}(\eta-(\xi\cdot\eta)\xi) \tag{9.142}$$

and

$$\mathrm{L}^*_\xi G^{(N)}(\Delta^*;\xi\cdot\eta) = -\frac{1}{4\pi}\frac{1}{1-\xi\cdot\eta}(\xi\wedge\eta) + \frac{1}{4\pi}\frac{P_N(\xi\cdot\eta)+P_{N+1}(\xi\cdot\eta)}{1-(\xi\cdot\eta)^2}(\xi\wedge\eta). \tag{9.143}$$

Integral Formulas. Let ξ be a fixed point of a regular region $\Gamma \subset \mathbb{S}^2$. Assume that F is continuously differentiable on $\overline{\Gamma} = \Gamma \cup \partial\Gamma$. Then, for sufficiently small $\rho > 0$, integration by parts, i.e., the first Green's surface theorem (9.28) gives

$$\begin{aligned}
&\int_{\substack{\xi\cdot\eta\le 1-\rho\\ \eta\in\Gamma}} \left\{F(\eta)\Delta^*_\eta G(\Delta^*;\xi\cdot\eta) + \nabla^*_\eta F(\eta)\cdot\nabla^*_\eta G(\Delta^*;\xi\cdot\eta)\right\}\, dS(\eta)\\
&= \int_{\substack{\xi\cdot\eta= 1-\rho\\ \eta\in\Gamma}} F(\eta)\frac{\partial}{\partial\nu_\eta}G(\Delta^*;\xi\cdot\eta)\, d\sigma(\eta)\\
&\quad + \int_{\substack{\xi\cdot\eta\le 1-\rho\\ \eta\in\partial\Gamma}} F(\eta)\frac{\partial}{\partial\nu_\eta}G(\Delta^*;\xi\cdot\eta)\, d\sigma(\eta),
\end{aligned} \tag{9.144}$$

where ν is the unit normal to the circle consisting of all points $\eta\in\mathbb{S}^2$ with $\xi\cdot\eta = 1-\rho$, tangential to $\mathbb{S}^2$, and directed exterior to the set of all points $\eta\in\mathbb{S}^2$ with $\xi\cdot\eta\le 1-\rho$. Explicitly, written out, we have

$$\nu_\eta = -(1-(\xi\cdot\eta)^2)^{-\frac{1}{2}}\eta\wedge(\eta\wedge\xi). \tag{9.145}$$

In the identity (9.144), we first observe the differential equation of the $\mathbb{S}^2$-sphere function

$$\int_{\substack{\xi\cdot\eta\le 1-\rho\\ \eta\in\Gamma}} F(\eta)\Delta^*_\eta G(\Delta^*;\xi\cdot\eta)\, dS(\eta) = -\frac{1}{4\pi}\int_{\substack{\xi\cdot\eta\le 1-\rho\\ \eta\in\Gamma}} F(\eta)\, dS(\eta). \tag{9.146}$$

By virtue of the logarithmic singularity of the $\mathbb{S}^2$-sphere function $G(\Delta^*;\cdot)$, we get in analogy to well-known results of potential theory (cf. O.D. Kellogg [1929])

$$\begin{aligned}
&\int_{\substack{\xi\cdot\eta=1-\rho\\ \eta\in\Gamma}} F(\eta)\frac{\partial}{\partial\nu_\eta}G(\Delta^*;\xi\cdot\eta)\,d\sigma(\eta) \qquad (9.147)\\
&= \int_{\substack{\xi\cdot\eta=1-\rho\\ \eta\in\Gamma}} F(\eta)\frac{\xi-(1-\rho)\eta}{\sqrt{1-(1-\rho)^2}}\cdot\left(-\frac{1}{4\pi\rho}(\xi-(1-\rho)\eta)\right)\,d\sigma(\eta)\\
&= -\frac{1}{4\pi}\int_{\substack{\xi\cdot\eta=1-\rho\\ \eta\in\Gamma}} F(\eta)\frac{\sqrt{1-(1-\rho)^2}}{\rho}\,d\sigma(\eta).
\end{aligned}$$

From the mean value theorem, we are able to deduce that

$$\begin{aligned}
&-\frac{1}{4\pi}\frac{\sqrt{1-(1-\rho)^2}}{\rho}\int_{\substack{\xi\cdot\eta=1-\rho\\ \eta\in\Gamma}} F(\eta)\,d\sigma(\eta) \qquad (9.148)\\
&= -\frac{1}{4\pi}\frac{\sqrt{1-(1-\rho)^2}}{\rho}2\pi\sqrt{1-(1-\rho)^2}F(\eta_\rho) = -\frac{1}{2}(2-\rho)F(\eta_\rho)
\end{aligned}$$

for some η_ρ lying on the circle $\{\eta\in\mathbb{S}^2 : 1-\xi\cdot\eta=\rho\}$. The continuity of F yields $F(\eta_\rho)\to F(\xi)$ as $\eta_\rho\to\xi$ for $\rho\to 0$ such that

$$\lim_{\rho\to 0}\int_{\substack{1-\xi\cdot\eta=\rho\\ \eta\in\Gamma}} F(\eta)\frac{\partial}{\partial\nu_\eta}G(\Delta^*;\xi\cdot\eta)\,d\sigma(\eta) = -F(\xi). \qquad (9.149)$$

Summarizing our results, we therefore obtain the following result.

Theorem 9.10. *Let Γ be a regular region with boundary $\partial\Gamma$. Suppose that $F:\overline{\Gamma}\to\mathbb{R}$ is a continuously differentiable function on $\overline{\Gamma}$, i.e., $F\in \mathrm{C}^{(1)}(\overline{\Gamma})$. Then, for every $\xi\in\Gamma$,*

$$\begin{aligned}
F(\xi) = &\frac{1}{4\pi}\int_\Gamma F(\eta)dS(\eta) - \int_\Gamma \nabla^*_\eta G(\Delta^*;\xi\cdot\eta)\cdot\nabla^*_\eta F(\eta)\,dS(\eta)\\
&+\int_{\partial\Gamma}(\nu_\eta\cdot\nabla^*_\eta G(\Delta^*;\xi\cdot\eta))F(\eta)\,d\sigma(\eta). \qquad (9.150)
\end{aligned}$$

Theorem 9.10 admits canonical extensions.

Theorem 9.11 (Fundamental Theorem for ∇^* on $\overline{\Gamma}$)**.** *Let Γ be a regular region with continuously differentiable boundary $\partial\Gamma$. Suppose that $F:\overline{\Gamma}\to\mathbb{R}$ is a continuously differentiable function on $\overline{\Gamma}$. Then, for every $\xi\in\overline{\Gamma}$,*

$$\begin{aligned}
\alpha_\Gamma(\xi)F(\xi) = &\frac{1}{4\pi}\int_\Gamma F(\eta)\,dS(\eta) - \int_\Gamma \nabla^*_\eta G(\Delta^*;\xi\cdot\eta)\cdot\nabla^*_\eta F(\eta)\,dS(\eta)\\
&+\int_{\partial\Gamma}(\nu_\eta\cdot\Delta^*_\eta G(\Delta^*;\xi\cdot\eta))F(\eta)\,d\sigma(\eta), \qquad (9.151)
\end{aligned}$$

where

$$\alpha_\Gamma(\xi) = \begin{cases} 1, & \xi \in \Gamma \\ \frac{1}{2}, & \xi \in \partial\Gamma \\ 0, & \xi \in \mathbb{S}^2 \setminus \overline{\Gamma}. \end{cases} \tag{9.152}$$

Proof. The same arguments can be used as for the Laplace operator in classical potential theory (see, e.g., O.D. Kellogg [1929]). □

In analogous way we obtain the following integral formula by aid of the first Green's surface theorem in the form of (9.29).

Theorem 9.12 (Fundamental Theorem for L^* on $\overline{\Gamma}$)**.** *Let Γ be a regular region with continuously differentiable boundary $\partial\Gamma$. Suppose that $F : \overline{\Gamma} \to \mathbb{R}$ is a continuously differentiable function on $\overline{\Gamma}$. Then*

$$\alpha_\Gamma(\xi)F(\xi) = \frac{1}{4\pi}\int_\Gamma F(\eta)\, dS(\eta) - \int_\Gamma \mathrm{L}^*_\eta G(\Delta^*;\xi\cdot\eta)\cdot \mathrm{L}^*_\eta F(\eta)\, dS(\eta)$$

$$+ \int_{\partial\Gamma} (\tau_\eta \cdot \mathrm{L}^*_\eta G(\Delta^*;\xi\cdot\eta))F(\eta)\, d\sigma(\eta). \tag{9.153}$$

Proof. Applying (9.29) for the operator L^*, we get for every (sufficiently small) $\rho > 0$ and $F \in \mathrm{C}^{(1)}(\overline{\Gamma})$,

$$\int_{\substack{\xi\cdot\eta\le 1-\rho \\ \eta\in\Gamma}} \mathrm{L}^*_\eta F(\eta)\cdot \mathrm{L}^*_\eta G(\Delta^*;\xi\cdot\eta)\, dS(\eta) + \int_{\substack{\xi\cdot\eta\le 1-\rho \\ \eta\in\Gamma}} F(\eta)\Delta^*_\eta G(\Delta^*;\xi\cdot\eta)\, dS(\eta)$$

$$= \int_{\substack{\xi\cdot\eta=1-\rho \\ \eta\in\Gamma}} F(\eta)\ \tau_\eta \cdot \mathrm{L}^*_\eta G(\Delta^*;\xi\cdot\eta)\, d\sigma(\eta)$$

$$+ \int_{\substack{\eta\in\partial\Gamma \\ \xi\cdot\eta\le 1-\rho}} F(\eta)\ \tau_\eta \cdot \mathrm{L}^*_\eta G(\Delta^*;\xi\cdot\eta)\, d\sigma(\eta), \tag{9.154}$$

where τ is defined as the (unit) surface vector on $\mathbb{S}^2$ tangential to the circle $\{\eta \in \mathbb{S}^2 \mid 1 - \xi\cdot\eta = \rho\}$. Explicitly written out, for $\eta \in \mathbb{S}^2$ with $1 - \xi\cdot\eta = \rho$, we have

$$\tau_\eta = (1 - (\xi\cdot\eta)^2)^{-1/2}\xi \wedge \eta. \tag{9.155}$$

Moreover, we know that

$$\mathrm{L}^*_\eta G(\Delta^*;\xi\cdot\eta) = -\frac{1}{4\pi}\frac{\eta\wedge\xi}{1-\xi\cdot\eta}. \tag{9.156}$$

In other words, the same reasoning as in Theorems 9.10 and 9.11 guarantees Theorem 9.12. □

From the second Green's surface theorem (9.30) we get for each sufficiently small $\rho > 0$ that

$$\int_{\substack{\xi\cdot\eta\leq 1-\rho\\ \eta\in\Gamma}} G(\Delta^*;\xi\cdot\eta)\Delta^*_\eta F(\eta) - F(\eta)\Delta^*_\eta G(\Delta^*;\xi\cdot\eta)\, dS(\eta) \tag{9.157}$$

$$= \int_{\substack{\xi\cdot\eta=1-\rho\\ \eta\in\Gamma}} G(\Delta^*;\xi\cdot\eta)\frac{\partial}{\partial\nu_\eta}F(\eta) - F(\eta)\frac{\partial}{\partial\nu_\eta}G(\Delta^*;\xi\cdot\eta)\, d\sigma(\eta)$$

$$+ \int_{\substack{\xi\cdot\eta\leq 1-\rho\\ \eta\in\partial\Gamma}} G(\Delta^*;\xi\cdot\eta)\frac{\partial}{\partial\nu_\eta}F(\eta) - F(\eta)\frac{\partial}{\partial\nu_\eta}G(\Delta^*;\xi\cdot\eta)\, d\sigma(\eta),$$

provided that F is twice continuously differentiable on $\overline{\Gamma}$. Observing the defining properties of the $\mathbb{S}^2$-sphere function for Δ^*, we can use the same arguments as known from potential theory (cf. O.D. Kellogg [1929]). In fact, the continuous differentiability of F on $\overline{\Gamma}$ leads us to

$$\lim_{\rho\to 0}\int_{\substack{\xi\cdot\eta=1-\rho\\ \eta\in\Gamma}} G(\Delta^*;\xi\cdot\eta)\frac{\partial}{\partial\nu_\eta}F(\eta)\, d\sigma(\eta) = 0. \tag{9.158}$$

Altogether, observing that $\Delta^* = \nabla^*\cdot\nabla^* = \mathrm{L}^*\cdot\mathrm{L}^*$ and applying the second Green's surface theorem (9.30) we obtain the following integral formula (see W. Freeden [1979]).

Theorem 9.13 (Fundamental Theorem for Δ^* on $\overline{\Gamma}$). *Let Γ be a regular surface with continuously differentiable boundary $\partial\Gamma$. Suppose that $F : \overline{\Gamma}\to\mathbb{R}$ is a twice continuously differentiable function on $\overline{\Gamma}$. Then, for every $\xi\in\overline{\Gamma}$,*

$$\alpha_\Gamma(\xi)F(\xi) = \frac{1}{4\pi}\int_\Gamma F(\eta)\, dS(\eta) + \int_\Gamma G(\Delta^*;\xi\cdot\eta)\Delta^*_\eta F(\eta)\, dS(\eta)$$

$$+ \int_{\partial\Gamma} (\nu_\eta\cdot\nabla^*_\eta G(\Delta^*;\xi\cdot\eta))F(\eta)\, d\sigma(\eta)$$

$$- \int_{\partial\Gamma} G(\Delta^*;\xi\cdot\eta)(\nu_\eta\cdot\nabla^*_\eta F(\eta))\, d\sigma(\eta) \tag{9.159}$$

and

$$\alpha_\Gamma(\xi)F(\xi) = \frac{1}{4\pi}\int_\Gamma F(\eta)\, dS(\eta) + \int_\Gamma G(\Delta^*;\xi\cdot\eta)\Delta^*_\eta F(\eta)\, dS(\eta)$$

$$+ \int_{\partial\Gamma} (\tau_\eta\cdot\mathrm{L}^*_\eta G(\Delta^*;\xi\cdot\eta))F(\eta)\, d\sigma(\eta)$$

$$- \int_{\partial\Gamma} G(\Delta^*;\xi\cdot\eta)(\tau_\eta\cdot\mathrm{L}^*_\eta F(\eta))\, d\sigma(\eta). \tag{9.160}$$

Choosing $F = 1$ we find

$$\alpha_\Gamma(\xi) = \frac{\|\Gamma\|}{4\pi} + \int_{\partial\Gamma} \nu_\eta \cdot \nabla^*_\eta G(\Delta^*; \xi \cdot \eta) \, d\sigma(\eta), \tag{9.161}$$

and

$$\alpha_\Gamma(\xi) = \frac{\|\Gamma\|}{4\pi} + \int_{\partial\Gamma} \tau_\eta \cdot \mathrm{L}^*_\eta G(\Delta^*; \xi \cdot \eta) \, d\sigma(\eta), \tag{9.162}$$

where

$$\|\Gamma\| = \int_\Gamma dS. \tag{9.163}$$

If $\Gamma \subset \mathbb{S}^2$ is regular, then $\mathbb{S}^2/\overline{\Gamma}$ is regular. Thus, all integral formulas remain valid for $\mathbb{S}^2/\Gamma$. If we add associated integral formulas for $\overline{\Gamma}$ and $\mathbb{S}^2/\Gamma$, then the sum of the boundary integrals vanishes (because of their opposite orientation). This leads us to the following results.

Theorem 9.14 (Green's Surface Theorem for ∇^* on $\mathbb{S}^2$). *Let ξ be a fixed point of the unit sphere $\mathbb{S}^2$. Suppose that F is a continuously differentiable function on $\mathbb{S}^2$. Then*

$$F(\xi) = \frac{1}{4\pi} \int_{\mathbb{S}^2} F(\eta) \, dS(\eta) - \int_{\mathbb{S}^2} \left(\nabla^*_\eta G(\Delta^*; \xi \cdot \eta)\right) \cdot \left(\nabla^*_\eta F(\eta)\right) \, dS(\eta). \tag{9.164}$$

In the same way, we are able to formulate the following corollary.

Corollary 9.15 (Green's Surface Theorem for L^* on $\mathbb{S}^2$). *Under the assumptions of Theorem 9.14*

$$F(\xi) = \frac{1}{4\pi} \int_{\mathbb{S}^2} F(\eta) \, dS(\eta) - \int_{\mathbb{S}^2} (\mathrm{L}^*_\eta G(\Delta^*; \xi \cdot \eta)) \cdot (\mathrm{L}^*_\eta F(\eta)) \, dS(\eta). \tag{9.165}$$

Theorem 9.16 (Green's Surface Theorem for Δ^* on $\mathbb{S}^2$). *Let ξ be a fixed point of the unit sphere $\mathbb{S}^2$. Suppose that F is a twice continuously differentiable function on $\mathbb{S}^2$. Then*

$$F(\xi) = \frac{1}{4\pi} \int_{\mathbb{S}^2} F(\eta) \, dS(\eta) + \int_{\mathbb{S}^2} G(\Delta^*; \xi \cdot \eta)(\Delta^*_\eta F(\eta)) \, dS(\eta). \tag{9.166}$$

In other words, the Green's theorems as stated above compare the value of a function at a point $\xi \in \mathbb{S}^2$ with the integral mean of F relative to the unit sphere $\mathbb{S}^2$ under explicit representation of the error term in integral form. Essential tool is the $\mathbb{S}^2$-sphere function for Δ^*.

Differential Equations Involving Sphere Functions. Combining Theorem 9.14 and observing the surface gradient of the $\mathbb{S}^2$-sphere function $G(\Delta^*; \cdot)$, we obtain the following theorem.

Theorem 9.17 (Differential Equation for ∇^* on $\mathbb{S}^2$). *Let $g : \mathbb{S}^2 \to \mathbb{R}^3$ be a continuously differentiable vector field on $\mathbb{S}^2$ with $\xi \cdot g(\xi) = 0$, $\mathrm{L}^*_\xi \cdot g(\xi) = 0$, $\xi \in \mathbb{S}^2$. Then*

$$F(\xi) = \frac{1}{4\pi} \int_{\mathbb{S}^2} \frac{1}{1 - \xi \cdot \eta} (\xi - (\xi \cdot \eta)\eta) \cdot g(\eta) \, dS(\eta) \tag{9.167}$$

is the uniquely determined solution of the differential equation

$$\nabla^*_\xi F(\xi) = g(\xi), \quad \xi \in \mathbb{S}^2, \tag{9.168}$$

satisfying

$$\frac{1}{4\pi} \int_{\mathbb{S}^2} F(\eta) \, dS(\eta) = 0. \tag{9.169}$$

Analogously, we get the following result on the differential equation of the surface curl gradient.

Theorem 9.18 (Differential Equation for L^* on $\mathbb{S}^2$). *Let $g : \mathbb{S}^2 \to \mathbb{R}^3$ be a continuously differentiable vector field on $\mathbb{S}^2$ with $\xi \cdot g(\xi) = 0$ $\nabla^*_\xi \cdot g(\xi) = 0$, $\xi \in \mathbb{S}^2$. Then*

$$F(\xi) = \frac{1}{4\pi} \int_{\mathbb{S}^2} \frac{1}{1 - \xi \cdot \eta} (\eta \wedge \xi) \cdot g(\eta) \, dS(\eta) \tag{9.170}$$

is the uniquely determined solution of the differential equation

$$\mathrm{L}^*_\xi F(\xi) = g(\xi), \quad \xi \in \mathbb{S}^2, \tag{9.171}$$

satisfying

$$\frac{1}{4\pi} \int_{\mathbb{S}^2} F(\xi) \, dS(\eta) = 0. \tag{9.172}$$

From Theorem 9.16 we are able to verify the following result on the Beltrami differential equation.

Theorem 9.19 (Differential Equations for Δ^* on $\mathbb{S}^2$). *Let H be a continuous function on $\mathbb{S}^2$ with*

$$\frac{1}{4\pi} \int_{\mathbb{S}^2} H(\xi) \, dS(\xi) = 0. \tag{9.173}$$

Let $F \in \mathrm{C}^{(2)}(\mathbb{S}^2)$ satisfy the Beltrami differential equation

$$\Delta^* F = H \tag{9.174}$$

such that

$$\frac{1}{4\pi} \int_{\mathbb{S}^2} F(\xi) \, dS(\xi) = 0. \tag{9.175}$$

Then

$$F(\xi) = \int_{\mathbb{S}^2} G(\Delta^*; \xi \cdot \eta) H(\eta) \, dS(\eta), \quad \xi \in \mathbb{S}^2. \tag{9.176}$$

9.3 Best Approximate Integration by Splines

As preparation of more general spline settings we are now concerned with spherical (Beltrami) splines of degree 0 providing best approximate integration formulas via spline interpolation. By virtue of the Cauchy–Schwarz inequality, we get from Theorem 9.16

$$\begin{aligned} &\left| F(\xi) - \frac{1}{4\pi} \int_{\mathbb{S}^2} F(\eta)\, dS(\eta) \right| \\ &\leq \left(G\left((\Delta^*)^2; \xi \cdot \xi\right)\right)^{1/2} \left(\int_{\mathbb{S}^2} \left|\Delta^* F(\eta)\right|^2 \, dS(\eta) \right)^{1/2} \\ &= \left(G((\Delta^*)^2; 1)\right)^{1/2} \left(\int_{\mathbb{S}^2} |\Delta^* F(\eta)|^2 \, dS(\eta) \right)^{1/2} \end{aligned} \tag{9.177}$$

for all $\xi \in \mathbb{S}^2$, where $G((\Delta^*)^2; \cdot) : (\xi, \eta) \mapsto G((\Delta^*)^2; \xi \cdot \eta)$, $\xi, \eta \in \mathbb{S}^2$, is defined by convolution as in (9.63), i.e.,

$$G((\Delta^*)^2; \xi \cdot \eta) = \int_{\mathbb{S}^2} G(\Delta^*; \xi \cdot \zeta) G(\Delta^*; \zeta \cdot \eta) \, dS(\zeta). \tag{9.178}$$

Obviously, due to (9.64) the bilinear series reads

$$G((\Delta^*)^2; \xi \cdot \eta) = \sum_{n=1}^{\infty} \frac{2n+1}{4\pi} \frac{1}{(-n(n+1))^2} P_n(\xi \cdot \eta). \tag{9.179}$$

Moreover, it follows that, for all $\xi, \eta \in \mathbb{S}^2$ with $-1 \leq \xi \cdot \eta < 1$

$$\Delta^*_\eta G((\Delta^*)^2; \xi \cdot \eta) = G(\Delta^*; \xi \cdot \eta). \tag{9.180}$$

Next, we are concerned with the explicit calculation of the iterated $\mathbb{S}^2$-sphere function $G((\Delta^*)^2; \cdot) : t \mapsto G((\Delta^*)^2; t)$, $t \in [-1, 1]$ given by

$$G((\Delta^*)^2; t) = \frac{1}{4\pi} \sum_{n=1}^{\infty} \frac{2n+1}{(-n(n+1))^2} P_n(t). \tag{9.181}$$

First, it is not hard to see

$$\frac{2n+1}{n^2(n+1)^2} = \frac{1}{n^2} - \frac{1}{(n+1)^2}. \tag{9.182}$$

This gives us for all $\xi \in \mathbb{S}^2$

$$G((\Delta^*)^2; \xi \cdot \xi) = G((\Delta^*)^2; 1) = \frac{1}{4\pi} \left(\sum_{n=1}^{\infty} \frac{1}{n^2} - \sum_{n=1}^{\infty} \frac{1}{(n+1)^2} \right) = \frac{1}{4\pi}. \tag{9.183}$$

Moreover, from the well-known result $\sum_{n=1}^{\infty}(-1)^n n^{-2} = -\pi^2/12$, we obtain

$$G((\Delta^*)^2;-1) = \frac{1}{4\pi} - \frac{\pi}{24}. \tag{9.184}$$

For the explicit calculation of $G((\Delta^*)^2;t)$ $t \in (-1,1)$, we use the differential equation

$$\begin{aligned} \frac{d}{dt}(1-t^2)\frac{d}{dt}G((\Delta^*)^2;t) &= G(\Delta^*;t) \\ &= \frac{1}{4\pi}\ln(1-t) + \frac{1}{4\pi}(1-\ln 2). \end{aligned} \tag{9.185}$$

First, for all $t \in (-1,1)$, we get by elementary manipulations

$$\begin{aligned} \int_{-1}^{t} \frac{d}{dx}(1-x^2)\frac{d}{dx}G((\Delta^*)^2;x)\,dx &= (1-t^2)\frac{d}{dt}G((\Delta^*)^2;t) \\ &= -\frac{1}{4\pi}(1-t)\ln(1-t) + \frac{\ln 2}{4\pi}(1-t). \end{aligned} \tag{9.186}$$

Second, for all $t \in (-1,1)$, we find

$$\begin{aligned} \int_{t}^{1} \frac{d}{dx}G((\Delta^*)^2;x)\,dx &= G((\Delta^*)^2;1) - G((\Delta^*)^2;t) \\ &= -\frac{1}{4\pi}\int_t^1 \frac{1}{1+x}\ln(1-x)\,dx + \frac{\ln 2}{4\pi}\int_t^1 \frac{1}{1+x}\,dx. \end{aligned} \tag{9.187}$$

Substituting $1-x=u$ we are led to

$$-\frac{1}{4\pi}\int_t^1 \frac{1}{1+x}\ln(1-x)\,dx = \frac{1}{4\pi}\int_{1-t}^{0}\frac{1}{2-u}\ln u\,du. \tag{9.188}$$

From a table of integrals (see, e.g., W. Gröbner, N. Hofreiter [1975]), we borrow

$$\int \frac{1}{2-u}\ln u\,du = -\ln u \ln\frac{2-u}{2} - \mathcal{L}_2\left(\frac{u}{2}\right) \quad (+C), \tag{9.189}$$

where $\mathcal{L}_2$ defines the so-called *dilogarithm*

$$\mathcal{L}_2(u) = -\int_0^u \frac{\ln(1-v)}{v}\,dv = \sum_{n=1}^{\infty}\frac{u^n}{n^2}. \tag{9.190}$$

It follows that the iterated $\mathbb{S}^2$-sphere function is expressible by means of the dilogarithm

$$\begin{aligned} \underbrace{G((\Delta^*)^2;1)}_{=\frac{1}{4\pi}} - G((\Delta^*)^2;t) &= \frac{1}{4\pi}\left(\ln(1-t)(\ln(1+t)-\ln 2)\right) \\ &\quad + \frac{1}{4\pi}\mathcal{L}_2\left(\frac{1-t}{2}\right) + \frac{(\ln 2)^2}{4\pi} - \frac{\ln 2}{4\pi}\ln(1+t). \end{aligned} \tag{9.191}$$

Note that

$$\lim_{\substack{t\to 1\\ t<1}} (\ln(1-t)\,(\ln(1+t)-\ln 2)) = 0 \tag{9.192}$$

and

$$\lim_{\substack{t\to -1\\ t>-1}} (\ln(1-t)\,(\ln(1+t)-\ln 2) - \ln 2\ln(1+t)) = -(\ln 2)^2. \tag{9.193}$$

This finally gives us the following representation (cf. W. Freeden, M. Schreiner [2009]).

Lemma 9.20. *For* $t \in (-1,1)$,

$$\begin{aligned} G((\Delta^*)^2;t) = &\frac{1}{4\pi} - \frac{1}{4\pi}\ln(1-t)\,(\ln(1+t)-\ln 2) \\ &- \frac{1}{4\pi}\mathcal{L}_2\left(\frac{1-t}{2}\right) - \frac{(\ln 2)^2}{4\pi} + \frac{\ln 2}{4\pi}\ln(1+t), \end{aligned} \tag{9.194}$$

where

$$\mathcal{L}_2\left(\frac{1-t}{2}\right) = \sum_{n=1}^{\infty}\left(\frac{1-t}{2}\right)^n \frac{1}{n^2} \tag{9.195}$$

and

$$\lim_{\substack{t\to 1\\ t<1}} G((\Delta^*)^2;t) = G((\Delta^*)^2;1) = \frac{1}{4\pi}, \tag{9.196}$$

$$\lim_{\substack{t\to -1\\ t>-1}} G((\Delta^*)^2;t) = G((\Delta^*)^2;-1) = \frac{1}{4\pi} - \frac{\pi}{24}. \tag{9.197}$$

Summarizing our results we therefore obtain the following result (see Figure 9.7).

Corollary 9.21. *The* $\mathbb{S}^2$*-sphere function* $G((\Delta^*)^2;\cdot) : (\xi,\eta) \mapsto G((\Delta^*)^2;\xi\cdot\eta)$, $\xi,\eta \in \mathbb{S}^2$, *with respect to the operator* $(\Delta^*)^2$ *is continuous, and we have*

$$G((\Delta^*)^2;\xi\cdot\eta) = \sum_{n=1}^{\infty}\frac{2n+1}{4\pi}\frac{1}{(-n(n+1))^2}P_n(\xi\cdot\eta) \tag{9.198}$$

$$= \begin{cases} \frac{1}{4\pi}, & 1-\xi\cdot\eta = 0, \\ -\frac{1}{4\pi}\ln(1-\xi\cdot\eta)\ln(1+\xi\cdot\eta) & \\ +\frac{\ln 2}{4\pi}\ln(1-(\xi\cdot\eta)^2) - \frac{1}{4\pi}\mathcal{L}_2\left(\frac{1-\xi\cdot\eta}{2}\right) & \\ +\frac{1}{4\pi}(1-(\ln 2)^2), & 1\pm\xi\cdot\eta \neq 0, \\ \frac{1}{4\pi} - \frac{\pi}{24}, & 1+\xi\cdot\eta = 0. \end{cases}$$

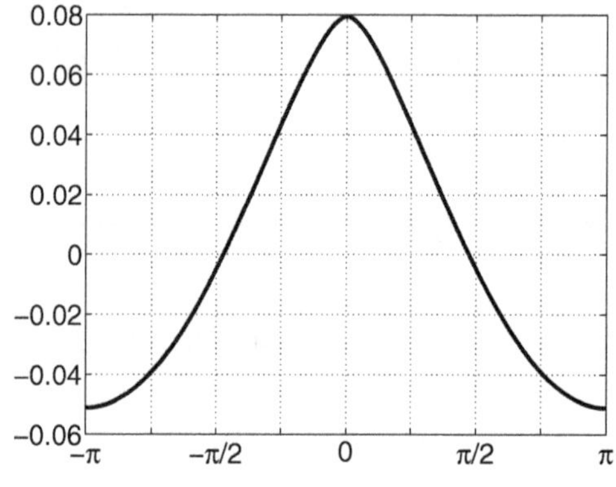

FIGURE 9.7: The iterated $\mathbb{S}^2$-sphere function $\vartheta \mapsto G((\Delta^*)^2; \cos\vartheta)$.

It should be mentioned that the integral formulas with respect to Δ^* can be used for *approximate integration methods* on the sphere $\mathbb{S}^2$. We give a concrete application: If $a_1, \ldots, a_N \in \mathbb{R}$ satisfy $\sum_{k=1}^N a_k = 1$, we have for all nodes $\eta_1, \ldots, \eta_N \in \mathbb{S}^2$,

$$\sum_{k=1}^N a_k F(\eta_k) - \frac{1}{4\pi}\int_{\mathbb{S}^2} F(\eta)\, dS(\eta) = \int_{\mathbb{S}^2} \sum_{k=1}^N a_k G(\Delta^*; \eta_k \cdot \eta)\Delta^*_\eta F(\eta)\, dS(\eta), \tag{9.199}$$

provided that F is twice continuously differentiable on $\mathbb{S}^2$. Using the Cauchy–Schwarz inequality, we get from (9.199)

$$\left| \frac{1}{4\pi}\int_{\mathbb{S}^2} F(\eta)\, dS(\eta) - \sum_{k=1}^N a_k F(\eta_k) \right| \tag{9.200}$$
$$\leq \left(\sum_{p=1}^N \sum_{q=1}^N a_p a_q G((\Delta^*)^2; \eta_p \cdot \eta_q) \right)^{1/2} \left(\int_{\mathbb{S}^2} (\Delta^*_\eta F(\eta))^2\, dS(\eta) \right)^{1/2}.$$

Thus, the "*best approximation*" formula corresponding to the given nodes $\eta_1, \ldots, \eta_N$

$$\sum_{k=1}^N \mathring{a}_k F(\eta_k) \tag{9.201}$$

to the integral

$$\frac{1}{4\pi}\int_{\mathbb{S}^2} F(\eta)\, dS(\eta) \tag{9.202}$$

is the solution of the quadratic optimization problem:

$$\sum_{p=1}^N \sum_{q=1}^N \mathring{a}_p \mathring{a}_q G\left((\Delta^*)^2; \eta_p \cdot \eta_q\right) \to \min. \tag{9.203}$$

under the constraints

$$\sum_{k=1}^{N} \mathring{a}_k = 1. \tag{9.204}$$

Therefore, it is not difficult to show in accordance with Lagrange's method of multipliers, that the solution $(\mathring{a}_1, \ldots, \mathring{a}_N)^T$ of the best approximation formula can be obtained by solving the linear system

$$\begin{array}{ccccccc} G((\Delta^*)^2; \eta_1 \cdot \eta_1)\mathring{a}_1 & + \ldots + & G((\Delta^*)^2; \eta_N \cdot \eta_1)\mathring{a}_N & - & \mathring{\lambda} & = & 0, \\ \vdots & \vdots & \vdots & & \vdots & & \\ G((\Delta^*)^2; \eta_1 \cdot \eta_N)\mathring{a}_1 & + \ldots + & G((\Delta^*)^2; \eta_N \cdot \eta_N)\mathring{a}_N & - & \mathring{\lambda} & = & 0, \\ \mathring{a}_1 & + \ldots + & \mathring{a}_N & & & = & 1 \end{array} \tag{9.205}$$

such that

$$\mathring{\lambda} = \sum_{p=1}^{N} \sum_{q=1}^{N} G((\Delta^*)^2, \eta_p \cdot \eta_q)\mathring{a}_p \mathring{a}_q. \tag{9.206}$$

In other words, the linear system does not only provide the coefficients $\mathring{a}_1, \ldots, \mathring{a}_N$ of the best approximation, but also the accuracy of the integration formula via the Lagrange multiplier (9.206)

$$\left| \frac{1}{4\pi} \int_{\mathbb{S}^2} F(\eta)\, dS(\eta) - \sum_{k=1}^{N} \mathring{a}_k F(\eta_k) \right| \leq \mathring{\lambda}^{1/2} \left(\int_{\mathbb{S}^2} \left(\Delta^*_\eta F(\eta)\right)^2\, dS(\eta) \right)^{1/2}. \tag{9.207}$$

Next, we deal with *spherical splines* (of degree 0): If $a_1, \ldots, a_N \in \mathbb{R}$ satisfy $\sum_{k=1}^{N} a_k = 0$ and if $\eta_1, \ldots, \eta_N$ are the prescribed nodes on $\mathbb{S}^2$, then

$$\int_{\mathbb{S}^2} \Delta^*_\eta S(\eta)\; \Delta^*_\eta F(\eta)\, dS(\eta) = \sum_{k=1}^{N} a_k F(\eta_k), \tag{9.208}$$

where $S : \mathbb{S}^2 \to \mathbb{R}$ is given by

$$S(\xi) = C_{0,1} Y_{0,1}(\xi) + \sum_{k=1}^{N} a_k G((\Delta^*)^2; \eta_k \cdot \xi). \tag{9.209}$$

Moreover, we have

$$\int_{\mathbb{S}^2} \Delta^*_\eta S(\eta)\; \Delta^*_\eta S(\eta)\, dS(\eta) = \sum_{k=1}^{N} a_k S(\eta_k). \tag{9.210}$$

Let $\alpha_1, \ldots, \alpha_N$ be given real values. Then, there exists one and only one function S of type (9.209) satisfying $\Sigma_{k=1}^{N} a_k = 0$ such that $S(\eta_i) = \alpha_i$, $i = 1, \ldots, N$. We denote this function by S_N (for more detailed information the reader is referred to Chapter 13).

Now, for all $F \in C^{(2)}(\mathbb{S}^2)$ with $F(\eta_i) = \alpha_i$, $i = 1, \ldots, N$, we find

$$\int_{\mathbb{S}^2} \Delta^*_\eta (S_N(\eta) - F(\eta)) \Delta^*_\eta (S_N(\eta) - F(\eta)) \, dS(\eta) \tag{9.211}$$
$$= \sum_{k=1}^{N} a_k \alpha_k - 2 \sum_{k=1}^{N} a_k \alpha_k + \int_{\mathbb{S}^2} (\Delta^*_\eta F(\eta))^2 \, dS(\eta).$$

Therefore we find

$$\int_{\mathbb{S}^2} (\Delta^*_\eta F(\eta))^2 \, dS(\eta) = \int_{\mathbb{S}^2} (\Delta^*_\eta S_N(\eta))^2 \, dS(\eta) + \int_{\mathbb{S}^2} (\Delta^*_\xi (S_N(\eta) - F(\eta)))^2 \, dS(\eta). \tag{9.212}$$

Expressed in terms of the "bending energy"

$$\|\Delta^* F\|^2_{L^2(\mathbb{S}^2)} = \int_{\mathbb{S}^2} (\Delta^*_\xi F(\eta))^2 \, dS(\eta) \tag{9.213}$$

we obtain

$$\|\Delta^* S_N\|^2_{L^2(\mathbb{S}^2)} = \|\Delta^* F\|^2_{L^2(\mathbb{S}^2)} - \|\Delta^* (S_N - F)\|^2_{L^2(\mathbb{S}^2)} \tag{9.214}$$

for all $F \in C^{(2)}(\mathbb{S}^2)$ satisfying $F(\eta_i) = \alpha_i$, $i = 1, \ldots, N$.

Remark 9.22. *The integral (9.213) may be physically interpreted (at least in linearized sense under some simplifying assumptions) as the bending energy of a (thin) membrane spanned wholly over the (unit) sphere, F denotes the deflection normal to the rest position supposed, of course, to be spherical. This physical model is suggested by the classical interpretation of the one-dimensional integral $\int_a^b |F''(x)|^2 \, dx$ as the potential energy of a statically deflected thin beam which indeed is proportional to the integral taken over the square of the (linearized) curvature of the beam.*

Next, we explain the intimate relationship between best approximate and spline integration. In fact, if $\mathring{a}_1, \ldots, \mathring{a}_N$ solve the linear system (9.205), we see that

$$\frac{1}{4\pi} \int_{\mathbb{S}^2} S(\xi) \, dS(\xi) = \frac{1}{\sqrt{4\pi}} \sum_{k=1}^{N} \mathring{a}_k C_{0,1} \tag{9.215}$$
$$= \sum_{k=1}^{N} \mathring{a}_k (C_{0,1} Y_{0,1}(\eta_k) + \sum_{r=1}^{N} \mathring{a}_r G((\Delta)^2; \eta_r \cdot \eta_k))$$
$$= \sum_{k=1}^{N} \mathring{a}_k S(\eta_k)$$

holds for all splines S of the form (9.209). In other words, the best approximation to the integral is precisely the unique approximation that is obtained by exact integration of spline functions.

"Mollified" Integration by Splines of Degree 0. We start our mollified spline settings from the identity

$$F(\xi) \approx \frac{1}{4\pi}\int_{\mathbb{S}^2} F(\eta)\, dS(\eta) - \int_{\mathbb{S}^2} \left(\nabla^*_\eta G^\rho(\Delta^*;\xi\cdot\eta)\right)\cdot\left(\nabla^*_\eta F(\eta)\right)\, dS(\eta) \quad (9.216)$$

provided that ρ is sufficiently small and F is continuously differentiable on $\mathbb{S}^2$. If $a_1,\ldots,a_N \in \mathbb{R}$ satisfy $\sum_{k=1}^N a_k = 0$, we have for all nodes $\eta_1,\ldots,\eta_N \in \mathbb{S}^2$,

$$\sum_{k=1}^N a_k F(\eta_k) \approx \int_{\mathbb{S}^2} \nabla^*_\eta \sum_{k=1}^N a_k G^\rho(\Delta^*;\eta_k\cdot\eta)\cdot\nabla^*_\eta F(\eta)\, dS(\eta), \qquad (9.217)$$

provided that F is continuously differentiable on $\mathbb{S}^2$. In other words, if $a_1,\ldots,a_N \in \mathbb{R}$ satisfy $\sum_{k=1}^N a_k = 0$ and if $\eta_1,\ldots,\eta_N$ are the prescribed nodes on $\mathbb{S}^2$, then

$$\int_{\mathbb{S}^2} \nabla^*_\eta S^\rho(\eta)\ \cdot\nabla^*_\eta F(\eta)\, dS(\eta) \approx \sum_{k=1}^N a_k F(\eta_k), \qquad (9.218)$$

where $S^\rho : \mathbb{S}^2 \to \mathbb{R}$ is given by

$$S^\rho(\xi) = \underbrace{C_{0,1}Y_{0,1}(\xi)}_{=\text{const.}} + \sum_{k=1}^N a_k G^\rho(\Delta^*;\eta_k\cdot\xi). \qquad (9.219)$$

Moreover, we have

$$\int_{\mathbb{S}^2} \nabla^*_\eta S^\rho(\eta)\ \cdot\nabla^*_\eta S^\rho(\eta)\, dS(\eta) \approx \sum_{k=1}^N a_k S^\rho(\eta_k). \qquad (9.220)$$

Let $\alpha_1,\ldots,\alpha_N$ be given real values. Then, there exists one and only one function S of type (9.219) satisfying $\Sigma_{k=1}^N a_k^{\rho,N} = 0$ such that $S(\eta_i) = \alpha_i$, $i = 1,\ldots,N$. We denote this function by S_N^ρ, more explicitly,

$$S_N^\rho(\xi) = C_{0,1}^{\rho;N} Y_{0,1}(\xi) + \sum_{k=1}^N a_k^{\rho;N} G^\rho(\Delta^*;\eta_k\cdot\xi), \quad \xi\in\mathbb{S}^2. \qquad (9.221)$$

Now, for all $F \in \mathrm{C}^{(2)}(\mathbb{S}^2)$ with $F(\eta_i) = \alpha_i$, $i = 1,\ldots,N$, we find

$$\begin{aligned}&\int_{\mathbb{S}^2} \nabla^*_\eta(S_N^\rho(\eta) - F(\eta))\cdot\nabla^*_\eta(S_N^\rho(\eta) - F(\eta))\, dS(\eta) \qquad (9.222)\\ &\approx \sum_{k=1}^N a_k^{\rho;N}\alpha_k - 2\sum_{k=1}^N a_k^{\rho;N}\alpha_k + \int_{\mathbb{S}^2} (\nabla^*_\eta F(\eta))^2\, dS(\eta).\end{aligned}$$

Therefore we find

$$\int_{\mathbb{S}^2} (\nabla^*_\eta F(\eta))^2 \, dS(\eta) \approx \int_{\mathbb{S}^2} (\nabla^*_\eta S^\rho_N(\eta))^2 \, dS(\eta) + \int_{\mathbb{S}^2} (\nabla^*_\xi (S^\rho_N(\eta) - F(\eta)))^2 \, dS(\eta). \tag{9.223}$$

Expressed in terms of the "bending energy"

$$\|\nabla^* F\|^2_{\mathrm{L}^2(\mathbb{S}^2)} = \int_{\mathbb{S}^2} (\nabla^*_\xi F(\eta))^2 \, dS(\eta) \tag{9.224}$$

we obtain

$$\|\nabla^* S^\rho_N\|^2_{\mathrm{L}^2(\mathbb{S}^2)} \approx \|\nabla^* F\|^2_{\mathrm{L}^2(\mathbb{S}^2)} - \|\nabla^*(S^\rho_N - F)\|^2_{\mathrm{L}^2(\mathbb{S}^2)} \tag{9.225}$$

for all $F \in \mathrm{C}^{(2)}(\mathbb{S}^2)$ satisfying $F(\eta_i) = \alpha_i$, $i = 1, \ldots, N$. Due to (9.124) integrating the interpolating spline leads to

$$\begin{aligned} \frac{1}{4\pi} \int_{\mathbb{S}^2} F(\eta) \, dS(\eta) \approx \int_{\mathbb{S}^2} S^\rho_N(\eta) \, dS(\eta) &= \frac{1}{\sqrt{4\pi}} \, C^{\rho;N}_{0,1} + \frac{\rho}{4} \sum_{k=1}^N a^{\rho;N}_k \\ &= \frac{1}{\sqrt{4\pi}} \, C^{\rho;N}_{0,1}. \end{aligned} \tag{9.226}$$

9.4 Integral Formulas under Boundary Conditions

Local purposes in geosciences require the integration over subregions of $\mathbb{S}^2$. In this respect, boundary curve information must be handled appropriately. In what follows we discuss Dirichlet and Neumann boundary values.

Homogeneous Boundary Conditions. Let $\Gamma \subset \mathbb{S}^2$ be a regular region, i.e., a region that allows the surface variant of the Gaussian theorem. Suppose that $F : \overline{\Gamma} \to \mathbb{R}$ is twice continuously differentiable. Furthermore, let ξ be a point of Γ. Under these assumptions we know that

$$\begin{aligned} F(\xi) = \frac{1}{4\pi} \int_\Gamma F(\eta) \, dS(\eta) &+ \int_\Gamma \Delta^*_\eta G((\Delta^*)^2; \xi \cdot \eta) \Delta^*_\eta F(\eta) \, dS(\eta) \\ &+ \int_{\partial\Gamma} (\nu_\eta \cdot \nabla^*_\eta \Delta^*_\eta G((\Delta^*)^2; \xi \cdot \eta)) F(\eta) \, d\sigma(\eta) \\ &- \int_{\partial\Gamma} \Delta^*_\eta G((\Delta^*)^2; \xi \cdot \eta)(\nu_\eta \cdot \nabla^*_\eta F(\eta)) \, d\sigma(\eta). \end{aligned} \tag{9.227}$$

First we are interested in an integral formula corresponding to the *prescribed boundary condition* $F|\partial\Gamma = B$. For that purpose we consider a function $G^{(2)}(\xi, \cdot) \in \mathrm{C}^{(4)}(\overline{\Gamma})$ solving the boundary-value problem

$$(\Delta^*)^2 G^{(2)}(\xi, \cdot)|\Gamma = 0 \tag{9.228}$$

such that

$$G^{(2)}(\xi,\cdot)|\partial\Gamma = G((\Delta^*)^2;\xi\cdot\,)|\partial\Gamma, \tag{9.229}$$

$$\Delta^* G^{(2)}(\xi,\cdot)|\partial\Gamma = G(\Delta^*;\xi\cdot\,)|\partial\Gamma. \tag{9.230}$$

it should be remarked that, for $\xi \in \Gamma$, $G((\Delta^*)^2;\xi\cdot)$ as well as $G(\Delta^*;\xi\cdot)$ are infinitely often differentiable functions on the curve $\partial\Gamma$. Furthermore, the boundary-value problem is equivalent to the following system of boundary-value problems

$$\Delta^* Q_\xi|\Gamma = 0, \tag{9.231}$$

$$Q_\xi|\partial\Gamma = G(\Delta^*;\xi\cdot\,)|\partial\Gamma, \tag{9.232}$$

and

$$\Delta^* G^{(2)}(\xi,\cdot)|\Gamma = Q_\xi, \tag{9.233}$$

$$G^{(2)}(\xi,\cdot)|\partial\Gamma = G((\Delta^*)^2;\xi\cdot\,)|\partial\Gamma. \tag{9.234}$$

Obviously, each of the problems can be solved separately. The second Green's surface theorem (9.30) then yields

$$\begin{aligned} 0 &= \int_\Gamma (\Delta^*_\eta)^2 G^{(2)}(\xi,\eta)F(\eta)\, dS(\eta) \\ &= \int_\Gamma \Delta^*_\eta G^{(2)}(\xi,\eta)(\Delta^*_\eta F(\eta))\, dS(\eta) \\ &\quad + \int_{\partial\Gamma} (\nu_\eta \cdot \nabla^*_\eta \Delta^*_\eta G^{(2)}(\xi,\eta))F(\eta)\, d\sigma(\eta) \\ &\quad - \int_{\partial\Gamma} (\Delta^*_\eta G^{(2)}(\xi,\eta))(\nu_\eta \cdot \nabla^*_\eta F(\eta))\, d\sigma(\eta). \end{aligned} \tag{9.235}$$

In connection with (9.227) this leads us to the following result.

Theorem 9.23. *Let $\Gamma \subset \mathbb{S}^2$ be a regular region. Suppose that $x \in \Gamma$. Let B be a given function of class $\mathrm{C}^{(2)}(\partial\Gamma)$. Let $F \in \mathrm{C}^{(2)}(\overline{\Gamma})$, $\overline{\Gamma} = \Gamma \cup \partial\Gamma$, satisfy the boundary condition*

$$F|\partial\Gamma = B. \tag{9.236}$$

Furthermore, suppose that $G^{(2)}(\xi,\cdot) \in \mathrm{C}^{(4)}(\overline{\Gamma})$ solves the boundary-value problem

$$(\Delta^*)^2 G^{(2)}(\xi,\cdot) = 0, \tag{9.237}$$

and

$$G^{(2)}(\xi,\cdot)|\partial\Gamma = G((\Delta^*)^2;\xi\cdot)|\partial\Gamma, \tag{9.238}$$

$$\Delta^* G^{(2)}(\xi,\cdot)|\partial\Gamma = G(\Delta^*;\xi\cdot)|\partial\Gamma. \tag{9.239}$$

Then

$$F(\xi) = \frac{1}{4\pi}\int_\Gamma F(\eta)\, dS(\eta) + \int_\Gamma \Delta^*_\eta D^{(2)}(\xi,\eta)\Delta^*_\eta F(\eta)\, dS(\eta) \qquad (9.240)$$
$$+ \int_{\partial\Gamma} (\nu_\eta \cdot \nabla^*_\eta \Delta^*_\eta D^{(2)}(\xi,\eta))B(\eta)\, d\sigma(\eta),$$

where

$$D^{(2)}(\xi,\cdot) = G((\Delta^*)^2;\xi\cdot) - G^{(2)}(\xi,\cdot). \qquad (9.241)$$

Even more, Theorem 9.23 can be used to "homogenize" the boundary conditions on $\partial\Gamma$. For that purpose, we introduce a function $W \in \mathrm{C}^{(2)}(\overline{\Gamma})$ such that

$$\Delta^* W|\Gamma = 0 \qquad (9.242)$$

and

$$W|\partial\Gamma = B. \qquad (9.243)$$

This gives us

$$W(\xi) = \frac{1}{4\pi}\int_\Gamma W(\eta)\, dS(\eta) + \int_{\partial\Gamma} (\nu_\eta \cdot \nabla^*_\eta \Delta^*_\eta D^{(2)}(\xi,\eta))B(\eta)\, d\sigma(\eta). \quad (9.244)$$

This leads to the following result.

Corollary 9.24. *Under the assumptions of Theorem 9.23 we have*

$$F(\xi) - W(\xi) = \frac{1}{4\pi}\int_\Gamma (F(\eta) - W(\eta))\, dS(\eta) \qquad (9.245)$$
$$+ \int_\Gamma \Delta^*_\eta D^{(2)}(\xi\cdot\eta)\Delta^*_\eta F(\eta)\, dS(\eta).$$

Corollary 9.24 give rise to introduce spherical splines on the regular region $\Gamma \in \mathbb{S}^2$. To this end we suppose there a given a set of points $\xi_1,\ldots,\xi_n \in \Gamma$ such that for the matrix $\mathbf{D}$ with $\mathbf{D}_{ij} = D^{(2)}(\xi_j\cdot\xi_i)$, $i,j = 1,\ldots,n$, holds that

$$\det(\mathbf{D}) \neq 0. \qquad (9.246)$$

Then, for all choices $a = (a_1,\ldots,a_N)^T$ of coefficients $a_i \in \mathbb{R}$, $i = 1,\ldots,N$, with

$$\sum_{i=1}^N a_i = 0 \qquad (9.247)$$

we get the *spline integration formula*

$$\sum_{i=1}^N a_i(F(\xi_i) - W(\xi_i)) = \int_\Gamma \Delta^*_\eta \sum_{i=1}^N a_i D^{(2)}(\xi_i,\eta)\Delta^*_\eta F(\eta)\, dS(\eta). \qquad (9.248)$$

Theorem 9.25. *Given $y_1, \ldots, y_N \in \mathbb{R}$. The spline function*

$$S = C_{0,1}Y_{0,1} + \sum_{i=1}^{N} a_i D^{(2)}(\xi_i, \cdot), \tag{9.249}$$

uniquely determined by

$$\sum_{i=1}^{N} a_i D^{(2)}(\xi_i, \xi_j) + C_{0,1}Y_{0,1}(\xi_j) = \alpha_j, \quad j = 1, \ldots, N \tag{9.250}$$

and

$$\sum_{i=1}^{N} a_i = 0, \tag{9.251}$$

where $\alpha_j = y_j - W(\eta_j)$ for $j = 1, \ldots, N$ and W is given by (9.244), satisfies the "smoothest interpolation property"

$$\int_\Gamma (\Delta^*_\eta F(\eta))^2 \, dS(\eta) = \int_\Gamma (\Delta^*_\eta S(\eta))^2 \, dS(\eta) + \int_\Gamma (\Delta^*_\eta (F(\eta) - S(\eta))^2) \, dS(\eta) \tag{9.252}$$

for all $F \in \mathrm{C}^{(2)}(\overline{\Gamma})$ with $F(\xi_i) = y_i$, $i = 1, \ldots, N$.

Proof. For all $F \in \mathrm{C}^{(2)}(\overline{\Gamma})$ with $F(\xi_i) = y_i$, $i = 1, \ldots, N$, we have

$$\int_\Gamma \Delta^*_\eta S(\eta) \Delta^*_\eta F(\eta) \, dS(\eta) = \sum_{i=1}^{N} a_i \underbrace{(y_i - W(\xi_i))}_{=\alpha_i}. \tag{9.253}$$

Moreover,

$$\int_\Gamma \Delta^*_\eta S(\eta) \Delta^*_\eta S(\eta) \, dS(\eta) = \sum_{i=1}^{N} a_i \underbrace{(y_i - W(\xi_i))}_{=\alpha_i}. \tag{9.254}$$

Therefore it follows that

$$\begin{aligned} &\int_\Gamma (\Delta^*_\eta (F(\eta) - S(\eta)))(\Delta^*_\eta (F(\eta) - S(\eta))) \, dS(\eta) \qquad (9.255) \\ &= \int_\Gamma (\Delta^*_\eta F(\eta))^2 \, dS(\eta) + \int_\Gamma (\Delta^*_\eta S(\eta))^2 \, dS(\eta) - 2 \sum_{i=1}^{N} a_i \alpha_i \\ &= \int_\Gamma (\Delta^*_\eta F(\eta))^2 \, dS(\eta) - \int_\Gamma (\Delta^*_\eta S(\eta))^2 \, dS(\eta). \end{aligned}$$

This is the desired result. □

More general formulas can be found in the PhD-thesis of T. Fehlinger [2009]. By integration of S as determined by (9.250) and (9.251) we are led to a spline integration rule over regular regions $\Gamma \subset \mathbb{S}^2$.

9.5 Sphere Functions and Shannon Kernels

A basic question when approximating a function on the sphere $\mathbb{S}^2$ by its truncated Fourier expansion (orthogonal expansion) in terms of spherical harmonics is the existence of a manageable error term. An adequate answer is the construction of integral formulas. It will be shown that the error term between a function and its truncated Fourier series expansion is explicitly available in integral form involving the Shannon kernel provided that sufficient smoothness is imposed on the function under consideration. An essential tool for the integral formulas is the theory of Green's function with respect to an operator

$$\partial_n = \Delta^* + n(n+1), \quad n \in \mathbb{N}_0 \tag{9.256}$$

and its iterations (which also turn out to play a fundamental role in the Helmholtz decomposition (cf. W. Freeden, M. Schreiner [2009]) of spherical vector as well as tensor fields.

Our considerations start with the introduction of Green's functions with respect to the operator ∂_n, $n \in \mathbb{N}_0$ (cf. W. Freeden [1978a, 1979]).

Definition 9.26. $G(\partial_n;\cdot,\cdot) : (\xi,\eta) \mapsto G(\partial_n;\xi,\eta)$, $-1 \leq \xi\cdot\eta < 1$, *is called* Green's function for the operator $\partial_n = \Delta^* + n(n+1)$, $n \in \mathbb{N}_0$, *if it satisfies the following properties:*

(i) (Differential equation) *for every point* $\xi \in \mathbb{S}^2, \eta \mapsto G(\partial_n;\xi,\eta)$ *is twice continuously differentiable on* $\{\eta \in \mathbb{S}^2 : -1 \leq \xi\cdot\eta < 1\}$, *and we have*

$$(\partial_n)_\eta G(\partial_n;\xi,\eta) = -\frac{2n+1}{4\pi} P_n(\xi\cdot\eta), \quad -1 \leq \xi\cdot\eta < 1, \tag{9.257}$$

where $(\partial_n)_\eta$ *means that the operator* ∂_n *is applied to the variable* η.

(ii) (Characteristic singularity) *for every* $\xi \in \mathbb{S}^2$, *the function*

$$\eta \mapsto G(\partial_n;\xi,\eta) - \frac{1}{4\pi} P_n(\xi\cdot\eta)\ln(1-\xi\cdot\eta) \tag{9.258}$$

is continuously differentiable on $\mathbb{S}^2$.

(iii) (Rotational symmetry) for every $\xi \in \mathbb{S}^2$ *and for all orthogonal transformations* $\mathbf{t} \in O(3)$,

$$G(\partial_n;\mathbf{t}\xi,\mathbf{t}\eta) = G(\partial_n;\xi,\eta). \tag{9.259}$$

(iv) (Normalization) for every $\xi \in \mathbb{S}^2$,

$$\int_{\mathbb{S}^2} G(\partial_n;\xi,\eta) P_n(\xi\cdot\eta)\, dS(\eta) = 0. \tag{9.260}$$

We prove the uniqueness of the Green's function with respect to the operator ∂_n.

Lemma 9.27. *$G(\partial_n; \cdot, \cdot)$ is uniquely determined by its defining properties (i)–(iv).*

Proof. Denote by $D(\partial_n; \cdot, \cdot)$ the difference between two Green's functions satisfying (i)–(iv). Then we have the following properties:

(i) for every $\xi \in \mathbb{S}^2$, $D(\partial_n; \xi, \cdot)$ is twice continuously differentiable for all points $\eta \in \mathbb{S}^2$ with $-1 \leq \xi \cdot \eta < 1$, and we have

$$(\partial_n)_\eta D(\partial_n; \xi, \eta) = (\Delta^*_\eta + n(n+1))D(\partial_n; \xi, \eta) = 0, \tag{9.261}$$

(ii) for every $\xi \in \mathbb{S}^2$, $D(\partial_n; \xi, \cdot)$ is continuously differentiable for all $\eta \in \mathbb{S}^2$,

(iii) For every orthogonal transformations $\mathbf{t} \in O(3)$,

$$D(\partial_n; \mathbf{t}\xi, \mathbf{t}\eta) = D(\partial_n; \xi, \eta), \tag{9.262}$$

(iv) For every $\xi \in \mathbb{S}^2$

$$\int_{\mathbb{S}^2} D(\partial_n; \xi, \eta) P_n(\xi \cdot \eta)\, dS(\eta) = 0. \tag{9.263}$$

The properties (i)–(iii) show that $D(\partial_n; \xi, \cdot)$ is an everywhere on the unit sphere $\mathbb{S}^2$ twice continuously differentiable function satisfying the differential equation (i). Therefore, $D(\partial_n; \xi, \cdot)$ must be a spherical harmonic of order n. $D(\partial_n; \xi, \eta)$ depends only on the scalar product of ξ and η, i.e.,

$$D(\partial_n; \xi, \eta) = \alpha_n P_n(\xi \cdot \eta). \tag{9.264}$$

From (iv) we obtain

$$\int_{\mathbb{S}^2} D(\partial_n; \xi, \eta) P_n(\xi \cdot \eta)\, dS(\eta) = \alpha_n \int_{\mathbb{S}^2} P_n(\xi \cdot \eta) P_n(\xi \cdot \eta)\, dS(\eta) = 0. \tag{9.265}$$

Hence, $\alpha_n = 0$. But this means that the Green's function $G(\partial_n; \cdot, \cdot)$ is uniquely determined by the defining properties (i)–(iv). □

Remark 9.28. *Because of (iii), $G(\partial_n; \cdot)$ is a zonal function such that $G(\partial_n; \xi, \eta)$ can be identified with $G(\partial_n; \xi \cdot \eta)$.*

Graphical impressions of the Green's functions $G(\partial_n; \cdot)$ can be found in Figure 9.8. Observing the characteristic singularity of the Green's function, we see by applying the Green's surface theorem that the spherical harmonics of degree n, i.e., the eigenfunctions Y_n of the Beltrami operator Δ^* with respect

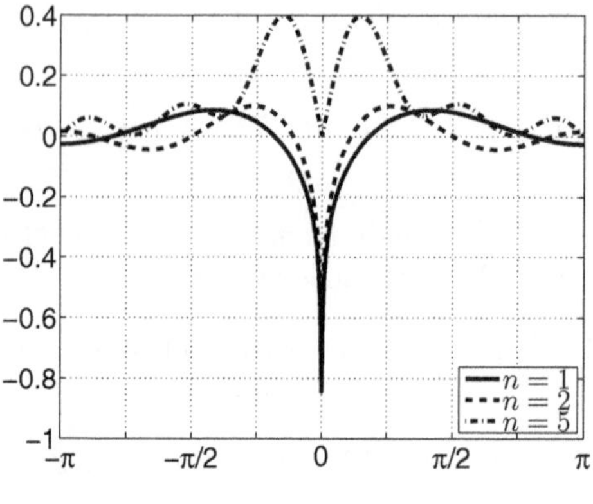

FIGURE 9.8: The functions $\vartheta \mapsto G(\partial_n; \cos\vartheta)$ for various values n.

to the eigenvalues $(\Delta^*)^\wedge(n)$, $n = 0, 1, \ldots$, are eigenfunctions of the Green's (kernel) function $G(\partial_n; \cdot)$ in the sense of the integral equation

$$\big(n(n+1) - k(k+1)\big) \int_{\mathbb{S}^2} G(\partial_n; \xi \cdot \eta) Y_k(\eta)\, dS(\eta) = (1 - \delta_{n,k}) Y_k(\xi). \quad (9.266)$$

Furthermore, if $\xi, \eta \in \mathbb{S}^2$ with $-1 \leq \xi \cdot \eta < 1$, $G(\partial_n; \xi \cdot \eta)$ allows the bilinear expansion

$$G(\partial_n; \xi \cdot \eta) = \sum_{(\partial_n)^\wedge(k) \neq 0} \frac{1}{(\partial_n)^\wedge(k)} \sum_{j=1}^{2k+1} Y_{k,j}(\xi) Y_{k,j}(\eta), \quad (9.267)$$

where $(\partial_n)^\wedge(k)$ is given by

$$(\partial_n)^\wedge(k) = (\Delta^*)^\wedge(n) - (\Delta^*)^\wedge(k) = n(n+1) - k(k+1). \quad (9.268)$$

The symbol $\sum_{(\partial_n)^\wedge(k)\neq 0}$ means that the sum is to be extended over all non-negative integers k for which the denominator $(\partial_n)^\wedge(k)$ is different from zero, i.e., $(\Delta^*)^\wedge(k) \neq (\Delta^*)^\wedge(n)$ such that $k \neq n$. Using the addition theorem, we are able to rewrite the bilinear expansion of $G(\partial_n; \xi \cdot \eta)$ in the form

$$G(\partial_n; \xi \cdot \eta) = \sum_{(\partial_n)^\wedge(k)\neq 0} \frac{2k+1}{4\pi} \frac{1}{(\partial_n)^\wedge(k)} P_k(\xi \cdot \eta). \quad (9.269)$$

This leads to the formulation of the following result.

Lemma 9.29. *For $n = 1, 2, \ldots$ and $\xi, \eta \in \mathbb{S}^2$ with $-1 \leq \xi \cdot \eta < 1$*

$$G(\partial_n; \xi \cdot \eta) = \frac{1}{4\pi} P_n(\xi \cdot \eta) \ln(1 - \xi \cdot \eta) - \frac{1}{2\pi} \sum_{k=0}^{n-1} \frac{2k+1}{(\partial_n)^\wedge(k)} P_k(\xi \cdot \eta)$$
$$- \frac{1}{4\pi} \Big(\frac{2n+1}{2} \int_{-1}^{1} P_n^2(t) \ln(1-t)\, dt \Big) P_n(\xi \cdot \eta). \quad (9.270)$$

In particular,

$$G(\partial_0;\xi\cdot\eta) = G(\Delta^*;\xi\cdot\eta) = \frac{1}{4\pi}\ln(1-\xi\cdot\eta) + \frac{1}{4\pi}(1-\ln(2)). \qquad (9.271)$$

Proof. We have to show that $G(\partial_n;\cdot)$ satisfies the defining properties (i)–(iv). An easy calculation yields

$$\left((1-t^2)\left(\frac{d}{dt}\right)^2 - 2t\frac{d}{dt} + n(n+1)\right)P_n(t)\ln(1-t) \qquad (9.272)$$
$$= -2(1+t)P_n'(t) - P_n(t).$$

Consequently we find

$$\left((1-t^2)\left(\frac{d}{dt}\right)^2 - 2t\frac{d}{dt} + n(n+1)\right)(-P_n(t)\ln(1-t)) \qquad (9.273)$$
$$+\left((1-t^2)(\frac{d}{dt})^2 - 2t\frac{d}{dt} + n(n+1)\right)\left(-2\sum_{k=0}^{n-1}\frac{2k+1}{n(n+1)-k(k+1)}P_k(t)\right)$$
$$= 2(1+t)P_n'(t) + P_n(t) - 2\sum_{k=0}^{n-1}(2k+1)P_k(t)$$
$$= (2n+1)P_n(t).$$

This shows that condition (i) of Definition 9.26 is valid. Because of $P_n(1) = 1$, $n = 1, 2, \ldots$, condition (ii) is certainly satisfied. Finally, it is not difficult to see that

$$\int_{\mathbb{S}^2} G(\partial_n;\xi\cdot\eta)P_n(\xi\cdot\eta)\, dS(\eta) = 0. \qquad (9.274)$$

□

Now we are in the position to formulate the Green's surface theorem for the operator $\partial_n = \Delta^* + n(n+1)$, $n \in \mathbb{N}_0$.

Theorem 9.30. *Let ξ be an arbitrary point of the unit sphere $\mathbb{S}^2$. Suppose that F is of class $\mathrm{C}^{(2)}(\mathbb{S}^2)$. Then, for $n = 0, 1, \ldots$,*

$$F(\xi) = \frac{2n+1}{4\pi}\int_{\mathbb{S}^2} F(\eta)P_n(\xi\cdot\eta)\, dS(\eta) \qquad (9.275)$$
$$+\int_{\mathbb{S}^2} G(\Delta^* + n(n+1);\xi\cdot\eta)(\Delta^*_\eta + n(n+1))F(\eta)\, dS(\eta).$$

Remark 9.31. *In particular with $n = 0$, for each $\eta \in \mathbb{S}^2, \xi \mapsto G(\partial_0^2;\xi\cdot\eta)$ is a continuous function on $\mathbb{S}^2$ and an infinitely often differentiable function in $\mathbb{S}^2\backslash\{\eta\}$ such that $(\partial_0)_\xi G(\partial_0^2;\xi\cdot\eta)$ is an infinitely often differentiable function in $\mathbb{S}^2\backslash\{\eta\}$ possessing a logarithmic singularity at $\eta \in \mathbb{S}^2$. Clearly, in distributional sense, we have*

$$(\partial_0^2)_\xi G(\partial_0^2;\xi\cdot\eta) = \delta_{1,\ldots,\infty}(\xi\cdot\eta) = \delta(\xi\cdot\eta) - \delta_0(\xi\cdot\eta), \quad \xi,\eta\in\mathbb{S}^2, \quad (9.276)$$

where the Dirac function(al) (Delta distribution) *is formally given by the zonal kernel function expansion*

$$\delta(\xi \cdot \eta) = \delta_{0,\ldots,\infty}(\xi \cdot \eta) = \sum_{n=0}^{\infty} \frac{2n+1}{4\pi} P_n(\xi \cdot \eta), \quad \xi, \eta \in \mathbb{S}^2, \tag{9.277}$$

and

$$\delta_0(\xi \cdot \eta) = \frac{1}{4\pi}, \quad \xi, \eta \in \mathbb{S}^2. \tag{9.278}$$

This observation explains the particular role of $G(\partial_0^2; \xi \cdot\ )$ in spherical spline theory as "Beltrami antiderivative" of a function with an isolated logarithmic singularity in ξ (seen in comparison to the one-dimensional spline context, where the characteristic singularity for the operator of the second derivative is known to be a finite jump).

The Shannon kernel $(\xi, \eta) \mapsto SH_{0,\ldots,m}(\xi \cdot \eta)$, $\xi, \eta \in \mathbb{S}^2$, of degree $m \in \mathbb{N}_0$ *is given by*

$$SH_{0,\ldots,m}(\xi \cdot \eta) = \sum_{n=0}^{m} \frac{2n+1}{4\pi} P_n(\xi \cdot \eta) = \sum_{n=0}^{m} \sum_{j=1}^{2n+1} Y_{n,j}(\xi) Y_{n,j}(\eta). \tag{9.279}$$

Obviously, it is the m-th truncated spherical harmonic expansion of the Dirac function(al)

$$\delta(\xi \cdot \eta) = SH_{0,\ldots,\infty}(\xi \cdot \eta), \quad \xi, \eta \in \mathbb{S}^2. \tag{9.280}$$

Furthermore, the Funk–Hecke formula (9.43), in connection with the orthogonality of the spherical harmonics (9.41), yields

$$\int_{\mathbb{S}^2} (\partial_0^2)_\xi G(\partial_0^2; \xi \cdot \eta) SH_{0,\ldots,m}(\zeta \cdot \eta)\, dS(\eta) = SH_{1,\ldots,m}(\xi \cdot \zeta) \tag{9.281}$$

for $m \in \mathbb{N}$ and all $\xi, \zeta \in \mathbb{S}^2$, where

$$SH_{1,\ldots,m}(\xi \cdot \zeta) = \delta_{1,\ldots,m}(\xi \cdot \zeta) = \sum_{n=1}^{m} \frac{2n+1}{4\pi} P_n(\xi \cdot \zeta) = \sum_{n=1}^{m} \sum_{j=1}^{2n+1} Y_{n,j}(\xi) Y_{n,j}(\zeta). \tag{9.282}$$

According to the classical Fredholm–Hilbert theory of linear integral equations (cf. D. Hilbert [1912]), we inductively define the Green's function with respect to iterated operators

$$\partial_{0,\ldots,m} = \partial_0 \ldots \partial_m, \quad m \in \mathbb{N}_0, \tag{9.283}$$

where

$$\partial_n = \Delta^* + n(n+1), \quad n = 0, \ldots, m, \tag{9.284}$$

by the following convolutions for $m = 1, 2 \ldots$

$$G(\partial_{0,\ldots,m}; \xi \cdot \eta) = \int_{\mathbb{S}^2} G(\partial_{0,\ldots,m-1}; \xi \cdot \zeta) G(\partial_m; \zeta \cdot \eta)\, dS(\zeta), \tag{9.285}$$

and $G(\partial_0; \xi \cdot \eta) = G(\Delta^*; \xi \cdot \eta)$. $G(\partial_{0,\ldots,m}; \cdot, \cdot) : \mathbb{S}^2 \times \mathbb{S}^2 \to \mathbb{R}$ is called *Green's function with respect to the operator* $\partial_{0,\ldots,m}$. An illustration of $G(\partial_{0,\ldots,m}, \cdot, \cdot)$ is given in Figure 9.9.

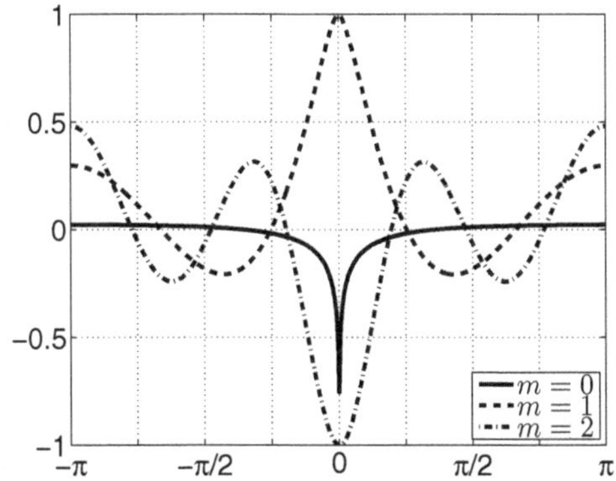

FIGURE 9.9: The functions $\vartheta \mapsto G(\partial_{0,\ldots,m}; \cos\vartheta)$ for $m = 0, 1, 2$ (normalized).

We are particularly interested in the Green's function with respect to the operator $\partial_0\partial_1 = \Delta^*(\Delta^* + 2)$ (cf. W. Freeden, M. Schreiner [2009]). First, it is clear that

$$G(\partial_0\partial_1; \xi \cdot \eta) = \sum_{n=2}^{\infty} \frac{2n+1}{4\pi} \frac{1}{-n(n+1)(-n(n+1)+2)} P_n(\xi \cdot \eta)$$

$$= \frac{1}{4\pi} \sum_{n=2}^{\infty} \frac{2n+1}{n(n+1)(n-1)(n+2)} P_n(\xi \cdot \eta). \tag{9.286}$$

Considering the derivatives of the sum

$$G(t) = \sum_{n=2}^{\infty} \frac{2n+1}{(n-1)n(n+1)(n+2)} P_n(t) \tag{9.287}$$

we obtain in a first step by (7.25)

$$G'(t) = \sum_{n=2}^{\infty} \frac{2n+1}{(n-1)n(n+1)(n+2)} P_n'(t) \tag{9.288}$$

$$= \frac{1}{t^2-1} \sum_{n=2}^{\infty} \frac{2n+1}{(n-1)n(n+1)(n+2)} \frac{n(n+1)}{2n+1} \left(P_{n+1}(t) - P_{n-1}(t)\right)$$

$$= \frac{1}{t^2-1} \sum_{n=2}^{\infty} \frac{1}{(n-1)(n+2)} \left(P_{n+1}(t) - P_{n-1}(t)\right).$$

By index shifts we are able to see that

$$G'(t) = \frac{1}{t^2-1}\left(\sum_{n=3}^{\infty}\frac{1}{(n-2)(n+1)}P_n(t) - \sum_{n=1}^{\infty}\frac{1}{n(n+3)}P_n(t)\right) \tag{9.289}$$

$$= \frac{1}{t^2-1}\left(\sum_{n=3}^{\infty}\frac{2(2n+1)}{(n-2)n(n+1)(n+3)}P_n(t) - \frac{1}{4}t - \frac{1}{10}\left(\frac{3}{2}t^2 - \frac{1}{2}\right)\right).$$

Another differentiation yields the expression

$$G''(t) = \frac{1}{t^2-1}\left(\sum_{n=3}^{\infty}\frac{2(2n+1)}{(n-2)n(n+1)(n+3)}P'_n(t) - \frac{1}{4} - \frac{3}{10}t\right)$$

$$- \frac{2t}{(t^2-1)^2}\left(\sum_{n=3}^{\infty}\frac{2(2n+1)}{(n-2)n(n+1)(n+3)}P_n(t)\right. \tag{9.290}$$

$$\left. - \frac{1}{4}t - \frac{1}{10}\left(\frac{3}{2}t^2 - \frac{1}{2}\right)\right).$$

The second sum can be transformed by use of (7.25) as follows

$$\sum_{n=3}^{\infty}\frac{2(2n+1)}{(n-2)n(n+1)(n+3)}P'_n(t) \tag{9.291}$$

$$= \frac{1}{t^2-1}\sum_{n=3}^{\infty}\frac{2}{(n-2)(n+3)}\left(P_{n+1}(t) - P_{n-1}(t)\right)$$

$$= \frac{1}{t^2-1}\left(\sum_{n=4}^{\infty}\frac{4(2n+1)}{(n-3)(n-1)(n+2)(n+4)}P_n(t) - \frac{1}{3}P_2(t) - \frac{1}{7}P_3(t)\right).$$

This provides us with the following representation of the second derivative

$$G''(t) = \frac{1}{(t^2-1)^2}\left(\sum_{n=3}^{\infty}\frac{-4(2n+1)}{(n-2)n(n+1)(n+3)}tP_n(t) + \frac{1}{2}t^2 + \frac{1}{5}tP_2(t)\right)$$

$$+ \frac{1}{(t^2-1)^2}\left(\sum_{n=4}^{\infty}\frac{4(2n+1)}{(n-3)(n-1)(n+2)(n+4)}P_n(t) - \frac{1}{3}P_2(t)\right.$$

$$\left. - \frac{1}{7}P_3(t) - \frac{1}{4}(t^2-1) - \frac{3}{10}t(t^2-1)\right). \tag{9.292}$$

Now, in the first sum, the three-term-recurrence relation (7.24) can be applied leading to the series

$$\begin{aligned}
&\sum_{n=3}^{\infty} \frac{2n+1}{(n-2)n(n+1)(n+3)} tP_n(t) \qquad (9.293)\\
&= \sum_{n=3}^{\infty} \frac{1}{(n-2)n(n+1)(n+3)} \left((n+1)P_{n+1}(t) + nP_{n-1}(t)\right)\\
&= \sum_{n=4}^{\infty} \frac{2n+1}{(n-3)(n-1)(n+2)(n+4)} P_n(t) + \frac{1}{24}P_2(t) + \frac{1}{70}P_3(t).
\end{aligned}$$

Collecting our results we therefore get as second derivative

$$\begin{aligned}
G''(t) = \frac{1}{(t^2-1)^2} \Bigg(&-4 \sum_{n=4}^{\infty} \frac{2n+1}{(n-3)(n-1)(n+2)(n+4)} P_n(t)\\
&- \frac{1}{6}P_2(t) - \frac{2}{35}P_3(t) + \frac{1}{2}t^2 + \frac{1}{5}tP_2(t)\\
&+ 4 \sum_{n=4}^{\infty} \frac{(2n+1)}{(n-3)(n-1)(n+2)(n+4)} P_n(t)\\
&- \frac{1}{3}P_2(t) - \frac{1}{7}P_3(t) - \frac{1}{4}(t^2-1) - \frac{3}{10}t(t^2-1) \Bigg)\\
= \frac{1}{(t^2-1)^2} &\left(-\frac{1}{2}(t-1)(t+1)^2 \right) = \frac{1}{2}\frac{1}{1-t}. \qquad (9.294)
\end{aligned}$$

This enables us to establish an elementary representation of $G(t)$ by integration. To this end, we need certain values of $G(t)$ to determine the constants of integration. In fact, we have

$$\begin{aligned}
G(-1) &= \sum_{n=2}^{\infty} \frac{2n+1}{(n-1)n(n+1)(n+2)} P_n(-1) \qquad (9.295)\\
&= \sum_{n=2}^{\infty} \frac{2n+1}{(n-1)n(n+1)(n+2)} (-1)^n = \frac{1}{6}
\end{aligned}$$

and

$$\begin{aligned}
G'(-1) &= \sum_{n=2}^{\infty} \frac{2n+1}{(n-1)n(n+1)(n+2)} P_n'(-1) \qquad (9.296)\\
&= \sum_{n=2}^{\infty} \frac{2n+1}{(n-1)n(n+1)(n+2)} \frac{n(n+1)}{2} (-1)^{n+1}\\
&= \sum_{n=2}^{\infty} \frac{2n+1}{2(n-1)(n+2)} (-1)^{n+1} = -\frac{5}{12}.
\end{aligned}$$

Thus, ordinary integration shows us that

$$G'(t) = -\frac{1}{2}\ln(1-t) + \frac{1}{2}\ln(2) - \frac{5}{12} \tag{9.297}$$

such that

$$\begin{aligned} G(t) = &\frac{1}{2}\ln(1-t)(1-t) - \frac{1}{2} + \frac{1}{2}t + \frac{1}{2}\ln(2)t - \frac{5}{12}t \\ &- \frac{1}{2}\ln(2) + 1 - \frac{5}{12} + \frac{1}{6}. \end{aligned} \tag{9.298}$$

Altogether we obtain the following result.

Lemma 9.32. *The Green's function $G(\partial_0\partial_1;\cdot)$ with respect to the operator $\partial_0\partial_1$ is continuous for all $(\xi,\eta) \in \mathbb{S}^2 \times \mathbb{S}^2$. Its explicit representation reads as follows*

$$\begin{aligned} G(\partial_0\partial_1;\xi\cdot\eta) = &\frac{1}{8\pi}(1-\xi\cdot\eta)\ln(1-\xi\cdot\eta) + \left(\frac{1}{12} + \frac{\ln(2)}{2}\right)\frac{\xi\cdot\eta}{4\pi} \\ &+ \frac{1}{4\pi}\left(\frac{1}{4} - \frac{\ln(2)}{2}\right). \end{aligned} \tag{9.299}$$

For positive integers m, $G(\partial_{0,\ldots,m};\xi\cdot\eta)$ is continuous on the whole sphere $\mathbb{S}^2$ as a function of η with ξ fixed, or as a function of ξ with η fixed. On the other hand, the bilinear expansion of $G(\partial_{0,\ldots,m};\xi\cdot\eta), m \in \mathbb{N}$,

$$\sum_{n=m+1}^{\infty}\sum_{j=1}^{2n+1} \frac{1}{(\partial_{0,\ldots,m})^\wedge(n)} Y_{n,j}(\xi)Y_{n,j}(\eta) = \frac{1}{4\pi}\sum_{n=m+1}^{\infty} \frac{2n+1}{(\partial_{0,\ldots,m})^\wedge(n)} P_n(\xi\cdot\eta) \tag{9.300}$$

with

$$(\partial_{0,\ldots,m})^\wedge(n) = (\partial_0)^\wedge(n)\cdot\ldots\cdot(\partial_m)^\wedge(n) \tag{9.301}$$

is absolutely and uniformly convergent both in ξ and η, respectively, and uniformly in ξ and η together. Hence,

$$G(\partial_{0,\ldots,m};\xi\cdot\eta) = \frac{1}{4\pi}\sum_{n=m+1}^{\infty} \frac{2n+1}{4\pi}\frac{1}{(\partial_{0,\ldots,m})^\wedge(n)} P_n(\xi\cdot\eta). \tag{9.302}$$

Remark 9.33. *Let m be an integer with $m \geq 2$. Then the derivative*

$$(\partial_2\cdot\ldots\cdot\partial_m)_\eta G(\partial_{0,\ldots,m};\xi\cdot\eta) = G(\partial_0\partial_1;\xi\cdot\eta), \quad -1 \leq \xi\cdot\eta \leq 1, \tag{9.303}$$

as a function of η for fixed ξ is a continuous function on $\mathbb{S}^2$. For integers $m \geq 1$, the derivative

$$(\partial_1\cdot\ldots\cdot\partial_m)_\eta G(\partial_{0,\ldots,m};\xi\cdot\eta) = G(\partial_0;\xi\cdot\eta), \quad -1 \leq \xi\cdot\eta < 1, \tag{9.304}$$

as a function of η possesses a logarithmic singularity in $\xi \in \mathbb{S}^2$.

Observing the fact that the p-th convolution of the Green's function with respect to ∂_m coincides with the Green's function with respect to ∂_m^p, i.e.,

$$G^{(p)}(\partial_m; \xi \cdot \eta) = G(\partial_m^p; \xi \cdot \eta), \quad p \geq 1, \ m \geq 0, \tag{9.305}$$

we obtain more generally for a function F is of class $\mathrm{C}^{(2p)}(\mathbb{S}^2)$,

$$F(\xi) = \frac{2m+1}{4\pi} \int_{\mathbb{S}^2} F(\eta) P_m(\xi \cdot \eta) \, dS(\eta) + \int_{\mathbb{S}^2} G(\partial_m^p; \xi \cdot \eta)((\partial_m^p)_\eta F(\eta)) \, dS(\eta), \tag{9.306}$$

where

$$G(\partial_m^p; \xi \cdot \eta) = \frac{1}{4\pi} \sum_{(\partial_m)^\wedge(n) \neq 0} \frac{2n+1}{4\pi} \frac{1}{((\partial_m)^\wedge(n))^p} P_n(\xi \cdot \eta). \tag{9.307}$$

Hence, we are able to compare a function $F \in \mathrm{C}^{(2p)}(\mathbb{S}^2)$ with the n-th degree term of its orthogonal expansion in terms of spherical harmonics.

By use of the Green's function $G(\partial_0 \partial_1; \cdot, \cdot)$ with respect to the operator $\partial_0 \partial_1$, we are able to generalize the integral formulas and to derive error estimates for truncated spherical harmonic expansions. Observing the recursion property

$$(\partial_1)_\eta G(\partial_0 \partial_1; \xi \cdot \eta) = G(\partial_0; \xi \cdot \eta) - \frac{3}{4\pi (\partial_0)^\wedge(1)} P_1(\xi \cdot \eta) \tag{9.308}$$

we obtain

$$\begin{aligned} &\int_{\mathbb{S}^2} G(\partial_0; \xi \cdot \eta)\big((\partial_0)_\eta F(\eta)\big) \, dS(\eta) \\ &= \int_{\mathbb{S}^2} (\partial_1)_\eta G(\partial_0 \partial_1; \xi \cdot \eta)\big((\partial_0)_\eta F(\eta)\big) \, dS(\eta) \\ &\quad + \frac{3}{4\pi (\partial_0)^\wedge(1)} \int_{\mathbb{S}^2} P_1(\xi \cdot \eta)\big((\partial_0)_\eta F(\eta)\big) \, dS(\eta). \end{aligned} \tag{9.309}$$

Integration by parts, i.e., application of the Green's surface theorem (9.33), yields for a function $F \in \mathrm{C}^{(4)}(\mathbb{S}^2)$

$$\begin{aligned} &\int_{\mathbb{S}^2} (\partial_1)_\eta G(\partial_0 \partial_1; \xi \cdot \eta)\big((\partial_0)_\eta F(\eta)\big) \, dS(\eta) \\ &\quad = \int_{\mathbb{S}^2} G(\partial_0 \partial_1; \xi \cdot \eta)\big((\partial_0 \partial_1)_\eta F(\eta)\big) \, dS(\eta). \end{aligned} \tag{9.310}$$

In the same way we get

$$\frac{3}{4\pi (\partial_0)^\wedge(1)} \int_{\mathbb{S}^2} P_1(\xi \cdot \eta) \left((\partial_0)_\eta F(\eta)\right) \, dS(\eta) = \frac{3}{4\pi} \int_{\mathbb{S}^2} P_1(\xi \cdot \eta) F(\eta) \, dS(\eta). \tag{9.311}$$

Therefore, we have

$$\int_{\mathbb{S}^2} G(\partial_0;\xi\cdot\eta)\big((\partial_0)_\eta F(\eta)\big)\,dS(\eta) \tag{9.312}$$
$$= \int_{\mathbb{S}^2} G(\partial_0\partial_1;\xi\cdot\eta)\big((\partial_0\partial_1)_\eta F(\eta)\big)\,dS(\eta) + \frac{3}{4\pi}\int_{\mathbb{S}^2} F(\eta)P_1(\xi\cdot\eta)\,dS(\eta)$$

provided that F is a four times continuously differentiable function on $\mathbb{S}^2$. Thus, via (9.312), we have for all functions $F \in \mathrm{C}^{(4)}(\mathbb{S}^2)$

$$F(\xi) = \sum_{n=0}^{1} \frac{2n+1}{4\pi}\int_{\mathbb{S}^2} F(\eta)P_n(\xi\cdot\eta)\,dS(\eta) \tag{9.313}$$
$$+ \int_{\mathbb{S}^2} G(\partial_0\partial_1;\xi\cdot\eta)\big((\partial_0\partial_1)_\eta F(\eta)\big)\,dS(\eta).$$

More generally, by successive integration by parts, we obtain in connection with the definition of $G(\partial_{0,\ldots,m};\cdot,\cdot)$ the following integral formulas.

Theorem 9.34. *Let m be a non-negative integer and ξ be a fixed point of the unit sphere $\mathbb{S}^2$. Let F be a $(2m+2)$-times continuously differentiable function on $\mathbb{S}^2$. Then*

$$F(\xi) = \int_{\mathbb{S}^2} SH_{0,\ldots,m}(\xi\cdot\eta)F(\eta)\,dS(\eta)$$
$$+ \int_{\mathbb{S}^2} G(\partial_{0,\ldots,m};\xi\cdot\eta)\big((\partial_{0,\ldots,m})_\eta F(\eta)\big)\,dS(\eta). \tag{9.314}$$

Theorem 9.34 gives a comparison between the m-th partial sum of the Fourier expansion of F in spherical harmonics

$$\int_{\mathbb{S}^2} SH_{0,\ldots,m}(\xi\cdot\eta)F(\eta)\,dS(\eta) = \sum_{n=0}^{m} \underbrace{\int_{\mathbb{S}^2} F(\eta)Y_{n,j}(\eta)\,dS(\eta)}_{=F^\wedge(n,j)}\,Y_{n,j}(\xi) \tag{9.315}$$

and the functional value $F(\xi)$ of $F \in \mathrm{C}^{(2m+2)}(\mathbb{S}^2)$ with explicit knowledge of the remainder term in integral form involving Green's function

$$\xi \mapsto \int_{\mathbb{S}^2} G(\partial_{0,\ldots,m};\xi\cdot\eta)\big((\partial_{0,\ldots,m})_\eta F(\eta)\big)\,dS(\eta), \quad \xi \in \mathbb{S}^2. \tag{9.316}$$

From Theorem 9.34 we immediately obtain

$$SH_{0,\ldots,m}(\xi\cdot\eta) = \int_{\mathbb{S}^2} SH_{0,\ldots,m}(\xi\cdot\zeta)SH_{0,\ldots,m}(\xi\cdot\eta)\,dS(\zeta), \tag{9.317}$$

since

$$(\partial_{0,\ldots,m})_\zeta SH_{0,\ldots,m}(\zeta\cdot\eta) = 0 \quad \zeta,\eta \in \mathbb{S}^2. \tag{9.318}$$

It should be noted that the identity (9.314) can be rewritten as follows

$$F(\xi) - \int_{\mathbb{S}^2} SH_{0,\ldots,m}(\xi\cdot\eta)F(\eta)\, dS(\eta) \tag{9.319}$$
$$= \int_{\mathbb{S}^2} (\partial_{0,\ldots,m})_\eta G(\partial^2_{0,\ldots,m};\xi\cdot\eta)\big((\partial_{0,\ldots,m})_\eta F(\eta)\big)\, dS(\eta),$$

where $\partial^2_{0,\ldots,m} = \partial_{0,\ldots,m}\partial_{0,\ldots,m}$ and $G(\partial^2_{0,\ldots,m};\cdot,\cdot)$ is given by the convolution integral (for a graphical impression see Figure 9.10)

$$G(\partial^2_{0,\ldots,m};\xi\cdot\eta) = G^{(2)}(\partial_{0,\ldots,m};\xi\cdot\eta) \tag{9.320}$$
$$= \int_{\mathbb{S}^2} G(\partial_{0,\ldots,m};\xi\cdot\zeta)G(\partial_{0,\ldots,m},\zeta\cdot\eta)\, dS(\zeta).$$

Clearly, $G(\partial^2_{0,\ldots,m};\cdot,\cdot)$ allows the bilinear expansion

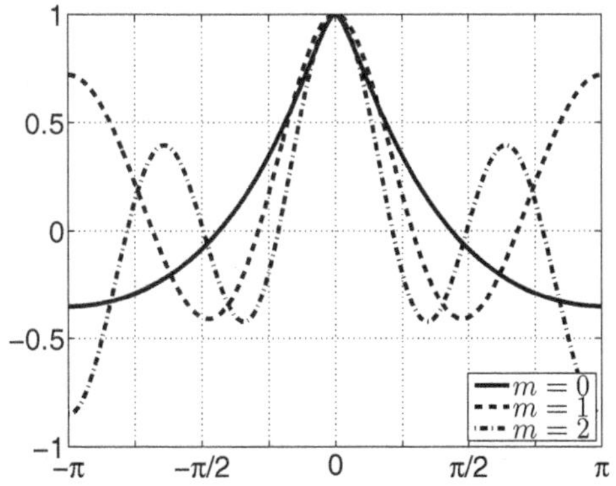

FIGURE 9.10: The functions $\vartheta \mapsto G(\partial^2_{0,\ldots,m};\cos\vartheta)$ for $m = 0,1,2$ (normalized).

$$G(\partial^2_{0,\ldots,m};\xi\cdot\eta) = \sum_{n=m+1}^{\infty} \frac{2n+1}{4\pi}\frac{1}{(\partial^2_{0,\ldots,m})^\wedge(n)} P_n(\xi\cdot\eta), \quad \xi,\eta\in\mathbb{S}^2, \tag{9.321}$$

with

$$(\partial^2_{0,\ldots,m})^\wedge(n) = ((\partial_{0,\ldots,m})^\wedge(n))^2. \tag{9.322}$$

From (9.319) we immediately obtain by application of the Cauchy–Schwarz inequality

$$\left\| F - \int_{\mathbb{S}^2} SH_{0,\ldots,m}(\cdot\eta)F(\eta)\, dS(\eta)\right\|_{\mathrm{L}^2(\mathbb{S}^2)}$$
$$\leq \left(\frac{1}{4\pi}\sum_{n=m+1}^{\infty}\frac{2n+1}{(\partial^2_{0,\ldots,m})^\wedge(n)}\right)^{1/2} \|(\partial_{0,\ldots,m})F\|_{\mathrm{L}^2(\mathbb{S}^2)}. \tag{9.323}$$

We omit $L^p(\mathbb{S}^2)$-estimates for $p \neq 2$ and $C^{(0)}(\mathbb{S}^2)$-estimates, since they are straightforward.

Throughout this work, let us denote by $\mathrm{Harm}_{0,\ldots,m}$, $m \in \mathbb{N}_0$, the space of all spherical harmonics of degree $\leq m$ so that in the sense of the inner product $\langle \cdot, \cdot \rangle_{L^2(\mathbb{S}^2)}$

$$\mathrm{Harm}_{0,\ldots,m} = \bigoplus_{l=0}^{m} \mathrm{Harm}_l \,. \tag{9.324}$$

Consequently, $\mathrm{Harm}_{0,\ldots,m}$ possesses the dimension

$$M = \dim(\mathrm{Harm}_{0,\ldots,m}) = \sum_{j=0}^{m} \dim(\mathrm{Harm}_j) = \sum_{j=0}^{m} (2j+1) = (m+1)^2. \tag{9.325}$$

The space $\mathrm{Harm}_{0,\ldots,m}$ equipped with the inner product $\langle \cdot, \cdot \rangle_{L^2(\mathbb{S}^2)}$ is an M-dimensional Hilbert space with uniquely determined reproducing kernel $K_{\mathrm{Harm}_{0,\ldots,m}}(\cdot,\cdot) : \mathbb{S}^2 \times \mathbb{S}^2 \to \mathbb{R}$ given by

$$K_{\mathrm{Harm}_{0,\ldots,m}}(\xi \cdot \eta) = SH_{0,\ldots,m}(\xi \cdot \eta) = \sum_{n=0}^{m} \frac{2n+1}{4\pi} P_n(\xi \cdot \eta). \tag{9.326}$$

Note that the recursion relation

$$(n+1)(P_{n+1}(t) - P_n(t)) - n(P_n(t) - P_{n-1}(t)) = (2n+1)(t-1)P_n(t) \tag{9.327}$$

implies the identity

$$(\xi \cdot \eta - 1) K_{\mathrm{Harm}_{0,\ldots,m}}(\xi \cdot \eta) = \frac{m+1}{4\pi} \left(P_{m+1}(\xi \cdot \eta) - P_m(\xi \cdot \eta) \right) \tag{9.328}$$

for all $(\xi, \eta) \in \mathbb{S}^2 \times \mathbb{S}^2$, such that the *Shannon kernel of degree* m is available in the explicit form

$$SH_{0,\ldots,m}(\xi \cdot \eta) = \begin{cases} \frac{(m+1)^2}{4\pi}, & \xi \cdot \eta = 1, \\ \frac{m+1}{4\pi} \frac{P_{m+1}(\xi \cdot \eta) - P_m(\xi \cdot \eta)}{\xi \cdot \eta - 1}, & \xi \cdot \eta \neq 1. \end{cases} \tag{9.329}$$

$G(\partial_0 \cdot \ldots \cdot \partial_m; \cdot, \cdot)$, $m \in \mathbb{N}$ can be defined by convolution in the form

$$G(\partial_0 \cdot \ldots \cdot \partial_m; \xi \cdot \eta) = \int_{\mathbb{S}^2} G(\partial_0 \cdot \ldots \cdot \partial_{m-1}; \xi, \zeta) \; G(\partial_m; \zeta, \eta) \; dS(\zeta). \tag{9.330}$$

For $m \in \mathbb{N}$ and all $\xi, \eta \in \mathbb{S}^2$, $G(\partial_{0,\ldots,m}; \cdot, \cdot)$ admits a uniformly convergent bilinear expansion

$$G(\partial_{0,\ldots,m}; \xi \cdot \eta) = \sum_{n=m+1}^{\infty} \frac{2n+1}{4\pi} \frac{1}{(\partial_{0,\ldots,m})^\wedge(n)} P_n(\xi \cdot \eta), \quad (\xi, \eta) \in \mathbb{S}^2. \tag{9.331}$$

with

$$(\partial_{0,\ldots,m})^\wedge(n) = (\partial_0)^\wedge(n) \cdots (\partial_m)^\wedge(n). \tag{9.332}$$

Let Y be an element of class $\mathrm{Harm}_{0,\ldots,m}$ of spherical harmonics of degree $\leq m$:

$$Y(\eta) = \sum_{n=0}^{m} \sum_{j=1}^{2n+1} Y^{\wedge}(n,j) Y_{n,j}(\eta) = \int_{\mathbb{S}^2} SH_{0,\ldots,m}(\,\cdot\eta) Y(\eta)\, dS(\eta), \quad \eta \in \mathbb{S}^2. \tag{9.333}$$

Then, observing the differential equation

$$(\partial_n)_\eta Y_{n,j}(\eta) = 0, \quad \eta \in \mathbb{S}^2, \tag{9.334}$$

for $n = 0, \ldots, m$, $j = 1, \ldots, 2n+1$ we have

$$\begin{aligned}(\partial_{0,\ldots,m})_\eta Y(\eta) &= (\partial_0 \cdot \ldots \cdot \partial_m)_\eta Y(\eta) \\ &= \sum_{n=0}^{m} \sum_{j=1}^{2n+1} Y^{\wedge}(n,j) (\partial_0 \cdot \ldots \cdot \partial_m)_\eta Y_{n,j}(\eta) = 0\end{aligned} \tag{9.335}$$

for all $\eta \in \mathbb{S}^2$. On the other hand, we know from Theorem 9.34 that any solution $Y \in \mathrm{C}^{(2m+2)}(\mathbb{S}^2)$ of the homogeneous differential equation

$$(\partial_{0,\ldots,m})_\eta Y(\eta) = 0, \quad \eta \in \mathbb{S}^2, \tag{9.336}$$

is representable in the following form for $\eta \in \mathbb{S}^2$:

$$Y(\eta) = \sum_{n=0}^{m} \sum_{j=1}^{2n+1} Y^{\wedge}(n,j) Y_{n,j}(\eta) = \int_{\mathbb{S}^2} SH_{0,\ldots,m}(\,\cdot\eta) Y(\eta)\, dS(\eta). \tag{9.337}$$

This means that $\mathrm{Harm}_{0,\ldots,m}$ is the null space of the operator $\partial_{0,\ldots,m}$.

For given $H \in \mathrm{C}^{(0)}(\mathbb{S}^2)$, the integral formula (Theorem 9.34) can be used to discuss the general differential equation

$$(\partial_{0,\ldots,m})_\eta F(\eta) = H(\eta), \quad \eta \in \mathbb{S}^2. \tag{9.338}$$

By virtue of the Green's surface identity, we first see that

$$\int_{\mathbb{S}^2} ((\partial_{0,\ldots,m})_\eta F(\eta)) Y(\eta)\, dS(\eta) = \int_{\mathbb{S}^2} ((\partial_{0,\ldots,m})_\eta Y(\eta)) F(\eta)\, dS(\eta) = 0 \tag{9.339}$$

for all elements $Y \in \mathrm{Harm}_{0,\ldots,m}$. From the considerations given above, it is clear that any function $Y \in \mathrm{Harm}_{0,\ldots,m}$ can be added to F without changing the differential equation (9.338). However, if we require that F is orthogonal to the null space $\mathrm{Harm}_{0,\ldots,m}$ of the operator $\partial_{0,\ldots,m}$, then the differential equation is uniquely solvable. This finally leads us to the following result.

Theorem 9.35. *Let H be a continuous function on $\mathbb{S}^2$ orthogonal to $\mathrm{Harm}_{0,\ldots,m}$, i.e.,*

$$\int_{\mathbb{S}^2} H(\eta) Y_{n,j}(\eta)\, dS = 0 \tag{9.340}$$

for $n = 0, \ldots, m$, $j = 1, \ldots, 2n+1$. *Then the function* F *given by*

$$F(\xi) = \int_{\mathbb{S}^2} G(\partial_{0,\ldots,m}; \xi \cdot \eta) H(\eta) \, dS(\eta), \quad \xi \in \mathbb{S}^2, \tag{9.341}$$

represents the only $(2m+2)$*-times continuously differentiable solution of the differential equation (9.338), i.e.,* $(\partial_{0,\ldots,m})_\eta F(\eta) = H(\eta)$, $\eta \in \mathbb{S}^2$ *which is orthogonal to* $\mathrm{Harm}_{0,\ldots,m}$.

Within the distributional Dirac framework Theorem 9.35 can be formally interpreted in the following way: Starting from

$$(\partial_{0,\ldots,m})_\eta G(\partial_{0,\ldots,m}; \xi \cdot \eta) = \delta_{m+1,\ldots,\infty}(\xi \cdot \eta) = SH_{m+1,\ldots,\infty}(\xi \cdot \eta) \tag{9.342}$$

we obtain by use of Green's theorem and the orthogonality of spherical harmonics

$$\begin{aligned} F(\xi) = F_{m+1,\ldots,\infty}(\xi) &= \int_{\mathbb{S}^2} SH_{m+1,\ldots,\infty}(\xi \cdot \eta) F(\eta) \, dS(\eta) \\ &= \int_{\mathbb{S}^2} (\partial_{0,\ldots,m})_\eta G(\partial_{0,\ldots,m}; \xi \cdot \eta) F_{m+1,\ldots,\infty}(\eta) \, dS(\eta) \\ &= \int_{\mathbb{S}^2} G(\partial_{0,\ldots,m}; \xi \cdot \eta) (\partial_{0,\ldots,m})_\eta F_{m+1,\ldots,\infty}(\eta) \, dS(\eta) \\ &= \int_{\mathbb{S}^2} G(\partial_{0,\ldots,m}; \xi \cdot \eta) H_{m+1,\ldots,\infty}(\eta) \, dS(\eta) \\ &= \int_{\mathbb{S}^2} G(\partial_{0,\ldots,m}; \xi \cdot \eta) H(\eta) \, dS(\eta). \end{aligned} \tag{9.343}$$

9.6 Peano's Theorem Involving Beltrami Operators

The integral formula (Theorem 9.34) enables us to formulate a spherical analog (cf. W. Freeden [1981b]) of the well-known Peano theorem. Later on, an extension will be given for rotation-invariant pseudodifferential operators.

Let us consider linear functionals on $\mathrm{C}^{(2m+2)}(\mathbb{S}^2)$ of the following structure:

$$\mathcal{I}F = \int_{\mathbb{S}^2} G(\eta) F(\eta) \, dS(\eta) + \sum_{k=1}^{L} b_k F(\xi_k), \tag{9.344}$$

where the function $G : \mathbb{S}^2 \to \mathbb{R}$ is assumed to be piecewise continuous on $\mathbb{S}^2$, the weights $b_1, \ldots, b_L$ are real numbers, and the points $\xi_1, \ldots, \xi_L$ lie on $\mathbb{S}^2$. Though (9.344) is not the most general class of functionals that might be considered (on the Sobolev space $\mathrm{H}^{(2m+2)}(\mathbb{S}^2) = \{F \in \mathrm{L}^2(\mathbb{S}^2) : \partial_{0,\ldots,m} F \in \mathrm{L}^2(\mathbb{S}^2)\}$), it is adequate for purposes of numerical integration. Obviously included as special cases are

(i) the integral over $\mathbb{S}^2$

$$\int_{\mathbb{S}^2} F(\eta)\, dS(\eta) \tag{9.345}$$

or any subdomain

$$\int_{\Gamma} F(\eta)\, dS(\eta), \tag{9.346}$$

(ii) the Fourier coefficients of a (continuous) function F on $\mathbb{S}^2$

$$F^{\wedge}(n,j) = \int_{\mathbb{S}^2} F(\eta) Y_{n,j}(\eta)\, dS(\eta), \tag{9.347}$$

(iii) any functional value of F at $\xi \in \mathbb{S}^2$

$$F(\xi), \tag{9.348}$$

(iv) any finite linear combination of functional values of F at prescribed points $\xi_1, \ldots, \xi_L$ on $\mathbb{S}^2$

$$\sum_{k=1}^{L} b_k F(\xi_k). \tag{9.349}$$

Theorem 9.36 (Peano's Theorem). *Let $\mathcal{I} : \mathrm{H} \to \mathbb{R}$ be a linear functional on $\mathrm{C}^{(2m+2)}(\mathbb{S}^2)$ of type (9.344) such that $\mathcal{I}(Y) = 0$ whenever $Y \in \mathrm{Harm}_{0,\ldots,m}$. Then, for $F \in \mathrm{H}$,*

$$\mathcal{I}F = \int_{\mathbb{S}^2} (\partial_{0,\ldots,m} K(\eta))(\partial_{0,\ldots,m} F(\eta))\, dS(\eta), \tag{9.350}$$

where

$$K(\eta) = \mathcal{I}_\xi G(\partial^2_{0,\ldots,m}; \xi \cdot \eta) \tag{9.351}$$

$\eta \in \mathbb{S}^2$. The function $K : \mathbb{S}^2 \to \mathbb{R}$ is called the Peano kernel for the linear functional $\mathcal{I}$. (As usual $\mathcal{I}_\xi$ means that the linear functional is applied to the ξ-variable.)

Proof. We apply $\mathcal{I}$ to both sides of the integral formula and use the fact that $\mathcal{I}$ vanishes on elements of class $\mathrm{Harm}_{0,\ldots,m}$:

$$\mathcal{I}F = \mathcal{I}_\xi \int_{\mathbb{S}^2} (\partial_{0,\ldots,m} G(\partial^2_{0,\ldots,m}; \xi \cdot \eta)(\partial_{0,\ldots,m} F(\eta))\, dS(\eta). \tag{9.352}$$

Now, the linear functional allows the interchange of $\mathcal{I}$ and the integral. This yields the desired result. □

Consider now an approximation of a linear functional $\mathcal{I}$ of the type (9.344). By a linear functional $\mathcal{L}$ (cf. A. Sard [1949]) of the form

$$\mathcal{L}F = \sum_{k=1}^{N} a_k F(\eta_k), \tag{9.353}$$

where $a_1, \ldots, a_N$ are real numbers, and $X_N = \{\eta_1, \ldots, \eta_N\}$ is a unisolvent system relative to $\mathrm{Harm}_{0,\ldots,m}$, i.e., a point set which contains $M = (m+1)^2$ points $Z_M = \{\eta_1, \ldots, \eta_M\} \subset X_N$ such that the $M \times M$-matrix

$$\mathbf{A}_{Z_M}(Y_{0,1}, \ldots, Y_{m,2m+1}) = \begin{pmatrix} Y_{0,1}(\eta_1) & \ldots & Y_{0,1}(\eta_M) \\ \vdots & & \vdots \\ Y_{m,2m+1}(\eta_1) & \ldots & Y_{m,2m+1}(\eta_M) \end{pmatrix} \tag{9.354}$$

is regular. For more details on unisolvence the reader is referred to Section 10.1.

$\mathcal{L}$ is called *exact for the degree* m, if $\mathcal{I}F = \mathcal{L}F$ whenever $F \in \mathrm{Harm}_{0,\ldots,m}$. The remainder, when $\mathcal{L}$ is used to approximate $\mathcal{I}$, is a linear functional $\mathcal{E}$ defined by $\mathcal{E} = \mathcal{I} - \mathcal{L}$. If the approximation of $\mathcal{I}$ by $\mathcal{L}$ is exact for the degree m and F is a function of class H, then $\mathcal{E}$ satisfies the assumptions of Peano's theorem. Hence, it follows that

$$\mathcal{E}F = \int_{\mathbb{S}^2} (\partial_{0,\ldots,m} K(\eta))(\partial_{0,\ldots,m} F(\eta))\, dS(\eta), \tag{9.355}$$

where

$$K(\eta) = \mathcal{E}_\xi G(\partial^2_{0,\ldots,m}; \xi \cdot \eta), \quad \eta \in \mathbb{S}^2. \tag{9.356}$$

Applying the Cauchy–Schwarz inequality to (9.355) we obtain

$$|\mathcal{E}F| \leq \left(\int_{\mathbb{S}^2} |\partial_{0,\ldots,m} K(\eta)|^2\, dS(\eta) \right)^{1/2} \left(\int_{\mathbb{S}^2} |\partial_{0,\ldots,m} F(\eta)|^2\, dS(\eta) \right). \tag{9.357}$$

We note that the quantity J given by

$$J = \int_{\mathbb{S}^2} |\partial_{0,\ldots,m} K(\eta)|^2\, dS(\eta) \tag{9.358}$$

depends on the knots $\eta_1, \ldots, \eta_N$ and the operator $\mathcal{I}$, but not on the function $F \in$ H. Inserting (9.356) into (9.358) we have

$$J = \mathcal{E}_\xi \mathcal{E}_\zeta \int_{\mathbb{S}^2} \partial_{0,\ldots,m} G(\partial^2_{0,\ldots,m}; \zeta \cdot \eta) \partial_{0,\ldots,m} G(\partial^2_{0,\ldots,m}; \xi \cdot \eta)\, dS(\eta), \tag{9.359}$$

i.e.,

$$J = \mathcal{E}_\xi \mathcal{E}_\zeta G(\partial^2_{0,\ldots,m}; \xi \cdot \zeta). \tag{9.360}$$

Collecting our results we therefore obtain the following *a priori* estimate.

Theorem 9.37. *Let $\mathcal{I}$ be a linear functional of type (9.344), and let $\mathcal{L}$ be an approximation to $\mathcal{I}$ of the form (9.353), exact for the degree m. Then, for each $F \in C^{(2m+2)}(\mathbb{S}^2)$,*

$$|\mathcal{E}(F)| \tag{9.361}$$

$$\leq \left((\mathcal{I}_\xi \mathcal{I}_\zeta - 2\mathcal{I}_\xi \mathcal{L}_\zeta + \mathcal{L}_\xi \mathcal{L}_\zeta) G(\partial^2_{0,\ldots,m}; \xi \cdot \zeta) \right)^{1/2} \left(\int_{\mathbb{S}^2} |\partial_{0,\ldots,m} F(\eta)|^2\, dS(\eta) \right)^{1/2}.$$

Chapter 10

Integration Rules with Polynomial Accuracy

Polynomial integration on the sphere is an intensively studied research area in mathematics (see, e.g., F. Filbir, W. Themistaclakis [2008], G.B. Folland [2001], W. Freeden et al. [1998], K. Hesse et al. [2010], M. Reimer [2003] and the references therein for points of departure). In our work we start with spherical counterparts to Lagrangian polynomial integration in terms of spherical harmonics. The difficulty in contrast to the one-dimensional theory is that pointwise interpolation cannot always be carried out automatically for any prescribed nodal system of pairwise different points because of Haar's theorem (see, e.g., P.J. Davis [1963]). As a consequence, the concept of unisolvence for point sets has to be taken int account (cf. W. Freeden et al. [1998]). The chapter ends with an overview on extremal point systems and designs. It culminates in the statement that Gaussian integration cannot be realized adequately on the unit sphere $\mathbb{S}^2$.

10.1 Lagrangian Integration

A system of N linearly independent functions $F_1, \ldots, F_N$ defined on a point set $X_N = \{\eta_1, \ldots, \eta_N\}$ is called a *fundamental system* on X_N if for the matrix $\mathbf{F}$ with $\mathbf{F}_{ij} = F_i(\eta_j)$, $i, j = 1, \ldots, N$, holds that

$$\det(\mathbf{F}) \neq 0. \tag{10.1}$$

In the one-dimensional theory, pointwise interpolation can always be carried out uniquely, i.e., the condition (10.1) is automatically satisfied because of the separation property of a point on the real line into right and left parts. In higher dimensions and on the sphere, the situation is vastly different, and the concept of unisolvence comes into play: Let us consider the $M \times N$-matrix (with $M = (m+1)^2 = \dim(\mathrm{Harm}_{0,\ldots,m})$)

$$\mathbf{A}_{X_N}(Y_{0,1}, \ldots, Y_{m,2m+1}) = \begin{pmatrix} Y_{0,1}(\eta_1) & \ldots & Y_{0,1}(\eta_N) \\ \vdots & & \vdots \\ Y_{m,2m+1}(\eta_1) & \ldots & Y_{m,2m+1}(\eta_N) \end{pmatrix} \tag{10.2}$$

corresponding to an $\mathrm{L}^2(\mathbb{S}^2)$-orthonormal system $\{Y_{n,j}\}$ of spherical harmonics. According to Haar's theorem (see, e.g., P.J. Davis [1963]), this matrix is not of maximal rank for *all* systems $X_N = \{\eta_1, \ldots, \eta_N\}$ of distinct points $\eta_1, \ldots, \eta_N \in \mathbb{S}^2$, $N \geq M$. However, we are able to conclude (see W. Freeden, M. Schreiner [2009], W. Freeden, B. Witte [1982], C. Müller [1966]) that there exists a large class of systems X_M having a non-degenerate matrix $\mathbf{A}_{X_M}(Y_{0,1}, \ldots, Y_{m,2m+1})$.

Definition 10.1. *Let X_N be a set of N distinct points $\eta_1, \ldots, \eta_N$, $N \geq M$, i.e., $X_N = \{\eta_1, \ldots, \eta_N\} \subset \mathbb{S}^2$ with $N = \#X_N$. For $m \in \mathbb{N}_0$, we let $\mathbf{A}_{X_N}(Y_{0,1}, \ldots, Y_{m,2m+1})$ be defined by (10.2). A system X_N of N distinct points $\eta_1, \ldots, \eta_N$, $N \geq M$, $m \geq 0$, is called* unisolvent *relative to the space* $\mathrm{Harm}_{0,\ldots,m}$, *if X_N contains a subset Z_M, $M = \#Z_M$, for which $\mathbf{A}_{Z_M}(Y_{0,1}, \ldots, Y_{m,2m+1})$ is a regular matrix.*

Remark 10.2. *Throughout this work, in case of a unisolvent system $X_N = \{\eta_1, \ldots, \eta_N\}$ relative to* $\mathrm{Harm}_{0,\ldots,m}$, *we assume that $X_M \subset X_N$ is a fundamental system relative to* $\mathrm{Harm}_{0,\ldots,m}$. *Of course, this property is always achievable by reordering a unisolvent system.*

It should be noted that, via the addition theorem, we have

$$\begin{aligned} &(\mathbf{A}_{X_M}(Y_{0,1}, \ldots, Y_{m,2m+1}))^T \mathbf{A}_{X_M}(Y_{0,1}, \ldots, Y_{m,2m+1}) \\ &= \mathbf{A}_{X_M}\left(\sum_{n=0}^{m} \frac{2n+1}{4\pi} P_n(\eta_1 \cdot\ ), \ldots, \sum_{n=0}^{m} \frac{2n+1}{4\pi} P_n(\eta_M \cdot\ )\right) \\ &= \mathbf{A}_{X_M}\left(SH_{0,\ldots,m}(\eta_1 \cdot\ ), \ldots, SH_{0,\ldots,m}(\eta_M \cdot\ )\right), \end{aligned} \tag{10.3}$$

where $SH_{0,\ldots,m}$ is the *Shannon kernel* (9.279). Consequently, the property of X_M being a fundamental system relative to the set $\mathrm{Harm}_{0,\ldots,m}$ is independent of the choice of the orthonormal basis $\{Y_{0,1}, \ldots, Y_{m,2m+1}\}$.

Let $X_N = \{\eta_1, \ldots, \eta_N\}$ be a unisolvent system relative to $\mathrm{Harm}_{0,\ldots,m}$. Assume that $F : \mathbb{S}^2 \to \mathbb{R}$ is a function of class $\mathrm{C}^{(2m+2)}(\mathbb{S}^2)$, $m \in \mathbb{N}_0$. Let Y be an element of $\mathrm{Harm}_{0,\ldots,m}$ of the decomposition

$$Y = \sum_{n=0}^{m} \sum_{j=1}^{2n+1} c_{n,j} Y_{n,j}, \quad c_{n,j} = Y^{\wedge}(n,j). \tag{10.4}$$

Then, for all solutions $a \in \mathbb{R}^N, a = (a_1, \ldots, a_N)^T$, of the linear system

$$\mathbf{A}_{X_N}(Y_{0,1,\ldots,}Y_{m,2m+1})a = c, \tag{10.5}$$

i.e., for all $a \in \mathbb{R}^N$ satisfying

$$\sum_{k=1}^{N} a_k Y_{n,j}(\eta_k) = c_{n,j}, \quad n = 0, \ldots, m, \ j = 1, \ldots, 2n+1, \tag{10.6}$$

we have

$$Y = \sum_{k=1}^{N} a_k \sum_{n=0}^{m} \sum_{j=1}^{2n+1} Y_{n,j}(\eta_k) Y_{n,j} = \sum_{k=1}^{N} a_k SH_{0,\ldots,m}(\,\cdot \eta_k). \tag{10.7}$$

Observing this fact and the reproducing property of the Shannon kernel in $\mathrm{Harm}_{0,\ldots,m}$ we obtain the following corollaries.

Corollary 10.3. *If* $F \in \mathrm{Harm}_{0,\ldots,m}$ *and* $Y_{n,j} \in \mathrm{Harm}_{0,\ldots,m}$, *then*

$$F^{\wedge}(n,j) = \int_{\mathbb{S}^2} Y_{n,j}(\eta) F(\eta)\, dS(\eta) = \sum_{k=1}^{N} a_k^{n,j} F(\eta_k) \tag{10.8}$$

holds for all $a_1^{n,j}, \ldots, a_N^{n,j} \in \mathbb{R}$ *with*

$$\sum_{k=1}^{N} a_k^{n,j} Y_{p,q}(\eta_k) = \delta_{p,n}\delta_{q,j}, \quad p = 0, \ldots, m, \; q = 1, \ldots, 2p+1. \tag{10.9}$$

Thus, already at this stage, we are led to a spherical counterpart to *Shannon's sampling rule in* $\mathrm{Harm}_{0,\ldots,m}$ based on the linear equations (10.9).

Corollary 10.4 (Shannon Sampling Theorem in $\mathrm{Harm}_{0,\ldots,m}$). *Let* F *be in* $\mathrm{Harm}_{0,\ldots,m}$ *and* $X_N = \{\eta_1, \ldots, \eta_N\}$ *be a unisolvent system relative to* $\mathrm{Harm}_{0,\ldots,m}$. *Then* F *can be reconstructed from its samples at points out of* X_N *by the following interpolation formula*

$$F(\xi) = \sum_{k=1}^{N} F(\eta_k) L_k(\xi), \quad \xi \in \mathbb{S}^2, \tag{10.10}$$

where $L_k \in \mathrm{Harm}_{0,\ldots,m}$ *is given by*

$$L_k(\xi) = \sum_{n=0}^{m} \sum_{j=1}^{2n+1} a_k^{n,j} Y_{n,j}(\xi) \tag{10.11}$$

and the coefficients $a_k^{n,j}$ *are determined by (10.9).*

Proof. From Corollary 10.3 we are able to deduce that

$$\begin{aligned} \sum_{k=1}^{N} F(\eta_k) L_k(\xi) &= \sum_{n=0}^{m} \sum_{j=1}^{2n+1} \left(\sum_{k=1}^{N} a_k^{n,j} F(\eta_k) \right) Y_{n,j}(\xi) \qquad (10.12) \\ &= \sum_{n=0}^{m} \sum_{j=1}^{2n+1} F^{\wedge}(n,j) Y_{n,j}(\xi) = F(\xi), \end{aligned}$$

since $F \in \mathrm{Harm}_{0,\ldots,m}$. This guarantees Corollary 10.7. □

From Corollary 10.4 in connection with (10.9) we are immediately able to derive $\mathbb{S}^2$-integrated versions of (10.10).

Lemma 10.5. *Let F be of class* $\mathrm{Harm}_{0,\ldots,m}$. *Suppose that $X_N = \{\eta_1, \ldots, \eta_N\}$ is a unisolvent system relative to* $\mathrm{Harm}_{0,\ldots,m}$. *Then*

$$\frac{1}{\sqrt{4\pi}} \int_{\mathbb{S}^2} F(\xi)\, dS(\xi) = \sum_{k=1}^{N} a_k^{0,1} F(\eta_k). \tag{10.13}$$

Proof. From (10.10) we get

$$\int_{\mathbb{S}^2} F(\xi)\, dS(\xi) = \sum_{k=1}^{N} F(\eta_k) \int_{\mathbb{S}^2} L_k(\xi)\, dS(\xi). \tag{10.14}$$

The integrals on the right side of (10.14) can be calculated explicitly by (10.11)

$$\begin{aligned} \frac{1}{\sqrt{4\pi}} \int_{\mathbb{S}^2} L_k(\xi)\, dS(\xi) &= \int_{\mathbb{S}^2} L_k(\xi) Y_{0,1}(\xi)\, dS(\xi) \qquad (10.15) \\ &= \int_{\mathbb{S}^2} \sum_{n=0}^{m} \sum_{j=1}^{2n+1} a_k^{n,j} Y_{n,j}(\xi) Y_{0,1}(\xi)\, dS(\xi) = a_k^{0,1}. \end{aligned}$$

□

Lemma 10.6. *Let $F \in$* $\mathrm{Harm}_{0,\ldots,m}$. *Suppose that $X_N = \{\eta_1, \ldots, \eta_N\}$ is a unisolvent system relative to* $\mathrm{Harm}_{0,\ldots,m}$. *Then*

$$\sum_{k=1}^{N} \sum_{l=1}^{N} F(\eta_k) F(\eta_l) \int_{\mathbb{S}^2} L_k(\xi) L_l(\xi)\, dS(\xi) = \sum_{n=0}^{m} \sum_{j=1}^{2n+1} \left(\sum_{k=1}^{N} a_k^{n,j} F(\eta_k) \right)^2. \tag{10.16}$$

Proof. It is clear that

$$\int_{\mathbb{S}^2} |F(\xi)|^2\, dS(\xi) = \sum_{k=1}^{N} \sum_{l=1}^{N} F(\eta_k) F(\eta_l) \int_{\mathbb{S}^2} L_k(\xi) L_l(\xi)\, dS(\xi). \tag{10.17}$$

From Corollary 10.3 we get

$$\sum_{n=0}^{m} \sum_{j=1}^{2n+1} (F^\wedge(n,j))^2 = \sum_{n=0}^{m} \sum_{j=1}^{2n+1} \left(\sum_{k=1}^{N} a_k^{n,j} F(\eta_k) \right)^2. \tag{10.18}$$

The Parseval identity in $\mathrm{Harm}_{0,\ldots,m}$ yields the desired result. □

Later we discuss the conditions, under which the Lagrangians are orthonormal in $\mathrm{L}^2(\mathbb{S}^2)$-sense.

Corollary 10.7. *The summation formula*

$$\sum_{k=1}^{N} a_k \int_{\mathbb{S}^2} \partial_{0,\ldots,m} G(\partial^2_{0,\ldots,m};\eta\cdot\eta_k)(\partial_{0,\ldots,m} F(\eta))\, dS(\eta) = \sum_{k=1}^{N} a_k F(\eta_k) \quad (10.19)$$

holds for every selection of vectors $a = (a_1, \ldots, a_N)^T \in \mathbb{R}^N$ *satisfying*

$$\mathbf{A}_{X_N}(Y_{0,1,\ldots,}Y_{m,2m+1})a = 0 \quad (10.20)$$

and every function $F \in \mathrm{C}^{(2m+2)}(\mathbb{S}^2)$.

Corollary 10.8. *The formula*

$$\int_{\mathbb{S}^2} (\partial_{0,\ldots,m} S(\eta))\,(\partial_{0,\ldots,m} F(\eta))\; dS(\eta) = \sum_{k=1}^{N} a_k F(\eta_k) \quad (10.21)$$

holds for all $S : \mathbb{S}^2 \to \mathbb{R}$ *of the form*

$$S(\eta) = Y(\eta) + \sum_{k=1}^{N} a_k G(\partial^2_{0,\ldots,m};\xi\cdot\eta), \quad (10.22)$$

$\eta \in \mathbb{S}^2$, $Y \in \mathrm{Harm}_{0,\ldots,m}$, *satisfying the linear system*

$$\sum_{k=1}^{N} a_k Y_{n,j}(\eta_k) = 0, \quad n = 0, \ldots, m, \ j = 1, \ldots, 2n+1. \quad (10.23)$$

In particular, for $N = M$, *we have* $a_1 = \ldots = a_m = 0$, *so that*

$$S(\eta) = Y(\eta), \quad \eta \in \mathbb{S}^2, \quad Y \in \mathrm{Harm}_{0,\ldots,m}. \quad (10.24)$$

In Chapter 13 the functions S occurring in Corollary 10.8 will be recognized as spherical spline functions of order m relative to X_N, and the identity (10.21) is the so-called spline integration formula which enables us to justify minimum norm properties in spherical spline interpolation (see W. Freeden [1981b] for more details).

Theorem 10.9. *For a given function* $W \in \mathrm{C}^{(0)}(\mathbb{S})$ *orthogonal to* $\mathrm{Harm}_{0,\ldots,m}$, *let* V *be the unique function of the class* $\mathrm{C}^{(2m+2)}(\mathbb{S}^2)$ *orthogonal to* $\mathrm{Harm}_{0,\ldots,m}$ *that satisfies* $\partial_{0,\ldots,m} V = W$. *For a given unisolvent system* $X_N = \{\eta_1, \ldots, \eta_N\}$ *and weights* $a_1, \ldots, a_N \in \mathbb{R}$ *with*

$$\mathbf{A}_{X_N}(Y_{0,1,\ldots,}Y_{m,2m+1})a = c, \quad (10.25)$$

let $K : \mathbb{S}^2 \to \mathbb{R}$ *designate a function of the form*

$$K(\eta) = S(\eta) - V(\eta), \quad \eta \in \mathbb{S}^2, \quad (10.26)$$

where

$$S(\eta) = \sum_{n=0}^{m} \sum_{j=1}^{2n+1} c_{n,j} Y_{n,j}(\eta) + \sum_{k=1}^{N} a_k G(\partial_{0,\ldots,m}; \eta_k \cdot \eta) \tag{10.27}$$
$$= Y(\eta) + \sum_{k=1}^{N} a_k G(\partial_{0,\ldots,m}; \eta_k \cdot \eta), \quad \eta \in \mathbb{S}^2.$$

Then, for every $F \in \mathrm{C}^{(2m+2)}(\mathbb{S}^2)$, the following identity holds true:

$$\sum_{k=1}^{N} a_k F(\eta_k) = \int_{\mathbb{S}^2} (W(\eta) + Y(\eta)) F(\eta) \, dS(\eta) + \int_{\mathbb{S}^2} K(\eta) \partial_{0,\ldots,m} F(\eta) \, dS(\eta). \tag{10.28}$$

In analogy to one-dimensional spline theory, a function K of the form (10.26) is called a *monospline* with respect to X_N and the operator $\partial_{0,\ldots,m}$. This notation will become more transparent later in this work.

The truncation error

$$\mathcal{E}^W F = \int_{\mathbb{S}^2} K(\eta) \partial_{0,\ldots,m} F(\eta) \, dS(\eta). \tag{10.29}$$

annihilates all functions F with $\partial_{0,\ldots,m} F = 0$. By specializing we obtain a variety of remainder estimates. We only mention an estimate in $\mathrm{L}^2(\mathbb{S}^2)$-topology:

$$|\mathcal{E}^W F| \leq \left(\int_{\mathbb{S}^2} |K(\eta)|^2 \, dS(\eta) \right)^{1/2} \left(\int_{\mathbb{S}^2} |\partial_{0,\ldots,m} F(\eta)|^2 \, dS(\eta) \right)^{1/2}. \tag{10.30}$$

The estimate (10.30) can be exploited to establish best integration formulas in the sense of Sard (for more details the reader is referred to W. Freeden et al. [1998]).

Now we study the *properties of the Lagrangians in* $\mathrm{Harm}_{0,\ldots,m}$ in the framework of the theory of determinants (cf. W. Freeden et al. [1998]):

Let $X_M = \{\eta_1, \ldots, \eta_M\}$, $M = (m+1)^2$, be a fundamental system relative to $\mathrm{Harm}_{0,\ldots,m}$. Then, the matrix $\mathbf{H} = \mathbf{A}_{X_M}(Y_{0,1}, \ldots, Y_{m,2m+1})$ is regular. The addition theorem of spherical harmonics yields

$$\mathbf{H}^T \mathbf{H} = \mathbf{K} = \mathbf{A}_{X_M}(SH_{0,\ldots,m}(\cdot \eta_1), \ldots SH_{0,\ldots,m}(\cdot \eta_M)). \tag{10.31}$$

Thus, $\mathbf{K}$ is a positive definite, symmetric matrix. Hence there exists an orthogonal matrix $\mathbf{Q}$ such that

$$\mathbf{Q}\mathbf{K}\mathbf{Q}^T = \mathbf{D} = \mathrm{diag}(\delta_1, \ldots, \delta_M) = \begin{pmatrix} \delta_1 & & 0 \\ & \ddots & \\ 0 & & \delta_M \end{pmatrix}, \tag{10.32}$$

where

$$\delta_{\min} = \delta_1 \leq \delta_2 \leq \ldots \leq \delta_M = \delta_{\max}. \tag{10.33}$$

For a fundamental system X_M relative to $\mathrm{Harm}_{0,\ldots,m}$, the *Lagrangians* $L_1, \ldots, L_m \in \mathrm{Harm}_{0,\ldots,m}$ defined by

$$L_1(\xi) = \frac{\begin{vmatrix} Y_{0,1}(\xi) & \ldots & Y_{m,1}(\xi) & \ldots & Y_{m,2m+1}(\xi) \\ Y_{0,1}(\eta_2) & \ldots & Y_{m,1}(\eta_2) & \ldots & Y_{m,2m+1}(\eta_2) \\ \vdots & & \vdots & & \vdots \\ Y_{0,1}(\eta_M) & \ldots & Y_{m,1}(\eta_M) & \ldots & Y_{m,2m+1}(\eta_M) \end{vmatrix}}{\begin{vmatrix} Y_{0,1}(\eta_1) & \ldots & Y_{m,1}(\eta_1) & \ldots & Y_{m,2m+1}(\eta_1) \\ Y_{0,1}(\eta_2) & \ldots & Y_{m,1}(\eta_2) & \ldots & Y_{m,2m+1}(\eta_2) \\ \vdots & & \vdots & & \vdots \\ Y_{0,1}(\eta_M) & \ldots & Y_{m,1}(\eta_M) & \ldots & Y_{m,2m+1}(\eta_M) \end{vmatrix}}, \tag{10.34}$$

$$L_2(\xi) = \frac{\begin{vmatrix} Y_{0,1}(\eta_1) & \ldots & Y_{m,1}(\eta_1) & \ldots & Y_{m,2m+1}(\eta_1) \\ Y_{0,1}(\xi) & \ldots & Y_{m,1}(\xi) & \ldots & Y_{m,2m+1}(\xi) \\ \vdots & & \vdots & & \vdots \\ Y_{0,1}(\eta_M) & \ldots & Y_{m,1}(\eta_M) & \ldots & Y_{m,2m+1}(\eta_M) \end{vmatrix}}{\begin{vmatrix} Y_{0,1}(\eta_1) & \ldots & Y_{m,1}(\eta_1) & \ldots & Y_{m,2m+1}(\eta_1) \\ Y_{0,1}(\eta_2) & \ldots & Y_{m,1}(\eta_2) & \ldots & Y_{m,2m+1}(\eta_2) \\ \vdots & & \vdots & & \vdots \\ Y_{0,1}(\eta_M) & \ldots & Y_{m,1}(\eta_M) & \ldots & Y_{m,2m+1}(\eta_M) \end{vmatrix}}, \tag{10.35}$$

etc. satisfy the property $L_j(\eta_k) = \delta_{j,k}$, $j, k = 1, \ldots, M$. Moreover, we have

$$|L_1(\xi)|^2 = \tag{10.36}$$

$$\frac{\begin{vmatrix} SH_{0,\ldots,m}(\xi \cdot \xi) & SH_{0,\ldots,m}(\xi \cdot \eta_2) & \ldots & SH_{0,\ldots,m}(\xi \cdot \eta_M) \\ \vdots & \vdots & & \vdots \\ SH_{0,\ldots,m}(\eta_M \cdot \xi) & SH_{0,\ldots,m}(\eta_M \cdot \eta_2) & \ldots & SH_{0,\ldots,m}(\eta_M \cdot \eta_M) \end{vmatrix}}{\begin{vmatrix} SH_{0,\ldots,m}(\eta_1, \cdot \eta_1) & SH_{0,\ldots,m}(\eta_1 \cdot \eta_2) & \ldots & SH_{0,\ldots,m}(\eta_1 \cdot \eta_M) \\ \vdots & \vdots & & \vdots \\ SH_{0,\ldots,m}(\eta_M \cdot \eta_1) & SH_{0,\ldots,m}(\eta_M \cdot \eta_2) & \ldots & SH_{0,\ldots,m}(\eta_M \cdot \eta_M) \end{vmatrix}},$$

$$|L_2(\xi)|^2 = \tag{10.37}$$

$$\frac{\begin{vmatrix} SH_{0,\ldots,m}(\eta_1 \cdot \eta_1) & SH_{0,\ldots,m}(\eta_1 \cdot \xi) & \ldots & SH_{0,\ldots,m}(\eta_1 \cdot \eta_M) \\ \vdots & \vdots & & \vdots \\ SH_{0,\ldots,m}(\eta_M \cdot \eta_1) & SH_{0,\ldots,m}(\eta_M \cdot \xi) & \ldots & SH_{0,\ldots,m}(\eta_M \cdot \eta_M) \end{vmatrix}}{\begin{vmatrix} SH_{0,\ldots,m}(\eta_1 \cdot \eta_1) & SH_{0,\ldots,m}(\eta_1 \cdot \eta_2) & \ldots & SH_{0,\ldots,m}(\eta_1 \cdot \eta_M) \\ \vdots & \vdots & & \vdots \\ SH_{0,\ldots,m}(\eta_M \cdot \eta_1) & SH_{0,\ldots,m}(\eta_M \cdot \eta_2) & \ldots & SH_{0,\ldots,m}(\eta_M \cdot \eta_M) \end{vmatrix}},$$

etc.

10.2 Lebesgue Functions

We are interested in the interpolation operator (cf. M. Reimer [1994, 2003])

$$\mathcal{T}_{\mathrm{Harm}_{0,\ldots,m}} : \mathrm{C}^{(0)}(\mathbb{S}^2) \to \mathrm{Harm}_{0,\ldots,m} \tag{10.38}$$

defined by (cf. Corollary 10.4)

$$\mathcal{T}_{\mathrm{Harm}_{0,\ldots,m}} F = \sum_{j=1}^{M} F(\eta_j) L_j \tag{10.39}$$

for $F \in \mathrm{C}^{(0)}(\mathbb{S}^2)$ and a unisolvent system $X_M = \{\eta_1, \ldots, \eta_M\} \subset \mathbb{S}^2$. $\mathrm{Harm}_{0,\ldots,m}$ is M-dimensional, so there is some member $Y \in \mathrm{Harm}_{0,\ldots,m}$ with

$$\|F - Y\|_{\mathrm{C}^{(0)}(\mathbb{S}^2)} = \inf_{P \in \mathrm{Harm}_{0,\ldots,m}} \|F - P\|_{\mathrm{C}^{(0)}(\mathbb{S}^2)}. \tag{10.40}$$

Furthermore, it is not difficult to see that $\mathcal{T}_{\mathrm{Harm}_{0,\ldots,m}} Y = Y$. Now, assume without loss of generality, that $\|F\|_{\mathrm{C}^{(0)}(\mathbb{S}^2)} \leq 1$. Then we obtain

$$|(\mathcal{T}_{\mathrm{Harm}_{0,\ldots,m}} F)(\xi)| \leq \sum_{j=1}^{M} |L_j(\xi)|, \quad \xi \in \mathbb{S}^2, \tag{10.41}$$

where the so-called *Lebesgue function* $L \in \mathrm{C}^{(0)}(\mathbb{S}^2)$ is given by

$$L(\xi) = \sum_{j=1}^{M} |L_j(\xi)|, \quad \xi \in \mathbb{S}^2. \tag{10.42}$$

L attains its maximum value, i.e., there exists a point $\xi_0 \in \mathbb{S}^2$ with

$$|(\mathcal{T}_{\mathrm{Harm}_{0,\ldots,m}} F)(\xi_0)| \leq L(\xi_0) = \max_{\xi \in \mathbb{S}^2} |L(\xi)|. \tag{10.43}$$

We let

$$\varepsilon_j = \mathrm{sign}\,(L_j(\xi_0)) \in \{-1, 0, +1\} \tag{10.44}$$

for $j = 1, \ldots, M$ and choose $d \in \mathbb{R}$ such that

$$0 < d < \frac{1}{2} \min\left\{|\eta_j - \eta_k| : j, k \in \{1, \ldots, M\}, j \neq k\right\}. \tag{10.45}$$

Then there exists a function $H \in \mathrm{C}^{(0)}(\mathbb{S}^2)$ satisfying

$$H(\xi) = \begin{cases} \varepsilon_j(1 - \frac{1}{d}|\xi - \eta_j|), & |\xi - \eta_j| \leq d, \quad j = 1, \ldots, M, \\ 0, & \text{otherwise}. \end{cases} \tag{10.46}$$

Clearly, $\|H\|_{C^{(0)}(\mathbb{S}^2)} \leq 1$. Thus it follows that

$$\|\mathcal{T}_{\mathrm{Harm}_{0,\ldots,m}}\|_{C^{(0)}(\mathbb{S}^2)} \geq (\mathcal{T}_{\mathrm{Harm}_{0,\ldots,m}} H)(\xi_0) = \sum_{j=1}^{M} \varepsilon_j L_j(\xi_0) = L(\xi_0). \quad (10.47)$$

This tells us that

$$\|\mathcal{T}_{\mathrm{Harm}_{0,\ldots,m}}\|_{C^{(0)}(\mathbb{S}^2)} = \max_{\xi \in \mathbb{S}^2} \sum_{j=1}^{M} |L_j(\xi)|. \quad (10.48)$$

It follows that for $Y \in \mathrm{Harm}_{0,\ldots,m}$ from (10.40)

$$\begin{aligned} \|F - \mathcal{T}_{\mathrm{Harm}_{0,\ldots,m}} F\|_{C^{(0)}(\mathbb{S}^2)} &= \|(F - Y) + \mathcal{T}_{\mathrm{Harm}_{0,\ldots,m}}(Y - F)\|_{C^{(0)}(\mathbb{S}^2)} \\ &\leq \left(1 + \|\mathcal{T}_{\mathrm{Harm}_{0,\ldots,m}}\|_{C^{(0)}(\mathbb{S}^2)}\right) \|F - Y\|_{C^{(0)}(\mathbb{S}^2)}. \end{aligned} \quad (10.49)$$

Definition 10.10. $\|\mathcal{T}_{\mathrm{Harm}_{0,\ldots,m}}\|_{C^{(0)}(\mathbb{S}^2)}$ *is called the* Lebesgue constant.

The determination of the *Lebesgue constant* is an important task in interpolation: In general, the explicit calculation of $\|\mathcal{T}_{\mathrm{Harm}_{0,\ldots,m}}\|_{C^{(0)}(\mathbb{S}^2)}$ becomes a severe problem. Therefore it is of high significance to know upper bounds. A first step is an easy, but reasonable estimate obtained by the Cauchy–Schwarz inequality, namely

$$\|\mathcal{T}_{\mathrm{Harm}_{0,\ldots,m}}\|^2_{C^{(0)}(\mathbb{S}^2)} \leq M \max_{\xi \in \mathbb{S}^2} \sum_{j=1}^{M} |L_j(\xi)|^2. \quad (10.50)$$

In other words, we are led to the question of finding an estimate for the Lagrangian square sum.

Lemma 10.11. *Let X_M be a fundamental system relative to* $\mathrm{Harm}_{0,\ldots,m}$. *Then*

$$\frac{M}{4\pi} \cdot \delta_{\max}^{-1} \leq \sum_{j=1}^{M} |L_j(\xi)|^2 \leq \frac{M}{4\pi} \cdot \delta_{\min}^{-1} \quad (10.51)$$

holds for all $\xi \in \mathbb{S}^2$ with $\delta_{\min}$, $\delta_{\max}$ from (10.33).

Proof. Let $l(\xi) = (L_1(\xi), \ldots, L_M(\xi))^T$, $\xi \in \mathbb{S}^2$. Furthermore, define

$$u(\xi) = (U_1(\xi), \ldots, U_M(\xi))^T = \mathbf{Q}\, l(\xi), \quad \xi \in \mathbb{S}^2 \quad (10.52)$$

with the orthogonal matrix $\mathbf{Q}$ from (10.32). Then, for $j = 1, \ldots, M$, U_j is an element of $\mathrm{Harm}_{0,\ldots,m}$, and it follows that

$$\sum_{j=1}^{M} |U_j(\xi)|^2 = \sum_{j=1}^{M} |L_j(\xi)|^2, \quad \xi \in \mathbb{S}^2. \quad (10.53)$$

Moreover, for any pair $(\xi, \eta) \in \mathbb{S}^2 \times \mathbb{S}^2$, we have

$$SH_{0,\ldots,m}(\xi \cdot \eta) = \sum_{j=1}^{M} \sum_{k=1}^{M} L_j(\xi) SH_{0,\ldots,m}(\eta_j \cdot \eta_k)\ L_k(\eta) \tag{10.54}$$

so that by (10.32)

$$SH_{0,\ldots,m}(\xi \cdot \eta) = l^T(\xi)\mathbf{K}\ l(\eta) = u^T(\xi)\mathbf{D}u(\eta). \tag{10.55}$$

But this means that

$$SH_{0,\ldots,m}(\xi \cdot \eta) = \sum_{j=1}^{M} \delta_j U_j(\xi) U_j(\eta), \quad \xi, \eta \in \mathbb{S}^2 \tag{10.56}$$

and

$$SH_{0,\ldots,m}(\xi \cdot \xi) = \sum_{j=1}^{M} \delta_j |U_j(\xi)|^2 = \frac{M}{4\pi}, \quad \xi \in \mathbb{S}^2. \tag{10.57}$$

Consequently, in connection with (10.32), we obtain

$$\delta_1 \sum_{j=1}^{M} |L_j(\xi)|^2 \leq \frac{M}{4\pi} \leq \delta_M \sum_{j=1}^{M} |L_j(\xi)|^2, \quad \xi \in \mathbb{S}^2. \tag{10.58}$$

This is the required result due to (10.33). □

From the estimate (10.50) we get as immediate consequences the following results.

Lemma 10.12. *Suppose that X_M is a fundamental system relative to* $\mathrm{Harm}_{0,\ldots,m}$. *Then*

$$\|\mathcal{T}_{\mathrm{Harm}_{0,\ldots,m}}\|_{\mathrm{C}^{(0)}(\mathbb{S}^2)} \leq \frac{M^2}{4\pi} \delta_{\min}^{-1}. \tag{10.59}$$

Lemma 10.13. *Suppose that X_M is a fundamental system relative to* $\mathrm{Harm}_{0,\ldots,m}$. *Then $\delta_{\min} = \delta_{\max}$ (i.e., $\delta_1 = \ldots = \delta_M$) in (10.33) if and only if*

$$\max_{\eta \in \mathbb{S}^2} \sum_{j=1}^{M} |L_j(\eta)|^2 = \sum_{j=1}^{M} |L_j(\xi)|^2 = 1, \quad \xi \in \mathbb{S}^2. \tag{10.60}$$

An easy consequence is the following estimate.

Lemma 10.14. *Let $\delta_{\min} = \delta_{\max}$ (i.e., $\delta_1 = \ldots = \delta_M$). Then*

$$\|\mathcal{T}_{\mathrm{Harm}_{0,\ldots,m}}\|_{\mathrm{C}^{(0)}(\mathbb{S}^2)} \leq \sqrt{M}. \tag{10.61}$$

Next we give a characterization of $\mathbf{K}^{-1}$ and its associated greatest eigenvalue $\delta_{\min}^{-1}$ in the context of the Lagrangians $L_1, \ldots, L_m$. We start with the fact that

$$\begin{aligned}\delta_{j,k} = L_k(\eta_j) &= \int_{\mathbb{S}^2} SH_{0,\ldots,m}(\eta_j \cdot \xi) L_k(\xi)\, dS(\xi) \\ &= \int_{\mathbb{S}^2} \sum_{p=1}^{M} L_p(\xi) SH_{0,\ldots,m}(\eta_j \cdot \eta_p) L_k(\xi)\, dS(\xi),\end{aligned} \tag{10.62}$$

i.e.,

$$\sum_{p=1}^{M} SH_{0,\ldots,m}(\eta_j \cdot \eta_p) \langle L_p, L_k \rangle_{\mathrm{L}^2(\mathbb{S}^2)} = \delta_{j,k}. \tag{10.63}$$

Hence, we have verified that

$$\mathbf{K}^{-1} = \left(\langle L_j, L_k \rangle_{\mathrm{L}^2(\mathbb{S}^2)} \right)_{j,k=1,\ldots,M}. \tag{10.64}$$

Especially, if $\delta_{\max} = \delta_{\min}$, we obtain

$$\langle L_j, L_j \rangle_{\mathrm{L}^2(\mathbb{S}^2)} = \frac{4\pi}{M} \tag{10.65}$$

for $j = 1, \ldots, M$. Moreover, from (10.64), we have

$$\delta_{\min}^{-1} = \max \left\{ \frac{\int_{\mathbb{S}^2} |F(\eta)|^2\, dS(\eta)}{\sum_{j=1}^{M} |F(\eta_j)|^2} : F \in \mathrm{Harm}_{0,\ldots,m},\ \sum_{j=1}^{M} |F(\eta_j)|^2 > 0 \right\}. \tag{10.66}$$

We know that

$$\mathbf{Q} = \mathbf{A}_{X_M}(U_1, \ldots, U_M), \tag{10.67}$$

and the functions $U_1, \ldots, U_M$ of (10.52) constitute a basis of $\mathrm{Harm}_{0,\ldots,m}$. The reproducing property of $SH_{0,\ldots,m}$ implies that

$$U_k(\eta) = \int_{\mathbb{S}^2} SH_{0,\ldots,m}(\xi \cdot \eta) U_k(\xi)\, dS(\xi), \quad \eta \in \mathbb{S}^2. \tag{10.68}$$

Inserting (10.56) in (10.68) we get

$$U_k(\eta) = \sum_{j=1}^{M} \delta_j \langle U_j, U_k \rangle_{\mathrm{L}^2(\mathbb{S}^2)} U_j(\eta), \quad \eta \in \mathbb{S}^2. \tag{10.69}$$

This shows that

$$\delta_j \langle U_j, U_k \rangle_{\mathrm{L}^2(\mathbb{S}^2)} = \delta_{j,k} \tag{10.70}$$

for $j, k = 1, \ldots, M$. Thus, the functions

$$V_j = \sqrt{\delta_j} U_j, \quad j = 1, \ldots, M, \tag{10.71}$$

constitute an orthonormal basis in $\mathrm{Harm}_{0,\ldots,m}$ such that the matrix

$$\mathbf{V} = \mathbf{A}_{X_M}(V_1, \ldots, V_M) \tag{10.72}$$

satisfies the properties

$$\mathbf{V}^T\mathbf{V} = \mathbf{K}, \qquad \mathbf{V}\mathbf{V}^T = \mathbf{D}. \tag{10.73}$$

Finally, the case of equal eigenvalues of $\mathbf{K}$ (i.e., $\delta_{\max} = \delta_{\min} = M/4\pi$) should be discussed in more detail. Under this assumption we obtain

$$SH_{0,\ldots,m}(\xi \cdot \eta) = \frac{M}{4\pi}\sum_{j=1}^{M} L_j(\xi)L_j(\eta), \tag{10.74}$$

so that

$$L_k(\eta) = \int_{\mathbb{S}^2} SH_{0,\ldots,m}(\xi \cdot \eta)L_k(\xi)\, dS(\xi) = \frac{M}{4\pi}\sum_{j=1}^{M}\langle L_j, L_k\rangle_{\mathrm{L}^2(\mathbb{S})}L_j(\eta). \tag{10.75}$$

Therefore it follows that

$$\frac{M}{4\pi}\langle L_j, L_k\rangle_{\mathrm{L}^2(\mathbb{S})} = \delta_{j,k} \tag{10.76}$$

for $j, k = 1, \ldots, M$. Moreover,

$$\langle L_j, SH_{0,\ldots,m}(\eta_k\cdot)\rangle_{\mathrm{L}^2(\mathbb{S})} = L_j(\eta_k) = 0 \tag{10.77}$$

for $k \neq j$. This leads us to the following result.

Theorem 10.15. *The identity*

$$\sum_{j=1}^{M} |L_j(\xi)|^2 = 1, \quad \xi \in \mathbb{S}^2 \tag{10.78}$$

holds if and only if X_M is a fundamental system relative to $\mathrm{Harm}_{0,\ldots,m}$*, and the corresponding Lagrangians constitute an orthogonal basis for* $\mathrm{Harm}_{0,\ldots,m}$*, where*

$$L_j = \frac{4\pi}{M}SH_{0,\ldots,m}(\,\cdot\eta_j) \tag{10.79}$$

and

$$\langle L_j, L_k\rangle_{\mathrm{L}^2(\mathbb{S})} = \int_{\mathbb{S}^2} L_j(\eta)L_k(\eta)\, dS(\eta) = \frac{4\pi}{M}\delta_{j,k}. \tag{10.80}$$

Let X_M be a fundamental system relative to $\mathrm{Harm}_{0,\ldots,m}$. Consider the approximate formula

$$\int_{\mathbb{S}^2} Y(\xi)F(\xi)\, dS(\xi) = \sum_{j=1}^{M} a_j F(\eta_j) + \mathcal{E}F, \tag{10.81}$$

where Y is an element of $\mathrm{Harm}_{0,\ldots,m}$ and $a_1, \ldots, a_M$ are given by

$$a_j = \int_{\mathbb{S}^2} Y(\eta) L_j(\eta)\, dS(\eta), \quad j = 1, \ldots, M. \tag{10.82}$$

Then we obtain for all $F \in \mathrm{C}^{(2m+2)}(\mathbb{S}^2)$, $m \geq 0$, that

$$\begin{aligned} \mathcal{E}F = & -\sum_{k=1}^{M} \int_{\mathbb{S}^2} Y(\eta) L_k(\eta)\, dS(\eta) \\ & \times \int_{\mathbb{S}^2} \partial_{0,\ldots,m} G(\partial^2_{0,\ldots,m}; \eta \cdot \eta_k)(\partial_{0,\ldots,m} F(\eta))\, dS(\eta). \end{aligned} \tag{10.83}$$

Consequently, we have $\mathcal{E}Y = 0$ for all $Y \in \mathrm{Harm}_{0,\ldots,m}$. The Cauchy–Schwarz inequality yields the following result.

Lemma 10.16. *For all* $Y \in \mathrm{Harm}_{0,\ldots,m}$ *and* $F \in \mathrm{C}^{(2m+2)}(\mathbb{S}^2)$

$$\begin{aligned} |\mathcal{E}F| \leq & \left(\max_{\xi \in \mathbb{S}^2} \sum_{k=1}^{M} |L_k(\xi)|^2 \right)^{1/2} \left(\int_{\mathbb{S}^2} |\partial_{0,\ldots,m} F(\eta)|^2 \, dS(\eta) \right)^{1/2} \\ & \times \left(\sum_{n=m+1}^{\infty} \frac{2n+1}{4\pi} \frac{1}{(\partial^{\wedge}_{0,\ldots,m}(n))^2} \|Y\|^2_{\mathrm{L}^2(\mathbb{S}^2)} \right)^{1/2}. \end{aligned} \tag{10.84}$$

In particular, for $Y = 1$ in (10.81), we obtain

$$\sum_{j=1}^{M} a_j Y_{n,k}(\eta_j) = \int_{\mathbb{S}^2} Y_{n,k}(\xi)\, dS(\xi) = \sqrt{4\pi}\,\delta_{n,0} \tag{10.85}$$

for $n = 0, \ldots, m$, $k = 1, \ldots, 2n+1$. Furthermore, we have by $\sum_{k=1}^{M} L_k(\eta) = 1$ for all $\eta \in \mathbb{S}^2$ due to (10.10) that

$$a_j = \int_{\mathbb{S}^2} L_j(\eta)\, dS(\eta) = \sum_{k=1}^{M} \langle L_j, L_k \rangle_{\mathrm{L}^2(\mathbb{S}^2)} \tag{10.86}$$

for $j = 1, \ldots, M$, hence, with (10.80)

$$\sum_{j=1}^{M} a_j = \sum_{j=1}^{M} \sum_{k=1}^{M} (L_j, L_k)_{\mathrm{L}^2(\mathbb{S}^2)} = 4\pi. \tag{10.87}$$

The weights a_j are the row sums in the matrix $\mathbf{K}^{-1}$. In the special case $\delta_{\max} = \delta_{\min}$ we obtain from Theorem 10.15 that

$$a_j = \frac{4\pi}{M} \int_{\mathbb{S}^2} SH_{0,\ldots,m}(\eta_j \cdot \eta)\, dS(\eta) = \frac{4\pi}{M} > 0, \quad j = 1, \ldots, M. \tag{10.88}$$

10.3 Spherical Geometry and Polynomial Cubature Rules

The distance between any two points ξ and η on the unit sphere $\mathbb{S}^2$ can be measured by the *geodesic distance* $\text{dist}(\xi,\eta) \in [0,2]$, where

$$\text{dist}(\xi,\eta) = 1 - \xi \cdot \eta, \tag{10.89}$$

which is the canonical metric on $\mathbb{S}^2$. It is related to the Euclidean distance $|\xi - \eta| \in [0,2]$ by

$$|\xi - \eta| = \sqrt{2(1 - \xi \cdot \eta)}, \quad \xi, \eta \in \mathbb{S}^2, \tag{10.90}$$

i.e.,

$$\text{dist}(\xi,\eta) = \frac{1}{2}|\xi - \eta|^2, \quad \xi, \eta \in \mathbb{S}^2. \tag{10.91}$$

The *spherical cap* with center $\zeta \in \mathbb{S}^2$ and radius $\rho \in [0,2]$ is given by

$$\Gamma_\rho(\zeta) = \left\{\xi \in \mathbb{S}^2 : 1 - \xi \cdot \zeta < \rho\right\}. \tag{10.92}$$

Clearly, with $\rho = 2$, $\overline{\Gamma_\rho(\zeta)}$ becomes the whole sphere $\mathbb{S}^2$.

Geometric Quantities. The quality of the geometric distribution of a set $X_N = \{\xi_1, \ldots, \xi_N\}$ of N distinct points of any point set on the sphere $\mathbb{S}^2$ is often characterized by the following two quantities and their ratio:

The *mesh norm*

$$h_{X_N} = \sup_{\eta \in \mathbb{S}^2} \min_{\xi_j \in X_N} \text{dist}(\eta, \xi_j) \tag{10.93}$$

and the *separation distance*

$$\delta_{X_N} = \min\left\{\text{dist}(\xi_j, \xi_k) : \xi_j, \xi_k \in X_N,\ j \neq k\right\}. \tag{10.94}$$

The mesh norm h_{X_N} is the covering radius for covering the sphere with spherical caps of the smallest possible equal radius centered at the points in X_N, while the separation distance δ_{X_N} is twice the packing radius, so $h_{X_N} \geq \delta_{X_N}/2$. (The packing radius of a point set X_N is the largest radius ρ such that different caps $\Gamma_\rho(\xi_j)$, $\xi_j \in X_N$, at most touch at the boundary.) Well known bounds for h_{X_N} and δ_{X_N}, the first simply coming from the condition that the N caps of radius h_{X_N} cover the sphere and the second from known bounds for best packing (see, e.g., E.B. Saff, A.B.J. Kuijlaars [1997], K. Hesse et al. [2010]), are given by

$$h_{X_N} \geq 2N^{-1/2}, \qquad \delta_{X_N} \leq \left(\frac{8\pi}{\sqrt{3}}\right)^{1/2} N^{-1/2}. \tag{10.95}$$

The mesh ratio

$$\sigma_{X_N} = \frac{2h_{X_N}}{\delta_{X_N}} \geq 1 \tag{10.96}$$

is a good measure for the quality of the geometric distribution of X_N: the smaller σ_{X_N} is, the more uniformly are the points distributed on $\mathbb{S}^2$. If the mesh ratio is bounded above by say σ then it follows from (10.95) that

$$h_{X_N} = \frac{1}{2}\sigma_{X_N}\delta_{X_N} \leq \left(\frac{2\pi}{\sqrt{3}}\right)^{1/2} \sigma N^{-1/2}. \tag{10.97}$$

Cubature Rules with Certain Geometric Properties. There is a large palette of approximate integration formulas on the sphere obeying certain geometric obligations. Although these rules are of high interest mathematically, they are of practical value only if measurements can be done in prescribed grids. This is the reason why we do not go into further details (see, e.g., K. Hesse et al. [2010] for more information). Nevertheless it should be mentioned that the theory of cubature rules on the sphere that are invariant under all the rotations in a particular finite subgroup G of the rotation group in $\mathbb{R}^3$ may be led back to a short paper by S. Sobolev [1962]. It has been used to develop concrete cubature rules by V.I. Lebedev [1975], V.I. Lebedev, D.N. Laikov [1999] (see also S. Sobolev, V. Vaskevich [1997]), and is therefore sometimes called Lebedev cubature (or quadrature).

Cubature Rules with Polynomial Accuracy. Consider a set $X_N = \{\xi_1, \ldots, \xi_N\}$ of N distinct points on the sphere $\mathbb{S}^2$. The quality of the geometric distribution of this point set is usually measured by considering the mesh norm h_{X_N} (see (10.93)), the separation distance δ_{X_N} (see (10.94)), and the mesh ratio $\sigma_{X_N} = 2h_{X_N}/\delta_{X_N} \geq 1$ (cf. (10.96)).

In H.N. Mhaskar et al. [2001] the existence of positive-weight cubature rules with a given set of points is proved, under the assumption that the points have a sufficiently uniform point distribution. The results in H.N. Mhaskar et al. [2001] are improved in F.J. Narcowich et al. [2006] as follows: There exists a constant $c > 0$ such that for every $\sigma \geq 1$ and every $0 < \beta < \frac{1}{2}$ the following statement holds: For every point set $X_N = \{\xi_1, \ldots, \xi_N\} \subset \mathbb{S}^2$ whose mesh ratio satisfies $\sigma_{X_N} \leq \sigma$ and for every $m \geq 1$ such that

$$h_{X_N} \leq \beta c^{-1} m^{-1}, \tag{10.98}$$

there exist positive weights $a_1, \ldots, a_N$ such that the cubature rule

$$\mathcal{L}_{X_N} F = \sum_{j=1}^{N} a_j F(\xi_j) \tag{10.99}$$

is exact on $\mathrm{Harm}_{0,\ldots,m}$, that is

$$\mathcal{L}_{X_N} Y = \sum_{j=1}^{N} a_j Y(\xi_j) = \int_{\mathbb{S}^2} Y(\xi)\, dS(\xi), \quad Y \in \mathrm{Harm}_{0,\ldots,m}. \tag{10.100}$$

The weights $w_1, \ldots, w_N$ satisfy the estimate

$$4\left(\frac{1-2\beta}{1-\beta}\right)\sigma^{-2}h_{X_N}^2 \leq w_j \leq \frac{4\pi}{(\lfloor m/2 \rfloor + 1)^2}, \qquad j = 1, \ldots, N. \tag{10.101}$$

If for a given point set X_N the polynomial degree m is chosen as large as possible, then from (10.98) $h_{X_N} \asymp m^{-1}$, and from (10.101) the weights satisfy $a_j \asymp m^{-2} \asymp h_{X_N}^2$.

Remark 10.17. *The notation $a_m \asymp b_m$ involving the two sequences $\{a_m\}$ and $\{b_m\}$ means that there exist positive constants c_1 and c_2, independent of m, such that $c_1 b_m \leq a_m \leq c_2 b_m$ for all m.*

Note that, from the boundedness of the mesh ratio (10.95), it follows that the number of points satisfies the property

$$N \asymp h_{X_N}^{-2} \asymp m^2 \asymp \dim(\mathrm{Harm}_{0,\ldots,m}). \tag{10.102}$$

10.4 Interpolatory Rules Based on Extremal Point Systems and Designs

Motivated by the work presented in W. Freeden et al. [1998] and M. Reimer [2003], I.H. Sloan, R.S. Womersley [2004], R. Womersley [2007] seek an interpolatory rule (that is, $N = M = (m+1)^2$) for which the polynomial basis matrix $\mathbf{H} = \mathbf{A}_{X_M}(Y_{0,1}, \ldots, Y_{m,2m+1})$ has the largest possible determinant. More precisely, when $N = M = (m+1)^2$ the corresponding symmetric positive semidefinite Gram matrix is $\mathbf{K} = \mathbf{H}^T\mathbf{H}$ (see (10.31)), and $\mathbf{K}$ is non-singular (and positive definite) if and only if $\mathbf{H}$ is non-singular, that is, if and only if X_M is a fundamental system.

Extremal Point Systems. *Extremal points* (see also, e.g., M. Reimer [1994, 2003], I.H. Sloan, R.S. Womersley [2004, 2009]) are chosen to maximize the determinant of $\mathbf{K}$ (or $\mathbf{H}$ or any other basis matrix). The cubature weights $a_1, \ldots, a_M$ are the solution to the linear system $\mathbf{K}a = e$, where $e = (1, 1, \ldots, 1)^T \in \mathbb{R}^M$. From the addition theorem, the elements of the matrix $\mathbf{K}$ are given by

$$SH_{0,\ldots,m}(\xi_i \cdot \xi_j) = \sum_{l=0}^{m}\sum_{k=1}^{2l+1} Y_{l,k}(\xi_i)Y_{l,k}(\xi_j) = \sum_{l=0}^{m} \frac{2l+1}{4\pi} P_l(\xi_i \cdot \xi_j), \tag{10.103}$$

where $i, j = 1, \ldots, M$. In particular, the diagonal elements of $\mathbf{K}$ are $SH_{0,\ldots,m}(\xi_i \cdot \xi_i) = (m+1)^2/(4\pi)$, so that

$$\zeta = \mathrm{trace}(\mathbf{K}) = \sum_{i=1}^{M} \lambda_i(\mathbf{K}) = \sum_{i=1}^{M} (\sigma_i(\mathbf{H}))^2 = \frac{M(m+1)^2}{4\pi} = \frac{(m+1)^4}{4\pi}, \tag{10.104}$$

where the $\lambda_i(\mathbf{K})$ are the eigenvalues of $\mathbf{K}$ and the $\sigma_i(\mathbf{H})$ are the singular values of $\mathbf{H}$. Maximizing the determinant is equivalent to either of the following two constrained optimization problems:

$$\max_{X_M} \prod_{j=1}^{M} \lambda_j(\mathbf{K}) \tag{10.105}$$

subject to $\sum_{j=1}^{M} \lambda_j(\mathbf{K}) = \zeta$, $\lambda_j(\mathbf{K}) \geq 0$, $j = 1, \ldots, M$, and

$$\max_{X_M} \prod_{j=1}^{M} \sigma_j(\mathbf{H}) \tag{10.106}$$

subject to $\sum_{j=1}^{M} \left(\sigma_j(\mathbf{H})\right)^2 = \zeta$.

Note that as $\mathbf{K}$ is positive semi-definite for any point set X_M, $\lambda_j(\mathbf{K}) \geq 0$, $j = 1, \ldots, M$, and by definition $\sigma_j(\mathbf{H}) \geq 0$, $j = 1, \ldots, M$. As $\zeta > 0$ the maximizer will have all eigenvalues (singular values) strictly positive. If it were possible to ignore the dependence of the eigenvalues on X_M, then both problems would be solved by taking all the eigenvalues $\lambda_j(\mathbf{K})$ of $\mathbf{K}$, and all the squares of the singular values $(\sigma_j(\mathbf{H}))^2$ of $\mathbf{H}$, respectively, equal, that is,

$$(\sigma_j(\mathbf{H}))^2 = \lambda_j(\mathbf{K}) = \lambda_{\text{avg}} = \frac{(m+1)^2}{4\pi}, \quad j = 1, \ldots, M, \tag{10.107}$$

corresponding to $\mathbf{K} = \lambda_{\text{avg}}\mathbf{I}$, where $\mathbf{I}$ is the $M \times M$ identity matrix. But this is not possible, except for $m = 1$ and $N = M = 4$ with the points forming the vertices of the regular tetrahedron, since otherwise it would contradict the theorem on the non-existence of tight spherical $2m$-designs (see E. Bannai, R.M. Damerell [1979] and Section 10.5). In summary, extremal points are attractive for both interpolation and cubature, but they are difficult to generate at least when M is increasing.

Spherical Designs. Cubature rules with specified degree of precision m and equal weights $a_j = 4\pi/N$ have attracted special interest (see, e.g., Ei. Bannai, Et. Bannai [2009] and the references therein). Following P. Delsarte et al. [1977], a *spherical m-design* on $\mathbb{S}^2$ is a set $X_N = \{\xi_1, \ldots, \xi_N\}$ of N points on $\mathbb{S}^2$ such that

$$\frac{4\pi}{N} \sum_{j=1}^{N} Y(\xi_j) = \int_{\mathbb{S}^2} Y(\xi) \, dS(\xi), \quad Y \in \text{Harm}_{0,\ldots,m}\,. \tag{10.108}$$

P. Delsarte et al. [1977] gave the lower bound for the number of points

$$N \geq \begin{cases} \frac{(m+1)(m+3)}{4}, & m \text{ odd}, \\ \frac{(m+2)^2}{4}, & m \text{ even}, \end{cases} \tag{10.109}$$

to be required to constitute an m-design. A spherical m-design that achieves these lower bounds is known as a *tight spherical design.* However, it is also known that for the unit sphere $\mathbb{S}^2$ tight spherical designs do not exist except for $m = 1, 2, 3$ or 5 (cf. E. Bannai, R.M. Damerell [1979]). On the other hand, it is known P.D. Seymour, T. Zaslavsky [1984] that for any $m \in \mathbb{N}$, spherical m-designs always exist if N is sufficiently large.

Based on equal weight quadrature for $[-1, 1]$ with degree of precision m, B. Bajnok [1991], J. Korevaar, J. Meyers [1993] give tensor product constructions of spherical m-designs with $N = O(m^3)$. For tabulated designs with modest numbers of points, see, e.g., R.H. Hardin, N. Sloane [1996], N. Sloane [2000]. Numerically there is evidence that spherical m-designs with $N = (m+1)^2 = M$ points exist. X. Chen, R. Womersley [2006], X. Chen et al. [2009] have used interval methods to *prove* existence of spherical m-designs with $N = (m + 1)^2 = M$ for all values of m up to 100. For more details, the reader is referred to K. Hesse et al. [2010].

10.5 Non-Existence of Spherical Gaussian Rules

An often asked natural question is: What is the analog for the sphere $\mathbb{S}^2$ of the Gaussian quadrature rules for $[-1, 1]$ of Chapter 8? The answer (see, e.g., K. Hesse et al. [2010]) is that, on the sphere $\mathbb{S}^2$, cubature rules with the "double-degree-of-precision" property do not exist. The following result states this precisely and is a consequence of the theorem on the non-existence of tight spherical $2m$-designs (cf. Ei. Bannai, Et. Bannai [2009], K. Hesse et al. [2010] and the references therein)

Corollary 10.18. *Let $m \geq 2$. There exists no positive-weight cubature rule on $\mathbb{S}^2$ with $M = (m + 1)^2$ points and polynomial degree of precision $2m$.*

In brief, the result follows by noting that if $M = (m+1)^2$ and the rule has the degree of precision $2m$, then, as the $Y_{n,k}$, $k = 1, \ldots, 2n+1$, $n = 0, \ldots, m$, form an $\mathrm{L}^2(\mathbb{S}^2)$-orthonormal system, the square matrix $\mathbf{H}$ satisfies

$$\mathbf{HAH}^T = \mathbf{I}, \tag{10.110}$$

where $\mathbf{I}$ is again the $M \times M$ identity matrix and $\mathbf{A} = \mathrm{diag}(a_1, \ldots, a_M)$ is the $M \times M$ diagonal matrix with the cubature weights as diagonal elements. From this it follows that $\mathbf{H}$ is invertible and $\mathbf{H}^T = \mathbf{A}^{-1}\mathbf{H}^{-1}$, so

$$\mathbf{K} = \mathbf{H}^T\mathbf{H} = \mathbf{A}^{-1}. \tag{10.111}$$

In particular, the j-th diagonal element of $\mathbf{K}$ gives $a_j^{-1} = M/(4\pi)$, by (10.103). Hence, the weights are independent of j and the cubature rule, if it exists, is a spherical $2m$-design that achieves the lower bound (10.109). However, tight spherical $2m$-designs on $\mathbb{S}^2$ do not exist for $m \geq 2$.

The negative result that Gaussian rules do no exist has unfortunate consequences for interpolatory cubature rules. Whereas interpolation of periodic polynomials of degree $\leq m$ at equally spaced points on the circle can be expressed by an explicit formula (because inner products are done exactly), no such formula exists for polynomial interpolation on $\mathbb{S}^2$.

Chapter 11

Latitude-Longitude Cubature

For computational tasks, a standard approach is to use latitude-longitude coordinates. In doing so, computations with functions on the sphere can be conveniently related to obligations with functions on a rectangular domain. The resulting latitude-longitude rules are useful whenever data sets are available in all nodes on the sphere that are originated by the rectangular grid points in accordance with the coordinate transform. However, this procedure also shows many calamities due to the artificial pole singularities introduced by the coordinate transform and the non-equidistribution of the data system generated on the sphere.

In what follows we explain the essential ingredients of latitude-longitude rules. We restrict ourselves to two types of underlying grids, namely equiangular latitude-longitude grid (ELLG) and the Gaussian grid (GG).

11.1 Associated Legendre Functions

Consider the function $L_n : \mathbb{R}^3 \to \mathbb{R}$, $n = 0, 1, \ldots$ defined by

$$L_n(x) = \frac{1}{2\pi} \int_0^{2\pi} (x_3 + ix_1 \cos\alpha + ix_2 \sin\alpha)^n \, d\alpha, \quad x = (x_1, x_2, x_3)^T. \tag{11.1}$$

Clearly, L_n is a homogeneous polynomial of degree n which is symmetric with respect to the x_3-axis. An easy calculation shows that $\Delta_x L_n(x) = 0$ for all $x \in \mathbb{R}^3$, i.e., L_n is harmonic in $\mathbb{R}^3$. Moreover, $L_n(\varepsilon^3) = 1$. Furthermore, by use of the coordinates (9.1) and the Laplace representation of the Legendre polynomial (7.41) we see that

$$\begin{aligned} L_n(x) &= \frac{r^n}{2\pi} \int_0^{2\pi} \left(t + i\sqrt{1-t^2}\cos(\alpha - \varphi)\right)^n \, d\alpha \\ &= \frac{r^n}{2\pi} \int_0^{2\pi} \left(t + i\sqrt{1-t^2}\cos\alpha\right)^n \, d\alpha \\ &= r^n P_n(t). \end{aligned} \tag{11.2}$$

This implies that

$$L_n\left(\frac{x}{|x|}\right) = P_n(t), \quad x = |x|\left(t\varepsilon^3 + \sqrt{1-t^2}(\cos\varphi\varepsilon^1 + \sin\varphi\varepsilon^2)\right) \tag{11.3}$$

for all $x \in \mathbb{R}^3 \setminus \{0\}$.

More specifically, we are able to conclude the following result (see W. Freeden, M. Schreiner [2009]): Let H_n be a homogeneous, harmonic polynomial of degree n with the following properties:

(i) $H_n(\mathbf{t}x) = H_n(x)$ for all orthogonal transformations $\mathbf{t} \in O(3)$ with $H_n(\mathbf{t}\varepsilon^3) = H_n(\varepsilon^3)$,

(ii) $H_n(\varepsilon^3) = 1$.

Then H_n is uniquely determined, and H_n coincides with the Legendre harmonic L_n of degree n, i.e.,

$$H_n(x) = L_n(x) = r^n P_n(t), \quad x = r\left(t\varepsilon^3 + \sqrt{1-t^2}(\cos\varphi\varepsilon^1 + \sin\varphi\varepsilon^2)\right). \tag{11.4}$$

In other words, the only function $Y_n \in \mathrm{Harm}_n$ satisfying

(i) $Y_n(\mathbf{t}\xi) = Y_n(\xi)$, $\xi \in \mathbb{S}^2$, for all orthogonal transformations $\mathbf{t} \in O(3)$ which leave η fixed (i.e., $Y_n(\mathbf{t}\eta) = Y_n(\eta)$)

(ii)) $Y_n(\eta) = 1$,

is given by $\xi \mapsto Y_n(\xi) = P_n(\xi \cdot \eta)$, $\xi \in \mathbb{S}^2$. By a straightforward calculation, we obtain from the Laplace representation of the Legendre polynomials (see Lemma 7.8)

$$(1-t^2)^{\frac{m}{2}}\left(\frac{d}{dt}\right)^m P_n(t) = \frac{(n+m)!\, i^m}{n!}\frac{}{\pi}\int_0^\pi \left(t + i\sqrt{1-t^2}\cos\varphi\right)^n \cos(m\varphi)\, d\varphi, \tag{11.5}$$

for $n = 0, 1, \ldots, m = 0, \ldots, n$.

Definition 11.1. *For* $n = 0, 1, \ldots, m = 0, \ldots, n$, $P_{n,m} : [-1,1] \to \mathbb{R}$ *given by*

$$P_{n,m}(t) = \frac{(n+m)!\, i^m}{n!\,\pi}\int_0^\pi \left(t + i\sqrt{1-t^2}\cos\varphi\right)^n \cos(m\varphi)\, d\varphi \tag{11.6}$$

is called the associated Legendre function of degree n and order m.

Obviously,

$$P_{n,m}(t) = (1-t^2)^{m/2}\left(\frac{d}{dt}\right)^m P_n(t), \quad t \in [-1,1]. \tag{11.7}$$

Observe that, in the sense of (11.7), $P_{n,m}(t) = 0$ for $m > n$. By use of the associated Legendre functions, we are immediately able to determine the expansion coefficients of the Fourier series of $(t + \sqrt{t^2-1}\cos\varphi)^n$, $t \in [-1,1]$,

$$\left(t + \sqrt{t^2-1}\cos\varphi\right)^n = \frac{a_0}{2} + \sum_{m=1}^{n} a_m \cos(m\varphi), \tag{11.8}$$

where

$$a_m = \frac{2}{\pi}\int_0^{\pi} (t + \sqrt{t^2-1}\cos\varphi)^n \cos(m\varphi)\, d\varphi = \frac{(2n)!}{(n+m)!} i^m P_{n,m}(t), \tag{11.9}$$

such that, for $t \in [-1,1]$, $\varphi \in [0, 2\pi)$, $n = 0, 1, \ldots$, and $m = 0, \ldots, n$,

$$\left(t + \sqrt{t^2-1}\cos\varphi\right)^n = P_{n,0}(t) + (2n)! \sum_{m=1}^{n} \frac{i^m}{(n+m)!} P_{n,m}(t)\cos(m\varphi). \tag{11.10}$$

Using the Rodriguez formula (cf. (7.16)) we are led to the identity

$$P_{n,m}(t) = \frac{1}{2^n n!}(1-t^2)^{m/2}\left(\frac{d}{dt}\right)^{n+m}(t^2-1)^n, \quad t \in [-1,1]. \tag{11.11}$$

Moreover, $P_{n,0}(t) = P_n(t)$, $t \in [-1,1]$. Furthermore, we have

$$P_{n,m}(t) = \frac{1}{2^n}(1-t^2)^{m/2}\sum_{k=0}^{\lfloor \frac{n}{2}\rfloor}(-1)^k \frac{(2n-2k)!}{k!(n-k)!(n-2k)!}\left(\frac{d}{dt}\right)^m t^{n-2k}. \tag{11.12}$$

Note that the m-th derivative of the power t^{n-2k} reads

$$\left(\frac{d}{dt}\right)^m t^{n-2k} = \frac{(n-2k)!}{(n-m-2k)!} t^{n-m-2k}. \tag{11.13}$$

This leads to the following explicit formula for any Legendre function.

Lemma 11.2. *For $n = 0, 1, \ldots$, $m = 0, \ldots, n$, and $t \in [-1,1]$ we have the representation*

$$P_{n,m}(t) = (1-t^2)^{m/2} \sum_{k=0}^{\lfloor \frac{n-m}{2}\rfloor} (-1)^k \frac{(2n-2k)!}{2^n k!(n-k)!(n-m-2k)!} t^{n-m-2k}. \tag{11.14}$$

Some graphical impressions of Legendre functions can be found in Figure 11.1. The Legendre polynomials P_n are known to satisfy the Legendre differential equation of (7.8). We differentiate (7.8) m-times with respect to t:

$$\left(\frac{d}{dt}\right)^m \left(-2t\frac{d}{dt}P_n(t) + (1-t^2)\left(\frac{d}{dt}\right)^2 P_n(t)\right) + n(n+1)\left(\frac{d}{dt}\right)^m P_n(t) = 0. \tag{11.15}$$

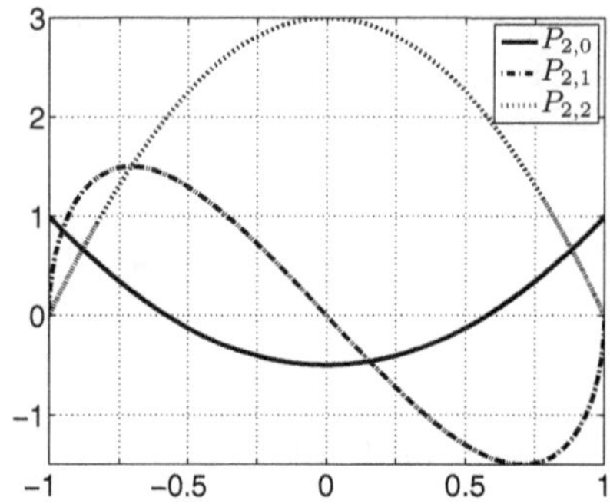

FIGURE 11.1: Legendre functions $t \mapsto P_{2,m}(t)$, $t \in [-1,1]$, $m = 0,1,2$.

An elementary calculation starting from (11.15) guarantees the validity of the differential equation for $P_{n,m}$, $n = 0,1,\ldots$, $m = 1,\ldots,n$:

$$(1-t^2)\left(\frac{d}{dt}\right)^2 P_{n,m}(t) - 2t\frac{d}{dt}P_{n,m}(t) + \left(n(n+1) - \frac{m^2}{1-t^2}\right)P_{n,m}(t) = 0. \tag{11.16}$$

We notice that the equation

$$\begin{aligned} &P_{r,m}(t)\left(\frac{d}{dt}(1-t^2)\frac{d}{dt}P_{n,m}(t) + \left(n(n+1) - \frac{m^2}{1-t^2}\right)P_{n,m}(t)\right) \qquad (11.17)\\ &- P_{n,m}(t)\left(\frac{d}{dt}(1-t^2)\frac{d}{dt}P_{r,m}(t) + \left(r(r+1) - \frac{m^2}{1-t^2}\right)P_{r,m}(t)\right) = 0 \end{aligned}$$

is equivalent to

$$\begin{aligned} &P_{r,m}(t)\frac{d}{dt}(1-t^2)\frac{d}{dt}P_{n,m}(t) - P_{n,m}(t)\frac{d}{dt}(1-t^2)\frac{d}{dt}P_{r,m}(t)\\ &+ P_{n,m}(t)P_{r,m}(t)\left(n(n+1) - r(r+1)\right) = 0. \end{aligned} \tag{11.18}$$

Integration with respect to t over the interval $[-1,1]$ yields

$$(n-r)(n+r+1)\int_{-1}^{1} P_{n,m}(t)P_{r,m}(t)\,dt = 0. \tag{11.19}$$

As $n+r+1 > 0$ and $n \neq r$, this leads to the following orthogonality relation

$$\int_{-1}^{1} P_{n,m}(t)P_{r,m}(t)\,dt = 0. \tag{11.20}$$

for all n, r with $n \neq r$, It is not difficult to see the following *recurrence formula*

$$\begin{aligned} P_{n,m+1}(t) &= (1-t^2)^{\frac{m}{2}+\frac{1}{2}}\left(\frac{d}{dt}\right)^{m+1} P_n(t)\\ &= (1-t^2)^{1/2}\frac{d}{dt}P_{n,m}(t) + mt(1-t^2)^{-1/2}P_{n,m}(t). \end{aligned} \tag{11.21}$$

There is also a large palette of three-term recursion relations, including

$$(n-m+1)P_{n+1,m}(t) - (2n+1)tP_{n,m}(t) + (n+m)P_{n-1,m}(t) = 0. \quad (11.22)$$

The effect of a change in the sign of the argument is

$$P_{n,m}(-t) = (-1)^{n+m} P_{n,m}(t). \quad (11.23)$$

Note that Legendre functions for negative orders are defined by the following relation

$$P_{n,-m}(t) = (-1)^m \frac{(n-m)!}{(n+m)!} P_{n,m}(t), \quad m = 0, \dots, n. \quad (11.24)$$

The identity (11.21) and the differential equation of the associated Legendre function can be used to verify

$$\int_{-1}^{1} (P_{n,m}(t))^2 \, dt = \frac{2}{2n+1} \frac{(n+m)!}{(n-m)!}. \quad (11.25)$$

for $n = 0, 1, \dots, m = 1, \dots, n$. Finally, we mention that, for $n, l = 0, 1, \dots$ and $m = 0, \dots, n$, $k = 0, \dots, l$, we have

$$\begin{aligned} &\int_0^{2\pi} \int_0^{\pi} P_{n,m}(\cos\vartheta) P_{l,k}(\cos\vartheta) \cos(m\varphi) \cos(k\varphi) \sin\vartheta \, d\vartheta \, d\varphi \\ &= \int_0^{2\pi} \cos(m\varphi) \cos(k\varphi) \, d\varphi \int_0^{\pi} P_{n,m}(\cos\vartheta) P_{l,k}(\cos\vartheta) \sin\vartheta \, d\vartheta \\ &= \delta_{l,n} \delta_{k,m} \frac{2\pi}{2-\delta_{m,0}} \frac{2}{2n+1} \frac{(n+m)!}{(n-m)!}, \end{aligned} \quad (11.26)$$

where, for $m, k \in \mathbb{N}_0$, we have observed the identity

$$\int_0^{2\pi} \cos(m\varphi) \cos(k\varphi) \, d\varphi = \frac{2\pi}{2-\delta_{m,0}} \delta_{k,m}. \quad (11.27)$$

In the same way, we also obtain for $n, l \in \mathbb{N}$ and $m = 1, \dots, n$, $k = 1, \dots, l$ that

$$\begin{aligned} &\int_0^{2\pi} \int_0^{\pi} P_{n,m}(\cos\vartheta) P_{l,k}(\cos\vartheta) \sin(m\varphi) \sin(k\varphi) \sin\vartheta \, d\vartheta \, d\varphi \\ &= \int_0^{2\pi} \sin(m\varphi) \sin(k\varphi) \, d\varphi \int_0^{\pi} P_{n,m}(\cos\vartheta) P_{l,k}(\cos\vartheta) \sin\vartheta \, d\vartheta \\ &= \delta_{k,m} \delta_{l,n} \frac{2\pi}{2n+1} \frac{(n+m)!}{(n-m)!}. \end{aligned} \quad (11.28)$$

11.2 Legendre Spherical Harmonics

The functions $P_{n,j}$ and H_j, $j = -n, \ldots, n$, defined by

$$H_j : \varphi \mapsto H(\varphi) = \begin{cases} \cos(j\varphi), & j \leq 0, \\ \sin(j\varphi), & j > 0, \end{cases}, \quad \varphi \in [0, 2\pi), \tag{11.29}$$

respectively, satisfy the differential equations

$$(1-t^2)P''_{n,j}(t) - 2tP'_{n,j}(t) + \left(n(n+1) - \frac{j^2}{1-t^2}\right) P_{n,j}(t) = 0, \tag{11.30}$$

$$H''(\varphi) + j^2 H(\varphi) = 0. \tag{11.31}$$

Therefore, the functions $L_{n,1}, \ldots, L_{n,2n+1} \in \mathrm{C}^{(\infty)}(\mathbb{S}^2)$ given by

$$L_{n,(n+1)+j}(\xi) = \begin{cases} P_{n,|j|}(t)\cos(j\varphi), & j = -n, \ldots, 0, \\ P_{n,j}(t)\sin(j\varphi), & j = 1, \ldots, n, \end{cases} \tag{11.32}$$

satisfy the differential equation

$$(\Delta^*_\xi + n(n+1))L_{n,(n+1)+j}(\xi) = 0, \quad \xi \in \mathbb{S}^2, j = -n, \ldots, n. \tag{11.33}$$

Note that, as always, $\xi = \sqrt{1-t^2}(\cos\varphi\varepsilon^1 + \sin\varphi\varepsilon^2) + t\varepsilon^3$ (cf. (9.1)). In addition, the functions $L_{n,(n+1)+j} \in \mathrm{C}^{(\infty)}(\mathbb{R}^3)$, $j = -n, \ldots, n$, given by

$$L_{n,(n+1)+j}(x) = |x|^n L_{n,(n+1)+j}(\xi), \quad x = |x|\xi,\ \xi \in \mathbb{S}^2, \tag{11.34}$$

form homogeneous harmonic polynomials of degree n in $\mathbb{R}^3$, i.e., they are members of $\mathrm{Harm}_n(\mathbb{R}^3)$.

Definition 11.3. *Let $L_{n,(n+1)+j}, j = -n, \ldots, n$, be defined by (11.32). Then $L_{n,(n+1)+j}$ is called* associated Legendre (spherical) harmonic of degree n and order j. *Correspondingly, the system $\{L^*_{n,(n+1)+j}\}_{j=-n,\ldots,n}$ given by*

$$L^*_{n,(n+1)+j} = C_{n,j}L_{n,(n+1)+j}, \quad j = -n, \ldots, n, \tag{11.35}$$

with

$$C_{n,j} = \sqrt{(2-\delta_{j,0})\frac{2n+1}{4\pi}\frac{(n-|j|)!}{(n+|j|)!}} \tag{11.36}$$

is called (fully) $\mathrm{L}^2(\mathbb{S}^2)$-orthonormal system of associated Legendre (spherical) harmonics *in* $\mathrm{Harm}_n(\mathbb{S}^2)$.

In terms of associated Legendre harmonics, the addition theorem (9.42) allows the following reformulation (that is standard in most geosciences):

Suppose that $\xi, \eta \in \mathbb{S}^2$ are given by

$$\xi = \sqrt{1-t_\xi^2}\cos\varphi_\xi \varepsilon^1 + \sqrt{1-t_\xi^2}\sin\varphi_\xi \varepsilon^2 + t_\xi \varepsilon^3, \tag{11.37}$$

$$\eta = \sqrt{1-t_\eta^2}\cos\varphi_\eta \varepsilon^1 + \sqrt{1-t_\eta^2}\sin\varphi_\eta \varepsilon^2 + t_\eta \varepsilon^3, \tag{11.38}$$

where $t_\xi, t_\eta \in [-1,1]$ and $\varphi_\xi, \varphi_\eta \in [0, 2\pi)$, respectively, so that

$$\begin{aligned}\xi \cdot \eta &= t_\xi t_\eta + \sqrt{1-t_\xi^2}\sqrt{1-t_\eta^2}(\cos\varphi_\xi \cos\varphi_\eta + \sin\varphi_\xi \sin\varphi_\eta) \\ &= t_\xi t_\eta + \sqrt{1-t_\xi^2}\sqrt{1-t_\eta^2}\cos(\varphi_\xi - \varphi_\eta). \end{aligned} \tag{11.39}$$

Then

$$\begin{aligned} &\frac{2n+1}{4\pi} P_n\left(t_\xi t_\eta + \sqrt{1-t_\xi^2}\sqrt{1-t_\eta^2}\cos(\varphi_\xi - \varphi_\eta)\right) \\ &\quad = \frac{2n+1}{4\pi} P_n(t_\xi) P_n(t_\eta) \\ &\qquad + \frac{2n+1}{2\pi} \sum_{m=1}^{n} \frac{(n-m)!}{(n+m)!} P_{n,m}(t_\xi) P_{n,m}(t_\eta) \cos(m(\varphi_\xi - \varphi_\eta)) \\ &\quad = \sum_{j=-n}^{n} L^*_{n,(n+1)+j}(\xi) L^*_{n,(n+1)+j}(\eta) \\ &\quad = \sum_{r=1}^{2n+1} L^*_{n,r}(\xi) L^*_{n,r}(\eta). \end{aligned} \tag{11.40}$$

In other words, summing up all spherical harmonics involving associated Legendre functions via the addition theorem leads (apart from a multiplicative factor) to the orthogonally invariant Legendre (kernel) functions. The geometrical interpretation of the spherical harmonics defined via associated Legendre functions is particularly useful. The harmonics with $j = 0$ – that is, the Legendre polynomials – are polynomials of degree n. They have n zeros. These n zeros are all real, different, and situated in the interval $(-1, 1)$. In consequence, the spherical harmonics with $j = 0$ change their sign n times in this interval; furthermore, they do not depend on the variable φ. Since they divide the sphere into zones, they are also called *zonal harmonics* (see Figure 11.2).

The associated Legendre functions $\vartheta \mapsto P_{n,|j|}(\cos\vartheta)$, $\vartheta \in [0, \pi]$, change their sign $n - |j|$ times in the interval $[0, \pi)$. The trigonometric functions $\varphi \mapsto \cos(j\varphi), j = -n, \ldots, 0$, have $2|j|$ zeros in the interval $[0, 2\pi)$, the functions $\varphi \mapsto \sin(j\varphi), j = 1, \ldots, n$, have $2j$ zeros in the interval $[0, 2\pi)$. Consequently, the geometrical representation of the harmonics for the case $|j| \neq n$ is similar to that of Figure 11.3. They divide the sphere into compartments in which they are alternately positive and negative, and are called *tesseral harmonics*. "Tesseral" means a square or rectangle.

In particular, for $|j| = n$, they degenerate into functions that divide the sphere into positive and negative sectors, in which case they are called *sectorial harmonics* (see Figure 11.4).

FIGURE 11.2: Zonal ($j = 0$) spherical harmonics $L^*_{n,(n+1)+j}$ of different degrees $1, \ldots, 6$ (from left to right). The black and white color indicate the zones of different signs of the function, respectively.

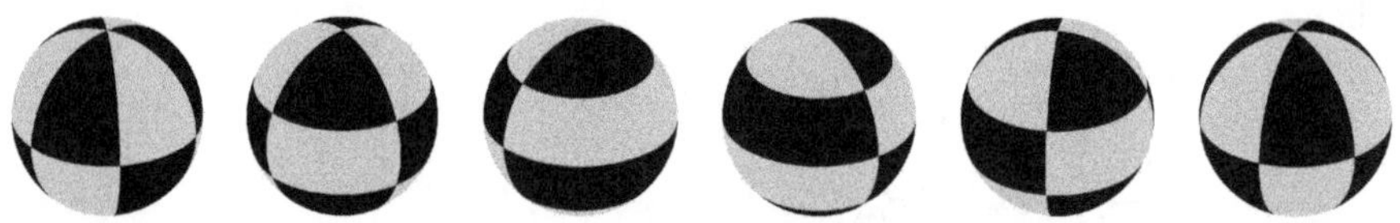

FIGURE 11.3: Tesseral ($j \neq \pm n$) spherical harmonics $L^*_{n,(n+1)+j}$ of degree $n = 4$, i.e., $L^*_{4,2}$, $L^*_{4,3}$, $L^*_{4,4}$, $L^*_{4,6}$, $L^*_{4,7}$, $L^*_{4,8}$ (from left to right). The black and white colors indicate the zones of different signs of the function, respectively.

FIGURE 11.4: Sectorial ($j = \pm n$) spherical harmonics $L^*_{n,(n+1)+j}$ of different degrees: $L^*_{1,3}$, $L^*_{2,5}$, $L^*_{3,7}$, $L^*_{4,9}$, $L^*_{5,11}$, $L^*_{6,13}$ (from left to right). The black and white colors indicate the zones of different signs of the function, respectively.

11.3 Latitude-Longitude Integration

It is well-known that latitude-longitude point sets on the unit sphere $\mathbb{S}^2 \subset \mathbb{R}^3$ enable approximate integration whose weights are available in a comparatively easy and fast way. They establish spherical harmonics exact cubature formulas which are of tremendous importance. Unfortunately, they are not equidistributed in the sense of Weyl (cf. Chapter 14).

In what follows we are first concerned with the arrangement of equiangular point sets. Systems of this type can be gained by suitably dividing the rectangle $[0,\pi] \times [0,2\pi)$ into latitude-longitude grids. Clearly, there are numerous possibilities for a choice of locations concerning the type and number of latitude as well as longitude distributions. Nonetheless, these choices are essential for the quality of the resulting integration formulas.

For application of latitude-longitude grid integration, suppose that the spherical harmonics are represented in the following way using (11.29), (11.32), and (11.35) for $n \in \mathbb{N}_0$

$$\begin{aligned} Y_{n,j}(\xi) &= L^*_{n,(n+1)+j}(\xi) = C_{n,j}P_{n,|j|}(\cos(\vartheta))H_j(\varphi) \\ &= C_{n,j}P_{n,|j|}(\cos(\vartheta)) \begin{cases} \cos(j\varphi), & j=-n,\ldots,0, \\ \sin(j\varphi), & j=1,\ldots,n, \end{cases} \end{aligned} \tag{11.41}$$

where we use spherical coordinates (9.1), i.e., $\xi = t\varepsilon^3 + \sqrt{1-t^2}(\cos\varphi\varepsilon^1 + \sin\varphi\varepsilon^2)$, $t = \cos\vartheta$, $t \in [-1,1]$, $\varphi \in [0,2\pi)$, $\vartheta \in [0,\pi]$ and $C_{n,j}$ is given by (11.36).

On the one hand, taking into account that $Y_{0,0} = \frac{1}{\sqrt{4\pi}}$, we obtain

$$\int_{\mathbb{S}^2} Y_{n,j}(\xi)\, dS(\xi) = \begin{cases} \sqrt{4\pi} & \text{if } n=j=0, \\ 0 & \text{else.} \end{cases} \tag{11.42}$$

On the other hand, it can be readily seen that

$$\int_{\mathbb{S}^2} Y_{n,j}(\xi)\, dS(\xi) = \int_0^\pi C_{n,j}P_{n,|j|}(\cos(\vartheta))\sin(\vartheta)\, d\vartheta \int_0^{2\pi} H_j(\varphi))\, d\varphi \tag{11.43}$$

with H_j from (11.29). The combination of (11.42) and (11.43) leads to the equation

$$\int_0^\pi C_{n,j}P_{n,|j|}(\cos(\vartheta))\sin(\vartheta)\, d\vartheta \int_0^{2\pi} H_j(\varphi)\, d\varphi = \begin{cases} \sqrt{4\pi} & \text{if } n=j=0, \\ 0 & \text{else.} \end{cases} \tag{11.44}$$

This identity is the point of departure for the derivation of spherical harmonics exact integration formulas. Together with $C_{0,0} = \frac{1}{\sqrt{4\pi}}$ we easily see that

$$\int_0^\pi P_{n,|j|}(\cos(\vartheta))\sin(\vartheta)\, d\vartheta \int_0^{2\pi} H_j(\varphi)\, d\varphi = \begin{cases} 4\pi & \text{if } n=j=0, \\ 0 & \text{else.} \end{cases} \tag{11.45}$$

Cubature Based on Periodic Integration. In accordance with (11.43) we are able to replace the integral on the whole unit sphere by latitude-longitude integration. As a consequence, the desired cubature rule on the sphere can be based on already known approaches to numerical integration in one dimension. In order to establish spherical harmonics exact integration formulas we pursue the following three-step strategy:

(i) Determine a number $N_\varphi \in \mathbb{N}$, the nodes $\varphi_0, \ldots, \varphi_{N_\varphi - 1} \in [0, 2\pi)$, and the weights $a_0^\varphi, \ldots, a_{N_\varphi-1}^\varphi \in \mathbb{R}$, so that the integration formula

$$\int_0^{2\pi} H_j(\varphi)\, d\varphi = \sum_{k=0}^{N_\varphi - 1} a_k^\varphi H_j(\varphi_k) = \begin{cases} 2\pi & \text{if } j = 0, \\ 0 & \text{else,} \end{cases} \tag{11.46}$$

is exact for $j = 0, \ldots, m$, where H_j is a sine or cosine (cf. (11.29)). Evidently, (11.46) is fulfilled for $j \neq 0$ if the longitude φ is discretized by N_φ different nodes in the described way. This is why the integral $\int_0^\pi P_{n,|j|}(\cos(\vartheta)) \sin(\vartheta)\, d\vartheta$ is only significant for $j = 0$ and $n = 0, \ldots, m$.

(ii) Concerning the discretization of the latitude ϑ it is sufficient to consider the following integration rule: Determine a number $N_\vartheta \in \mathbb{N}$, the nodes $\vartheta_0, \ldots, \vartheta_{N_\vartheta - 1} \in [0, \pi]$, and the weights $a_0^\vartheta, \ldots, a_{N_\vartheta - 1}^\vartheta \in \mathbb{R}$, so that

$$\int_0^\pi \underbrace{P_{n,0}(\cos(\vartheta))}_{=P_n(\cos(\vartheta))} \sin(\vartheta) d\vartheta = \sum_{l=0}^{N_\vartheta - 1} a_l^\vartheta P_n(\cos(\vartheta_l)) \tag{11.47}$$

is exact for $n = 0, \ldots, m$.

(iii) Combine the latitude-longitude nodal systems to the product set

$$(\vartheta_l, \varphi_k) \in [0, \pi) \times [0, 2\pi), \quad l = 0, \ldots, N_\vartheta - 1,\ k = 0, \ldots, N_\varphi - 1, \tag{11.48}$$

and collect the integration weights $a_0^\varphi, \ldots, a_{N_\varphi - 1}^\varphi$ and $a_0^\vartheta, \ldots a_{N_\vartheta - 1}^\vartheta$.

We begin with the discretization with respect to the longitude that is well-known from (6.66).

Lemma 11.4 (Longitude Integration). *For $N_\varphi \in \mathbb{N}$, assume that the nodes are given by $\varphi_k = k\frac{2\pi}{N_\varphi} \in [0, 2\pi)$, $k = 0, \ldots, N_\varphi - 1$. Then, the integration rule*

$$\int_0^{2\pi} T(\varphi)\, d\varphi = \sum_{k=0}^{N_\varphi - 1} \frac{2\pi}{N_\varphi} T\left(\frac{2\pi k}{N_\varphi}\right) \tag{11.49}$$

holds true for all trigonometric functions T of degree $\leq N_\varphi - 1$.

In order to discretize the latitude ϑ, however, it is evident that the integration weights cannot be expected to be equal for all samples. In fact, there are several possibilities to determine point sets and weights which fulfill (11.46). The methods presented here reflect straightforward numerical application.

In accordance with (11.47) we are interested in constructing an integral formula of the form

$$\int_0^\pi P_n(\cos(\vartheta))\sin(\vartheta)\, d\vartheta = \sum_{l=0}^{N_\vartheta - 1} a_l^\vartheta P_n(\cos(\vartheta_l)) \tag{11.50}$$

for $n = 0, \ldots, m$. Since we already know that the set $\{P_0, \ldots, P_{N_\vartheta}\}$, $N_\vartheta \in \mathbb{N}$, represents a *Tschebyscheff-system* in the univariate sense (see also P.J. Davis, P. Rabinowitz [1975]), integration formulas of the type (11.50) exist if the angles $\vartheta_l, l = 0, \ldots, N_\vartheta - 1$, are chosen to be pairwise distinct. In particular, we can think of an equiangular distribution of the samples on latitudes. As a matter of fact, we briefly explain an integration technique whose origin dates back to J.R. Driscoll, R.M. Healy [1994]. For that purpose we notice that, for an even $N_\vartheta \in \mathbb{N}$ and for $n = 0, \ldots, N_\vartheta - 1$

$$\int_0^\pi P_n(\cos(\vartheta))\sin(\vartheta)\, d\vartheta = \frac{1}{2}\int_{-\pi}^\pi P_n(\cos(\vartheta))\sin(\vartheta)\,\mathrm{sign}(\vartheta)\, d\vartheta. \tag{11.51}$$

Remembering the well-known sign expansion (see, e.g., W. Magnus et al. [1966])

$$\mathrm{sign}(\vartheta) = \frac{4}{\pi}\sum_{s=0}^\infty \frac{\sin((2s+1)\vartheta)}{2s+1}, \quad \vartheta \in [-\pi, \pi], \tag{11.52}$$

we obtain for $n = 0, \ldots, N_\vartheta - 1$

$$\int_0^\pi P_n(\cos(\vartheta))\sin(\vartheta)\, d\vartheta = \frac{2}{\pi}\sum_{s=0}^\infty \int_{-\pi}^\pi P_n(\cos(\vartheta))\sin(\vartheta)\frac{\sin((2s+1)\vartheta)}{2s+1}\, d\vartheta. \tag{11.53}$$

It is clear that, for $n = 0, \ldots, N_\vartheta - 1$, the term $P_n(\cos(\vartheta))\sin(\vartheta)$ represents a trigonometric function of degree $\leq N_\vartheta$. Due to the orthogonality relations we readily find that

$$\int_{-\pi}^\pi P_n(\cos(\vartheta))\sin\vartheta\sin((2s+1)\vartheta)\, d\vartheta = 0, \quad s > \frac{N_\vartheta}{2} - 1. \tag{11.54}$$

Hence, the series (11.53) reduces to

$$\int_0^\pi P_n(\cos(\vartheta))\sin(\vartheta)\, d\vartheta = \frac{2}{\pi}\sum_{s=0}^{\frac{N_\vartheta}{2}-1} \int_{-\pi}^\pi P_n(\cos(\vartheta))\sin(\vartheta)\frac{\sin((2s+1)\vartheta)}{2s+1}\, d\vartheta. \tag{11.55}$$

The integrand on the right side is a trigonometric function of degree $2N_\vartheta - 1$, hence, it is not difficult to show that the identity

$$\frac{2}{\pi} \sum_{s=0}^{\frac{N_\vartheta}{2}-1} \int_{-\pi}^{\pi} P_n(\cos(\vartheta)) \sin(\vartheta) \frac{\sin((2s+1)\vartheta)}{2s+1} \, d\vartheta \tag{11.56}$$

$$= \frac{2}{N_\vartheta} \sum_{l=-N_\vartheta}^{N_\vartheta - 1} P_n\left(\cos\left(l\frac{\pi}{N_\vartheta}\right)\right) \sin\left(l\frac{\pi}{N_\vartheta}\right) \sum_{s=0}^{\frac{N_\vartheta}{2}-1} \frac{1}{2s+1} \sin\left((2s+1)\, l\frac{\pi}{N_\vartheta}\right)$$

holds true for $n = 0, \ldots, N_\vartheta - 1$. Moreover, the symmetry of the trigonometric functions enables us to rewrite the first sum on the right-hand side of (11.56) as follows

$$\int_0^\pi P_n(\cos(\vartheta)) \sin(\vartheta) \, d\vartheta = \frac{2}{N_\vartheta} \sum_{l=0}^{N_\vartheta - 1} P_n\left(\cos\left(l\frac{\pi}{N_\vartheta}\right)\right) \sin\left(l\frac{\pi}{N_\vartheta}\right) \times \sum_{s=0}^{\frac{N_\vartheta}{2}-1} \frac{1}{2s+1} \sin\left((2s+1) l\frac{\pi}{N_\vartheta}\right) \tag{11.57}$$

for $n = 0, \ldots, N_\vartheta - 1$. Finally, taking into account that the system of Legendre polynomials P_n, for $n = 0, \ldots, N_\vartheta - 1$, forms a basis of the space of all polynomials of degree less or equal to $N_\vartheta - 1$ we arrive at following result.

Theorem 11.5 (Latitude Integration). *Let $N_\vartheta \in \mathbb{N}$ be an even number. Suppose that $\vartheta_l = l\frac{\pi}{N_\vartheta}$, $l = 0, \ldots, N_\vartheta - 1$. Then, the integration rule*

$$\int_0^\pi P(\cos(\vartheta)) \sin(\vartheta) \, d\vartheta = \int_{-1}^{+1} P(t) \, dt = \sum_{l=0}^{N_\vartheta - 1} a_l^\vartheta P(\cos(\vartheta_l)) \tag{11.58}$$

with the weights

$$a_l^\vartheta = \frac{4}{N_\vartheta} \sin\left(l\frac{\pi}{N_\vartheta}\right) \sum_{s=0}^{\frac{N_\vartheta}{2}-1} \frac{1}{2s+1} \sin\left((2s+1) l\frac{\pi}{N_\vartheta}\right) \tag{11.59}$$

holds true for all polynomials P of degree $\leq N_\vartheta - 1$.

Now, especially for $N_\vartheta = N_\varphi = m+1$ with $m \in \mathbb{N}$ odd, the combination of the above schemes for latitude-longitude distribution enables us to arrange an *equiangular latitude-longitude distributed grid* (ELLG) integration rule, where the nodes and weights, respectively, are given by

$$(\vartheta_l, \varphi_k) \in [0, \pi) \times [0, 2\pi), \quad \vartheta_l = l\frac{\pi}{m+1}, \; l = 0, \ldots, m, \tag{11.60}$$

$$\varphi_k = k\frac{2\pi}{m+1}, \; k = 0, \ldots, m,$$

and

$$a_l^\vartheta = \frac{4}{m+1} \sin\left(l\frac{\pi}{m+1}\right) \sum_{s=0}^{\frac{m+1}{2}-1} \frac{1}{2s+1} \sin\left((2s+1)l\frac{\pi}{m+1}\right), \; l = 0, \ldots, m,$$
$$a_k^\varphi = \frac{2\pi}{m+1}, \quad k = 0, \ldots, m. \tag{11.61}$$

Figure 11.5 (left column) gives a graphical illustration of the equiangular latitude-longitude grid (ELLG).

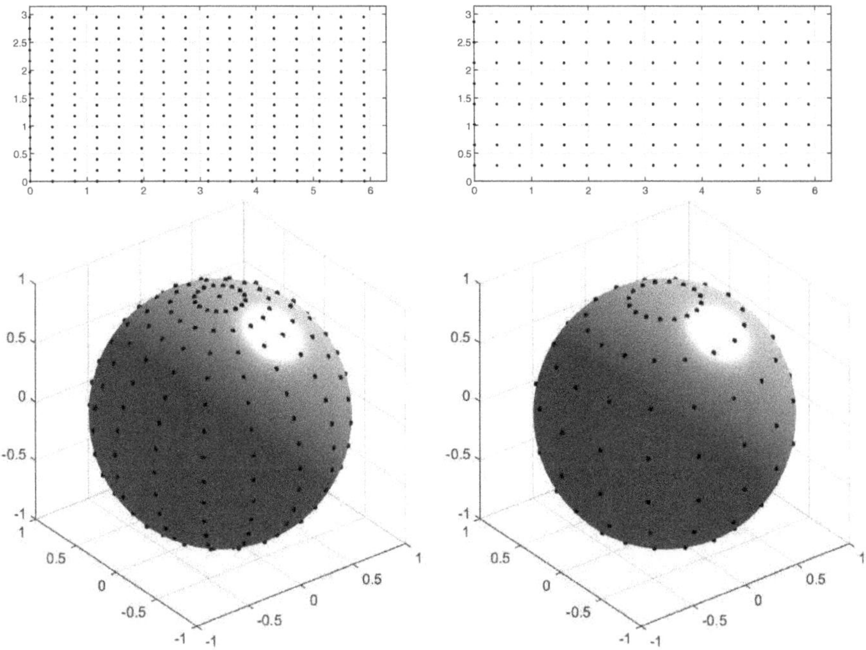

FIGURE 11.5: The ELLG latitude-longitude grid with $m = 15$ (left column) and the Gaussian Grid (GG) with $m = 15$ (right column).

As a consequence, any $Y \in Harm_{0,\ldots,m}$ can be integrated exactly in the form

$$\int_{\mathbb{S}^2} Y(\eta)\, dS(\eta) = \sum_{l=0}^{m} \sum_{k=0}^{m} a_l^\vartheta \, a_k^\varphi \, Y(\vartheta_l, \varphi_k). \tag{11.62}$$

In particular, if a product FY is of class $Harm_{0,\ldots,m}$, then the inner product $\langle F, Y \rangle_{\mathrm{L}^2(\mathbb{S}^2)}$ can be evaluated exactly by means of a proposed latitude-longitude grid based on $(m+1)^2$ points.

Cubature Based on Gaussian Integration. From the identity (11.45) we know that integration involving spherical harmonics along spherical longitudes and latitudes, respectively, can be treated independently from each other. Thus, classical Gaussian quadrature can also be used to construct an efficient cubature formula for latitude approximation.

We first choose the longitude $\varphi \in [0, 2\pi)$ to be discretized in an equiangular way. Our purpose is to reduce the number N_ϑ of the required latitudes in relation to the already discussed equiangular latitude-longitude grid (ELLG) by use of Gaussian quadrature. As a consequence, let N_ϑ and N_φ, as usual, denote the number of spherical latitudes and longitudes, respectively, to be involved. Assume, especially, that $m \in \mathbb{N}$ is odd, and N_ϑ as well as N_φ are chosen (see Figure 11.5 (right column) for an example of a Gaussian grid (GG)) as follows:

$$N_\varphi = m + 1, \tag{11.63}$$

$$N_\vartheta = \frac{1}{2}(m + 1). \tag{11.64}$$

Let $t^\vartheta_{1,N_\vartheta} < \ldots < t^\vartheta_{N_\vartheta,N_\vartheta}$ be the zeros of the Legendre polynomial of degree N_ϑ, and suppose that the grid (GG) is given by

$$(\vartheta_l, \varphi_k) \in [0, \pi] \times [0, 2\pi), \tag{11.65}$$

with

$$\vartheta_l = \arccos(t^\vartheta_{l,N_\vartheta}), \quad l = 1, \ldots, N_\vartheta = \frac{1}{2}(m+1), \tag{11.66}$$

$$\varphi_k = k\frac{2\pi}{m+1}, \quad k = 0, \ldots, m. \tag{11.67}$$

Then, for $Y \in Harm_{0,\ldots,m}$ we have

$$\begin{aligned} \int_{\mathbb{S}^2} Y(\eta)\, dS(\eta) &= \sum_{l=0}^{\frac{1}{2}(m+1)} \sum_{k=0}^{m} a^\vartheta_{l,N_\vartheta} a^\varphi_k Y(\vartheta_l, \varphi_k) \\ &= \frac{2\pi}{m+1} \sum_{l=0}^{\frac{1}{2}(m+1)} a^\vartheta_{l,N_\vartheta} \sum_{k=0}^{m} Y(\vartheta_l, \varphi_k), \end{aligned} \tag{11.68}$$

where the weights $a^\vartheta_{l,N_\vartheta}$ are given by the Gaussian integration (see Chapter 8).

Thus, for $m \in \mathbb{N}$ odd and $F, Y \in \mathrm{Harm}_{0,\ldots,\frac{m-1}{2}}$, the product FY is of class $\mathrm{Harm}_{0,\ldots,m}$, hence, the inner product of F and Y given by

$$\langle Y, F\rangle_{\mathrm{L}^2(\mathbb{S}^2)} = \int_{\mathbb{S}^2} Y(\eta)F(\eta)\, dS(\eta) \tag{11.69}$$

can be calculated in an exact way by use of the Gaussian grid (GG) with $\frac{1}{2}(m+1)^2$ points. In particular, this result can be applied to the determination of the Fourier coefficients of an arbitrary function $F \in \mathrm{L}^2(\mathbb{S}^2)$. In more detail, if the Fourier series expansion of a function $F \in \mathrm{L}^2(\mathbb{S}^2)$, i.e., in the sense of $\mathrm{L}^2(\mathbb{S}^2)$ we have

$$F = \sum_{n=0}^{\infty} \sum_{j=-n}^{n} \langle F, Y_{n,j} \rangle_{\mathrm{L}^2(\mathbb{S}^2)} \, Y_{n,j} \tag{11.70}$$

with the Fourier coefficients given by the usual projection

$$\langle F, Y_{n,j} \rangle_{\mathrm{L}^2(\mathbb{S}^2)} = \int_{\mathbb{S}^2} F(\eta) Y_{n,j}(\eta) \, dS(\eta), \tag{11.71}$$

is replaced by its bandlimited version of degree m (for sufficiently large odd m)

$$F_{0,\ldots,m} = \sum_{n=0}^{m} \sum_{j=-n}^{n} \langle Y_{n,j}, F \rangle_{\mathrm{L}^2(\mathbb{S}^2)} \, Y_{n,j}, \tag{11.72}$$

we are allowed to conclude that

$$\langle F, Y_{n,j}, \rangle_{\mathrm{L}^2(\mathbb{S}^2)} = \langle F_{0,\ldots,m}, Y_{n,j} \rangle_{\mathrm{L}^2(\mathbb{S}^2)}, \quad n = 0, \ldots, m, \; j = -n, \ldots, n. \tag{11.73}$$

In connection with (11.68) we then obtain with $N_\vartheta = \frac{1}{2}(m+1)$ the following discrete version

$$F_{0,\ldots,m} = \sum_{n=0}^{m} \sum_{j=-n}^{n} \left(\frac{2\pi}{m+1} \sum_{l=0}^{\frac{1}{2}(m+1)} w_{l,N_\vartheta}^{\vartheta} \sum_{k=0}^{m} Y_{n,j}(\vartheta_l, \varphi_k) \, F(\vartheta_l, \varphi_k) \right) Y_{n,j}. \tag{11.74}$$

Remark 11.6. *Recently, a novel cubature formula based on the relation of spherical harmonics to Wigner functions has been proposed by J.D. McEwen, Y. Wiaux [2011], that achieves the same degree of accuracy as the Gaussian formula with (approximately) half as many points as the aforementioned approach of J.R. Driscoll, R.M. Healy [1994] (the details will not be discussed here).*

Once more, it should be pointed out that both the equiangular latitude-longitude grid (ELLG) and the Gaussian grid (GG) are constructed by a uniform distribution of points in the two-dimensional interval $[0, \pi] \times [0, 2\pi)$. Unfortunately, this means that these point sets are not equidistributed on the sphere in the sense of Weyl (see, e.g., W. Freeden et al. [1998] for more details). In fact, the nodal width of points around the poles of $\mathbb{S}^2$ is comparatively small, while larger distances between the nodes can be observed around the equator. In the case of (ELLG), even a certain number of points coincides with the poles, but their corresponding weight is 0. The Gaussian grid (GG) achieves the same degree of accuracy with (approximately) half as many points as the system (ELLG). However, the zeros of the Legendre polynomials have to be included which is not as simple as for the grid (ELLG). Besides the prescription of the location of nodal points where the measurements must be done, the observation that the integrand is usually not sampled uniformly on the sphere $\mathbb{S}^2$ is of disadvantage except, e.g., for certain satellite campaigns, where the data are acquired in a similar way. No doubt, the facts that the approximate integration rules are explicitly known, easily implementable, and economical even for a higher degree m, make latitude-longitude sampling attractive. All in all, the cubature method presented here is only useful for regular grids generated by partitioning of the two-dimensional interval $[0, \pi) \times [0, 2\pi)$. For scattered data distributions this integration technique is not applicable. In this case, other cubature formulas (e.g., best approximate integration as proposed in Section 9.3 and Chapter 13) have to come into play.

A. Townsend et al. [2016] are concerned with extensions of the latitude-longitude grids, the so-called "double Fourier sphere" technique, and a collection of resulting algorithms. The *double Fourier sphere (DFS) method* is a procedure for which new algorithmic tools (software packages) alleviate many of the drawbacks inherent with standard polar coordinate transform, namely no artificial pole singularities, representations resulting in fast algorithms, structural features that make differentiation stable, and an underlying interpolation grid that rarely oversamples functions near poles.

Instead of latitude-longitude integration one may think of a function on $\mathbb{S}^2$ as a periodic function of latitude and longitude, so that *bivariate trigonometric interpolation* comes into play to fit the data. One problem besides the non-equidistribution is that not all bivariate trigonometric polynomials are continuous on $\mathbb{S}^2$. Several contributions have considered approximation by special classes of bivariate trigonometric polynomials (see, e.g., J. P. Boyd [2001] and the references therein). In M. Ganesh et al. [1998], a subspace of bivariate trigonometric polynomials was introduced to facilitate interpolation of functions on $\mathbb{S}^2$. Following M. Ganesh, H.N. Mhaskar [2006] the interpolation is called matrix-free, if a system of interpolation nodes and basis functions is explicitly given, so that a formula analogous to the FFT-type defines an interpolation operator. As a consequence, the authors are able to show that the uniform norm of the interpolation operator is of the order $(\log N)^2$, where N is the number of interpolation points.

M. Ganesh, H.N. Mhaskar [2006] also point out that their constructions are different from the classical theory of spherical harmonics interpolation, where the resulting spherical polynomial interpolation operator as well as the computationally interesting condition number of the interpolation matrix are a challenge for future work.

M. Ganesh, V. Dominguez [2013] discuss and implement interpolatory approximation and Filon-type cubature for a class of very low to very high frequency Fourier integrals. The paper is a canonical extension of the construction of matrix-free interpolation operators on the sphere. In addition to the contribution of M. Ganesh, H.N. Mhaskar [2006] Sobolev norm error estimates of FFT-type interpolation operators are studied. Numerical examples demonstrate the "wideband" cubature. Furthermore, an optimal order convergence analysis is presented.

Chapter 12

Remainder Terms Involving Pseudodifferential Operators

In Chapter 9, we developed integral formulas showing explicit remainder term expressions involving Green's functions with respect to (iterated) Beltrami operators. The difficulty of using Green's functions for Beltrami operators is their explicit representation in terms of numerically manageable expressions, so that implementations often turn out to be laborious. This is the reason why we are interested in generalizing these results to rotation-invariant pseudodifferential operators showing much more accessible realizations of "Green's functions" and integral formulas for computational purposes, at least for a large class of pseudodifferential operators. Our considerations given in this chapter represent an extension of results presented in W. Freeden et al. [1998].

12.1 Sobolev Spaces

The point of departure for our considerations (cf. W. Freeden [1990], W. Freeden et al. [1996]) is the linear space $\mathfrak{A}$ consisting of all sequences $\{A_n\}$ of real numbers A_n, $n \in \mathbb{N}_0$:

$$\mathfrak{A} = \{\{A_n\} : \ A_n \in \mathbb{R}, \ n \in \mathbb{N}_0\}. \tag{12.1}$$

Let $\{A_n\} \in \mathfrak{A}$ be a sequence with $A_n \neq 0$ for all n. Consider the set $\mathrm{E} = \mathrm{E}(\{A_n\}; \mathbb{S}^2)$ of all infinitely differentiable functions F on $\mathbb{S}^2$ satisfying

$$\sum_{n=0}^{\infty} \sum_{j=1}^{2n+1} |A_n|^2 \langle F, Y_{n,j} \rangle^2_{\mathrm{L}^2(\mathbb{S}^2)} < \infty. \tag{12.2}$$

From the Cauchy–Schwarz inequality it follows that

$$\left| \sum_{n=0}^{\infty} \sum_{j=1}^{2n+1} |A_n|^2 \langle F, Y_{n,j} \rangle_{\mathrm{L}^2(\mathbb{S}^2)} \langle G, Y_{n,j} \rangle_{\mathrm{L}^2(\mathbb{S}^2)} \right|^2 \tag{12.3}$$

$$\leq \left(\sum_{n=0}^{\infty} \sum_{j=1}^{2n+1} |A_n|^2 \langle F, Y_{n,j} \rangle^2_{\mathrm{L}^2(\mathbb{S}^2)} \right) \left(\sum_{n=0}^{\infty} \sum_{j=1}^{2n+1} |A_n|^2 \langle G, Y_{n,j} \rangle^2_{\mathrm{L}^2(\mathbb{S}^2)} \right)$$

for all $F, G \in \mathrm{E}$, hence, the left-hand side of (12.3) is finite whenever each member of the right-hand side is finite. We can therefore define an inner product $\langle \cdot, \cdot \rangle_{\mathrm{H}(\{A_n\};\mathbb{S}^2)}$ on the space E by letting

$$\langle F, G \rangle_{\mathrm{H}(\{A_n\};\mathbb{S}^2)} = \sum_{n=0}^{\infty} \sum_{j=1}^{2n+1} |A_n|^2 \langle F, Y_{n,j} \rangle_{\mathrm{L}^2(\mathbb{S}^2)} \langle G, Y_{n,j} \rangle_{\mathrm{L}^2(\mathbb{S}^2)}. \tag{12.4}$$

The associated norm is given by

$$\|F\|_{\mathrm{H}(\{A_n\};\mathbb{S}^2)} = \langle F, F \rangle^{1/2}_{\mathrm{H}(\{A_n\};\mathbb{S}^2)}. \tag{12.5}$$

Definition 12.1. *The* Sobolev space $\mathrm{H}(\{A_n\};\mathbb{S}^2)$ *is the completion of the class* $\mathrm{E}(\{A_n\};\mathbb{S}^2)$ *under the norm (12.5):*

$$\mathrm{H}(\{A_n\};\mathbb{S}^2) = \overline{\mathrm{E}(\{A_n\};\mathbb{S}^2)}^{\|\cdot\|_{\mathrm{H}(\{A_n\};\mathbb{S}^2)}}. \tag{12.6}$$

$\mathrm{H}(\{A_n\};\mathbb{S}^2)$ equipped with the inner product corresponding to the norm (12.5) is a Hilbert space. It follows from the Cauchy–Schwarz inequality (12.3) that $\langle F, G \rangle_{\mathrm{H}(\{1\};\mathbb{S}^2)}$ exists if $F \in \mathrm{H}(\{A_n\};\mathbb{S}^2)$ and $G \in \mathrm{H}(\{A_n^{-1}\};\mathbb{S}^2)$. In addition,

$$|\langle F, G \rangle_{\mathrm{H}(\{1\};\mathbb{S}^2)}| \leq \|F\|_{\mathrm{H}(\{A_n\};\mathbb{S}^2)} \|G\|_{\mathrm{H}(\{A_n^{-1}\};\mathbb{S}^2)}. \tag{12.7}$$

Hence, $\langle \cdot, \cdot \rangle_{\mathrm{H}(\{1\};\mathbb{S}^2)}$ defines a duality of $\mathrm{H}(\{A_n\};\mathbb{S}^2)$ and $\mathrm{H}(\{A_n^{-1}\};\mathbb{S}^2)$.

Remark 12.2. $\mathrm{H}(\{A_n\};\mathbb{S}^2)$ *may be identified with a subspace of* $\mathfrak{A}$ *by associating with each element* $F \in \mathrm{H}(\{A_n\};\mathbb{S}^2)$ *its sequence* $\{F^\wedge(n,j)\}_{(n,j)}$, *where* $F^\wedge(n,j) = \langle F, Y_{n,j} \rangle_{\mathrm{L}^2(\mathbb{S}^2)}$.

As already mentioned, $\Delta^* Y_{n,j} = -n(n+1) Y_{n,j}$ holds true for $n = 0, 1, \ldots$ and $j = 1, \ldots, 2n+1$. Therefore, we are formally able to set

$$\left(\left(-\Delta^* + \frac{1}{4} \right)^{s/2} F \right)^\wedge (n,j) = \left(n + \frac{1}{2} \right)^s F^\wedge(n,j). \tag{12.8}$$

For brevity, we let

$$\mathrm{H}^s(\mathbb{S}^2) = \mathrm{H}\left(\left\{ \left(n + \frac{1}{2} \right)^s \right\}; \mathbb{S}^2 \right) \tag{12.9}$$

for each $s \in \mathbb{R}$. Hence, the norm in $\mathrm{H}^s(\mathbb{S}^2)$ is given by

$$\|F\|_{\mathrm{H}^s(\mathbb{S}^2)} = \|F\|_{\mathrm{H}(\{(n+\frac{1}{2})^s\};\mathbb{S}^2)} = \left\| \left(-\Delta^* + \frac{1}{4} \right)^{s/2} F \right\|_{\mathrm{L}^2(\mathbb{S}^2)}. \tag{12.10}$$

In particular, $\mathrm{H}^0(\mathbb{S}^2) = \mathrm{L}^2(\mathbb{S}^2)$, and we define the norm $\| \cdot \|_{\mathrm{H}^0(\mathbb{S}^2)}$ to be, simply, the L^2-norm. Moreover, according to our construction, $\mathrm{C}^{(\infty)}(\mathbb{S}^2)$ is a dense subspace of $\mathrm{H}^s(\mathbb{S}^2)$ for each s. If $t < s$, then $\|F\|_{\mathrm{H}^t(\mathbb{S}^2)} \leq \|F\|_{\mathrm{H}^s(\mathbb{S}^2)}$, and $\mathrm{H}^s(\mathbb{S}^2) \subset \mathrm{H}^t(\mathbb{S}^2)$. Thus, the union of H^s-spaces over all integers s may be understood as a subspace of $\mathfrak{A}$. If $F \in \mathrm{H}^s(\mathbb{S}^2)$ and $G \in \mathrm{H}^{-s}(\mathbb{S}^2)$, then

$$\langle F, G\rangle_{\mathrm{H}(\{1\};\mathbb{S}^2)} = \int_{\mathbb{S}^2} \left(\left(-\Delta^*_\xi + \frac{1}{4}\right)^{s/2} F(\xi)\right)\left(\left(-\Delta^*_\xi + \frac{1}{4}\right)^{-s/2} G(\xi)\right) dS(\xi). \tag{12.11}$$

It should be noted that if F and G are both in $\mathrm{L}^2(\mathbb{S}^2)$, then we have

$$\langle F, G\rangle_{\mathrm{H}(\{1\};\mathbb{S}^2)} = \langle F, G\rangle_{\mathrm{L}^2(\mathbb{S}^2)} = \int_{\mathbb{S}^2} F(\xi)\ G(\xi)\ dS(\xi). \tag{12.12}$$

The operator $(-\Delta^* + \frac{1}{4})^{t/2}$ is a bounded operator from $\mathrm{H}^{s+t}(\mathbb{S}^2)$ to $\mathrm{H}^s(\mathbb{S}^2)$ for each s, indeed, we have

$$\left\| \left(-\Delta^* + \frac{1}{4}\right)^{t/2} F \right\|_{\mathrm{H}^s(\mathbb{S}^2)} = \|F\|_{\mathrm{H}^{s+t}(\mathbb{S}^2)} \tag{12.13}$$

for all $F \in \mathrm{H}^{s+t}(\mathbb{S}^2)$.

12.2 Pseudodifferential Operators

We restrict ourselves to rotation-invariant pseudodifferential operators, i.e., pseudodifferential operators A with symbol independent of j, i.e., $AY_{n,j} = A^\wedge(n)Y_{n,j}$ for all n, j. A more detailed discussion can be found in W. Freeden et al. [1998], W. Freeden, M. Schreiner [2009], S. L. Svensson [1983]. General references to the mathematical theory in Euclidean spaces are, e.g., G.I. Eskin [1981], L. Hörmander [1965], and many others.

Definition 12.3. *Let $\{A^\wedge(n)\}_{n=0,1,\ldots}$ be a sequence of real numbers $A^\wedge(n)$ satisfying*

$$\lim_{n\to\infty} \frac{|A^\wedge(n)|}{(n+\frac{1}{2})^t} = const. \neq 0 \tag{12.14}$$

for some $t \in \mathbb{R}$. Then the operator $A : \mathrm{H}^s(\mathbb{S}^2) \to \mathrm{H}^{s-t}(\mathbb{S}^2)$ defined by

$$AF = \sum_{n=0}^{\infty} \sum_{j=1}^{2n+1} A^\wedge(n) F^\wedge(n,j) Y_{n,j}, \quad F \in \mathrm{H}^s(\mathbb{S}^2), \tag{12.15}$$

is called (rotation-invariant) pseudodifferential operator of order t. $\{A^\wedge(n)\}$ *is called* spherical symbol *of A. Moreover, if*

$$\lim_{n\to\infty} \frac{|A^\wedge(n)|}{(n+\frac{1}{2})^t} = 0 \tag{12.16}$$

for all $t \in \mathbb{R}$*, then the operator* $A : \mathrm{H}^s \to \mathrm{C}^{(\infty)}(\mathbb{S}^2)$ *is called* pseudodifferential operator of order $-\infty$ *(note that Equation (12.15) is understood in the* $\mathrm{H}^{s-t}(\mathbb{S}^2)$*-topology).*

The spherical symbol has many appealing properties. It is easily seen that $(A' + A'')^\wedge(n) = (A')^\wedge(n) + (A'')^\wedge(n)$, $(A'A'')^\wedge(n) = (A')^\wedge(n)(A'')^\wedge(n)$ for all n. Obviously, we have in case of rotation-invariance

$$AY_{n,j} = A^\wedge(n)Y_{n,j}, \quad n = 0, 1, \ldots, j = 1, \ldots, 2n+1. \tag{12.17}$$

Moreover, we have

$$\|U\|_{\mathrm{H}(\{A^\wedge(n)B_n\})} = \|AU\|_{\mathrm{H}(\{B_n\};\mathbb{S}^2)}, \quad U \in \mathrm{H}(\{A^\wedge(n)B_n\};\mathbb{S}^2). \tag{12.18}$$

Hence, $\mathrm{H}(\{A^\wedge(n)B_n\};\mathbb{S}^2) = A^{-1}\mathrm{H}(\{B_n\};\mathbb{S}^2)$. Obviously, $\{r^n\}$, $0 \le r < 1$ (resp. $\{e^{-\frac{1}{2}n(n+1)h}\}$, $h > 0$) define pseudodifferential operators of order $-\infty$.

Some rotation-invariant examples, which are of significance in geosciences, should be listed below:

(i) *Beltrami-operator:* It is not invertible since $(\Delta^*)^\wedge(0) = 0$, but $\Delta^* + \frac{1}{4}$ has the symbol $\{(-\Delta^* + \frac{1}{4})^\wedge(n)\}$ with $(-\Delta^* + \frac{1}{4})^\wedge(n) = (n + \frac{1}{2})^2$, $n \in \mathbb{N}_0$. Hence, $(-\Delta^* + \frac{1}{4})$ has an inverse $(-\Delta^* + \frac{1}{4})^{-1}$ which is a rational pseudodifferential operator of order -2, i.e., $n \mapsto ((-\Delta^* + \frac{1}{4})^{-1})^\wedge(n)$, $n \in \mathbb{N}_0$, is a rational function of order -2.

More generally, $(-\Delta^* + \frac{1}{4})^s$ is a rational pseudodifferential operator of order 2s and has the symbol $\{((-\Delta^* + \frac{1}{4})^s)^\wedge(n)\}$ with

$$\left(\left(-\Delta + \frac{1}{4}\right)^s\right)^\wedge(n) = \left(n + \frac{1}{2}\right)^{2s}, \quad n = 0, 1, \ldots. \tag{12.19}$$

(ii) *Integral operator* A *of the single layer potential on* $\mathbb{S}^2$: The operator defined by

$$AU(\xi) = \frac{1}{2\pi}\int_{\mathbb{S}^2} U(\eta)\frac{1}{|\xi - \eta|}\, dS(\eta), \quad \xi \in \mathbb{S}^2, \tag{12.20}$$

is a rational pseudodifferential operator of order -1 possessing the symbol $\{A^\wedge(n)\}$ with $A^\wedge(n) = (n + \frac{1}{2})^{-1} = ((-\Delta^* + \frac{1}{4})^{-1/2})^\wedge(n)$. This can be easily seen by inserting the series expansion of $|\xi - \eta|^{-1}$ in terms of Legendre polynomials. A consequence is that A is invertible.

(iii) The operator $A = (-2\Delta^*)(-\Delta^* + \frac{1}{4})^{1/2}$ has the symbol $\{A^\wedge(n)\}$, where $A^\wedge(n)$ is given by $A^\wedge(n) = (2n+1)n(n+1)$, $n = 0, 1, \ldots$. A is not invertible, but the operator B with symbol $\{B^\wedge(n)\}$ given by

$$B^\wedge(n) = \begin{cases} 1, & n = 0, \\ (2n+1)(n(n+1)), & n = 1, 2, \ldots, \end{cases} \tag{12.21}$$

is invertible. Since (see, e.g., W. Magnus et al. [1966])

$$\sum_{n=1}^{\infty} \frac{1}{n(n+1)} P_n(t) = 1 - 2\ln\left(1 + \sqrt{\frac{1-t}{2}}\right), \quad t \in [-1,1], \tag{12.22}$$

we find

$$Q(\xi \cdot \eta) = \sum_{n=1}^{\infty} \frac{2n+1}{4\pi} \frac{1}{B^\wedge(n)} P_n(\xi \cdot \eta) = \frac{1}{4\pi} - \frac{1}{2\pi} \ln\left(1 + \sqrt{\frac{1-\xi\cdot\eta}{2}}\right), \tag{12.23}$$

such that

$$B^{-1}U(\xi) = \int_{\mathbb{S}^2} \left(Q(\xi\cdot\eta) - \frac{1}{4\pi^2}\right) U(\eta)\, dS(\eta), \quad \xi \in \mathbb{S}^2, \tag{12.24}$$

This pseudodifferential operator plays an important role in the characterization of equidistributions on the sphere (see Section 14.4).

(iv) $A^\wedge(n) = r^n$, $0 \le r < 1$ (or $A^\wedge(n) = e^{\frac{1}{2}n(n+1)h}$, $h > 0$) define pseudodifferential operators of order $-\infty$.

(v) *Integral operator A of the double layer potential on $\mathbb{S}^2$*: The operator defined by

$$AU(\xi) = \frac{1}{2\pi} \int_{\mathbb{S}^2} U(\eta) \frac{\partial}{\partial \nu_\eta} \frac{1}{|\xi - \eta|}\, dS(\eta), \quad \xi \in \mathbb{S}^2, \tag{12.25}$$

($\partial/\partial\nu$ is the derivative in the direction of the exterior normal of $\mathbb{S}^2$) can be written in the form $A = (-\frac{1}{2})(-\Delta^* + \frac{1}{4})^{-1/2}$.

(vi) *Stokes integral operator*: This operator is defined by

$$AU(\xi) = \frac{1}{4\pi} \int_{\mathbb{S}^2} St(\xi, \eta)\, U(\eta)\, dS(\eta), \tag{12.26}$$

where $St(\cdot,\cdot)$ is the *Stokes kernel.* A has the spherical symbol $\{A^\wedge(n)\}$:

$$A^\wedge(n) = \begin{cases} 0, & n = 1, \\ \frac{1}{n-1}, & n = 0, 2, 3, 4, \ldots. \end{cases} \tag{12.27}$$

The problem of determining the "disturbing potential" U with $AU = F$ from prescribed "gravity anomalies" F in spherical nomenclature is the "fundamental problem of classical physical geodesy" (see, e.g., W.A. Heiskanen, H. Moritz [1967], B. Hofmann-Wellenhof, H. Moritz [2006], M. Hotine [1969]).

(vii) *Green's integral operator with respect to ∂_0*: The operator given by

$$AU(\xi) = \frac{1}{4\pi} \int_{\mathbb{S}^2} G(\partial_0; \xi, \eta) U(\eta)\, dS(\eta) \tag{12.28}$$

has the spherical symbol $\{A^\wedge(n)\}$, where (cf. (9.331))

$$A^\wedge(n) = \begin{cases} 0, & n = 0, \\ \frac{1}{n(n+1)}, & n = 1, 2, \ldots. \end{cases} \tag{12.29}$$

We see that all these listed operators form mappings from $\mathrm{H}^s(\mathbb{S}^2)$ into some $\mathrm{H}^{s-t}(\mathbb{S}^2)$, where t is the order of the pseudodifferential operator A under consideration. For example, the Beltrami operator is an operator from $\mathrm{H}^s(\mathbb{S}^2)$ into $\mathrm{H}^{s-2}(\mathbb{S}^2)$, etc.

(vii) *Poisson operator*: The operator A given by

$$AU(\xi) = \frac{r^2 - 1}{4\pi} \int_{\mathbb{S}^2} \frac{U(\eta)}{|r\xi - \eta|^3}\, dS(\eta), \quad r > 1, \tag{12.30}$$

has the symbol $\{r^{-(n+1)}\}$. This operator serves as upward continuation operator of data from the Earth's surface to a satellite height.

(viii) *Operator of first order radial derivative at the height r:* This operator given by

$$AU(\xi) = \int_{\mathbb{S}^2} D_r(\xi \cdot \eta) U(\eta)\, dS(\eta), \quad r > 1, \tag{12.31}$$

with

$$\begin{aligned} D_r(\xi \cdot \eta) &= \frac{\partial}{\partial r} \frac{1}{4\pi} \frac{r^2 - 1}{(r^2 + 1 - 2r(\xi \cdot \eta))^{3/2}} \\ &= \frac{\partial}{\partial r} \sum_{n=0}^{\infty} \frac{2n+1}{4\pi} \left(\frac{1}{r}\right)^{n+1} P_n(t) \end{aligned} \tag{12.32}$$

has the symbol $\{-(n+1)r^{-(n+2)}\}$. Its inversion plays a particular role in satellite-to-satellite tracking (SST) problems, i.e., the downward continuation of first order radial derivative data at satellite height to the Earth's surface.

(ix) *Operator of second order radial derivative at the height r:* This operator given by

$$AU(\xi) = \int_{\mathbb{S}^2} \left(\frac{\partial}{\partial r} D_r(\xi \cdot \eta)\right) U(\eta)\, dS(\eta), \quad r > 1, \tag{12.33}$$

has the symbol $\{(n+1)(n+2)r^{-(n+3)}\}$. Its inversion is of fundamental importance in satellite gravity gradiometry (SGG), i.e., the downward continuation of second order radial derivative data at satellite height to the Earth's surface.

Sometimes an operator A satisfying $A^\wedge(n) \to 0$ is called a smoothing operator, because $A^\wedge(n) \to 0$ means that the higher order harmonics are subdued by the operator. The Beltrami operator has an opposite effect, because the higher order harmonics are amplified.

Spherical harmonic analysis, i.e., the computation of the Fourier coefficients of a function T

$$T^\wedge(n,j) = \langle T, Y_{n,j}\rangle_{\mathrm{L}^2(\mathbb{S}^2)} = \int_{\mathbb{S}^2} T(\eta) Y_{n,j}(\eta)\, dS(\eta), \tag{12.34}$$

and *spherical harmonic synthesis*, i.e., the reconstruction of a potential T by its Fourier coefficients

$$T = \sum_{n=0}^{\infty} \sum_{j=1}^{2n+1} T^\wedge(n,j) Y_{n,j}, \quad T \in \mathrm{L}^2(\mathbb{S}^2), \tag{12.35}$$

constantly attract attention, e.g., in physical geodesy, where T is the gravitational (disturbing) potential of the Earth. Nowadays reference models for the gravitational potential include spherical harmonic coefficients $T^\wedge(n,j)$ up to $m \geq 2000$ or even more (see, e.g., N.K. Pavlis et al. [2012]). If the table of the potential coefficients $T^\wedge(n,j)$ is known, it can be used to evaluate geodetic observables, i.e., linear functionals such as geoid undulation, gravity anomaly, radial derivative at Earth's surface or at satellite height, etc. The relation between T and its geodetic observables AT is usually provided by (rotation-invariant) pseudodifferential operators A (cf. S. L. Svensson [1983]), the table of spherical harmonic coefficients of $F = AT$ is given by

$$F^\wedge(n,j) = A^\wedge(n) T^\wedge(n,j), \tag{12.36}$$

where $\{A^\wedge(n)\}_{n=0,1,\ldots}$ is the spherical symbol of A. Note that $AY_{n,j} = A^\wedge(n) Y_{n,j}$ for all $n \in \mathbb{N}_0$, $j = 1, \ldots, 2n+1$.

In mathematical language, the spherical harmonics $Y_{n,j}$, may be interpreted as eigenfunctions of the operator A corresponding to the eigenvalues $A^\wedge(n)$, and it follows that

$$F = AT = \sum_{n=0}^{\infty} \sum_{j=1}^{2n+1} A^\wedge(n) T^\wedge(n,j) Y_{n,j}. \tag{12.37}$$

The list of corresponding spherical symbols $\{A^\wedge(n)\}_{n=0,1,\ldots}$ (given in Table 12.1) motivates to introduce the *Integrated Spherical Harmonic Model of Physical Geodesy* (sometimes called "Pocket Guide of Physical Geodesy", cf. P. Meissl [1971], H. Nutz [2002], R. Rummel [1992]).

In Table 12.1, μ denotes the Earth's mass, R is the Earth's mean radius, and k is the gravitational constant.

TABLE 12.1: Geodetic Operators in Spherical Nomenclature (cf. S. L. Svensson [1983]).

Operator/quantity A	$A^\wedge(n)$	Order
Gravity anomaly	$\frac{n-1}{R}$	1
Geoid undulations	$\frac{R^2}{k\mu}$	0
Stokes operator	$\frac{R}{n-1}$	-1
Single layer	$\frac{R}{n+\frac{1}{2}}$	-1
Double layer	$-\frac{R}{2n+1}$	-1
First radial derivative	$-\frac{n+1}{R}$	1
Second rad. derivative	$\frac{(n+1)(n+2)}{R^2}$	2
Upward continuation	$\left(\frac{R}{r}\right)^{n+1}$	$-\infty$

12.3 Reproducing Kernels and Remainder Terms

We consider elements of the Sobolev spaces $\mathrm{H}(\{A_n\};\mathbb{S}^2)$ as formal orthogonal expansions or "generalized functions" (distributions). A fundamental lemma due to Sobolev says that under certain circumstances the formal orthogonal expansion actually converges uniformly to a function in ordinary sense.

Next, this lemma will be discussed in more detail. We begin by introducing (cf. W. Freeden [1990], W. Freeden et al. [1998]) the following setting.

Definition 12.4. *A sequence* $\{A_n\} \in \mathfrak{A}$ *is said to be* $\{B_n\}$-summable of order m *if the following conditions are satisfied:*

(i) $A_n \neq 0$ *for all* $n \geq m+1$

(ii)

$$\Sigma_m(\{B_n\},\{A_n\}) = \sum_{n=m+1}^{\infty} \frac{2n+1}{4\pi}\frac{|B_n|^2}{|A_n|^2} < \infty, \quad \{B_n\} \in \mathfrak{A}. \tag{12.38}$$

Furthermore, $\{A_n\} \in \mathfrak{A}$ *is said to be* summable of order m *if*

(i) $A_n \neq 0$ *for all* $n \geq m+1$

(ii)

$$\Sigma_m(\{A_n\}) = \sum_{n=m+1}^{\infty} \frac{2n+1}{4\pi} \frac{1}{|A_n|^2} < \infty. \tag{12.39}$$

For simplicity, we call $\{A_n\} \in \mathfrak{A}$ summable *if it is summable of order* $m = -1$.

Remark 12.5. *$\{B_n\}$-summability is formally understood as $\{B_n\}$-summability of order -1, and we write*

$$\Sigma(\{B_n\}, \{A_n\}) = \Sigma_{-1}(\{B_n\}, \{A_n\}). \tag{12.40}$$

Associate with F the Fourier series $\sum_{n=0}^{\infty} \sum_{j=1}^{2n+1} F^{\wedge}(n,j) Y_{n,j}$. It will be of fundamental importance to know when this series converges uniformly so that its limit is continuous. This answer is supplied by the *Sobolev Lemma.*

Lemma 12.6. *Let the sequence $\{A_n\}$ be $\{B_n\}$-summable, $B_n \neq 0$, (more explicitly, $\{B_n\}$-summable of order -1). Then, each $F \in \mathrm{H}(\{B_n^{-1} A_n\}; \mathbb{S}^2)$ corresponds to a continuous function on $\mathbb{S}^2$.*

Proof. For each positive integer N, we have

$$\left| \sum_{n=0}^{N} \sum_{j=1}^{2n+1} F^{\wedge}(n,j) Y_{n,j}(\xi) \right|^2 \tag{12.41}$$

$$= \left| \sum_{n=0}^{N} \sum_{j=1}^{2n+1} B_n A_n^{-1} Y_{n,j}(\xi) A_n B_n^{-1} F^{\wedge}(n,j) \right|^2$$

$$\leq \left(\sum_{n=0}^{N} \sum_{j=1}^{2n+1} |B_n|^2 |A_n|^{-2} (Y_{n,j}(\xi))^2 \right) \left(\sum_{n=0}^{N} \sum_{j=1}^{2n+1} |B_n|^{-2} |A_n|^2 (F^{\wedge}(n,j))^2 \right)$$

$$\leq \Sigma(\{B_n\}, \{A_n\}) \|F\|^2_{\mathrm{H}(\{B_n^{-1} A_n\}; \mathbb{S}^2)}.$$

Due to (12.38) we find the Sobolev Lemma as desired. □

In particular, we mention the following consequence.

Lemma 12.7. *If $F \in \mathrm{H}^s(\mathbb{S}^2)$, where $s > k+1$, then F corresponds to a function of class $\mathrm{C}^{(k)}(\mathbb{S}^2)$.*

It is well-known that a continuously differentiable function is Lipschitz-continuous on $\mathbb{S}^2$. The following lemma yields an estimate of the Lipschitz constant.

Lemma 12.8. *If $F \in \mathrm{H}^s(\mathbb{S}^2)$, $s > 2$, then F is Lipschitz-continuous. More explicitly,*

$$|F(\xi) - F(\eta)| \leq C_F(s) |\xi - \eta|, \tag{12.42}$$

where

$$C_F(s) = \left(\frac{1}{2} \sum_{n=1}^{\infty} \frac{2n+1}{4\pi} \frac{n(n+1)}{(n+\frac{1}{2})^{2s}} \right)^{1/2} \|F\|_{\mathrm{H}^s(\mathbb{S}^2)}. \tag{12.43}$$

Proof. The Cauchy–Schwarz inequality shows us in connection with the addition theorem of spherical harmonics (9.42) that

$$\begin{aligned} |F(\xi)-F(\eta)|^2 &= \left|\sum_{n=0}^{\infty}\sum_{j=1}^{2n+1} F^\wedge(n,j)(Y_{n,j}(\xi)-Y_{n,j}(\eta))\right|^2 \\ &\leq 2\left(\sum_{n=0}^{\infty}\sum_{j=1}^{2n+1}\left(n+\frac{1}{2}\right)^{2s}(F^\wedge(n,j))^2\right) \\ &\quad\times\left(\sum_{n=0}^{\infty}\frac{2n+1}{4\pi}\left(n+\frac{1}{2}\right)^{-2s}(1-P_n(\xi\cdot\eta))\right) \\ &\leq \|F\|^2_{\mathrm{H}^s(\mathbb{S}^2)}\left(\sum_{n=0}^{\infty}\frac{2n+1}{4\pi}\frac{n(n+1)}{(n+1/2)^{2s}}\right)|\xi-\eta|^2. \end{aligned} \tag{12.44}$$

This proves Lemma 12.8. □

Let F be a member of the Sobolev space $\mathrm{H}^s(\mathbb{S}^2), s>1$. Then it follows that

$$\begin{aligned} \left|F(\xi)-\sum_{n=0}^{N}\sum_{j=1}^{2n+1}F^\wedge(n,j)Y_{n,j}(\xi)\right| &\leq \left(\sum_{n=N+1}^{\infty}\sum_{j=1}^{2n+1}(F^\wedge(n,j))^2(n+\frac{1}{2})^{2s}\right)^{1/2} \\ &\quad\times\left(\sum_{n=N+1}^{\infty}\sum_{j=1}^{2n+1}|Y_{n,j}(\xi)|^2(n+\frac{1}{2})^{-2s}\right)^{1/2} \end{aligned} \tag{12.45}$$

for each $\xi\in\mathbb{S}^2$. This enables us to derive an estimate between $F\in\mathrm{H}^s(\mathbb{S}^2)$ and its N-th truncated spherical harmonic expansion.

Lemma 12.9. *For $F\in\mathrm{H}^s(\mathbb{S}^2), s>1$,*

$$\begin{aligned} &\sup_{\xi\in\mathbb{S}^2}\left|F(\xi)-\sum_{n=0}^{N}\sum_{j=1}^{2n+1}F^\wedge(n,j)Y_{n,j}(\xi)\right| \\ &= \sup_{\xi\in\mathbb{S}^2}\left|F(\xi)-\int_{\mathbb{S}^2}SH_{0,\ldots,N}(\xi\cdot\eta)F(\eta)\,dS(\eta)\right| \leq \frac{C}{N^{s-1}}\|F\|_{\mathrm{H}^s(\mathbb{S}^2)}, \end{aligned} \tag{12.46}$$

where the constant C depends only on s.

Of great importance for our considerations are Sobolev spaces equipped with reproducing kernel structure.

Theorem 12.10. *Let $\{A_n\}$ be summable. Then the space* $\mathrm{H}=\mathrm{H}(\{A_n\};\mathbb{S}^2)$ *(from now on we write* H *instead of* $\mathrm{H}(\{A_n\};\mathbb{S}^2)$ *if no confusion is likely to* *arise) is a functional Hilbert subspace of* $\mathrm{C}^{(0)}(\mathbb{S}^2)$. H *has the reproducing kernel* $K_{\mathrm{H}} : \mathbb{S}^2 \times \mathbb{S}^2 \to \mathbb{R}$ *given by*

$$K_{\mathrm{H}}(\xi \cdot \eta) = K_{\mathrm{H}(\{A_n\};\mathbb{S}^2)}(\xi \cdot \eta) = \sum_{n=0}^{\infty} \sum_{j=1}^{2n+1} |A_n|^{-2} Y_{n,j}(\xi) Y_{n,j}(\eta), \qquad (12.47)$$

for $\xi, \eta \in \mathbb{S}^2$, *i.e.:*

(i) for every $\eta \in \mathbb{S}^2$, *the function* $K_{\mathrm{H}}(\ \cdot \eta) : \xi \mapsto K_{\mathrm{H}}(\xi \cdot \eta)$ *belongs to* H*;*

(ii) for all $F \in \mathrm{H}$ *and any* $\eta \in \mathbb{S}^2$, *the reproducing property*

$$F(\eta) = \langle F, K_{\mathrm{H}}(\ \cdot \eta) \rangle_{\mathrm{H}} \qquad (12.48)$$

is valid.

The system $\{A_n^{-1} Y_{n,j}\}_{n=0,1,\ldots,j=1,\ldots,2n+1}$, *represents a Hilbert-basis in* H.

Proof. That H is a subset of $\mathrm{C}^{(0)}(\mathbb{S}^2)$ follows from the Sobolev Lemma (cf. Lemma 12.6). A necessary and sufficient condition that H have a reproducing kernel function (see N. Aronszajn [1950]) is that for each $\xi \in \mathbb{S}^2$, the linear functional $\mathcal{L}_\xi : \mathrm{H} \to \mathbb{R}$ defined by $\mathcal{L}_\xi F = F(\xi), \xi \in \mathbb{S}^2$, is bounded, i.e., $|\mathcal{L}_\xi F| \leq C \|F\|_{\mathrm{H}}$ for some constant C and for all $F \in \mathrm{H}$. In fact, the Cauchy–Schwarz inequality shows that

$$|F(\xi)| \leq C \|F\|_{\mathrm{H}}, \qquad (12.49)$$

where

$$C^2 = K_{\mathrm{H}}(\xi \cdot \xi) \leq \Sigma(\{A_n\}) = \sum_{n=0}^{\infty} \frac{2n+1}{4\pi} |A_n|^{-2}. \qquad (12.50)$$

This means C is uniformly bounded. Suppose, conversely, that $|\mathcal{L}_\xi F| \leq C \|F\|_{\mathrm{H}}$ holds for each fixed $\xi \in \mathbb{S}^2$. Then one can find a function $G_\xi \in \mathrm{H}$ such that $F(\xi) = \langle F, G_\xi \rangle_{\mathrm{H}}$. Now set $K_{\mathrm{H}}(\ \cdot \xi) = G_\xi$, and it is clear that K_{H} is the reproducing kernel (cf. N. Aronszajn [1950]). Clearly, the reproducing kernel is unique in H. □

The space $\mathrm{H}_{0,\ldots,m} = \mathrm{Harm}_{0,\ldots,m}$, $m \in \mathbb{N}_0$, is a finite-dimensional Hilbert space ($\dim(\mathrm{Harm}_{0,\ldots,m}) = M = (m+1)^2$) with inner product

$$\langle F, G \rangle_{\mathrm{H}_{0,\ldots,m}} = \sum_{n=0}^{m} \sum_{j=1}^{2n+1} |A_n|^2 \langle F, Y_{n,j} \rangle_{\mathrm{L}^2(\mathbb{S}^2)} \langle G, Y_{n,j} \rangle_{\mathrm{L}^2(\mathbb{S}^2)} \qquad (12.51)$$

and reproducing kernel

$$K_{\mathrm{H}_{0,\ldots,m}}(\xi \cdot \eta) = \sum_{n=0}^{m} \sum_{j=1}^{2n+1} |A_n|^{-2} Y_{n,j}(\xi) Y_{n,j}(\eta). \qquad (12.52)$$

Let us denote by $\mathrm{H}^{\perp}_{0,\ldots,m} = \mathrm{H}^{\perp}_{0,\ldots,m}(\{A_n\};\mathbb{S}^2)$ the orthogonal complement of $\mathrm{Harm}_{0,\ldots,m}$ in $\mathrm{H}(\{A_n\},\mathbb{S}^2)$. The linear space $\mathrm{H}^{\perp}_{0,\ldots,m}(\{A_n\};\mathbb{S}^2)$ is a Hilbert space with inner product $\langle\cdot,\cdot\rangle_{\mathrm{H}^{\perp}_{0,\ldots,m}}$

$$\langle F,G\rangle_{\mathrm{H}^{\perp}_{0,\ldots,m}} = \sum_{n=m+1}^{\infty}\sum_{j=1}^{2n+1}|A_n|^2\langle F,Y_{n,j}\rangle_{\mathrm{L}^2(\mathbb{S}^2)}\langle G,Y_{n,j}\rangle_{\mathrm{L}^2(\mathbb{S}^2)} \tag{12.53}$$

and reproducing kernel

$$K_{\mathrm{H}^{\perp}_{0,\ldots,m}}(\xi\cdot\eta) = \sum_{n=m+1}^{\infty}\sum_{j=1}^{2n+1}|A_n|^{-2}Y_{n,j}(\xi)Y_{n,j}(\eta). \tag{12.54}$$

Hence, H is the orthogonal direct sum of $\mathrm{H}_{0,\ldots,m}$ and $\mathrm{H}^{\perp}_{0,\ldots,m}$ with the inner product

$$\langle\cdot,\cdot\rangle_{\mathrm{H}} = \langle\cdot,\cdot\rangle_{\mathrm{H}_{0,\ldots,m}} + \langle\cdot,\cdot\rangle_{\mathrm{H}^{\perp}_{0,\ldots,m}} \tag{12.55}$$

and the reproducing kernel

$$K_{\mathrm{H}} = K_{\mathrm{H}_{0,\ldots,m}} + K_{\mathrm{H}^{\perp}_{0,\ldots,m}}. \tag{12.56}$$

The null space of the semi-norm $\|\cdot\|_{\mathrm{H}^{\perp}_{0,\ldots,m}(\{A_n\};\mathbb{S}^2)}$ is the linear space $\mathrm{H}_{0,\ldots,m} = \mathrm{Harm}_{0,\ldots,m}$.

Let us consider a pseudodifferential operator A with summable symbol $\{A^\wedge(n)\}$. Then it follows that

$$\begin{aligned} F(\xi) &= \langle F, K_{\mathrm{H}(\{A^\wedge(n)\};\mathbb{S}^2)}(\,\cdot\xi)\rangle_{\mathrm{H}(\{A^\wedge(n)\};\mathbb{S}^2)} \\ &= \langle F, K_{\mathrm{H}_{0,\ldots,m}(\{A^\wedge(n)\};\mathbb{S}^2)}(\,\cdot\xi)\rangle_{\mathrm{H}_{0,\ldots,m(\{A^\wedge(n)\};\mathbb{S}^2)}} \\ &\quad + \langle F, K_{\mathrm{H}^{\perp}_{0,\ldots,m}(\{A^\wedge(n)\};\mathbb{S}^2)}(\,\cdot\xi)\rangle_{\mathrm{H}^{\perp}_{0,\ldots,m(\{A^\wedge(n)\};\mathbb{S}^2)}} \end{aligned} \tag{12.57}$$

holds true for $m\in\mathbb{N}_0$, $\xi\in\mathbb{S}^2$, and $F\in\mathrm{H}(\{A^\wedge(n)\};\mathbb{S}^2)$. By virtue of Parseval's identity it is not difficult to verify that

$$\langle F, K_{\mathrm{H}_{0,\ldots,m(\{A^\wedge(n)\};\mathbb{S}^2)}}(\,\cdot\xi)\rangle_{\mathrm{H}_{0,\ldots,m(\{A^\wedge(n)\};\mathbb{S}^2)}} = \int_{\mathbb{S}^2} SH_{0,\ldots,m}(\xi\cdot\eta)F(\eta)\,dS(\eta), \tag{12.58}$$

whereas

$$\begin{aligned} &\langle F, K_{\mathrm{H}^{\perp}_{0,\ldots,m}(\{A^\wedge(n)\};\mathbb{S}^2)}(\,\cdot\xi)\rangle_{\mathrm{H}^{\perp}_{0,\ldots,m}(\{A^\wedge(n)\};\mathbb{S}^2)} \\ &= \int_{\mathbb{S}^2} A_\eta K_{\mathrm{H}^{\perp}_{0,\ldots,m}(\{A^\wedge(n)\};\mathbb{S}^2)}(\xi\cdot\eta)(A_\eta F(\eta))\,dS(\eta). \end{aligned} \tag{12.59}$$

This leads us to the following *integral formula relative to pseudodifferential operators* with summable symbol.

Theorem 12.11. *Let A be a pseudodifferential operator with summable symbol $\{A^\wedge(n)\}$. Then*

$$F(\xi) = \int_{\mathbb{S}^2} SH_{0,\ldots,m}(\xi \cdot \eta) F(\eta)\, dS(\eta) \qquad (12.60)$$
$$+ \int_{\mathbb{S}^2} A_\eta K_{\mathrm{H}^\perp_{0,\ldots,m}(\{A^\wedge(n)\};\mathbb{S}^2)}(\xi \cdot \eta)(A_\eta F(\eta))\, dS(\eta)$$

holds for every $\xi \in \mathbb{S}^2$ and every $F \in \mathrm{H}(\{A^\wedge(n)\};\mathbb{S}^2)$.

This leads to the following remainder term estimate in problems of approximate integration by setting $m = 0$.

Theorem 12.12. *Let A be a pseudodifferential operator with summable symbol $\{A^\wedge(n)\}$. Then*

$$F(\xi) = \frac{1}{4\pi} \int_{\mathbb{S}^2} F(\eta)\, dS(\eta) \qquad (12.61)$$
$$+ \int_{\mathbb{S}^2} A_\eta K_{\mathrm{H}^\perp_0(\{A^\wedge(n)\};\mathbb{S}^2)}(\xi \cdot \eta)(A_\eta F(\eta))\, dS(\eta)$$

holds for every $\xi \in \mathbb{S}^2$ and every $F \in \mathrm{H}(\{A^\wedge(n)\};\mathbb{S}^2)$.

In other words, the integral mean can be compared with a functional value at a point $\xi \in \mathbb{S}^2$ under explicit knowledge of the remainder term involving the pseudodifferential operator with summable symbol $\{A^\wedge(n)\}$. Choosing $A = \Delta^*$ leads back to Theorem 9.16.

Corollary 12.13. *Assume that $X_N = \{\eta_1, \ldots, \eta_N\}$ is a unisolvent system relative to* $\mathrm{Harm}_{0,\ldots,m}$. *Then, for $n \leq m$,*

$$F^\wedge(n,j) = \sum_{k=1}^{N} a_k^{n,j} F(\eta_k) \qquad (12.62)$$
$$- \sum_{k=1}^{N} a_k^{n,j} \int_{\mathbb{S}^2} A_\eta K_{\mathrm{H}^\perp_{0,\ldots,m}(\{A^\wedge(n)\};\mathbb{S}^2)}(\eta \cdot \eta_k)(A_\eta F(\eta))\, dS(\eta)$$

holds for every $F \in \mathrm{H}(\{A^\wedge(n)\};\mathbb{S}^2)$ and for all $a_1^{n,j}, \ldots, a_N^{n,j}$ satisfying

$$\sum_{k=1}^{N} a_k^{n,j} Y_{p,q}(\eta_k) = \delta_{p,n}\delta_{q,j}, \quad p = 0, \ldots, m,\ q = 1, \ldots, 2n+1. \qquad (12.63)$$

In consequence we are led to the following sampling result involving the pseudodifferential operator A and the Lagrangians $L_k \in \mathrm{Harm}_{0,\ldots,m}$ given by (10.11)

$$\underbrace{\int_{\mathbb{S}^2} SH_{0,\ldots,m}(\xi \cdot \eta) F(\eta)\, dS(\eta)}_{=F_{0,\ldots,m}(\xi)} - \sum_{k=1}^{N} F(\eta_k) L_k(\xi) \tag{12.64}$$
$$= -\sum_{k=1}^{N} L_k(\xi) \int_{\mathbb{S}^2} A_\eta K_{\mathrm{H}^{\perp}_{0,\ldots,m}(\{A^\wedge(n)\};\mathbb{S}^2)}(\eta, \eta_k)(A_\eta F(\eta))\, dS(\eta).$$

Summarizing our considerations we come to the following conclusion: Assume that $X_N = \{\eta_1, \ldots, \eta_N\}$ is a unisolvent system relative to $\mathrm{Harm}_{0,\ldots,m}$. Then, for every $F \in \mathrm{H}(\{A^\wedge(n)\}, \mathbb{S}^2)$, the error between the m-th truncated Fourier expansion $F_0, \ldots, m$ of F and the sampling sum $\sum_{k=1}^{N} F(\eta_k) L_k$ of F in X_N relative to $\mathrm{Harm}_{0,\ldots,m}$ is given by the right side of (12.64), i.e.,

$$F_{0,\ldots,m}(\xi) \tag{12.65}$$
$$= \sum_{k=1}^{N} \left(F(\eta_k) - \int_{\mathbb{S}^2} A_\eta K_{\mathrm{H}^{\perp}_{0,\ldots,m}(\{A^\wedge(n)\};\mathbb{S}^2)}(\eta \cdot \eta_k)(A_\eta F(\eta))\, dS(\eta) \right) L_k(\xi)$$

for all $\xi \in \mathbb{S}^2$.

12.4 Particular Types of Kernel Functions

Up to this point, our theoretical conclusions have been given under predefined Hilbert space structure. However, the choice of the norm is an important problem in constructive approximation. Next we give a list of possible choices of kernel functions. The list is divided into four parts:

(1) Classical Green's functions with respect to (iterated) Beltrami differential operators,

(2) Exponential kernels, i.e., Abel-Poisson kernel as well as Gauss–Weierstrass kernel,

(3) Kernels related to Euclidean radial basis functions,

(4) Locally supported kernels (see Section 12.5).

All choices of this section, i.e., the radial basis functions of (1)–(3) share the property of having a global support, i.e., their support covers the whole sphere $\mathbb{S}^2$. For numerical purposes, locally supported kernels of (4) as listed in Section 12.5 are of particular significance.

Green's Functions Associated to Beltrami Operators. Let us construct the Hilbert space $\mathrm{H} = \mathrm{H}(\{A_n\};\mathbb{S}^2)$ by choosing the inner product $\langle\cdot,\cdot\rangle_{\mathrm{H}(\{A_n\};\mathbb{S}^2)}$ corresponding to the norm $\|\cdot\|_{\mathrm{H}(\{A_n\};\mathbb{S}^2)}$, given by

$$\|F\|^2_{\mathrm{H}(\{A_n\};\mathbb{S}^2)} = \left|\int_{\mathbb{S}^2} F(\xi)Y_{0,1}(\xi)\, dS(\xi)\right|^2 + \sum_{n=1}^{\infty}\sum_{j=1}^{2n+1} ((\Delta^*)^\wedge(n))^2 \left|\int_{\mathbb{S}^2} F(\xi)Y_{n,j}(\xi)\, dS(\xi)\right|^2. \tag{12.66}$$

That means the sequence $\{A_n\}$ is given by $A_0 = 1$, $A_n = (\Delta^*)^\wedge(n) = (\partial_0)^\wedge(n)$, $n \in \mathbb{N}$, i.e., the sequence $\{(\Delta^*)^\wedge(n)\}$ is the spherical symbol of the Beltrami operator $\partial_0 = \Delta^*$. Obviously, $\{A_n\}$ is $\{(n+\frac{1}{2})^\tau\}$-summable for all $\tau < 1$. Observing the definition of the spherical harmonics (9.38) we get via Green's surface identity (9.33)

$$\begin{aligned}(\Delta^*)^\wedge(n)\int_{\mathbb{S}^2} F(\xi)Y_{n,j}(\xi)\, dS(\xi) &= \int_{\mathbb{S}^2} F(\xi)(\Delta^*_\xi Y_{n,j}(\xi))\, dS(\xi) \\ &= \int_{\mathbb{S}^2} (\Delta^*_\xi F(\xi))Y_{n,j}(\xi)\, dS(\xi).\end{aligned} \tag{12.67}$$

Consequently, according to Parseval's identity, we obtain

$$\|F\|^2_{\mathrm{H}(A;\mathbb{S}^2)} = \frac{1}{4\pi}\left|\int_{\mathbb{S}^2} F(\xi)\, dS(\xi)\right|^2 + \int_{\mathbb{S}^2} |\Delta^*_\xi F(\xi)|^2\, dS(\xi). \tag{12.68}$$

For the reproducing kernel in $\mathrm{H} = \mathrm{H}(\{A_n\};\mathbb{S}^2)$ we find

$$K_{\mathrm{H}}(\xi\cdot\eta) = \frac{1}{4\pi} + G(\partial_0^2;\xi\cdot\eta) = \frac{1}{4\pi} + \frac{1}{4\pi}\sum_{n=1}^{\infty}\frac{2n+1}{(n(n+1))^2}P_n(\xi\cdot\eta). \tag{12.69}$$

Apart from an additive constant the series on the right-hand side coincides with Green's function on the unit sphere corresponding to the differential operator ∂_0^2.

Another example is obtained by considering the sequence $\{A_n\}$ with $A_n = 1$ for $n = 0,\ldots,m$ and $A_n = (\partial_{0,\ldots,m})^\wedge(n)$ for $n = m+1, m+2, \ldots$, where the sequence $\{(\partial_{0,\ldots,m})^\wedge(n)\}$ is the spherical symbol of the operator $\partial_{0,\ldots,m} = \partial_0, \cdot \ldots \cdot \partial_m$ where $\partial_n = \Delta^* + n(n+1)$. The kernel of the operator $\partial_{0,\ldots,m}$ is the space $\mathrm{Harm}_{0,\ldots,m}$. The norm is given by

$$\|F\|^2_{\mathrm{H}(\{A_n\};\mathbb{S}^2)} = \sum_{n=0}^{m}\sum_{j=1}^{2n+1}\left|\int_{\mathbb{S}^2} F(\xi)Y_{n,j}(\xi)\, dS(\xi)\right|^2 + \int_{\mathbb{S}^2} |(\partial_{0,\ldots,m})_\xi F(\xi)|^2\, dS(\xi), \tag{12.70}$$

whereas the reproducing kernel in $\mathrm{H}(\{A_n\};\mathbb{S}^2)$ is the zonal function of the form

$$K_{\mathrm{H}}(\xi\cdot\eta) = \underbrace{\sum_{n=0}^{m} \frac{2n+1}{4\pi} P_n(\xi\cdot\eta)}_{=SH_{0,\ldots,m}(\xi\cdot\eta)} + \frac{1}{4\pi} \sum_{k=m+1}^{\infty} \frac{2k+1}{((\partial_{0,\ldots,m})^\wedge(n))^2} P_k(\xi\cdot\eta). \quad (12.71)$$

The infinite series on the right side of (12.71) is the bilinear expansion of Green's function of the operator $\partial^2_{0,\ldots,m}$ (cf. W. Freeden [1981a]). The sequence $\{A_n\}$ is $\{(n+\frac{1}{2})^\tau\}$-summable for all $\tau < 2m+1$.

Exponential Kernels. Next we list a variety of kernels corresponding to pseudodifferential operators of exponential order.

Let $\{A_n\} = \{A^\wedge(n)\}$, where A is a pseudodifferential operator on $\mathbb{S}^2$ with spherical symbol $\{A^\wedge(n)\}$ such that

(i) $n \mapsto |A^\wedge(n)|^2$, $n \in \mathbb{N}_0$, is a real rational function

(ii) There exist two positive constants C, C' with

$$C\left(n+\frac{1}{2}\right)^{2+\varepsilon} \leq (A^\wedge(n))^2 \leq C'\left(n+\frac{1}{2}\right)^{\sigma}, \quad (12.72)$$

for some $\varepsilon > 0, \sigma \geq 2+\varepsilon$.

Via the Parseval identity the norm is given by

$$\|F\|^2_{\mathrm{H}(\{A_n\};\mathbb{S}^2)} = \int_{\mathbb{S}^2} |A_\eta F(\eta)|^2 \, dS(\eta) = \sum_{n=0}^{\infty} \sum_{j=1}^{2n+1} (A^\wedge(n))^2 (F^\wedge(n,j))^2. \quad (12.73)$$

For the reproducing kernel in $\mathrm{H} = \mathrm{H}(\{A^\wedge(n)\};\mathbb{S}^2)$ we find

$$K_{\mathrm{H}}(\xi\cdot\eta) = \sum_{n=0}^{\infty} \frac{2n+1}{4\pi} \frac{1}{(A^\wedge(n))^2} P_n(\xi\cdot\eta). \quad (12.74)$$

The sequence $\{A_n\}$ is $\{(n+\frac{1}{2})^\tau\}$-summable for all $\tau < \varepsilon/2$. As a matter of fact, examples can be found for which the kernel representations are expressible as elementary functions. An example is $A^\wedge(0) = 1$, $A^\wedge(n) = ((2n+1)n(n+1))^{1/2}$, $n = 1, 2, \ldots$ (cf. W. Freeden [1990]).

Note that the Legendre-transform of the function $t \mapsto 1 - 2\ln(1 + \sqrt{(1-t)/2}$, $t \in [-1,1])$ is given by the sequence $\{4\pi(2n+1)n(n+1))^{-1}\}_{n=1,2,\ldots}$. This function turns out to be useful in the theory of equidistribution (cf. J. Cui, W. Freeden [1997]) for establishing explicitly evaluable discrepancies (see Chapter 14).

An alternative concept to come to a good candidate for numerical purposes is to replace $\{A^\wedge(n)\}$ by the sequence $\{h^{-n/2}A^\wedge(n)\}$, $0 < h < 1$, for which the corresponding kernel is available as elementary function and to use this kernel for parameters h close to the value 1. More concretely, let $\{A_n\} = \{h^{-n/2}R^\wedge(n)\}$, $0 < h < 1$, where $n \mapsto |R^\wedge(n)|^2$, $n \in \mathbb{N}_0$, is a real rational function with

$$0 < |R^\wedge(n)|^2 \leq C' \left(n + \frac{1}{2}\right)^\tau \tag{12.75}$$

for some τ. The parameter h gives additional flexibility in choosing the norm, and also results in more general sequences $\{A_n\}$, since the factor h^n appears in the infinite sum for the reproducing kernel. In fact, the norm reads as follows

$$\|F\|^2_{\mathrm{H}(\{A_n\};\mathbb{S}^2)} = \sum_{n=0}^{\infty} \sum_{j=1}^{2n+1} \frac{1}{h^n} \left| \int_{\mathbb{S}^2} (R_\xi F(\xi)) Y_{n,j}(\xi)\, dS(\xi) \right|^2, \tag{12.76}$$

while we find as series expansion for the reproducing kernel in $\mathrm{H}(\{A_n\};\mathbb{S}^2)$

$$K_{\mathrm{H}(\{A_n\};\mathbb{S}^2)}(\xi \cdot \eta) = \sum_{n=0}^{\infty} \frac{2n+1}{4\pi} (R^\wedge(n))^{-2} h^n P_n(\xi \cdot \eta). \tag{12.77}$$

The sequence $\{A_n\}$ is $\{(n+\frac{1}{2})^\tau\}$-summable for all τ. Moreover, different types of kernel representations are available in terms of elementary functions. We only mention here (cf. W. Freeden [1987], M. Gutting [2008]):

(i) *Abel-Poisson kernel:* $R^\wedge(n) = 1$, $n = 0, 1, \ldots$

$$K_{\mathrm{H}(\{A_n\};\mathbb{S}^2)}(\xi \cdot \eta) = Q_h(\xi \cdot \eta) = \frac{1}{4\pi} \frac{1-h^2}{(L_h(\xi \cdot \eta))^{3/2}} \tag{12.78}$$

(ii) *Singularity kernel:* $R^\wedge(n) = (n + \frac{1}{2})^{\frac{1}{2}}, n = 0, 1, \ldots$

$$K_{\mathrm{H}(\{A_n\};\mathbb{S}^2)}(\xi \cdot \eta) = \frac{1}{2\pi} \frac{1}{(L_h(\xi \cdot \eta))^{1/2}} \tag{12.79}$$

(iii) *Logarithmic kernel:* $R^\wedge(n) = ((2n+1)(n+1))^{\frac{1}{2}}$, $n = 0, 1, \ldots$

$$K_{\mathrm{H}(\{A_n\};\mathbb{S}^2)}(\xi \cdot \eta) = \frac{1}{2\pi h} \ln\left(1 + \frac{2h}{(L_h(\xi \cdot \eta))^{1/2} + 1 - h}\right), \tag{12.80}$$

where we have used the abbreviation $L_h(\xi \cdot \eta) = 1 + h^2 - 2h(\xi \cdot \eta)$. The *Abel-Poisson kernel* $Q_h : [-1, 1] \to \mathbb{R}$, $h \in (0, 1)$,

$$Q_h(t) = \frac{1}{4\pi} \frac{1-h^2}{(1+h^2-2ht)^{\frac{3}{2}}} = \frac{1}{4\pi} \sum_{n=0}^{\infty} (2n+1) h^n P_n(t) \tag{12.81}$$

is characterized by a large number of significant properties (cf. W. Freeden et al. [1998]). For example, from $1 + h^2 - 2ht = (1-h)^2 + 2h(1-t)$ it follows that $Q_h(t) \geq 0$ for all $t \in [-1,1]$. By virtue of the Funk–Hecke formula in connection with (12.81) we get

$$\int_{\mathbb{S}^2} Q_h(\xi \cdot \eta)\, dS(\eta) = 1 \tag{12.82}$$

for all $\xi \in \mathbb{S}^2$ and $h < 1$.

Remark 12.14. *It should be pointed out that the kernels (12.78) and (12.79) are of high importance in numerical computation because of their efficiency in fast multipole methods (cf. M. Gutting [2008, 2012, 2014, 2017]).*

Set $\{A_n\} = \{e^{\frac{1}{2}n(n+1)\rho} R^\wedge(n)\}$, $\rho > 0$, where R is a pseudodifferential operator satisfying the property that $n \mapsto (R^\wedge(n))^2, n = 0, 1, \ldots$ is a real rational function with $0 < |R^\wedge(n)|^2 \leq C'(n + (\frac{1}{2}))^\tau$ for some τ. The simplest example is the *Gauss–Weierstrass kernel* where $R^\wedge(n) = 1$, $n = 0, 1, \ldots$,

$$K_{\mathrm{H}(\{A_n\};\mathbb{S}^2)}(\xi \cdot \eta) = W_\rho(\xi \cdot \eta) = \sum_{n=0}^{\infty} e^{-n(n+1)\rho} \frac{2n+1}{4\pi} P_n(\xi \cdot \eta). \tag{12.83}$$

$W_\rho(t) \geq 0$ for all $t \in [-1,1]$. Clearly, $(W_\rho)^\wedge(n) = e^{-n(n+1)\rho}, n = 0, 1, \ldots$. Furthermore,

$$\int_{\mathbb{S}^2} W_\rho(\xi \cdot \eta)\, dS(\eta) = 1 \tag{12.84}$$

for all $\xi \in \mathbb{S}^2$.

Counterparts to Euclidean Radial Basis Functions. Until now, the sequence $\{A_n\}$ defining the corresponding reproducing kernel Hilbert spaces $\mathrm{H}(\{A_n\};\mathbb{S}^2)$ has been considered under the assumption that $A_n \neq 0$ for all $n \in \mathbb{N}_0$. In practice, however, one often has to deal with the more general case, where some A_n are allowed to be zero. This situation should be studied in more detail. We start with the following splitting.

Definition 12.15. *Let $\{A_n\} \in \mathfrak{A}$ be a sequence. We split $\mathbb{N}_0$ into two parts such that*

$$\mathbb{N}_0 = \mathcal{N} \cup \mathcal{N}_0, \qquad \mathcal{N} \cap \mathcal{N}_0 = \emptyset, \tag{12.85}$$

where

$$\mathcal{N} = \{n \in \mathbb{N}_0 : \ A_n \neq 0\}, \qquad \mathcal{N}_0 = \{n \in \mathbb{N}_0 : \ A_n = 0\}. \tag{12.86}$$

*The sequence $\{A_n\}$ is called $\{B_n\}$-*summable with respect to $\mathcal{N}$ *if*

$$\Sigma^{\mathcal{N}}(\{B_n\}, \{A_n\}) = \sum_{n \in \mathcal{N}} \frac{2n+1}{4\pi} \frac{|B_n|^2}{|A_n|^2} < \infty. \tag{12.87}$$

A $\{1\}$-summable sequence with respect to $\mathcal{N}$ is simply called summable with respect to $\mathcal{N}$*:*

$$\Sigma^{\mathcal{N}}(\{A_n\}) = \sum_{n \in \mathcal{N}} \frac{2n+1}{4\pi} \frac{1}{|A_n|^2} < \infty. \tag{12.88}$$

For a given sequence $\{A_n\}$, consider the linear space $\mathrm{E}^{\mathcal{N}}(\{A_n\};\mathbb{S}^2)$ of all functions $F \in \mathrm{C}^{(\infty)}(\mathbb{S}^2)$ satisfying $\langle F, Y_{n,j}\rangle_{\mathrm{L}^2(\mathbb{S}^2)} = 0$ for all $n \in \mathcal{N}$ such that (cf. (12.2))

$$\sum_{n\in\mathcal{N}} \sum_{j=1}^{2n+1} |A_n|^2 \langle F, Y_{n,j}\rangle_{\mathrm{L}^2(\mathbb{S}^2)} < \infty. \tag{12.89}$$

On the space $\mathrm{E}^{\mathcal{N}}(\{A_n\};\mathbb{S}^2)$ we are able to introduce an inner product by

$$\langle F, G\rangle_{\mathrm{H}^{\mathcal{N}}(\{A_n\};\mathbb{S}^2)} = \sum_{n\in\mathcal{N}} |A_n|^2 \sum_{j=1}^{2n+1} \langle F, Y_{n,j}\rangle_{\mathrm{L}^2(\mathbb{S}^2)} \langle G, Y_{n,j}\rangle_{\mathrm{L}^2(\mathbb{S}^2)}, \tag{12.90}$$

Accordingly we define the space $\mathrm{H}^{\mathcal{N}}(\{A_n\};\mathbb{S}^2)$ to be the completion of $\mathrm{E}^{\mathcal{N}}(\{A_n\};\mathbb{S}^2)$ with respect to the topology induced by $\langle\cdot,\cdot\rangle_{\mathrm{H}^{\mathcal{N}}(\{A_n\};\mathbb{S}^2)}$. Then we end up with a Hilbert space. If there is no confusion likely to arise we simply write in the following H instead of $\mathrm{H}^{\mathcal{N}}(\{A_n\};\mathbb{S}^2)$. Note that in this situation not all Harm_n are contained in H. Indeed, we have $\mathrm{Harm}_n \subset \mathrm{H}$ if and only if $n \in \mathcal{N}$.

The Hilbert space H corresponding to a summable sequence $\{A_n\}$ with respect to $\mathcal{N}$ possesses a reproducing kernel K_{H}, since the evaluation functional $F \mapsto F(\xi)$, $F \in \mathrm{H}$, is continuous for every $\xi \in \mathbb{S}^2$ (see, e.g., N. Aronszajn [1950]). It can be even seen, that K_{H} admits a uniformly convergent series representation of the form

$$K_{\mathrm{H}}(\xi\cdot\eta) = \sum_{n\in\mathcal{N}} \frac{2n+1}{4\pi}\frac{1}{|A_n|^2} P_n(\xi\cdot\eta), \quad (\xi,\eta)\in\mathbb{S}^2\times\mathbb{S}^2. \tag{12.91}$$

Its Legendre coefficients are given by

$$K_{\mathrm{H}}^{\wedge}(n) = \begin{cases} |A_n|^{-2}, & n \in \mathcal{N}, \\ 0, & n \in \mathcal{N}_0. \end{cases} \tag{12.92}$$

Assume that the summable sequence $\{A_n\}$ with respect to $\mathcal{N}$ satisfies, in addition, $A_n \neq 0$ for $n = 0,\ldots,m$, i.e., $\{0,\ldots,m\} \subset \mathcal{N}$. Then the space $\mathrm{H} = \mathrm{H}(\{A_n\};\mathbb{S}^2)$ can be decomposed into the orthogonal direct sum

$$\mathrm{H} = \mathrm{H}_{0,\ldots,m} \oplus \mathrm{H}_{0,\ldots,m}^{\perp}, \tag{12.93}$$

where $\mathrm{H}_{0,\ldots,m} = \mathrm{Harm}_{0,\ldots,m}$. $\mathrm{H}_{0,\ldots,m}$ and $\mathrm{H}_{0,\ldots,m}^{\perp}$ defined in canonical way are Hilbert spaces possessing the reproducing kernels

$$K_{\mathrm{H}_{0,\ldots,m}}(\xi\cdot\eta) = \sum_{n=0}^{m} \frac{2n+1}{4\pi}\frac{1}{|A_n|^2} P_n(\xi\cdot\eta), \quad (\xi,\eta)\in\mathbb{S}^2\times\mathbb{S}^2, \tag{12.94}$$

$$K_{\mathrm{H}_{0,\ldots,m}^{\perp}}(\xi\cdot\eta) = \sum_{\substack{n\in\mathcal{N}\\ n>m}} \frac{2n+1}{4\pi}\frac{1}{|A_n|^2} P_n(\xi\cdot\eta), \quad (\xi,\eta)\in\mathbb{S}^2\times\mathbb{S}^2, \tag{12.95}$$

respectively. In the same way as described above we have the decomposition of the kernel $K_{\mathrm{H}} = K_{\mathrm{H}_{0,\ldots,m}} + K_{\mathrm{H}_{0,\ldots,m}^{\perp}}$ and $\langle\cdot,\cdot\rangle_{\mathrm{H}} = \langle\cdot,\cdot\rangle_{\mathrm{H}_{0,\ldots,m}} + \langle\cdot,\cdot\rangle_{\mathrm{H}_{0,\ldots,m}^{\perp}}$. The norm $\|\cdot\|_{\mathrm{H}_{0,\ldots,m}^{\perp}}$ of the space $\mathrm{H}_{0,\ldots,m}^{\perp}$ can be understood as a semi-norm in H implying the null space $\mathrm{H}_{0,\ldots,m} = \mathrm{Harm}_{0,\ldots,m}$. This fact will be of importance for the forthcoming definition of spherical splines. The reproducing kernels (12.94) and (12.95) can be seen to satisfy the condition of positive definiteness.

Definition 12.16. *Let $K : [-1,1] \to \mathbb{R}$, given by*

$$K(t) = \sum_{n=0}^{\infty} \frac{2n+1}{4\pi} K^\wedge(n) P_n(t), \tag{12.96}$$

satisfy

$$\sum_{n=0}^{\infty} \frac{2n+1}{4\pi} K^\wedge(n) < \infty. \tag{12.97}$$

K is said to be conditionally positive definite of order m, *if for any unisolvent system $X_N = \{\eta_1, \ldots, \eta_N\}$ on $\mathbb{S}^2$ and all real scalars $a_1, \ldots, a_N$ satisfying*

$$\mathbf{A}_{X_N}(Y_{0,1}, \ldots, Y_{m,2m+1})a = 0 \tag{12.98}$$

the quadratic form

$$\sum_{k=1}^{N} \sum_{l=1}^{N} a_k a_l K(\eta_k \cdot \eta_l) \tag{12.99}$$

is non-negative. K is said to be conditionally strictly positive definite of order m *on $\mathbb{S}^2$ if the quadratic form is positive.* (Strict) positive definiteness of K *is formally understood as conditionally (strict) positive definiteness of order* -1.

It is a remarkable fact that positive definite functions can be characterized by their Legendre expansion. Important results go back to I.J. Schoenberg [1942].

Lemma 12.17. *Each Legendre polynomial P_n is positive definite.*

Proof. It follows from the addition theorem that

$$\sum_{k=1}^{N} \sum_{l=1}^{N} a_k a_l P_n(\eta_k \cdot \eta_l) = \frac{2n+1}{4\pi} \sum_{j=1}^{2n+1} \left| \sum_{k=1}^{N} a_k Y_{n,j}(\eta_k) \right|^2 \geq 0 \tag{12.100}$$

holds true for all systems $\{\eta_1, \ldots, \eta_N\} \subset \mathbb{S}^2$ and coefficients $(a_1, \ldots, a_N)^T \in \mathbb{R}^N \setminus \{0\}$. This proves the assertion of Lemma 12.17. □

Now we come to a more detailed characterization of positive definiteness.

Theorem 12.18. *Assume that the Legendre coefficients of $K : [-1,1] \to \mathbb{R}$ satisfy (12.97). Then the following statements are equivalent:*

(i) for all $\Phi \in \mathrm{C}^{(\infty)}(\mathbb{S}^2)$

$$\int_{\mathbb{S}^2} \int_{\mathbb{S}^2} K(\xi \cdot \eta) \Phi(\xi) \Phi(\eta) \, dS(\xi) \, dS(\eta) \geq 0 \tag{12.101}$$

(ii) for all $Y_{n,j}$ with $n = 0, 1, \ldots; j = 1, \ldots, 2n+1$

$$\int_{\mathbb{S}^2}\int_{\mathbb{S}^2} K(\xi\cdot\eta)Y_{n,j}(\xi)Y_{n,j}(\eta)\, dS(\xi)\, dS(\eta) \geq 0 \tag{12.102}$$

(iii) $K^\wedge(n) \geq 0$ for all $n \geq 0$

(iv) K is positive definite.

Proof. That (i) implies (ii) is clear. We therefore turn to (ii) $\Rightarrow$ (iii). (iii) follows immediately from (ii) since for all $n = 0, 1, \ldots$, $j = 1, \ldots, 2n+1$

$$K^\wedge(n) = \int_{\mathbb{S}^2}\int_{\mathbb{S}^2} K(\xi\cdot\eta)Y_{n,j}(\xi)Y_{n,j}(\eta)\, dS(\xi)\, dS(\eta). \tag{12.103}$$

In order to develop (iv) from (iii) suppose that $\eta_1, \ldots, \eta_N \in \mathbb{S}^2$ are pairwise distinct and assume that $(a_1, \ldots, a_n)^T$ is different from $(0, \ldots, 0)^T$. Then

$$\begin{aligned} 0 &\leq \sum_{n=0}^{\infty} h^{2n} \sum_{i=1}^{2n+1} \left(\sum_{k=1}^{N} a_k Y_{n,j}(\eta_k)\right)^2 K^\wedge(n) \qquad (12.104)\\ &= \int_{\mathbb{S}^2}\int_{\mathbb{S}^2} K(\xi\cdot\eta)\Phi_h(\xi)\Phi_h(\eta)\, dS(\xi)\, dS(\eta), \end{aligned}$$

where $\Phi_h(\xi) = \sum_{k=1}^{N} a_k Q_h(\eta_k\cdot\xi)$, $\xi \in \mathbb{S}^2$, and Q_h is the Abel–Poisson kernel (cf. (12.78)). Letting $h \to 1$ we see that

$$\sum_{k=1}^{N}\sum_{l=1}^{N} a_k a_l K(\eta_k\cdot\eta_l) \geq 0, \tag{12.105}$$

as required.

Finally, assume that (iv) is true and suppose that $\Phi \in \mathrm{C}^{(\infty)}(\mathbb{S}^2)$. Since K is continuous and, therefore, Riemann-integrable there exists a sequence of numbers r_k^N, $N = 1, 2, \ldots$, $k = 1, \ldots, N$, and a sequence of point sets $X_N = \{\eta_1^N, \ldots, \eta_N^N\} \subset \mathbb{S}^2$, $N \in \mathbb{N}$, with pairwise different points $\eta_1^N, \ldots, \eta_N^N$ such that

$$\int_{\mathbb{S}^2}\int_{\mathbb{S}^2} K(\xi\cdot\eta)\Phi(\xi)\Phi(\eta)\, dS(\xi)\, dS(\eta) = \lim_{N\to\infty}\sum_{k=1}^{N}\sum_{l=1}^{N} r_k^N r_l^N K(\eta_k\cdot\eta_l)\Phi(\eta_k)\Phi(\eta_l). \tag{12.106}$$

By assumption,

$$\sum_{k=1}^{N}\sum_{l=1}^{N} K(\eta_k\cdot\eta_l)(r_k^N\Phi(\eta_k))(r_l^N\Phi(\eta_l)) \tag{12.107}$$

is non-negative. Thus, assertion (i) is verified. □

Corollary 12.19. *Assume that $\{A_n\}$ is summable with respect to $\mathcal{N}$. Then $K_{\mathrm{H}^{\mathcal{N}}(\{A_n\};\mathbb{S}^2)}$ satisfies (i)–(iv) of Theorem 12.18.*

For purposes of interpolation the property of strictly conditionally positive definiteness is of importance. As a first result we obtain from Definition 12.15.

Lemma 12.20. *If K of the form (12.96) satisfying (12.96) is conditionally strictly positive definite of order m, then $K^{\perp}_{0,\ldots,m}$ defined by*

$$K^{\perp}_{0,\ldots,m}(t) = K(t) - \sum_{n=0}^{m} \frac{2n+1}{4\pi} K^{\wedge}(n) P_n(t), \quad t \in [-1,1] \tag{12.108}$$

is conditionally strictly positive definite of order m.

Remark 12.21. *Conventionally, in the case that $m = -1$, the sum $\Sigma_{n=0}^{m}$ is assumed to be zero.*

Unfortunately, we do not know equivalent conditions comparable to those of Theorem 12.18 for the case of strictly positive definiteness. That is the reason why we are interested in sufficient conditions for a kernel to be strictly positive definite. A first result is due to Y. Xu, E.W. Cheney [1992], where it is proved that if all the Legendre coefficients of a kernel are positive, then the kernel is strictly positive definite. By using locally supported kernels we are able to improve this result (cf. Theorem 12.32). An equivalent condition for strict positive definiteness is provided by the following lemma (cf. W. Freeden et al. [1998], M. Schreiner [1995]).

Lemma 12.22. *Let $K : [-1,1] \to \mathbb{R}$ of the form (12.96) satisfy*

$$\sum_{n=0}^{\infty} \frac{2n+1}{4\pi} K^{\wedge}(n) < \infty. \tag{12.109}$$

Then the kernel K is strictly positive definite if and only if the functions $K(\,\cdot\eta_1), \ldots, K(\,\cdot\eta_N)$ are linearly independent for any choice of pairwise distinct points $\eta_1, \ldots, \eta_N \in \mathbb{S}^2$.

Proof. According to our assumptions K is the reproducing kernel of a Hilbert space H. Therefore, $K(\eta_k \cdot \eta_l) = \langle K(\,\cdot\eta_k), K(\,\cdot\eta_l)\rangle_{\mathrm{H}}$. Hence,

$$\mathbf{A}_{X_N}(K(\,\cdot\eta_1), \ldots, K(\,\cdot\eta_N)) = \begin{pmatrix} K(\eta_1 \cdot \eta_1) & \ldots & K(\eta_1 \cdot \eta_N) \\ \vdots & & \vdots \\ K(\eta_N \cdot \eta_1) & \ldots & K(\eta_N \cdot \eta_N) \end{pmatrix} \tag{12.110}$$

is as a Gram matrix positive definite if and only if $K(\,\cdot\eta_1), \ldots, K(\,\cdot\eta_N)$ are linearly independent (cf., e.g., P.J. Davis [1963]). □

Next we continue our list of kernel functions. In doing so we drop the condition that all the Legendre coefficients of a kernel have to be positive. We first come to kernels related to radial basis functions in Euclidean space $\mathbb{R}^3$ and concentrate afterwards on locally supported kernels (i.e., axisymmetric "finite elements") on the sphere.

Definition 12.23. *A function $G \in \mathrm{C}^{(0)}(0,\infty)$ is termed* conditionally strictly positive definite of order m in $\mathbb{R}^3$, *if for any unisolvent system $X_N = \{x_1,\ldots,x_N\} \subset \mathbb{R}^3$ and scalars $a_1,\ldots,a_N$ satisfying $\sum_{i=1}^N a_i P(x_i) = 0$ for all polynomials P of degree $\leq m$ in $\mathbb{R}^3$ the quadratic form*

$$\sum_{k=1}^{N}\sum_{i=1}^{N} a_k a_i G(|x_k - x_i|) \tag{12.111}$$

is positive.

Comparing this definition (in Euclidean space $\mathbb{R}^3$) with Definition 12.16 (cf. N. Dyn [1989]) we obtain the following statement: if $G \in \mathrm{C}^{(0)}[0,\infty)$ is conditionally strictly positive definite of order m in $\mathbb{R}^3$, then $t \mapsto K(t) = G(\sqrt{2-2t}), t \in [-1,1]$ is continuous on $[-1,+1]$ and conditionally strictly positive definite of order m on $\mathbb{S}^2$ (in the sense of Definition 12.16).

A result in I.J. Schoenberg [1938] shows us that $t \mapsto K(t) = G(\sqrt{2-2t})$, $t \in [-1,1]$, is strictly positive definite whenever $G \in \mathrm{C}^{(\infty)}[0,\infty)$ satisfies $(-1)^j \frac{d^j}{d\tau^j} G(\sqrt{\tau}) > 0$, $\tau > 0$, $j = 0,1,\ldots$. An extension of this result is proved in J.J. Benedetto, P.J.S.G. Ferreira [2001]: Let $G \in \mathrm{C}^{(\infty)}(0,\infty) \cap \mathrm{C}^{(0)}[0,\infty)$ satisfy $(-1)^j \frac{d^j}{d\tau^j} G(\sqrt{\tau}) \geq 0$, $\tau > 0$, $j > m$ and $\frac{d^{m+1}}{d\tau^{m+1}} G(\sqrt{\tau}) \neq$ const. . Then $t \mapsto K(t) = G(\sqrt{2-2t})$, $t \in [-1,1]$, is conditionally strictly positive definite of order m.

A list of functions satisfying one of these conditions includes (cf. M.D. Buhmann [2003], N. Dyn [1989]):

(i) $G(\tau) = (c^2+\tau^2)^{-\frac{1}{2}}$, $\quad c > 0$, (inverse multiquadric)

(ii) $G(\tau) = (-1)^{m+1}\tau^{\beta}$, $\quad 2m < \beta < 2m+2$, $m \geq 0$,

(iii) $G(\tau) = (-1)^{m+1}\tau^{2m}\ln\tau$, $\quad m \geq 1$,

(iv) $G(\tau) = (-1)^{m+1}(c^2+\tau^2)^{\frac{\beta}{2}}$, $\quad 2m < \beta < 2m+2$, $c > 0$, $m \geq 0$,

(v) $G(\tau) = (-1)^{m+1}(c^2+\tau^2)^m \ln(c^2+\tau^2)^{\frac{1}{2}}$, $\quad c > 0$, $m \geq 0$.

Another criterion that guarantees the regularity of the matrix

$$\mathbf{G} = \begin{pmatrix} G(\sqrt{2-2\eta_1\cdot\eta_1}) & \ldots & G(\sqrt{2-2\eta_N\cdot\eta_1}) \\ \vdots & & \vdots \\ G(\sqrt{2-2\eta_1\cdot\eta_N}) & \ldots & G(\sqrt{2-2\eta_N\cdot\eta_N}) \end{pmatrix} \tag{12.112}$$

was proved in the note C. Miccelli [1986]: Let $G \in \mathrm{C}^{(\infty)}(0,\infty) \cap \mathrm{C}^{(0)}[0,\infty)$ satisfying the conditions $G(\tau) > 0$ and $(-1)^{j+1}\frac{d^j}{d\tau^j}G(\sqrt{\tau}) \geq 0$, $\tau > 0$, $j \geq 1$, then

$$(-1)^{N-1} \det(\mathbf{G}) > 0. \tag{12.113}$$

A function satisfying (12.113) is the well-known multiquadric (see R. Hardy [1971]).

$$K(t) = G(\sqrt{2-2t}) = (c^2 + 2 - 2t)^{\nu}, \quad 0 < \nu < 1,\ c > 0. \tag{12.114}$$

By calculating a simple example it can be shown that the multiquadric without approximation order (i.e., $m = -1$) is not conditionally strictly positive definite. However if we take the approximation order 0 (cf., e.g., examples of (v)), the corresponding kernel defines a reproducing kernel in the space $\mathrm{H}_0^{\perp}$.

12.5 Locally Supported Kernels

For numerical purposes, locally supported kernels are of particular significance. Two types of locally supported kernels are discussed, namely (isotropic) piecewise polynomial generators (cf. W. Freeden, J.C. Mason [1990], M. Schreiner [1997], S.L. Svensson [1984]) and zonal counterparts to so-called Euclidean Wendland functions (cf. H. Wendland [2005]).

Consider the piecewise polynomial function $B_h^{(k)} : [-1,1] \to \mathbb{R}$, $k = 0, 1, \ldots$, $h \in (0,1)$, given by

$$B_h^{(k)}(t) = \begin{cases} 0, & -1 \leq t \leq h, \\ \frac{(t-h)^k}{(1-h)^k}, & h < t \leq 1. \end{cases} \tag{12.115}$$

Let $\eta \in \mathbb{S}^2$ be fixed. Then the η-zonal function $B_h^{(k)}(\eta\cdot\ ) : \mathbb{S}^2 \to \mathbb{R}$ has a local support. The case $k = 0$ is a rotationally invariant counterpart of the Haar kernel (cf. A. Haar [1910]). More explicitly, a spherical cap given by

$$\begin{aligned} \operatorname{supp}\left(B_h^{(k)}(\ \cdot\eta)\right) &= \{\xi \in \mathbb{S}^2 : h \leq \xi\cdot\eta \leq 1\} \\ &= \{\xi \in \mathbb{S}^2 : 1 - \xi\cdot\eta \leq 1 - h\} = \overline{\Gamma_{1-h}(\eta)}. \end{aligned} \tag{12.116}$$

An easy calculation shows that the iterated η-zonal function $(B_h^{(k)})^{(2)}(\ \cdot\eta)$ has the support

$$\operatorname{supp}\left((B_h^{(k)})^{(2)}(\ \cdot\eta)\right) = \{\xi \in \mathbb{S}^2 : 2h^2 - 1 \leq \xi\cdot\eta \leq 1\} = \overline{\Gamma_{2-2h^2}(\eta)}. \tag{12.117}$$

It follows from the Funk–Hecke formula (9.43) that

$$(B_h^{(k)})^{\wedge}(n) = 2\pi \int_h^1 \frac{(t-h)^k}{(1-h)^k} P_n(t)\,dt. \tag{12.118}$$

A recursion formula can be derived by considering the integral in (12.118).

Lemma 12.24. *For $k \geq 0$, $n \geq 1$ we have*

$$(B_h^{(k)})^\wedge(0) = 2\pi \frac{1-h}{k+1} \tag{12.119}$$

$$(k+2)(B_h^{(k)})^\wedge(1) = (k+1+h)(B_h^{(k)})^\wedge(0) \tag{12.120}$$

$$\begin{aligned}(n+k+2)(B_h^{(k)})^\wedge(n+1) &= (2n+1)h(B_h^{(k)})^\wedge(n) \\ &\quad + (k+1-n)(B_h^{(k)})^\wedge(n-1).\end{aligned} \tag{12.121}$$

Next we consider the asymptotic behavior of $(B_h^{(k)})^\wedge(n)$ for fixed $h \in (0,1)$ as $n \to \infty$. For $k = 0$ and $n \geq 1$ we get from (7.21)

$$I_{n,0} = \int_h^1 P_n(t)dt = \frac{1}{2n+1}(P_{n+1}(h) - P_{n-1}(h)). \tag{12.122}$$

From the estimates known for the Legendre polynomials we see that $|(B_h^{(0)})^\wedge(n)| = O(n^{-\frac{3}{2}})$ as n tends to infinity. We conclude recursively that $|(B_h^{(k)})^\wedge(n)| = O(n^{-\frac{3}{2}-k})$. Hence, for all $k \geq 0$, we have a summable sequence with respect to $\mathcal{N}$ given by

$$\left\{ \left((B_h^{(k)})^\wedge(n) \right)^{-1} \right\} \quad \text{where } \mathcal{N} = \{n : (B_h^{(k)})^\wedge(n) \neq 0\}. \tag{12.123}$$

Note that if $(B_h^{(k)})^\wedge(n) = 0$ the n-th entry in this sequence has to be substituted by zero. We conclude that the kernel $(B_h^{(k)})^{(2)}$ can be regarded as reproducing kernel of the space $\mathrm{H} = \mathrm{H}(\{1/(B_h^{(k)})^\wedge(n)\}; \mathbb{S}^2)$. For a detailed characterization of H it is important to know which numbers $(B_h^{(k)})^\wedge(n)$ are zero. This investigation can be found in W. Freeden et al. [1998].

Lemma 12.25. *For $h \in (0,1)$ and $k = 0, 1, \ldots$, the following statements are valid.*

(i) $(B_h^{(k)})^\wedge(n) \neq 0$ *for* $n = 0, 1, \ldots, k+2$.

(ii) For $n \geq k+2$, $(B_h^{(k)})^\wedge(n) = 0$ *if and only if* $C_{n-k-1}^{(k+3/2)}(h) = 0$, *where* $C_{n-k-1}^{(k+3/2)}$ *denotes the Gegenbauer polynomial of degree* $n-k-1$ *(for more details on Gegenbauer polynomials see, e.g., M. Abramowitz, I.A. Stegun [1972], W. Freeden, M. Gutting [2013], W. Gautschi [2004], W. Magnus et al. [1966]).*

Thus we are able to formulate the following result.

Corollary 12.26. *We assume again that $k \geq 0$. The following statements are valid.*

(i) There exist parameters $h \in (0,1)$ such that $(B_h^{(k)})^\wedge(n) \neq 0$ for all $n \in \mathbb{N}_0$.

(ii) Let $m \in \mathbb{N}_0$ be given. Then there exists a number $h_0 \in (0,1)$ such that for all $h \in (h_0, 1)$ and all $n \leq m$, $(B_h^{(k)})^\wedge(n) \neq 0$.

Proof. Part (i) follows immediately from Lemma 12.25, since there exist only countably infinite points being a zero of a Gegenbauer polynomial.

It can be deduced from the theory of orthogonal polynomials (cf., e.g., L.C. Andrews [1985], W. Gautschi [2004], W. Freeden, M. Gutting [2013]) that the largest number $h \in (0,1)$ for which $C_{n-k-1}^{k+3/2}(h) = 0$ for an integer n with $k+2 \leq n \leq m$ is the largest zero of $C_{m-k-1}^{k+3/2}$. So just take this value as h_0. This proves part (ii). □

The first statement of Corollary 12.26 is of more theoretical concern than practical use, since it is also known that the set of all zeros of the Gegenbauer polynomials $C_n^{k+3/2}$, $n = 0, \ldots$, is dense in $[-1,1]$. But the second statement is very useful since it shows that we can ensure by a suitable choice of the parameter $h \in (0,1)$ that all spherical harmonics up to a certain order are contained in the Hilbert space H, possessing $(B_h^{(k)})^{(2)}$ as reproducing kernel. Summarizing our results we finally obtain

Theorem 12.27. *Assume that $k \geq 0$ and $h \in (0,1)$. Suppose that the sets $\mathcal{N} = \{n \in \mathbb{N}_0 : (B_h^{(k)})(n) \neq 0\}$ and $\mathcal{N}_0 = \{n \in \mathbb{N}_0 : (B_h^{(k)})^\wedge(n) = 0\}$ are determined by Lemma 12.25. Then the sequence*

$$A_n = \begin{cases} \frac{1}{(B_h^{(k)})^\wedge(n)}, & n \in \mathcal{N}, \\ 0, & n \in \mathcal{N}_0, \end{cases} \tag{12.124}$$

is summable and defines a reproducing kernel Hilbert space $\mathrm{H} = \mathrm{H}(\{A_n\}; \mathbb{S}^2)$. The reproducing kernel of H is given by

$$K_{\mathrm{H}}(\xi \cdot \eta) = (B_h^{(k)})^{(2)}(\xi \cdot \eta), \quad \xi, \eta \in \mathbb{S}^2, \tag{12.125}$$

where the support is the spherical cap

$$\operatorname{supp} K_{\mathrm{H}}(\,\cdot \eta) = \{\xi \in \mathbb{S}^2 : \; 2h^2 - 1 \leq \xi \cdot \eta\} = \overline{\Gamma_{2-2h^2}(\eta)}. \tag{12.126}$$

This theorem shows how reproducing kernel Hilbert spaces with locally supported reproducing kernels can be constructed.

However, taking $m \geq 0$, the modified kernels $K_{\mathrm{H}^{\perp}_{0,\ldots,m}}$, are now globally supported. Thus, we are now interested in developing a method to construct also kernels $K_{\mathrm{H}^{\perp}_{0,\ldots,m}}$ that still remain locally supported. The idea is simple (cf. W. Freeden et al. [1998]): Let $m \in \mathbb{N}_0$ be fixed. Choose $m+2$ values $h_1, \ldots, h_{m+2} \in (0,1)$ such that $h_1 < \ldots < h_{m+2}$. Then determine real numbers $k_1, \ldots, k_{m+2}$ in such a way that

$$K(t) = \sum_{i=1}^{m+2} k_i B_{h_i}^{(k)}(t), \quad t \in [-1,1], \tag{12.127}$$

satisfies

$$K^\wedge(n) = 0, \quad n = 0, \ldots, m. \tag{12.128}$$

If we define

$$K_{\mathrm{H}}(\xi \cdot \eta) = \sum_{n=0}^{m} \frac{2n+1}{4\pi} P_n(\xi \cdot \eta) + K_{\mathrm{H}_{0,\ldots,m}^{\perp}}(\xi \cdot \eta), \quad \xi, \eta \in \mathbb{S}^2, \tag{12.129}$$

we end up with the reproducing kernel of a space $\mathrm{H} = \mathrm{H}^{\mathcal{N}}(\{A_n\}; \mathbb{S}^2)$, where the sequence $\{A_n\}$ is given by

$$A_n = \begin{cases} \frac{1}{k_1 (B_{h_1}^{(k)})^\wedge(n) + \ldots + k_{m+2} (B_{h_{m+2}}^{(k)})^\wedge(n)}, & n \in \mathcal{N}, n \geq m+1, \\ 1, & n = 0, \ldots, m, \\ 0, & n \in \mathcal{N}_0, \end{cases} \tag{12.130}$$

with

$$\mathcal{N} = \left\{ n \in \mathbb{N}_0 : \sum_{i=1}^{m+2} k_i (B_{h_i}^{(k)})^\wedge(n) \neq 0 \right\} \tag{12.131}$$

and

$$\mathcal{N}_0 = \left\{ n \geq m+1 : \sum_{i=1}^{m+2} k_i (B_{h_i}^{(k)})^\wedge(n) = 0 \right\}. \tag{12.132}$$

Furthermore, it follows that $K_{\mathrm{H}_{0,\ldots,m}^{\perp}} = K^{(2)}$, i.e., the iterated kernel corresponding to K of (12.127). Hence,

$$\operatorname{supp} K^{(2)}(\,\cdot \eta) = \{\xi \in \mathbb{S}^2 : \; 2h_1^2 - 1 \leq \xi \cdot \eta\} = \overline{\Gamma_{2-2h_1^2}(\eta)} \tag{12.133}$$

for all $\eta \in \mathbb{S}^2$. Note that we have used in definition (12.129) the iterated kernel $K^{(2)}$. This ensures that all the Legendre coefficients of K_{H} are non-negative. So it remains to show that we can find $k_1, \ldots, k_{m+2} \in \mathbb{R}$ such that (12.128) is fulfilled:

Theorem 12.28. *Suppose that $m \in \mathbb{N}_0$, $0 < h_1 < \ldots < h_{m+2} < 1$. Then there exists $k_1, \ldots, k_{m+2} \in \mathbb{R}$ such that (12.127) satisfies (12.128). If we require, in addition, that*

$$k_1 + \ldots + k_{m+2} = 1, \tag{12.134}$$

then the numbers k_i are uniquely determined. The support of $K^{(2)}(\,\cdot \eta)$ is, for fixed $\eta \in \mathbb{S}^2$, given by (12.133).

Proof. We only have to show that the system of linear equations defined by (12.128) and (12.134) is uniquely solvable. It can be written down as follows:

$$\begin{pmatrix} 1 & \cdots & 1 \\ (B_{h_1}^{(k)})^\wedge(0) & \cdots & (B_{h_{m+2}}^{(k)})^\wedge(0) \\ \vdots & \ddots & \vdots \\ (B_{h_1}^{(k)})^\wedge(m) & \cdots & (B_{h_{m+2}}^{(k)})^\wedge(m) \end{pmatrix} \begin{pmatrix} k_1 \\ k_2 \\ \vdots \\ k_{m+2} \end{pmatrix} = \begin{pmatrix} 1 \\ 0 \\ \vdots \\ 0 \end{pmatrix}. \tag{12.135}$$

But we have seen in Lemma 12.24 that $\{1, (B_h^{(k)})^\wedge(0), \ldots, (B_h^{(k)})^\wedge(m)\}$ represents a system of polynomials in h of degree $0, \ldots, m+1$, respectively. Thus, they form a unisolvent system for interpolation on the interval $[-1,1]$ (cf. P.J. Davis [1963]). Therefore, the system (12.135) is uniquely solvable, as required. The support property is obvious from our previous discussion. □

Finally, we shall prove that the locally supported kernels as introduced above are strictly positive definite.

Theorem 12.29. *Let $k \geq 0$. Then the following statements are valid:*

(i) If $h \in (0,1)$, then $(B_h^{(k)})^{(2)}$ is strictly positive definite.

(ii) Let $m \in \mathbb{N}_0$ and $0 < h_1 < \ldots < h_{m+2} < 1$, and assume that the kernel $K : [-1,1] \to \mathbb{R}$ is defined as in Theorem 12.28. Then $K^{(2)} = K * K$ is strictly positive definite.

Proof. Assertion (i) is equivalent to the positive definiteness of the symmetric matrix

$$\begin{pmatrix} (B_h^{(k)})^{(2)}(\eta_1 \cdot \eta_1) & \cdots & (B_h^{(k)})^{(2)}(\eta_1 \cdot \eta_N) \\ \vdots & \ddots & \vdots \\ (B_h^{(k)})^{(2)}(\eta_N \cdot \eta_1) & \cdots & (B_h^{(k)})^{(2)}(\eta_N \cdot \eta_N) \end{pmatrix} \tag{12.136}$$

for any point set $X_N = \{\eta_1, \ldots, \eta_N\}$ of distinct points. However, since

$$(B_h^{(k)})^{(2)}(\eta_i \cdot \eta_k) = \int_{\mathbb{S}^2} B_h^{(k)}(\eta_i \cdot \zeta) B_h^{(k)}(\zeta \cdot \eta_k)\, dS(\zeta), \tag{12.137}$$

the matrix (12.136) can be recognized as a Gram matrix, and is therefore positive definite if and only if the functions $B_h^{(k)}(\,\cdot \eta_i)$, $i = 1, \ldots, N$, are linearly independent. But this can be concluded from the discontinuity of the k-th derivative of $B_h^{(k)}$ in the following way:

Assume that $F = \sum_{i=1}^N a_i B_h^{(k)}(\,\cdot \eta_i) = 0$. Thus, F is trivially of class $\mathrm{C}^{(\infty)}$. On the other hand side, the directional derivative of order k of the function $\xi \mapsto B_h^{(k)}(\eta_i \cdot \xi)$ in direction $\eta_i - (\eta_i \cdot \xi)\xi$ does not exist for all $\xi \in \partial\Gamma_{1-h}(\eta_i)$ with $\partial\Gamma_{1-h}(\eta_i) = \{\xi \in \mathbb{S}^2 : \ \xi \cdot \eta_i = h\}$. Suppose that an a_i (without loss of generality, say a_1) is different from zero. Since due to H.S.M. Coxeter [1961] there exists a point

$$\xi_1 \in \partial\Gamma_{1-h}(\eta_1) \setminus \bigcup_{j=2}^{N} \partial\Gamma_{1-h}(\eta_j), \tag{12.138}$$

it follows that the described k-th derivative of F at the point ξ_1 does not exist, a contradiction. Thus, $a_1 = \ldots = a_N = 0$ is proved, and therefore part (i) is shown. The second statement follows by similar arguments. □

Next we turn to strict positive definiteness: It is known from Y. Xu, E.W. Cheney [1992] that a sufficient condition for a positive definite kernel K to be strictly positive definite is $K^\wedge(n) > 0$ for all $n = 0, 1, \ldots$. Now we prove a more general result by use of locally supported kernels and approximation properties of the Abel-Poisson kernel.

Let Q_h be the Abel-Poisson kernel, i.e.,

$$Q_h(\xi \cdot \eta) = \frac{1}{4\pi} \frac{1 - h^2}{(1 + h^2 - 2\xi \cdot \eta)^{3/2}} = \sum_{n=0}^{\infty} \frac{2n+1}{4\pi} h^n P_n(\xi \cdot \eta). \tag{12.139}$$

Then we define for $m \in \mathbb{N}$

$$(Q_h)^{\perp}_{0,\ldots,m}(t) = Q_h(t) - \sum_{n=0}^{m} \frac{2n+1}{4\pi} h^n P_n(t), \quad t \in [-1, 1]. \tag{12.140}$$

The following slight generalization of Poisson integral formula is well-known (cf. W. Freeden, M. Schreiner [2009]).

Theorem 12.30. *Let $F \in \mathrm{C}^{(0)}(\mathbb{S}^2)$ satisfy $\langle F, Y_{n,j}\rangle_{\mathrm{L}^2(\mathbb{S}^2)} = 0$ for all $n = 0, \ldots, m$, $j = 1, \ldots, 2n+1$. Then*

$$\lim_{\substack{h \to 1 \\ h < 1}} \sup_{\xi \in \mathbb{S}^2} \left| F(\xi) - \int_{\mathbb{S}^2} (Q_h)^{\perp}_{0,\ldots,m}(\xi \cdot \eta) F(\eta) \, dS(\eta) \right| = 0. \tag{12.141}$$

This leads us to the following results (cf. W. Freeden et al. [1998], M. Schreiner [1995, 1997]).

Theorem 12.31. *Suppose that $h \in (0, 1)$, $m \in \mathbb{N}_0$. Then the function $(Q_h)^{\perp}_{0,\ldots,m}$ is strictly positive definite.*

Proof. Let $\eta_1, \ldots, \eta_N \in \mathbb{S}^2$ be pairwise distinct. According to Lemma 12.22 we shall show that the functions $(Q_h)^{\perp}_{0,\ldots,m}(\,\cdot \eta_N)$ are linearly independent. Assume, therefore, that $a_1, \ldots, a_N \in \mathbb{R}$ satisfy

$$\sum_{i=1}^{N} a_i (Q_h)^{\perp}_{0,\ldots,m}(\,\cdot \eta_i) = 0. \tag{12.142}$$

Since the series of $(Q_h)^{\perp}_{0,\dots,m}$ in terms of the Legendre polynomials is uniformly convergent, it follows for all $\xi \in \mathbb{S}^2$ and $n = 0, \dots, m$ that

$$0 = \left\langle \sum_{i=1}^{N} a_i (Q_h)^{\perp}_{0,\dots,m}(\eta_i \cdot\ ), \frac{2n+1}{4\pi} P_n(\ \cdot \xi) \right\rangle_{\mathrm{L}^2(\mathbb{S}^2)} \tag{12.143}$$
$$= \sum_{i=1}^{N} a_i h^n P_n(\eta_i \cdot \xi) = h^n \sum_{i=1}^{N} a_i P_n(\eta_i \cdot \xi).$$

Thus, $\sum_{i=1}^{N} a_i P_n(\ \cdot \eta_i) = 0$. The same calculation performed in the backward direction implies then

$$\sum_{i=1}^{N} a_i (Q_h)^{\perp}_{0,\dots,m}(\eta_i \cdot\ ) = 0 \tag{12.144}$$

for all $h \in (0,1)$. Now, let $i^* \in \{1, \dots, N\}$ be fixed, and choose $\varepsilon < \Theta_N$, where Θ_N is given by $\Theta_N = \max_{j=1,\dots,N} \sup_{j \neq i} |\eta_i - \eta_j|$. We deduce from Theorem 12.28 that there exists a continuous function $H \in \mathrm{C}^{(0)}(\mathbb{S}^2)$ with the properties

(i) $H(\eta_{i^*}) = 1$,

(ii) $\mathrm{supp}(H) \subset \{\xi \in \mathbb{S}^2\ : |\xi - \eta| \leq \varepsilon\}$,

(iii) $\langle H, Y_n \rangle_{\mathrm{L}^2(\mathbb{S}^2)} = 0$ for all $Y_n \in \mathrm{Harm}_n$, $n = 0, \dots, m$.

Thus, we are able to deduce from Theorem 12.30 and the fact that $H(\eta_i) = 0$ if $i \neq i^*$,

$$a_{i^*} = \sum_{i=1}^{N} a_i H(\eta_i) = \lim_{\substack{h \to 1 \\ h < 1}} \left\langle \sum_{i=1}^{N} a_i (Q_h)^{\perp}_{0,\dots,m}(\ \cdot \eta_i), H \right\rangle_{\mathrm{L}^2(\mathbb{S}^2)}. \tag{12.145}$$

On the other hand, it follows from (12.144) for all $h \in (0,1)$ that

$$\left\langle \sum_{i=1}^{N} a_i (Q_h)^{\perp}_{0,\dots,m}(\ \cdot \eta_i), H \right\rangle = 0. \tag{12.146}$$

Thus, $a_{i^*} = 0$ and hence the functions $(Q_h)^{\perp}_{0,\dots,m}(\ \cdot \eta_1), \dots, (Q_h)^{\perp}_{0,\dots,m}(\ \cdot \eta_N)$ are linearly independent, as required. □

This theorem finally allows us to prove (cf. M. Schreiner [1995]) the following result.

Theorem 12.32. *Let $K : [-1,1] \to \mathbb{R}$ of the form (12.96) be a positive definite kernel. Assume that only finitely many $K^\wedge(n)$ are zero. Then K is strictly positive definite.*

Proof. Let m denote the index of the largest integer n for that $K^\wedge(n) = 0$. Then $K^\wedge(n) > 0$ for all $n > m$. Assume that $\eta_1 \ldots, \eta_N \in \mathbb{S}^2$ are pairwise distinct. We shall show that $K(\,\cdot \eta_1), \ldots, K(\,\cdot \eta_N)$ are linearly independent.

If $\sum_{i=1}^N a_i K(\,\cdot \eta_i) = 0$, it follows as in the proof of Theorem 12.31 that

$$\sum_{i=1}^{N} a_i P_n(\,\cdot \eta_i) = 0 \tag{12.147}$$

for all n for which $K^\wedge(n) > 0$. In particular, (12.147) is true for all $n > m$. But then we obtain that for every $h \in (0,1)$

$$\sum_{i=1}^{N} a_i (Q_h)^{\perp}_{0,\ldots,m}(\,\cdot \eta_i) = 0. \tag{12.148}$$

Since we know from Theorem 12.31 that $(Q_h)^{\perp}_{0,\ldots,m}$ is strictly positive definite, $a_1 = \ldots = a_N = 0$. Hence, K is strictly positive definite. This is the desired result. □

Next we introduce a class of locally supported radial basis functions, called *Euclidean Wendland functions* (see H. Wendland [1995, 2005]), from which zonal counterparts on $\mathbb{S}^2$ are canonically derivable. For that purpose we start with the explanation of the following operators:

(1) Let ϕ be a function such that $\rho \mapsto \rho\phi(\rho)$, $\rho \geq 0$, is of class $\mathrm{L}^1[0,\infty)$. Then $\mathcal{I}$ is defined by

$$\mathcal{I}\phi(\rho) = \int_\rho^\infty r\phi(r)\, dr, \quad \rho \geq 0. \tag{12.149}$$

(2) Let ϕ be a function of class $\mathrm{C}^{(2)}(\mathbb{R})$. Then $\mathcal{D}$ is defined by

$$\mathcal{D}\phi(\rho) = -\frac{1}{\rho}\frac{d}{d\rho}\phi(\rho), \quad \rho > 0. \tag{12.150}$$

There exist a palette of properties between $\mathcal{I}$ and $\mathcal{D}$ (cf. H. Wendland [1995]) from which we mention without proof:

(i) If $\rho \mapsto \rho\phi(\rho)$ is of class $\mathrm{L}^1[0,\infty)$, then $\mathcal{D}\mathcal{I}\phi = \phi$.

(ii) If $\phi \in \mathrm{C}^{(2)}(\mathbb{R})$ with derivative $\phi' \in \mathrm{L}^2[0,\infty)$, then $\mathcal{I}\mathcal{D}\phi = \phi$.

Indeed, the operators $\mathcal{I}$ and $\mathcal{D}$ are inverse in the sense of (i) and (ii). Moreover, the operators walk through the space dimension in steps of width 2. To be more precise, we are able to verify the following results:

(iii) Let $\phi_{(q)}$ be the radial basis function given by

$$\phi_{(q)}(x) = \phi(|x|), \quad x \in \mathbb{R}^q. \tag{12.151}$$

Then the Fourier transform (cf. (17.73) for more details)

$$(\phi_{(q)})^{\wedge}_{\mathbb{R}^q}(y) = \int_{\mathbb{R}^q} \phi_{(q)}(|x|)e^{-2\pi i y\cdot x}\, dx = \frac{\Gamma(\frac{q}{2})}{2\pi^{\frac{q}{2}}} \int_0^\infty \phi(r) r^{q-1} J_0(q; 2\pi|y|r)\, dr \tag{12.152}$$

satisfies the identity

$$(\phi_{(q)})^{\wedge}_{\mathbb{R}^q} = \mathcal{I}(\phi_{(q-2)})^{\wedge}_{\mathbb{R}^{q-2}}, \quad q \geq 3, \tag{12.153}$$

where $J_0(q;\cdot)$ is the Bessel function of order 0 and dimension q (see, e.g., W. Freeden, M. Gutting [2013]).

(iv) If $\phi \in \mathrm{C}^{(2)}(\mathbb{R})$ with $\rho^q \phi' \in \mathrm{L}^1[0,\infty)$, then

$$(\phi_{(q)})^{\wedge}_{\mathbb{R}^q} = \left((\mathcal{D}\phi)^{\wedge}_{\mathbb{R}^q}\right)_{(q-2)}, \quad q \geq 3. \tag{12.154}$$

(v) If $(\phi_{(q)})^{\wedge}_{\mathbb{R}^q} \in \mathrm{L}^1(\mathbb{R}^q)$, $q \geq 3$, then $\phi_{(q)}$ is positive definite in $\mathbb{R}^q$ if and only if $(\mathcal{I}\phi)_{(q-2)}$ is positive definite in $\mathbb{R}^{q-2}$.

(vi) If $\phi \in \mathrm{C}^{(2)}(\mathbb{R})$ with $\rho^q \phi' \in \mathrm{L}^1[0,\infty)$, then $\phi_{(q)}$ is positive definite on $\mathbb{R}^q$ if and only if $\mathcal{D}\phi_{(q-2)}$ is positive in $\mathbb{R}^{q+2}$.

For $k \in \mathbb{N}_0$, let the function ϕ_k be defined as follows

$$\rho \mapsto \phi_k(\rho) = \mathcal{I}^k (1-\rho)_+^{k+2}, \quad \rho \geq 0, \tag{12.155}$$

where, as usual, the truncated power function is given by (2.1). From H. Wendland [1995, 2005] we borrow the following properties:

(i) ϕ_k admits the representation

$$\phi_k(\rho) = \begin{cases} P_{3k+2}(\rho), & 0 \leq \rho \leq 1, \\ 0, & 1 < \rho, \end{cases} \tag{12.156}$$

where P_{3k+2} is a univariate polynomial of degree $3k+2$.

(ii) There exist constants $A, B > 0$ depending only on k such that the Fourier transform of the associated function $(\phi_k)_{(3)}$, $k \in \mathbb{N}_0$, of class $C^{(2k)}(\mathbb{R}^3)$ satisfies the bounds:

$$\frac{A}{(1+|x|^2)^{k+2}} \leq \left((\phi_k)_{(3)}\right)^{\wedge}_{\mathbb{R}^3}(x) \leq \frac{B}{(1+|x|)^{k+2}}, \quad x \in \mathbb{R}^3. \tag{12.157}$$

The counterpart of $(\phi_k)_{(3)}$ on the unit sphere $\mathbb{S}^2$ in $\mathbb{R}^3$ defined by

$$\Phi_k(\xi\cdot\eta) = (\phi_k)_{(3)}\left(\sqrt{2-2\xi\cdot\eta}\right), \quad \xi,\eta \in \mathbb{S}^2, \tag{12.158}$$

provides a zonal function possessing the support $[\frac{1}{2}, 1]$. Especially, we have

$$\Phi_0(\xi \cdot \eta) = \begin{cases} 3 - 2\xi \cdot \eta - 2\sqrt{2 - 2\xi \cdot \eta}, & \frac{1}{2} \leq \xi \cdot \eta \leq 1 \\ 0, & -1 \leq \xi \cdot \eta < \frac{1}{2}. \end{cases} \tag{12.159}$$

$(\phi_k)_{(3)}$, $k \in \mathbb{N}_0$, is positive definite on $\mathbb{R}^3$. Its Fourier transform is a radial basis function. Therefore, by well-known results (see H. Wendland [1995]), it follows that Φ_k, $k \in \mathbb{N}_0$, is strictly positive definite on the unit sphere $\mathbb{S}^2$. Finally, it should be mentioned that we are able to conclude from (12.157), that

$$\mathrm{H}^{k+\frac{3}{2}}(\mathbb{S}^2) = \mathrm{H}(\{\Phi_k^\wedge(n)\}; \mathbb{S}^2), \quad k \in \mathbb{N}_0. \tag{12.160}$$

12.6 Zonal Function Exact Integration

W. Freeden et al. [1998], A. Sommariva, R. Womersley [2005] and others propose estimating the integral by integrating exactly a *zonal function* approximation of the form

$$F(\xi) \approx \sum_{j=1}^{N} a_j K_{\mathrm{H}}(\xi \cdot \eta_j), \quad \xi \in \mathbb{S}^2, \tag{12.161}$$

so that

$$\int_{\mathbb{S}^2} F(\xi)\, dS(\xi) \approx \sum_{j=1}^{N} a_j \int_{\mathbb{S}^2} K_{\mathrm{H}}(\xi \cdot \eta_j)\, dS(\xi). \tag{12.162}$$

To be more concrete, given arbitrary points $\eta_1, \ldots, \eta_N$, the problem that we address is the construction of a "smooth" function defined over $\mathbb{S}^2$ which interpolates the given data points $(\eta_i, \alpha_i) \in \mathbb{S}^2 \times \mathbb{R}$, $i = 1, \ldots, N$. Applications include modeling closed surfaces in computer aided geometric design and representing functions which estimate physical quantities such as temperature, pressure, gravitational or magnetic potentials at all points on the surface of the spherical Earth based on a discrete sample of values taken at arbitrary locations.

Let $\{A_n\} = \{A^\wedge(n)\}$ be the symbol of a pseudodifferential operator A being summable with respect to $\mathcal{N}$. From the summability we know that $\mathrm{H} = \mathrm{H}^{\mathcal{N}}(\{A_n\}; \mathbb{S}^2)$ admits the reproducing kernel K_{H} given by (12.91). We assume that K_{H} is strictly positive definite. As a consequence we know that a sufficient condition for K_{H} to be strictly positive definite is if $\mathcal{N}_0$ consists of only finitely many elements.

Let there be given N data points $(\eta_1, \alpha_1), \ldots, (\eta_N, \alpha_N) \in \mathbb{S}^2 \times \mathbb{R}$ corresponding to a system $X_N = \{\eta_1, \cdots, \eta_N\}$, $\eta_i \neq \eta_k$ for $i \neq k$. We first consider the interpolation problem of finding the H-*smallest norm interpolant* to the prescribed data:

$$\inf_{F\in\mathcal{I}_N(\alpha)} \|F\|_{\mathrm{H}} = \inf_{F\in\mathcal{I}_N(\alpha)} \|A(F)\|_{\mathrm{L}^2(\mathbb{S}^2)}, \tag{12.163}$$

where

$$\mathcal{I}_N(\alpha) = \{F \in \mathrm{H} : F(\eta_i) = \alpha_i,\ i = 1,\ldots,N\}. \tag{12.164}$$

The smallest norm problem is standard in interpolation theory:

Theorem 12.33. *The interpolation problem*

$$\|S_N^F\|_{\mathrm{H}} = \inf_{G\in\mathcal{I}_N(\alpha)} \|G\|_{\mathrm{H}} \tag{12.165}$$

is well-posed in the sense that its solution exists, is unique, and depends continuously on the data $\alpha_1,\ldots,\alpha_N$. *The uniquely determined solution is given in the explicit form by*

$$S_N^F(\xi) = \sum_{i=1}^{N} a_i K_{\mathrm{H}}(\xi\cdot\eta_i), \quad \xi \in \mathbb{S}^2, \tag{12.166}$$

where the coefficients $a_i \in \mathbb{R}$ *satisfy the system of linear equations*

$$S_N^F(\eta_k) = \sum_{i=1}^{N} a_i K_{\mathrm{H}}(\eta_k\cdot\eta_i) = \alpha_k, \quad k = 1,\ldots,N, \tag{12.167}$$

whose matrix is positive definite.

Once the coefficients a_i in the interpolant (12.166) are known, an approximation to the integral follows easily by integrating the interpolant,

$$\int_{\mathbb{S}^2} S_N^F(\xi)\, dS(\xi) = \frac{1}{A_0^2}\sum_{i=1}^{N} a_i, \tag{12.168}$$

where

$$\frac{1}{A_0^2} = \int_{\mathbb{S}^2} K_{\mathrm{H}}(\xi\cdot\eta_i)\, dS(\xi), \quad i = 1,\ldots,N, \tag{12.169}$$

which is independent of η_i. The integration rule in (12.168) is not presented in the canonical form of a sum of weights times function values, but it can be easily rewritten in that form. There is no practical necessity to do this, but to understand the connection and get formulas for the weights, we use as usual the "Lagrangians".

Corollary 12.34 (Shannon Sampling Theorem)**.** *Let* F *be a member of*

$$\mathrm{Spline}(\{A_n\}; X_N) = \mathrm{span}\,\{K_{\mathrm{H}}(\,\cdot\eta_i) : \ j = 1,\ldots,N\}\,. \tag{12.170}$$

Then F can be reconstructed from its samples at points of X_N by the following interpolation formula

$$F(\xi) = \sum_{k=1}^{N} F(\eta_k) L_k(\xi), \quad \xi \in \mathbb{S}^2, \tag{12.171}$$

where $L_k \in \mathrm{Spline}(\{A_n\}; X_N)$ is given by

$$L_k(\xi) = \sum_{l=1}^{N} a_l^k K_{\mathrm{H}}(\xi \cdot \eta_l) \tag{12.172}$$

and the coefficients a_l^k satisfy the linear equations

$$\sum_{l=1}^{N} a_l^k K_{\mathrm{H}}(\eta_i \cdot \eta_l) = \delta_{i,k}. \tag{12.173}$$

If we represent now the interpolant (12.171) in the form

$$S_N^F = \sum_{j=1}^{N} F(\eta_j) L_j, \tag{12.174}$$

then integration over $\mathbb{S}^2$ yields the standard rule already known from one-dimensional theory

$$\int_{\mathbb{S}^2} S_N^F(\xi)\, dS(\xi) = \sum_{j=1}^{N} w_j F(\eta_j), \tag{12.175}$$

with

$$w_j = \int_{\mathbb{S}^2} L_j(\xi)\, dS(\xi). \tag{12.176}$$

It is readily seen that the weights w_j, $j = 1, \ldots, N$, are the solutions of the linear system

$$\sum_{j=1}^{N} w_j K_{\mathrm{H}}(\eta_k \cdot \eta_j) = \int_{\mathbb{S}^2} K_{\mathrm{H}}(\eta_k \cdot \xi)\, dS(\xi) = \frac{1}{A_0^2}, \qquad k = 1, \ldots, N, \tag{12.177}$$

which expresses the obvious fact that the rule is exact for all functions in $\mathrm{Spline}(\{A_n\}; X_N)$. The weights w_j for this rule are not necessarily positive.

In the above construction, interpolation can be replaced by smoothing, the most natural approach being to replace the interpolant S_N^F of the integrand F in (12.168) by a smoothed approximation of F, that respects but does not interpolate the data (see, e.g., W. Freeden [1981a,b], W. Freeden, B. Witte [1982]. Such an approach is especially relevant for applications with noisy data, but we do not consider it further.

The convergence of zonal kernel interpolation with decreasing mesh norm is known from W. Freeden [1984a,b]. K. Hesse et al. [2010], A. Sommariva, R. Womersley [2005], Q. Le Gia et al. [2009] and others consider a number of different possibilities for kernels $K_{\mathrm{H}}(\cdot)$, in particular convergence of zonal kernel exact integration using compactly supported zonal kernel functions. There are obvious computational attractions in using compactly supported zonal kernel functions: the coefficient matrix is then sparse (because of the compactness of the support). Most of the kernels are explicitly known in terms of elementary expressions, so that an efficient and economical computation can be realized. On the other hand, the uncertainty principle (cf. W. Freeden et al. [1998], W. Freeden, M. Schreiner [2009], N. Lain Fernández [2003]) tells us about deficiencies in the frequency domain, for example, an infinite number of Legendre coefficients vanishes, so that the corresponding spherical harmonics degrees are not reflected for purposes of approximation, in particular, approximate integration.

Chapter 13

Spline Exact Integration

Spline theory is strongly related to zonal kernel exact approximation, where no polynomial precision is of interest. It is canonically based on a variational approach under certain polynomial accuracy (cf. W. Freeden [1978a, 1981a], G. Wahba [1981], W. Freeden [1990], W. Freeden et al. [1998], K.Hesse, I.H. Sloan [2005, 2006]) that minimizes a weighted Sobolev norm of the interpolant, with a large class of spline manifestations provided by pseudodifferential operators being at the disposal of the user. In fact, the Sobolev space framework involving rotation-invariant pseudodifferential operators as realized in Chapter 12 yields some important benefits of spline integration as preparatory tool for spherical sampling. Global spherical spline approximation has its roots in physically motivated problems. It was independently developed by W. Freeden [1978a, 1981a], G. Wahba [1981]. A spline survey is given e.g., by W. Freeden et al. [1996, 1998], G. Whaba [1990]. Generalizations to multi-dimensional hyperspheres are due to W. Freeden, P. Hermann [1985], W. Freeden, M. Gutting [2013] and many others, while generalizations to invariant vectorial and tensorial problems of geosciences and satellite-based disciplines have been discussed by J. Cui, W. Freeden [1997], W. Freeden [1982b, 1987, 1990], R. Reuter [1982], L. Shure et al. [1982], and many others. Related work in multivariate periodic theory can be found in Chapter 21 (see also the references therein).

13.1 Spline Interpolation

Let m be a non-negative integer. We write $\mathrm{H} = \mathrm{H}^{\mathcal{N}}(\{A^\wedge(n)\};\mathbb{S}^2)$ and assume, in addition, that $\{0,\ldots,m\} \subset \mathcal{N}$. This leads to an orthogonal decomposition

$$\mathrm{H} = \mathrm{H}^{\mathcal{N}}(\{A_n\};\mathbb{S}^2) = \mathrm{Harm}_{0,\ldots,m} \oplus \mathrm{H}^{\perp}_{0,\ldots,m}. \tag{13.1}$$

The corresponding reproducing kernels are denoted by $K_{\mathrm{Harm}_{0,\ldots,m}}$ and $K_{\mathrm{H}^{\perp}_{0,\ldots,m}}$, and $\|\cdot\|_{\mathrm{H}^{\perp}_{0,\ldots,m}}$ is a semi-norm in H with null space $\mathrm{Harm}_{0,\ldots,m}$. We assume that K_{H} is conditionally strictly positive definite of order m.

Let there be given N data points $(\eta_1,\alpha_1),\ldots,(\eta_N,\alpha_N)$ corresponding to a unisolvent system $X_N = \{\eta_1,\cdots,\eta_N\}$ relative to $\mathrm{Harm}_{0,\ldots,m}$ and a function $F \in \mathrm{H}$, so that $F(\eta_k) = \alpha_k, k = 1,\ldots,N$. We consider the interpolation problem of finding the $\mathrm{H}^{\perp}_{0,\ldots,m}$-*smallest norm interpolant* to the prescribed data:

$$\inf_{F\in\mathcal{I}_N(\alpha)} \|F\|_{\mathrm{H}^{\perp}_{0,\ldots,m}} = \inf_{F\in\mathcal{I}_N(\alpha)} \left(\sum_{\substack{n\geq m+1\\ n\in\mathcal{N}}} \sum_{j=1}^{2n+1} |A_n|^2 \langle F, Y_{n,j}\rangle^2_{\mathrm{L}^2(\mathbb{S}^2)} \right)^{1/2}. \quad (13.2)$$

in the set $\mathcal{I}_N(\alpha)$ of all H-interpolants to the given data:

$$\mathcal{I}_N(\alpha) = \{G \in \mathrm{H} : G(\eta_i) = F(\eta_i) = \alpha_i,\ i = 1, ..., N\}. \quad (13.3)$$

Definition 13.1. *Let* $X_N = \{\eta_1, \ldots, \eta_N\}$ *be a unisolvent system relative to the space* $\mathrm{Harm}_{0,\ldots,m}$. *Then any function* $S \in \mathrm{H}(\{A_n\}; \mathbb{S}^2)$ *of the form*

$$S(\xi) = Y(\xi) + \sum_{i=1}^{N} a_i K_{\mathrm{H}^{\perp}_{0,\ldots,m}}(\eta_i \cdot \xi), \quad \xi \in \mathbb{S}^2, \quad Y \in \mathrm{Harm}_{0,\ldots,m}, \quad (13.4)$$

is called spherical spline in H of order m *relative to the system* $X_N = \{\eta_1, ..., \eta_N\} \subset \mathbb{S}^2$, *if* $a \in \mathbb{R}^N$, $a = (a_1, ..., a_N)^T$, *satisfies the linear system*

$$\mathbf{A}_{X_N}(Y_{0,1}, \ldots, Y_{m,2m+1})a = 0. \quad (13.5)$$

Correspondingly, the class of all spherical splines of order m in H is denoted by $\mathrm{Spline}_{0,\ldots,m}(\{A_n\}; X_N)$. Evidently, $\mathrm{Spline}_{0,\ldots,m}(\{A_n\}; X_N)$ contains the class $\mathrm{Harm}_{0,\ldots,m}$ as subset. More details can be found in W. Freeden et al. [1998].

Theorem 13.2. *The spline interpolation problem*

$$\|S_N^F\|_{\mathrm{H}^{\perp}_{0,\ldots,m}} = \inf_{G\in\mathcal{I}_N(\alpha)} \|G\|_{\mathrm{H}^{\perp}_{0,\ldots,m}} \quad (13.6)$$

is well-posed in the sense that its solution exists, is unique, and depends continuously on the data $\alpha_1, \ldots, \alpha_N$. *The uniquely determined solution* S_N *is given in the explicit form*

$$S_N^F(\xi) = \sum_{n=0}^{m} \sum_{j=1}^{2n+1} c_{n,j} Y_{n,j}(\xi) + \sum_{i=1}^{N} a_i K_{\mathrm{H}^{\perp}_{0,\ldots,m}}(\eta_i \cdot \xi), \quad \xi \in \mathbb{S}^2, \quad (13.7)$$

where the coefficients $c_{n,j}, a_i$ *satisfy the linear equations*

$$\sum_{i=1}^{N} a_i K_{\mathrm{H}^{\perp}_{0,\ldots,m}}(\eta_k \cdot \eta_i) + \sum_{n=0}^{m} \sum_{j=1}^{2n+1} c_{n,j} Y_{n,j}(\eta_k) = \alpha_k, \quad k = 1, \ldots, N \quad (13.8)$$

and

$$\sum_{i=1}^{N} a_i Y_{n,j}(\eta_i) = 0, \quad n = 0, \ldots, m, \ j = 1, \ldots, 2n+1. \quad (13.9)$$

Next we mention the corresponding Shannon sampling theorem.

Theorem 13.3 (Shannon sampling theorem in $\text{Spline}_{0,\ldots,m}(\{A_n\};X_N)$). *Let F be an element of* $\text{Spline}_{0,\ldots,m}(\{A_n\};X_N)$. *Then F can be reconstructed from its samples at points of X_N by the following interpolation formula (cf. W. Freeden [1981a])*

$$F(\xi) = \sum_{k=1}^{N} F(\eta_k) L_k(\xi), \quad \xi \in \mathbb{S}^2, \tag{13.10}$$

where $L_k \in \text{Spline}_{0,\ldots,m}(\{A_n\};X_N)$ *is given by*

$$L_k(\xi) = \sum_{n=0}^{m} \sum_{j=1}^{2n+1} c_{n,j}^k Y_{n,j}(\xi) + \sum_{l=1}^{N} a_l^k K_{\mathrm{H}_{0,\ldots,m}^{\perp}}(\xi \cdot \eta_l) \tag{13.11}$$

and the coefficients $c_{n,j}^k$, a_l^k satisfy the linear equations

$$\sum_{l=1}^{N} a_l^k K_{\mathrm{H}_{0,\ldots,m}^{\perp}}(\eta_i \cdot \eta_l) + \sum_{n=0}^{m} \sum_{j=1}^{2n+1} c_{n,j}^k Y_{n,j}(\eta_i) = \delta_{i,k}, \tag{13.12}$$

$$\sum_{l=1}^{N} a_l^k Y_{p,q}(\eta_l) = 0, \tag{13.13}$$

where $i, k = 1, \ldots, N$, $p = 0, 1, \ldots, m$, $q = 1, \ldots, 2p+1$.

Numerical experiences with the linear systems of equations stated in Theorem 13.2 have shown that the systems tend to be ill-conditioned unless m and N are not too large. However, it turns out that the condition is strongly dependent on the reproducing kernel function used for the interpolation problem. Moreover, R. Reuter [1982] proposes a combined Lagrange and Newton technique, which also opens the advantage of a permanence property in spline interpolation.

Integration over $\mathbb{S}^2$ yields

$$\int_{\mathbb{S}^2} S_N^F(\xi)\, dS(\xi) = \int_{\mathbb{S}^2} \sum_{n=0}^{m} \sum_{j=1}^{2n+1} c_{n,j} Y_{n,j}(\xi) + \sum_{i=1}^{N} a_i K_{\mathrm{H}_{0,\ldots,m}^{\perp}}(\eta_i \cdot \xi)\, dS(\xi) \tag{13.14}$$

so that

$$\int_{\mathbb{S}^2} S_N^F(\xi)\, dS(\xi) = \sqrt{4\pi} c_{0,1} = \sum_{k=1}^{N} w_k F(\eta_k), \tag{13.15}$$

with

$$w_k = \int_{\mathbb{S}^2} L_k(\xi)\, dS(\xi), \quad k = 1, \ldots, N. \tag{13.16}$$

The weights w_k, $k = 1, \ldots, N$, are given in such a way that

$$\sum_{k=1}^{N} w_k S(\eta_k) = \int_{\mathbb{S}^2} S(\xi)\, dS(\xi), \tag{13.17}$$

which expresses the obvious fact that the rule is exact for all functions S in $\mathrm{Spline}_{0,\ldots,m}(\{A_n\}; X_N)$.

For an approximation order $m \geq 0$ there is also the possibility of establishing a Lagrange expansion of the interpolant, in combination with a Newton part of the approximation when additional points are being added over time (see W. Freeden [1982b, 1984a], W. Freeden et al. [1998]). To this end we remember that there exists in $\mathrm{Harm}_{0,\ldots,m}$ a unique Lagrange basis $\{L_1, \ldots, L_M\}$ with $M = (m+1)^2$ of the form

$$L_r(\xi) = \sum_{k=1}^{M} b_k^r \sum_{n=0}^{m} \sum_{j=1}^{2n+1} Y_{n,j}(\eta_k) Y_{n,j}(\xi), \quad \xi \in \mathbb{S}^2, \tag{13.18}$$

satisfying $L_r(\eta_k) = \delta_{r,k}$, $r, k = 1, \ldots, M$, provided that $X_M = \{\eta_1, \ldots, \eta_M\}$ is a fundamental system on $\mathbb{S}^2$ relative to $\mathrm{Harm}_{0,\ldots,m}$. Therefore, for every $F \in \mathrm{H}$, the unique $\mathrm{Harm}_{0,\ldots,m}$-interpolant $\mathrm{Proj}_{\mathrm{Harm}_{0,\ldots,m}} F$ of F for the $\mathrm{Harm}_{0,\ldots,m}$-fundamental system X_M is given by the "Lagrange form"

$$\mathrm{Proj}_{\mathrm{Harm}_{0,\ldots,m}} F = \sum_{i=1}^{M} F(\eta_i) L_i. \tag{13.19}$$

The mapping $\mathrm{Proj}_{\mathrm{Harm}_{0,\ldots,m}} : \mathrm{H} \to \mathrm{H}$ is a continuous linear projector of $\mathrm{H} \subset \mathrm{C}^{(0)}(\mathbb{S}^2)$ onto $\mathrm{Harm}_{0,\ldots,m}$. Hence, $\mathrm{Proj}_{\mathrm{Harm}_{0,\ldots,m}}$ determines the following direct sum decomposition of H:

$$\mathrm{H} = \mathrm{Harm}_{0,\ldots,m} \oplus \mathring{\mathrm{H}}^{\perp}_{0,\ldots,m}, \tag{13.20}$$

where

$$\mathring{\mathrm{H}}^{\perp}_{0,\ldots,m} = \mathring{\mathrm{H}}^{\perp}_{0,\ldots,m}(\{A_n\}; \mathbb{S}^2) = \{F \in \mathrm{H} : F(\eta_j) = 0,\ j = 1, \ldots, M\}. \tag{13.21}$$

Consequently, any $F \in \mathrm{H}$ can be written uniquely as

$$F = \mathrm{Proj}_{\mathrm{Harm}_{0,\ldots,m}} F + \mathring{F}, \qquad \mathring{F} \in \mathring{\mathrm{H}}^{\perp}_{0,\ldots,m}. \tag{13.22}$$

$\mathring{\mathrm{H}}^{\perp}_{0,\ldots,m}$, as defined by (13.21) equipped with $\langle \cdot, \cdot \rangle_{\mathrm{H}^{\perp}_{0,\ldots,m}}$ is a Hilbert subspace of $\mathrm{C}^{(0)}(\mathbb{S}^2)$. Furthermore, the reproducing structure in H enables us to show (cf. W. Freeden [1984a,b]) that

$$\langle \mathring{F}, K_{\mathring{\mathrm{H}}^{\perp}_{0,\ldots,m}}(\cdot, \eta) \rangle_{\mathrm{H}^{\perp}_{0,\ldots,m}} = \mathring{F}(\eta) \tag{13.23}$$

for all $\eta \in \mathbb{S}^2$ and all $\mathring{F} \in \mathring{\mathrm{H}}^{\perp}_{0,\ldots,m}$, where the reproducing kernel $K_{\mathring{\mathrm{H}}^{\perp}_{0,\ldots,m}}$: $\mathbb{S}^2 \times \mathbb{S}^2 \to \mathbb{R}$ is given by

$$K_{\mathring{\mathrm{H}}^{\perp}_{0,\ldots,m}}(\xi, \eta) = K_{\mathrm{H}^{\perp}_{0,\ldots,m}}(\xi \cdot \eta) - \sum_{i=1}^{M} K_{\mathrm{H}^{\perp}_{0,\ldots,m}}(\xi \cdot \eta_i) L_i(\eta) \tag{13.24}$$

$$- \sum_{i=1}^{M} L_i(\xi) K_{\mathrm{H}^{\perp}_{0,\ldots,m}}(\eta \cdot \eta_i) + \sum_{i=1}^{M} \sum_{j=1}^{M} L_i(\xi) K_{\mathrm{H}^{\perp}_{0,\ldots,m}}(\eta_i \cdot \eta_j) L_j(\eta).$$

Obviously,

$$\left\| \sum_{k=1}^{M} \alpha_k L_k \right\|_{\mathrm{H}_{0,\ldots,m}^{\perp}} = 0. \tag{13.25}$$

This means that the spline interpolation problem (13.2) reduces to the problem

$$\left\| \mathring{S}_{M+1,N} \right\|_{\mathrm{H}_{0,\ldots,m}^{\perp}} = \inf_{\mathring{F} \in \mathcal{I}_{M+1,N}(\alpha)} \left\| \mathring{F} \right\|_{\mathrm{H}_{0,\ldots,m}^{\perp}}, \tag{13.26}$$

where

$$\mathcal{I}_{M+1,N}(\alpha) = \{F \in \mathring{\mathrm{H}}_{0,\ldots,m}^{\perp} : F(\eta_k) = \alpha_k,\ k = M+1,\ldots,N\}. \tag{13.27}$$

From the same arguments leading to Theorem 12.33, however, we know that

$$\mathring{S}_{M+1,N}(\xi) = \sum_{k=M+1}^{N} d_k K_{\mathring{\mathrm{H}}_{0,\ldots,m}^{\perp}}(\xi, \eta_k), \quad \xi \in \mathbb{S}^2, \tag{13.28}$$

where the coefficients $d_{M+1}, \ldots, d_N$ satisfy the linear equations

$$\sum_{k=M+1}^{N} d_k K_{\mathring{\mathrm{H}}_{0,\ldots,m}^{\perp}}(\eta_i, \eta_k) = \alpha_i - \sum_{k=1}^{M} \alpha_k L_k(\eta_i), \quad i = M+1, \ldots, N. \tag{13.29}$$

Theorem 13.4. *The uniquely determined solution S_N of the spline interpolation problem*

$$\|S_N^F\|_{\mathrm{H}_{0,\ldots,m}^{\perp}} = \inf_{G \in \mathcal{I}_N(\alpha)} \|G\|_{\mathrm{H}_{0,\ldots,m}^{\perp}} \tag{13.30}$$

is given in the explicit form

$$S_N^F(\xi) = \sum_{k=1}^{M} \alpha_k L_k(\xi) + \sum_{k=M+1}^{N} d_k K_{\mathring{\mathrm{H}}_{0,\ldots,m}^{\perp}}(\xi, \eta_k), \quad \xi \in \mathbb{S}^2, \tag{13.31}$$

where the coefficients $d_{M+1}, \ldots, d_N$ satisfy the linear equations (13.29).

The coefficient matrix of the linear system (13.29) of $N - M$ equations in the unknowns d_k is, in fact, a Gram matrix consisting of $N - M$ linearly independent elements out of $\mathring{\mathrm{H}}_{0,\ldots,m}^{\perp}$. Thus it is symmetric and positive definite. This implies the unique solvability of the linear system (13.29).

The convergence of the spline technique as proposed above is known from W. Freeden [1984a,b], W. Freeden et al. [1998]. Integration over $\mathbb{S}^2$ shows that

$$\int_{\mathbb{S}^2} S_N^F(\xi)\, dS(\xi) = \sum_{j=1}^{M} w_j F(\eta_j) + \sum_{j=M+1}^{N} d_j \int_{\mathbb{S}^2} K_{\mathring{\mathrm{H}}_{0,\ldots,m}^{\perp}}(\xi, \eta_j)\, dS(\xi), \tag{13.32}$$

i.e.,

$$\int_{\mathbb{S}^2} S_N^F(\xi)\, dS(\xi) = \sum_{j=1}^{M} w_j F(\eta_j) \tag{13.33}$$

$$- \sum_{s=M+1}^{N} d_j \left(\sum_{i=1}^{M} w_i K_{\mathrm{H}^{\perp}_{0,\dots,m}}(\eta_s \cdot \eta_i) - \sum_{i=1}^{M}\sum_{j=1}^{M} w_i K_{\mathrm{H}^{\perp}_{0,\dots,m}}(\eta_i \cdot \eta_j) L_j(\eta_s) \right),$$

where

$$w_j = \int_{\mathbb{S}^2} L_j(\xi)\, dS(\xi), \quad j = 1, \dots, M. \tag{13.34}$$

13.2 Peano's Theorem in Terms of Pseudodifferential Operators

Again, we write $\mathrm{H} = \mathrm{H}(\{A_n\}; \mathbb{S}^2)$ throughout this section. A is a rotation-invariant pseudodifferential operator with symbol $\{A^\wedge(n)\}$ given by $A^\wedge(n) = A_n$ (cf. Section 12.2). The concept of pseudodifferential operators A enable us to formulate a spherical analog (cf. W. Freeden [1981a]) of the well-known Peano theorem. In fact, by choosing

$$A^\wedge(n) = \begin{cases} 1, & n = 2, \dots, m \\ (\partial_{0,\dots,m})^\wedge(n), & n = m+1, \dots \end{cases} \tag{13.35}$$

we are led back to the iterated Beltrami derivatives of Section 9.6.

Theorem 13.5 (Peano's Theorem). *Let $\mathcal{I} : \mathrm{H} \to \mathbb{R}$ be a bounded linear functional on* H *such that $\mathcal{I}Y = 0$ whenever $Y \in \mathrm{Harm}_{0,\dots,m}$. Then, for $F \in \mathrm{H}$,*

$$\mathcal{I}F = \int_{\mathbb{S}^2} (A_\eta K(\eta))(A_\eta F(\eta))\, dS(\eta), \tag{13.36}$$

where

$$K(\eta) = \mathcal{I}_\xi K_{\mathrm{H}^{\perp}_{0,\dots,m}}(\xi, \eta), \quad \eta \in \mathbb{S}^2. \tag{13.37}$$

The function $K : \mathbb{S}^2 \to \mathbb{R}$ is called the Peano kernel for the linear functional $\mathcal{I}$. As usual, $\mathcal{I}_\xi$ means that the linear functional is applied to the ξ-variable.

Proof. We apply $\mathcal{I}$ to both sides of the integral formula (Theorem 12.11) and use the fact that $\mathcal{I}$ vanishes on elements of class $\mathrm{Harm}_{0,\dots,m}$:

$$\mathcal{I}F = \mathcal{I}_\xi \int_{\mathbb{S}^2} (A_\eta K_{\mathrm{H}^{\perp}_{0,\dots,m}}(\xi, \eta))(A_\eta F(\eta))\, dS(\eta). \tag{13.38}$$

The linear functional allows the interchange of $\mathcal{I}$ and the integral. This yields the desired result. □

Consider now an approximation of a linear functional $\mathcal{I}$ of the type (9.344). By a linear functional $\mathcal{L}$ (cf. A. Sard [1949]) of the form

$$\mathcal{L}F = \sum_{k=1}^{N} a_k F(\eta_k), \tag{13.39}$$

where $a_1, \ldots, a_N$ are real numbers, and $X_N = \{\eta_1, \ldots, \eta_N\}$ is a unisolvent system relative to $\mathrm{Harm}_{0,\ldots,m}$. $\mathcal{L}$ is called *exact for the degree* m, if $\mathcal{I}F = \mathcal{L}F$ whenever $F \in \mathrm{Harm}_{0,\ldots,m}$. The remainder, when $\mathcal{L}$ is used to approximate $\mathcal{I}$, is a linear functional $\mathcal{E}$ defined by $\mathcal{E} = \mathcal{I} - \mathcal{L}$. If the approximation of $\mathcal{I}$ by $\mathcal{L}$ is exact for the degree m and F is a function of class H, then $\mathcal{E}$ satisfies the assumptions of Peano's theorem. Hence it follows that

$$\mathcal{E}F = \int_{\mathbb{S}^2} (AK(\eta))(AF(\eta))\, dS(\eta), \tag{13.40}$$

where

$$K(\eta) = \mathcal{E}_\xi K_{\mathrm{H}^{\perp}_{0,\ldots,m}}(\xi, \eta), \quad \eta \in \mathbb{S}^2. \tag{13.41}$$

Applying the Cauchy–Schwarz inequality to (13.40) we obtain

$$|\mathcal{E}F| \leq \left(\int_{\mathbb{S}^2} |AK(\eta)|^2 \, dS(\eta) \right)^{1/2} \left(\int_{\mathbb{S}^2} |AF(\eta)|^2 \, dS(\eta) \right)^{1/2}. \tag{13.42}$$

We note that the quantity J given by

$$J = \int_{\mathbb{S}^2} |AK(\eta)|^2 \, dS(\eta) \tag{13.43}$$

depends on the knots $\eta_1, \ldots, \eta_N$ and the functional $\mathcal{I}$, but not on the function $F \in \mathrm{H}$. Inserting (13.41) into (13.43) we have

$$J = \mathcal{E}_\xi \mathcal{E}_\zeta \int_{\mathbb{S}^2} AK_{\mathrm{H}^{\perp}_{0,\ldots,m}}(\xi, \eta) AK_{\mathrm{H}^{\perp}_{0,\ldots,m}}(\zeta, \eta)\, dS(\eta), \tag{13.44}$$

i.e.

$$J = \mathcal{E}_\xi \mathcal{E}_\zeta K_{\mathrm{H}^{\perp}_{0,\ldots,m}}(\xi, \zeta). \tag{13.45}$$

Collecting our results we therefore obtain the following *a priori* estimate.

Theorem 13.6. *Let $\mathcal{I}$ be a linear functional of type (9.344), and let $\mathcal{L}$ be an approximation to $\mathcal{I}$ of the form (13.39), exact for the degree m. Then, for each $F \in \mathrm{C}^{(2m+2)}(\mathbb{S}^2)$,*

$$|\mathcal{E}(F)| \tag{13.46}$$
$$\leq \left((\mathcal{I}_\xi \mathcal{I}_\zeta - 2\mathcal{I}_\xi \mathcal{L}_\zeta + \mathcal{L}_\xi \mathcal{L}_\zeta) K_{\mathrm{H}^{\perp}_{0,\ldots,m}}(\xi, \zeta) \right)^{1/2} \left(\int_{\mathbb{S}^2} |AF(\eta)|^2 \, dS(\eta) \right)^{1/2}.$$

13.3 Best Approximations

Let us assume that $X_N = \{\eta_1, \ldots, \eta_N\}$ is a unisolvent system relative to $\text{Harm}_{0,\ldots,m}$. Let $\mathcal{I}$ be a linear functional of type (9.344). An approximation $\mathcal{L}$ of the form

$$\mathcal{L}F = \sum_{k=1}^{N} a_k F(\eta_k), \tag{13.47}$$

exact for the degree m, is called a *best approximation* to $\mathcal{I}$, if, within the class of such approximations, there is one for which the quantity J (cf. Theorem 13.6) given by

$$J = (\mathcal{I}_\xi \mathcal{I}_\zeta - 2\mathcal{I}_\xi \mathcal{L}_\zeta + \mathcal{L}_\xi \mathcal{L}_\zeta) K_{\text{H}^{\perp}_{0,\ldots,m}}(\xi, \zeta) \tag{13.48}$$

assumes its minimum. The minimum of J can be obtained by solving the uniquely determined quadratic optimization problem

$$\begin{aligned} &\frac{1}{2}\mathcal{I}_\xi \mathcal{I}_\zeta K_{\text{H}^{\perp}_{0,\ldots,m}}(\xi \cdot \zeta) - \sum_{k=1}^{N} a_k \mathcal{I}_\xi K_{\text{H}^{\perp}_{0,\ldots,m}}(\xi \cdot \eta_k) \\ &+ \frac{1}{2}\sum_{k=1}^{N}\sum_{r=1}^{N} a_k a_r K_{\text{H}^{\perp}_{0,\ldots,m}}(\eta_r \cdot \eta_l) \to \min . \end{aligned} \tag{13.49}$$

under the M-constraints

$$\sum_{k=1}^{N} a_k Y_{p,q}(\eta_k) = \mathcal{I}(Y_{p,q}), \quad p = 0, \ldots, m, \; q = 1, \ldots, 2p+1. \tag{13.50}$$

According to Lagrange's method of multipliers there exists a vector $v \in \mathbb{R}^M$, $v = (v_{0,1}, \ldots, v_{m,2m+1})^T$, such that

$$\begin{aligned} &\frac{\partial}{\partial a_k}\left(\frac{1}{2}\mathcal{I}_\xi \mathcal{I}_\zeta K_{\text{H}^{\perp}_{0,\ldots,m}}(\xi \cdot \zeta)\right) \\ &- \frac{\partial}{\partial a_k}\left(\sum_{k=1}^{N} a_k \mathcal{I}_\xi K_{\text{H}^{\perp}_{0,\ldots,m}}(\xi \cdot \eta_k)\right) \\ &+ \frac{\partial}{\partial a_k}\left(\frac{1}{2}\sum_{k=1}^{N}\sum_{r=1}^{N} K_{\text{H}^{\perp}_{0,\ldots,m}}(\eta_r \cdot \eta_k) a_k a_r\right) \\ &+ \frac{\partial}{\partial a_k}\left(\sum_{p=0}^{m}\sum_{q=1}^{2p+1} v_{p,q}\left(\sum_{k=1}^{N} a_k Y_{p,q}(\eta_k) - \mathcal{I}Y_{p,q}\right)\right) = 0, \quad k = 1, \ldots, N. \end{aligned} \tag{13.51}$$

In connection with the constraints (13.50) this leads us to the $(N+M) \times (N+M)$ linear system

$$\begin{pmatrix} \mathbf{K} & \mathbf{H}^T \\ \mathbf{H} & \mathbf{0} \end{pmatrix} \begin{pmatrix} a \\ v \end{pmatrix} = \begin{pmatrix} j \\ i \end{pmatrix} \tag{13.52}$$

and $\mathbf{K}$, $\mathbf{H}$, i, j are given as follows:

$$\mathbf{K} = \mathbf{A}_{X_N}(K_{\mathrm{H}^{\perp}_{0,\ldots,m}}(\cdot\ \eta_1), \ldots, K_{\mathrm{H}^{\perp}_{0,\ldots,m}}(\cdot\ \eta_N)), \tag{13.53}$$

$$\mathbf{H} = \mathbf{A}_{X_N}(Y_{0,1}, \ldots, Y_{m,2m+1}) \tag{13.54}$$

and

$$j = (j_1, \ldots, j_N)^T, \quad j_k = \mathcal{I}_\xi K_{\mathrm{H}^{\perp}_{0,\ldots,m}}(\xi \cdot \eta_k), \quad k = 1, \ldots, N, \tag{13.55}$$

$$i = (i_{0,1}, \ldots, i_{m,2m+1})^T, \quad i_{p,q} = \mathcal{I}Y_{p,q}, \quad p = 0, \ldots, m, \ q = 1, \ldots, 2p+1. \tag{13.56}$$

Observe that

$$K_{\mathrm{H}^{\perp}_{0,\ldots,m}}(\eta_r \cdot \eta_s) = \int_{\mathbb{S}^2} AK_{\mathrm{H}^{\perp}_{0,\ldots,m}}(\eta_r \cdot \eta) AK_{\mathrm{H}^{\perp}_{0,\ldots,m}}(\eta \cdot \eta_s)\ dS(\eta) \tag{13.57}$$

for $r, s = 1, \ldots, N$. Hence, $\mathbf{K}$ is a $N \times N$ matrix of Gram type. Since the functions $K_{\mathrm{H}^{\perp}_{0,\ldots,m}}(\ \cdot\eta_r)$, $r = 1, \ldots, N$ are linearly independent, it follows that $\det(K) > 0$ (cf. P.J. Davis, P. Rabinowitz [1975]), and the matrix $\mathbf{K}$ is positive definite. It is a remarkable fact that the $(N+M) \times (N+M)$ matrix

$$\begin{pmatrix} \mathbf{K} & \mathbf{H}^T \\ \mathbf{H} & \mathbf{0} \end{pmatrix} \tag{13.58}$$

has already been considered in spline interpolation of order m. We know that (13.58) is non-singular. The unique solution of the linear system (13.52) is given by

$$a = \mathbf{K}^{-1}(j + \mathbf{H}^T v), \tag{13.59}$$

$$v = -(\mathbf{H}\mathbf{K}^{-1}\mathbf{H}^T)^{-1}(\mathbf{H}\mathbf{K}^{-1}j - i). \tag{13.60}$$

The vector $a \in \mathbb{R}^N$, $a = (a_1, \ldots, a_N)^T$, in (13.59) is the only N-tuple for which the quadratic optimization problem (13.49) is minimal under the constraints (13.50). Summarizing our results developed above we obtain the following result.

Theorem 13.7. *Let $\mathcal{I}$ be a linear functional of type (9.344). Let $(a_1, \ldots, a_N)^T$, $(v_{0,1}, \ldots, v_{m,2m+1})^T$ be the solution of the linear system (13.52). Then, for every $F \in \mathrm{H}$, the linear functional $\mathcal{L}$ of type (13.47) represents the best approximation to $\mathcal{I}$. The approximation formula*

$$\mathcal{I}F \approx \mathcal{L}F = \sum_{k=1}^{N} a_k F(\eta_k) \tag{13.61}$$

possesses the a posteriori estimate

$$\left|\mathcal{I}F - \sum_{k=1}^{N} a_k F(\eta_k)\right| \tag{13.62}$$

$$\leq \left(\mathcal{I}_\xi \mathcal{I}_\zeta K_{\mathrm{H}^\perp_{0,\ldots,m}}(\xi\cdot\zeta) - \sum_{k=1}^{N} a_k \mathcal{I}_\xi K_{\mathrm{H}^\perp_{0,\ldots,m}}(\xi\cdot\eta_k) + \sum_{n=0}^{m}\sum_{j=1}^{2n+1} \upsilon_{n,j}\mathcal{I}Y_{n,j}\right)^{1/2} \left(\int_{\mathbb{S}^2} |AF(\eta)|^2 dS(\eta)\right)^{1/2}.$$

In particular, for the linear functional $\mathcal{I}$ given by

$$\mathcal{I}F = F^\wedge(n,j) = \int_{\mathbb{S}^2} F(\eta)Y_{n,j}(\eta)\, dS(\eta), \quad n,j \text{ fixed}, \quad n \leq m, \tag{13.63}$$

we have the *a posteriori* estimate

$$\left|\int_{\mathbb{S}^2} F(\eta)Y_{n,j}(\eta)\, dS(\eta) - \sum_{k=1}^{N} a_k F(\eta_k)\right| \leq \sqrt{\upsilon_{n,j}}\sqrt{\int_{\mathbb{S}^2} |AF(\eta)|^2\, dS(\eta)}. \tag{13.64}$$

For $n = 0$ and therefore $j = 1$, we obtain

Corollary 13.8.

$$\left|\frac{1}{\sqrt{4\pi}}\int_{\mathbb{S}^2} F(\eta)\, dS(\eta) - \sum_{k=1}^{N} a_k F(\eta_k)\right| \leq \sqrt{\upsilon_{0,1}}\sqrt{\int_{\mathbb{S}^2} |AF(\eta)|^2\, dS(\eta)}. \tag{13.65}$$

13.4 Spline Exact Integration Formulas

Once more, we write $\mathrm{H} = \mathrm{H}^{\mathcal{N}}(\{A_n\};\mathbb{S}^2)$ and assume, in addition, that $\{0,\ldots,m\} \subset \mathcal{N}$, and that the kernel K_H is strictly positive definite of order m. Within the space H we look for the solution S_N^F of the spline interpolation problem (13.6) in the set $\mathcal{I}_N(\alpha)$ of all H-interpolants to the given data α defined in (13.3).

As is well-known, this procedure is reasonable because it produces an interpolating function S_N^F which is as close to being a member of $\mathrm{Harm}_{0,\ldots,m}$ as can be achieved for the data. In addition, the spline interpolant S_N^F is the "smoothest" in the sense of (13.6) thereby avoiding that oscillations become too severe for interpolating the data. Moreover, ordinary interpolation by spherical harmonics (i.e., the previous case when $N = M$) may also be included since the uniquely defined interpolant in $\mathrm{Harm}_{0,\ldots,m}$ is evidently a solution of the problem (13.6).

For a given bounded linear functional $\mathcal{I}$ on H and a unisolvent system $X_N = \{\eta_1, \ldots, \eta_N\}$ relative to $\mathrm{Harm}_{0,\ldots,m}$, an approximation $\mathcal{L}$ of the form

$$\mathcal{L}F = \sum_{k=1}^{N} a_k F(\eta_k), \quad F \in \mathrm{H}, a_k \in \mathbb{R}, k = 1, \ldots, N, \tag{13.66}$$

is not uniquely determined by the requirement that $\mathcal{L}$ be exact for the degree m provided that $N > M$. However, as we shall see, $\mathcal{L}$ is in fact uniquely determined by the stronger requirement that $\mathcal{I}F = \mathcal{L}F$ whenever $F \in \mathrm{Spline}_{0,\ldots,m}(\{A_n\}; X_N)$.

Theorem 13.9. *Let X_N be a unisolvent system relative to* $\mathrm{Harm}_{0,\ldots,m}$ *and $\mathcal{I}$ be a linear functional of type (9.344). Then there is a unique functional $\mathcal{L}$ of the form (13.66) satisfying the properties*

(i) $\mathcal{I}Y = \mathcal{L}Y$ whenever $Y \in \mathrm{Harm}_{0,\ldots,m}$,

(ii) there exists a unique $Y \in \mathrm{Harm}_{0,\ldots,m}$ such that

$$(\mathcal{I}_\xi - \mathcal{L}_\xi) K_{\mathrm{H}^{\perp}_{0,\ldots,m}}(\xi \cdot \eta_k) = Y(\eta_k), \quad k = 1, \ldots, N. \tag{13.67}$$

For a given function $F \in \mathrm{H}$, let S_N^F denote the unique function of class $\mathrm{Spline}_{0,\ldots,m}(\{A_n\}; X_N)$ *that interpolates the data points $(\eta_i, F(\eta_i)) \in \mathbb{S}^2 \times \mathbb{R}, i = 1, \ldots, N$. Then $\mathcal{L}$ is also uniquely determined by the property that $\mathcal{L}F = \mathcal{I}S_N^F$ whenever $F \in \mathrm{H}$. Finally, $\mathcal{L}$ is uniquely determined by the requirement that $\mathcal{L}F = \mathcal{I}F$ whenever $F \in$* $\mathrm{Spline}_{0,\ldots,m}(\{A_n\}; X_N)$.

Remark 13.10. *The proof of Theorem 13.9 is similar to one-dimensional results (of Theorem 2.12). An analogous proof appeared in W. Freeden [1981b].*

Proof. If $\mathcal{L}$ is a bounded linear functional on H, the conditions (i), (ii) stated in the first part of Theorem 13.9 are equivalent to the system of linear equations (13.52) for determining the polynomial Y and the coefficients $a_1, \ldots, a_N$. We know from Section 13.3 that the matrix of the linear system is non-singular. Thus Y and $\mathcal{L}$ are uniquely determined by the system of equations (13.52).

We now prove that, for the approximation L so determined, the assumption $F \in \mathrm{Spline}_{0,\ldots,m}(\{A_n\}; X_N)$ implies $\mathcal{L}F = \mathcal{I}F$. Indeed, $F \in \mathrm{Spline}_{0,\ldots,m}(\{A_n\}; X_N)$ implies that F is of the form

$$F = Y + \sum_{k=1}^{N} v_k K_{\mathrm{H}^{\perp}_{0,\ldots,m}}(\cdot \eta_k), \quad Y \in \mathrm{Harm}_{0,\ldots,m} \tag{13.68}$$

with coefficients $v_1, \ldots, v_N$ satisfying

$$\sum_{k=1}^{N} v_k Y_{n,j}(\eta_k) = 0, \quad n = 0, \ldots, m, \quad j = 1, \ldots, 2n+1. \tag{13.69}$$

Hence,

$$\mathcal{I}F = \mathcal{I}Y + \sum_{k=1}^{N} v_k j_k. \tag{13.70}$$

Then, by (13.52), it follows that

$$\mathcal{I}F = \mathcal{L}Y + \sum_{r=1}^{N} v_r \left(\sum_{k=1}^{N} a_k K_{\mathrm{H}^{\perp}_{0,\ldots,m}}(\eta_r \cdot \eta_k) + Y(\eta_r) \right). \tag{13.71}$$

Reversing the order of summation and observing (13.69) we obtain

$$\mathcal{I}F = \mathcal{L}Y + \sum_{k=1}^{N} a_k \sum_{r=1}^{N} v_r K_{\mathrm{H}^{\perp}_{0,\ldots,m}}(\eta_r \cdot \eta_k), \tag{13.72}$$

i.e.,

$$\mathcal{I}F = \mathcal{L}F. \tag{13.73}$$

Conversely, let $\mathcal{L}'$ be any functional of the form (13.66) having the property that $\mathcal{L}'S = \mathcal{I}S$ whenever $S \in \mathrm{Spline}_{0,\ldots,m}(\{A_n\}; X_N)$. Then for $F \in \mathrm{H}$, $\mathcal{L}'S_N^F = \mathcal{I}S_N^F$ and, from the definition of S_N^F, $\mathcal{L}'S_N^F = \mathcal{L}'F$. Thus it follows that $\mathcal{L}'F = \mathcal{I}S_N^F$.

In particular, for $\mathcal{L}' = \mathcal{L}$, we have $\mathcal{L}F = \mathcal{I}S_N^F$, and therefore $\mathcal{L}'F = \mathcal{I}S_N^F = \mathcal{L}F$ whenever $F \in \mathrm{H}$. But this means that $\mathcal{L}' = \mathcal{L}$ on H. Thus, L is uniquely determined either by the assumption that $\mathcal{L}S = \mathcal{I}S$ whenever $S \in \mathrm{Spline}_{0,\ldots,m}(\{A_n\}; X_N)$, or by the property $\mathcal{L}F = \mathcal{I}S_N^F$ whenever $F \in \mathrm{H}$. □

Finally we are able to prove that the best approximation $\mathcal{L}$ of the form (13.66), exact for the degree m, to a bounded linear functional $\mathcal{I}$ on H is precisely the unique approximation given by Theorem 13.9 that is exact for functions $F \in \mathrm{Spline}_{0,\ldots,m}(\{A_n\}; X_N)$.

Theorem 13.11 (Schoenberg's Theorem). *Suppose that $\mathcal{I}$ is a bounded linear functional of type (9.344). Let $\mathcal{L}'$ be any approximation to $\mathcal{I}$ of the form (13.66), exact for the degree m, and let $\mathcal{L}$ denote the approximation given by Theorem 13.9. Let J and J' be the quantities given by (13.43) corresponding to $\mathcal{E} = \mathcal{I} - \mathcal{L}$ and $\mathcal{E}' = \mathcal{I} - \mathcal{L}'$, respectively. Then $J \le J'$ with equality only if $\mathcal{L}' = \mathcal{L}$.*

Proof. Denote by $\mathcal{L}''$ a linear functional defined by $\mathcal{L}'' = \mathcal{L} - \mathcal{L}'$. By virtue of Theorem 13.9 and the assumption on $\mathcal{L}'$

$$\mathcal{L}F = \mathcal{I}F = \mathcal{L}'F \tag{13.74}$$

holds whenever $F \in \mathrm{Harm}_{0,\ldots,m}$. Thus $\mathcal{L}''$ satisfies the assumption of Peano's theorem (Theorem 13.5), and for every $F \in \mathrm{H}$ we have

$$\mathcal{L}''F = \int_{\mathbb{S}^2} (AK''(\eta))(AF(\eta)) \, dS(\eta), \tag{13.75}$$

where

$$K''(\eta) = \mathcal{L}''_\xi K_{\mathrm{H}^\perp_{0,\ldots,m}}(\xi \cdot \eta_k), \quad \eta \in \mathbb{S}^2. \tag{13.76}$$

As difference of $\mathcal{L}$ and $\mathcal{L}'$ the functional $\mathcal{L}''$ is of the form (13.66). Hence, with coefficients $a_1, \ldots, a_N$ satisfying $\sum_{k=1}^N a_k Y_{n,j}(\eta_k) = 0$, $n = 0, \ldots, m$, $j = 1, \ldots, 2m+1$, it follows that

$$K''(\eta) = \sum_{k=1}^N a_k K_{\mathrm{H}^\perp_{0,\ldots,m}}(\eta_k \cdot \eta), \quad \eta \in \mathbb{S}^2. \tag{13.77}$$

Now we consider the functional $\mathcal{E}$ given by $\mathcal{I} - \mathcal{L}$ and the functional $\mathcal{E}' = \mathcal{I} - \mathcal{L}'$ associated with $\mathcal{L}'$. Correspondingly, let K and K' be the Peano kernels with respect to $\mathcal{E}$ and $\mathcal{E}'$. Then, because of $\mathcal{L}'' = \mathcal{L} - \mathcal{L}'$, it follows that

$$\mathcal{L}'' = (\mathcal{I} - \mathcal{E}) - (\mathcal{I} - \mathcal{E}') = \mathcal{E}' - \mathcal{E} \tag{13.78}$$

and by (13.76) and (13.78)

$$K'' = K' - K. \tag{13.79}$$

Furthermore, in light of (13.78) and (13.79),

$$\int_{\mathbb{S}^2} (AK(\eta))(AK''(\eta))\; dS(\eta) = \mathcal{E}_\xi \sum_{k=1}^N a_k K_{\mathrm{H}^\perp_{0,\ldots,m}}(\xi \cdot \eta_k). \tag{13.80}$$

Thus, by Theorem 13.9 we get

$$\int_{\mathbb{S}^2} (AK(\eta))(AK''(\eta))\; dS(\eta) = \sum_{k=1}^N a_k Y(\eta_k) = 0. \tag{13.81}$$

In view of (13.79) and (13.81) it is clear that

$$\int_{\mathbb{S}^2} |AK'(\eta)|^2 \; dS(\eta) = \int_{\mathbb{S}^2} |AK(\eta)|^2 \; dS(\eta) + \int_{\mathbb{S}^2} |AK''(\eta)|^2 \; dS(\eta). \tag{13.82}$$

Therefore we have $J \leq J'$, where equality holds only if

$$\int_{\mathbb{S}^2} |AK''(\eta)|^2 \; dS(\eta) = \sum_{k=1}^N \sum_{r=1}^N a_r a_k K_{\mathrm{H}^\perp_{0,\ldots,m}}(\eta_k \cdot \eta_r) = 0. \tag{13.83}$$

This, however, means $a_1 = \ldots = a_N = 0$. □

Chapter 14

Equidistributions and Discrepancy Methods

The idea of the low discrepancy method is simple: approximate the integral of a function $F : \mathbb{S}^2 \to \mathbb{R}$ by a mean of functional values at prescribed points. This is a reasonable approach if the data set is somehow "equidistributed" over the sphere and if the function shows "good" behavior. These statements can be made rigorous by the concept of equidistribution and the notion of discrepancy. The starting point of the theory of equidistributed point sets is the work of H. Weyl [1916]. Results for Euclidean spaces $\mathbb{R}^n$ can be found in L. Kuipers, H. Niederreiter [1974], I.H. Sloan, S. Joe [1994], S.K. Zaremba [1968], and many others. An extension of the Koksma–Hlawka formula to the spherical context has been proved by W. Freeden [1978c]. More detailed work about equidistribution on the sphere is due to E. Hlawka [1981], W. Freeden, M. Schreiner [1995], J. Cui, W. Freeden [1997].

14.1 Equidistributions

Assume that $X_N = \{\xi_1^N, \ldots, \xi_N^N\}$ is a system of points $\xi_k^N \in \mathbb{S}^2$, $k = 1, \ldots, N$.

Definition 14.1. $\mathcal{P}(X_N) = \{\mathcal{U}(\xi_1^N), \mathcal{U}(\xi_2^N), \ldots, \mathcal{U}(\xi_N^N)\}$ *is called an* associated partition of $\mathbb{S}^2$ to X_N, *if the following four conditions are satisfied:*

(i) any subset $\mathcal{U}(\xi_k^N)$ of $\mathbb{S}^2$ is measurable and

$$\|\mathcal{U}(\xi_k^N)\| = \int_{\mathcal{U}(\xi_k^N)} dS > 0, \quad k = 1, 2, \ldots, N, \tag{14.1}$$

(ii) $\xi_k^N \in \mathcal{U}(\xi_k^N)$ for all $k = 1, \ldots, N$,

(iii) $\bigcup_{k=1}^N \mathcal{U}(\xi_k^N) = \mathbb{S}^2$, and

(iv) $\mathcal{U}(\xi_i^N) \cap \mathcal{U}(\xi_k^N) = \emptyset$ for all $i, k = 1, 2, \ldots, N$, $i \neq k$.

Definition 14.2. *An associated partition* $\mathcal{P}(X_N) = \{\mathcal{U}(\xi_1^N), \ldots, \mathcal{U}(\xi_N^N)\}$ *of* $\mathbb{S}^2$ *to the point set* X_N *is called* equidistributed, *if*

$$\|\mathcal{U}(\xi_k^N)\| = \int_{\mathcal{U}(\xi_k^N)} dS(\xi) = \frac{4\pi}{N}, \quad k \in \{1, \ldots, N\}. \tag{14.2}$$

The set of all equidistributed partitions $\mathcal{P}(X_N)$ *of* $\mathbb{S}^2$ *to* X_N *is denoted by* $\Pi(X_N)$. *The* partition size of X_N *is defined by*

$$\sigma(X_N) = \inf\left\{ \max_{k=1,\ldots,N} \sup_{\xi \in \mathcal{U}(\xi_k^N)} |\xi - \xi_k^N| : \mathcal{P}(X_N) \in \Pi(X_N) \right\}. \tag{14.3}$$

It follows that, for every equidistributed partition $\mathcal{P}(X_N) = \{\mathcal{U}(\xi_1^N), \ldots, \mathcal{U}(\xi_N^N)\}$ of the above type,

$$\begin{aligned} \frac{1}{4\pi} \int_{\mathbb{S}^2} F(\eta)\, dS(\eta) &= \frac{1}{4\pi} \sum_{k=1}^{N} \int_{\mathcal{U}(\xi_k^N)} F(\eta)\, dS(\eta) \qquad (14.4) \\ &= \frac{1}{N} \sum_{k=1}^{N} F(\xi_k^N) + \frac{1}{4\pi} \sum_{k=1}^{N} \int_{\mathcal{U}(\xi_k^N)} F(\eta) - F(\xi_k^N)\, dS(\eta). \end{aligned}$$

Thus we obtain the following result.

Lemma 14.3. *Let* $X_N = \{\xi_1^N, \ldots, \xi_N^N\}$ *be a system of points on* $\mathbb{S}^2$ *and let* F *be Lipschitz-continuous on* $\mathbb{S}^2$ *with Lipschitz constant* C_F. *Then*

$$\left| \frac{1}{4\pi} \int_{\mathbb{S}^2} F(\eta)\, dS(\eta) - \frac{1}{N} \sum_{k=1}^{N} F(\xi_k^N) \right| \leq C_F \sigma(X_N). \tag{14.5}$$

In other words, the "*integral mean*"

$$\mathcal{I}F = \frac{1}{4\pi} \int_{\mathbb{S}^2} F(\eta)\, dS(\eta) \tag{14.6}$$

may be replaced by the "*approximate sum*" (*arithmetic mean*)

$$\mathcal{L}_{X_N} F = \frac{1}{N} \sum_{k=1}^{N} F(\xi_k^N) \tag{14.7}$$

provided that the partition size $\sigma(X_N)$ is sufficiently small.

Let us consider sequences $\{X_N\}$ of systems $X_N = \{\xi_1^N, \ldots, \xi_N^N\} \subset \mathbb{S}^2$.

Definition 14.4. *A sequence* $\{X_N\}$ *is called a* hierarchical sequence, *whenever* $X_N \subset X_{N+1}$ *for all* N. *One can always assume that* $\xi_k^N = \xi_k^{N+1}$ *holds for all* $k = 1, \ldots, N$ *and all positive integers* N.

A sequence (or hierarchical sequence) $\{X_N\}$ *is called* equidistribution *(or* hierarchical equidistribution*), if*

$$\mathcal{I}F = \frac{1}{4\pi}\int_{\mathbb{S}^2} F(\xi)\, dS(\xi) = \lim_{N\to\infty}\frac{1}{N}\sum_{k=1}^{N} F(\xi_k^N) = \lim_{N\to\infty}\mathcal{L}_{X_N}F \tag{14.8}$$

holds for all $F \in \mathrm{C}^{(2)}(\mathbb{S}^2)$.

This is obviously the case for sequences $\{X_N\}$ with partition size $\sigma(X_N) \to 0$ for N tending to infinity. Note that, in Section 15.4, Definition 15.15 and Definition 15.16 provide a generalization to subsets $\mathcal{U} \subset \mathbb{S}^2$.

14.2 Discrepancy Variants

Extending the well-known Erdös–Turan inequality (cf. L. Kuipers, H. Niederreiter [1974], E. Hlawka [1975]), P.J. Grabner [1991] establishes the following bound for the so-called *spherical cap discrepancy*

$$\begin{aligned} D_\Gamma(X_N) &= \sup_\Gamma \left| \frac{1}{N}\sum_{k=1}^{N} \chi_\Gamma(\xi_k^N) - \|\Gamma\| \right| \qquad (14.9)\\ &= \sup_{\xi\in\mathbb{S}^2}\ \sup_{\rho\in[-1,1]} \left| \frac{1}{N}\sum_{k=1}^{N} \chi_{\Gamma_\rho(\xi)}(\xi_k^N) - \|\Gamma_\rho(\xi)\| \right| \end{aligned}$$

where $\Gamma_\rho(\xi) = \{\eta \in \mathbb{S}^2 : 1 - \xi\cdot\eta < \rho\}$ and $\chi_{\Gamma_\rho(\xi)}$ denotes the characteristic function of the spherical cap. The supremum in the first line of (14.9) is taken over all spherical caps $\Gamma \subset \mathbb{S}^2$. In other words, the discrepancy G_Γ measures the maximal deviation between the empirical distribution of the point set X_N and uniform distribution. More explicitly, for any positive integer L and constants $C_1, C_2, C_3 > 0$, the inequality of P.J. Grabner [1991]

$$D_\Gamma(X_N) \le \frac{C_1}{L+1} + \sum_{n=1}^{L}\left(\frac{C_2}{n} + \frac{C_3}{L+1}\right)\sum_{j=1}^{2n+1}\left|\frac{1}{N}\sum_{k=1}^{N} Y_{n,j}(\xi_k^N)\right| \tag{14.10}$$

holds true. Vaaler's approximation kernel (cf. J.D. Vaaler [1985]), which is very suitable to approximate step functions by periodic polynomials, is the main ingredient for the proof of this estimate.

The above upper bound for the cap discrepancy in terms of spherical harmonics suggests that a notion of discrepancy based on spherical harmonics might be fruitful. In analogy to the so-called polynomial discrepancy for sequences in $[0,1)^q$ (cf. E. Hlawka [1975], R.F. Tichy [1984]) we introduce a spherical polynomial discrepancy defined for $q = 2$ by

$$D_{\mathrm{Harm}}(X_N) = \sup_{n\geq 1} \left(\frac{1}{n^2} \max_{1\leq j\leq 2n+1} \left| \frac{1}{N} \sum_{k=1}^{N} Y_{n,j}(x_k^N) \right| \right). \tag{14.11}$$

It is a natural question to ask whether uniform distribution of infinite point sequences can be defined via cap discrepancy as well as via the *spherical polynomial discrepancy.* By well-known arguments (cf. L. Kuipers, H. Niederreiter [1974]) it can be shown that this is equivalent to

$$\lim_{N\to\infty} D_\Gamma(X_N) = 0 \quad \text{or} \quad \lim_{N\to\infty} D_{\mathrm{Harm}}(X_N) = 0. \tag{14.12}$$

B. Klinger, R.F. Tichy [1996] provides the following relation between these two concepts of discrepancy using two constants $C_1, C_2 > 0$

$$C_1 D_{\mathrm{Harm}}(X_N) \leq D_\Gamma(X_N) \leq C_2 \left(D_{\mathrm{Harm}}(X_N)\right)^{1/4}. \tag{14.13}$$

An upper bound for the spherical cap discrepancy in terms of Legendre polynomials P_n was proved by P.J. Grabner [1991] with constants $C, \widetilde{C} > 0$

$$D_\Gamma(X_N) \leq \frac{C}{L+1} + \widetilde{C} \sum_{n=1}^{L} n^{-\frac{1}{2}} \frac{1}{N} \left(\sum_{i=1}^{N} \sum_{k=1}^{N} P_n(\xi_i \cdot \xi_k) \right)^{1/2}. \tag{14.14}$$

This leads P.J. Grabner et al. [1997] to define a discrepancy based on Legendre polynomials

$$D_{\mathrm{Leg}}(X_N) = \sup_{n\geq 1} \frac{1}{n^2} \frac{1}{N} \left(\sum_{i=1}^{N} \sum_{k=1}^{N} P_n(\xi_i \cdot \xi_k) \right)^{1/2}. \tag{14.15}$$

which is compatible with the aforementioned concepts (cf. B. Klinger, R.F. Tichy [1996]), i.e., with two constants $C_1, C_2 > 0$

$$C_1 D_{\mathrm{Leg}}(X_N) \leq D_\Gamma(X_N) \leq C_2 (D_{\mathrm{Leg}}(X_N))^{\frac{2}{7}}. \tag{14.16}$$

An immediate consequence of (9.180) and (9.199) is the following integral formula (cf. W. Freeden [1978c]).

Theorem 14.5 (Koksma–Hlawka Formula). *Assume $X_N = \{\xi_1^N, \ldots, \xi_N^N\}$ is a system of points $\xi_1^N, \ldots, \xi_N^N$ on $\mathbb{S}^2$. Then*

$$\begin{aligned} \frac{1}{4\pi} \int_{\mathbb{S}^2} F(\eta)\, dS(\eta) &= \frac{1}{N} \sum_{k=1}^{N} F(\xi_k^N) \\ &\quad - \frac{1}{N} \sum_{k=1}^{N} \int_{\mathbb{S}^2} (\Delta_\eta^* G((\Delta^*)^2; \xi_k^N \cdot \eta))(\Delta_\eta^* F(\eta))\, dS(\eta) \end{aligned} \tag{14.17}$$

holds true for all $F \in \mathrm{C}^{(2)}(\mathbb{S}^2)$.

Remark 14.6. *Remember* $G((\Delta^*)^2;\xi,\eta)$, $\xi,\eta \in \mathbb{S}^2$, *is explicitly available both as bilinear expansion* (9.179) *in terms of spherical harmonics and as explicit function in Corollary 9.21.*

Using the abbreviation

$$\mathcal{E}_{X_N} F = \mathcal{I}F - \mathcal{L}_{X_N} F \tag{14.18}$$

we get by virtue of Theorem 14.5 that

$$\mathcal{E}_{X_N} F = \int_{\mathbb{S}^2} (\mathcal{E}_{X_N} \Delta^*_\eta G((\Delta^*)^2;\ \cdot\eta))(\Delta^*_\eta F(\eta))\ dS(\eta) \tag{14.19}$$

provided that F is a member of $\mathrm{C}^{(2)}(\mathbb{S}^2)$. Observing the expression (14.19) we obtain from the Cauchy–Schwarz inequality

$$\begin{aligned} &|\mathcal{E}_{X_N} F| \\ &\leq \frac{1}{N}\left(\sum_{k=1}^{N}\sum_{i=1}^{N} G((\Delta^*)^2;\xi_k^N\cdot\xi_i^N)\right)^{1/2}\left(\int_{\mathbb{S}^2}|\Delta^*_\eta F(\eta)|^2\ dS(\eta)\right)^{1/2}. \end{aligned} \tag{14.20}$$

The expression

$$D(X_N;\Delta^*) = \frac{1}{N}\left(\sum_{k=1}^{N}\sum_{i=1}^{N} G((\Delta^*)^2;\xi_k^N\cdot\xi_i^N)\right)^{1/2} \tag{14.21}$$

is called *discrepancy of* X_N (more accurately *discrepancy of* X_N *to the operator* Δ^*). This allows the following reformulation of (14.20).

Lemma 14.7. *Assume that the function* F *is of class* $\mathrm{C}^{(2)}(\mathbb{S}^2)$. *Then the following estimate is valid:*

$$|\mathcal{E}_{X_N} F| \leq D(X_N;\Delta^*)V(F), \tag{14.22}$$

where the $\mathrm{L}^2(\mathbb{S}^2)$-variance $V(F)$ *is given by*

$$V(F) = \left(\int_{\mathbb{S}^2}|\Delta^*_\eta F(\eta)|^2\ dS(\eta)\right)^{1/2}. \tag{14.23}$$

An easy consequence using Lemma 14.3 is the following result.

Lemma 14.8. *If* $\sigma(X_N)\to 0$ *as* $N\to\infty$, *then* $\{X_N\}$ *is an equidistribution.*

Moreover, we have the following lemma expanding (14.8) to continuous functions F instead of functions of class $\mathrm{C}^{(2)}(\mathbb{S}^2)$.

Lemma 14.9. *Suppose that* $\{X_N\}$ *is an equidistribution. Then (14.8) holds for all* $F \in \mathrm{C}^{(0)}(\mathbb{S}^2)$.

Proof. We know already that (14.8) is true for all functions of $\mathrm{C}^{(2)}(\mathbb{S}^2)$. Take now a continuous function F on $\mathbb{S}^2$. For $\varepsilon > 0$ arbitrary, the closure of the system of spherical harmonics tells us that there exists a function $P \in \mathrm{C}^{(2)}(\mathbb{S}^2)$, such that

$$\sup_{\xi \in \mathbb{S}^2} |F(\xi) - P(\xi)| \leq \varepsilon. \tag{14.24}$$

Therefore it follows via the triangle inequality that

$$\begin{aligned} &\left| \frac{1}{4\pi} \int_{\mathbb{S}^2} F(\eta)\, dS(\eta) - \frac{1}{N} \sum_{k=1}^{N} F(\eta_k^N) \right| \qquad (14.25)\\ &\leq \frac{1}{4\pi} \int_{\mathbb{S}^2} |F(\eta) - P(\eta)|\, dS(\eta) + \left| \frac{1}{4\pi} \int_{\mathbb{S}^2} P(\eta)\, dS(\eta) - \frac{1}{N} \sum_{k=1}^{N} P(\eta_k^N) \right| \\ &\quad + \frac{1}{N} \sum_{k=1}^{N} |P(\eta_k^N) - F(\eta_k^N)| \\ &\leq 2\varepsilon + \left| \frac{1}{4\pi} \int_{\mathbb{S}^2} P(\eta)\, dS(\eta) - \frac{1}{N} \sum_{k=1}^{N} P(\eta_k^N) \right|. \end{aligned}$$

For N sufficiently large we thus obtain $|\mathcal{E}_{X_N} F| \leq 3\varepsilon$, as required. □

It is not very complicated to extend Lemma 14.9 to all Riemann-integrable functions. We omit these considerations.

Next we want to adopt a famous criterion due to H. Weyl [1916]. We consider the system $\{\mathcal{L}_{X_N} Y_{n,j}\}$ of "*Weyl-sums*" $\mathcal{L}_{X_N} Y_{n,j}$, $n = 0, 1, \ldots,$ $j = 1, \ldots, 2n+1$.

Theorem 14.10. *Let $\{X_N\}$ be a sequence of point sets $X_N = \{\xi_1^N, \ldots, \xi_N^N\}$. Then the condition*

$$\lim_{N\to\infty} \frac{1}{N} \sum_{k=1}^{N} Y_{n,j}(\xi_k^N) = 0 \quad \textit{for } n = 1, 2, \ldots,\; j = 1, \ldots, 2n+1, \tag{14.26}$$

is equivalent to

$$\lim_{N\to\infty} \frac{1}{N} \sum_{k=1}^{N} F(\xi_k^N) = \frac{1}{4\pi} \int_{\mathbb{S}^2} F(\eta)\, dS(\eta) \quad \textit{for all } F \in \mathrm{C}^{(0)}(\mathbb{S}^2). \tag{14.27}$$

Proof. Assume that (14.26) is true. Then we know from Theorem 14.5 that for all $F \in \mathrm{C}^{(2)}(\mathbb{S}^2)$

$$\lim_{N\to\infty} |\mathcal{E}_{X_N} F| = \lim_{N\to\infty} \left| \int_{\mathbb{S}^2} \frac{1}{N} \sum_{k=1}^{N} G((\Delta^*)^2; \xi_k^N \cdot \eta) Y_{n,j}(\eta)\, dS(\eta) \right| = 0 \tag{14.28}$$

for all $n = 1, 2, \ldots, j = 1, \ldots, 2n + 1$. Hence,

$$\lim_{N\to\infty} \frac{1}{N} \sum_{k=1}^{N} G((\Delta^*)^2; \xi_k^N \cdot\ ) = 0. \tag{14.29}$$

Then, by virtue of Lemma 14.7, condition (14.27) is true for all $F \in \mathrm{C}^{(0)}(\mathbb{S}^2)$. That condition (14.27) implies (14.26) is clear from the orthogonality of spherical harmonics. □

As an immediate consequence we obtain the following result.

Corollary 14.11. *$\{X_N\}$ is an equidistribution if and only if*

$$\lim_{N\to\infty} D(X_N; \Delta^*) = 0. \tag{14.30}$$

14.3 Examples of Equidistributions on the Sphere

Finally we list some examples of *sequences generating equidistributions* which are of practical importance. Thereby, any point of the below described ensembles is given in terms of the standard representation (cf. (9.1)) using the longitude $\varphi \in [0, 2\pi)$ and the latitude $\vartheta \in [0, \pi]$.

Example 14.12. *Depending on a parameter $\gamma \in \mathbb{N}$ the spherical coordinates of the* Reuter grid *are given by*

$$(\vartheta_0, \varphi_{0,1}) = (0, 0) \qquad \textit{(North Pole)}, \tag{14.31}$$

$$(\vartheta_i, \varphi_{i,j}) = \left(i\frac{\pi}{\gamma}, \left(j - \frac{1}{2}\right)\frac{2\pi}{\gamma_i}\right), \quad i = 1, \ldots, \gamma - 1,\ j = 1, \ldots, \gamma_i, \tag{14.32}$$

$$(\vartheta_\gamma, \varphi_{\gamma,1}) = (\pi, 0) \qquad \textit{(South Pole)}, \tag{14.33}$$

where

$$\gamma_i = \left\lfloor 2\pi \left(\arccos \left(\frac{\cos(\pi/\gamma) - \cos^2(\vartheta_i)}{\sin^2(\vartheta)} \right) \right)^{-1} \right\rfloor. \tag{14.34}$$

This construction due to R. Reuter [1982] leads to a system of points $X_{N(\gamma)}$ (see Figure 14.1 (left column) for an illustration). Notice that we get an equidistribution but not a hierarchical equidistribution by increasing the parameter γ.

Lemma 14.13. *The number of points $N(\gamma)$ for a given $\gamma \in \mathbb{N}$ can be estimated by*

$$N(\gamma) \leq 2 + \frac{4}{\pi}\gamma^2. \tag{14.35}$$

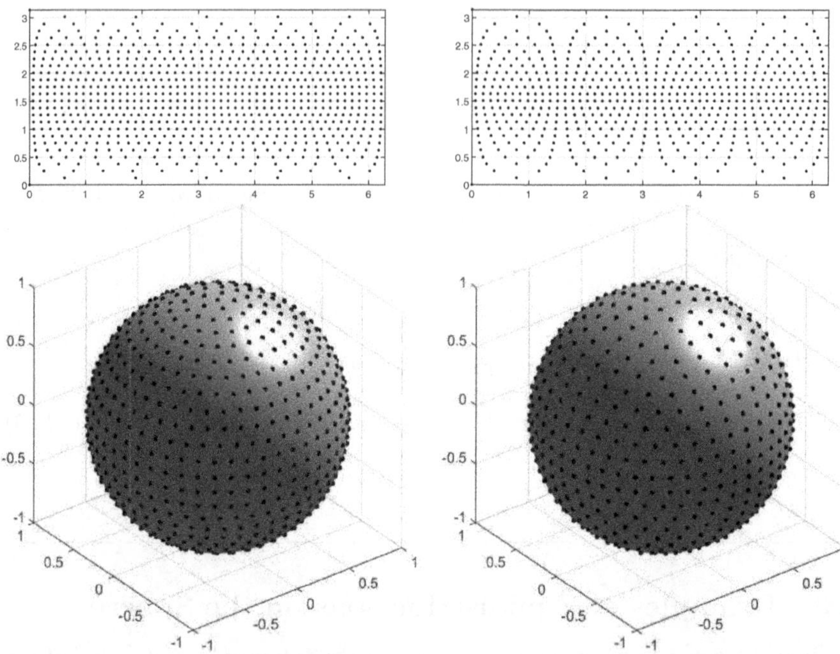

FIGURE 14.1: Coordinates and points of the Reuter grid of Example 14.12 for $\gamma = 25$ with $N(\gamma) = 786$ (left column) and of the grid of Example 14.14 also for $\gamma = 25$ with $N(\gamma) = 626$ (right column).

Proof. For $1 \le i \le \gamma - 1$ we have the estimate

$$\frac{\gamma_i}{\gamma} = \frac{1}{\gamma}\left\lfloor 2\pi/\arccos\left(\frac{\cos(\pi/\gamma) - \cos^2(i\pi/\gamma)}{\sin^2(i\pi/\gamma)}\right)\right\rfloor \le 2\sin\left(\frac{i\pi}{\gamma}\right). \tag{14.36}$$

Hence, remembering the trapezoidal rule (cf. Chapter 5) and observing that $\sin(\pi t)$ is concave for $t \in (-1, 1)$ we obtain

$$\begin{aligned} N(\gamma) &= 2 + \sum_{i=1}^{\gamma-1} \gamma_i = 2 + \gamma^2 \sum_{i=1}^{\gamma-1} \frac{1}{\gamma}\frac{\gamma_i}{\gamma} \qquad (14.37)\\ &\le 2 + 2\gamma^2 \sum_{i=1}^{\gamma-1} \frac{1}{\gamma}\sin\left(\frac{i\pi}{\gamma}\right) \le 2 + 2\gamma^2 \int_0^1 \sin(\pi t)\, dt = 2 + \frac{4}{\pi}\gamma^2, \end{aligned}$$

as required. □

Example 14.14. *Again dependent on a parameter $\gamma \in \mathbb{N}$, we define the ensemble $X_{N(\gamma)}$ by the following relations* (14.31)–(14.33), *but now*

$$\gamma_i = \begin{cases} 4i, & i \le \gamma/2, \\ 4(\gamma - i), & i > \gamma/2, \end{cases} \tag{14.38}$$

instead of (14.34) *(see Figure 14.1 (right column) for an illustration).*

Remark 14.15. *The total number $N(\gamma)$ of points is determined by the choice of the value γ. We have*

$$N(\gamma) = 2 + 4\left\lfloor \frac{\gamma+1}{2} \right\rfloor \left\lfloor \frac{\gamma}{2} \right\rfloor . \tag{14.39}$$

By doubling the parameter γ a hierarchical equidistribution is constructed (for more details see W. Freeden et al. [1998] and the references therein).

The generation of the following two systems is based on the theory of uniform distributions in Euclidean spaces $\mathbb{R}^q$ (cf. L. Kuipers, H. Niederreiter [1974]). More explicitly, the idea is to use a sequence in a rectangle which is uniformly distributed and then to transform it via spherical coordinates to the unit sphere $\mathbb{S}^2$. This approach yields more "irregular" point distributions for numerical integration. Based on the idea of J.G. van der Corput [1935a,b] we consider for $p \in \{2, 3, \ldots\}$ the unique expansion of $n-1$, where $n \in \mathbb{N}$, of the form

$$n - 1 = \sum_{j=0}^{s} a_j p^j, \quad a_j \in \{0, \ldots, p-1\} \tag{14.40}$$

and define

$$\Phi_p(n) = \sum_{j=0}^{s} a_j p^{-j-1}. \tag{14.41}$$

Clearly, $\Phi_p(n) \in [0, 1)$ for all $n \in \mathbb{N}$. The mapping Φ_p is essentially involved in the next two examples.

Example 14.16. *For a given $N \in \mathbb{N}$ define the system of points by their spherical coordinates (cf.* (9.1)*)*

$$(\vartheta_n, \varphi_n) = \left(\arccos(2\Phi_2(n) - 1), 2\pi \frac{n-1}{N} \right), \quad n = 1, \ldots, N, \tag{14.42}$$

The sequence of these point systems is called Hammersley sequence. *See Figure 14.2 (left column) for an illustration.*

It is known for this Hammersley sequence (cf. H. Niederreiter [1992]) that, for all $F \in \mathrm{C}^{(0)}(\mathbb{S}^2)$,

$$\lim_{N\to\infty} \frac{1}{N} \sum_{n=1}^{N} F^*(\varphi_k, t_k) = \frac{1}{4\pi} \int_0^{2\pi} \int_{-1}^{1} F^*(\varphi, t)\, dt\, d\varphi = \frac{1}{4\pi} \int_{\mathbb{S}^2} F(\eta)\, dS(\eta), \tag{14.43}$$

where $t_k = \cos(\vartheta_k)$ and $F^*(\varphi, t) = F(t\varepsilon^3 + \sqrt{1-t^2}(\cos\varphi\varepsilon^1 + \sin\varphi\varepsilon^2))$. Hence, according to Theorem 14.10, we have the limit relations

$$\lim_{N\to\infty} D(X_N) = 0. \tag{14.44}$$

It is easy to see that we do not get a hierarchical sequence of systems by increasing N. In order to get a hierarchical sequence of ensembles we change the last definition slightly.

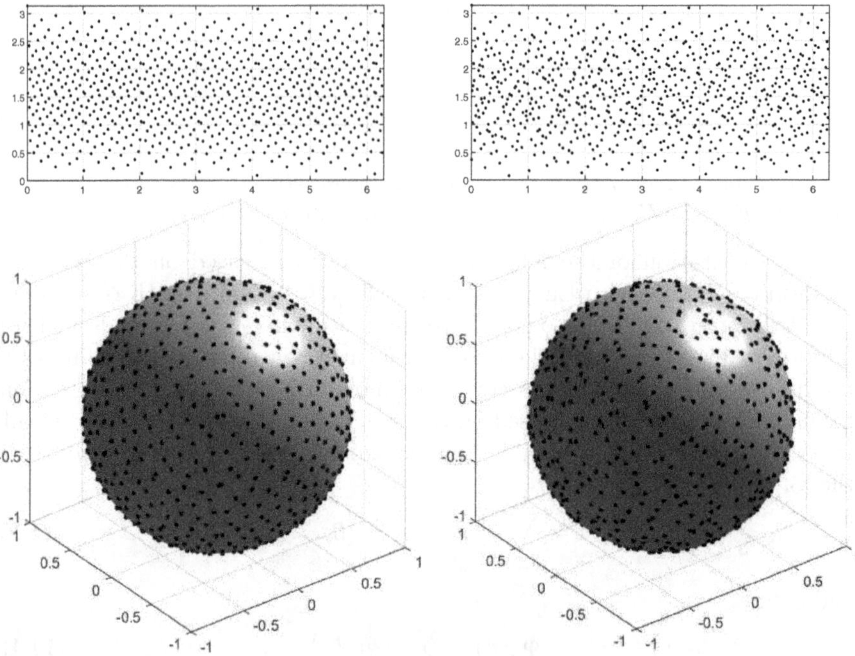

FIGURE 14.2: Coordinates and points of the Hammersley sequence of Example 14.16 (left column) and of the Corput–Halton sequence of Example 14.17 (right column). The number of points is $N = 786$ in both cases.

Example 14.17. *Let for* $k = 1, \ldots, N$,

$$(\varphi_k, \vartheta_k) = (2\pi\Phi_2(k), \arccos(2\Phi_3(k) - 1)) \,. \tag{14.45}$$

This Corput–Halton sequence *also satisfies (14.43), (14.44). It is obvious that a hierarchical sequence of systems is generated for increasing values of* N. *An illustration is given in Figure 14.2 (right column).*

Example 14.18. *Given* $N \in \mathbb{N}$, $N > 1$, *E. Rakhmanov et al. [1994], E.B. Saff, A.B.J. Kuijlaars [1997] define the* spiral grid X_N *using the spherical coordinates*

$$(\vartheta_1, \varphi_1) = (\pi, 0) \qquad \textit{(South Pole)}, \tag{14.46}$$

$$(\vartheta_j, \varphi_j) = \left(\arccos(h_j), \left(\varphi_{j-1} + \frac{3.6}{\sqrt{N}}(1 - h_j^2)^{-1/2}\right) (\mathrm{mod}\ (2\pi))\right), \tag{14.47}$$

$$(\vartheta_N, \varphi_N) = (0, 0) \qquad \textit{(North Pole)}, \tag{14.48}$$

where $j = 2, \ldots, N-1$ *and* $h_j = -1 + \frac{2j-2}{N-1}$.

K. Thomson [2007] uses this construction and introduces the improved spiral grid X_N *using the spherical coordinates*

$$(\vartheta_1, \varphi_1) = (\pi, 0) \qquad \textit{(South Pole)}, \tag{14.49}$$

$$(\vartheta_j, \varphi_j) = \left(\arccos(h_j), \left(\varphi_{j-1} + \frac{3.6}{\sqrt{N}} \frac{2}{r_j + r_{j-1}}\right) (\text{mod } (2\pi))\right), \tag{14.50}$$

$$(\vartheta_N, \varphi_N) = (0, 0) \qquad \textit{(North Pole)}, \tag{14.51}$$

where $j = 2, \ldots, N-1$ *and*

$$h_j = -1 + \frac{2k(j) - 2}{N - 1} \quad \textit{with } k(j) = (1 - \frac{1}{N-3})(j+1) + \frac{1}{2}\frac{N+1}{N-3}, \tag{14.52}$$

$$r_j = \sqrt{1 - h_j^2}, \quad r_1 = 0. \tag{14.53}$$

See Figure 14.3 (left column) for an illustration.

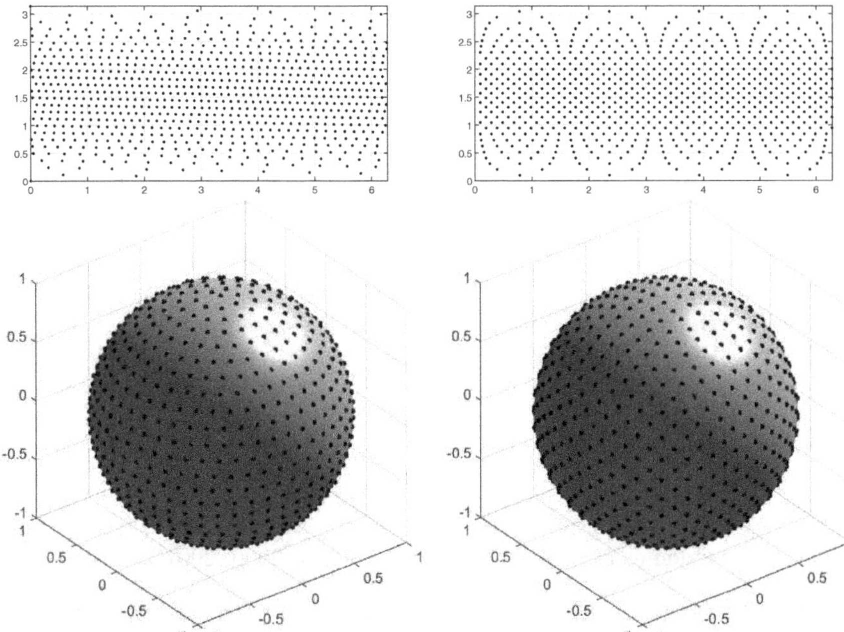

FIGURE 14.3: Coordinates and points of the improved spiral grid from Example 14.18 (left column) and the HEALPix grid of Example 14.19 for $N_s = 8$ (right column). The number of points is $N = 786$ for both the improved spiral and the HEALPix case.

Example 14.19. *Depending on the parameter* $N_s \in \mathbb{N}$ *we construct the so-called* HEALPix grid *consisting of* $N = 12N_s^2$ *points on the sphere. HEALPix* *is an acronym for Hierarchical Equal Area isoLatitude Pixelization of a sphere. As suggested in the name, this pixelization produces a subdivision of a spherical surface in which each pixel covers the same surface area as every other pixel. For further details see K.M. Górski et al. [2005]. The construction consists of three steps:*

At first the northern polar cap is considered:

$$(\vartheta_j, \varphi_j) = \left(\arccos\left(1 - \frac{k^2}{3N_s^2}\right), \frac{\pi}{2k}\left(j - 2k(k-1) - \frac{1}{2}\right)\right), \tag{14.54}$$

where $j = 1, \ldots, N_{cap} = 2N_s(N_s - 1)$ *and*

$$k = \left\lfloor \left(\frac{j}{2} - \left(\left\lfloor \frac{j}{2} \right\rfloor \right)^{1/2} \right)^{1/2} \right\rfloor + 1. \tag{14.55}$$

For the second step the remaining northern hemisphere and the equator are considered:

$$\begin{aligned} &(\vartheta_{N_{cap}+j}, \varphi_{N_{cap}+j}) \qquad (14.56) \\ &= \left(\arccos\left(\frac{4}{3} - \frac{2k}{3N_s}\right), \frac{\pi}{2N_s}\left((j-1)(\mathrm{mod}\,(4N_s)) + 1 - \frac{m}{2}\right)\right), \end{aligned}$$

where $j = 1, \ldots, N_{belt} = 4N_s(N_s + 1)$, $m = (k - N_s + 1)(\mathrm{mod}\, 2)$, *and*

$$k = \left\lfloor \frac{j-1}{4N_s} \right\rfloor + N_s. \tag{14.57}$$

Finally, the points of the southern hemisphere are constructed by

$$(\vartheta_{N-j+1}, \varphi_{N-j+1}) = (\pi - \vartheta_j, \varphi_j), \tag{14.58}$$

where $j = 1, \ldots, N_{south} = N_{cap} + N_{belt} - 4N_s$. *See Figure 14.3 (right column) for an illustration.*

Besides these point systems there are many other examples (see, for example, W. Freeden et al. [1998, 2010], P. Leopardi [2006, 2007]).

Remark 14.20. *Remembering the grid ELLG* (11.60) *we deal with points in terms of the spherical coordinates* $(\vartheta_l, \varphi_k) \in [0, \pi) \times [0, 2\pi)$ *given by*

$$\begin{aligned} \vartheta_l &= l\frac{\pi}{m+1}, \quad l = 0, \ldots, m, \qquad (14.59) \\ \varphi_k &= k\frac{2\pi}{m+1}, \quad k = 0, \ldots, m, \end{aligned}$$

for odd $m \in \mathbb{N}$. *Obviously, this point system* does not generate *an equidistribution on the sphere* $\mathbb{S}^2$. *Nevertheless, the trapezoidal rule (see Chapter 5) plays a particular role in numerical integration. Together with such a point system it enables us to compute "discrete convolutions" on the sphere, for example, discretized wavelet transforms by use of fast Fourier transform techniques (see for example, W. Freeden, U. Windheuser [1996, 1997]), as far as the data set is available in the required points. Another field of interest is partial differential equations on the sphere, e.g., the Navier-Stokes equations (cf. M.J. Fengler, W. Freeden [2005], M. Ganesh et al. [2011]).*

Lattice Point Generated Equidistribution. Next we come to the problem of generating an equidistributed point system, that is obtained by projection of lattice points of $\mathbb{Z}^3 \setminus \{0\}$ to $\mathbb{S}^2$ (cf. E. Hlawka [1981], A.V. Malysev [1957] for this construction principle).

Example 14.21. *For $n \in \mathbb{N}$, let us consider the number $r(n)$ of integer solutions $g = (n_1, n_2, n_3)^T \in \mathbb{Z}^3$ with $n_i = g \cdot \varepsilon^i$, $i = 1, 2, 3$, of the equation $|g|^2 = n_1^2 + n_2^2 + n_3^2 = n$, i.e., the number $r(n)$ of representations of $n \in \mathbb{N}$ as* sum of three squares.

There exists a sequence $\{m_j\}_{j=1,2,\ldots} \subset \mathbb{N}$ (see, e.g., W. Freeden [2011]) such that $N_j = r(m_j) > 0$ and $\lim_{j\to\infty} m_j = \infty$.

Keeping this sequence $\{m_j\}_{j=1,2,\ldots} \subset \mathbb{N}$ in mind we let $\widetilde{X}_{N_j} = \left\{x_1^{N_j}, \ldots, x_{N_j}^{N_j}\right\}$ be the set of points of $\mathbb{Z}^3$ on the sphere $\mathbb{S}^2_{\sqrt{m_j}}$ around the origin 0 with radius $\sqrt{m_j}$, i.e.,

$$\widetilde{X}_{N_j} = \mathbb{Z}^3 \cap \mathbb{S}^2_{\sqrt{m_j}}. \tag{14.60}$$

Furthermore, we understand by $X_{N_j} \subset \mathbb{S}^{q-1}$ the set

$$X_{N_j} = \left\{\xi_1^{N_j}, \ldots, \xi_{N_j}^{N_j}\right\}, \tag{14.61}$$

where

$$\xi_i^{N_j} = \frac{1}{\sqrt{m_j}} x_i^{N_j}, \quad j = 1, \ldots, N_j. \tag{14.62}$$

In other words, X_{N_j} is the projection of the point set $\widetilde{X}_{N_j}$ to the unit sphere $\mathbb{S}^2 \subset \mathbb{R}^3$. For more theoretical background we refer to R. Arenstorf, D. Johnson [1979].

From the Green's surface theorem (Theorem 9.16) on $\mathbb{S}^2$, in particular (9.306) for the operator $\partial_0^m = (\Delta^*)^m$, as well as the property $1 = (N_j)^{-1} \sum_{\xi \in X_{N_j}} 1$, it follows that

$$\begin{aligned} \frac{1}{4\pi} \int_{\mathbb{S}^2} F(\eta)\, dS(\eta) = {} & \frac{1}{N_j} \sum_{\xi \in X_{N_j}} F(\xi) \\ & - \frac{1}{N_j} \sum_{\xi \in X_{N_j}} \int_{\mathbb{S}^2} G((\Delta^*)^m; \xi \cdot \zeta)(\Delta^*_\zeta)^m F(\zeta)\, dS(\zeta) \end{aligned} \tag{14.63}$$

holds true for $j = 1, 2, \ldots$, provided that F is a function of class $C^{(2m)}(\mathbb{S}^2)$, $m \in \mathbb{N}$. The identity (14.63) can be understood as a *spherical counterpart of the Hlawka-Koksma formula* (for the classical approach in Euclidean spaces $\mathbb{R}^q$ see, e.g., L. Kuipers, H. Niederreiter [1974], and the references therein) formulated for the "projected" lattice point set (14.61) to $\mathbb{S}^2$.

We are interested in the difference $\mathcal{E}_{N_j} F$ of the integral mean

$$\mathcal{I}F = \frac{1}{4\pi} \int_{\mathbb{S}^2} F(\eta)\, dS(\eta) \tag{14.64}$$

and the "approximating sum"

$$\mathcal{L}_{N_j} F = \frac{1}{N_j} \sum_{\xi \in X_{N_j}} F(\xi), \tag{14.65}$$

i.e., the "*remainder term*" $\mathcal{E}_{N_j} F$ is given by

$$\mathcal{E}_{N_j}(F) = \mathcal{I}F - \mathcal{L}_{N_j} F \tag{14.66}$$

within the reference space $C^{(2m)}(\mathbb{S}^2)$. Note that $\mathcal{E}_{N_j} Y_{0,1} = 0$ for $Y_{0,1} \in \mathrm{Harm}_0$. In this notational framework, the *Hlawka-Koksma formula* (14.63) can be rewritten as follows

Lemma 14.22. *For $F \in C^{(2m)}(\mathbb{S}^2)$, $m \in \mathbb{N}$,*

$$\mathcal{E}_{N_j} F = \int_{\mathbb{S}^2} \left(\mathcal{E}_{N_j} G((\Delta^*)^m;\ \cdot\eta)\right) (\Delta^*_\eta)^m F(\eta)\, dS(\eta). \tag{14.67}$$

The error term (14.66) can be estimated from above in different norms: For $m \geq 1$ we obtain from the Cauchy–Schwarz inequality

$$|\mathcal{E}_{N_j} F| \leq D(X_{N_j}; (\Delta^*)^m) \left(\int_{\mathbb{S}^2} |(\Delta^*_\eta)^m F(\eta)|^2\, dS(\eta) \right)^{1/2}, \tag{14.68}$$

where the expression $D(X_{N_j}; (\Delta^*)^m)$ given by

$$\begin{aligned}
\left(D(X_{N_j}; (\Delta^*)^m)\right)^2 &= \mathcal{L}_{N_j} \mathcal{L}_{N_j} G((\Delta^*)^{2m};\ \cdot\ ) \qquad (14.69)\\
&= \frac{1}{N_j^2} \sum_{\xi \in X_{N_j}} \sum_{\eta \in X_{N_j}} G((\Delta^*)^{2m}; \xi \cdot \eta)\\
&= \frac{1}{N_j^2} \sum_{\xi \in X_{N_j}} \sum_{\eta \in X_{N_j}} \int_{\mathbb{S}^2} G((\Delta^*)^m; \eta \cdot \zeta) G((\Delta^*)^m; \xi \cdot \zeta)\, dS(\zeta)
\end{aligned}$$

is called the *discrepancy of X_{N_j} of order m*, $m \geq 1$ (more accurately the *discrepancy of X_{N_j} to the operator $(\Delta^*)^m$*).

As already known, for $m > 1$, the function $\eta \mapsto G\big((\Delta^*)^m; \xi \cdot \eta\big)$, $\xi \in \mathbb{S}^2$, is continuous and we are able to formulate an estimate of the error term (14.66) in the $C^{(0)}$-norm (see, e.g., E. Hlawka [1981]) as follows.

Lemma 14.23. *Assume that F is of class $\mathrm{C}^{(2m)}(\mathbb{S}^2)$, $m > 1$. Then*

$$\left|\mathcal{E}_{N_j} F\right| \leq D_{\mathrm{C}^{(0)}(\mathbb{S}^2)}(X_{N_j}) \int_{\mathbb{S}^2} \left|(\Delta^*_\eta)^m F(\eta)\right| \, dS(\eta). \tag{14.70}$$

The expression

$$D_{\mathrm{C}^{(0)}(\mathbb{S}^2)}(X_{N_j}) = \sup_{\eta \in \mathbb{S}^2} \left| \frac{1}{N_j} \sum_{\xi \in X_{N_j}} G((\Delta^*)^m; \xi \cdot \eta) \right| \tag{14.71}$$

is called the $\mathrm{C}^{(0)}\left(\mathbb{S}^2\right)$-*discrepancy* of X_{N_j} of order m, $m > 1$.

Obviously, the sequence $\{X_{N_j}\}_{j=1,2,\ldots}$ is equidistributed in the sense that $\lim_{j\to\infty} \mathcal{E}_{N_j} F = 0$ if the discrepancy of X_{N_j} tends to 0. R. Arenstorf, D. Johnson [1979] verified (under certain circumstances not specified here) that

$$D_{\mathrm{C}^{(0)}(\mathbb{S}^2)}(X_{N_j}) \leq \frac{C_m}{\ln\left(\ln(m_j)\right)}, \quad m > 1, \tag{14.72}$$

where $\{m_j\}$ is the sequence of Example 14.21. The proof is are rather technical, it will be omitted here.

Remark 14.24. *A more detailed overview on the activities in the theory of equidistribution by projection of lattice points of $\mathbb{Z}^q \setminus \{0\}$ from certain spheres $\mathbb{S}^{q-1}_{\sqrt{m_j}}$ to the (unit) sphere $\mathbb{S}^{q-1} \subset \mathbb{R}^q$ is due to E. Hlawka [1981].*

14.4 Sobolev Space Based Generalized Discrepancy

A concept of generalized discrepancy, which involves pseudodifferential operators to give a quantifying criterion of equidistributed point sets, is of great interest. We develop an explicit formula in terms of elementary functions for the generalized discrepancy corresponding to a specific operator following J. Cui, W. Freeden [1997]. Essential tools are Sobolev space structures introduced in Section 12.1 and pseudodifferential operator techniques of Section 12.2. It should be mentioned that an optimal point set may be obtained by minimizing the generalized discrepancy.

Our investigations show that there are many promising ways to generate point systems on the sphere such that the generalized discrepancy becomes small. Furthermore, a quantitative comparison between the different point systems is given. Nevertheless, for certain numbers of points there are special other point sets such as spherical designs (cf. Section 10.4) which provide better results.

Next we discuss the problem of equidistribution in a Sobolev context. This problem can be stated as follows: find a point set $X_N = \{\xi_1, \ldots, \xi_N\}$ such that the remainder term

$$\left| \frac{1}{4\pi} \int_{\mathbb{S}^2} F(\eta)\, dS(\eta) - \frac{1}{N} \sum_{i=1}^{N} F(\xi_i) \right|, \tag{14.73}$$

for any function $F \in \mathrm{H}^s(\mathbb{S}^2)$ converges to 0 as $N \to \infty$.

We begin with an *a priori* estimate of the remainder which is an analog to the Koksma-Hlawka inequality in Euclidean spaces (see also Theorem 14.5).

Theorem 14.25. *Let A be a pseudodifferential operator of order s, $s > 1$, with symbol $\{A^\wedge(n)\}$ satisfying $A^\wedge(n) \neq 0$, $n \geq 1$. Then, for any function $AF \in \mathrm{L}^2(\mathbb{S}^2)$, we have the estimate*

$$\begin{aligned} &\left| \frac{1}{4\pi} \int_{\mathbb{S}^2} F(\eta)\, dS(\eta) - \frac{1}{N} \sum_{i=1}^{N} F(\xi_i) \right| \\ &\leq \frac{1}{N} \left(\sum_{k=1}^{N} \sum_{i=1}^{N} \sum_{n=1}^{\infty} \frac{2n+1}{4\pi (A^\wedge(n))^2} P_n(\xi_i \cdot \xi_k) \right)^{1/2} \|AF\|_{\mathrm{L}^2(\mathbb{S}^2)}. \end{aligned} \tag{14.74}$$

Proof. From the definition of the Sobolev space $\mathrm{H}^s(\mathbb{S}^2)$ (cf. (12.9)) we know that $AF \in \mathrm{L}^2(\mathbb{S}^2)$ is equivalent to $F \in \mathrm{H}^s(\mathbb{S}^2)$. Because of the assumption $s > 1$ it follows from Lemma 12.6 that the spherical harmonic expansion

$$F(\xi) = \sum_{n=0}^{\infty} \sum_{j=1}^{2n+1} F^\wedge(n,j) Y_{n,j}(\xi), \quad \xi \in \mathbb{S}^2, \tag{14.75}$$

converges absolutely and uniformly. We rewrite the spherical harmonic expansion of F as follows:

$$F(\xi) = \frac{1}{4\pi} \int_{\mathbb{S}^2} F(\eta)\, dS(\eta) + \sum_{n=1}^{\infty} \sum_{j=1}^{2n+1} \int_{\mathbb{S}^2} \frac{(AF)(\eta) Y_{n,j}(\eta)}{A^\wedge(n)}\, dS(\eta) Y_{n,j}(\xi). \tag{14.76}$$

By setting $\xi = \xi_i$ and taking the sum over all indices i, $1 \leq i \leq N$, we obtain

$$\begin{aligned} \frac{1}{N} \sum_{i=1}^{N} F(\xi_i) &= \frac{1}{4\pi} \int_{\mathbb{S}^2} F(\eta)\, dS(\eta) \\ &\quad + \sum_{n=1}^{\infty} \sum_{j=1}^{2n+1} \int_{\mathbb{S}^2} \frac{(AF)(\eta) Y_{n,j}(\eta)}{A^\wedge(n)}\, dS(\eta) \left(\frac{1}{N} \sum_{i=1}^{N} Y_{n,j}(\xi_i) \right). \end{aligned} \tag{14.77}$$

Hence, the integral error can then be estimated as follows

$$\begin{aligned}
&\left| \frac{1}{4\pi} \int_{\mathbb{S}^2} F(\eta)\, dS(\eta) - \frac{1}{N} \sum_{i=1}^{N} F(\eta_i) \right| \qquad (14.78)\\
&\leq \left| \frac{1}{N} \sum_{n=1}^{\infty} \sum_{j=1}^{2n+1} \sum_{i=1}^{N} \frac{Y_{n,j}(\xi_i)}{A^\wedge(n)} \int_{\mathbb{S}^2} (AF)(\eta) Y_{n,j}(\eta)\, dS(\eta) \right| \\
&= \left| \frac{1}{N} \int_{\mathbb{S}^2} (AF)(\eta) \sum_{n=1}^{\infty} \sum_{j=1}^{2n+1} \sum_{i=1}^{N} \frac{Y_{n,j}(\xi_i) Y_{n,j}(\eta)}{A^\wedge(n)}\, dS(\eta) \right|.
\end{aligned}$$

By using the Cauchy–Schwarz inequality, Parseval identity and the addition theorem of the spherical harmonics (9.42) we obtain

$$\begin{aligned}
&\left| \frac{1}{4\pi} \int_{\mathbb{S}^2} F(\eta)\, dS(\eta) - \frac{1}{N} \sum_{i=1}^{N} F(\xi_i) \right| \qquad (14.79)\\
&\leq \frac{1}{N} \|AF\|_{\mathrm{L}^2(\mathbb{S}^2)} \left\| \sum_{n=1}^{\infty} \sum_{j=1}^{2n+1} \sum_{i=1}^{N} \frac{Y_{n,j}(\xi_i)}{A^\wedge(n)} Y_{n,j} \right\|_{\mathrm{L}^2(\mathbb{S}^2)} \\
&= \frac{1}{N} \|AF\|_{\mathrm{L}^2(\mathbb{S}^2)} \left(\sum_{n=1}^{\infty} \sum_{j=1}^{2n+1} \left(\frac{1}{A^\wedge(n)} \sum_{i=1}^{N} Y_{n,j}(\xi_i) \right)^2 \right)^{1/2} \\
&= \frac{1}{N} \|AF\|_{\mathrm{L}^2(\mathbb{S}^2)} \left(\sum_{i=1}^{N} \sum_{k=1}^{N} \sum_{n=1}^{\infty} \sum_{j=1}^{2n+1} \frac{Y_{n,j}(\xi_i) Y_{n,j}(\xi_k)}{(A^\wedge(n))^2} \right)^{1/2} \\
&= \frac{1}{N} \|AF\|_{\mathrm{L}^2(\mathbb{S}^2)} \left(\sum_{i=1}^{N} \sum_{k=1}^{N} \sum_{n=1}^{\infty} \frac{2n+1}{4\pi (A^\wedge(n))^2} P_n(\xi_i \cdot \xi_k) \right)^{1/2}.
\end{aligned}$$

This completes the proof. □

Theorem 14.25 extends a result given for the Beltrami operator in W. Freeden [1978c, 1980b] to the context of pseudodifferential operators. It is obvious that the right-hand side of the estimate (14.74) consists of two parts. The first part depends only on the function F under consideration, while the second part depends merely on the point set. This gives rise to the definition of generalized discrepancy.

Definition 14.26 (Generalized Discrepancy). *Let A be a pseudodifferential operator of order s, $s > 1$, with symbol $\{A^\wedge(n)\}$, $A^\wedge(n) \neq 0$ for $n \geq 1$. Then the* generalized discrepancy *associated to a pseudodifferential operator A for a point set $X_N = \{\xi_1, \ldots, \xi_N\}$ is defined by*

$$D(X_N; A) = \frac{1}{N} \left(\sum_{i=1}^{N} \sum_{k=1}^{N} \sum_{n=1}^{\infty} \frac{2n+1}{4\pi (A^\wedge(n))^2} P_n(\xi_i \cdot \xi_k) \right)^{1/2}. \qquad (14.80)$$

The generalized discrepancy also characterizes "how well a point set is equidistributed" which can be explained by the following heuristic geometrical interpretation: for any subset $\Gamma \subset \mathbb{S}^2$ and any point set $X_N = \{\xi_1, \ldots, \xi_N\}$ on $\mathbb{S}^2$ we consider the counting function,

$$\#\Gamma = \sum_{i=1}^{N} \chi_\Gamma(\xi_i), \tag{14.81}$$

where χ_Γ is the characteristic function of Γ. Thus $\#\Gamma$ indicates the number n, $1 \leq n \leq N$, of points ξ_i contained in Γ. Let $\chi_\Gamma^\varepsilon \in \mathrm{H}^s(\mathbb{S}^2)$, $s > 1$, be a "smooth" approximation of the characteristic function χ_Γ. Then, for any non-empty Lebesgue-measurable $\Gamma \subset \mathbb{S}^2$, we have an integral inequality of the following form

$$\begin{aligned}\left|\frac{\#\Gamma}{N} - \frac{1}{4\pi}\int_{\mathbb{S}^2} \chi_\Gamma(\eta)\, dS(\eta)\right| &= \left|\frac{1}{N}\sum_{i=1}^{N} \chi_\Gamma(\xi_i) - \frac{1}{4\pi}\int_{\mathbb{S}^2} \chi_\Gamma(\eta)\, dS(\eta)\right| \qquad (14.82)\\ &\approx \left|\frac{1}{N}\sum_{i=1}^{N} \chi_\Gamma^\varepsilon(\xi_i) - \frac{1}{4\pi}\int_{\mathbb{S}^2} \chi_\Gamma^\varepsilon(\eta)\, dS(\eta)\right|\\ &\leq D(X_N; A)\|A\chi_\Gamma^\varepsilon\|_{\mathrm{L}^2(\mathbb{S}^2)}.\end{aligned}$$

Therefore, the generalized discrepancy gives a quantitative description of how well a point set is equidistributed (cf. Section 14.2). For different pseudodifferential operators we have different generalized discrepancies. It is of interest to understand the relationship between different generalized discrepancies associated to different pseudodifferential operators.

Lemma 14.27. *Let A, B be two pseudodifferential operators of order s_1, $s_2 > 1$, and with symbols $\{A^\wedge(n)\}$, $\{B^\wedge(n)\}$ satisfying $A^\wedge(n) \neq 0$, $B^\wedge(n) \neq 0$ for $n \geq 1$, respectively. Then, the following relations are valid for the generalized discrepancies associated to the pseudodifferential operators A and B:*

(i) If the two pseudodifferential operators A and B are of the same order, i.e., $s_1 = s_2$, then there exist two constants $C_1 > 0$ and $C_2 > 0$ such that

$$C_1 D(X_N; A) \leq D(X_N; B) \leq C_2 D(X_N; A) \tag{14.83}$$

for all point sets $X_N = \{\xi_1, \ldots, \xi_N\} \subset \mathbb{S}^2$ and all $N \in \mathbb{N}$.

(ii) If $s_1 > s_2$, then there exists a constant $C_3 > 0$ such that

$$D(X_N; A) < C_3 D(X_N; B). \tag{14.84}$$

for all point sets $X_N = \{\xi_1, \ldots, \xi_N\} \subset \mathbb{S}^2$ and all $N \in \mathbb{N}$.

Proof. From (14.79) it follows that the generalized discrepancy can be rewritten as follows:

$$D(X_N; A) = \left(\sum_{n=1}^{\infty}\sum_{j=1}^{2n+1}\left(\frac{1}{A^\wedge(n)}\sum_{i=1}^{N} Y_{n,j}(\xi_i)\right)^2\right)^{1/2}. \tag{14.85}$$

If $s_1 = s_2$, then it follows from our definition of pseudodifferential operators (cf. Definition 12.3) that there exist two constants C_1 and C_2 such that

$$C_1 \left(\frac{1}{A^\wedge(n)}\right)^2 \leq \left(\frac{1}{B^\wedge(n)}\right)^2 \leq C_2 \left(\frac{1}{A^\wedge(n)}\right)^2, \quad n \geq 1. \tag{14.86}$$

Therefore, it is clear that

$$C_1 D(X_N; A) \leq D(X_N; B) \leq C_2 D(X_N; A). \tag{14.87}$$

The second part can be proved in the same manner. □

From Lemma 14.27 we deduce that the generalized discrepancy is dependent only on the order of the pseudodifferential operators. Using the generalized discrepancy we can study the property of point systems in quantitative way.

Definition 14.28 (Equidistribution in $\mathrm{H}^s(\mathbb{S}^2)$). *A point system* $X_N = \{\xi_1^N, \ldots, \xi_N^N\}$, $N = 1, 2, \ldots$, *is called* A-equidistributed *in* $\mathrm{H}^s(\mathbb{S}^2)$, $s > 1$, *if the generalized discrepancy associated to a pseudodifferential operator A of order s, $s > 1$, satisfies*

$$\lim_{N \to \infty} D(X_N; A) = 0. \tag{14.88}$$

It should be noted that if a point system is equidistributed in $\mathrm{H}^{s_0}(\mathbb{S}^2)$, $s_0 > 1$, then the point system is equidistributed in $\mathrm{H}^s(\mathbb{S}^2)$, $s > s_0$. This shows us that we should use s_0, $s_0 > 1$, as small as possible for our quantitative investigations. However, in order to compute the generalized discrepancy as illustrated above, one has to evaluate a series expansion in terms of Legendre polynomials. This summation is very time consuming, in particular, for relatively small s. Fortunately, for certain pseudodifferential operators, the series is representable in terms of elementary functions (see also Section 12.2). One example should be discussed in more detail.

Consider the pseudodifferential operator $A = (-2\Delta^*)^{\frac{1}{2}}(-\Delta^* + \frac{1}{4})^{\frac{1}{4}}$ (cf. W. Freeden [1990]). Its symbol is $\{A^\wedge(n)\}$ with $A^\wedge(n) = \sqrt{(2n+1)n(n+1)}$, its order is $\frac{3}{2}$. Thus, the condition $AF \in \mathrm{L}^2(\mathbb{S}^2)$ is equivalent to $F \in \mathrm{H}^{\frac{3}{2}}(\mathbb{S}^2)$, more precisely,

$$\|AF\|_{\mathrm{L}^2(\mathbb{S}^2)} \leq \sqrt{6}\|F\|_{\mathrm{H}^{\frac{3}{2}}(\mathbb{S}^2)}. \tag{14.89}$$

It is worth noting that the functions in the Sobolev space $\mathrm{H}^{\frac{3}{2}}(\mathbb{S}^2)$ are "slightly smoother" than the continuous functions, and its elements generally are not Lipschitz-continuous. Observing our particular choice and using the addition theorem of spherical harmonics (9.42) the generalized discrepancy is then given by

$$D(X_N; A) = \frac{1}{N}\left(\sum_{i=1}^{N}\sum_{k=1}^{N}\sum_{n=1}^{\infty} \frac{1}{4\pi n(n+1)} P_n(\xi_i \cdot \xi_k)\right)^{1/2}. \tag{14.90}$$

As mentioned above, the series on the right-hand side is expressible in terms of elementary functions by (12.22) (cf., e.g., W. Magnus et al. [1966])

$$D(X_N; A) = \frac{1}{2\sqrt{\pi}N} \left(\sum_{i=1}^{N} \sum_{k=1}^{N} \left(1 - 2\ln\left(1 + \sqrt{\frac{1-\xi_i\cdot\xi_k}{2}}\right)\right)\right)^{1/2}. \quad (14.91)$$

P.J. Grabner et al. [1997] guarantee the existence of a constant C so that for the discrepancy $D_{\mathrm{Leg}}(X_N)$ of (14.15) holds that

$$D_{\mathrm{Leg}}(X_N) \le D(X_N; A) \le C D_{\mathrm{Leg}}(X_N). \quad (14.92)$$

To keep an appropriate balance between computation and the requirement that the order should be as small as possible, we simply speak in the following of equidistribution on the sphere, when $X_N = \{\xi_1^N, \ldots, \xi_N^N\}$, $N = 1, 2, \ldots$, is A-equidistributed, i.e.,

$$\lim_{N\to\infty} D(X_N; A) = 0, \quad (14.93)$$

where $D(X_N; A)$ is the discrepancy of (14.91). In connection with Theorem 14.25 and (14.89) we then obtain

$$\left| \frac{1}{4\pi} \int_{\mathbb{S}^2} F(\eta)\, dS(\eta) - \frac{1}{N}\sum_{i=1}^{N} F(\xi_i) \right| \le \sqrt{6} D(X_N; A) \|F\|_{\mathrm{H}^{\frac{3}{2}}(\mathbb{S}^2)}, \quad (14.94)$$

where $X_N = \{\xi_1, \ldots, \xi_N\}$, $D(X_N; A)$ is given by (14.91) and $F \in \mathrm{H}^{\frac{3}{2}}(\mathbb{S}^2)$.

To get an optimal point system on the sphere we have to minimize the expression $D(X_N; A)$ within a set of N points $\xi_1, \ldots, \xi_N \in \mathbb{S}^2$. In spite of the elementary representation of the generalized discrepancy, this is a non-linear optimization problem similar to the search for extremal points (cf. Section 10.4).

Remark 14.29. *Note that, recently, A. Ishtiaq, V. Michel [2017] have generalized this idea to the three-dimensional ball.*

Another possibility of construction for point sets is to require that sum of some low degree spherical harmonics vanish, that is to say, moments up to a particular degree should vanish. This leads back to the spherical designs of Section 10.4. We call a point set $X_N = \{\xi_1, \ldots, \xi_N\}$ an m-design if

$$\frac{1}{4\pi} \int_{\mathbb{S}^2} Y_n(\xi)\, dS(\xi) = \frac{1}{N}\sum_{i=1}^{N} Y_n(\eta_i), \quad (14.95)$$

holds for all spherical harmonics Y_n of degree $n \le m$. Equivalently, the following definition can be used.

Definition 14.30 (m-Design). *A point set $\{\eta_1, \ldots, \eta_N\}$ is called an m-design if the Weyl sums (cf. Theorem 14.10) vanish for $n = 1, \ldots, m$. To be specific,*

$$\sum_{i=1}^{N} Y_{n,j}(\xi_i) = 0 \quad \textit{for } n = 1, 2, \ldots m;\ j = 1, \ldots 2n+1. \quad (14.96)$$

In this book we only mention the possibility of an estimate of the generalized discrepancy in case of an m-design X_N.

Lemma 14.31. *Let A be a pseudodifferential operator of order s, $s > 1$, with symbol $\{A^\wedge(n)\}$, $A^\wedge(n) \neq 0$ for $n \geq 1$. If a point set $X_N = \{\xi_1, \ldots, \xi_N\}$ is an m-design, then the generalized discrepancy satisfies*

$$D(X_N; A) \leq \frac{C}{m^{s-1}}, \tag{14.97}$$

where the constant C depends only on s.

Proof. From the definition of m-design and generalized discrepancy it follows that

$$\begin{aligned} D(\{\eta_1, \ldots, \eta_N\}; A) &= \frac{1}{N}\left(\sum_{i=1}^{N}\sum_{k=1}^{N}\sum_{n=1}^{\infty} \frac{2n+1}{4\pi(A^\wedge(n))^2} P_n(\eta_i \cdot \eta_k)\right)^{1/2} \qquad (14.98) \\ &= \frac{1}{N}\left(\sum_{n>m}^{\infty}\sum_{i=1}^{N}\sum_{k=1}^{N} \frac{2n+1}{4\pi(A^\wedge(n))^2} P_n(\eta_i \cdot \eta_k)\right)^{1/2} \\ &\leq \frac{1}{N}\left(\sum_{n>m}^{\infty}\sum_{i=1}^{N}\sum_{k=1}^{N} \frac{2n+1}{4\pi(A^\wedge(n))^2}\right)^{1/2}. \end{aligned}$$

By using the well-known inequality

$$\sum_{n=m+1}^{\infty} \frac{1}{n^s} \leq \frac{\zeta(s)}{m^{s-1}}, \tag{14.99}$$

where the zeta function $\zeta(s)$ is given by (5.12), we have the desired estimate. □

For large s the m-design gives a good rate of convergence. Suppose that the number of points is $N = m^t$. For $t \geq 4.5$ the existence of an m-design is known (cf. P.D. Seymour, T. Zaslavsky [1984]), but the minimal value of t is unknown. Then the rate of convergence amounts to $N^{(s-1)/t}$. However, for $s = \frac{3}{2}$ and $t = 1$, we obtain a rate of convergence of order $N^{\frac{1}{2}}$. Therefore this type of an m-design is not well-suited.

We numerically compute the generalized discrepancies of several point systems of Section 14.3. More precisely, we compare using log-log-scale plots:

- The Reuter grid of Example 14.12 (see Figure 14.4 (left side)),
- The ELLG of Remark 14.20 (see Figure 14.4 (right side)),
- The HEALPix grid of Example 14.19 (see Figure 14.5 (left side)),
- The Halton points of Example 14.17 (see Figure 14.5 (right side)),

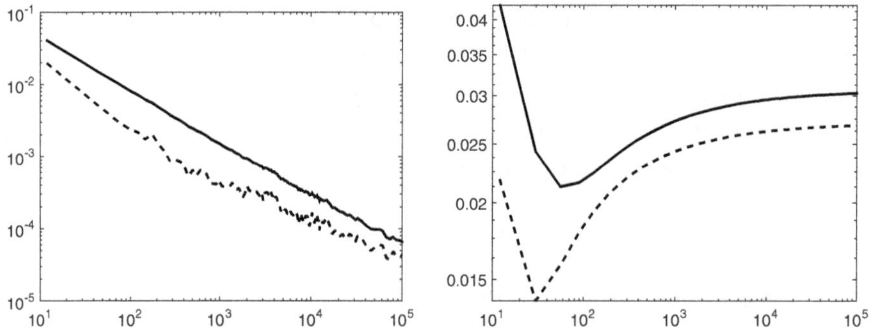

FIGURE 14.4: Discrepancy of the Reuter grid of Example 14.12 (left) and the ELLG of Remark 14.20 (right). The solid line is $D(X_N; A)$ with $A^\wedge(n) = \sqrt{(2n+1)n(n+1)}$ (14.91), the dashed line is $D(X_N; \Delta^*)$ (14.21), i.e. $A = \Delta^*$ and $A^\wedge(n) = -n(n+1)$.

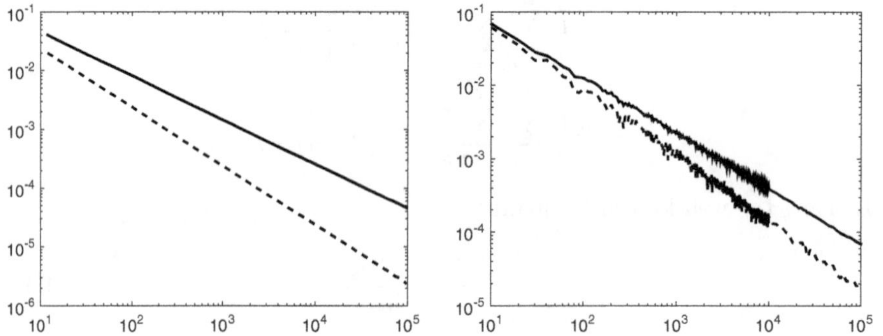

FIGURE 14.5: Discrepancy of the HEALPix grid of Example 14.19 (left) and the Halton points of Example 14.17 (right). The solid line is $D(X_N; A)$ with $A^\wedge(n) = \sqrt{(2n+1)n(n+1)}$ (14.91), the dashed line is $D(X_N; \Delta^*)$ (14.21), i.e. $A = \Delta^*$ and $A^\wedge(n) = -n(n+1)$.

- The Hammersley points of Example 14.16 (see Figure 14.6 (left side)),
- The improved spiral grid of Example 14.18 (see Figure 14.6 (right side)).

Thereby, we use $D(X_N; A)$ with $A^\wedge(n) = \sqrt{(2n+1)n(n+1)}$, i.e., the discrepancy is given by (14.91), as well as $D(X_N; \Delta^*)$ of (14.21), i.e. $A = \Delta^*$ and $A^\wedge(n) = -n(n+1)$. Note that the ELLG is not equidistributed as indicated by Figure 14.4 (right side).

In particular for the point numbers $N_1 = 30,000$ and $N_2 = 100,000$, we provide the values of $D(X_N; A)$ and $D(X_N; \Delta^*)$ in Table 14.1 for the equidistributed point sets. Note that the improvement of the spiral grid (cf. Example 14.18) can be seen in the values, but plays a smaller role for larger numbers of points.

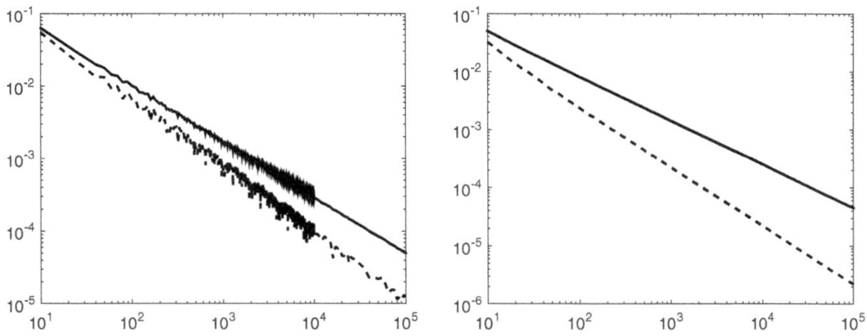

FIGURE 14.6: Discrepancy of the Hammersley points of Example 14.16 (left) and the improved spiral grid of Example 14.18 (right). The solid line is $D(X_N; A)$ with $A^\wedge(n) = \sqrt{(2n+1)n(n+1)}$ (14.91), the dashed line is $D(X_N; \Delta^*)$ (14.21), i.e. $A = \Delta^*$ and $A^\wedge(n) = -n(n+1)$.

TABLE 14.1: Discrepancies for $N_1 = 30,000$ and $N_2 = 100,000$ points (note that for the HEALPix grid we use $N_2 = 97{,}200$ and for the Reuter grid we use $N_1 = 29{,}333$ and $N_2 = 99{,}677$).

Point system	$D(X_{N_1}; \Delta^*)$	$D(X_{N_2}; \Delta^*)$	$D(X_{N_1}; A)$	$D(X_{N_2}; A)$
Spiral grid	8.2984e-06	2.4857e-06	1.1124e-04	4.5068e-05
Imp. spiral grid	7.2937e-06	2.1867e-06	1.1112e-04	4.5041e-05
HEALPix grid	7.8274e-06	2.3495e-06	1.1194e-04	4.6347e-05
Reuter grid	7.2157e-05	4.1556e-05	1.3903e-04	6.5137e-05
Hammersley points	3.6891e-05	1.1081e-05	1.2482e-04	4.9802e-05
Halton points	5.4757e-05	1.8146e-05	1.6755e-04	6.9253e-05

In what follows we show some examples for spline exact integration as well as numerical integration using equal weights for the following functions (see also the examples in J. Fliege, U. Maier [1999], I.H. Sloan, R.S. Womersly [2001]):

$$F_0(x) = |x_1| + |x_2| + |x_3|, \tag{14.100}$$

$$F_1(x) = (x_1 - 0.9)_+^{\frac{3}{4}} + (x_3 - 0.9)_+^{\frac{3}{4}}, \tag{14.101}$$

$$F_2(x) = \exp(x_1 + x_2 + x_3) + (0.01 - x_1^2 - x_2^2 - (x_3 - 1)^2)_+, \tag{14.102}$$

$$F_3(x) = (101 - 100x_3)^{-1}, \tag{14.103}$$

$$F_4(x) = (|x_1| + |x_2| + |x_3|)^{-1}, \tag{14.104}$$

$$F_5(x) = \begin{cases} \cos^2\left(\frac{3}{2}\pi d(x)\right), & d(x) = \left|x - (-1/2, -1/2, 1/\sqrt{2})^T\right| < \frac{1}{3} \\ 0, & \text{else} \end{cases} \tag{14.105}$$

Example 14.32. *We use spline exact integration to approximate the integrals $\mathcal{I}F_k$ and compare the absolute errors $\mathcal{E}F_k$, $k = 0,\ldots,5$, of the results for the improved spiral grid of Example 14.18 in Table 14.2 and for the HEALPix grid of Example 14.19 in Table 14.3.*

TABLE 14.2: Error for different grid sizes of the improved spiral grid with N points using spline exact integration (of order 0).

N	$\mathcal{E}F_0$	$\mathcal{E}F_1$	$\mathcal{E}F_2$	$\mathcal{E}F_3$	$\mathcal{E}F_4$	$\mathcal{E}F_5$
1,200	3.6120e-04	3.9776e-05	6.7977e-06	1.0006e-04	5.3088e-04	1.5244e-04
2,700	4.3322e-05	5.2317e-05	4.9850e-06	1.7534e-05	8.6417e-05	4.1182e-05
4,800	3.2044e-05	1.8310e-05	7.2686e-07	8.0616e-06	7.3765e-06	1.1382e-05
10,800	2.0218e-05	4.5929e-05	8.0715e-08	1.2247e-06	1.6298e-05	4.4924e-07
19,200	3.7675e-05	6.6810e-06	7.2129e-09	2.5913e-07	2.1187e-05	6.8907e-07
30,000	2.9301e-06	2.4804e-06	1.2888e-09	7.3703e-08	2.7509e-06	6.5598e-09

TABLE 14.3: Error for different grid sizes of HEALPix with N points using spline exact integration (of order 0).

N	$\mathcal{E}F_0$	$\mathcal{E}F_1$	$\mathcal{E}F_2$	$\mathcal{E}F_3$	$\mathcal{E}F_4$	$\mathcal{E}F_5$
1,200	8.3153e-04	9.1344e-04	4.3995e-07	5.3651e-04	6.9147e-04	2.2466e-04
2,700	3.8814e-04	5.5379e-04	9.9363e-06	6.9437e-05	3.4026e-04	5.5985e-06
4,800	2.2486e-04	6.3452e-04	9.1449e-06	1.4264e-05	1.8350e-04	8.3164e-06
10,800	1.0324e-04	1.9986e-04	1.9654e-06	1.3794e-06	8.3816e-05	6.4331e-06
19,200	5.9089e-05	1.5679e-04	2.1752e-06	2.5576e-07	4.7889e-05	1.0324e-07
30,000	3.8226e-05	8.2342e-05	8.6565e-07	6.9073e-08	3.0961e-05	5.9136e-07

Note that the values of the integrals are approximately $\mathcal{I}F_0 = 18.8495$, $\mathcal{I}F_1 = 0.1278$, $\mathcal{I}F_2 = 19.8623$, $\mathcal{I}F_3 = 0.3332$, $\mathcal{I}F_4 = 8.4720$, $\mathcal{I}F_5 = 0.1038$, *respectively.*

Example 14.33. *Now we use equal weights for the numerical integration of the six functions (14.100),...,(14.105), respectively. We compare the absolute errors $\mathcal{E}F_k$, $k = 0,\ldots,5$, of the results for the improved spiral grid of Example 14.18 in Table 14.4 and for the HEALPix grid of Example 14.19 in Table 14.5.*

Note again that the values of the integrals are approximately $\mathcal{I}F_0 = 18.8495$, $\mathcal{I}F_1 = 0.1278$, $\mathcal{I}F_2 = 19.8623$, $\mathcal{I}F_3 = 0.3332$, $\mathcal{I}F_4 = 8.4720$, $\mathcal{I}F_5 = 0.1038$, *respectively.*

TABLE 14.4: Error for different grid sizes of the improved spiral grid with N points using equidistribution based integration (of order 0).

N	$\mathcal{E}F_0$	$\mathcal{E}F_1$	$\mathcal{E}F_2$	$\mathcal{E}F_3$	$\mathcal{E}F_4$	$\mathcal{E}F_5$
1,200	6.5038e-04	3.2093e-05	1.0880e-03	7.2254e-04	7.8097e-04	1.5508e-04
2,700	1.1658e-04	5.4037e-05	4.8771e-04	1.4958e-04	1.5449e-04	4.1077e-05
4,800	9.5279e-06	1.7883e-05	1.2563e-04	4.8137e-05	4.4769e-05	1.1393e-05
10,800	2.8483e-05	4.5772e-05	1.4117e-05	9.6259e-06	2.4737e-05	4.0654e-07
19,200	3.2631e-05	6.6571e-06	2.4477e-05	3.0589e-06	1.6535e-05	6.7962e-07
30,000	5.5022e-06	2.5050e-06	9.6873e-06	1.2554e-06	5.3483e-06	5.0432e-09
43,200	6.4913e-06	1.6815e-06	1.3374e-06	6.0610e-07	5.4398e-06	1.8830e-07
67,500	8.2343e-06	3.6297e-06	3.5870e-06	2.4858e-07	6.5586e-06	6.5395e-08
97,200	1.8292e-06	6.5716e-07	1.2598e-06	1.1989e-07	1.0134e-06	5.6143e-08
172,800	1.3679e-06	4.7543e-07	4.4024e-07	3.7952e-08	8.7191e-07	1.5205e-09

TABLE 14.5: Error for different grid sizes of HEALPix with N points using equidistribution based integration (of order 0).

N	$\mathcal{E}F_0$	$\mathcal{E}F_1$	$\mathcal{E}F_2$	$\mathcal{E}F_3$	$\mathcal{E}F_4$	$\mathcal{E}F_5$
1,200	2.9377e-03	2.9097e-04	3.6102e-04	3.5315e-03	1.4427e-03	4.6355e-04
2,700	1.3614e-03	2.7756e-04	1.4142e-04	1.5298e-03	6.7932e-04	9.9990e-05
4,800	7.8098e-04	7.8981e-04	9.4085e-05	8.5402e-04	4.1092e-04	5.0902e-05
10,800	3.5371e-04	1.3087e-04	3.9827e-05	3.7764e-04	1.8994e-04	3.2688e-05
19,200	2.0079e-04	1.9560e-04	1.9120e-05	2.1206e-04	1.0886e-04	1.4858e-05
30,000	1.2920e-04	5.7510e-05	1.2763e-05	1.3561e-04	7.0428e-05	1.0030e-05
43,200	9.0041e-05	8.1575e-05	1.0216e-05	9.4133e-05	4.9256e-05	6.7404e-06
67,500	5.7830e-05	3.7715e-05	5.9761e-06	6.0224e-05	3.1709e-05	4.2991e-06
97,200	4.0254e-05	1.3787e-05	3.4602e-06	4.1815e-05	2.2145e-05	2.7649e-06
172,800	2.2709e-05	2.7479e-06	2.4830e-06	2.3516e-05	1.2527e-05	1.6376e-06

14.5 Statistics for Equidistributions

Next, we closely follow C. Choirat, R. Seri [2013] to derive the asymptotic statistical properties of a class of generalized discrepancies to test equidistribution on the sphere. We show that the generalized discrepancies have highly desirable properties and encompass several statistics already proposed in the literature. In particular, it turns out that the limiting distribution is an (infinite) weighted sum of chi-squared random variables. Issues concerning the approximation of this distribution are considered in detail and explicit bounds for the approximation error are given. In C. Choirat, R. Seri [2013] the statistics are applied to assess the equidistribution of Hammersley low discrepancy sequences on the sphere (cf. Example 14.16) and the uniformity of a dataset.

Let $X_N = \{\xi_1, \xi_2, \ldots, \xi_N\}$ be a set of points on $\mathbb{S}^2$. The generalized discrepancy $D(X_N; A)$ associated with a pseudodifferential operator A with symbol $\{A^\wedge(n)\}$, $A^\wedge(n) \neq 0$ for $n \geq 1$, is defined by (14.80) in Definition 14.26. A difference with respect to the case considered in the previous section is the fact that there it is supposed that $A^\wedge(n)$ is finite for every $n \geq 1$, while in statistical applications there is some interest in allowing for $(A^\wedge(n))^{-1} = 0$ for some n as the examples below demonstrate.

Test Statistics. The computation of $D^2(X_N; A)$ for a specific point set X_N is generally much more difficult than it seems. The Legendre polynomials appearing in (14.80) require the use of stable numerical formulas (cf. C.W. Clenshaw [1955], W. Freeden, M. Gutting [2013]) and the series has to be truncated suitably. Therefore, it is better and easier to have a closed form expression for the series in (14.80).

Some tests used in statistics are characterized by explicit forms (see, e.g., C. Choirat, R. Seri [2013] and the references therein). Closed formulas trivially exist for the Rayleigh statistic

$$\frac{1}{N^2} \sum_{i=1}^{N} \sum_{j=1}^{N} P_1(\xi_i \cdot \xi_j) \tag{14.106}$$

obtained taking $(A^\wedge(1))^2 = \frac{3}{4\pi}$ and $(A^\wedge(n))^{-1} = 0$ for $n \neq 1$ (cf. E. Giné [1975]) and for the Bingham statistic

$$\frac{1}{N^2} \sum_{i=1}^{N} \sum_{j=1}^{N} P_2(\xi_i \cdot \xi_j) \tag{14.107}$$

obtained taking $(A^\wedge(2))^2 = \frac{5}{4\pi}$ and $(A^\wedge(n))^{-1} = 0$ for $n \neq 2$ (cf. E. Giné [1975]). Other, more complicated cases are described as follows:

- Beran's form of Ajne's statistic (see, e.g., M.J. Prentice [1978])

$$\sum_{n=1}^{\infty} \frac{2n+1}{4\pi (A^\wedge(n))^2} P_n(t) = \frac{1}{4} - \frac{1}{2\pi} \arccos(t), \tag{14.108}$$

  obtained taking (see, e.g., I.A. Gradshteyn, I.M. Ryzhik [1965])

$$(A^\wedge(n))^2 = \begin{cases} +\infty, & n \text{ even}, \\ n^2 \left(\Gamma(\frac{n+3}{2})\right)^2 \left(\Gamma(\frac{n+2}{2})\right)^{-2}, & n \text{ odd}. \end{cases} \tag{14.109}$$

- Giné's statistic (see E. Giné [1975]):

$$\sum_{n=1}^{\infty} \frac{2n+1}{4\pi (A^\wedge(n))^2} P_n(t) = \frac{1}{2} - \frac{2}{\pi} \sin(\arccos(t)), \tag{14.110}$$

obtained taking (see, e.g., I.A. Gradshteyn, I.M. Ryzhik [1965])

$$(A^\wedge(n))^2 = \begin{cases} \frac{n-1}{n+2}\left(\Gamma(\frac{n}{2})\right)^2 \left(\Gamma(\frac{n+1}{2})\right)^{-2}, & n \text{ even}, \\ +\infty, & n \text{ odd}. \end{cases} \tag{14.111}$$

- Pycke's statistic (see J.-R. Pycke [2007a,b]):

$$\begin{aligned} \sum_{n=1}^{\infty} \frac{2n+1}{4\pi(A^\wedge(n))^2} P_n(t) &= -\frac{1}{4\pi} \ln\left(\frac{e}{2}(1-t)\right) \\ &= -\frac{1}{4\pi}\left(\ln(1-t) + 1 - \ln 2\right), \end{aligned} \tag{14.112}$$

obtained taking $A^\wedge(n) = \sqrt{n(n+1)}$. Note that this statistic does not respect the condition $\sum_{n=1}^{\infty} \frac{2n+1}{(A^\wedge(n))^2} < \infty$, and has to be analyzed as in Theorem 14.34 (iv). Obviously, this is just the negative sphere function $-G(\Delta^*, t)$ (see Lemma 9.4).

- A prominent case arises when $A^\wedge(n) = \sqrt{(2n+1)n(n+1)}$ which leads to (see (14.91) or (12.22))

$$D^2(X_N; A) = \frac{1}{N^2} \sum_{i=1}^{N} \sum_{j=1}^{N} \frac{1}{4\pi}\left(1 - 2\ln\left(1 + \sqrt{\frac{1-\xi_i\cdot\xi_j}{2}}\right)\right). \tag{14.113}$$

This is the case considered in J. Cui, W. Freeden [1997], R.J. Hinde [2000], J.-T. Zhang [2005].

- When $A^\wedge(n) = -n(n+1)$ we also obtain a closed form as described in detail in Corollary 9.21, i.e.,

$$\sum_{n=1}^{\infty} \frac{2n+1}{4\pi(A^\wedge(n))^2} P_n(t) = \begin{cases} \frac{1}{4\pi}, & t = 1, \\ -\frac{1}{4\pi}\ln(1-t)\ln(1+t) \\ +\frac{\ln 2}{4\pi}\ln(1-t^2) - \frac{1}{4\pi}\mathcal{L}_2\left(\frac{1-t}{2}\right) \\ +\frac{1}{4\pi}(1-(\ln 2)^2), & t \neq \pm 1, \\ \frac{1}{4\pi} - \frac{\pi}{24}, & t = -1, \end{cases} \tag{14.114}$$

where $\mathcal{L}_2(\cdot)$ is the so-called dilogarithm (cf. (9.190)). Obviously, this is the Green's function $G((\Delta^*)^2; \cdot)$ of Section 9.3.

- The case of Riesz kernels (see, e.g., S.B. Damelin [2008]) appearing in the analysis of s-energies has attracted a lot of interest (cf. A. Kuijlaars [1993], S.B. Damelin, P.J. Grabner [2003]). For $\xi_1, \xi_2 \in \mathbb{S}^2$, we define

$$K(\xi_1, \xi_2) = \begin{cases} \text{sign}(s)|\xi_1 - \xi_2|^{-s}, & s \neq 0 \\ -\ln|\xi_1 - \xi_2|^2, & s = 0. \end{cases} \tag{14.115}$$

The choices $s = 0$ and $s = 1$ lead, respectively, to the logarithmic and Coulombic potential. We can write this kernel as $K(\xi_1, \xi_2) = K_s(\xi_1 \cdot \xi_2)$, where

$$K_s(t) = \begin{cases} \operatorname{sign}(s)|2(1-t)|^{-\frac{s}{2}}, & s \neq 0 \\ -\ln 2(1-t), & s = 0, \end{cases} \tag{14.116}$$

for $t \in [-1, 1)$. When $s \neq 0$, the choice

$$(A^\wedge(n))^2 = \frac{2^{s-2}\Gamma\left(\frac{s}{2}\right)\Gamma\left(-\frac{s}{2}+n+2\right)}{\pi\Gamma\left(\frac{s}{2}+n\right)\Gamma\left(1-\frac{s}{2}\right)} \tag{14.117}$$

for $s < 2$ yields

$$\sum_{n=1}^{\infty} \frac{2n+1}{4\pi(A^\wedge(n))^2} P_n(t) = |2(1-t)|^{-\frac{s}{2}} - \frac{2^{-s}}{1-\frac{s}{2}} = \operatorname{sign}(s) K_s(t) - \frac{2^{-s}}{1-\frac{s}{2}}. \tag{14.118}$$

When $s = 0$, the choice $(A^\wedge(n))^2 = \frac{n(n+1)}{4\pi}$ yields

$$\sum_{n=1}^{\infty} \frac{2n+1}{4\pi(A^\wedge(n))^2} P_n(t) = -\ln 2(1-t) - \ln \frac{e}{4} = K_0(t) - 1 + 2\ln 2. \tag{14.119}$$

This case is simply a version of Pycke's statistic.

Note that $(A^\wedge(n))^2 \sim n^{-s+2}$, i.e., $\lim_{n\to\infty} (A^\wedge(n))^2 / n^{-s+2} = 1$, so that $\sum_{n=1}^{\infty} \frac{2n+1}{4\pi(A^\wedge(n))^2} \sim \sum_{n=1}^{\infty} n^{s-1}$, which converges if $s < 0$. Therefore, the $D^2(X_N; A)$ form of the discrepancy can be used only when $s < 0$. On the other hand, $d^2(X_N; A)$ (see Theorem 14.34 (iv) below) can be used if

$$\sum_{n=1}^{\infty} \left(\frac{2n+1}{4\pi(A^\wedge(n))^2}\right)^2 \sim \sum_{n=1}^{\infty} n^{2s-2} \tag{14.120}$$

converges which holds if $s < \frac{1}{2}$. The requirement that $s < 2$ is quite unfortunate. Generalized Riesz kernels remove it (see, e.g., S.B. Damelin [2008]). In J.S. Brauchart [2005], some computations are available that allow one to recover $(A^\wedge(n))^2$ for $s > 0$ in the case of generalized Riesz kernels.

- Some further cases with closed forms can be obtained from Section 12.4. For $(A^\wedge(n))^2 = (2n+1)h^{-n}$ with $h \in (0,1)$ the generating function of the Legendre polynomials (see Lemma 7.6) yields the singularity kernel (12.79), i.e.,

$$\sum_{n=1}^{\infty} \frac{2n+1}{4\pi(A^\wedge(n))^2} P_n(t) = \frac{1}{4\pi} \frac{1}{(1-2ht+h^2)^{1/2}} - \frac{1}{4\pi}, \tag{14.121}$$

while for $(A^\wedge(n))^2 = h^{-n}$ with $h \in (0,1)$ we obtain the Abel-Poisson kernel (12.78), i.e.,

$$\sum_{n=1}^{\infty} \frac{2n+1}{4\pi(A^\wedge(n))^2} P_n(t) = \frac{1}{4\pi} \frac{1-h^2}{(1-2ht+h^2)^{3/2}} - \frac{1}{4\pi}, \tag{14.122}$$

Other situations in which closed form expressions exist are given in M. Englis, J. Peetre [1998], G.G. Gregory [1977], but not all of them respect the condition $\sum_{n=1}^{\infty} \frac{2n+1}{(A^\wedge(n))^2} < \infty$ of Theorem 14.34.

Asymptotic Properties. The following theorem of C. Choirat, R. Seri [2013] provides the most relevant asymptotic properties of the generalized discrepancies under the null hypothesis of uniform distribution of the sample of points X_N.

Theorem 14.34. *Let X_N be a sample of iid uniform random variables on $\mathbb{S}^2$, and suppose that the spherical symbol $\{A^\wedge(n)\}$ of the pseudodifferential operator A is such that $A^\wedge(n) \neq 0$ for any n. Then, if $\sum_{n=1}^{\infty} \frac{2n+1}{(A^\wedge(n))^2} < \infty$, the first three properties hold.*

(i) $D^2(X_N; A)$ converges almost surely to 0.

(ii) The following convergence in distribution holds:

$$N\, D^2(X_N; A) \xrightarrow{\mathcal{D}} \sum_{n=1}^{\infty} \frac{1}{4\pi (A^\wedge(n))^2} \chi_n^2(2n+1), \tag{14.123}$$

where $\chi^2(k)$ denotes a χ^2 random variable with k degrees of freedom, and $\{\chi_n^2(2n+1) : n \in \mathbb{N}\}$ is a sequence of independent χ^2 random variables.

(iii) The following uniform bound holds:

$$\left| \mathbb{P}\{N\, D^2(X_N; A) \le y\} - \mathbb{P}\left\{ \sum_{n=1}^{\infty} \frac{\chi_n^2(2n+1)}{4\pi (A^\wedge(n))^2} \le y \right\} \right| \tag{14.124}$$

$$\le \frac{\exp\left(\frac{c}{\lambda_{13}} \sqrt{\sum_{n=1}^{\infty} \frac{2n+1}{(A^\wedge(n))^4}} \right)}{N} \left(1 + \left(\sum_{n=1}^{\infty} \frac{2n+1}{(A^\wedge(n))^2} \right)^{\frac{3}{2}} \left(\sum_{n=1}^{\infty} \frac{2n+1}{(A^\wedge(n))^4} \right)^{-\frac{3}{4}} \right),$$

where λ_{13} is defined in the proof of the theorem in C. Choirat, R. Seri [2013].

(iv) If $\sum_{n=1}^{\infty} \left(\frac{2n+1}{(A^\wedge(n))^2} \right)^2 < \infty$, the modified statistic

$$d^2(X_N; A) = \frac{1}{N^2} \sum_{\substack{i,j=1 \\ i \neq j}}^{N} \sum_{n=1}^{\infty} \frac{2n+1}{4\pi (A^\wedge(n))^2} P_n(\xi_i \cdot \xi_j) \tag{14.125}$$

is such that $d^2(X_N; A)$ almost surely converges to 0 and the following convergence in distribution holds true:

$$N\, d^2(X_N; A) \xrightarrow{\mathcal{D}} \sum_{n=1}^{\infty} \frac{1}{4\pi (A^\wedge(n))^2} (\chi_n^2(2n+1) - (2n+1)). \tag{14.126}$$

The reader is referred to C. Choirat, R. Seri [2013] for the proof.

Remark 14.35. *We are able to investigate several special cases as follows:*

(i) For $A^\wedge(n) = \sqrt{(2n+1)n(n+1)}$, *we obtain*

$$\sum_{n=1}^{\infty} \frac{2n+1}{(A^\wedge(n))^2} = \sum_{n=1}^{\infty} \frac{1}{n(n+1)} = 1 < \infty \tag{14.127}$$

and

$$N\, D^2(X_N; A) \xrightarrow{\mathcal{D}} \sum_{n=1}^{\infty} \frac{\chi_n^2(2n+1)}{4\pi(2n+1)n(n+1)}. \tag{14.128}$$

(ii) When $A^\wedge(n) = h^{-n/2}$, *then* $N\, D^2(X_N; A) \xrightarrow{\mathcal{D}} \frac{1}{4\pi} \sum_{n=1}^{\infty} h^n \chi_n^2(2n+1)$.

(iii) For $A^\wedge(n) = \sqrt{n(n+1)}$, *i.e., Pycke's statistic, we have to use the statistic* d^2 *and we find that*

$$N\, d^2(X_N; A) \xrightarrow{\mathcal{D}} \sum_{n=1}^{\infty} \frac{\chi_n^2(2n+1) - (2n+1)}{4\pi n(n+1)}. \tag{14.129}$$

(iv) If A is such that $k_1 \cdot n^{1+\alpha} \leq A^\wedge(n) \leq k_2 \cdot n^{1+\alpha}$ *for* $\alpha > 0$ *and* $0 < k_1 \leq k_2 < +\infty$, *then the bound (14.124) has the following behavior in terms of N and* α:

$$\left| \mathbb{P}\{N\, D^2 \leq y\} - \mathbb{P}\left\{ \sum_{n=1}^{\infty} \frac{\chi_n^2(2n+1)}{4\pi(A^\wedge(n))^2} \leq y \right\} \right| \tag{14.130}$$

$$\leq \frac{\exp\left(c\, 9^\alpha \sqrt{\zeta(3+4\alpha)}\right)}{N} \left(1 + \frac{\zeta^{\frac{3}{2}}(1+2\alpha)}{\zeta^{\frac{3}{4}}(3+4\alpha)}\right)$$

where ζ *is the Riemann zeta function (5.12). The only practical implication of* α *is in the exponent. This shows that a large value of* α *implies a worse convergence rate.*

Remark 14.36. *Due to the results in E. Giné [1975], it is possible to show that the condition* $(A^\wedge(n))^2 < +\infty$ *for any n is equivalent to the fact that* $D^2(X_N; A) \xrightarrow{\mathbb{P}} 0$ *if and only if the sample* X_N *comes from a uniform distribution. This implies that a test based on a generalized discrepancy may converge to 0 also for nonuniform distributions if* $(A^\wedge(n))^2 = +\infty$ *for some n. As an example, Gine's statistic converges to 0 for every probability measure uniform on the projective space and not only on the sphere (see E. Giné [1975]).*

As concerns the asymptotic properties under the alternative hypothesis that the points are independent realizations of a nonuniform distribution, we just quote them, together with references to results in the literature that allow for deriving them. Consider an iid sample X_N^* drawn from a distribution $\mathbb{P}^*$ on the sphere different from the uniform distribution. Note that, according to Remark 14.36, if $(A^\wedge(n))^2 = +\infty$ for some n, then $D^2(X_N^*; A)$ can converge in probability to 0 even if the sample X_N^* does not come from a uniform distribution. The following properties hold if $D^2(X_N^*; A)$ does not converge in probability to 0. Due to a corollary of I.B. Alberink, V. Bentkus [2001], it is possible to show that when $D^2(X_N^*; A)$ is suitably centered and scaled, it converges in distribution to a standard normal random variable. A uniform bound on the distance between the finite sample and the asymptotic distribution can be obtained through I.B. Alberink, V. Bentkus [2001] and decreases as $1/\sqrt{N}$. The asymptotic distribution of $N\, D^2(X_N^*; A)$ under a local alternative converging to the null, i.e., under a Pitman drift, is given by a weighted sum of noncentral χ^2 random variables (see G.G. Gregory [1977]).

Computation of the Asymptotic Distribution. Now we consider approximations of the asymptotic distribution of $ND^2(X_N; A)$, denoted, in the following as $X = \sum_{n=1}^{\infty} \frac{\chi_n^2(2n+1)}{4\pi(A^\wedge(n))^2}$ through three different methods. This topic is relevant for any distribution given by a weighted sum of χ^2 random variables, but is particularly critical here. Indeed, contributions to X corresponding to large values of n that is $\frac{\chi_n^2(2n+1)}{4\pi(A^\wedge(n))^2}$ can be quite relevant since the degrees of freedom of the χ^2 random variable depend on n. Note that this problem does not arise with the most common test statistics on the real line, such as Cramér-von Mises and Anderson-Darling, and is a facet of the curse of dimensionality problem.

The first approximation amounts to considering a scaled and translated Γ distributed random variable X_Γ that matches the first three moments of X:

$$X_\Gamma = \mu + \frac{\Gamma(\frac{\nu}{2}, 2) - \nu}{\sqrt{2\nu}} \sigma, \tag{14.131}$$

where $\Gamma(\alpha, \theta)$ denotes a Γ random variable with shape parameter α and scale parameter θ,

$$\mu = \sum_{n=1}^{\infty} \frac{2n+1}{4\pi(A^\wedge(n))^2}, \quad \sigma^2 = 2\sum_{n=1}^{\infty} \frac{2n+1}{(4\pi(A^\wedge(n))^2)^2} \tag{14.132}$$

and

$$\nu = \left(\sum_{n=1}^{\infty} \frac{2n+1}{(A^\wedge(n))^4}\right)^3 \left(\sum_{n=1}^{\infty} \frac{2n+1}{(A^\wedge(n))^6}\right)^{-2}. \tag{14.133}$$

This approximation for linear combinations of χ^2 random variables has been considered in M.J. Buckley, G.K. Eagleson [1988], H. Leeb [2002], J.-T. Zhang [2005], where uniform bounds are also derived.

We propose also the following truncated version of X:

$$X_\nu = \sum_{n=1}^{\nu} \frac{\chi_n^2(2n+1)}{4\pi(A^\wedge(n))^2}. \tag{14.134}$$

C. Choirat, R. Seri [2013] show that a better alternative is to use a centered version of X_ν:

$$X_\nu^* = \sum_{n=1}^{\nu} \frac{\chi_n^2(2n+1)}{4\pi(A^\wedge(n))^2} + \sum_{n=\nu+1}^{\infty} \frac{2n+1}{4\pi(A^\wedge(n))^2}. \tag{14.135}$$

The computational of X_ν can then be easily performed using the algorithm described in R.B. Davies [1973] and C. Choirat, R. Seri [2006]. We majorize the errors implied by the replacement of X with X_ν or X_ν^* using the following bounds.

Theorem 14.37. *We have with the help of Euler's Beta function $B(\cdot,\cdot)$ (see, e.g., W. Freeden, M. Gutting [2013])*

$$\begin{aligned} |\mathbb{P}\{X \le x\} - \mathbb{P}\{X_\nu \le x\}| \le \left(2 \sum_{n=\nu+1}^{\infty} \frac{2n+1}{(A^\wedge(n))^4} + \left(\sum_{n=\nu+1}^{\infty} \frac{2n+1}{(A^\wedge(n))^2}\right)^2\right)^{1/2} \\ \times \left(\sum_{n=1}^{\nu} \frac{2n+1}{(A^\wedge(n))^4}\right)^{-1/2} \frac{2}{\pi\sqrt{3}} B\left(\frac{1}{2}, \frac{1}{4}\right) \end{aligned} \tag{14.136}$$

and

$$|\mathbb{P}\{X \le x\} - \mathbb{P}\{X_\nu^* \le x\}| \le \frac{5}{\pi}\left(\sum_{n=\nu+1}^{\infty} \frac{2n+1}{(A^\wedge(n))^4}\right)\left(\sum_{n=2}^{\nu} \frac{2n+1}{(A^\wedge(n))^4}\right)^{-1}. \tag{14.137}$$

If A is such that $(A^\wedge(n))^{-1} \sim Cn^{-1-\alpha}$ for $\alpha > 0$ (so that A is a pseudodifferential operator of order $t = 1 + \alpha$), we have:

$$|\mathbb{P}\{X \le x\} - \mathbb{P}\{X_\nu \le x\}| \le \frac{C^2}{\alpha\sigma} \frac{\sqrt{3}}{16\pi^2} B\left(\frac{1}{2}, \frac{1}{4}\right) \nu^{-2\alpha}(1 + o(1)), \tag{14.138}$$

$$\begin{aligned} |\mathbb{P}\{X \le x\} - \mathbb{P}\{X_\nu^* \le x\}| \le \frac{C^4}{(1+2\alpha)\left(\sigma^2 - \frac{3}{8\pi^2}(A^\wedge(1))^{-4}\right)} \\ \times \frac{5}{8\pi^3} \nu^{-2-4\alpha}(1 + o(1)), \end{aligned} \tag{14.139}$$

where σ^2 has been defined in (14.132).

Again, the reader is referred to C. Choirat, R. Seri [2013] for the proof.

Remark 14.38. *Note that C. Choirat, R. Seri [2013] also show that both bounds are tight. The previous bounds assert that the approximation of X through X_ν and X_ν^* is better when the order of the pseudodifferential operator A is higher. This is in line with J. Cui, W. Freeden [1997], where it is stated that computation of $D^2(X_N; A)$ is easier the larger t is. When using the statistic $d^2(X_N; A)$, centering is automatic and only the second bound applies.*

Remark 14.39. *For the interesting case* $A^\wedge(n) = \sqrt{(2n+1)n(n+1)}$, *we have* $k = 1/\sqrt{2}$ *and* $\alpha = 1/2$, *so that*

$$|\mathbb{P}\{X \le x\} - \mathbb{P}\{X_\nu \le x\}| \le 6.41233\,\nu^{-1}(1+o(1)), \tag{14.140}$$

$$|\mathbb{P}\{X \le x\} - \mathbb{P}\{X_\nu^* \le x\}| \le 28.33329\,\nu^{-4}(1+o(1)), \tag{14.141}$$

where $X_\nu^* = X_\nu + \frac{1}{4\pi(\nu+1)}$. *Even if the constant of the second bound is quite large, the bound shows that the approximation is very good even for small values of* ν.

When $(A^\wedge(n))^2 = (2n+1)h^{-n}$, $h \in (0,1)$, *the first bound is* $O(h^\nu)$ *while the second is* $O(h^{2\nu})$. *For* $(A^\wedge(n))^2 = h^{-n}$, $h \in (0,1)$, *the former is* $O(\nu h^\nu)$ *while the latter is* $O(\nu h^{2\nu})$.

Simulation Results and Applications. We briefly summarize some simulation results of C. Choirat, R. Seri [2013], where further details can be found and additionally the rate of convergence to the asymptotic distribution is investigated. Only the Cui–Freeden test statistic $N\,D^2(X_N; A)$ for $A^\wedge(n) = \sqrt{(2n+1)n(n+1)}$ is considered, since in this case a simple closed form expression exists.

The derivation $\sup_x |\mathbb{P}\{X \le x\} - \mathbb{P}\{X_\nu \le x\}|$ between the asymptotic distribution of $N\,D^2(X_N; A)$ and the proposed approximation is denoted by UNC-Real, while CEN-Real is $\sup_x |\mathbb{P}\{X \le x\} - \mathbb{P}\{X_\nu^* \le x\}|$. UNC-Comp and CEN-Comp are the bounds computed through Theorem 14.37. It is possible to see that both UNC-Comp and CEN-Comp are precise upper bounds respectively on UVC-Real and CEN-Real.

In C. Choirat, R. Seri [2013] there is a dramatic improvement in the performance of the approximation method that can be obtained simply through centering, even if this operation involves a shift of the distribution through a small constant such as $\sum_{n=\nu+1}^{\infty} \frac{2n+1}{4\pi(A^\wedge(n))^2} = \frac{1}{4\pi(\nu+1)}$. Centering allows for a reduction of the error behavior from $\Theta(\nu^{-1})$ to $\Theta(\nu^{-4})$ as shown by Theorem 14.37. It also seems that the Γ approximation cannot compete with the other methods, but the use of a uniform bound hides the fact that for most quantities of interest (confined to the upper tail) the Γ approximation is extremely good.

Next we investigate how generalized discrepancies can be used in the assessment of low-discrepancy (often called quasi-Monte Carlo) sequences. We recall the definition of the Hammersley sequence on the sphere (see Example 14.16). In T.T. Wong et al. [1997], the Hammersley sequence has been proposed for the solution of some problems in computer graphics. In particular, the authors remark that when p increases, the uniformity of the sequence seems to get worse.

The results in C. Choirat, R. Seri [2013] show that the sequence with $p = 2$ is uniformly the best one. When p increases, the generalized discrepancy $N\,D^2(X_N; A)$ shows an oscillating behavior but remains consistently above the curve corresponding to $p = 2$. By the way, this is coherent with the

TABLE 14.6: Computed test statistics on the data due to C. Choirat, R. Seri [2013]

Name	$(A^\wedge(n))^2$	$\sum_{n=1}^{\infty} \frac{2n+1}{4\pi(A^\wedge(n))^2} P_n(t)$	Asymptotic distribution
Rayleigh	$\begin{cases} \frac{3}{4\pi}, & n=1, \\ +\infty, & n \neq 1, \end{cases}$	t	$\chi_n^2(3)/3$
Bingham	$\begin{cases} \frac{5}{4\pi}, & n=1, \\ +\infty, & n \neq 1, \end{cases}$	$-\frac{1}{2}(1-3t^2)$	$\chi_n^2(5)/5$
Beran/Ajne	$\begin{cases} +\infty, & n \text{ even}, \\ n^2\left(\Gamma\left(\frac{n+3}{2}\right)\right)^2\left(\Gamma\left(\frac{n+2}{2}\right)\right)^{-2}, & n \text{ odd}, \end{cases}$	$\frac{1}{4}-(2\pi)^{-1}\arccos(t)$	$\sum_{n=1}^{\infty} \frac{(B(\frac{2n-1}{2},\frac{3}{2}))^2 \chi_n^2(4n-1)}{16\pi(\Gamma(\frac{3}{2}))^2}$
Giné	$\begin{cases} +\infty, & n \text{ odd}, \\ \frac{n-1}{n+2}\left(\Gamma\left(\frac{n}{2}\right)\right)^2\left(\Gamma\left(\frac{n+1}{2}\right)\right)^{-2}, & n \text{ even}, \end{cases}$	$\frac{1}{2}-\frac{2}{\pi}\sin(\arccos(t))$	$\sum_{n=1}^{\infty} \frac{(2n-1)(B(\frac{2n-1}{2},\frac{3}{2}))^2 \chi_n^2(4n-1)}{8\pi(n+1)(\Gamma(\frac{3}{2}))^2}$
Cui-Freeden	$(2n+1)n(n+1)$	$\frac{1}{4\pi}\left(1-2\ln\left(1+\sqrt{\frac{1-t}{2}}\right)\right)$	$\sum_{n=1}^{\infty} \frac{\chi_n^2(2n+1)}{4\pi(2n+1)n(n+1)}$
Freeden-Schreiner	$n^2(n+1)^2$	$G((\Delta^*)^2,t)$ (cf. (9.198))	$\sum_{n=1}^{\infty} \frac{\chi_n^2(2n+1)}{4\pi n^2(n+1)^2}$
Pycke	$n(n+1)$	$-\frac{1}{4\pi}\ln\frac{e}{2}(l-t)$	$\sum_{n=1}^{\infty} \frac{\chi_n^2(2n+1)-(2n+1)}{4\pi n(n+1)}$
Name	$N\,D^2(X_n;A)$	p-value UNC	p-value CEN
Rayleigh	0.7539653	0.5198599 (–)	0.5198599 (–)
Bingham	1.325083	0.2500217 (–)	0.2500217 (–)
Beran/Ajne	0.2226953	0.4652019 (0.002812024)	0.4656521 (8.269991e-11)
Giné	0.7638484	0.1023675 (0.008622067)	0.1027497 (7.9393300e-12)
Cui-Freeden	0.0881051	0.3134682 (0.006405284)	0.3142182 (2.822021e-11)
Freeden-Schreiner	0.0744169	0.4369813 (4.392691e-06)	0.436982 (1.234307e-16)
Pycke	0.1531135	– (–)	0.09439549 (6.3335351e-06)

behavior of the star discrepancy of Hammersley sequences that can be shown to worsen when p increases (see H. Niederreiter [1992]).

A Statistical Example. C. Choirat, R. Seri [2013] considered a classical dataset concerning 52 measurements of magnetic remanence from specimens of red beds from Bowen Basin, Queensland, after thermal demagnetisation to 670°C (see N.I. Fisher et al. [1993]). They computed seven statistics as

$$D^2(X_N; A) = \frac{1}{N^2} \sum_{i=1}^{N} \sum_{j=1}^{N} \sum_{n=1}^{\infty} \frac{2n+1}{4\pi(A^\wedge(n))^2} P_n(\xi_i \cdot \xi_j), \tag{14.142}$$

where Pycke's statistic is the only one computed removing the elements with $i = j$ in the double sum over i and j. The values of N, $D^2(X_N; A)$ are summarized in Table 14.6 along with the other values of interest. The values of the cumulative distribution function have been computed through X_ν and X_ν^*, where $\nu = 1000$. The seven statistics yield similar results in the sense that all of them fail to reject the null hypothesis of uniformity, even if one of them (Giné's statistic) comes very near to reject it at 90%. The respective error bounds are also displayed in parentheses. The difference in the bounds for the centered and the noncentered cases is eloquent. It should also be remarked that the bounds for the Cui–Freeden statistic are coherent with the rates of decrease derived in Remark 14.39.

Chapter 15

Multiscale Approximate Integration

As already mentioned in the Introduction, spherical wavelet approximation dates back to many sources. Each method has is own specific characteristic. In our approach here (cf. W. Freeden, K. Hesse [2002]), we are essentially interested in rotation-invariant (numerically) locally supported wavelet variants for use in geoscientific problems and the resulting multiscale integration.

15.1 Singular Integrals and Approximate Identities

In this section approximate identities for continuous functions as well as square-integrable functions are presented (see also H. Berens et al. [1968], W. Freeden et al. [1998], W. Freeden, K. Hesse [2002]), i.e., families of operators $\{\mathcal{I}_h\}_{h\in(-1,1)}$, $\mathcal{I}_h : X(\mathbb{S}^2) \to X(\mathbb{S}^2)$, $F \mapsto \mathcal{I}_h F = K_h * F$, where $\{K_h\}_{h\in(-1,1)} \subset \mathrm{L}^1[-1,1]$ for $X(\mathbb{S}^2) = \mathrm{C}^{(0)}(\mathbb{S}^2)$, and $\{K_h\}_{h\in(-1,1)} \subset \mathrm{L}^2[-1,1]$ for $X(\mathbb{S}^2) = \mathrm{L}^2(\mathbb{S}^2)$, which fulfill $\lim_{h\to 1,\, h<1} \|F - \mathcal{I}_h F\|_{X(\mathbb{S}^2)} = 0$. Of special interest are approximate identities generated by so-called $[h,1]$-locally supported
kernels.

First the terminology of singular integrals, approximate identities, and scaling functions is explained and several equivalent characterizations of an approximate identity will be formulated. Then we are concerned with approximate identities generated by so-called $[h,1]$-locally supported kernels, i.e., kernels $\{K_h\}_{h\in(-1,1)}$ showing the additional property $\mathrm{supp}(K_h) = [h,1]$ for all $h \in (-1,1)$. Error estimates for $\|F - \mathcal{I}_h F\|_{\mathrm{C}^{(0)}(\mathbb{S}^2)}$ will be presented in explicit form. Finally, two particularly important examples of non-negative $[h,1]$-locally supported kernels that generate an approximate identity in $\mathrm{C}^{(0)}(\mathbb{S}^2)$ (and also in $\mathrm{L}^2(\mathbb{S}^2)$) will be discussed in more detail.

Definition 15.1. *Let $\{K_h\}_{h\in(-1,1)}$ be a family of functions in $\mathrm{L}^1[-1,1]$ or in $\mathrm{L}^2[-1,1]$ satisfying the condition $K_h^\wedge(0) = 1$ for all $h \in (-1,1)$. Then the family of bounded linear operators $\{\mathcal{I}_h\}_{h\in(-1,1)}$, $\mathcal{I}_h : X(\mathbb{S}^2) \to X(\mathbb{S}^2)$, $F \mapsto \mathcal{I}_h F$, given by*

$$(\mathcal{I}_h F)(\xi) = (K_h * F)(\xi) = \int_{\mathbb{S}^2} K_h(\xi\cdot\eta) F(\eta)\, dS(\eta), \qquad (15.1)$$

where $X(\mathbb{S}^2) = \mathrm{C}^{(0)}(\mathbb{S}^2)$ *for* $\{K_h\}_{h\in(-1,1)} \subset \mathrm{L}^1[-1,1]$ *and* $X(\mathbb{S}^2) = \mathrm{L}^2(\mathbb{S}^2)$ *for* $\{K_h\}_{h\in(-1,1)} \subset \mathrm{L}^2[-1,1]$, *is called a* (spherical) singular integral. *The family* $\{K_h\}_{h\in(-1,1)}$ *is called the* kernel *of the (spherical) singular integral.*

The singular integral $\{\mathcal{I}_h\}_{h\in(-1,1)}$ *is said to be an* approximate identity *in* $X(\mathbb{S}^2)$ *corresponding to the so-called* scaling function $\{K_h\}_{h\in(-1,1)}$, *if the following limit relation holds true:*

$$\lim_{\substack{h\to 1,\\ h<1}} \|F - \mathcal{I}_h F\|_{X(\mathbb{S}^2)} = 0 \tag{15.2}$$

for all $F \in X(\mathbb{S}^2)$.

Obviously, the assumption that a kernel $\{K_h\}_{h\in(-1,1)}$ is a scaling function implies that $\{\mathcal{I}_h\}_{h\in(-1,1)}$ is an approximate identity. The scaling function $\{K_h\}_{h\in(-1,1)}$ is said to generate the approximate identity $\{\mathcal{I}_h\}_{h\in(-1,1)}$.

The kernel $\{K_h\}_{h\in(-1,1)}$ is called continuous kernel, L^1-kernel, L^2-kernel, non-negative kernel, etc., if all members K_h have this property.

Theorem 15.2. *Let* $\{K_h\}_{h\in(-1,1)}$ *be a family of functions in* $\mathrm{L}^1[-1,1]$ *or in* $\mathrm{L}^2[-1,1]$, *which satisfies* $K_h^\wedge(0) = 1$ *for all* $h \in (-1,1)$ *and which is* L^1*-uniformly bounded, i.e.,*

$$2\pi \int_{-1}^{1} |K_h(t)|\, dt \le M, \qquad h \in (-1,1), \tag{15.3}$$

with some constant M *independent of* h. *Then the spherical singular integral* $\{\mathcal{I}_h\}_{h\in(-1,1)}$ *defined in* (15.1) *is an approximate identity in* $X(\mathbb{S}^2)$ *if and only if*

$$\lim_{\substack{h\to 1,\\ h<1}} K_h^\wedge(n) = 1, \quad n \in \mathbb{N}_0. \tag{15.4}$$

Proof. Our considerations have to be performed separately for the two cases $\{K_h\}_{h\in(-1,1)} \subset \mathrm{L}^1[-1,1]$ and $\{K_h\}_{h\in(-1,1)} \subset \mathrm{L}^2[-1,1]$. We start with $\{K_h\}_{h\in(-1,1)} \subset \mathrm{L}^1[-1,1]$.

Let $\{\mathcal{I}_h\}_{h\in(-1,1)}$ be an approximate identity in $\mathrm{C}^{(0)}(\mathbb{S}^2)$, i.e., (15.2) holds for all $F \in \mathrm{C}^{(0)}(\mathbb{S}^2)$. Particularly, this limit relation holds for all spherical harmonics Y_n of degree n. The Funk–Hecke formula (9.43) implies that

$$(\mathcal{I}_h Y_n)(\xi) = K_h^\wedge(n) Y_n(\xi), \quad \xi \in \mathbb{S}^2. \tag{15.5}$$

Thus,

$$0 = \lim_{\substack{h\to 1,\\ h<1}} \|Y_n - \mathcal{I}_h Y_n\|_{\mathrm{C}^{(0)}(\mathbb{S}^2)} = \lim_{\substack{h\to 1,\\ h<1}} |1 - K_h^\wedge(n)|\, \|Y_n\|_{\mathrm{C}^{(0)}(\mathbb{S}^2)}, \tag{15.6}$$

and (15.4) follows because of $\|Y_n\|_{C^{(0)}(\mathbb{S}^2)} \neq 0$ for all spherical harmonics $Y_n \neq 0$ of degree $n \in \mathbb{N}_0$.

Let now (15.4) be satisfied. Let $Y_n \in \mathrm{Harm}_n(\mathbb{S}^2)$, $n \in \mathbb{N}_0$, be arbitrary. We know that $(\mathcal{I}_h Y_n)(\xi) = K_h^\wedge(n) Y_n(\xi)$ for all $\xi \in \mathbb{S}^2$, for all $n \in \mathbb{N}_0$, and all $h \in (-1, 1)$. Furthermore, $\lim_{h \to 1, h<1} K_h^\wedge(n) = 1$ for all $n \in \mathbb{N}_0$. Thus we get

$$\lim_{\substack{h \to 1,\\ h<1}} \|Y_n - \mathcal{I}_h Y_n\|_{C^{(0)}(\mathbb{S}^2)} = \lim_{\substack{h \to 1,\\ h<1}} |1 - K_h^\wedge(n)| \, \|Y_n\|_{C^{(0)}(\mathbb{S}^2)} = 0. \tag{15.7}$$

Thus the assertion $\lim_{h \to 1, h<1} \|F - \mathcal{I}_h F\|_{C^{(0)}(\mathbb{S}^2)} = 0$ is true for all spherical harmonics Y_n of degree $n \in \mathbb{N}_0$.

Let $F \in C^{(0)}(\mathbb{S}^2)$ be arbitrary. We have to show, that for every $\varepsilon > 0$ there exists a value h_0 so that for all $h \in [h_0, 1)$ the estimate $\|F - \mathcal{I}_h F\|_{C^{(0)}(\mathbb{S}^2)} \leq \varepsilon$ holds. This task can be realized in the following way:

Let $\varepsilon > 0$ be arbitrary. Choose a complete orthonormal system $\{Y_{n,j}\}$ of spherical harmonics in $L^2(\mathbb{S}^2)$. Then F can be approximated arbitrarily well (with respect to $\|\cdot\|_{C^{(0)}(\mathbb{S}^2)}$) by finite linear combinations of the spherical harmonics $Y_{n,j}$. Let L_F be such a linear combination, so that $\|F - L_F\|_{C^{(0)}(\mathbb{S}^2)} \leq \min\{\varepsilon/3, \varepsilon/(3M)\}$, where M is the constant given in (15.3). By the triangle inequality we than get

$$\begin{aligned} &\|F - \mathcal{I}_h F\|_{C^{(0)}(\mathbb{S}^2)} \\ &\leq \|F - L_F\|_{C^{(0)}(\mathbb{S}^2)} + \|L_F - \mathcal{I}_h L_F\|_{C^{(0)}(\mathbb{S}^2)} + \|\mathcal{I}_h L_F - \mathcal{I}_h F\|_{C^{(0)}(\mathbb{S}^2)}. \end{aligned} \tag{15.8}$$

The first summand in (15.8) can be estimated by $\varepsilon/3$. For the second summand we observe that

$$L_F(\xi) = \sum_{n=0}^{\infty} \sum_{j=1}^{2n+1} l_{n,j} Y_{n,j}(\xi), \tag{15.9}$$

where only a finite number of the coefficients $l_{n,j}$ is different from zero. By virtue of the linearity of $\mathcal{I}_h$, we find that

$$\begin{aligned} |L_F(\xi) - (\mathcal{I}_h L_F)(\xi)| &= \left| \sum_{n=0}^{\infty} \sum_{j=1}^{2n+1} l_{n,j} \Big(Y_{n,j}(\xi) - (\mathcal{I}_h Y_{n,j})(\xi) \Big) \right| \\ &\leq \sum_{n=0}^{\infty} \sum_{j=1}^{2n+1} |l_{n,j}| \, |Y_{n,j}(\xi) - (\mathcal{I}_h Y_{n,j})(\xi)|. \end{aligned} \tag{15.10}$$

Let N be the number of coefficients $l_{n,j}$, which are not zero. All these coefficients can be estimated by some constant C which depends only on L_F. Choose h_0 so, that for all $h \in [h_0, 1)$, and for all tuples (n, j) with $l_{n,j} \neq 0$ the norm $\|\mathcal{I}_h Y_{n,j} - Y_{n,j}\|_{C^{(0)}(\mathbb{S}^2)} \leq \varepsilon/(3NC)$. Then, for all $h \in [h_0, 1)$, $\|L_F - \mathcal{I}_h L_F\|_{C^{(0)}(\mathbb{S}^2)} \leq NC(\varepsilon/(3NC)) = \varepsilon/3$.

In order to estimate the last summand in (15.8), the L^1-uniform boundedness of the kernels K_h may be used such that

$$\begin{aligned}
|(\mathcal{I}_h L_F)(\xi) - (\mathcal{I}_h F)(\xi)| &= \left| \int_{\mathbb{S}^2} K_h(\xi\cdot\eta)\,(L_F(\eta) - F(\eta))\; dS(\eta)\right| \\
&\leq \int_{\mathbb{S}^2} |K_h(\xi\cdot\eta)|\,|L_F(\eta) - F(\eta)|\; dS(\eta) \\
&\leq \|L_F - F\|_{C^{(0)}(\mathbb{S}^2)} 2\pi \int_{-1}^{1} |K_h(t)|\; dt \qquad (15.11)\\
&\leq \frac{\varepsilon}{3M} M = \frac{\varepsilon}{3}.
\end{aligned}$$

Hence, (15.4) implies $\lim_{h\to 1, h<1} \|F - \mathcal{I}_h F\|_{C^{(0)}(\mathbb{S}^2)} = 0$ for all $F \in C^{(0)}(\mathbb{S}^2)$.

In case of $\{K_h\}_{h\in(-1,1)} \subset L^2[-1,1]$ the proof can be given as follows:

Let $\{\mathcal{I}_h\}_{h\in(-1,1)}$ be an approximate identity in $L^2(\mathbb{S}^2)$, i.e., (15.2) holds for all $F \in L^2(\mathbb{S}^2)$. This part of the proof is identical to the one for $\{K_h\}_{h\in(-1,1)} \subset L^1[-1,1]$, only the $\|\cdot\|_{C^{(0)}(\mathbb{S}^2)}$-norm has to be replaced by the $\|\cdot\|_{L^2(\mathbb{S}^2)}$-norm.

Let now (15.4) be satisfied. The L^1-uniform boundedness (15.3) of the functions K_h, $h \in (-1,1)$, and $|P_n(t)| \leq 1$ for all $t \in [-1,1]$ and all $n \in \mathbb{N}_0$ imply that

$$K_h^\wedge(n) \leq 2\pi \int_{-1}^{1} |K_h(t)|\,|P_n(t)|\; dt \leq 2\pi \int_{-1}^{1} |K_h(t)|\; dt \leq M, \qquad (15.12)$$

$$K_h^\wedge(n) \geq -2\pi \int_{-1}^{1} |K_h(t)|\,|P_n(t)|\; dt \geq -2\pi \int_{-1}^{1} |K_h(t)|\; dt \geq -M. \qquad (15.13)$$

Hence, $K_h^\wedge(n) \in [-M, M]$ for all $n \in \mathbb{N}_0$ and all $h \in (-1,1)$. Moreover,

$$\|F - \mathcal{I}_h F\|^2_{L^2(\mathbb{S}^2)} = \sum_{n=0}^{\infty}\sum_{j=1}^{2n+1} \left(1 - K_h^\wedge(n)\right)^2 \left(F^\wedge(n,j)\right)^2 \leq (M+1)^2 \|F\|^2_{L^2(\mathbb{S}^2)} \qquad (15.14)$$

for all $h \in (-1,1)$ and all $F \in L^2(\mathbb{S}^2)$. As the upper bound $(M+1)$ of the expression $|1 - K_h^\wedge(n)|$ is independent of $h \in (-1,1)$, the limit for $h \to 1$ and the sum may be interchanged. Hence,

$$\begin{aligned}
&\lim_{\substack{h\to 1,\\ h<1}} \|F - \mathcal{I}_h F\|_{L^2(\mathbb{S}^2)} \qquad (15.15)\\
&= \left(\sum_{n=0}^{\infty}\sum_{j=1}^{2n+1} \lim_{\substack{h\to 1,\\ h<1}} \left(1 - K_h^\wedge(n)\right)^2 \left(F^\wedge(n,j)\right)^2 \right)^{1/2} = 0
\end{aligned}$$

for all $F \in L^2(\mathbb{S}^2)$. □

Restricting our attention to non-negative kernels $\{K_h\}_{h\in(-1,1)}$, i.e., all K_h, $h \in (-1,1)$, satisfy $K_h(t) \geq 0$ for almost all $t \in [-1,1]$, more equivalent characterizations of an approximate identity are deducible. The main advantage of non-negative kernels $\{K_h\}_{h\in(-1,1)}$ is that the property $K_h^\wedge(0) = 1$ implies

$$1 = K_h^\wedge(0) = 2\pi \int_{-1}^{1} K_h(t)\, dt = 2\pi \int_{-1}^{1} |K_h(t)|\, dt = 2\pi \|K\|_{\mathrm{L}^1[-1,1]}, \quad (15.16)$$

i.e., the L^1-uniformly boundedness condition (15.3) is valid with the sharp bound $M = 1$.

Theorem 15.3. *Let $\{K_h\}_{h\in(-1,1)}$ be a family of functions in $\mathrm{L}^1[-1,1]$ or in $\mathrm{L}^2[-1,1]$, which satisfy $K_h^\wedge(0) = 1$ and which are non-negative. Let $\{\mathcal{I}_h\}_{h\in(-1,1)}$ be the spherical singular integral as defined in Definition 15.2. Then the following properties are equivalent:*

(i) $\{K_h\}_{h\in(-1,1)}$ *is a non-negative scaling function,*

(ii) $\{\mathcal{I}_h\}_{h\in(-1,1)}$ *is an approximate identity,*

(iii) $\lim_{h\to 1,\, h<1} K_h^\wedge(n) = 1$ *for all* $n \in \mathbb{N}_0$,

(iv) $\lim_{h\to 1,\, h<1} K_h^\wedge(1) = 1$,

(v) $\{K_h\}_{h\in(-1,1)}$ *satisfies the localization property* $\lim_{h\to 1,\, h<1} \int_{-1}^{\delta} K_h(t)\, dt = 0$ *for all* $\delta \in (-1,1)$.

Proof. The following considerations are valid for both cases, $\{K_h\}_{h\in(-1,1)} \subset \mathrm{L}^1[-1,1]$ as well as $\{K_h\}_{h\in(-1,1)} \subset \mathrm{L}^2[-1,1]$. The statements (i) and (ii) are equivalent by definition and the equivalence of (ii) and (iii) was verified in Theorem 15.2. Obviously, (iii) implies (iv). It remains to show, that (v) follows from (iv) and that (v) implies (ii).

"(iv) $\Longrightarrow$ (v)": Let $\delta \in (-1,1)$ be arbitrary. Because of the non-negativity of K_h,

$$\begin{aligned} 0 \leq \int_{-1}^{\delta} K_h(t)\, dt &\leq \frac{1}{(1-\delta)} \int_{-1}^{\delta} (1-t) K_h(t)\, dt \\ &\leq \frac{1}{(1-\delta)} \int_{-1}^{1} (1-t) K_h(t)\, dt \\ &= \frac{1}{2\pi} \frac{1}{(1-\delta)} \Big(K_h^\wedge(0) - K_h^\wedge(1) \Big). \end{aligned} \quad (15.17)$$

Taking the limit for $h \to 1$ the localization property follows from (iv).

"(v) $\Longrightarrow$ (iii)": Property (iii) is equivalent to the following assertion: For every $n \in \mathbb{N}$, and for every $\varepsilon > 0$, there exists $h_0 = h_0(\varepsilon, n) \in (-1,1)$ such that $1 - \varepsilon \leq K_h^\wedge(n)) \leq 1$ for all $h \in [h_0, 1)$. By the non-negativity of K_h and the estimate $|P_n(t)| \leq 1$ for all $n \in \mathbb{N}_0$, we obtain

$$K_h^\wedge(n) = 2\pi \int_{-1}^{1} K_h(t) P_n(t)\, dt \leq 2\pi \int_{-1}^{1} K_h(t)\, dt = K_h^\wedge(0) = 1. \quad (15.18)$$

Let $n \in \mathbb{N}$ and $\varepsilon > 0$ be arbitrary. For $\delta \in (-1,1)$ it is clear that

$$K_h^\wedge(n) = 2\pi \int_{-1}^{\delta} K_h(t)P_n(t)\, dt + 2\pi \int_{\delta}^{1} K_h(t)P_n(t)\, dt. \tag{15.19}$$

As $P_n(1) = 1$, $\delta \in (-1,1)$ can be chosen so close to 1, that $P_n(t) \geq \sqrt{1-\frac{\varepsilon}{2}}$ for all $t \in [\delta, 1]$. Thus,

$$K_h^\wedge(n) \geq 2\pi \int_{-1}^{\delta} K_h(t)P_n(t)\, dt + 2\pi\sqrt{1-\frac{\varepsilon}{2}} \int_{\delta}^{1} K_h(t)\, dt. \tag{15.20}$$

As $|P_n(t)| \leq 1$, for all $\delta \in (-1,1)$

$$-2\pi \int_{-1}^{\delta} K_h(t)\, dt \leq 2\pi \int_{-1}^{\delta} K_h(t)P_n(t)\, dt \leq 2\pi \int_{-1}^{\delta} K_h(t)\, dt. \tag{15.21}$$

On the one hand the localization property (v) implies that there exists $h_1 \in (-1,1)$, such that the estimate $2\pi \int_{-1}^{\delta} K_h(t)P_n(t)\, dt \geq -\frac{\varepsilon}{2}$ is valid for all $h \in [h_1, 1)$.

On the other hand, $K_h^\wedge(0) = 1$ for all $h \in (-1,1)$, and the localization property implies the identity

$$\begin{aligned} \frac{1}{2\pi} &= \lim_{\substack{h\to 1,\\ h<1}} \int_{-1}^{1} K_h(t)\, dt \\ &= \lim_{\substack{h\to 1,\\ h<1}} \int_{-1}^{\delta} K_h(t)\, dt + \lim_{\substack{h\to 1,\\ h<1}} \int_{\delta}^{1} K_h(t)\, dt \\ &= \lim_{\substack{h\to 1,\\ h<1}} \int_{\delta}^{1} K_h(t)\, dt. \end{aligned} \tag{15.22}$$

Hence, there exists $h_2 \in (-1,1)$ such that $2\pi \int_{\delta}^{1} K_h(t)\, dt \geq \sqrt{1-\frac{\varepsilon}{2}}$ for all $h \in [h_2, 1)$. Equation (15.20) implies $1 - \varepsilon \leq K_h^\wedge(n) \leq 1$ for all $h \in [h_0, 1)$ where $h_0 = \max\{h_1, h_2\}$. □

Finally, it is worth mentioning that a non-negative scaling function $\{K_h\}_{h\in(-1,1)} \subset \mathrm{L}^1[-1,1]$, and $\{K_h\}_{h\in(-1,1)} \subset \mathrm{L}^2[-1,1]$, respectively, which fulfills the assumptions of Theorem 15.3, satisfies the estimate $\|K_h * F\|_{X(\mathbb{S}^2)} \leq \|F\|_{X(\mathbb{S}^2)}$ for all $h \in (-1,1)$ and for all $F \in X(\mathbb{S}^2)$, where $X(\mathbb{S}^2) = \mathrm{C}^{(0)}(\mathbb{S}^2)$, and $X(\mathbb{S}^2) = \mathrm{L}^2(\mathbb{S}^2)$, respectively.

15.2 Locally Supported Scaling Functions

Next approximate identities generated by so-called (non-negative) $[h,1]$-locally supported kernels $\{K_h\}_{h\in(-1,1)}$ are investigated (cf. W. Freeden,

K. Hesse [2002]). Due to their non-negativity and local support, i.e., $\operatorname{supp}(K_h) = [h, 1]$, such kernels generate an approximate identity, and therefore, are scaling functions. The approximation error $\|F - \mathcal{I}_h F\|_{\mathrm{C}^{(0)}(\mathbb{S}^2)}$ will be estimated for approximate identities generated by such non-negative $[h, 1]$-locally supported scaling functions $\{K_h\}_{h\in(-1,1)} \subset \mathrm{L}^1[-1, 1]$. Finally we shall present some examples, namely the Haar scaling function and smoothed Haar scaling functions. We note that all the examples are at least piecewise continuous and consequently, then also are L^2-scaling functions.

Our considerations start with the definition of $[h, 1]$-locally supported kernels.

Definition 15.4. *Let $\{K_h\}_{h\in(-1,1)}$ be a family of functions in $\mathrm{L}^1[-1, 1]$ or in $\mathrm{L}^2[-1, 1]$ that satisfy $K_h^\wedge(0) = 1$ for all $h \in (-1, 1)$ and $\operatorname{supp}(K_h) = [h, 1]$ for all $h \in (-1, 1)$. Then $\{K_h\}_{h\in(-1,1)}$ is called an $[h, 1]$-locally supported kernel.*

We mention the following important property of non-negative $[h, 1]$-locally supported kernels.

Theorem 15.5. *Suppose $\{K_h\}_{h\in(-1,1)} \subset \mathrm{L}^1[-1, 1]$ or $\{K_h\}_{h\in(-1,1)} \subset \mathrm{L}^2[-1, 1]$ is a non-negative $[h, 1]$-locally supported kernel. Then the family $\{\mathcal{I}_h\}_{h\in(-1,1)}$ as defined by(15.1) is an approximate identity corresponding to the $[h, 1]$-locally supported non-negative scaling function $\{K_h\}_{h\in(-1,1)}$.*

Proof. The family $\{K_h\}_{h\in(-1,1)}$ satisfies the assumptions of Theorem 15.3 and obeys the localization property because of $\operatorname{supp}(K_h) = [h, 1]$. Thus, the assertion immediately follows from Theorem 15.3. □

Lemma 15.6. *Let $\{K_h\}_{h\in(-1,1)} \subset \mathrm{L}^1[-1, 1]$ be a non-negative $[h, 1]$-locally supported scaling function. Suppose that $\{\mathcal{I}_h\}_{h\in(-1,1)}$, $\mathcal{I}_h : \mathrm{C}^{(0)}(\mathbb{S}^2) \to \mathrm{C}^{(0)}(\mathbb{S}^2)$, $F \mapsto \mathcal{I}_h F = K_h * F$ is the approximate identity generated by $\{K_h\}_{h\in(-1,1)}$. Assume that F is of class $\mathrm{C}^{(0)}(\mathbb{S}^2)$. Then,*

$$|F(\xi) - (\mathcal{I}_h F)(\xi)| \leq \left(2\pi \int_h^{\tilde{h}} K_h(t)\, dt\right) \sup_{\substack{\eta\in\mathbb{S}^2,\\ h\leq\xi\cdot\eta\leq\tilde{h}}} |F(\xi) - F(\eta)| + \left(2\pi \int_{\tilde{h}}^1 K_h(t)\, dt\right) \sup_{\substack{\eta\in\mathbb{S}^2,\\ \tilde{h}\leq\xi\cdot\eta\leq 1}} |F(\xi) - F(\eta)| \tag{15.23}$$

for every $\tilde{h} \in [h, 1]$ and for all $\xi \in \mathbb{S}^2$. If $F \in \mathrm{C}^{(0)}(\mathbb{S}^2)$ is additionally Lipschitz-continuous with Lipschitz-constant C_F, then

$$\|F - \mathcal{I}_h F\|_{\mathrm{C}^{(0)}(\mathbb{S}^2)} \leq 2\pi\sqrt{2}C_F \left(\int_h^{\tilde{h}} K_h(t)\, dt\ (1-h)^{1/2} + \int_{\tilde{h}}^1 K_h(t)\, dt\ (1-\tilde{h})^{1/2}\right) \tag{15.24}$$

for every $\tilde{h} \in [h, 1]$.

Proof. Let F be an arbitrary member of $\in \mathrm{C}^{(0)}(\mathbb{S}^2)$. For $\xi \in \mathbb{S}^2$ fixed but arbitrary,

$$\|K_h(\xi \cdot \eta)\|_{\mathrm{L}^1(\mathbb{S}^2)} = K_h^{\wedge}(0) = 1. \tag{15.25}$$

This implies

$$\begin{aligned} |F(\xi) - (\mathcal{I}_h F)(\xi)| &= \left| \int_{\substack{h \le \xi\cdot\eta \le 1,\\ \eta\in\mathbb{S}^2}} K_h(\xi \cdot \eta)(F(\xi) - F(\eta))\, dS(\eta) \right| \\ &\le \int_{\substack{h \le \xi\cdot\eta \le \tilde{h},\\ \eta\in\mathbb{S}^2}} K_h(\xi \cdot \eta)|F(\xi) - F(\eta)|\, dS(\eta) \\ &\quad + \int_{\substack{\tilde{h} \le \xi\cdot\eta \le 1,\\ \eta\in\mathbb{S}^2}} K_h(\xi \cdot \eta)|F(\xi) - F(\eta)|\, dS(\eta) \\ &\le \left(2\pi \int_h^{\tilde{h}} K_h(t)\, dt\right) \sup_{\substack{h \le \xi\cdot\eta \le \tilde{h},\\ \eta\in\mathbb{S}^2}} |F(\xi) - F(\eta)| \\ &\quad + \left(2\pi \int_{\tilde{h}}^1 K_h(t)\, dt\right) \sup_{\substack{\tilde{h} \le \xi\cdot\eta \le 1,\\ \eta\in\mathbb{S}^2}} |F(\xi) - F(\eta)| \end{aligned} \tag{15.26}$$

for every $\tilde{h} \in [h, 1]$ and for all $\xi \in \mathbb{S}^2$. This proves (15.23). To verify (15.24) note that $|\xi - \eta|^2 = (\xi - \eta) \cdot (\xi - \eta) = 2(1 - \xi \cdot \eta)$ for all $\xi, \eta \in \mathbb{S}^2$. If the function F is additionally Lipschitz-continuous, then

$$|F(\xi) - F(\eta)| \le C_F\sqrt{2}|1 - \xi \cdot \eta|^{1/2} \le C_F\sqrt{2}(1 - h)^{1/2} \tag{15.27}$$

for all $\xi, \eta \in \mathbb{S}^2$ satisfying $h \le \xi \cdot \eta \le 1$, and

$$|F(\xi) - F(\eta)| \le C_F\sqrt{2}(1 - \tilde{h})^{1/2} \tag{15.28}$$

for all $\xi, \eta \in \mathbb{S}^2$ satisfying $\tilde{h} \le \xi\cdot\eta \le 1$. Hence, (15.24) follows from (15.23). □

The reason for the introduction of the parameter $\tilde{h} \in [h, 1]$ in the previous lemma is the following: In many cases the kernel K_h will assume only very small values on a certain subset $[h, \tilde{h}]$ of $[h, 1]$ and grow extremely fast on $[\tilde{h}, 1]$. Therefore, the subdivision of the interval $[h, 1]$ into $[h, \tilde{h}]$ and $[\tilde{h}, 1]$, $\tilde{h} \ne 1, h$ chosen suitably, will yield much better results than the cases $\tilde{h} = 1$ or $\tilde{h} = h$ in estimate (15.24).

Haar Scaling Functions. In what follows, we present some examples of non-negative $[h, 1]$-locally supported scaling functions. The Legendre coefficients, the supremum norm, and Lipschitz-constants of the examples are calculated.

Example 15.7. *The* Haar scaling function $\{H_h\}_{h\in(-1,1)} \subset \mathrm{L}^2[-1,1]$, *where* $H_h : [-1,1] \to \mathbb{R}$, $t \mapsto H_h(t)$, *is given by*

$$H_h(t) = \begin{cases} 0, & t \in [-1,h), \\ \frac{1}{2\pi}\frac{1}{(1-h)}, & t \in [h,1]. \end{cases} \tag{15.29}$$

Obviously, $H_h(t) \geq 0$ for all $t \in [-1,1]$ and $H_h^\wedge(0) = 2\pi\|H_h\|_{\mathrm{L}^1[-1,1]} = 1$ are fulfilled. Thus $\{H_h\}_{h\in(-1,1)}$ generates an approximate identity in $\mathrm{C}^{(0)}(\mathbb{S}^2)$ (as well as in $\mathrm{L}^2(\mathbb{S}^2)$). Further properties of the Haar scaling function follow in the next example by specialization.

Example 15.8. *Let* $k \in \mathbb{N}_0$. *The* smoothed Haar scaling function $\{L_h^{(k)}\}_{h\in(-1,1)} \subset \mathrm{C}^{(k-1)}[-1,1]$ *is defined by* $L_h^{(k)} : [-1,1] \to \mathbb{R}$, $t \mapsto L_h^{(k)}(t)$,

$$L_h^{(k)}(t) = ((B_h^{(k)})^\wedge(0))^{-1} B_h^{(k)}(t) \tag{15.30}$$

with $B_h^{(k)}$ *from* (12.115).

By definition, $L_h^{(k)}$ is non-negative, has the support $[h,1]$, and satisfies $(L_h^{(k)})^\wedge(0) = 1$. Hence it is a non-negative $[h,1]$-locally supported scaling function. The function $L_h^{(0)}$, $h \in (-1,1)$, coincides with the Haar function H_h. The Legendre coefficients of $B_h^{(k)}$ and, hence, $L_h^{(k)}$, $h \in (-1,1)$, $k \in \mathbb{N}_0$, can be calculated recursively using Lemma 12.24. It can be shown that

$$|(L_h^{(k)})^\wedge(n)| = O\left((n(1-h))^{-(3/2)-k}\right) \tag{15.31}$$

for $n \to \infty$. The functions $L_h^{(k)}$, $h \in (-1,1)$, $k \in \mathbb{N}_0$, assume their maximum in $t = 1$. For $k > 2$ the Lipschitz-constant $C_h^{(k)}$ for $L_h^{(k)}$ can be chosen as the maximum of the first derivative, which is taken in the point $t = 1$, too. Thus, we obtain

$$\|L_h^{(k)}\|_{\mathrm{C}^{(0)}[-1,1]} = L_h^{(k)}(1) = \frac{1}{2\pi}\frac{(k+1)}{(1-h)}, \qquad k \in \mathbb{N}_0. \tag{15.32}$$

Furthermore,

$$C_h^{(k)} = \|(L_h^{(k)})'\|_{\mathrm{C}^{(0)}[-1,1]} = (L_h^{(k)})'(1) = \frac{1}{2\pi}\frac{k(k+1)}{(1-h)^2}, \qquad k \geq 2. \tag{15.33}$$

The function $L_h^{(0)}$ is constant on its support. Consequently, Equation (15.33) is also valid for $k = 0$ on $\mathrm{supp}(L_h^{(0)}) = [h,1]$. For $k = 1$ the function $L_h^{(k)}$ is continuous and piecewise linear, thus the Lipschitz-constant $C_h^{(1)}$ can be chosen as the first derivative of $L_h^{(1)}$ on $\mathrm{supp}(L_h^{(1)})$. Hence, Equation (15.33) is also true for $k = 1$.

15.3 Locally Supported Difference Wavelets

A scaling function $\{K_h\}_{h\in(-1,1)}$ generates an approximate identity, which provides nothing else than a sequence of low-pass filters. The so-called spherical difference wavelets are understood to describe the difference between such low-pass filters. In other words, spherical difference wavelets act as band-pass filters as the difference of two smoothings.

Definition 15.9. *Let $\{\mathcal{I}_h\}_{h\in(-1,1)}$ be an approximate identity in $\mathrm{C}^{(0)}(\mathbb{S}^2)$ or $\mathrm{L}^2(\mathbb{S}^2)$, generated by the scaling function $\{K_h\}_{h\in(-1,1)} \subset \mathrm{L}^1[-1,1]$, and $\{K_h\}_{h\in(-1,1)} \subset \mathrm{L}^2[-1,1]$, respectively. Let $\{h_j\}_{j\in\mathbb{N}_0} \subset (-1,1]$ be a strict monotonically increasing sequence with $\lim_{j\to\infty} h_j = 1$. Let there be given the sequence $\{\mathcal{T}_j\}_{j\in\mathbb{N}_0}$ of bounded linear operators*

$$\mathcal{T}_j : X(\mathbb{S}^2) \to X(\mathbb{S}^2), \quad F \mapsto \mathcal{T}_j F = \mathcal{I}_{h_j} F = K_{h_j} * F, \tag{15.34}$$

where $X(\mathbb{S}^2) = \mathrm{C}^{(0)}(\mathbb{S}^2)$ for $\{K_h\}_{h\in(-1,1)} \subset \mathrm{L}^1[-1,1]$, and $X(\mathbb{S}^2) = \mathrm{L}^2(\mathbb{S}^2)$ for $\{K_h\}_{h\in(-1,1)} \subset \mathrm{L}^2[-1,1]$, respectively. The family $\{\Psi_j\}_{j\in\mathbb{N}_0} \subset \mathrm{L}^1[-1,1]$, and $\{\Psi_j\}_{j\in\mathbb{N}_0} \subset \mathrm{L}^2[-1,1]$, respectively, given by

$$\Psi_j = K_{h_{j+1}} - K_{h_j} \tag{15.35}$$

is called spherical difference wavelet *corresponding to the scaling function $\{K_{h_j}\}_{j\in\mathbb{N}_0}$. Furthermore, a family $\{\mathcal{R}_j\}_{j\in\mathbb{N}_0}$ of bounded linear operators is defined by*

$$\mathcal{R}_j : X(\mathbb{S}^2) \to X(\mathbb{S}^2), \quad F \mapsto \mathcal{R}_j F = \Psi_j * F. \tag{15.36}$$

Note that the spherical difference wavelet $\{\Psi_j\}_{j\in\mathbb{N}_0}$ corresponding to a scaling function $\{K_h\}_{h\in(-1,1)}$ satisfies

$$\int_{-1}^{1} \Psi_j(t)\, dt = 0, \quad j \in \mathbb{N}_0, \tag{15.37}$$

because of the condition $K_h^\wedge(0) = 1$ for all $h \in (-1,1)$.

Definition 15.10. *Let $\{K_h\}_{h\in(-1,1)} \subset \mathrm{L}^1[-1,1]$ or $\{K_h\}_{h\in(-1,1)} \subset \mathrm{L}^2[-1,1]$ be an $[h,1]$-locally supported scaling function, and let $\{h_j\}_{j\in\mathbb{N}_0} \subset (-1,1]$ be a strict monotonically increasing sequence with $\lim_{j\to\infty} h_j = 1$. Then $\{\Psi_j\}_{j\in\mathbb{N}_0}$ is called an* $[h,1]$-locally supported spherical difference wavelet *corresponding to the $[h,1]$-locally supported scaling function $\{K_{h_j}\}_{j\in\mathbb{N}_0}$).*

Throughout this work, the spherical difference wavelets are usually called wavelets. The next theorem shows, that the low-pass filter $\mathcal{T}_J$, $J \in \mathbb{N}_0$, can be decomposed into a sum of the low-pass filter $\mathcal{T}_{J_0}$ and the band-pass filters $\mathcal{R}_j$, $j \in \{J_0, J_0+1, \ldots, J-1\}$ and, thus, be reconstructed as a sum of the latter.

Theorem 15.11. *Let the assumptions and the notation be as in Definition 15.9. Then,*

$$\lim_{J\to\infty} \|F - \mathcal{T}_J F\|_{X(\mathbb{S}^2)} = 0, \quad F \in X(\mathbb{S}^2), \tag{15.38}$$

where

$$\mathcal{T}_J F = \mathcal{T}_{J-1} F + \mathcal{R}_{J-1} F = \mathcal{T}_{J_0} F + \sum_{j=J_0}^{J-1} \mathcal{R}_j F \tag{15.39}$$

for all $J, J_0 \in \mathbb{N}_0$, $0 \le J_0 \le J-1$. Equation (15.39) is called a reconstruction of the approximation $\mathcal{T}_J F$. Particularly, we have

$$F = \mathcal{T}_{J_0} F + \sum_{j=J_0}^{\infty} \mathcal{R}_j F \tag{15.40}$$

in $X(\mathbb{S}^2)$-sense.

Proof. Equation (15.39) is an immediate consequence of the definitions of the operators $\mathcal{T}_j$ and $\mathcal{R}_j$:

$$\mathcal{T}_J F = K_{h_J} * F = (K_{h_{J-1}} + \Psi_{J-1}) * F = \mathcal{T}_{J-1} F + \mathcal{R}_{J-1} F. \tag{15.41}$$

This statement proves the first equality. The second equality analogously follows by repeating this process for $\mathcal{T}_{J-1}F, \ldots, \mathcal{T}_{J_0+1}F$. Equation (15.40) is a consequence of Equation (15.39) and the fact that $\mathcal{T}_J F$ converges to F in $X(\mathbb{S}^2)$ for $J \to \infty$. □

Haar Wavelets. Next the spherical difference wavelets are computed for the examples of $[h,1]$-locally supported scaling functions. In all the examples the same strict monotonically increasing sequence $\{h_j\}_{j\in\mathbb{N}_0}$, $h_j = 1 - 2^{-j}$, is used. Additionally (an estimate of) the supremum norm $\|\Psi_j\|_{\mathrm{C}^{(0)}[-1,1]}$ and Lipschitz-constants for Ψ_j are deduced explicitly.

Example 15.12. *For the sequence $\{h_j\}_{j\in\mathbb{N}_0}$, $h_j = 1 - 2^{-j}$, the* Haar scaling function $\{H_{h_j}\}_{j\in\mathbb{N}_0} \subset \mathrm{L}^2[-1,1]$ *is given by*

$$H_{h_j}(t) = \begin{cases} 0, & t \in [-1, 1-2^{-j}), \\ (2\pi)^{-1} 2^j, & t \in [1-2^{-j}, 1]. \end{cases} \tag{15.42}$$

The corresponding Haar wavelet *is the family $\{\Psi_j^H\}_{j\in\mathbb{N}_0} \subset \mathrm{L}^2[-1,1]$ of functions $\Psi_j^H : [-1,1] \to \mathbb{R}$, $t \mapsto \Psi_j^H(t)$, given by*

$$\Psi_j^H(t) = \begin{cases} 0, & t \in [-1, 1-2^{-j}), \\ -(2\pi)^{-1} 2^j, & t \in [1-2^{-j}, 1-2^{-(j+1)}), \\ (2\pi)^{-1} 2^j, & t \in [1-2^{-(j+1)}, 1]. \end{cases} \tag{15.43}$$

Obviously, we have

$$\|\Psi_j^H\|_{\mathrm{C}^{(0)}[-1,1]} = (2\pi)^{-1}2^j = \|H_{h_j}\|_{\mathrm{C}^{(0)}[-1,1]}. \tag{15.44}$$

The function Ψ_j^H is not Lipschitz-continuous on its support, because of the jump at $t = 1-2^{-(j+1)}$, but it is Lipschitz-continuous with Lipschitz-constant $\tilde{C}_j^{(0)} = 0$ on each subset of its support, where Ψ_j^H is constant, i.e., on the intervals $[1-2^{-j}, 1-2^{-(j+1)})$ and $[1-2^{-(j+1)}, 1]$.

The Haar wavelet additionally shows the property that

$$\int_{-1}^{1} \Psi_j^H(t)\Psi_k^H(t)\,dt = 0, \quad j,k \in \mathbb{N}_0,\ j \neq k, \tag{15.45}$$

$$\int_{-1}^{1} H_{h_0}(t)\Psi_j^H(t)\,dt = 0, \quad j \in \mathbb{N}_0. \tag{15.46}$$

This implies that $H_{h_0}(\xi\cdot\eta), \Psi_0^H(\xi\cdot\eta), \Psi_1^H(\xi\cdot\eta), \ldots$ as functions of $\eta \in \mathbb{S}^2$, where $\xi \in \mathbb{S}^2$ is kept fixed, are orthogonal. Equations (15.45) and (15.46) easily follow from the definition $\Psi_j^H(t) = H_{h_{j+1}}(t) - H_{h_j}(t)$, $t \in [-1,1]$, and calculation of the integrals.

Example 15.13. *For the sequence* $\{h_j\}_{j\in\mathbb{N}_0}$, $h_j = 1-2^{-j}$, *the* smoothed Haar scaling functions $\{L_{h_j}^{(k)}\}_{j\in\mathbb{N}_0} \subset \mathrm{C}^{(k-1)}[-1,1]$, $k \in \mathbb{N}_0$, *are defined by*

$$L_{h_j}^{(k)}(t) = \begin{cases} 0, & t \in [-1, 1-2^{-j}), \\ (2\pi)^{-1}(k+1)2^{j(k+1)}(t-1+2^{-j})^k, & t \in [1-2^{-j}, 1]. \end{cases} \tag{15.47}$$

The corresponding smoothed Haar wavelets *are the families* $\{\Psi_j^{(k)}\}_{j\in\mathbb{N}_0} \subset \mathrm{C}^{(k-1)}[-1,1]$, $k \in \mathbb{N}_0$, *of functions* $\Psi_j^{(k)} : [-1,1] \to \mathbb{R}$, $t \mapsto \Psi_j^{(k)}(t)$, *given by*

$$\Psi_j^{(k)}(t) \tag{15.48}$$
$$= \begin{cases} 0, & t \in [-1, 1-2^{-j}), \\ -(2\pi)^{-1}(k+1)2^{j(k+1)}(t-1+2^{-j})^k, & t \in [1-2^{-j}, 1-2^{-(j+1)}), \\ (2\pi)^{-1}(k+1)2^{j(k+1)}\left[2^{k+1}(t-1+2^{-(j+1)})^k \right. \\ \qquad\qquad \left. -(t-1+2^{-j})^k\right], & t \in [1-2^{-(j+1)}, 1]. \end{cases}$$

The supremum norm and a Lipschitz-constant $C_j^{(k)}$ of $L_{h_j}^{(k)}$ are given by

$$\|L_{h_j}^{(k)}\|_{\mathrm{C}^{(0)}[-1,1]} = \frac{1}{2\pi}(k+1)2^j, \quad k \in \mathbb{N}_0, \tag{15.49}$$

and

$$C_j^{(k)} = \frac{1}{2\pi}k(k+1)2^{2j}, \quad k \in \mathbb{N}. \tag{15.50}$$

An estimate for the supremum norm of $\Psi_j^{(k)}$ and a Lipschitz-constant $\tilde{C}_j^{(k)}$ for $\Psi_j^{(k)}$, $k \in \mathbb{N}$, can be obtained by using that $\Psi_j^{(k)}$ and its (piecewise defined) first derivative $(\Psi_j^{(k)})'$ are each the difference of two non-negative functions. Thus, $\|\Psi_j^{(k)}\|_{C^{(0)}[-1,1]}$, and $\|(\Psi_j^{(k)})'\|_{C^{(0)}[-1,1]}$ respectively, are not larger than the maximum of the supremum norms of the two non-negative functions. Hence,

$$\|\Psi_j^{(k)}\|_{C^{(0)}[-1,1]} \leq \|L_{h_{j+1}}^{(k)}\|_{C^{(0)}[-1,1]} = \frac{1}{2\pi}(k+1)2^{j+1}, \tag{15.51}$$

We obtain

$$\tilde{C}_j^{(k)} = \|(L_{h_{j+1}}^{(k)})'\|_{C^{(0)}[-1,1]} = \left(L_{h_{j+1}}^{(k)}\right)'(1) = \frac{1}{2\pi}k(k+1)2^{2j+2} = C_{j+1}^{(k)}. \tag{15.52}$$

Note that $\{\Psi_j^{(0)}\}_{j\in\mathbb{N}_0}$ is the Haar wavelet of Example 15.12.

15.4 Integration for Large Equidistributed Data

Next we focus our attention on numerical aspects of approximate identities in $C^{(0)}(\mathbb{S}^2)$ and their wavelet reconstruction. The approximation $\mathcal{T}_J F$ of a function $F \in C^{(0)}(\mathbb{S}^2)$ and its reconstruction from $\mathcal{T}_{J_0}F$ by means of $\mathcal{R}_j F$, $j = J_0, J_0+1, \ldots, J-1$, are discussed from numerical point of view. An approximate integration rule for the evaluation of the convolution integrals $(\mathcal{T}_j F)(\xi)$, and $(\mathcal{R}_j F)(\xi)$, respectively, will be derived, where the integrands are supposed to be at least continuous. This is no serious restriction as the functions K_h can be chosen continuous and the approximated functions F can be assumed to be continuous in the practical applications we are interested in.

As mentioned above, the theory as proposed here is meant, e.g., for satellite problems, where huge amounts of data are given, i.e., F is known at a very large set of points on the (orbital) sphere, and where the data are equidistributed on the sphere (in the sense of H. Weyl [1916]) (cf. Chapter 14, in particular Definition 14.4). There are various integration rules for the numerical evaluation of the convolution integrals $(K_{h_j} * F)(\xi)$ and $(\Psi_j * F)(\xi)$, but most of these integration techniques are not applicable to problems, where millions of data occur (for example, when satellite data come into play). In such cases summing up the values of the integrand in the data points with identical integration weights seems to be a simple and suitable way to cope with the large amount of data.

First, some terminology connected with point sets and partitions will be introduced. After that the approximate integration rule, based on Weyl's law of equidistribution, (cf. H. Weyl [1916]) will be presented including some error estimates. Finally, the numerical computation of $(\mathcal{T}_J F)(\xi)$ reconstructed as the sum of $(\mathcal{T}_{J_0}F)(\xi)$ and $(\mathcal{R}_{J_0}F)(\xi), (\mathcal{R}_{J_0+1}F)(\xi), \ldots, (\mathcal{R}_{J-1}F)(\xi)$

(see (15.39)) will be outlined and the approximation error will be estimated. In order to discuss the efficiency of our approximate integration rule we will determine the number of elementary operations that are needed. An elementary operation consists of one multiplication and one addition.

In what follows (cf. W. Freeden, K. Hesse [2002]) we remember the concept of equidistribution known from Chapter 14 involving Lipschitz continuous functions. Our purpose is to combine the idea of multiscale recovery of signals with the essentials of approximate integration based on equidistributed data sets. Our considerations start with

Definition 15.14. *Let $\mathcal{U} \subset \mathbb{S}^2$ be a measurable subset of $\mathbb{S}^2$. Suppose that $X_N^{\mathcal{U}} = \{\eta_1^N, \eta_2^N, \ldots, \eta_N^N\} \subset \mathcal{U}$ and $\tilde{X}_N^{\mathcal{U}} = \{\tilde{\eta}_1^N, \tilde{\eta}_2^N, \ldots, \tilde{\eta}_N^N\} \subset \mathcal{U}$ are two subsets of N points in the set $\mathcal{U}$. $X_N^{\mathcal{U}}$ and $\tilde{X}_N^{\mathcal{U}}$ are called* equivalent, *if there exists a permutation p of $\{1, 2, \ldots, N\}$ such that $\tilde{\eta}_{p(k)}^N = \eta_k^N$ for all $k = 1, \ldots, N$.*

An equivalence class with respect to this equivalence relation is called an ensemble of $\mathcal{U}$. *An ensemble of $\mathcal{U}$ is identified with its elements (it is a set of N points in $\mathcal{U}$ without any ordering).*

We now generalize Definition 14.1 and Definition 14.2 to subsets $\mathcal{U} \subset \mathbb{S}^2$.

Definition 15.15. *Let $X_N^{\mathcal{U}} = \{\eta_1^N, \eta_2^N, \ldots, \eta_N^N\}$ be an ensemble of a measurable subset $\mathcal{U} \subset \mathbb{S}^2$. $\mathcal{P}(X_N^{\mathcal{U}}) = \{\mathcal{U}(\eta_1^N), \mathcal{U}(\eta_2^N), \ldots, \mathcal{U}(\eta_N^N)\}$ is called an* associated partition of $\mathcal{U}$ to the ensemble $X_N^{\mathcal{U}}$, *if the following four conditions are satisfied:*

(i) any subset $\mathcal{U}(\eta_k^N)$ of $\mathcal{U}$ is measurable and

$$\|\mathcal{U}(\eta_k^N)\| = \int_{\mathcal{U}(\eta_k^N)} dS > 0, \quad k = 1, 2, \ldots, N, \tag{15.53}$$

(ii) $\eta_k^N \in \mathcal{U}(\eta_k^N)$ for all $k = 1, \ldots, N$,

(iii) $\bigcup_{k=1}^N \mathcal{U}(\eta_k^N) = \mathcal{U}$, and

(iv) $\mathcal{U}(\eta_i^N) \cap \mathcal{U}(\eta_k^N) = \emptyset$ for all $i, k = 1, 2, \ldots, N$, $i \neq k$.

Definition 15.16. *An associated partition $\mathcal{P}(X_N^{\mathcal{U}}) = \{\mathcal{U}(\eta_1^N), \ldots, \mathcal{U}(\eta_N^N)\}$ of $\mathcal{U}$ to the ensemble $X_N^{\mathcal{U}}$ is called* equidistributed, *if*

$$\|\mathcal{U}(\eta_k^N)\| = \int_{\mathcal{U}(\eta_k^N)} dS(\eta) = \frac{\|\mathcal{U}\|}{N}, \quad k \in \{1, \ldots, N\}. \tag{15.54}$$

The set of all equidistributed partitions $\mathcal{P}(X_N^{\mathcal{U}})$ of $\mathcal{U}$ to $X_N^{\mathcal{U}}$ is denoted by $\Pi(X_N^{\mathcal{U}})$. The partition size of the ensemble $X_N^{\mathcal{U}}$ *of $\mathcal{U}$ is defined by*

$$\sigma(X_N^{\mathcal{U}}) = \inf \left\{ \max_{k=1,\ldots,N} \sup_{\xi \in \mathcal{U}(\eta_k^N)} |\xi - \eta_k^N| : \mathcal{P}(X_N^{\mathcal{U}}) \in \Pi(X_N^{\mathcal{U}}) \right\}. \tag{15.55}$$

Theorem 15.17. *Let $\mathcal{U}$ be a measurable subset of $\mathbb{S}^2$, let $X_N^{\mathcal{U}} = \{\eta_1^N, \ldots, \eta_N^N\} \subset \mathcal{U}$ be an ensemble of $\mathcal{U}$, and let $\mathcal{P}(X_N^{\mathcal{U}}) = \{\mathcal{U}(\eta_1^N), \ldots, \mathcal{U}(\eta_N^N)\}$ be an arbitrary equidistributed, associated partition of $\mathcal{U}$ to the ensemble $X_N^{\mathcal{U}}$. Suppose that F is a continuous, real valued function on $\mathcal{U}$ which is bounded on $\mathcal{U}$. Then*

$$\left| \frac{1}{\|\mathcal{U}\|} \int_{\mathcal{U}} F(\eta) dS(\eta) - \frac{1}{N} \sum_{k=1}^{N} F(\eta_k^N) \right| \tag{15.56}$$

$$\leq \inf \left\{ \max_{k=1,\ldots,N} \sup_{\eta \in \mathcal{U}(\eta_k^N)} |F(\eta) - F(\eta_k^N)| : \mathcal{P}(X_N^{\mathcal{U}}) \in \Pi(X_N^{\mathcal{U}}) \right\}.$$

Proof. The definition of an equidistributed associated partition of $\mathcal{U}$ to $X_N^{\mathcal{U}}$ implies that

$$\frac{1}{\|\mathcal{U}\|} \int_{\mathcal{U}} F(\eta)\, dS(\eta) - \frac{1}{N} \sum_{k=1}^{N} F(\eta_k^N) = \frac{1}{\|\mathcal{U}\|} \sum_{k=1}^{N} \int_{\mathcal{U}(\eta_k^N)} \Big(F(\eta) - F(\eta_k^N)\Big) dS(\eta). \tag{15.57}$$

The estimate (15.56) follows from (15.57) by estimating the integrals by the supremum of the integrand multiplied with $\|\mathcal{U}(\eta_k^N)\| = \|\mathcal{U}\|/N$, and by taking the infimum over all equidistributed associated partitions $\mathcal{P}(X_N^{\mathcal{U}})$ of $\mathcal{U}$ to $X_N^{\mathcal{U}}$. □

Corollary 15.18. *Let $\mathcal{U} \subset \mathbb{S}^2$ be a measurable subset. Suppose that $\{N_j\}_{j\in\mathbb{N}_0} \subset \mathbb{N}$ is a strictly monotonically increasing sequence. Assume that $\{X_{N_j}^{\mathcal{U}}\}_{j\in\mathbb{N}_0}$ is a sequence of ensembles $X_{N_j}^{\mathcal{U}} = \{\eta_1^{N_j}, \eta_2^{N_j}, \ldots, \eta_{N_j}^{N_j}\}$ of $\mathcal{U}$ satisfying $\lim_{j\to\infty} \sigma(X_{N_j}^{\mathcal{U}}) = 0$. Then, for $F \in \mathrm{C}^{(0)}(\overline{\mathcal{U}})$,*

$$\lim_{j\to\infty} \left| \frac{1}{\|\mathcal{U}\|} \int_{\mathcal{U}} F(\eta)\, dS(\eta) - \frac{1}{N_j} \sum_{k=1}^{N_j} F(\eta_k^{N_j}) \right| = 0. \tag{15.58}$$

If $F \in \mathrm{C}^{(0)}(\overline{\mathcal{U}})$ additionally is Lipschitz-continuous with Lipschitz-constant C_F, then

$$\left| \frac{1}{\|\mathcal{U}\|} \int_{\mathcal{U}} F(\eta)\, dS(\eta) - \frac{1}{N_j} \sum_{k=1}^{N_j} F(\eta_k^{N_j}) \right| \leq C_F \sigma(X_{N_j}^{\mathcal{U}}). \tag{15.59}$$

Proof. For a Lipschitz-continuous function F the statement is a consequence of the estimate (15.56) in Theorem 15.17. If F is assumed to be continuous only, Equation (15.58) follows from the estimate (15.56) in Theorem 15.17 and the fact that F is uniformly continuous on $\overline{\mathcal{U}}$. □

Corollary 15.18 shows that the mean value of the function values of a continuous function $F \in \mathrm{C}^{(0)}(\overline{\mathcal{U}})$ in the points of a suitable ensemble $X_N^{\mathcal{U}}$ of $\mathcal{U} \subset \mathbb{S}^2$ yields an approximation of the integral $\frac{1}{\|\mathcal{U}\|}\int_{\mathcal{U}} F(\eta)\, dS(\eta)$, i.e., an *approximate integration rule for continuous integrands*
$F \in \mathrm{C}^{(0)}(\overline{\mathcal{U}})$

$$\frac{1}{\|\mathcal{U}\|}\int_{\mathcal{U}} F(\eta)\, dS(\eta) \approx \frac{1}{N}\sum_{k=1}^{N} F(\eta_k^N). \tag{15.60}$$

This approximate integration rule is useful for very large and equidistributed point sets, because all the data can be used and the number of elementary operations is just N for a dataset $\{(\eta_1^N, F(\eta_1^N)), (\eta_2^N, F(\eta_2^N)), \ldots, (\eta_N^N, F(\eta_N^N))\}$.

Now we are prepared to apply this approximate integration rule for the evaluation of the approximation $\mathcal{T}_J F$, $F \in \mathrm{C}^{(0)}(\mathbb{S}^2)$.

Theorem 15.19. *Assume that $\{N_j\}_{j\in\mathbb{N}_0} \subset \mathbb{N}$ is a strict monotonically increasing sequence. Let $\{X_{N_j}^{\mathbb{S}^2}\}_{j\in\mathbb{N}_0}$, $X_{N_j}^{\mathbb{S}^2} = \{\eta_1^{N_j}, \eta_2^{N_j}, \ldots, \eta_{N_j}^{N_j}\}$, be a sequence of ensembles of $\mathbb{S}^2$ that is hierarchical, i.e., $\eta_i^{N_j} = \eta_i^{N_{j+1}}$ for all $i = 1, 2, \ldots, N_j$ and for all $j \in \mathbb{N}_0$, and that satisfies $\lim_{j\to\infty}\sigma(X_{N_j}^{\mathbb{S}^2}) = 0$. Let $\{K_h\}_{h\in(-1,1)} \subset \mathrm{C}^{(0)}[-1,1]$ be a continuous scaling function. Suppose $\{h_j\}_{j\in\mathbb{N}_0} \subset (-1,1]$ is a strict monotonically increasing sequence with $\lim_{j\to\infty} h_j = 1$. Let $\mathcal{T}_j$ and $\mathcal{R}_j$, $j \in \mathbb{N}_0$, denote the operators defined in Definition 15.9 corresponding to the sequence $\{h_j\}_{j\in\mathbb{N}_0}$ and the scaling function $\{K_h\}_{h\in(-1,1)}$. Set $\mathcal{U}_j(\xi) = \overline{\{\eta \in \mathbb{S}^2 \mid K_{h_j}(\xi\cdot\eta) \neq 0\}}$, $\tilde{\mathcal{U}}_j(\xi) = \overline{\{\eta \in \mathbb{S}^2 \mid \Psi_j(\xi\cdot\eta) \neq 0\}}$, $j \in \mathbb{N}_0$, and define $\hat{X}_{N_j}^{\mathcal{U}} = X_{N_j}^{\mathbb{S}^2} \cap \mathcal{U}$, $j \in \mathbb{N}_0$, for an arbitrary measurable subset $\mathcal{U}$ of $\mathbb{S}^2$ and regard $\hat{X}_{N_j}^{\mathcal{U}}$ as an ensemble of $\mathcal{U}$. Suppose F is of class in $\mathrm{C}^{(0)}(\mathbb{S}^2)$.*

Then $\mathcal{T}_j F$ of (15.34), and $\mathcal{R}_j F$ of (15.36), $j \in \mathbb{N}_0$, respectively, can be evaluated approximately by the formulas

$$(\mathcal{T}_j F)(\xi) \approx \frac{\|\mathcal{U}_j(\xi)\|}{\#(\hat{X}_{N_j}^{\mathcal{U}_j(\xi)})} \sum_{\zeta\in\hat{X}_{N_j}^{\mathcal{U}_j(\xi)}} F(\zeta) K_{h_j}(\xi\cdot\zeta) \tag{15.61}$$

and

$$(\mathcal{R}_j F)(\xi) \approx \frac{\|\tilde{\mathcal{U}}_j(\xi)\|}{\#(\hat{X}_{N_{j+1}}^{\tilde{\mathcal{U}}_j(\xi)})} \sum_{\zeta\in\hat{X}_{N_{j+1}}^{\tilde{\mathcal{U}}_j(\xi)}} F(\zeta)\Psi_j(\xi\cdot\zeta) \tag{15.62}$$

for all $\xi \in \mathbb{S}^2$. An arbitrary prescribed accuracy of the right-hand side of (15.61) and (15.62) can be obtained uniformly in $\xi \in \mathbb{S}^2$ by choosing the sequence $\{X_{N_j}^{\mathbb{S}^2}\}_{j\in\mathbb{N}_0}$ in an appropriate way. The function $F \in \mathrm{C}^{(0)}(\mathbb{S}^2)$ can *be approximated by* $F_J : \mathbb{S}^2 \to \mathbb{R}$, $\xi \mapsto F_J(\xi)$, *as follows:*

$$F_J(\xi) = \frac{\|\mathcal{U}_{J_0}(\xi)\|}{\#(\hat{X}_{N_{J_0}}^{\mathcal{U}_{J_0}(\xi)})} \sum_{\zeta \in \hat{X}_{N_{J_0}}^{\mathcal{U}_{J_0}(\xi)}} F(\zeta) K_{h_{J_0}}(\xi \cdot \zeta) + \sum_{j=J_0}^{J-1} \frac{\|\tilde{\mathcal{U}}_j(\xi)\|}{\#(\hat{X}_{N_{j+1}}^{\tilde{\mathcal{U}}_j(\xi)})} \sum_{\zeta \in \hat{X}_{N_{j+1}}^{\tilde{\mathcal{U}}_j(\xi)}} F(\zeta) \Psi_j(\xi \cdot \zeta), \quad J > J_0. \tag{15.63}$$

If F *and the functions* K_h *are additionally Lipschitz-continuous, the error of the approximation* F_J *can be estimated as follows:*

$$\begin{aligned} &\|F - F_J\|_{\mathrm{C}^{(0)}(\mathbb{S}^2)} \\ &\leq \|F - \mathcal{T}_J F\|_{\mathrm{C}^{(0)}(\mathbb{S}^2)} \\ &\quad + \|\mathcal{U}_{J_0}(\xi)\| \left(C_{J_0} \|F\|_{\mathrm{C}^{(0)}(\mathbb{S}^2)} + C_F \|K_{h_{J_0}}\|_{\mathrm{C}^{(0)}[-1,1]} \right) \sigma\left(\hat{X}_{N_{J_0}}^{\mathcal{U}_{J_0}(\xi)} \right) \\ &\quad + \sum_{j=J_0}^{J-1} \|\tilde{\mathcal{U}}_j(\xi)\| \left(\tilde{C}_j \|F\|_{\mathrm{C}^{(0)}(\mathbb{S}^2)} + C_F \|\Psi_j\|_{\mathrm{C}^{(0)}[-1,1]} \right) \sigma\left(\hat{X}_{N_{j+1}}^{\tilde{\mathcal{U}}_j(\xi)} \right), \end{aligned} \tag{15.64}$$

where C_F, C_{J_0}, *and* $\tilde{C}_j$, $j = J_0, J_0+1, \ldots, J-1$, *respectively, are Lipschitz-constants for the functions* F, $K_{h_{J_0}}$, *and* Ψ_j, $j = J_0, J_0+1, \ldots, J-1$, *respectively.*

Proof of Theorem 15.19. The approximation formulas (15.61) and (15.62) are immediate consequences of the numerical integration rule (15.60).

That every prescribed accuracy of the right-hand side of (15.61) can be obtained uniformly in $\xi \in \mathbb{S}^2$ by choosing the sequence $\{X_{N_j}^{\mathbb{S}^2}\}_{j \in \mathbb{N}_0}$ in an appropriate way follows from the uniform continuity of F and K_{h_j} and from the estimate

$$\begin{aligned} |K_{h_j}(\xi \cdot \eta) F(\eta) - K_{h_j}(\xi \cdot \zeta) F(\zeta)| &\leq \|K_{h_j}\|_{\mathrm{C}^{(0)}[-1,1]} |F(\eta) - F(\zeta)| \\ &\quad + \|F\|_{\mathrm{C}^{(0)}(\mathbb{S}^2)} |K_{h_j}(\xi \cdot \eta) - K_{h_j}(\xi \cdot \zeta)| \end{aligned} \tag{15.65}$$

for all $\xi, \eta, \zeta \in \mathbb{S}^2$. The argumentation for (15.62) is similar. These formulas combined with Theorem 15.11 imply that $F_J(\xi)$ is a suitable approximation for $(\mathcal{T}_J F)(\xi)$ and consequently for $F(\xi)$, $\xi \in \mathbb{S}^2$, provided the sequence $\{X_{N_j}^{\mathbb{S}^2}\}_{j \in \mathbb{N}_0}$ is chosen in adaptation to the growth of the functions K_{h_j}.

In order to prove the estimate of the approximation error in case F, K_h, and Ψ_j, $j \in \mathbb{N}_0$ are Lipschitz-continuous we first determine Lipschitz-constants for the integrands in $(\mathcal{T}_j F)(\xi)$ and $(\mathcal{R}_j F)(\xi)$, $\xi \in \mathbb{S}^2$ fixed but arbitrary:

$$\begin{aligned} &|F(\eta) K_{h_j}(\xi \cdot \eta) - F(\zeta) K_{h_j}(\xi \cdot \zeta)| \\ &\qquad \leq \left(C_j \|F\|_{\mathrm{C}^{(0)}(\mathbb{S}^2)} + C_F \|K_{h_j}\|_{\mathrm{C}^{(0)}[-1,1]} \right) |\eta - \zeta|, \end{aligned} \tag{15.66}$$

$$\begin{aligned} &|F(\eta) \Psi_j(\xi \cdot \eta) - F(\zeta) \Psi_j(\xi \cdot \zeta)| \\ &\qquad \leq \left(\tilde{C}_j \|F\|_{\mathrm{C}^{(0)}(\mathbb{S}^2)} + C_F \|\Psi_j\|_{\mathrm{C}^{(0)}[-1,1]} \right) |\eta - \zeta| \end{aligned} \tag{15.67}$$

for all $\eta, \zeta \in \mathbb{S}^2$, where C_F, C_j, and $\tilde{C}_j$ respectively, are the Lipschitz-constants of the functions F, K_{h_j}, and Ψ_j respectively. The triangle inequality yields

$$|F(\xi) - F_J(\xi)| \leq |F(\xi) - (\mathcal{T}_J F)(\xi)| + |(\mathcal{T}_J F)(\xi) - F_J(\xi)|. \tag{15.68}$$

The second term can be estimated with the help of the triangle inequality, the reconstruction theorem (Theorem 15.11), and Corollary 15.18.

$$\begin{aligned}
&|(\mathcal{T}_J F)(\xi) - F_J(\xi)| \qquad (15.69)\\
&\leq \left| (\mathcal{T}_{J_0} F)(\xi) - \frac{\|\mathcal{U}_{J_0}(\xi)\|}{\#(\hat{X}_{N_{J_0}}^{\mathcal{U}_{J_0}(\xi)})} \sum_{\zeta \in \hat{X}_{N_{J_0}}^{\mathcal{U}_{J_0}(\xi)}} F(\zeta) K_{h_{J_0}}(\xi \cdot \zeta) \right| \\
&\quad + \sum_{j=J_0}^{J-1} \left| (\mathcal{R}_j F)(\xi) - \frac{\|\tilde{\mathcal{U}}_j(\xi)\|}{\#(\hat{X}_{N_{j+1}}^{\tilde{\mathcal{U}}_j(\xi)})} \sum_{\zeta \in \hat{X}_{N_{j+1}}^{\tilde{\mathcal{U}}_j(\xi)}} F(\zeta) \Psi_j(\xi \cdot \zeta) \right| \\
&\leq \|\mathcal{U}_{J_0}(\xi)\| \left(C_{J_0} \|F\|_{\mathrm{C}^{(0)}(\mathbb{S}^2)} + C_F \|K_{h_{J_0}}\|_{\mathrm{C}^{(0)}[-1,1]} \right) \sigma(\hat{X}_{N_{J_0}}^{\mathcal{U}_{J_0}(\xi)}) \\
&\quad + \sum_{j=J_0}^{J-1} \|\tilde{\mathcal{U}}_j(\xi)\| \left(\tilde{C}_j \|F\|_{\mathrm{C}^{(0)}(\mathbb{S}^2)} + C_F \|\Psi_j\|_{\mathrm{C}^{(0)}[-1,1]} \right) \sigma(\hat{X}_{N_{j+1}}^{\tilde{\mathcal{U}}_j(\xi)}).
\end{aligned}$$

□

The estimate (15.64) in Theorem 15.19 implies that an approximation F_J with ε-accuracy can be obtained, if J is chosen so large, that

$$\|F - \mathcal{T}_J F\|_{\mathrm{C}^{(0)}(\mathbb{S}^2)} \leq \frac{\varepsilon}{2} \tag{15.70}$$

and if the hierarchical sequence $\{X_{N_j}^{\mathbb{S}^2}\}_{j \in \mathbb{N}_0}$ is chosen such that the $\sigma(X_{N_j}^{\mathbb{S}^2})$, $j = J_0, \ldots, J$, get so small, that each of the other summands in (15.64) becomes smaller than $\varepsilon/(2(J - J_0 + 1))$.

The same can be done in the case of continuous (but not Lipschitz-continuous) integrands, but in this case the theorem yields no quantitative error estimate. The partition sizes $\sigma(X_{N_j}^{\mathbb{S}^2})$, $j = J_0, \ldots, J$, that are needed for a prescribed accuracy depend on the growth behavior of the kernel and on the undulations of the integrand. Generally it can be said that, the larger the index j the faster the growth of the kernel K_{h_j}, and consequently a smaller partition size, i.e., more discretization points, is needed.

In practical applications a function F will be known on a large (in the sense of Weyl) equidistributed point set $X_{N_J}^{\mathbb{S}^2}$. The continuous scaling function $\{K_h\}_{h \in (-1,1)}$, as well as the sequence $\{h_j\}_{j \in \mathbb{N}_0}$ and the hierarchical point sets $X_{N_j}^{\mathbb{S}^2} \subset X_{N_J}^{\mathbb{S}^2}$, $j = J_0, \ldots, J-1$, have then to be chosen in adaptation to the problem and the given dataset. It is desirable to choose the numbers N_j so that they satisfy the relations $N_j \approx N_{j+1}/2$ for $J_0 \le j \le J-1$, because in this case the approximation $\mathcal{T}_J F$ of F can be reconstructed with at most $2N_J$ elementary operations.

One difficulty in this approach is the choice of the hierarchical point sets $X_{N_j}^{\mathbb{S}^2}$, $j = J_0, \ldots, J-1$, from the given set of points $X_{N_J}^{\mathbb{S}^2}$, where the function F is known.

Improvements for Locally Supported Kernels. All results from above can be applied to our examples of approximate identities generated by non-negative $[h,1]$-locally supported scaling functions. In this case the number of needed elementary operations is even much smaller because of the local support of the integrand in the convolution integrals. This enables the calculation of F_J in Theorem 15.19 for $\{h_j\}_{j\in\mathbb{N}_0}$ chosen as $h_j = 1 - 2^{-j}$ with approximately $2^{-J_0+1}N_J$ elementary operations (for N_J given data points) without hierarchical point sets, as it will be explained in detail below.

In the examples of non-negative $[h,1]$-locally supported scaling functions the sequence $\{h_j\}_{j\in\mathbb{N}_0}$, given by $h_j = 1 - 2^j$, was chosen. Thus, $\mathcal{U}_j(\xi) = \Gamma_{1-h_j}(\xi)$, and $\tilde{\mathcal{U}}_j(\xi) = \Gamma_{1-h_j}(\xi)$ in Theorem 15.19, where $\Gamma_{1-h}(\xi) = \{\eta \in \mathbb{S}^2 : 1 - \xi\cdot\eta < 1-h\}$ is the spherical cap with center ξ and radius $1-h$.

Obviously $\|\Gamma_{1-h}(\xi)\| = 2\pi(1-h)$, hence, $\#(\hat{X}_{N_j}^{\mathcal{U}_j(\xi)}) \approx 2^{-j-1}\#(X_{N_j}^{\mathbb{S}^2})$, and $\#(\hat{X}_{N_{j+1}}^{\tilde{\mathcal{U}}_j(\xi)}) \approx 2^{-j-1}\#(X_{N_{j+1}}^{\mathbb{S}^2})$, $j = J_0, \ldots, J$. This means that the total amount of elementary operations in Theorem 15.19 is approximately

$$2^{-J_0-1}\#(X_{N_{J_0}}^{\mathbb{S}^2}) + \sum_{j=J_0}^{J-1} 2^{-j-1}\#(X_{N_{j+1}}^{\mathbb{S}^2}) = 2^{-J_0-1}N_{J_0} + \sum_{j=J_0}^{J-1} 2^{-j-1}N_{j+1} \tag{15.71}$$

in case of an $[h,1]$-locally supported scaling function and the sequence $\{h_j\}_{j\in\mathbb{N}_0}$ given by $h_j = 1 - 2^j$.

To use this advantage in practical applications it is necessary to find a quick strategy to decide whether a point $\eta \in \mathbb{S}^2$ lies in the spherical cap $\Gamma_{1-h}(\xi)$. We will not elaborate on this point here further and refer to V. Michel [2011] for similar problems.

Using such a strategy the approximation F_J given in Theorem 15.19 can also be calculated with approximately $2^{-J_0+1}N_J$ elementary operations without the use of hierarchical point sets for $[h,1]$-locally supported kernels and the sequence $\{h_j\}_{j\in\mathbb{N}_0}$, $h_j = 1 - 2^j$: To achieve this, the different sets $X_{N_j}^{\mathbb{S}^2}$ in Theorem 15.19 are all replaced by $X_{N_J}^{\mathbb{S}^2}$, and the formula (15.71) yields at most $2^{-J_0+1}N_J$ elementary operations. This means on the one hand that some of the convolution integrals are calculated with unnecessary accuracy, but on the other hand the problem of the choice of the hierarchical point sets does no longer occur. Additionally the error estimate gets much simpler.

15.5 Error Discussion

The estimates of the approximation error in case of hierarchical point sets given in Theorem 15.19 are applied to the approximate identities generated by the smoothed Haar scaling functions of Example 15.8 (with $k \geq 1$). The case of the Haar scaling function (cf. Example 15.7) is treated separately because of the discontinuity.

In this section we make the following assumptions: Let $\{N_j\}_{j\in\mathbb{N}} \subset \mathbb{N}$ be a strictly monotonically increasing sequence, and let $\{X_{N_j}^{\mathbb{S}^2}\}_{j\in\mathbb{N}}$, where $X_{N_j}^{\mathbb{S}^2} = \{\eta_1^{N_j}, \eta_2^{N_j}, \ldots, \eta_{N_j}^{N_j}\}$, be a hierarchical sequence of ensembles of $\mathbb{S}^2$, such that $\lim_{j\to\infty} \sigma(X_{N_j}^{\mathbb{S}^2}) = 0$. Moreover, the notation from Theorem 15.19 will be adopted.

Lemma 15.20. *Let $\{L_h^{(k)}\}_{h\in(-1,1)}$, $k \in \mathbb{N}$, be the smoothed Haar scaling functions introduced in Example 15.8, and define $\{h_j\}_{j\in\mathbb{N}}$ as usual by $h_j = 1 - 2^{-j}$. Assume that $F \in \mathrm{C}^{(0)}(\mathbb{S}^2)$ is a Lipschitz-continuous function with Lipschitz-constant C_F. Then the error of the approximation $F_J^{(k)}$ of F defined in Theorem 15.19 by (15.63) can be estimated by*

$$\begin{aligned} \|F - F_J^{(k)}\|_{\mathrm{C}^{(0)}(\mathbb{S}^2)} \leq \Bigg(&\sum_{j=J_0-1}^{J-1} (k+1)\left(2^{j+2}k\|F\|_{\mathrm{C}^{(0)}(\mathbb{S}^2)} + 2C_F\right)\sigma(\hat{X}_{N_{j+1}}^{\tilde{\mathcal{U}}_j(\xi)})\Bigg) \\ &+ \sqrt{2}C_F \min_{\tilde{h}\in[1-2^{-J},1]} \Big(2^{J(k+(1/2))}(\tilde{h} - (1-2^{-J}))^{k+1} \\ &+ \left(1 - 2^{J(k+1)}(\tilde{h} - (1-2^{-J}))^{k+1}\right)(1-\tilde{h})^{1/2}\Big), \end{aligned} \tag{15.72}$$

where $\tilde{\mathcal{U}}_j(\xi) = \Gamma_{1-h_j}(\xi)$ for $j = J_0, \ldots, J-1$, $\tilde{\mathcal{U}}_{J_0-1}(\xi) = \Gamma_{1-h_{J_0}}(\xi)$, and $\Gamma_{1-h}(\xi) = \{\eta \in \mathbb{S}^2 : \xi \cdot \eta < 1-h\}$.

The second summand in the estimate of the approximation error is the estimate of $\|F - \mathcal{T}_J F\|_{\mathrm{C}^{(0)}(\mathbb{S}^2)}$ developed in Lemma 15.6.

Lemma 15.21. *Let $\{H_h\}_{h\in(-1,1)}$ be the Haar scaling function introduced in Example 15.7, and define $\{h_j\}_{j\in\mathbb{N}}$ by $h_j = 1 - 2^{-j}$. Assume that $F \in \mathrm{C}^{(0)}(\mathbb{S}^2)$ is a Lipschitz-continuous function with Lipschitz-constant C_F.*

Then the error of the approximation $F_J^H : \mathbb{S}^2 \to \mathbb{R}$, $\xi \mapsto F_J^H(\xi)$ of F, defined by

$$F_J^H(\xi) = \frac{1}{\#(\hat{X}_{N_{J_0}}^{\tilde{\mathcal{U}}_{J_0}(\xi)})} \sum_{\zeta \in \hat{X}_{N_{J_0}}^{\tilde{\mathcal{U}}_{J_0}(\xi)}} F(\zeta) \tag{15.73}$$

$$+ \frac{1}{2} \sum_{j=J_0}^{J-1} \left(\frac{1}{\#(\hat{X}_{N_{j+1}}^{\tilde{\mathcal{U}}_{j+1}(\xi)})} \sum_{\zeta \in \hat{X}_{N_{j+1}}^{\tilde{\mathcal{U}}_{j+1}(\xi)}} F(\zeta) \right.$$

$$\left. - \frac{1}{\#(\hat{X}_{N_{j+1}}^{\tilde{\mathcal{U}}_j(\xi)\setminus\tilde{\mathcal{U}}_{j+1}(\xi)})} \sum_{\zeta \in \hat{X}_{N_{j+1}}^{\tilde{\mathcal{U}}_j(\xi)\setminus\tilde{\mathcal{U}}_{j+1}(\xi)}} F(\zeta) \right),$$

can be estimated by

$$\|F - F_J^H\|_{\mathrm{C}^{(0)}(\mathbb{S}^2)} \tag{15.74}$$

$$\leq C_F \left(\sigma(\hat{X}_{N_{J_0}}^{\tilde{\mathcal{U}}_{J_0}(\xi)}) + \frac{1}{2} \sum_{j=J_0}^{J-1} \left(\sigma(\hat{X}_{N_{j+1}}^{\tilde{\mathcal{U}}_{j+1}(\xi)}) + \sigma(\hat{X}_{N_{j+1}}^{\tilde{\mathcal{U}}_j(\xi)\setminus\tilde{\mathcal{U}}_{j+1}(\xi)}) \right) \right)$$

$$+ \min_{\tilde{h} \in [1-2^{-J},1]} \sqrt{2} C_F \left(2^{J/2}(\tilde{h} - (1 - 2^{-J})) + 2^J (1 - \tilde{h})^{3/2} \right),$$

where $\tilde{\mathcal{U}}_j(\xi) = \Gamma_{1-h_j}(\xi)$ for $j = J_0, \ldots, J$, and the spherical caps $\Gamma_{1-h}(\xi) = \{\eta \in \mathbb{S}^2 : \xi \cdot \eta < 1 - h\}$.

Proof. $F_J^H(\xi)$ can also be written in the following form

$$F_J^H(\xi) = \frac{\|\tilde{\mathcal{U}}_{J_0}(\xi)\|}{\#(\hat{X}_{N_{J_0}}^{\tilde{\mathcal{U}}_{J_0}(\xi)})} \sum_{\zeta \in \hat{X}_{N_{J_0}}^{\tilde{\mathcal{U}}_{J_0}(\xi)}} F(\zeta) H_{h_{J_0}}(\xi \cdot \zeta) \tag{15.75}$$

$$+ \sum_{j=J_0}^{J-1} \left(\frac{\|\tilde{\mathcal{U}}_{j+1}(\xi)\|}{\#(\hat{X}_{N_{j+1}}^{\tilde{\mathcal{U}}_{j+1}(\xi)})} \sum_{\zeta \in \hat{X}_{N_{j+1}}^{\tilde{\mathcal{U}}_{j+1}(\xi)}} F(\zeta) \Psi_j^H(\xi \cdot \zeta) \right.$$

$$\left. + \frac{\|\tilde{\mathcal{U}}_j(\xi) \setminus \tilde{\mathcal{U}}_{j+1}(\xi)\|}{\#(\hat{X}_{N_{j+1}}^{\tilde{\mathcal{U}}_j(\xi)\setminus\tilde{\mathcal{U}}_{j+1}(\xi)})} \sum_{\zeta \in \hat{X}_{N_{j+1}}^{\tilde{\mathcal{U}}_j(\xi)\setminus\tilde{\mathcal{U}}_{j+1}(\xi)}} F(\zeta) \Psi_j^H(\xi \cdot \zeta) \right).$$

Hence, F_J^H is a suitable approximation for F according to Theorem 15.17. To verify the error estimate we proceed similarly as in the proof of Theorem 15.19. By the triangle inequality we find

$$|F(\xi) - F_J^H(\xi)| \leq |F(\xi) - (\mathcal{T}_J^H F)(\xi)| + |(\mathcal{T}_J^H F)(\xi) - F_J^H(\xi)|. \tag{15.76}$$

The estimate of the first term is a consequence of Lemma 15.6. Applying the triangle inequality to the second term yields

$$\begin{aligned}
&|(\mathcal{T}_J^H F)(\xi) - F_J^H(\xi)| \qquad (15.77)\\
&\leq \left|(\mathcal{T}_{J_0}^H F)(\xi) - \frac{\|\tilde{\mathcal{U}}_{J_0}(\xi)\|}{\#(\hat{X}_{N_{J_0}}^{\tilde{\mathcal{U}}_{J_0}(\xi)})} \sum_{\zeta \in \hat{X}_{N_{J_0}}^{\tilde{\mathcal{U}}_{J_0}(\xi)}} F(\zeta) H_{h_{J_0}}(\xi\cdot\zeta)\right|\\
&\quad + \sum_{j=J_0}^{J-1} \left(\left| \int_{\tilde{\mathcal{U}}_{j+1}(\xi)} F(\eta)\Psi_j^H(\xi\cdot\eta)\, dS(\eta) \right.\right.\\
&\qquad \left. - \frac{\|\tilde{\mathcal{U}}_{j+1}(\xi)\|}{\#(\hat{X}_{N_{j+1}}^{\tilde{\mathcal{U}}_{j+1}(\xi)})} \sum_{\zeta \in \hat{X}_{N_{j+1}}^{\tilde{\mathcal{U}}_{j+1}(\xi)}} F(\zeta)\Psi_j^H(\xi\cdot\zeta) \right|\\
&\qquad + \left| \int_{\tilde{\mathcal{U}}_j(\xi)\setminus\tilde{\mathcal{U}}_{j+1}(\xi)} F(\eta)\Psi_j^H(\xi\cdot\eta)\, dS(\eta) \right.\\
&\qquad \left.\left. - \frac{\|\tilde{\mathcal{U}}_j(\xi)\setminus\tilde{\mathcal{U}}_{j+1}(\xi)\|}{\#(\hat{X}_{N_{j+1}}^{\tilde{\mathcal{U}}_j(\xi)\setminus\tilde{\mathcal{U}}_{j+1}(\xi)})} \sum_{\zeta \in \hat{X}_{N_{j+1}}^{\tilde{\mathcal{U}}_j(\xi)\setminus\tilde{\mathcal{U}}_{j+1}(\xi)}} F(\zeta)\Psi_j^H(\xi\cdot\zeta) \right|\right).
\end{aligned}$$

Note that $\|\tilde{\mathcal{U}}_j(\xi) \setminus \tilde{\mathcal{U}}_{j+1}(\xi)\| = \|\tilde{\mathcal{U}}_{j+1}(\xi)\|$ because of the definition of the sequence $\{h_j\}_{j\in\mathbb{N}}$ and the fact that the function H_{J_0} as well as the wavelet functions Ψ_j^H are constant in each term. Hence these constants can be put outside of the absolute value and each term can be estimated by applying Corollary 15.18.

$$\begin{aligned}
&|(\mathcal{T}_J^H F)(\xi) - F_J^H(\xi)| \qquad (15.78)\\
&\leq \frac{\|\tilde{\mathcal{U}}_{J_0}(\xi)\|}{2\pi(1-h_{J_0})} C_F \sigma(\hat{X}_{N_{J_0}}^{\tilde{\mathcal{U}}_{J_0}(\xi)})\\
&\quad + \sum_{j=J_0}^{J-1} \frac{\|\tilde{\mathcal{U}}_{j+1}(\xi)\|}{2\pi(1-h_j)} C_F \left(\sigma(\hat{X}_{N_{j+1}}^{\tilde{\mathcal{U}}_{j+1}(\xi)}) + \sigma(\hat{X}_{N_{j+1}}^{\tilde{\mathcal{U}}_j(\xi)\setminus\tilde{\mathcal{U}}_{j+1}(\xi)}) \right)
\end{aligned}$$

Inserting $\|\tilde{\mathcal{U}}_j(\xi)\| = 2\pi(1-h_j) = 2\pi 2^{-j}$ for $j = J_0, \ldots, J$ yields the desired assertion. □

Finally, it should be pointed out that the use of the concept of equidistribution in multiscale approximation by locally supported difference wavelets is appropriate for applications in geosciences, whenever large and dense data sets are available (for instance, in satellite technology, W. Freeden, K. Hesse [2002]). Another advantage is the simple structure of the approximation mechanisms. The disadvantage is the slow convergence of the integrationtechniques.

Remark 15.22. *W. Freeden, K. Hesse [2002] use the rotation-invariant, locally supported multiscale apparatus in its discrete realization based on equidistributed data sets as presented in this chapter to study "downward continuation" of SST and/or SGG satellite data in Earth's gravitational determination. In doing so, the spacelimited character of the wavelets enables a fast calculation of the gravitational potential at satellite orbit, so that the gravitational potential at the (spherical) Earth's surface can be obtained by truncated singular value or Tikhonov regularization.*

Part III

Integration on 2D-Surfaces

Chapter 16

Surface Integration

In this chapter we deal with the integration over regular surfaces Σ, i.e., boundary surfaces $\Sigma = \partial\mathcal{G}$ of regular regions $\mathcal{G} \subset \mathbb{R}^3$. Integration over regular surfaces (manifolds) has seen a renewed interest during the last years (see, e.g., I. Pesenson [2015] and the references therein). Different to integration over spheres is the problem that the system of eigenfunctions of the Beltrami operator on a general surface Σ is not explicitly known, which seriously limits the applicability in numerical practice. In addition, the determination of the null space of the Beltrami operator causes difficulties.

Accordingly, in our approach, we propose two variants where we are led to surface quantities such as "distortion function" or "mean curvature" that principally are unknown as well, at least when only discrete data of Σ are available. Nevertheless, if the discrete information about the geometry of the surface is sufficiently dense, the differential operators constituting the distortion function and the mean curvature may be replaced by difference operators, so that approximate values become obtainable. For example, when working with the actual Earth's surface based on data from global navigation satellite systems (GNSS), we are able to derive mean curvature values on the Earth's surface.

The layout of this chapter is as follows: In a first step, since most of the geoscientifically relevant surfaces (e.g., sphere, ellipsoidal, telluroidal, etc.) are star-shaped, we are concerned with transformation procedures back to the sphere (cf. Section 16.1), making use of the distortion function of the surface element. This approach closely follows W. Freeden, H. Kersten [1980] and takes advantage from the spherical integration methods already discussed in Part II. In a second step (cf. Section 16.2) we turn to not-necessarily star-shaped surfaces. Their study demands the knowledge of differential geometric tools, such as mean curvature. In both cases we end up with integration formulas (see Section 16.3) enabling various types of approximate integration rules, for instance, best approximate integration (cf. Section 16.4) and equidistributions in the sense of Weyl (cf. Section 16.5).

16.1 Transformation Back to the Sphere

Analogously to the spherical theory in Part II we start from the following definition.

Definition 16.1. *A bounded region $\mathcal{G} \subset \mathbb{R}^q$, $q \geq 2$, is called* regular, *if it allows the* Gauss theorem

$$\int_{\mathcal{G}} \nabla \cdot f(x)\, dx = \int_{\partial\mathcal{G}} f(x) \cdot \nu(x)\, dS(x) \tag{16.1}$$

for all continuously differentiable vector fields f on $\overline{\mathcal{G}}$, $\overline{\mathcal{G}} = \mathcal{G} \cup \partial\mathcal{G}$ (throughout this work, ν is the (unit) normal field on $\partial\mathcal{G}$ directed to the exterior $\mathcal{G}^c$ of $\overline{\mathcal{G}}$). The volume of a regular region $\mathcal{G} \subset \mathbb{R}^q$ *is given by* $\|\mathcal{G}\| = \int_{\mathcal{G}} dx$, *where $dx = dx_1 \ldots dx_q$ is the volume element.*

Examples of regular regions are cube, other polyhedra showing a non-smooth boundary $\partial\mathcal{G}$ with vertices and edges, ball, ellipsoid, geoid, (real) Earth's body, etc., showing a smooth boundary $\partial\mathcal{G}$ as discussed in Part III.

Definition 16.2. *The boundary $\partial\mathcal{G}$ of a regular region $\mathcal{G} \subset \mathbb{R}^q$ is called a* regular surface. *The* area of the boundary $\partial\mathcal{G}$ of a regular region $\mathcal{G} \subset \mathbb{R}^q$ *is given by* $\|\partial\mathcal{G}\| = \int_{\partial\mathcal{G}} dS(x)$.

Throughout this chapter, we restrict ourselves to the three-dimensional case, i.e., $q = 3$, which is of particular geoscientific relevance.

Definition 16.3. *Any boundary surface $\Sigma = \partial\mathcal{G}$ of a regular region $\mathcal{G} \subset \mathbb{R}^3$ consisting of points $x = r(\xi)\xi$, $\xi \in \mathbb{S}^2$, with $r(\xi)$ being a positive and (sufficiently) smooth function is called a* star-shaped surface*:*

$$\Sigma = \{x \in \mathbb{R}^3 : x = r(\xi)\xi,\ \xi \in \mathbb{S}^2\}. \tag{16.2}$$

Obviously, every point $x \in \Sigma$ of a star-shaped surface is the unique radial projection of $x \in \Sigma$ to a point $\xi \in \mathbb{S}^2$. In terms of polar coordinates

$$x = r(\xi)\xi, \quad \xi = \begin{pmatrix} \sqrt{1-t^2}\,\cos\varphi \\ \sqrt{1-t^2}\,\sin\varphi \\ t \end{pmatrix}, \quad t \in [-1,1],\ \varphi \in [0, 2\pi), \tag{16.3}$$

the surface element $dS(x)$ may be expressed as follows:

$$dS(x) = \left| \frac{\partial}{\partial\varphi} x \wedge \frac{\partial}{\partial t} x \right| d\varphi\, dt = \left| \frac{\partial}{\partial\varphi} x \wedge \frac{\partial}{\partial t} x \right| dS(\xi). \tag{16.4}$$

An elementary calculation yields

$$\begin{aligned} dS(x) &= \left| \frac{\partial}{\partial\varphi} r(\xi)\xi \wedge \frac{\partial}{\partial t} r(\xi)\xi \right| dS(\xi) \qquad (16.5) \\ &= \sqrt{\frac{1}{1-t^2} r(\xi)^2 \left(\frac{\partial}{\partial\varphi} r(\xi)\right)^2 + (1-t^2) r(\xi)^2 \left(\frac{\partial}{\partial t} r(\xi)\right)^2 + r(\xi)^4}\, dS(\xi). \end{aligned}$$

Looking at (16.5) we are led to introduce the *distortion function* $c(\xi)$ given by

$$c(\xi) = \sqrt{\frac{1}{1-t^2} r(\xi)^2 \left(\frac{\partial}{\partial \varphi} r(\xi) \right)^2 + (1-t^2) r(\xi)^2 \left(\frac{\partial}{\partial t} r(\xi) \right)^2 + r(\xi)^4} \quad (16.6)$$

so that

$$dS(x) = c(\xi)\ dS(\xi). \quad (16.7)$$

This result offers the opportunity to transform an integral over a star-shaped Σ to an integral over the unit sphere $\mathbb{S}^2$. More concretely, we are allowed to write

$$\int_\Sigma F(x)\ dS(x) = \frac{1}{4\pi} \int_{\mathbb{S}^2} \widetilde{F}(\xi)\ dS(\xi), \quad (16.8)$$

where $\widetilde{F}$ is given by

$$\widetilde{F}(\xi) = 4\pi\, c(\xi) F(r(\xi)\xi) = \widetilde{c}(x)\ F(x) \quad (16.9)$$

with

$$\widetilde{c}(x) = 4\pi c \left(\frac{x}{|x|} \right), \ \xi = \frac{x}{|x|}, \quad x \in \Sigma. \quad (16.10)$$

Summarizing our preparatory considerations we are able to deduce the following integral formula.

Theorem 16.4 (Integral Formula for Starshaped Surfaces). *Let F be of class $\mathrm{C}^{(2)}(\Sigma)$. Let $\widetilde{F}$ is given by (16.9). Then we have*

$$\widetilde{F}(\xi) = \int_\Sigma F(y)\ dS(y) + \int_{\mathbb{S}^2} G(\Delta^*; \xi \cdot \eta) \Delta_\eta^* \widetilde{F}(\eta)\ dS(\eta). \quad (16.11)$$

The integral formula (16.11) enables us to apply all spherical integration rules developed earlier in Part II. For example, we are able to investigate the difference between an integral over a star-shaped surface $\int_\Sigma F(y)\ dS(y)$ and the mean value $\frac{1}{N} \sum_{k=1}^N \widetilde{F}(y_k^N)$ based on data points $y_k^N \in \Sigma$, thereby taking advantage of the discrepancy method.

Lemma 16.5. *Let $Y_N = \{y_1^N, \ldots, y_N^N\}$ be a point set on Σ, $y_k^N = r(\eta_k^N)\eta_k^N$, and $X_N = \{\eta_1^N, \ldots, \eta_N^N\}$, $\eta_k^N \in \mathbb{S}^2$, $k = 1, \ldots, N$. Assume that the function F is of class $\mathrm{C}^{(2)}(\Sigma)$. Then the following estimate is valid:*

$$|\mathcal{E}_{Y_N} F| \leq D(X_N; \Delta^*)\ V(\widetilde{F}), \quad (16.12)$$

where $D(X_N; \Delta^*)$ is the discrepancy of X_N (cf. (14.21) for the spherical case), i.e.,

$$D(X_N; \Delta^*) = \left(\frac{1}{N^2} \sum_{k=1}^N \sum_{l=1}^N G((\Delta^*)^2; \eta_k^N \cdot \eta_l^N) \right)^{1/2}, \quad (16.13)$$

and the variance $V(\widetilde{F})$ *(cf.* (14.23)*) is given by*

$$V(\widetilde{F}) = \left(\int_{\mathbb{S}^2} |\Delta_\eta^* \widetilde{F}(\eta)|^2 \, dS(\eta) \right)^{1/2}. \tag{16.14}$$

In a straightforward way we are able to verify the following equivalence.

Lemma 16.6. *Let* $\{Y_N\}$ *be a sequence of systems* $Y_N = \{y_1^N, \ldots, y_N^N\}$ *on* Σ *with* $y_{k_1}^N = r(\eta_k^N)\eta_k^N$, $X_N = \{\eta_1^N, \ldots, \eta_N^N\}$, $\eta_k^N \in \mathbb{S}^2$, $k = 1, \ldots, N$. *Then the following statements are equivalent:*

(i) For all $F \in \mathrm{C}^{(2)}(\mathbb{S}^2)$ *and* $\widetilde{F}$ *given by (16.9) holds the limit*

$$\int_\Sigma F(y) \, dS(y) = \lim_{N \to \infty} \frac{1}{N} \sum_{k=1}^{N} \widetilde{F}(\eta_k^N). \tag{16.15}$$

(ii) The discrepancy of X_N *fulfills the limit*

$$\lim_{N \to \infty} D(X_N; \Delta^*) = \lim_{N \to \infty} \left(\frac{1}{N^2} \sum_{k=1}^{N} \sum_{l=1}^{N} G((\Delta^*)^2; \eta_k^N \cdot \eta_l^N) \right)^{1/2} = 0. \tag{16.16}$$

For numerical purposes, the "distortion function" c in (16.5) has to be known or numerically accessible (e.g., in case of suitable discrete information about the surface Σ).

16.2 Use of Differential Geometric Means

Let $\mathcal{G}$ be a regular region in $\mathbb{R}^3$. Throughout the next sections of this chapter we additionally assume that the boundary $\Sigma = \partial\mathcal{G}$ is (locally) twice continuously differentiable, in order to avoid technicalities with edges and corners.

Tangential-Normal Coordinate System. There exist (universal) constants $M, \delta > 0$ such that, for every $x \in \Sigma$, the set $\Sigma_\delta(x) = \Sigma \cap \mathbb{B}_\delta(x)$, where $\mathbb{B}_\delta(x)$ is a ball of radius δ around x, can be represented in a local (tangential-normal) coordinate system associated with x (cf. Figure 16.1) in the form

$$u_3 = F(u), \quad u = (u_1, u_2)^T \in \mathcal{U}, \tag{16.17}$$

where $\mathcal{U}$ is a subset of the tangent plane of Σ at x (such that the ε^3-axis coincides with the unit normal vector $\nu(x)$ in $x \in \Sigma$ pointing into the outer space $\mathcal{G}^c = \mathbb{R}^3 \setminus \overline{\mathcal{G}}$) and $F : \mathcal{U} \to \mathbb{R}$ is a twice continuously differentiable function satisfying

$$|F(u)| \leq M \; |u|^2, \tag{16.18}$$

$$|\nabla F(u)| \leq M \; |u|, \tag{16.19}$$

for all $u = (u_1, u_2)^T \in \mathcal{U}$.

In the tangential-normal system the point $x \in \Sigma$ represents the origin, i.e., it has the coordinates $(0,0,0)^T$ so that $F(0,0) = 0$. Furthermore, by rotating the coordinate system the plane $u_3 = 0$ can be transferred into the tangential plane at x such that the first partial derivatives of F vanish at the origin. Via Taylor's formula this observation explains the estimates (16.18) and (16.19) (for more details see, e.g., W. Freeden, C. Gerhards [2013]).

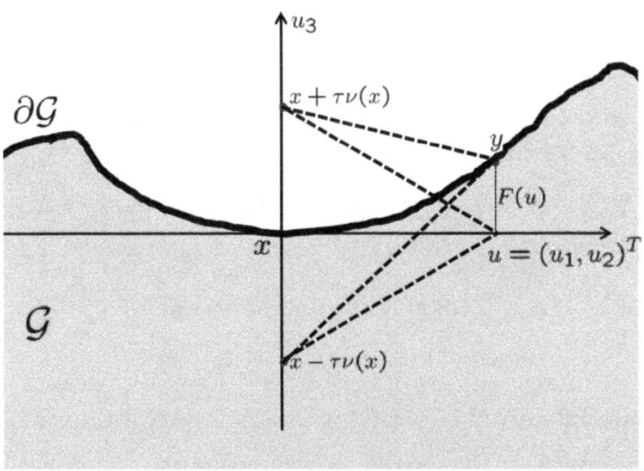

FIGURE 16.1: Sectional representation of a tangential-normal system (with $\nu(x) = \varepsilon^3$).

By virtue the representation (16.17) we are able to associate a point $y \in \Sigma$ to $y = (u_1, u_2, F(u_1, u_2))^T$ with x as origin of the local tangential-normal system. In the nomenclature of the local tangential-normal system (see Figure 16.1) we are able to write

$$|x \pm \tau\nu(x) - y| = \left(u_1^2 + u_2^2 + (\tau \mp F(u_1, u_2))^2\right)^{1/2}. \tag{16.20}$$

Furthermore, we find

$$\begin{aligned} |\nu(x) \cdot (y - x)| &= \left|\varepsilon^3 \cdot \left(u_1\varepsilon^1 + u_2\varepsilon^2 + F(u_1, u_2)\varepsilon^3\right)\right| = |F(u_1, u_2)| \qquad (16.21) \\ &\leq M(u_1^2 + u_2^2) \\ &\leq M(u_1^2 + u_2^2 + F(u_1, u_2)^2) \\ &\leq M|y - x|^2. \end{aligned}$$

Observing the representation

$$\nu(y) = \frac{1}{(1+|\nabla F(u_1,u_2)|^2)^{1/2}} \left(\varepsilon^3 - F_{|1}(u_1,u_2)\varepsilon^1 - F_{|2}(u_1,u_2)\varepsilon^2\right), \quad (16.22)$$

where we use the notation $\nabla F = (\frac{\partial F}{\partial u_1}, \frac{\partial F}{\partial u_2})^T = (F_{|1}, F_{|2})^T$, we find

$$\begin{aligned} &|\nu(x) - \nu(y)| \\ &= \frac{\left|F_{|1}(u_1,u_2)\varepsilon^1 + F_{|2}(u_1,u_2)\varepsilon^2 + \left((1+|\nabla F(u_1,u_2)|^2)^{1/2} - 1\right)\varepsilon^3\right|}{(1+|\nabla F(u_1,u_2)|^2)^{1/2}} \\ &\leq \frac{\left(|\nabla F(u_1,u_2)|^2 + \frac{1}{4}|\nabla F(u_1,u_2)|^4\right)^{1/2}}{(1+|\nabla F(u_1,u_2)|^2)^{1/2}} \\ &\leq |\nabla F(u_1,u_2)| \\ &\leq M\, |x-y|, \end{aligned} \quad (16.23)$$

where we have used the well-known inequality $(1+t)^{1/2} - 1 \leq \frac{t}{2}$ for $t > 0$. Summarizing we are able to formulate

Lemma 16.7. *Let $\mathcal{G}$ be a regular region. Then, for every $x \in \Sigma = \partial\mathcal{G}$ and all $y \in \Sigma_\delta(x)$,*

$$|\nu(x) - \nu(y)| \leq M|x-y|, \quad (16.24)$$

$$|\nu(x) \cdot (x-y)| \leq M|x-y|^2. \quad (16.25)$$

Next we are interested in verifying an additional estimate.

Lemma 16.8. *For every $x \in \Sigma$ and all $y \in \Sigma_\delta(x)$ we have*

$$|\nu(y) \cdot (y-x)| \leq 2M|x-y|^2. \quad (16.26)$$

Proof. For $x \in \Sigma$ and $y \in \Sigma_\delta(x)$ we find by virtue of (16.23) and Lemma 16.7

$$\begin{aligned} |\nu(y)\cdot(y-x)| &= |(\nu(y) - \nu(x))\cdot(y-x) + \nu(x)\cdot(y-x)| \quad (16.27) \\ &\leq |\nu(y) - \nu(x)|\; |y-x| + M|x-y|^2 \\ &\leq 2M|x-y|^2. \end{aligned}$$

This yields the desired result of Lemma 16.8. □

Fundamental Tensors and Curvature. The following concept is borrowed from C. Müller [1969, 1995]. We have already seen that a tangential-normal coordinate system may be introduced in any point x_0 on the surface Σ. Moreover, any $y \in \Sigma_\delta(x_0) = \Sigma \cap \mathbb{B}_\delta(x_0)$ is representable in the form

$$y = u_1\varepsilon^1 + u_2\varepsilon^2 + u_3\varepsilon^3 = u_1\varepsilon^1 + u_2\varepsilon^2 + F(u_1,u_2)\varepsilon^3, \quad (16.28)$$

so that any point $x \in \mathbb{R}^3$ in a neighborhood of $\Sigma_\delta(x_0)$ may be written as follows:

$$x = y + u_3\nu(y). \tag{16.29}$$

The normal $\nu(y)$ is given by

$$\nu(y) = \frac{y_{|1} \wedge y_{|2}}{|y_{|1} \wedge y_{|2}|}, \tag{16.30}$$

where $y_{|i} = \frac{\partial y}{\partial u_i}$, $i = 1, 2$. An easy calculation (see also (16.22)) shows that

$$\nu(y) = \frac{1}{(1 + F_{|1}^2 + F_{|2}^2)^{1/2}}(-F_{|1}\varepsilon^1 - F_{|2}\varepsilon^2 + \varepsilon^3). \tag{16.31}$$

The so-called *fundamental tensors* are defined by

$$g_{ik} = \frac{\partial x}{\partial u_i} \cdot \frac{\partial x}{\partial u_k} = x_{|i} \cdot x_{|k}, \quad i, k \in \{1, 2, 3\}, \tag{16.32}$$

and

$$\gamma_{\alpha\beta} = \frac{\partial y}{\partial u_\alpha} \cdot \frac{\partial y}{\partial u_\beta} = y_{|\alpha} \cdot y_{|\beta}, \quad \alpha, \beta \in \{1, 2\}. \tag{16.33}$$

In accordance with our approach we are able to represent the derivative $\nu_{|\alpha}$ as linear combination of $y_{|\beta}$, $\beta = 1, 2$, as follows

$$\nu_{|\alpha} = -\sum_{\kappa=1}^{2} L_\alpha^\kappa y_{|\kappa}, \quad \alpha = 1, 2. \tag{16.34}$$

The dual components $\gamma^{\alpha\beta}$ to $\gamma_{\alpha\beta}$ are uniquely determined by the requirement

$$\sum_{\kappa=1}^{2} \gamma^{\alpha\kappa}\gamma_{\kappa\beta} = \delta_{\alpha,\beta}, \quad \alpha, \beta \in \{1, 2\}. \tag{16.35}$$

Note that $\gamma\gamma^{11} = \gamma_{22}$, $\gamma\gamma^{12} = \gamma\gamma^{21} = -\gamma_{12}$, and $\gamma\gamma^{22} = \gamma_{11}$, where γ is the determinant of $\gamma_{\alpha\beta}$, i.e., $\gamma = \det(\gamma_{\alpha\beta})$. Analogously, we are able to introduce g^{ik} by letting

$$\sum_{l=1}^{3} g^{kl} g_{li} = \delta_{i,k}, \quad i, k \in \{1, 2, 3\}. \tag{16.36}$$

We let

$$\sum_{\kappa=1}^{2} \gamma_{\alpha\kappa} L_\beta^\kappa = L_{\alpha\beta}, \quad \alpha, \beta \in \{1, 2\}. \tag{16.37}$$

A straightforward calculation allows us to compare $g_{\alpha\beta}$ with $\gamma_{\alpha\beta}$, $\alpha, \beta \in \{1, 2\}$, as follows

$$g_{\alpha\beta} = \gamma_{\alpha\beta} - 2u_3 L_{\alpha\beta} + u_3^2 \sum_{\kappa=1}^{2} L_\alpha^\kappa L_{\kappa\beta}, \quad \alpha, \beta \in \{1, 2\}. \tag{16.38}$$

The following intrinsic curvature settings are in use:

$$H = \frac{1}{2} \sum_{\kappa=1}^{2} L_{\kappa}^{\kappa} \tag{16.39}$$

is called the *mean curvature*, while the *Gaussian curvature* is defined by

$$K = \det(L_{\alpha}^{\beta}). \tag{16.40}$$

H and K are invariants of the surface under coordinate transformations (cf. C. Müller [1969]), and we have

$$\sum_{\kappa=1}^{2} L_{\alpha}^{\kappa} L_{\kappa\beta} = 2HL_{\alpha\beta} - K\gamma_{\alpha\beta}, \quad \alpha, \beta \in \{1,2\}. \tag{16.41}$$

In fact, the last identity is valid in the origin x_0 of the coordinate system under consideration. We denote equality in the origin by $\overset{\circ}{=}$ and obtain

$$\begin{array}{lll} L_{11} \overset{\circ}{=} k_1, & L_{22} \overset{\circ}{=} k_2, & L_{12} \overset{\circ}{=} 0, \\ \gamma_{\alpha\beta} \overset{\circ}{=} \delta_{\alpha,\beta}, & 2H \overset{\circ}{=} k_1 + k_2, & K \overset{\circ}{=} k_1 k_2. \end{array} \tag{16.42}$$

Now it should be remarked that any point $x_0 \in \Sigma$ can be chosen as origin of a tangential-normal coordinate system in the same way. This means (cf. C. Müller [1969, 1995]) that the identity (16.41) holds true for every point on the surface Σ.

As a consequence we obtain the following results

$$g_{\alpha\beta} = (1 - u_3^2 K)\gamma_{\alpha\beta} - 2u_3(1 - u_3 H)L_{\alpha\beta}, \quad \alpha, \beta = 1, 2, \tag{16.43}$$

and

$$g_{33} = \nu \cdot \nu = 1, \tag{16.44}$$

$$g_{3,\alpha} = \nu \cdot (y_{|\alpha} + u_3 \nu_{|\alpha}) = 0, \quad \alpha = 1, 2. \tag{16.45}$$

In terms of the determinants $g = \det(g_{ik})$ and $\gamma = \det(\gamma_{\alpha\beta})$ we are able to verify that

$$\det\left(\sum_{\kappa=1}^{2} g^{\alpha\kappa} \gamma_{\kappa\beta} \right) = \frac{\gamma}{g}. \tag{16.46}$$

The identity (16.46) is an invariant of the surface (with u_3 fixed). It is not hard to see that $\gamma \overset{\circ}{=} 1$ and $g \overset{\circ}{=} (1 - 2u_3 H + u_3^2 K)^2$. This enables us to conclude that

$$g = \gamma(1 - 2u_3 H + u_3^2 K)^2 = \gamma G^2. \tag{16.47}$$

Note that $G^2 \neq 0$ for sufficiently small $|u_3|$, so that $g \neq 0$. We notice that

$$\gamma = y_{|1}^2 y_{|2}^2 - (y_{|1} \wedge y_{|2})^2 = |y_{|1} \wedge y_{|2}|^2, \tag{16.48}$$

so that the surface element dS is expressible in the form

$$dS(y) = \sqrt{\gamma}\, du_1\, du_2. \tag{16.49}$$

Surface Differential Operators. Let a be a sufficiently often differentiable vector field in the neighborhood of the surface Σ. Then

$$a_i = a \cdot \frac{\partial x}{\partial u_i}, \qquad i = 1, 2, 3, \tag{16.50}$$

are called the *covariant components* of a, while

$$a^i = \sum_{k=1}^{3} g^{ki} a_k, \qquad i = 1, 2, 3, \tag{16.51}$$

are called the *contravariant components* of a. It is not difficult to see that

$$a = \sum_{i=1}^{3} a^i \frac{\partial x}{\partial u_i}. \tag{16.52}$$

We let

$$x_{|i|k} = \frac{\partial^2 x}{\partial u_i \partial u_k} = \sum_{r=1}^{3} \Gamma^r_{ik} \frac{\partial x}{\partial u_r} = \sum_{r=1}^{3} \Gamma^r_{ik} x_{|r} \tag{16.53}$$

where Γ^r_{ik} are the so-called *Christoffel symbols.* Moreover, we introduce the notation $a^i_{\|s}$ by

$$a^i_{\|s} = a^i_{|s} + \sum_{r=1}^{3} \Gamma^i_{sr} a^r \tag{16.54}$$

which is known as *covariant derivative* of a. As a consequence we obtain

$$a = \sum_{j=1}^{3} a^j x_{|j}. \tag{16.55}$$

Furthermore,

$$\begin{aligned}
\nabla \cdot a &= \sum_{i=1}^{3} \sum_{j=1}^{3} \frac{\partial}{\partial x_i} a^j \frac{\partial x_i}{\partial u_j} \qquad (16.56)\\
&= \sum_{k=1}^{3} \sum_{i=1}^{3} \sum_{j=1}^{3} \frac{\partial}{\partial u_k} \left(a^j \frac{\partial x_i}{\partial u_j} \right) \frac{\partial u_k}{\partial x_i}\\
&= \sum_{j=1}^{3} a^j_{|j} + \sum_{j=1}^{3} \sum_{i=1}^{3} \sum_{k=1}^{3} a^j \frac{\partial^2 x_i}{\partial u_k \partial u_j} \frac{\partial u_k}{\partial x_i}\\
&= \sum_{i=1}^{3} a^i_{\|i}.
\end{aligned}$$

Going over to the surface, i.e., choosing (formally) $u_3 = 0$ we are led to the *surface divergence* applied to a tangential field a, i.e.,

$$\nabla^* \cdot a = \sum_{\alpha=1}^{2} \frac{\partial a^\alpha}{\partial u_\alpha} + \sum_{\alpha=1}^{2}\sum_{\beta=1}^{2} \Gamma^\alpha_{\alpha\beta} a^\beta. \tag{16.57}$$

The *gradient* ∇ applied to a sufficiently often differentiable function F on Σ (understood as imbedding in $\mathbb{R}^3$) is given by

$$\nabla F = \left(\frac{\partial F}{\partial x_1}, \frac{\partial F}{\partial x_2}, \frac{\partial F}{\partial x_3} \right)^T = f. \tag{16.58}$$

Clearly, $f_j = f \cdot x_{|j} = F_{|j}$. For the *Laplace operator* $\Delta = \sum_{i=1}^{3} \frac{\partial^2}{\partial x_i^2}$ applied to a function F an elementary calculation yields the representation

$$\begin{aligned} \Delta F &= \sum_{i=1}^{3}\sum_{j=1}^{3} \left(g^{ij} F_{|j}\right)_{\|i} \qquad (16.59)\\ &= \sum_{i=1}^{3}\sum_{j=1}^{3} \frac{\partial}{u_i} g^{ij} \frac{\partial F}{\partial u_j} + \sum_{i=1}^{3}\sum_{j=1}^{3}\sum_{k=1}^{3} \Gamma^k_{ki} g^{ij} \frac{\partial F}{\partial u_j}. \end{aligned}$$

In order to calculate the symbols Γ^k_{ki} we start from the well-known settings

$$g = \det(g_{ik}) = \det\left(\frac{\partial x}{\partial u_i} \cdot \frac{\partial x}{\partial u_k} \right) = |x_{|1}, x_{|2}, x_{|3}|^2, \tag{16.60}$$

where $|x_{|1}, x_{|2}, x_{|3}|$ denotes the triple product of the vectors $x_{|1}$, $x_{|2}$, $x_{|3}$. By elementary differentiation we obtain, in connection with (16.53),

$$\begin{aligned} \frac{\partial g}{\partial u_j} &= 2\sqrt{g}\,\big(|x_{|1|j}, x_{|2}, x_{|3}| \qquad (16.61)\\ &\qquad + |x_{|1}, x_{|2|j}, x_{|3}|\\ &\qquad + |x_{|1}, x_{|2}, x_{|3|j}|\big)\\ &= 2g \sum_{r=1}^{3} \Gamma^r_{rj}, \quad j = 1, 2, 3. \end{aligned}$$

Consequently, we are able to write (cf. C. Müller [1969]) for a sufficiently often differentiable function F on Σ (understood as imbedding in $\mathbb{R}^3$)

$$\begin{aligned} \Delta F &= \sum_{i=1}^{3}\sum_{j=1}^{3} \frac{\partial}{\partial u_i}\left(g^{ij} \frac{\partial}{\partial u_j} F \right) + \frac{1}{2g} \sum_{i=1}^{3}\sum_{j=1}^{3} \frac{\partial g}{\partial u_i} g^{ij} \frac{\partial F}{\partial u_j}\\ &= \frac{1}{\sqrt{g}} \sum_{i=1}^{3}\sum_{j=1}^{3} \frac{\partial}{\partial u_i}\left(\sqrt{g}\, g^{ij} \frac{\partial F}{\partial u_j} \right). \qquad (16.62) \end{aligned}$$

By formal expansion with respect to u_3 we obtain in connection with $g = \gamma G^2$

$$G\Delta = \frac{1}{\sqrt{\gamma}} \sum_{\alpha=1}^{2} \sum_{\beta=1}^{2} \frac{\partial}{\partial u_\alpha} \left(G\sqrt{\gamma} g^{\alpha\beta} \frac{\partial}{\partial u_\beta} \right) + \frac{\partial}{\partial u_3} \left(G \frac{\partial}{\partial u_3} \right)$$

$$= \sum_{n=0}^{\infty} \frac{u_3^n}{n!} \Delta_n + \frac{\partial}{\partial u_3} \left(G \frac{\partial}{\partial u_3} \right) \tag{16.63}$$

with the expansion

$$G g^{\alpha\beta} = \sum_{n=1}^{\infty} \frac{u_3^n}{n!} S_n^{\alpha\beta}, \quad \alpha, \beta \in \{1, 2\}, \tag{16.64}$$

and

$$\Delta_n = \frac{1}{\sqrt{\gamma}} \sum_{\alpha=1}^{2} \sum_{\beta=1}^{2} \frac{\partial}{\partial u_\alpha} \left(\sqrt{\gamma}\, S_n^{\alpha\beta} \frac{\partial}{\partial u_\beta} \right). \tag{16.65}$$

In particular, we have for the zeroth expansion coefficient

$$S_0^{\alpha\beta} = \gamma^{\alpha\beta}. \tag{16.66}$$

For harmonic functions F with $u_3 = 0$ and the *Beltrami operator* $\Delta_0 = \Delta^*$ we are led to the following statement.

Lemma 16.9. *Let F be harmonic, i.e., $\Delta F = 0$ in the neighborhood of $\Sigma = \partial\mathcal{G}$. Then, the differential equation*

$$\left(\Delta^* + \left(\frac{\partial}{\partial \nu} \right)^2 - 2H \frac{\partial}{\partial \nu} \right) F = 0 \tag{16.67}$$

holds true on Σ.

The nabla (gradient) operator expressed in terms of the system u_i, $i = 1, 2, 3$, is given by

$$\nabla F = \sum_{i=1}^{3} \sum_{k=1}^{3} g^{ki} x_{|i} F_{|k} = \sum_{\alpha=1}^{2} \sum_{\beta=1}^{2} g^{\alpha\beta} (y_{|\beta} + u_3 \nu_{|\beta}) F_{|\alpha} + F_{|3} \nu. \tag{16.68}$$

Formal series expansion now yields

$$\sum_{\beta=1}^{2} g^{\alpha\beta} (y_{|\beta} + \nu_{|\beta} u_3) = \sum_{\beta=1}^{2} \sum_{\kappa=1}^{2} g^{\alpha\beta} (\delta_{\kappa,\beta} - u_3 L_\beta^\kappa) y_{|\kappa} \tag{16.69}$$

$$= \sum_{\kappa=1}^{2} y_{|\kappa} \sum_{n=0}^{\infty} \frac{u_3^n}{n!} Q_n^{\alpha\kappa}. \tag{16.70}$$

This implies the formula for the expansion coefficients

$$\nabla_n = \sum_{\alpha=1}^{2}\sum_{\beta=1}^{2} Q_n^{\alpha\beta} y_{|\beta} \frac{\partial}{\partial u_\alpha}, \tag{16.71}$$

with the zeroth expansion coefficient

$$Q_0^{\alpha\beta} = y^{\alpha\beta}, \tag{16.72}$$

so that

$$\nabla = \sum_{n=0}^{\infty} \frac{u_3^n}{n!} \nabla_n + \nu \frac{\partial}{\partial u_3}. \tag{16.73}$$

In particular, choosing $u_3 = 0$ we obtain from $\Delta = \nabla \cdot \nabla$ the representation

$$\Delta^* F = \Delta_0 F = \nabla_0 \cdot \nabla_0 F = \nabla^* \cdot \nabla^* F, \tag{16.74}$$

where $\nabla^* = \nabla_0$ is the *surface gradient* and $\nabla^* \cdot = \nabla_0 \cdot$ is the *surface divergence.*

Gaussian Theorems. The differential geometric framework enables the formulation of the surface *Gaussian Theorems* (for more details see C. Müller [1969]).

Theorem 16.10. *Suppose that the functions $A, B : \Sigma \to \mathbb{R}$ are continuously differentiable. Then*

$$\int_\Sigma A(x)(\nabla_n B(x))\, dS(x) = -\int_\Sigma B(x)(\nabla_n A(x))\, dS(x) \tag{16.75}$$

$$- \int_\Sigma A(x)B(x)a_n\, dS(x),$$

where

$$a_0 = 2\nu H, \tag{16.76}$$

$$a_n = \frac{1}{\sqrt{\gamma}} \sum_{\alpha=1}^{2}\sum_{\beta=1}^{2} \frac{\partial}{\partial u_\alpha} \left(\sqrt{\gamma} Q_n^{\alpha\beta} y_{|\beta}\right). \tag{16.77}$$

Proof. We only give a sketch: For that purpose we subdivide $\Sigma = \partial\mathcal{G}$ into a finite set of surface elements Γ_k. We use integration by parts on each Γ_k. In doing so the integrals over the boundaries $\partial\Gamma_k$ do not contribute to the integral over the closed surface Σ. Consequently we obtain ($\triangleq$ means the equality up to boundary terms).

$$\int_{\Gamma_k} B(x)\nabla_n A(x)\, dS(x) \tag{16.78}$$

$$= \int_{\Gamma_k} B(x(u_1,u_2))\Big(\sum_{\alpha,\beta=1}^{2} \gamma^{\alpha\beta} \frac{\partial y}{\partial u_\beta} \frac{\partial A(x(u_1,u_2))}{\partial u_\alpha} \Big) \sqrt{\gamma}\, du_1\, du_2$$

$$\triangleq - \int_{\Gamma_k} A(x(u_1,u_2))\Big(\sum_{\alpha,\beta=1}^{2} \frac{\partial}{\partial u_\alpha} (B(x(u_1,u_2)))\gamma^{\alpha\beta} \frac{\partial y}{\partial u_\beta} \sqrt{\gamma} \Big)\, du_1\, du_2$$

$$= - \int_{\Gamma_k} A(x)\nabla_n B(x)\, dS(x)$$

$$- \int_{\Gamma_k} A(x(u_1,u_2))B(x(u_1,u_2)) \frac{1}{\sqrt{\gamma}} \sum_{\alpha,\beta=1}^{2} \frac{\partial}{\partial u_\alpha} \left(\sqrt{\gamma} Q_n^{\alpha\beta} \frac{\partial y}{\partial u_\beta} \right) \sqrt{\gamma}\, du_1\, du_2$$

$$= - \int_{\Gamma_k} A(x)\nabla_n B(x)\, dS(x)$$

$$- \int_{\Gamma_k} A(x(u_1,u_2))B(x(u_1,u_2)) a_n\, \sqrt{\gamma}\, du_1\, du_2.$$

In particular, for $n = 0$, by observing the identity

$$a_0 = \Delta^* y = \left(2H \frac{\partial x}{\partial \nu} - \frac{\partial^2 x}{\partial \nu^2} \right)\Bigg|_{u_3=0} = 2H\nu \tag{16.79}$$

and the harmonicity of $x = y + u_3 \nu$ we obtain the desired result from Lemma 16.9. □

As immediate consequences we obtain the following corollaries for $n = 0$.

Corollary 16.11. *Let the functions A, B be of class $C^{(1)}(\partial\mathcal{G})$. Then*

$$\int_{\Sigma} B(x)(\nabla_0 A(x))\, dS(x) \tag{16.80}$$

$$= - \int_{\Sigma} A(x)(\nabla_0 B(x))\, dS(x) - 2 \int_{\Sigma} A(x)B(x)H(x)\nu(x)\, dS(x).$$

Corollary 16.12. *Suppose that the function $A, B : \Sigma \to \mathbb{R}$ are twice continuously differentiable. Then*

$$\int_{\Sigma} A(x)\Delta_0 B(x)\, dS(x) = \int_{\Sigma} B(x)\Delta_0 A(x)\, dS(x). \tag{16.81}$$

Note that $\Delta^* = \Delta_0 = \nabla_0 \cdot \nabla_0 = \nabla^* \cdot \nabla^*$.

16.3 Integral Formulas

For the formulation of the third Green's theorem (sometimes also designated fundamental theorem) of surface theory we begin with the introduction of the so-called fundamental solution for the Beltrami operator.

Definition 16.13. *The function* $x \mapsto G(\Delta^*; |x-y|)$, $|x-y| > 0$, *given by*

$$G_\Sigma(\Delta^*; |x-y|) = \frac{1}{2\pi} \ln |x-y| - \frac{1}{4\pi}, \quad x, y \in \Sigma, \quad x \neq y, \tag{16.82}$$

is called the fundamental solution for the Beltrami operator Δ^* *on the regular surface* Σ *at a point* $y \in \Sigma$. *Note that we simply write* $G(\Delta^*; |x-y|)$ *if no confusion is likely to arise.*

Remark 16.14. *Clearly, for* $\Sigma = \mathbb{S}^2$,

$$\begin{aligned} G_{\mathbb{S}^2}(\Delta^*; |x-y|) &= \frac{1}{2\pi} \ln |x-y| - \frac{1}{4\pi} = \frac{1}{4\pi}(2\ln|x-y| - 1) \\ &= \frac{1}{4\pi}(\ln(|x|^2 + |y|^2 - 2(x \cdot y)) - 1), \end{aligned} \tag{16.83}$$

reduces to the standard setting (9.83) *from Lemma 9.4*

$$\begin{aligned} G(\Delta^*; \xi \cdot \eta) &= G_{\mathbb{S}^2}(\Delta^*; |\xi - \eta|) \\ &= \frac{1}{4\pi} \ln(2 - 2\xi \cdot \eta) - \frac{1}{4\pi} \\ &= \frac{1}{4\pi} \ln(1 - \xi \cdot \eta) + \frac{1}{4\pi} \ln 2 - \frac{1}{4\pi}. \end{aligned} \tag{16.84}$$

Using a local (u_1, u_2)-coordinate system, as indicated earlier in this chapter, we are able to rewrite the fundamental solution. More explicitly, in connection with the first order derivatives applied to the variable y

$$\begin{aligned} \frac{\partial}{\partial u_i} G(\Delta^*; |x-y|) &= \frac{1}{4\pi} \frac{\partial}{\partial u_i} (2 \ln |x-y| - 1) \\ &= \frac{1}{4\pi} \frac{2}{|x-y|^2} (x-y) \cdot \left(-\frac{\partial y}{\partial u_i} \right), \end{aligned} \tag{16.85}$$

the second order derivatives applied to the variable y can be obtained as follows:

$$\begin{aligned}
&\frac{\partial^2}{\partial u_i \partial u_j} G(\Delta^*; |x-y|) \qquad (16.86)\\
&= \frac{1}{4\pi}\frac{\partial}{\partial u_i}\left(\frac{2}{|x-y|^2}(x-y)\cdot\left(-\frac{\partial y}{\partial u_j}\right)\right)\\
&= \frac{1}{4\pi}\underbrace{\frac{\partial}{\partial u_i}\frac{2}{|x-y|^2}}_{=\frac{4(x-y)}{|x-y|^4}\cdot\frac{\partial y}{\partial u_i}}(x-y)\cdot\left(-\frac{\partial y}{\partial u_j}\right) + \frac{1}{4\pi}\frac{2}{|x-y|^2}\underbrace{\frac{\partial}{\partial u_i}(x-y)}_{=-\frac{\partial y}{\partial u_i}}\cdot\left(-\frac{\partial y}{\partial u_j}\right)\\
&\quad + \frac{1}{4\pi}\frac{2}{|x-y|^2}(x-y)\cdot\frac{\partial}{\partial u_i}\left(-\frac{\partial y}{\partial u_j}\right)\\
&= -\frac{1}{4\pi}4\frac{1}{|x-y|^4}(x-y)\cdot\frac{\partial y}{\partial u_i}(x-y)\cdot\frac{\partial y}{\partial u_j} + \frac{1}{4\pi}2\frac{1}{|x-y|^2}\frac{\partial y}{\partial u_i}\cdot\frac{\partial y}{\partial u_j}\\
&\quad - \frac{1}{4\pi}2\frac{1}{|x-y|^2}(x-y)\cdot\frac{\partial}{\partial u_i}\frac{\partial y}{\partial u_i}.
\end{aligned}$$

Now, it is helpful to remember the following relation

$$\sum_{i=1}^{2}\sum_{j=1}^{2} g^{ij}\frac{\partial y}{\partial u_i}\frac{\partial y}{\partial u_j} = \mathbf{I} - \nu(y)\otimes\nu(y) \qquad (16.87)$$

where $\mathbf{I} \in \mathbb{R}^{3\times 3}$ is the identity matrix and $\otimes$ designates the usual tensor product in $\mathbb{R}^3$. Moreover, it is not hard to see that

$$\sum_{i=1}^{2}\sum_{j=1}^{2} g^{ij}\frac{\partial y}{\partial u_i}\cdot\frac{\partial y}{\partial u_j} = 2. \qquad (16.88)$$

As a consequence, the Beltrami derivative of $G(\Delta^*; |x-y|)$ can be expressed in the form

$$\begin{aligned}
&\Delta^*_y G(\Delta^*; |x-y|) \qquad (16.89)\\
&= \sum_{i=1}^{2}\sum_{j=1}^{2} g^{ij}(y)\frac{\partial^2}{\partial u_i\partial u_j}G(\Delta^*; |x-y|)\\
&= -\frac{1}{4\pi}\sum_{i=1}^{2}\sum_{j=1}^{2} 4\frac{1}{|x-y|^4}(x-y)\cdot\left((x-y)\cdot g^{ij}(y)\frac{\partial y}{\partial u_i}\frac{\partial y}{\partial u_j}\right)\\
&\quad + \frac{1}{4\pi}\sum_{i=1}^{2}\sum_{j=1}^{2} 2\frac{1}{|x-y|^2}g^{ij}(y)\frac{\partial y}{\partial u_i}\cdot\frac{\partial y}{\partial u_j}\\
&\quad - \frac{1}{4\pi}\sum_{i=1}^{2}\sum_{j=1}^{2} 2\frac{1}{|x-y|^2}(x-y)\cdot g^{ij}(y)\frac{\partial}{\partial u_i}\frac{\partial y}{\partial u_i}.
\end{aligned}$$

Now we use (16.87) as well as (16.88) and it follows that

$$\begin{aligned}
&\Delta^*_y G(\Delta^*; |x-y|) \qquad (16.90)\\
&= -\frac{1}{4\pi}\sum_{i=1}^{2}\sum_{j=1}^{2} 4\frac{1}{|x-y|^4}(x-y)\cdot\Big((\mathbf{I}-\nu(y)\otimes\nu(y))\Big)(x-y)\\
&\quad+\frac{1}{4\pi}\sum_{i=1}^{2}\sum_{j=1}^{2} 2\frac{1}{|x-y|^2}2\\
&\quad-\frac{1}{4\pi}\sum_{i=1}^{2}\sum_{j=1}^{2} 2\frac{1}{|x-y|^2}(x-y)\cdot\Delta^* y,
\end{aligned}$$

so that due to (16.79)

$$\begin{aligned}
&\Delta^*_y G(\Delta^*; |x-y|) \qquad (16.91)\\
&= \frac{1}{4\pi}4\frac{1}{|x-y|^4}(x-y)\cdot\nu(y)(x-y)\cdot\nu(y)\\
&\quad-\frac{1}{4\pi}4\frac{1}{|x-y|^2}H(y)(x-y)\cdot\nu(y)\\
&= \frac{1}{\pi}\frac{\nu(y)\cdot(x-y)}{|x-y|^2}\left(H(y)-\frac{\nu(y)\cdot(x-y)}{|x-y|^2}\right).
\end{aligned}$$

Remark 16.15. *For $x = y$, we are able to deduce from (16.25) that the Beltrami derivative $\Delta^*_y G(\Delta^*; |x-y|)$ admits a bounded continuation, denoted by $Q(x,y)$ throughout the remaining part of this chapter.*

Based on these preliminaries the well-known technique of potential theory (see, e.g., O.D. Kellogg [1929]) leading to the third Green's formula provides the following integral formula.

Theorem 16.16 (Fundamental Theorem of Surface Theory)**.** *Let $\Sigma = \partial\mathcal{G}$ be the boundary surface of a regular region $\mathcal{G}$ (as indicated above). Assume that F is of class $\mathrm{C}^{(2)}(\Sigma)$. Then the integral formula*

$$F(x) = \int_\Sigma G(\Delta^*; |x-y|)\Delta^*_y F(y)\, dS(y) - \int_\Sigma (\Delta^*_y G(\Delta^*; |x-y|))F(y)\, dS(y). \qquad (16.92)$$

is valid for all $x \in \Sigma$. More explicitly, we have

$$\begin{aligned}
F(x) &= \int_\Sigma \left(\frac{1}{2\pi}\ln|x-y| - \frac{1}{4\pi}\right)\Delta^*_y F(y)\, dS(y) \qquad (16.93)\\
&\quad+\frac{1}{\pi}\int_\Sigma \frac{\nu(y)\cdot(x-y)}{|x-y|^2}\left(H(y)-\frac{\nu(y)\cdot(x-y)}{|x-y|^2}\right)F(y)\, dS(y).
\end{aligned}$$

Proof. Once more, because of the similarity to potential theoretic arguments, we only give a sketch: Writing out Corollary 16.12 for a (regular) subregion $\Gamma \subset \Sigma$ with continuously differentiable boundary curve $\partial\Gamma$ we are led to the second Green's formula of surface theory

$$\int_\Gamma A(y)\Delta_y^* B(y)\, dS(y) = \int_\Gamma B(y)\Delta_y^* A(y)\, dS(y) + \int_{\partial\Gamma} B(y)(\nu(y)\cdot\nabla_y^*)A(y)\, d\sigma(y) - \int_{\partial\Gamma} A(y)(\nu(y)\cdot\nabla_y^*)B(y)\, d\sigma(y) \tag{16.94}$$

where $d\sigma$ is the line element, ν is the unit normal field pointing outward of Γ, and A, B are twice continuously functions on $\overline{\Gamma} = \Gamma \cap \partial\Gamma$.

Suppose now that x is an arbitrary point of Σ. Choosing $\Gamma = \Sigma \setminus \overline{\mathbb{B}_\delta(x)}$ in (16.94), where δ is assumed to be positive and sufficiently small, $A(y) = G(\Delta^*; |x-y|$, and $B(y) = F(y)$ we obtain

$$\begin{aligned}&\int_{\substack{|x-y|\geq\delta\\ y\in\Sigma}} G(\Delta^*;|x-y|)\Delta_y^* F(y)\, dS(y) \\ &= \int_{\substack{|x-y|\geq\delta\\ y\in\Sigma}} \Delta_y^* G(\Delta^*;|x-y|)F(y)\, dS(y) \\ &\quad + \int_{\substack{|x-y|=\delta\\ y\in\Sigma}} G(\Delta^*;|x-y|)(\nu(y)\cdot\nabla_y^*)F(y)\, d\sigma(y) \\ &\quad - \int_{\substack{|x-y|=\delta\\ y\in\Sigma}} (\nu(y)\cdot\nabla_y^*)G(\nabla^*;|x-y|)F(y)\, d\sigma(y).\end{aligned} \tag{16.95}$$

In the usual manner as known from potential theory we are able to verify (see, e.g., K. Strubecker [1969]) that

$$\lim_{\delta\to 0}\int_{\substack{|x-y|=\delta\\ y\in\Sigma}} G(\Delta^*;|x-y|)(\nu(y)\cdot\nabla_y^*)F(y))\, d\sigma(y) = 0 \tag{16.96}$$

and

$$\lim_{\delta\to 0}\int_{\substack{|x-y|=\delta\\ y\in\Sigma}} (\nu(y)\cdot\nabla_y^*)G(\Delta;|x-y|)F(y) d\sigma(y) = -F(x). \tag{16.97}$$

These limit relations yield the desired result. □

Once more, it should be remarked that the occurrence of the mean curvature in the integral formula (16.93) strongly limits its numerical use if only discrete information about Σ is available and approximate substitutes have to come into play.

In terms of the bounded continuation Q the identity (16.93) admits the following reformulation.

Corollary 16.17. *Under the assumptions of Theorem 16.16 we have for $x \in \Sigma$ that*

$$F(x) = \int_\Sigma G(\Delta^*; |x-y|)\Delta^*_y F(y)\, dS(y) - \int_\Sigma Q(x,y)F(y)\, dS(y). \tag{16.98}$$

We notice that Corollary 16.17 in its general form does not give a comparison between the integral over F on the regular surface Σ and an arbitrary functional value $F(x)$, $x \in \Sigma$.

Choosing $F = 1$ in Corollary 16.17 we see that

$$\int_\Sigma Q(x,y)\, dS(y) = -1. \tag{16.99}$$

Introducing the kernel $K(x,y)$ by

$$K(x,y) = Q(x,y) + \frac{1}{\|\Sigma\|}, \quad x, y \in \Sigma, \tag{16.100}$$

we see that

$$\int_\Sigma K(x,y)\, dS(y) = 0. \tag{16.101}$$

Collecting our results we therefore arrive at the following integral formulas.

Corollary 16.18 (Integral Formula). *Under the assumption of Theorem 16.16 we have for $x \in \Sigma$ that*

$$\begin{aligned} F(x) - \frac{1}{\|\Sigma\|} \int_\Sigma F(y)\, dS(y) & \qquad\qquad (16.102) \\ = \int_\Sigma G(\Delta^*; |x-y|)\, \Delta^*_y F(y)\, dS(y) - \int_\Sigma K(x,y)F(y)\, dS(y), \end{aligned}$$

that is equivalent to

$$\begin{aligned} F(x) = \int_\Sigma (G(\Delta^*; |x-y|)) - K(x,y)\, \Delta^*_y F(y) dS(y) & \qquad (16.103) \\ - \int_\Sigma K(x,y)(F(y) - \Delta^*_y F(y))\, dS(y), \end{aligned}$$

where $K(x,y)$ is given by (16.100).

As a consequence, if $\Delta^* F - F = \text{const.}$ on Σ, then

$$F(x) - \frac{1}{\|\Sigma\|} \int_\Sigma F(y)\, dS(y) = \int_\Sigma (G(\Delta^*; |x-y|) - K(x,y))\Delta^*_y F(y) dS(y). \tag{16.104}$$

Obviously, Corollary 16.18 provides a comparison between a functional value at $x \in \Sigma$ and the mean integral value of F over the surface Σ under explicit knowledge of the remainder terms in integral form.

For later use we introduce the following iterated kernels

$$G^{(2)}(\Delta^*; x, y) = \int_\Sigma G(\Delta^*; |x-z|)\, G(\Delta^*; |y-z|)\, dS(z), \qquad (16.105)$$

and

$$K^{(2)}(x, y) = \int_\Sigma K(x, z)\, K(y, z)\, dS(z). \qquad (16.106)$$

From similar arguments of potential theory (see, e.g., O.D. Kellogg [1929], K. Strubecker [1969]) we know that $G^{(2)}(\Delta^*; x, y)$ defines a continuous function in x with respect to fixed y and in y with respect to fixed x. From Corollary 16.17 and Corollary 16.18 we therefore obtain

Lemma 16.19. *Under the assumption of Theorem 16.16 we have for $x \in \Sigma$ that*

$$\begin{aligned} \Delta_x^* \int_\Sigma G(\Delta^*; |x-y|) F(y)\, dS(y) &= \int_\Sigma G(\Delta^*; |x-y|) \Delta_y^* F(y)\, dS(y) \qquad (16.107) \\ &\quad + \int_\Sigma (Q(y,x) - Q(x,y)) F(y)\, dS(y) \\ &= F(x) + \int_\Sigma Q(y,x) F(y)\, dS(y), \end{aligned}$$

where, as usual, $Q(y,x)$ is a bounded continuation of $G^{(2)}(\Delta; x, y)$ in $x = y$. In addition, for $|x-y| > 0$, we have

$$\Delta_y^* G^{(2)}(\Delta; x, y) = G(\Delta^*; |x-y|) + \int_\Sigma Q(z,y) G(\Delta^*; |z-x|)\, dS(z). \qquad (16.108)$$

Consequently, we are led to

$$\begin{aligned} &\int_\Sigma G(\Delta^*; |x-y|) F(y)\, dS(y) \qquad (16.109) \\ &= \int_\Sigma \Delta_y^* G^{(2)}(\Delta^*; x, y) F(y)\, dS(y) \\ &\quad + \int_\Sigma \int_\Sigma Q(z,y) G(\Delta^*; |z-x|)\, dS(z) F(y)\, dS(y), \end{aligned}$$

such that

$$\begin{aligned} &\Delta_x^* \int_\Sigma G(\Delta^*; |x-y|) F(y)\, dS(y) \qquad (16.110) \\ &= \Delta_x^* \int_\Sigma \Delta_y^* G^{(2)}(\Delta^*; x, y) F(y)\, dS(y) \\ &\quad + \Delta_x^* \int_\Sigma \int_\Sigma Q(z,y) G(\Delta^*; |z-x|)\, dS(z)\; F(y)\, dS(y). \end{aligned}$$

Note that, for all $F \in \mathrm{C}^{(2)}(\Sigma)$ with $\Delta^* F = 0$, we arrive at

$$F(x) = -\int_\Sigma Q(x,y)F(y)\, dS(y), \tag{16.111}$$

which characterizes the *null space of the Beltrami operator* Δ^* on Σ.

Spherical Specialization. We consider now the spherical case, i.e., $\Sigma = \mathbb{S}^2$. Since the unit normal vector ν coincides with ε^r (in a tangential-normal system) the mixed terms satisfy $\mathrm{L}_{12} = \mathrm{L}_{\varphi t} = 0$ and the trace of the matrix L_{ik} can be written as

$$\mathrm{L}_{11} = \mathrm{L}_{\varphi\varphi} = t^2 - 1, \quad \mathrm{L}_{22} = \mathrm{L}_{tt} = \frac{1}{t^2-1}. \tag{16.112}$$

These facts inform us that

$$\left(g_{ik}\right)_{i,k=1,2} = -\left(L_{ik}\right)_{i,k=1,2}. \tag{16.113}$$

The Gaussian curvature H on the unit sphere $\mathbb{S}^2$ is given by

$$H = \frac{1}{2}\frac{g_{22}\mathrm{L}_{11} + g_{11}\mathrm{L}_{22}}{g_{11}g_{22} - g_{12}^2} = \frac{1}{2}\frac{\frac{1}{1-t^2}(t^2-1) + (1-t^2)\frac{1}{t^2-1}}{1} = -1. \tag{16.114}$$

By observing that

$$\begin{aligned}\Delta_y^* G(\Delta^*; |x-y|) &= -\frac{1}{\pi}\frac{-(1-(x\cdot y))}{2(1-(x\cdot y))}\left(-1 - \frac{-(1-(x\cdot y))}{2(1-(x\cdot y))}\right) \\ &= -\frac{1}{\pi}\frac{(-1)}{2}\left(-1 - \frac{(-1)}{2}\right) = -\frac{1}{4\pi}, \quad x \neq y, \end{aligned} \tag{16.115}$$

we see that $K(x,y) = 0$, $x, y \in \mathbb{S}^2$. In other words, the integral formula (16.102) coincides with its spherical analog (cf. Theorem 9.16) so that the discrepancy estimate can be handled in the already known way.

Ellipsoidal Specialization. Considering an ellipsoid of revolution

$$\mathcal{G}_{a,b} = \left\{ x = (x_1, x_2, x_3)^T \in \mathbb{R}^3 : \left(\frac{x_1}{a}\right)^2 + \left(\frac{x_2}{a}\right)^2 + \left(\frac{x_3}{b}\right)^2 \leq 1 \right\} \tag{16.116}$$

we are canonically led (cf. M. Klug [2014]) to ellipsoidal coordinates

$$x(s,\varphi,t) = \begin{pmatrix} \sqrt{s^2+E^2}\sqrt{1-t^2}\cos(\varphi) \\ \sqrt{s^2+E^2}\sqrt{1-t^2}\sin(\varphi) \\ st \end{pmatrix}, \tag{16.117}$$

where the *eccentricity* E is given by $E = \sqrt{a^2-b^2}$ and $s > 0$. Note that the semi-minor axis coincides with the x_3-axis. The geometric longitude is denoted by $\varphi \in [0, 2\pi)$, and $t = \cos(\vartheta) \in [-1, 1]$ denotes the polar distance, $\vartheta \in [0, \pi]$ is the corresponding (co)latitude. More explicitly,

$$s(x) = \left(\frac{1}{2}(|x|^2 - E^2) + \sqrt{\frac{1}{4}(|x|^2 - E^2)^2 + x_3^2 E^2} \right)^{1/2}, \tag{16.118}$$

$$\varphi(x) = \arctan\left(\frac{x_2}{x_1}\right), \quad x_1 \neq 0, \tag{16.119}$$

$$t(x) = \frac{x_3}{s(x)}, \quad x = (x_1, x_2, x_3)^T \in \mathbb{R}^3. \tag{16.120}$$

An orthonormal basis at a point $x = x(s, \varphi, t)$ is provided by

$$\varepsilon^s(s, \varphi, t) = \begin{pmatrix} \frac{s}{\sqrt{s^2+E^2t^2}} \sqrt{1-t^2} \cos(\varphi) \\ \frac{s}{\sqrt{s^2+E^2t^2}} \sqrt{1-t^2} \sin(\varphi) \\ \frac{\sqrt{s^2+E^2}}{\sqrt{s^2+E^2t^2}} t \end{pmatrix}, \tag{16.121}$$

$$\varepsilon^\varphi(s, \varphi, t) = \begin{pmatrix} -\sin(\varphi) \\ \cos(\varphi) \\ 0 \end{pmatrix},$$

$$\varepsilon^t(s, \varphi, t) = \begin{pmatrix} -\frac{\sqrt{s^2+E^2}}{\sqrt{s^2+E^2t^2}} t \cos(\varphi) \\ -\frac{\sqrt{s^2+E^2}}{\sqrt{s^2+E^2t^2}} t \sin(\varphi) \\ \frac{s}{\sqrt{s^2+E^2t^2}} \sqrt{1-t^2} \end{pmatrix}. \tag{16.122}$$

The unit normal vector ν at a point $x = x(s, \varphi, t)$ is given by

$$\nu(x) = \begin{pmatrix} \frac{s}{\sqrt{s^2+E^2t^2}} \sqrt{1-t^2} \cos(\varphi) \\ \frac{s}{\sqrt{s^2+E^2t^2}} \sqrt{1-t^2} \sin(\varphi) \\ \frac{\sqrt{s^2+E^2}}{\sqrt{s^2+E^2t^2}} t \end{pmatrix}. \tag{16.123}$$

Since s is constant for the surface of an ellipsoid defined by (16.117), we are allowed to write (φ, t) instead of (s, φ, t). Consequently, for the mean curvature H on $\Sigma = \partial\mathcal{G}_{a,b}$ we obtain

$$H(x) = -\frac{1}{2} \frac{1}{\sqrt{(s^2+E^2)(s^2+E^2t^2)}} \left(\frac{s(s^2+E^2)}{\sqrt{s^2+E^2t^2}} + s\sqrt{s^2+E^2t^2} \right). \tag{16.124}$$

The explicit representations of the operators ∇^* and Δ^* for the ellipsoid of revolution are easily derivable (see, e.g., E.W. Grafarend et al. [1999], M. Klug [2014]). The gradient expressed in ellipsoidal coordinates is given as follows

$$\nabla = \frac{\sqrt{s^2+E^2}}{\sqrt{s^2+E^2t^2}} \varepsilon^s \frac{\partial}{\partial s} + \frac{\sqrt{1-t^2}}{\sqrt{s^2+E^2t^2}} \varepsilon^t \frac{\partial}{\partial t} + \frac{1}{\sqrt{s^2+E^2}\sqrt{1-t^2}} \varepsilon^\varphi \frac{\partial}{\partial \varphi}, \tag{16.125}$$

so that the *ellipsoidal surface gradient* is given by

$$\nabla^* = \frac{\sqrt{1-t^2}}{\sqrt{s^2+E^2t^2}} \varepsilon^t \frac{\partial}{\partial t} + \frac{1}{\sqrt{s^2+E^2}\sqrt{1-t^2}} \varepsilon^\varphi \frac{\partial}{\partial \varphi}. \tag{16.126}$$

The Laplace operator expressed in ellipsoidal coordinates shows the representation

$$\Delta = \frac{1}{s^2+E^2t^2}\left((s^2+E^2)\frac{\partial^2}{\partial s^2} + 2s\frac{\partial}{\partial s} + (1-t^2)\frac{\partial^2}{\partial t^2} - 2t\frac{\partial}{\partial t} \right.$$
$$\left. \frac{s^2+E^2t^2}{(s^2+E^2)(1-t^2)}\frac{\partial^2}{\partial \varphi^2}\right) \tag{16.127}$$

while the *Beltrami operator* $\Delta^* = \nabla^* \cdot \nabla^*$ in ellipsoidal framework results in

$$\Delta^* = \frac{1}{(s^2+E^2)(1-t^2)}\frac{\partial^2}{\partial\varphi^2} - \frac{-2t(s^2+E^2t^2)+tE^2(1-t^2)}{(s^2+E^2t^2)^2}\frac{\partial}{\partial t} \tag{16.128}$$
$$+\frac{(1-t^2)}{s^2+E^2t^2}\frac{\partial^2}{\partial t^2}.$$

16.4 Best Approximate Integration

For $x \in \Sigma$ we start from the following identity (cf. Corollary 16.18)

$$F(x) = \frac{1}{\|\Sigma\|}\int_\Sigma F(y)\, dS(y) \tag{16.129}$$
$$+\int_\Sigma G(\Delta^*;|x-y|)\Delta^*_y F(y) dS(y) - \int_\Sigma K(x,y)F(y)\, dS(y),$$

where $K(x,y) = Q(x,y) + \frac{1}{\|\Sigma\|}$ and $Q(x,y)$ is the bounded continuation of $y \mapsto \Delta^*_y G(\Delta^*;|x-y|)$, $y \in \Sigma$.

Suppose that the system $X_N = \{x_1,\ldots,x_N\} \subset \Sigma$ is chosen in such a way that $x_i \neq x_j$, $i \neq j$. Assume that the coefficient system $\{a_1,\ldots,a_N\} \subset \mathbb{R}$ is arbitrary. Then, the Cauchy–Schwarz inequality yields the estimate

$$0 \le \left|\int_\Sigma \sum_{i=1}^N a_i K(x_i,z)\, dS(z)\right|^2 \le \|\Sigma\| \int_\Sigma \left|\sum_{i=1}^N a_i K(x_i,z)\right|^2 dS(z)$$
$$= \|\Sigma\| \sum_{i=1}^N \sum_{j=1}^N a_i a_j K^{(2)}(x_i,x_j). \tag{16.130}$$

Restricting ourselves to N-tuples $(a_1,\ldots,a_n)^T \in \mathbb{R}^N$ satisfying

$$\sum_{i=1}^N a_i K^{(2)}(x_i,x_j) = 0, \quad j = 1\ldots,N, \tag{16.131}$$

we are able to verify from (16.130) that

$$0 = \|\Sigma\| \sum_{i=1}^{N} \sum_{j=1}^{N} a_i a_j K^{(2)}(x_i, x_j) = \left| \sum_{i=1}^{N} a_i \underbrace{\int_\Sigma K(x_i, z)\, dS(z)}_{=0} \right|. \tag{16.132}$$

This result shows that (16.131) is non-trivially solvable with

$$\det(\mathbf{K}) = 0, \quad \mathbf{K} = \left(K^{(2)}(x_i, x_j) \right)_{i,j=1,\ldots,N}. \tag{16.133}$$

As a consequence, we are able to deduce that there exist non-trivial vectors $(a_1, \ldots, a_N)^T \in \mathbb{R}^N$ with $\Sigma_{i=1}^N a_i = 1$ satisfying (16.131).

From the integral formula (16.129) we obtain

$$\begin{aligned} \sum_{i=1}^{N} a_i F(x_i) = {} & \frac{1}{\|\Sigma\|} \int_\Sigma F(y)\, dS(y) \\ & + \int_\Sigma \sum_{i=1}^{n} a_i G(\Delta^*; |x_i - y) \Delta_y^* F(y)\, dS(y) \\ & - \int_\Sigma \sum_{i=1}^{N} a_i K(x_i, y) F(y)\, dS(y). \end{aligned} \tag{16.134}$$

From our aforementioned argument it immediately follows that

$$\begin{aligned} 0 & \leq \left| \int_\Sigma \sum_{i=1}^{N} a_i K(x_i, y) F(y) dS(y) \right|^2 \\ & \leq \underbrace{\int_\Sigma \left| \sum_{i=1}^{N} a_i K(x_i, y) \right|^2 dS(y)}_{=0} \int_\Sigma |F(y)|^2 dS(y) \end{aligned} \tag{16.135}$$

so that

$$\begin{aligned} \sum_{i=1}^{N} a_i F(x_i) = {} & \frac{1}{\|\Sigma\|} \int_\Sigma F(y)\, dS(y) \\ & + \int_\Sigma \sum_{i=1}^{N} a_i G(\Delta^*; |x_i - y|)\, \Delta_y^* F(y)\, dS(y) \end{aligned} \tag{16.136}$$

holds for all $(a_1, \ldots, a_N)^T \in \mathbb{R}^N$ satisfying the linear equations

$$\sum_{i=1}^{N} a_i = 1, \tag{16.137}$$

$$\sum_{i=1}^{N} a_i K^{(2)}(x_i, x_j) = 0, \quad j = 1, \ldots, N. \tag{16.138}$$

Therefore, best approximate integration (in the sense of Sard) can be realized by well-known arguments using Lagrange's method of multipliers. The details are left to the reader.

16.5 Equidistributions

Assume that a system $Y_N = \{y_1^N, \ldots, y_N^N\}$ of points $y_k^N \in \Sigma$, $k = 1, \ldots, N$, is associated with a partition $\mathcal{P}(Y_N) = \{\mathcal{U}(y_1^N), \ldots, \mathcal{U}(y_N^N)\}$ of the regular surface Σ into N mutually distinct parts $\mathcal{U}(y_k^N) \subset \Sigma$, $y_k^N \in \mathcal{U}(y_k^N)$, satisfying

$$0 < \|\mathcal{U}(y_k^N)\| = \int_{\mathcal{U}(y_k^N)} dS(x) = \frac{\|\Sigma\|}{N}, \quad k = 1, \ldots, N. \tag{16.139}$$

As in the spherical case (cf. Section 14.1) it makes sense to consider the quantity

$$\max_{k=1,\ldots,N} \sup_{x \in \mathcal{U}(y_k^N)} |x - y_k^N|. \tag{16.140}$$

So, the *partition size* of the system Y_N is defined by letting

$$\sigma(Y_N) = \inf \max_{k=1,\ldots,N} \sup_{x \in \mathcal{U}(y_k^N)} |x - y_k^N|, \tag{16.141}$$

where the infimum is taken over all partitions $\mathcal{P}(Y_N)$ of Σ satisfying (16.139).

For every partition $\mathcal{P}(Y_N) = \{\mathcal{U}(y_1^N), \ldots, \mathcal{U}(y_N^N)\}$ of the above type,

$$\begin{aligned} \frac{1}{\|\Sigma\|} \int_\Sigma F(y)\, dS(y) &= \frac{1}{\|\Sigma\|} \sum_{k=1}^N \int_{\mathcal{U}(y_k^N)} F(y)\, dS(y) \\ &= \frac{1}{N} \sum_{k=1}^N F(y_k^N) + \frac{1}{\|\Sigma\|} \sum_{k=1}^N \int_{\mathcal{U}(y_k^N)} (F(y) - F(y_k^N))\, dS(y). \end{aligned} \tag{16.142}$$

This identity immediately leads to the following result in analogy to Lemma 14.3.

Lemma 16.20. *Let $Y_N = \{y_1^N, \ldots, y_N^N\}$ be a system of points on Σ. Suppose that F is Lipschitz continuous on Σ with Lipschitz constant C_F. Then*

$$\left| \frac{1}{\|\Sigma\|} \int_\Sigma F(y)\, dS(y) - \frac{1}{N} \sum_{k=1}^N F(y_k^N) \right| \leq C_F\, \sigma(Y_N). \tag{16.143}$$

In other words, the "integral mean"

$$\mathcal{I}F = \frac{1}{\|\Sigma\|} \int_\Sigma F(y)\, dS(y) \tag{16.144}$$

may be replaced by the "arithmetical mean"

$$\mathcal{L}_{Y_N} F = \frac{1}{N} \sum_{k=1}^{N} F(y_k^N) \tag{16.145}$$

provided that the partition size of Y_N is sufficiently small. From Corollary 16.18 we immediately obtain the following integration formula.

Theorem 16.21. *Assume $Y_N = \{y_1^N, \ldots, y_N^N\}$ is a system of points $y_1^N, \ldots, y_N^N$ on Σ. Then the integration rule*

$$\begin{aligned} &\frac{1}{N} \sum_{k=1}^{N} F(y_k^N) - \frac{1}{\|\Sigma\|} \int_{\Sigma} F(y)\, dS(y) \\ &= \int_{\Sigma} \frac{1}{N} \sum_{i=1}^{N} G(\Delta^*; |y_k^N - y|) \Delta_y^* F(y)\, dS(y) - \int_{\Sigma} \frac{1}{N} \sum_{i=1}^{N} K(y_k^N, y) F(y)\, dS(y) \end{aligned} \tag{16.146}$$

holds true for all $F \in \mathrm{C}^{(2)}(\Sigma)$.

Using the abbreviation $\mathcal{E}_{Y_N}$ for the *remainder term*

$$\mathcal{E}_{Y_N} F = \mathcal{I} F - \mathcal{L}_{Y_N} F \tag{16.147}$$

we find by application of the Cauchy–Schwarz inequality, in connection with Theorem 16.21, that

$$\begin{aligned} |\mathcal{E}_{Y_N} F| \le &\left| \frac{1}{N} \int_{\Sigma} \sum_{k=1}^{N} K(y_k^N, y) F(y)\, dS(y) \right| \\ &+ \left| \frac{1}{N} \int_{\Sigma} \sum_{k=1}^{N} G(\Delta^*; |y_k^N - y|) \Delta_y^* F(y)\, dS(y) \right|, \end{aligned}$$

so that

$$\begin{aligned} &|\mathcal{E}_{Y_N} F| \\ &\le \left(\frac{1}{N^2} \sum_{k=1}^{N} \sum_{l=1}^{N} \int_{\Sigma} K(y_k^N, y) K(y_l^N, y)\, dS(y) \right)^{1/2} \left(\int_{\Sigma} |F(y)|^2\, dS(y) \right)^{1/2} \\ &+ \left(\frac{1}{N^2} \sum_{k=1}^{N} \sum_{l=1}^{N} \int_{\Sigma} G(\Delta^*; |y_k^N - y|) G(\Delta^*; |y_l^N - y|)\, dS(y) \right)^{1/2} \\ &\qquad \times \left(\int_{\Sigma} |\Delta_y^* F(y)|^2\, dS(y) \right)^{1/2}. \end{aligned} \tag{16.148}$$

We call the expression

$$D_{\Sigma}(Y_N; K) = \left(\frac{1}{N^2} \sum_{k=1}^{N} \sum_{l=1}^{N} K^{(2)}(y_k^N, y_l^N) \right)^{1/2} \tag{16.149}$$

the *non-sphericity discrepancy of* Y_N (more accurately, *non-sphericity* Σ-*discrepancy of* Y_N, while the expression

$$D_\Sigma(Y_N;\Delta^*) = \left(\frac{1}{N^2}\sum_{k=1}^{N}\sum_{l=1}^{N} G((\Delta^*)^2;|y_k^N - y_l^N|)\right)^{1/2} \tag{16.150}$$

is designated *discrepancy of* Y_N (more accurately, Σ-*discrepancy of* Y_N *to the operator* Δ^*).

Summarizing our results we obtain

Lemma 16.22. *Assume that the function* F *is of class* $\mathrm{C}^{(2)}(\Sigma)$. *Then the following estimate is valid:*

$$|\mathcal{E}_{Y_N}F| \le D_\Sigma(Y_N;K)\|F\|_{\mathrm{L}^2(\Sigma)} + D_\Sigma(Y_N;\Delta^*)V_\Sigma(F), \tag{16.151}$$

where $\|F\|_{\mathrm{L}^2(\Sigma)}$ *and the "*Σ*-variance"* $V_\Sigma(F)$ *are given by*

$$\|F\|_{\mathrm{L}^2(\Sigma)} = \left(\int_\Sigma |F(y)|^2\, dS(y)\right)^{1/2}, \tag{16.152}$$

$$V_\Sigma(F) = \left(\int_\Sigma |\Delta_y^* F(y)|^2\, dS(y)\right)^{1/2}. \tag{16.153}$$

The following definition is canonical in the concept of equidistributions.

Definition 16.23. *A sequence* $\{Y_N\}$ *of point sets* $Y_N \subset \Sigma$ *is called equidistribution, if*

$$\frac{1}{\|\Sigma\|}\int_\Sigma F(y)\, dS(y) = \lim_{N\to\infty}\frac{1}{N}\sum_{k=1}^{N} F(y_k^N) \tag{16.154}$$

holds for all $F \in \mathrm{C}^{(2)}(\Sigma)$.

Evidently, we have

Lemma 16.24. *If* $D_\Sigma(Y_N;K)\to 0$ *and* $D_\Sigma(Y_N;\Delta^*)\to 0$ *for* $N\to\infty$, *then* $\{Y_N\}$ *is an equidistribution.*

Similar to the spherical case (see Lemma 14.9) we are able to show that an equidistribution defined over the reference space $\mathrm{C}^{(2)}(\Sigma)$ can be extended to the space $\mathrm{C}^{(0)}(\Sigma)$.

Lemma 16.25. *Suppose that* $\{Y_N\}$ *is an equidistribution in the sense of Definition 16.23. Then (16.154) holds for all* $F \in \mathrm{C}^{(0)}(\Sigma)$.

We are interested in showing that an equidistribution is equivalent with the property that the non-sphericity discrepancy (16.149) vanishes for $N\to\infty$ with simultaneous extinction of the discrepancy (16.150) for $N\to\infty$. We start our considerations with the following result.

Lemma 16.26. *If $\sigma(Y_N) \to 0$ as $N \to \infty$, then $\{Y_N\}$ is an equidistribution.*

Proof. This is an immediate consequence of (16.143). □

In connection with (16.107) and Theorem 16.21 we find the following equivalences.

Theorem 16.27. *Let Σ be a regular surface and $\{Y_N\}$ a sequence of point sets $Y_N \subset \Sigma$. Assume that F is of class $\mathrm{C}^{(2)}(\Sigma)$. Then, the conditions (16.155)–(16.158) are equivalent:*

$$\lim_{N\to\infty} \frac{1}{N} \sum_{k=1}^{N} F(y_k^N) = \frac{1}{\|\Sigma\|} \int_\Sigma F(y)\, dS(y), \tag{16.155}$$

$$\lim_{N\to\infty} \int_\Sigma \frac{1}{N} \sum_{k=1}^{N} G(\Delta^*; |y_k^N - y|)\Delta_y^* F(y)\, dS(y) \tag{16.156}$$

$$= \lim_{N\to\infty} \int_\Sigma \frac{1}{N} \sum_{k=1}^{N} K(y_k^N, y) F(y)\, dS(y),$$

$$\lim_{N\to\infty} \left(\int_\Sigma \frac{1}{N} \sum_{k=1}^{N} G(\Delta^*; |y_k^N - y|)\Delta_y^* F(y)\, dS(y) \right. \tag{16.157}$$

$$\left. - \int_\Sigma \frac{1}{N} \sum_{k=1}^{N} Q(y_k^N, y) F(y)\, dS(y) \right) = \frac{1}{\|\Sigma\|} \int_\Sigma F(y)\, dS(y),$$

$$\lim_{N\to\infty} \left(\frac{1}{N} \sum_{k=1}^{N} F(y_k^N) + \int_\Sigma \frac{1}{N} \sum_{k=1}^{N} Q(y, y_k^N) F(y)\, dS(y) \right) \tag{16.158}$$

$$= \lim_{N\to\infty} \frac{1}{N} \sum_{k=1}^{N} \left(\Delta_x^* \int_\Sigma G(\Delta^*; |x - y|) F(y)\, dS(y) \right) \Bigg|_{x = y_k^N} = 0.$$

If F is a member of the null space of the Beltrami operator Δ^* on Σ, i.e.,

$$F(x) = - \int_\Sigma Q(x, y) F(y)\, dS(y), \tag{16.159}$$

then it follows that

$$\frac{1}{\|\Sigma\|} \int_\Sigma F(y)\, dS(y) = - \lim_{N\to\infty} \int_\Sigma \frac{1}{N} \sum_{k=1}^{N} Q(y_k^N, y) F(y)\, dS(y), \tag{16.160}$$

provided that $Y_N = \{y_1^N, \ldots, y_N^N\}$ is an equidistribution. In this case, we find

$$\int_\Sigma \frac{1}{N} \sum_{k=1}^{N} G(\Delta^*; |y_k^N - y|)\Delta_y^* F(y)\, dS(y) = 0, \tag{16.161}$$

such that

$$\lim_{N\to\infty} \int_\Sigma \frac{1}{N}\sum_{k=1}^{N} K(y_k^N, y)F(y)\, dS(y) = 0. \tag{16.162}$$

Finally, we discuss the integral

$$\frac{1}{\|\Sigma\|}\int_\Sigma V(y)\, dS(y), \tag{16.163}$$

where we use the abbreviation

$$V(y) = \frac{1}{N}\sum_{k=1}^{N} K(y, y_k^N)F(y), \quad y \in \Sigma, \quad F \in \mathrm{C}^{(2)}(\Sigma). \tag{16.164}$$

Certainly, V is a bounded function on Σ, that is Lipschitz continuous for all $x \in \Sigma \setminus \bigcup_{k=1}^{N} \mathbb{B}_\varepsilon(y_k^N)$ with $\varepsilon > 0$ arbitrary, small. Therefore, for every $\varepsilon > 0$, we can replace V by a function V_ε that coincides on $\Sigma \setminus \bigcup_{k=1}^{N} \mathbb{B}_\varepsilon(y_k^N)$ with V such that V_ε is Lipschitz-continuous on the entire surface Σ and

$$\left| \frac{1}{\|\Sigma\|}\int_\Sigma V(y)\, dS(y) - \frac{1}{\|\Sigma\|}\int_\Sigma V_\varepsilon(y)\, dS(y) \right| \leq C\varepsilon, \tag{16.165}$$

where $C > 0$ is a constant independent of ε. From (16.142) we know that

$$\frac{1}{\|\Sigma\|}\int_\Sigma V_\varepsilon(y)\, dS(y) = \frac{1}{N}\sum_{k=1}^{N} V_\varepsilon(y_k^N) + \frac{1}{\|\Sigma\|}\sum_{k=1}^{N}\int_{\mathcal{U}(y_k^N)} V_\varepsilon(y) - V_\varepsilon(y_k^N)\, dS(y), \tag{16.166}$$

such that

$$\left| \frac{1}{\|\Sigma\|}\int_\Sigma V_\varepsilon(y)\, dS(y) - \frac{1}{N}\sum_{k=1}^{N} V_\varepsilon(y_k^N) \right| \leq C_{V_\varepsilon}\sigma(Y_N). \tag{16.167}$$

Because of the properties of our construction the Lipschitz constant C_{V_ε} of V_ε can be estimated from above by a constant independent of ε. Clearly, $\sigma(Y_N) \to 0$ for $N \to \infty$, this implies that there is an $N(\varepsilon)$ such that for all $N > N(\varepsilon)$

$$\left| \frac{1}{\|\Sigma\|}\int_\Sigma V_\varepsilon(y)\, dS(y) - \frac{1}{N}\sum_{k=1}^{N} V_\varepsilon(y_k^N) \right| \leq C\varepsilon. \tag{16.168}$$

By the triangle inequality we obtain from (16.165) and (16.167)

$$\left| \frac{1}{\|\Sigma\|}\int_\Sigma V(y)\, dS(y) - \frac{1}{N}\sum_{k=1}^{N} V_\varepsilon(y_k^N) \right| \leq 2C\varepsilon. \tag{16.169}$$

This shows that

$$\frac{1}{\|\Sigma\|}\int_\Sigma V(y)\, dS(y) \tag{16.170}$$

can be approximated uniformly by

$$\frac{1}{N}\sum_{k=1}^{N} V_\varepsilon(y_k^N). \tag{16.171}$$

Note that V and V_ε coincide on $\Sigma \setminus \bigcup_{k=1}^{N} \mathbb{B}_\varepsilon(y_k^N)$.

Part IV

Integration over qD-Volumes

Chapter 17

Lattices, Periodic Polynomials, and Integral Formulas

In this chapter we present basic preparatory material which is needed for approximate integration over volumes, i.e., regular regions in Euclidean space $\mathbb{R}^q$. We start with the introduction of lattices in Section 17.1. Then we are concerned in Section 17.2 with multivariate periodic polynomials. Section 17.3 delivers vector analytic background in the Euclidean space $\mathbb{R}^q$. Green's theorems are recapitulated for regular regions. It follows the representation of the essentials for classical Fourier theory (in Section 17.4) and its weak variants in Gauss–Weierstrass context. The chapter ends with the concept of periodization in its manifestation involving the Poisson summation formula (in Section 17.5).

17.1 Lattices

We start with the standard introduction of lattices in Euclidean spaces.

Definition 17.1. *Let $g_1, \ldots, g_q$ be linearly independent vectors in q-dimensional Euclidean space $\mathbb{R}^q$. The set Λ (more precisely, $\Lambda_{(q)}$) of all points*

$$g = n_1 g_1 + \ldots + n_q g_q \tag{17.1}$$

with $n_i \in \mathbb{Z}$, $i = 1, \ldots, q$, is called a lattice in $\mathbb{R}^q$ with basis $g_1, \ldots, g_q$.

Clearly, the canonical basis $\{\varepsilon^1, \ldots, \varepsilon^q\}$ in $\mathbb{R}^q$ forms a lattice basis of $\mathbb{Z}^q$. A lattice basis $\{g_1, \ldots, g_q\}$ is related to the canonical basis $\{\varepsilon^1, \ldots, \varepsilon^q\}$ in $\mathbb{R}^q$ via the formula $g_i = \sum_{s=1}^{q} (g_i \cdot \varepsilon^s)\, \varepsilon^s$ (for more details on lattices see, e.g., J.W.S. Cassels [1968], C.G. Lekkerkerker [1969]).

Definition 17.2. *Let $\mathcal{F}$ (more precisely, $\mathcal{F}_\Lambda$) be the half-open parallelotope (cf. Figure 17.1) consisting of the points $x \in \mathbb{R}^q$ with*

$$x = t_1 g_1 + \ldots + t_q g_q, \quad -\frac{1}{2} \le t_i < \frac{1}{2}, \quad i = 1, \ldots, q. \tag{17.2}$$

$\mathcal{F}_\Lambda$ is called the fundamental cell of the lattice Λ.

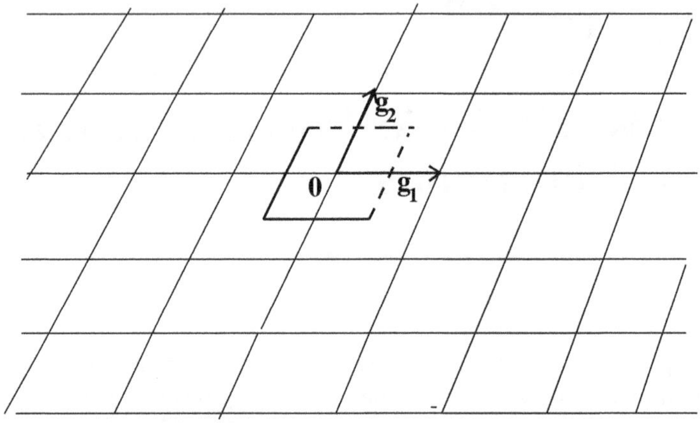

FIGURE 17.1: Two-dimensional lattice.

From linear algebra (see, e.g., P.J. Davis [1963]) it is well-known that the volume of $\mathcal{F}_\Lambda$ is equal to the quantity

$$\|\mathcal{F}_\Lambda\| = \int_{\mathcal{F}_\Lambda} dx = |\det(g_1, \ldots, g_q)| . \tag{17.3}$$

Since the vectors $g_1, \ldots, g_q$ are assumed to be linearly independent, there exists a system of vectors $h_1, \ldots, h_q$ in $\mathbb{R}^q$ such that

$$h_j \cdot g_i = \delta_{i,j} = \begin{cases} 0 & , i \neq j \\ 1 & , i = j. \end{cases} \tag{17.4}$$

($\delta_{i,j}$ is the Kronecker symbol). In more detail, for $i, j = 1, \ldots, q$, we let

$$g_i \cdot g_j = \gamma_{ij} \tag{17.5}$$

The scalars γ^{ij}, $i, j = 1, \ldots, q$, are defined by

$$\sum_{j=1}^{q} \gamma^{ij} \gamma_{jk} = \delta_{i,k}. \tag{17.6}$$

Then the vectors h_j, $j = 1, \ldots, q$, given by

$$h_j = \sum_{k=1}^{q} \gamma^{jk} g_k, \quad j = 1, \ldots, q, \tag{17.7}$$

satisfy the equations

$$h_j \cdot g_i = \sum_{k=1}^{q} \gamma^{jk} g_k \cdot g_i = \sum_{k=1}^{q} \gamma^{jk} \gamma_{ki} = \delta_{j,i}, \quad i, j = 1, \ldots, q. \tag{17.8}$$

Moreover, we find

$$h_i \cdot h_j = \sum_{k=1}^{q} \gamma^{ik} g_k \cdot \sum_{l=1}^{q} \gamma^{jl} g_l \tag{17.9}$$

$$= \sum_{l=1}^{q} \gamma^{jl} \sum_{k=1}^{q} \gamma^{ik} \gamma_{kl} = \gamma^{ji}, \quad i,j = 1,\ldots,q.$$

Definition 17.3. *The lattice with basis* $h_1, \ldots, h_q$ *is called the* inverse lattice Λ^{-1} *to* Λ.

The inverse lattice Λ^{-1} consists of all vectors $h \in \mathbb{R}^q$ such that the inner product $h \cdot g$ is an integer for all $g \in \Lambda$. Obviously, $\Lambda = (\Lambda^{-1})^{-1}$. Moreover, $\|\mathcal{F}_{\Lambda^{-1}}\| = \|\mathcal{F}_\Lambda\|^{-1}$.

Example 17.4. *Let* $\Lambda = \tau\mathbb{Z}^q$ *be the lattice generated by the "dilated" basis* $\tau\varepsilon^1, \ldots, \tau\varepsilon^q$, *where* τ *is a positive number and* $\varepsilon^1, \ldots, \varepsilon^q$ *forms the canonical orthonormal basis in* $\mathbb{R}^q$. *Then, the volume of the fundamental cell of* $\tau\mathbb{Z}^q$ *is* $\|\mathcal{F}_{\tau\mathbb{Z}^q}\| = \tau^q$. *Generating vectors of the inverse lattice* Λ^{-1} *are* $\tau^{-1}\varepsilon^1, \ldots, \tau^{-1}\varepsilon^q$. *The volume of the fundamental cell of the inverse lattice is given by* $\left\|\mathcal{F}_{(\tau\mathbb{Z}^q)^{-1}}\right\| = \tau^{-q} = \left\|\mathcal{F}_{\tau\mathbb{Z}^q}\right\|^{-1}$. *In particular, for* $\tau = 1$, *i.e., the lattice* $\Lambda = \mathbb{Z}^q$, *we have* $\Lambda^{-1} = \mathbb{Z}^q = \Lambda$ *such that* $\|\mathcal{F}_{(\mathbb{Z}^q)^{-1}}\| = 1 = \|\mathcal{F}_{\mathbb{Z}^q}\|$. *This fact has been permanently used in one-dimensional theory, and it is always obvious throughout this work whenever* $\Lambda = \mathbb{Z}^q$.

As preparation for number theoretic ingredients in our cubature approach we first give some basic results on the number of lattice points inside spheres $\mathbb{S}_N^{q-1}(0)$ of radii $N > 0$ around the origin 0

$$\mathbb{S}_N^{q-1}(0) = \{x \in \mathbb{R}^q : |x| = N\} \tag{17.10}$$

(for more background material the reader is referred, e.g., to F. Fricker [1982]). By $\#_\Lambda(\overline{\mathbb{B}_N^q(0)})$ we denote the *total number of lattice points* $g \in \Lambda$ such that g is a member of $\overline{\mathbb{B}_N^q(0)}$ with $\mathbb{B}_N^q(0) = \{x \in \mathbb{R}^q : |x| < N\}$. According to C.F. Gauss [1826] we consider the polyhedron

$$\mathbb{P}_N^q(0) = \bigcup_{g \in \overline{\mathbb{B}_N^q(0)}} \mathcal{F}_\Lambda(g), \tag{17.11}$$

where $\mathcal{F}_\Lambda(g)$ is the cell of the lattice Λ containing the lattice point $g \in \Lambda$:

$$\mathcal{F}_\Lambda(g) = \{x \in \mathbb{R}^q : x = y + g, \quad y \in \mathcal{F}_\Lambda\}. \tag{17.12}$$

Obviously, the volume of $\mathbb{P}_N^q(0)$ is equal to $\|\mathcal{F}_\Lambda\| \#_\Lambda(\overline{\mathbb{B}_N^q(0)})$ (see Figure 17.2).

Now, there exists a constant d such that the estimate $|x| \leq d$ holds true for all $x \in \mathcal{F}_\Lambda$ (for example, for $\Lambda = \mathbb{Z}^q$, $d = \frac{\sqrt{q}}{2}$, see the illustration of

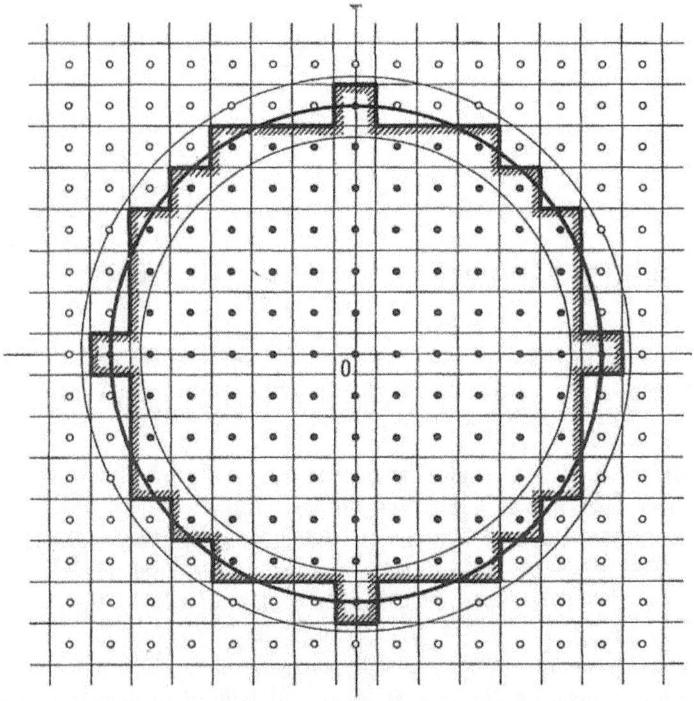

FIGURE 17.2: Gaussian polyhedron for $\Lambda = \mathbb{Z}^q$.

Figure 17.2). Thus, for the polyhedron $\mathbb{P}_N^q(0)$ the following inclusion is valid with $(N > d)$:

$$\overline{\mathbb{B}_{N-d}^q(0)} \subset \mathbb{P}_N^q(0) \subset \overline{\mathbb{B}_{N+d}^q(0)}. \tag{17.13}$$

Note that, for given $x \in \mathbb{P}_N^q(0)$, there exist $g \in \Lambda$ such that

$$|x| \leq |x - g + g| \leq |x - g| + |g| \leq d + N. \tag{17.14}$$

Moreover, for $x \in \overline{\mathbb{B}_0^q(N-d)} \cap \overline{\mathcal{F}_\Lambda(g)}$ we have

$$|g| \leq |x - g| + |x| \leq d + N, \tag{17.15}$$

i.e., $g \in \overline{\mathbb{B}_N^q(0)}$. This leads to the following estimate:

$$\|\overline{\mathbb{B}_{N-d}^q(0)}\| \leq \|\mathcal{F}_\Lambda\| \#_\Lambda(\overline{\mathbb{B}_{N+d}^q(0)}) \leq \|\overline{\mathbb{B}_{N-d}^q(0)}\| \tag{17.16}$$

This is equivalent to

$$\frac{\pi^{\frac{q}{2}}}{\Gamma(\frac{q}{2}+1)}(N-d)^q \leq \|\mathcal{F}_\Lambda\| \#_\Lambda(\overline{\mathbb{B}_N^q(0)}) \leq \frac{\pi^{\frac{q}{2}}}{\Gamma(\frac{q}{2}+1)}(N+d)^q. \tag{17.17}$$

Therefore, we obtain the following asymptotic relation.

Lemma 17.5. *For $N \to \infty$*

$$\#_\Lambda(\overline{\mathbb{B}_N^q(0)}) = \frac{1}{\|\mathcal{F}_\Lambda\|} \frac{\pi^{\frac{q}{2}}}{\Gamma(\frac{q}{2}+1)} N^q + O(N^{q-1}). \tag{17.18}$$

From Lemma 17.5 we are immediately able to verify the following result which will be used in Poisson summation procedures over Euclidean spaces later.

Corollary 17.6. *For $N \to \infty$*

$$\#_\Lambda\left(\overline{\mathbb{B}_{N+d}^q(0)}/\mathbb{B}_{N-d}^q(0)\right) = O(N^{q-1}). \tag{17.19}$$

It should be remarked that the problem of determining

$$\alpha_q = \inf\left\{\varepsilon : \ \#_\Lambda(\overline{\mathbb{B}_N^q(0)}) = \frac{\|\overline{\mathbb{B}_N^q(0)}\|}{\|\mathcal{F}_\Lambda\|} + O(N^\varepsilon)\right\} \tag{17.20}$$

is known as the *lattice point problem for the sphere* (for $q = 2$, the *lattice point problem for the circle*). From the approach due to C.F. Gauss [1826], i.e., Lemma 17.5, we know that $\alpha_q \le q-1$. Improvements and generalizations of this result form extensive studies of today's analytic number theory (see, e.g., W. Freeden [2011] and the references therein). Moreover we are able to conclude that $\alpha_q \ge q-2$ (see, e.g., F. Fricker [1975, 1982]).

17.2 Periodic Polynomials

The functions Φ_h, $h \in \Lambda^{-1}$, defined by

$$\Phi_h(x) = \frac{1}{\sqrt{\|\mathcal{F}_\Lambda\|}} e(h\cdot x) = \frac{1}{\sqrt{\|\mathcal{F}_\Lambda\|}} e^{2\pi i(h\cdot x)}, \quad x \in \mathbb{R}^q, \tag{17.21}$$

are Λ-*periodic*, i.e.,

$$\Phi_h(x+g) = \Phi_h(x) \tag{17.22}$$

for all $x \in \mathbb{R}^q$ and $g \in \Lambda$. The space of all $F \in \mathrm{C}^{(m)}(\mathbb{R}^q)$ that are Λ-periodic is denoted by $\mathrm{C}_\Lambda^{(m)}(\mathbb{R}^q)$, $0 \le m \le \infty$. $\mathrm{L}_\Lambda^p(\mathbb{R}^q)$, $1 \le p < \infty$ is the space of all $F : \mathbb{R}^q \to \mathbb{C}$ that are Λ-periodic and are Lebesgue measurable on $\mathcal{F}_\Lambda$ with

$$\|F\|_{\mathrm{L}_\Lambda^p(\mathbb{R}^q)} = \left(\int_{\mathcal{F}_\Lambda} |F(x)|^p \, dx\right)^{\frac{1}{p}} < \infty. \tag{17.23}$$

Clearly, $\mathrm{C}_\Lambda^{(0)}(\mathbb{R}^q) \subset \mathrm{L}_\Lambda^p(\mathbb{R}^q)$. As is well-known, $\mathrm{L}_\Lambda^2(\mathbb{R}^q)$ is the completion of $\mathrm{C}_\Lambda^{(0)}(\mathbb{R}^q)$ with respect to the norm $\|\cdot\|_{\mathrm{L}_\Lambda^2(\mathbb{R}^q)}$:

$$\mathrm{L}_\Lambda^2(\mathbb{R}^q) = \overline{\mathrm{C}_\Lambda^{(0)}(\mathbb{R}^q)}^{\|\cdot\|_{\mathrm{L}_\Lambda^2(\mathbb{R}^q)}}. \tag{17.24}$$

An easy calculation shows that

$$\langle \Phi_h, \Phi_{h'} \rangle_{\mathrm{L}^2_\Lambda(\mathbb{R}^q)} = \int_{\mathcal{F}_\Lambda} \Phi_h(x)\overline{\Phi_{h'}(x)}\, dx = \delta_{h,h'} = \begin{cases} 1 & , h = h', \\ 0 & , h \neq h'. \end{cases} \tag{17.25}$$

In other words, the system $\{\Phi_h\}_{h\in\Lambda^{-1}}$ is orthonormal with respect to the $\mathrm{L}^2_\Lambda(\mathbb{R}^q)$-inner product: $\langle \Phi_h, \Phi_{h'} \rangle_{\mathrm{L}^2_\Lambda(\mathbb{R}^q)} = \delta_{h,h'}$. For the Laplace operator Δ in $\mathbb{R}^q$ an elementary calculation shows that

$$(\Delta_x + \Delta^\wedge(h))e^{2\pi i h\cdot x} = (\Delta_x + 4\pi^2 h^2)e^{2\pi i h\cdot x} = 0, \quad x \in \mathbb{R}^q \tag{17.26}$$

holds true with $h^2 = h \cdot h$ and $\Delta^\wedge(h) = 4\pi^2 h^2$. We shall say that λ is an eigenvalue of the lattice Λ with respect to Δ if there is a non-trivial solution U of the differential equation $(\Delta+\lambda)U = 0$ satisfying the "boundary condition of periodicity" $U(x + g) = U(x)$ for all $x \in \mathbb{R}^q$ and $g \in \Lambda$. The function U is called an eigenfunction of the lattice Λ with respect to the eigenvalue λ and the operator Δ. In analogy to the one-dimensional case we are able to verify that the spectrum is given by

$$\mathrm{Spect}_\Delta(\Lambda) = \left\{\Delta^\wedge(h) : \Delta^\wedge(h) = 4\pi^2 h^2,\ h \in \Lambda^{-1}\right\}. \tag{17.27}$$

Clearly, an analog of Theorem 3.5 also holds true for the multidimensional case (see, e.g., P.J. Davis [1963]). In consequence, the system $\{\Phi_h\}_{h\in\Lambda^{-1}}$ is closed and complete in the pre-Hilbert space $(\mathrm{C}^{(0)}_\Lambda(\mathbb{R}^q), \|\cdot\|_{\mathrm{L}^2_\Lambda(\mathbb{R}^q)})$ as well as in the Hilbert space $(\mathrm{L}^2_\Lambda(\mathbb{R}^q), \|\cdot\|_{\mathrm{L}^2_\Lambda(\mathbb{R}^q)})$.

A fundamental result in Fourier analysis is that each $F \in \mathrm{L}^2_\Lambda(\mathbb{R}^q)$ can be represented by its Fourier series in the sense

$$\lim_{N\to\infty} \left\| F - \sum_{\substack{|h|\leq N \\ h\in\Lambda^{-1}}} F^\wedge_{\mathcal{F}_\Lambda}(h)\Phi_h \right\|_{\mathrm{L}^2_\Lambda(\mathbb{R}^q)} = 0. \tag{17.28}$$

where the Fourier coefficients $F^\wedge_{\mathcal{F}_\Lambda}(h)$ of F are given by

$$F^\wedge_{\mathcal{F}_\Lambda}(h) = \int_{\mathcal{F}_\Lambda} F(x)\overline{\Phi_h(x)}\, dx, \quad h \in \Lambda^{-1}. \tag{17.29}$$

The Parseval identity tells us that

$$\int_{\mathcal{F}_\Lambda} |F(x)|^2\, dx = \sum_{h\in\Lambda^{-1}} |F^\wedge_{\mathcal{F}_\Lambda}(h)|^2 \tag{17.30}$$

for all $F \in \mathrm{L}^2_\Lambda(\mathbb{R}^q)$. Suppose that $F \in \mathrm{C}^{(k)}_\Lambda(\mathbb{R}^q)$ with $k > \frac{q}{2}$. Then

$$\int_{\mathcal{F}_\Lambda} (\nabla^\alpha F)(x)\overline{\Phi_h(x)}\, dx = (-2\pi i h)^\alpha F^\wedge_{\mathcal{F}_\Lambda}(h), \tag{17.31}$$

whenever $F \in \mathrm{C}^{(k)}_{\Lambda}(\mathbb{R}^q)$ and $\alpha = (\alpha_1, \ldots, \alpha_q)^T \in \mathbb{N}_0^q$ is chosen such that $|\alpha| = \alpha_1 + \ldots + \alpha_q \le k$. Since $\nabla^\alpha F$ is continuous in $\mathcal{F}_\Lambda$ it must belong to $\mathrm{L}^2_\Lambda(\mathbb{R}^q)$. In other words,

$$\sum_{|\alpha|=n} \left(\sum_{h\in\Lambda^{-1}} \left|F^{\wedge}_{\mathcal{F}_\Lambda}(h)\right|^2 |(2\pi i h)^\alpha|^2 \right) < \infty. \tag{17.32}$$

Moreover, there exists a constant C (dependent on k, q) such that

$$C|h|^{2k} \le \sum_{|\alpha|=k} |(2\pi i h)^\alpha|^2. \tag{17.33}$$

From the Cauchy–Schwarz inequality we therefore obtain for all $N > 0$

$$\sum_{\substack{|h|\le N\\ h\in\Lambda^{-1}}} |F^{\wedge}_{\mathcal{F}_\Lambda}(h)| \le \sum_{\substack{|h|\le N\\ h\in\Lambda^{-1}}} |F^{\wedge}_{\mathcal{F}_\Lambda}(h)| \left(\sum_{|\alpha|=k} |(2\pi i h)^\alpha|^2 \right)^{1/2} C^{-1/2} |h|^{-k} \tag{17.34}$$

$$\le \left(\sum_{\substack{|h|\le N\\ h\in\Lambda^{-1}}} |F^{\wedge}_{\mathcal{F}_\Lambda}(h)|^2 \sum_{|\alpha|=k} |(2\pi i h)^\alpha|^2 \right)^{1/2} \left(\sum_{\substack{|h|\le N\\ h\in\Lambda^{-1}}} |h|^{-2k} \right)^{1/2} C^{-1/2}.$$

If $k > \frac{q}{2}$, the sum $\sum_{|h|\le N} |h|^{-2k}$ is finite, hence, the last expression must also be finite. This leads to the following statement.

Theorem 17.7. *If $F \in \mathrm{C}^{(k)}_{\Lambda}(\mathbb{R}^q)$ with $k > \frac{q}{2}$, then $\sum_{h\in\Lambda^{-1}} |F^{\wedge}_{\mathcal{F}_\Lambda}(h)| < \infty$, where $F^{\wedge}_{\mathcal{F}_\Lambda}(h)$ are the Fourier coefficients of F.*

17.3 Regular Regions and Green's Integral Theorems

First we replicate some key results of classical vector analysis and potential theory, which are useful for our work. To this end we remember the concept of a regular region in $\mathbb{R}^q$, $q \ge 2$, given in Definition 16.1. By letting $f = \nabla F$, $F \in \mathrm{C}^{(2)}(\overline{\mathcal{G}})$, $\mathcal{G} \subset \mathbb{R}^q$ regular, we obtain from the theorem of Gauss (16.1)

$$\int_{\mathcal{G}} \Delta F(x)\, dx = \int_{\partial\mathcal{G}} \frac{\partial F}{\partial \nu}(x)\, dS(x), \tag{17.35}$$

where $\frac{\partial}{\partial\nu}$ denotes the derivative in the direction of the outer (unit) normal field ν. Consequently, for all functions $F \in \mathrm{C}^{(2)}(\overline{\mathcal{G}})$ satisfying the Laplace equation $\Delta F = 0$ in $\mathcal{G}$, we have

$$\int_{\partial\mathcal{G}} \frac{\partial F}{\partial \nu}(x)\, dS(x) = 0. \tag{17.36}$$

For all vector fields $f = F\nabla G$, $F \in \mathrm{C}^{(1)}(\overline{\mathcal{G}})$, $G \in \mathrm{C}^{(2)}(\overline{\mathcal{G}})$, we obtain from the Gauss theorem

Theorem 17.8 (First Green's Theorem)**.** *Suppose that $\mathcal{G} \subset \mathbb{R}^q$ is a regular region. For $F \in \mathrm{C}^{(1)}(\overline{\mathcal{G}}), G \in \mathrm{C}^{(2)}(\overline{\mathcal{G}})$ we have*

$$\int_{\mathcal{G}} F(x)\Delta G(x) + \nabla F(x) \cdot \nabla G(x)\ dx = \int_{\partial\mathcal{G}} F(x)\frac{\partial G}{\partial \nu}(x)\ dS(x). \tag{17.37}$$

Taking $f = F\nabla G - G\nabla F$ with $F, G \in \mathrm{C}^{(2)}(\overline{\mathcal{G}})$ we obtain

Theorem 17.9 (Second Green's Theorem)**.** *Suppose that $\mathcal{G} \subset \mathbb{R}^q$ is a regular region. For $F, G \in \mathrm{C}^{(2)}(\overline{\mathcal{G}})$ we have*

$$\begin{aligned} &\int_{\mathcal{G}} G(x)\Delta F(x) - F(x)\Delta G(x)\ dx \\ &= \int_{\partial\mathcal{G}} G(x)\frac{\partial F}{\partial \nu}(x) - F(x)\frac{\partial G}{\partial \nu}(x)\ dS(x). \end{aligned} \tag{17.38}$$

Next we mention an extension of the second Green's Theorem.

Theorem 17.10 (Extended Second Green's Theorem)**.** *For a number $\lambda \in \mathbb{R}$ and a regular region $\mathcal{G} \subset \mathbb{R}^q$, $q \geq 2$, and for $F \in \mathrm{C}^{(2m)}(\overline{\mathcal{G}})$, $m \in \mathbb{N}$, we have*

$$\begin{aligned} \int_{\mathcal{G}} G(x)(\Delta+\lambda)^m F(x)\ dx &= \int_{\mathcal{G}} F(x)(\Delta+\lambda)^m G(x)\ dx \\ &+ \sum_{r=0}^{m-1} \int_{\partial\mathcal{G}} \Big((\Delta+\lambda)^{m-(r+1)} G(x)\Big)\Big(\frac{\partial}{\partial\nu}(\Delta+\lambda)^r F(x)\Big)\ dS(x) \\ &- \sum_{r=0}^{m-1} \int_{\partial\mathcal{G}} \Big(\frac{\partial}{\partial\nu}(\Delta+\lambda)^{m-(r+1)} G(x)\Big)\Big((\Delta+\lambda)^r F(x)\Big)\ dS(x). \end{aligned} \tag{17.39}$$

The simplest form of (17.39) is the case with vanishing boundary integrals such that (17.39) reduces to the formula

$$\int_{\mathcal{G}} G(x)(\Delta+\lambda)^m F(x) - F(x)(\Delta+\lambda)^m G(x)\ dx = 0. \tag{17.40}$$

Next we come to the well-known definition of harmonic functions.

Definition 17.11. *$U \in \mathrm{C}^{(2)}(\mathcal{G})$ is called a* harmonic function *(sometimes also called a* harmonic*) in a region $\mathcal{G} \subset \mathbb{R}^q$ if it satisfies the Laplace equation*

$$\Delta U(x) = \sum_{i=1}^{q} \left(\frac{\partial}{\partial x_i}\right)^2 U(x_1, \ldots, x_q) = 0, \quad x = (x_1, \ldots, x_q)^T \in \mathcal{G}. \tag{17.41}$$

$U \in \mathrm{C}^{(2m)}(\mathcal{G})$, $m \in \mathbb{N}$, is called a polyharmonic function of degree *m in $\mathcal{G} \subset \mathbb{R}^q$ if*

$$\Delta^m U(x) = 0, \quad x \in \mathcal{G}. \tag{17.42}$$

Fundamental Solutions of the Laplacian. Let $y \in \mathbb{R}^q$ be fixed. We are looking for a harmonic function U in $\mathbb{R}^q \setminus \{y\}$ such that

$$U(x) = F(|x-y|), \quad x \in \mathbb{R}^q \setminus \{y\}, \tag{17.43}$$

i.e., U depends only on the mutual distance of x and y. It is well-known from potential theory (see, e.g., O.D. Kellogg [1929]) that $F(|x-y|)$ can be written in the form

$$F(|x-y|) = \begin{cases} C_1 \ln|x-y| + C_2, & q = 2, \\ C_1 |x-y|^{2-q} + C_2, & q \geq 3, \end{cases} \tag{17.44}$$

with some constants C_1, C_2. By convention, the function

$$x \mapsto F_q(|x-y|) = \begin{cases} -\frac{1}{2\pi} \ln|x-y|, & q = 2, \\ \frac{|x-y|^{2-q}}{(q-2)\|\mathbb{S}^{q-1}\|}, & q \geq 3, \end{cases} \tag{17.45}$$

is called the *fundamental solution in* $\mathbb{R}^q$ *for the Laplace operator* Δ (later on, we will see that the choice (17.45) for the coefficients C_1, C_2 is very useful in our context).

Remark 17.12. *As already used in the introduction of one-dimensional Green's functions, the one-dimensional fundamental solution for the Laplacian, i.e., the operator of the second order derivative, is given by the continuous function*

$$x \mapsto F_1(|x-y|) = -\frac{1}{2}|x-y| = -\frac{1}{2}(x-y)\ \mathrm{sign}(x-y), \quad x, y \in \mathbb{R}. \tag{17.46}$$

This is a remarkable feature with nice applications – unfortunately only in the one-dimensional theory.

The fundamental solution of the Laplace operator possesses the following property (see, e.g., W. Freeden, C. Gerhards [2013]).

Lemma 17.13. *Let* $\mathcal{G} \subset \mathbb{R}^q$ *be a regular region with continuously differentiable boundary* $\partial\mathcal{G}$*. Then*

$$-\int_{\partial\mathcal{G}} \frac{\partial}{\partial\nu_x} F_q(|x-y|)\, dS(x) = \begin{cases} 1, & y \in \mathcal{G} \\ \frac{1}{2}, & y \in \partial\mathcal{G} \\ 0, & y \notin \overline{\mathcal{G}}. \end{cases} \tag{17.47}$$

In other words, the integral is a measure for the "solid angle" $\alpha_G(y)$ subtended by the boundary $\partial\mathcal{G}$ at the point $y \in \mathbb{R}^q$.

Remark 17.14. *From potential theory (see, e.g., O.D. Kellogg [1929]) it is known that Lemma 17.13 may be extended to regular regions* $\mathcal{G}$ *such as cube, simplex, polyhedron, more concretely, to regular regions with solid angle* $\alpha_{\mathcal{G}}(y)$ *at* $y \in \mathbb{R}^q$ *subtended by the surface* $\partial\mathcal{G}$ *(cf. Figures 17.3 and 17.4):*

$$-\int_{\partial\mathcal{G}} \frac{\partial}{\partial\nu_x} F_q(|x-y|)\, dS(x) = \alpha_{\mathcal{G}}(y), \quad y \in \mathbb{R}^q. \tag{17.48}$$

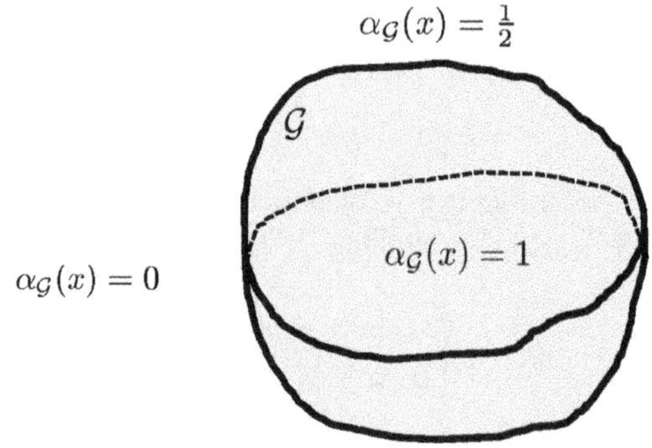

FIGURE 17.3: Solid angle subtended at $x \in \mathbb{R}^3$ by the surface $\partial\mathcal{G}$ of a regular region $\mathcal{G}$ with "smooth boundary".

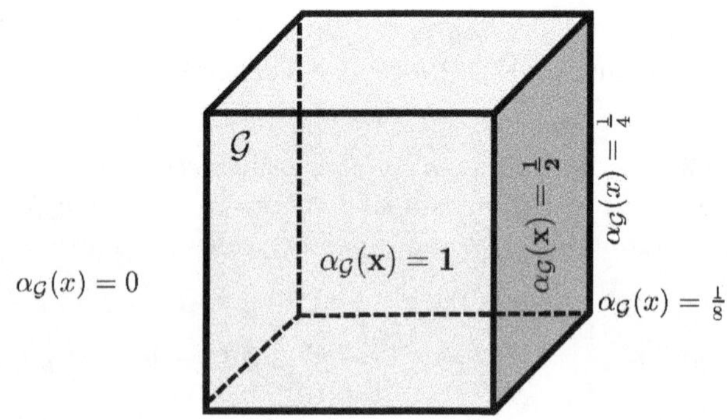

FIGURE 17.4: Solid angle subtended at $x \in \mathbb{R}^3$ by the surface $\partial\mathcal{G}$ of the "non-smooth" cube $\mathcal{G} = (-1,1)^3$.

Example 17.15. *For the cube (cf. Figure 17.4)*

$$\mathcal{G} = (-1,1)^3 \subset \mathbb{R}^3 \tag{17.49}$$

we have

(i) $\alpha_{\mathcal{G}}(y) = 1$ *if* y *is located in the open cube* $\mathcal{G}$,

(ii) $\alpha_{\mathcal{G}}(y) = \frac{1}{2}$ *if* y *is located on one of the six faces of the boundary* $\partial\mathcal{G}$ *of the cube* $\mathcal{G}$ *but not on an edge or in a vertex,*

(iii) $\alpha_{\mathcal{G}}(y) = \frac{1}{4}$ *if* y *is located on one of the eight edges of* $\partial\mathcal{G}$ *but not in a vertex,*

(iv) $\alpha_{\mathcal{G}}(y) = \frac{1}{8}$ *if* y *is located in one of the eight vertices of* $\partial\mathcal{G}$.

Lemma 17.13 is a special case of the third Green's theorem in $\mathbb{R}^q$ (see, e.g., O.D. Kellogg [1929], N. Aronszaijn et al. [1983]).

Theorem 17.16 (Third Green's Theorem). *Let $\mathcal{G}$ be a regular region. Suppose that $U : \overline{\mathcal{G}} \to \mathbb{R}$ is twice continuously differentiable, i.e., $U \in \mathrm{C}^{(2)}\left(\overline{\mathcal{G}}\right)$. Then we have*

$$\int_{\partial\mathcal{G}} F_q(|x-y|)\left(\frac{\partial U}{\partial\nu}(x)\right) - U(x)\left(\frac{\partial}{\partial\nu_x}F_q(|x-y|)\right)\, dS(x)$$
$$- \int_{\mathcal{G}} F_q(|x-y|)\Delta U(x)\, dx = \alpha_{\mathcal{G}}(y)U(y), \tag{17.50}$$

where $\alpha_{\mathcal{G}}(y)$, $y \in \mathbb{R}^q$, is the solid angle at y subtended by the surface $\partial\mathcal{G}$.

Finally we mention the *Poisson differential equation*, which is a classical result in potential theory (see, e.g., S.G. Michlin [1970] for the q-dimensional case or W. Freeden, M. Schreiner [2009] for an alternative proof in $\mathbb{R}^3$). We restrict our formulation to the case $q \geq 3$. The case $q = 2$ follows by obvious modification (cf. O.D. Kellogg [1929]).

Lemma 17.17. *Let F be of class $\mathrm{C}^{(0)}\left(\overline{\mathcal{G}}\right)$, $\overline{\mathcal{G}} = \mathcal{G} \cup \partial\mathcal{G} \subset \mathbb{R}^q$, $q \geq 3$. Then the* Newton potential *$U : \overline{\mathcal{G}} \to \mathbb{R}$ given by*

$$U(x) = \int_{\mathcal{G}} \frac{F(y)}{|x-y|^{q-2}}\, dy \tag{17.51}$$

is of class $\mathrm{C}^{(1)}\left(\overline{\mathcal{G}}\right)$, and we have

$$\nabla U(x) = \int_{\mathcal{G}} F(y)\nabla_x \frac{1}{|x-y|^{q-2}}\, dy. \tag{17.52}$$

If F is bounded in $\overline{\mathcal{G}} = \mathcal{G}\cup\partial\mathcal{G}$ satisfying a Hölder-condition in the neighborhood of the point $x \in \mathcal{G}$ (more precisely, $|F(y) - F(z)| \leq C_F|y-z|^\mu$ with $\mu \in (0,1]$ for all y, z in the neighborhood of the point $x \in \mathcal{G}$), then U as given by (17.51) is twice continuously differentiable in the neighborhood of $x \in \mathcal{G}$, and we have

$$\Delta U(x) = -(q-2)\|\mathbb{S}^{q-1}\|F(x). \tag{17.53}$$

Fundamental Solutions for Iterated Laplacians. By induction on m we obtain (see N. Aronszaijn et al. [1983])

$$\Delta^m |x|^k = A_{m,k}|x|^{k-2m}, \tag{17.54}$$

where we have used the abbreviations

$$A_{0,1} = 1, \quad A_{m,k} = \prod_{j=1}^{m}(k-2j+2)(k-2j+q). \tag{17.55}$$

An elementary calculation yields

$$A_{m,k} = 2^{2m}\frac{\Gamma\left(\frac{k}{2}+1\right)\Gamma\left(\frac{k+q}{2}\right)}{\Gamma\left(\frac{k}{2}-m+1\right)\Gamma\left(\frac{k+q}{2}-m\right)}, \tag{17.56}$$

except when k is even and $k \leq -2$ or $k+q$ is even and $k+q \leq 0$. Moreover,

$$\Delta\left(|x|^k \ln|x|\right) = k(k+q-2)|x|^{k-2}\ln|x| + (2k+q-2)|x|^{k-2}. \tag{17.57}$$

By induction on m we find (cf. N. Aronszaijn et al. [1983]) that

$$\begin{aligned}\Delta^m\left(|x|^k \ln|x|\right) = {} & A_{m,k}|x|^{k-2m}\ln|x| \\ & + |x|^{k-2m}\sum_{i=1}^{m} A_{i-1,k}A_{m-i,k-2i}(2k+q+2-4i).\end{aligned} \tag{17.58}$$

By virtue of (17.56) the preceding formula can be transformed into the identity

$$\Delta^m\left(|x|^k \ln|x|\right) = A_{m,k}|x|^{k-2m}\left(\ln|x| + \sum_{i=1}^{m}\left(\frac{1}{k-2i+2} + \frac{1}{k+q-2i}\right)\right). \tag{17.59}$$

For each operator Δ^l there corresponds a function $S_q^{(l)}$, $q \geq 2$, solving the equation $\Delta^l S_q^{(l)} = 0$ in $\mathbb{R}^q \setminus \{0\}$. To be more concrete, $S_q^{(l)}$ is given by

$$S_q^{(l)}(x) = \begin{cases} \frac{|x|^{2l-q}}{\gamma_{l-1}}, & q \text{ odd},\ l = 1,2,\ldots, \\ \frac{|x|^{2l-q}}{\tilde{\gamma}_{l-1}}, & q \text{ even},\ l = 1,2,\ldots,\frac{q}{2}-1, \\ \frac{-|x|^{2l-q}\ln|x|}{\tilde{\gamma}_{l-1}}, & q \text{ even},\ l = \frac{q}{2},\frac{q}{2}+1,\ldots. \end{cases} \tag{17.60}$$

The constants γ_l and $\tilde{\gamma}_l$ (in the notation of (17.55)) are given as follows:

- q odd:

$$\gamma_l = A_{l,2l+2-q} = 2^{2l}l!\frac{\Gamma\left(l+2-\frac{q}{2}\right)}{\Gamma\left(2-\frac{q}{2}\right)}, \quad l = 0,1,\ldots, \tag{17.61}$$

- $q = 2$:

$$\tilde{\gamma}_l = 2^{2l}(l!)^2, \quad l \in \mathbb{N}_0. \tag{17.62}$$

- q even and $q = 2r \geq 4$:

$$\tilde{\gamma}_l = A_{l,2l+2-q} = (-1)^l 2^{2l} l! \frac{(r-2)!}{(r-l-2)!}, \quad l = 0,\ldots,r-2, \tag{17.63}$$

$$\tilde{\gamma}_l = (-1)^{r+1}2^{2l-1}(r-2)!\ l!\ (l+1-r)!, \quad l = r-1, r, \ldots. \tag{17.64}$$

For brevity we let

$$\Omega_q = \begin{cases} \frac{1}{(q-2)\|\mathbb{S}^{q-1}\|}, & q \neq 2, \\ \frac{1}{\|\mathbb{S}^{q-1}\|}, & q = 2. \end{cases} \tag{17.65}$$

The *fundamental solution* $F_q^{(l)} : \mathbb{R}^q \setminus \{0\} \to \mathbb{R}$ for the equation $\Delta^l F_q^{(l)} = 0$ is defined by

$$F_q^{(l)}(|x|) = \begin{cases} \Omega_q \dfrac{|x|^{2l-q}}{\gamma_{l-1}}, & q \text{ odd}, \ l = 1, 2, \ldots, \\ \Omega_q \dfrac{|x|^{2l-q}}{\tilde{\gamma}_{l-1}}, & q = 2r, \ l = 1, \ldots, r-1 \\ \Omega_q \left(\dfrac{-|x|^{2l-q} \ln |x|}{\tilde{\gamma}_{l-1}} + \dfrac{C_{l-\frac{q}{2}}}{\tilde{\gamma}_{l-1}} |x|^{2l-q} \right), & q = 2r, \ l = r, r+1, \ldots, \end{cases} \tag{17.66}$$

where the constants C_l, $l = 0, 1, \ldots$, are given by

$$C_0 = 0, \tag{17.67}$$

$$C_l = \sum_{k=1}^{l} \frac{1}{2k} + \sum_{k=r}^{l+r-1} \frac{1}{2k}, \quad l > 0. \tag{17.68}$$

With this choice of constants C_l it immediately follows that, for all dimensions $q = 2, 3, \ldots$ and all degrees $l = 1, 2, \ldots$, we have

$$\Delta^m F_q^{(l)} = F_q^{(l-m)}, \quad m < l. \tag{17.69}$$

Remark 17.18. *The fundamental solution (17.66) plays an important role in the characterization of the singularity behavior of the lattice function which is needed in the establishment of multivariate Euler summation formulas (see Section 18.1).*

Now, suppose that $\mathcal{G}$ is a regular region. By standard arguments we obtain from the extended second Green's theorem (Theorem 17.10) the following extension of the third Green's theorem.

Theorem 17.19 (Extended Third Green's Theorem)**.** *Let $\mathcal{G} \subset \mathbb{R}^q$, $q \geq 2$, be a regular region. Let $F : \overline{\mathcal{G}} \to \mathbb{R}$ be of the class $\mathrm{C}^{(2m)}\left(\overline{\mathcal{G}}\right), \overline{\mathcal{G}} = \mathcal{G} \cup \partial\mathcal{G}$, $m \geq 1$. Then*

$$\begin{aligned} &\sum_{l=0}^{m-1} \int_{\partial\mathcal{G}} F_q^{(l+1)}(|x-y|) \left(\frac{\partial}{\partial\nu_y} \Delta_y^l F(y) \right) dS(y) \\ &- \sum_{l=0}^{m-1} \int_{\partial\mathcal{G}} \left(\frac{\partial F_q^{(l+1)}}{\partial\nu_y}(|x-y|) \right) \Delta_y^l F(y) \, dS(y) \\ &- \int_{\mathcal{G}} F_q^{(m)}(|x-y|) \Delta_y^m F(y) \, dy \\ &= \alpha_{\mathcal{G}}(x) F(x), \end{aligned} \tag{17.70}$$

where $\alpha_{\mathcal{G}}(x)$, $x \in \mathbb{R}^q$, is the solid angle at x subtended by the surface $\partial\mathcal{G}$.

Remark 17.20. *For $m = 1$, Theorem 17.19 is the usual* third Green's theorem for the Laplace operator, *i.e., Theorem 17.16.*

Observing the differential equation (17.69) we obtain the following reformulation of Theorem 17.19.

Corollary 17.21. *Under the assumptions of Theorem 17.19 we have*

$$\begin{aligned}
&\sum_{l=0}^{m-1} \int_{\partial\mathcal{G}} \left(\Delta_y^{2m-l-1} F_q^{(2m)}(|x-y|)\right) \left(\frac{\partial}{\partial\nu_y} \Delta_y^l F(y)\right) \, dS(y) \\
&- \sum_{l=0}^{m-1} \int_{\partial\mathcal{G}} \left(\frac{\partial}{\partial\nu_y} \Delta_y^{2m-l-1} F_q^{(2m)}(|x-y|)\right) \left(\Delta_y^l F(y)\right) dS(y) \qquad (17.71) \\
&- \int_{\mathcal{G}} \left(\Delta_y^m F_q^{(2m)}(|x-y|)\right) \left(\Delta_y^m F(y)\right) \, dy \\
&= \alpha_{\mathcal{G}}(x) F(x),
\end{aligned}$$

where, as usual, $\alpha_{\mathcal{G}}(x)$, $x \in \mathbb{R}^q$, is the solid angle at x subtended by the surface $\partial\mathcal{G}$.

17.4 Fourier Transform in Euclidean Spaces

Next we deal with various spaces of functions defined on $\mathbb{R}^q$ (for more details see, e.g., M.A. Pinsky [2002], E.M. Stein, G. Weiss [1971]). $\mathrm{L}^p(\mathbb{R}^q)$, $1 \le p < \infty$, are the spaces of all measurable functions $F : \mathbb{R}^q \to \mathbb{C}$ such that

$$\|F\|_{\mathrm{L}^p(\mathbb{R}^q)} = \left(\int_{\mathbb{R}^q} |F(x)|^p \, dx\right)^{1/p} < \infty. \qquad (17.72)$$

The number $\|F\|_{\mathrm{L}^p(\mathbb{R}^q)}$ is called the L^p-norm of F. If F is of class $\mathrm{L}^1(\mathbb{R}^q)$, the *Fourier transform* of F is the function $F^\wedge_{\mathbb{R}^q}$ defined by letting

$$F^\wedge_{\mathbb{R}^q}(x) = \int_{\mathbb{R}^q} F(y) e^{-2\pi i x \cdot y} \, dy \qquad (17.73)$$

for all $x \in \mathbb{R}^q$. If F is of class $\mathrm{L}^1(\mathbb{R}^q)$, then it is well-known that $F^\wedge_{\mathbb{R}^q}$ is uniformly continuous. Moreover, the Riemann–Lebesgue Theorem of Fourier analysis tells us that $F^\wedge_{\mathbb{R}^q}(x) \to 0$ as $|x| \to \infty$. In addition to the linear-space operations, $\mathrm{L}^1(\mathbb{R}^q)$ is endowed with a *convolution* defined in the following way: If F, G belong to $\mathrm{L}^1(\mathbb{R}^q)$, their convolution $H = F * G$ is the function whose value at $x \in \mathbb{R}^q$ is given by

$$H(x) = \int_{\mathbb{R}^q} F(x-y) G(y) \, dy. \qquad (17.74)$$

Clearly, $H \in \mathrm{L}^1(\mathbb{R}^q)$ and $\|H\|_{\mathrm{L}^1(\mathbb{R}^q)} \le \|F\|_{\mathrm{L}^1(\mathbb{R}^q)} \|G\|_{\mathrm{L}^1(\mathbb{R}^q)}$. The operation is commutative and associative. Even more, if $F \in \mathrm{L}^p(\mathbb{R}^q)$, $1 \le p < \infty$, and

$G \in \mathrm{L}^1(\mathbb{R}^q)$, then $H = F * G$ belongs to $\mathrm{L}^p(\mathbb{R}^q)$, and we have the estimate $\|H\|_{\mathrm{L}^p(\mathbb{R}^q)} \leq \|F\|_{\mathrm{L}^p(\mathbb{R}^q)}\|G\|_{\mathrm{L}^1(\mathbb{R}^q)}$. An essential feature is the fact that the Fourier transform of the convolution of two functions is the (pointwise) product of their Fourier transforms. More precisely, if F and G belong to $\mathrm{L}^1(\mathbb{R}^q)$, then

$$(F * G)^{\wedge}_{\mathbb{R}^q} = F^{\wedge}_{\mathbb{R}^q} G^{\wedge}_{\mathbb{R}^q}. \tag{17.75}$$

Many other important operations have simple relations in the context of the Fourier transform:

- If τ_d denotes the *translation* by $d \in \mathbb{R}^q$ (by this we mean the operator mapping the function $F \in \mathrm{L}^1(\mathbb{R}^q)$ into the function $\tau_d F = F(\cdot - d)$), then

$$(\tau_d F)^{\wedge}_{\mathbb{R}^q}(x) = e(-d \cdot x) F^{\wedge}_{\mathbb{R}^q}(x), \quad x \in \mathbb{R}^q \tag{17.76}$$

  and

$$(e(d \cdot\ ) F)^{\wedge}_{\mathbb{R}^q}(x) = (\tau_d F^{\wedge}_{\mathbb{R}^q})(x), \quad x \in \mathbb{R}^q. \tag{17.77}$$

- If δ_a, $a > 0$, denotes the *dilation* (by this we mean the operator mapping $F \in \mathrm{L}^1(\mathbb{R}^q)$ into the function $\delta_a F = F(a \cdot\ )$), then

$$a^q (\delta_a F)^{\wedge}_{\mathbb{R}^q}(x) = F^{\wedge}_{\mathbb{R}^q}(a^{-1}x), \quad x \in \mathbb{R}^q. \tag{17.78}$$

- If $x \mapsto P(x)$, $x \in \mathbb{R}^q$, is a polynomial in q variables $x = (x_1, \ldots, x_q)^T$ and $P(\nabla)$ is the associated differential operator (i.e., replace $x_i = x \cdot \varepsilon^i$ by $\frac{\partial}{\partial x_i} = \nabla \cdot \varepsilon^i$, $i \in \{1, \ldots, q\}$), then

$$P(\nabla) F^{\wedge}_{\mathbb{R}^q}(x) = (P(-2\pi i \cdot\ ) F)^{\wedge}_{\mathbb{R}^q}(x), \quad x \in \mathbb{R}^q, \tag{17.79}$$

  and

$$(P(\nabla) F)^{\wedge}_{\mathbb{R}^q}(x) = P(2\pi i x) F^{\wedge}_{\mathbb{R}^q}(x), \quad x \in \mathbb{R}^q. \tag{17.80}$$

Applying Fubini's Theorem we immediately have

$$\int_{\mathbb{R}^q} F^{\wedge}(x) \overline{G(x)}\, dx = \int_{\mathbb{R}^q} F(x) \overline{G^{\wedge}(x)}\, dx \tag{17.81}$$

provided that F, G belong to $\mathrm{L}^1(\mathbb{R}^q)$. Moreover, if $F \in \mathrm{L}^1(\mathbb{R}^q)$, $G \in \mathrm{L}^p(\mathbb{R}^q)$, $1 \leq p \leq 2$, then $H = F * G \in \mathrm{L}^p(\mathbb{R}^q)$ satisfies $H^{\wedge}_{\mathbb{R}^q}(x) = F^{\wedge}_{\mathbb{R}^q}(x) G^{\wedge}_{\mathbb{R}^q}(x)$ (for almost every x). Finally we mention the *inversion formula* (see, e.g., E.M. Stein, G. Weiss [1971])

$$F(x) = \int_{\mathbb{R}^q} F^{\wedge}_{\mathbb{R}^q}(y) e^{2\pi i x \cdot y} dy \tag{17.82}$$

(for almost every x) provided that $F, F^{\wedge}_{\mathbb{R}^q}$ belong to $\mathrm{L}^1(\mathbb{R}^q)$. It should be remarked that there are important examples in the theory of multivariate alternating functions where the last sufficient conditions (i.e., $F, F^{\wedge}_{\mathbb{R}^q} \in \mathrm{L}^1(\mathbb{R}^q)$) for the validity of the inversion formula cannot be guaranteed. In these cases we have to find a way out to specify a methodical strategy closely dependent on the properties of $F^{\wedge}_{\mathbb{R}^q}$. Another procedure is to understand the inversion formula in summability sense which we discuss next.

Gaussian Inversion Formula. Next we are concerned with a weighted variant of the inverse Fourier transform, namely the Gauss–Weierstrass integral transform in their particular interrelation to the classical Fourier transform. In the usual $L^1(\mathbb{R}^q)$-nomenclature, the theory of these integral transforms is well understood (see, e.g., E.M. Stein, G. Weiss [1971]). In the approach here, however, we are essentially inspired by C. Müller [1998], W. Freeden [2011] to transfer the theory of Fourier transforms (defined as spherical principal values) to so-called spherically continuous functions. In doing so we are able to deal with structures and settings in pointwise sense that are adequate in inversion procedures.

We begin with the *"Gaussian function"* $G : \mathbb{R}^q \to \mathbb{C}$ given by

$$G(x) = e^{2\pi i x\cdot z} e^{-\tau\pi^2 x^2}, \quad z \in \mathbb{R}^q, \quad \tau > 0 \tag{17.83}$$

and the *"Weierstrass function"* $W : \mathbb{R}^q \to \mathbb{C}$ given by

$$W(x) = \left(\frac{1}{\pi\tau}\right)^{q/2} e^{-\frac{(z-x)^2}{\tau}}, \quad z \in \mathbb{R}^q, \quad \tau > 0. \tag{17.84}$$

By elementary calculations (see, e.g., W. Freeden [2011]) we are led to

Lemma 17.22. *For all $x \in \mathbb{R}^q$ we have*

$$W^{\wedge}_{\mathbb{R}^q}(-x) = G(x) = \int_{\mathbb{R}^q} \underbrace{G^{\wedge}_{\mathbb{R}^q}(y)}_{=W(y)} \, e^{2\pi i x\cdot y} \, dy, \tag{17.85}$$

$$G^{\wedge}_{\mathbb{R}^q}(x) = W(x) = \int_{\mathbb{R}^q} \underbrace{W^{\wedge}_{\mathbb{R}^q}(y)}_{=G(-y)} \, e^{2\pi i x\cdot y} \, dy. \tag{17.86}$$

In similar way we obtain for the *Gauss transform* $\mathcal{G}_\tau(F)$ of $F : \mathbb{R}^q \to \mathbb{C}$ at $x \in \mathbb{R}^q$

$$\mathcal{G}_\tau(F)(x) = \left(\frac{1}{\pi\tau}\right)^{q/2} \int_{\mathbb{R}^q} e^{-\frac{(x-y)^2}{\tau}} F(y) \, dy, \tag{17.87}$$

and the *Weierstrass transform* $\mathcal{W}_\tau(F)$ of $F : \mathbb{R}^q \to \mathbb{C}$ at $x \in \mathbb{R}^q$

$$\mathcal{W}_\tau(F)(x) = \int_{\mathbb{R}^q} e^{-\tau\pi^2 y^2} e^{2\pi i x\cdot y} F(y) \, dy. \tag{17.88}$$

Lemma 17.23. *For $F \in L^1(\mathbb{R}^q)$, $\tau > 0$, and $z \in \mathbb{R}^q$*

$$\mathcal{W}_\tau(F^{\wedge}_{\mathbb{R}^q})(z) = \mathcal{G}_\tau(F)(z), \tag{17.89}$$

$$\mathcal{G}_\tau(F^{\wedge}_{\mathbb{R}^q})(z) = \mathcal{W}_\tau(F)(-z). \tag{17.90}$$

Next we mention that $\mathcal{W}_\tau(F)(x)$ and $\mathcal{G}_\tau(F)(x)$ tend to $F(x)$ as $\tau \to 0$, if suitable conditions are imposed on the position x as well as the function F under consideration. We begin with the Weierstrass transform.

Lemma 17.24. *Suppose that F is continuous and uniformly bounded in $\mathbb{R}^q$ such that, for $z \in \mathbb{R}^q$, the limit*

$$\lim_{N\to\infty} \int_{\substack{|x|\le N\\ x\in\mathbb{R}^q}} F(x)\ e^{-2\pi i z\cdot x}\ dx \tag{17.91}$$

exists. Then

$$\lim_{\substack{\tau\to 0\\ \tau>0}} \mathcal{W}_\tau(F)(z) = \lim_{N\to\infty} \int_{\substack{|x|\le N\\ x\in\mathbb{R}^q}} F(x)\ e^{-2\pi i x\cdot z}\ dx. \tag{17.92}$$

The next result (cf. W. Freeden [2011]) helps us to come closer to the ordinary Fourier inversion formula in $\mathbb{R}^q$ (under conditions of interest in our approach).

Lemma 17.25. *Suppose that $F \in \mathrm{L}^1(\mathbb{R}^q)$ is spherically continuous at $z \in \mathbb{R}^q$, i.e., the mean over the unit sphere $\mathbb{S}^{q-1}$ around the origin given by*

$$\Phi(r;z) = \frac{1}{\|\mathbb{S}^{q-1}\|} \int_{\mathbb{S}^{q-1}} F(z+r\xi)\ dS(\xi) \tag{17.93}$$

exists for all $r > 0$ and is continuous for $r \to 0$ with $\Phi(0;z) = F(z)$. Then

$$\lim_{\substack{\tau\to 0\\ \tau>0}} \mathcal{G}_\tau(F)(z) = \lim_{\substack{\tau\to 0\\ \tau>0}} \mathcal{W}_\tau(F^\wedge_{\mathbb{R}^q})(z) = F(z). \tag{17.94}$$

It should be pointed out that Lemma 17.25 allows the characterization as *pointwise Fourier inversion formula.*

Theorem 17.26. *Suppose that $F \in \mathrm{L}^1(\mathbb{R}^q)$ is spherically continuous at $z \in \mathbb{R}^q$. Furthermore, assume that the "spherical principal value"*

$$\lim_{N\to\infty} \int_{\substack{|y|\le N\\ y\in\mathbb{R}^q}} e^{2\pi i z\cdot y} F^\wedge_{\mathbb{R}^q}(y)\ dy \tag{17.95}$$

exists. Then we have

$$F(z) = \lim_{N\to\infty} \int_{\substack{|y|\le N\\ y\in\mathbb{R}^q}} e^{2\pi i z\cdot y} F^\wedge_{\mathbb{R}^q}(y)\ dy, \tag{17.96}$$

If the spherical principal value does not exist, the limit relation

$$F(z) = \lim_{\substack{\tau\to 0\\ \tau>0}} \int_{\mathbb{R}^q} e^{-\tau\pi^2 y^2} e^{2\pi i z\cdot y} F^\wedge_{\mathbb{R}^q}(y)\ dy \tag{17.97}$$

remains valid as a "mean representation" of F at the point z of spherical continuity.

The "spherical continuity" is of particular significance if F is a "discontinuous function".

Lemma 17.27. *Let $\mathcal{G}$ be a regular region in $\mathbb{R}^q$. Suppose that F is of class $\mathrm{C}^{(0)}(\overline{\mathcal{G}})$, $\overline{\mathcal{G}} = \mathcal{G} \cup \partial\mathcal{G}$. Then*

$$\lim_{\substack{\tau\to 0\\ \tau>0}} \int_{\mathbb{R}^q} e^{-\tau\pi^2 y^2} e^{2\pi i z\cdot y} \int_{\mathcal{G}} F(x)\, e^{-2\pi i x\cdot y}\, dx\, dy \tag{17.98}$$

$$= \lim_{\substack{\tau\to 0\\ \tau>0}} \int_{\mathbb{R}^q} e^{-\tau\pi^2 y^2}\, e^{2\pi i z\cdot y} F^{\wedge}_{\mathcal{G}}(y)\, dy = \alpha_{\mathcal{G}}(z) F(z),$$

where $\alpha_{\mathcal{G}}(z)$ denotes the solid angle subtended by $\partial\mathcal{G}$ at z and $F^{\wedge}_{\mathcal{G}}$ is given by

$$F^{\wedge}_{\mathcal{G}}(y) = \int_{\mathcal{G}} F(x)\, e^{-2\pi i x\cdot y}\, dx, \quad y \in \mathbb{R}^q. \tag{17.99}$$

Moreover,

$$\int_{\mathbb{R}^q} e^{2\pi i z\cdot y} F^{\wedge}_{\mathcal{G}}(y)\, dy = \alpha_{\mathcal{G}}(z) F(z), \tag{17.100}$$

where

$$\int_{\mathbb{R}^1} e^{2\pi i z\cdot y} F^{\wedge}_{\mathcal{G}}(y)\, dy = \lim_{N\to\infty} \int_{\substack{|x|\le N\\ x\in\mathbb{R}^q}} e^{2\pi i z\cdot y} F^{\wedge}_{\mathcal{G}}(y)\, dy. \tag{17.101}$$

17.5 Periodization and Poisson Summation Formula

Suppose we are given an (appropriate) function F in Euclidean space $\mathbb{R}^q$. The question is what its periodic analog is. Our purpose is to show that there are at least two ways of obtaining a periodization of the function under consideration. The "first periodization" is straightforward:

$$x \mapsto S(x) = \sum_{g\in\Lambda} F(x+g). \tag{17.102}$$

Since this (formal) sum is extended over all points of the lattice Λ it is obviously periodic. The "second periodization" is based on the Fourier expansion

$$x \mapsto \frac{1}{\sqrt{\|\mathcal{F}_\Lambda\|}} \sum_{h\in\Lambda^{-1}} S^{\wedge}_{\mathcal{F}_\Lambda}(h)\, \Phi_h(x). \tag{17.103}$$

The Fourier coefficients $S^{\wedge}_{\mathcal{F}_\Lambda}(h)$, $h \in \Lambda^{-1}$, of the series (17.103) are calculable as follows

$$S^{\wedge}_{\mathcal{F}_\Lambda}(h) = \int_{\mathcal{F}_\Lambda} \sum_{g\in\Lambda} F(x+g)\overline{\Phi_h(x)}\, dx = \sum_{g\in\Lambda} \int_{\mathcal{F}_\Lambda(g)} F(x)\overline{\Phi_h(x)}\, dx, \tag{17.104}$$

where $\mathcal{F}_\Lambda(g)$ denotes the translate of $\mathcal{F}_\Lambda$ containing the lattice point $g \in \Lambda$. Since the fundamental cell $\mathcal{F}_\Lambda$ of the lattice Λ defines translates $\mathcal{F}_\Lambda(g)$, $g \in \Lambda$, that are mutually disjoint, their union is $\mathbb{R}^q$. This observation implies the identity

$$\int_{\mathcal{F}_\Lambda} \sum_{g\in\Lambda} F(x+g)\overline{\Phi_h(x)}\, dx = \int_{\mathbb{R}^q} F(x)\overline{\Phi_h(x)}\, dx = \frac{1}{\sqrt{\|\mathcal{F}_\Lambda\|}} F^\wedge_{\mathbb{R}^q}(h). \quad (17.105)$$

The conclusion is that the first and second periodization are identical so that the *Poisson summation formula for the Euclidean space* $\mathbb{R}^q$

$$\sum_{g\in\Lambda} F(x+g) = \sum_{h\in\Lambda^{-1}} \int_{\mathbb{R}^q} F(y)\overline{\Phi_h(y)}\, dy\ \Phi_h(x) \quad (17.106)$$

holds true. The identity (17.106) is expressible in many ways. The simplest and standard manifestation (see E.M. Stein, G. Weiss [1971]) is the following.

Theorem 17.28. *Suppose that* F *and* $F^\wedge_{\mathbb{R}^q}$ *satisfy the conditions*

$$F(x) = O\left(\frac{1}{|x|^{q+\varepsilon}}\right), \quad \varepsilon > 0, \quad |x| \to \infty, \quad (17.107)$$

$$F^\wedge_{\mathbb{R}^q}(x) = O\left(\frac{1}{|x|^{q+\varepsilon}}\right), \quad \varepsilon > 0, \quad |x| \to \infty \quad (17.108)$$

with

$$F(x) = \int_{\mathbb{R}^q} F^\wedge_{\mathbb{R}^q}(y) e^{2\pi i x\cdot y}\, dy, \quad x \in \mathbb{R}^q. \quad (17.109)$$

Then, F *and* $F^\wedge_{\mathbb{R}^q}$ *are continuous in* $\mathbb{R}^q$ *such that*

$$\sum_{g\in\Lambda} F(x+g) = \sum_{h\in\Lambda^{-1}} \int_{\mathbb{R}^q} F(y)\overline{\Phi_h(y)}\, dy\ \Phi_h(x). \quad (17.110)$$

In particular,

$$\sum_{g\in\Lambda} F(g) = \frac{1}{\sqrt{\|\mathcal{F}_\Lambda\|}} \sum_{h\in\Lambda^{-1}} \int_{\mathbb{R}^q} F(y)\overline{\Phi_h(y)}\, dy. \quad (17.111)$$

The four series in (17.110) *and* (17.111) *converge absolutely.*

Later on, we shall provide sufficient conditions for the validity of the Poisson summation formula such that the series on the left-hand side of (17.111) involved in its representation do not necessarily show absolute convergence.

Chapter 18

Euler Summation Based Integration

This chapter first deals with multi-dimensional analogs of the Bernoulli polynomials related to iterated Laplace operators (cf. W. Freeden [2011]). As in the one-dimensional theory, the multi-dimensional "Bernoulli polynomials" may be understood as Green's functions for the Laplace operator Δ and the "boundary condition of Λ-lattice periodicity", so that multivariate Euler summation formulas become accessible for regular regions.

The layout of this chapter is as follows: In Section 18.1 we introduce the Green's function for the Laplace operator and the "boundary condition of Λ-periodicity" – briefly called Λ-lattice function $G(\Delta;\cdot)$ – by its constituting properties, i.e., differential equation, characteristic singularity, boundary condition, and normalization. We also mention the uniqueness of this function. Contrary to the one-dimensional case, the existence of the Λ-lattice function $G(\Delta;\cdot)$ cannot be guaranteed just by the representation in terms of its bilinear expansion. Instead we need as auxiliary mathematical tool the Fredholm-Hilbert theory of linear (weakly) singular integral equations. From this theory we are able to deduce that each iteration, i.e., each convolution over the fundamental cell, reduces the order of the singularity by two, such that $G(\Delta^m;\cdot)$, $m \geq \frac{q}{2}$, turns out to be a continuous function in $\mathbb{R}^q$. Based on the properties of the Λ-lattice function for iterated Laplace operators, Section 18.2 develops the Euler summation formula by use of integration by parts, i.e., the second Green's theorem in quite analogous way as in one-dimensional theory (cf. Section 3.3). Moreover, Euler summation formulas are formulated under Dirichlet/Neumann boundary conditions for regular regions. Furthermore, we discuss the theory of the zeta function within its q-dimensional framework, thereby taking special advantage of the Euler summation formula for establishing its functional equation. The chapter continues with the formulation of Euler cubature formulas on regular regions for iterated Laplace operators. Finally, in Section 18.4, multivariate Romberg extrapolation involving regular regions and based on Euler summation formulas for iterated Laplace operators is investigated in more detail.

18.1 Euler Summation Formulas for Laplace Operators

Our considerations start with the defining constituents of the Λ-lattice function. In accordance with the one-dimensional theory and seen from the point of mathematical physics, the Λ-lattice function as introduced in Definition 18.1 is nothing more than the Green's function for the Laplace operator Δ in Euclidean space $\mathbb{R}^q$ corresponding to the "boundary condition of periodicity with regard to the lattice Λ". Formally understood we are interested in a Λ-periodic function $G(\Delta;\cdot)$ possessing the Fourier expansion

$$\frac{1}{\sqrt{\|\mathcal{F}_\Lambda\|}} \sum_{\substack{h\neq 0 \\ h\in\Lambda^{-1}}} \frac{1}{-\Delta^\wedge(h)}\, \Phi_h \tag{18.1}$$

such that its Laplace derivative

$$\Delta G(\Delta;\cdot) = \delta - \frac{1}{\|\mathcal{F}_\Lambda\|} \tag{18.2}$$

leads over to the Λ-periodic Dirac function(al)

$$\delta = \frac{1}{\sqrt{\|\mathcal{F}_\Lambda\|}} \sum_{h\in\Lambda^{-1}} \Phi_h. \tag{18.3}$$

In contrary to the one-dimensional case, however, the Fourier series (18.1) is not convergent. Thus, for dimensions $q \geq 2$, there is a need to introduce $G(\Delta;\cdot)$ by defining properties.

Definition 18.1. $G(\Delta;\cdot) : \mathbb{R}^q \setminus \Lambda \to \mathbb{R}$ *is called the* Λ-lattice function for the operator Δ *if it satisfies the following properties:*

*(i) (*Periodicity*) For all $x \in \mathbb{R}^q \setminus \Lambda$ and $g \in \Lambda$*

$$G(\Delta; x+g) = G(\Delta; x) \tag{18.4}$$

*(ii) (*Differential equation*) $G(\Delta;\cdot)$ is twice continuously differentiable for all $x \notin \Lambda$, such that*

$$\Delta G(\Delta; x) = -\frac{1}{\|\mathcal{F}_\Lambda\|}, \tag{18.5}$$

holds true for $x \notin \Lambda$.

*(iii) (*Characteristic singularity*) In the neighborhood of the origin*

$$G(\Delta; x) + F_q(|x|) = \begin{cases} O(1) & , q=2, \\ O\left(|x|^{3-q}\ln|x|\right) & , q=4, \\ O\left(|x|^{3-q}\right) & , q\neq 2,4 \end{cases} \tag{18.6}$$

and

$$\nabla_x G(\Delta; x) + \nabla_x F_q(|x|) = \begin{cases} O(1) & , q=2, \\ O\left(|x|^{2-q}\ln|x|\right) & , q=4, \\ O\left(|x|^{2-q}\right) & , q\neq 2,4, \end{cases} \tag{18.7}$$

where $F_q : x \mapsto F_q(|x|), x \neq 0$, is the fundamental solution in $\mathbb{R}^q$ with respect to the Laplace operator Δ (see (17.45)).

(iv) *(*Normalization*)*

$$\int_{\mathcal{F}_\Lambda} G(\Delta; x)\overline{\Phi_h(x)}\, dx = 0. \tag{18.8}$$

The uniqueness of the Λ-lattice function for the Laplace operator Δ can be guaranteed by showing that the difference of two Λ-lattice functions is the zero function. The existence of the Λ-lattice function for the Laplace operator follows from well-known arguments based on Hilbert's theory of integral equations (for more details the reader is referred to W. Freeden [2011]).

The Λ-lattice functions for operators Δ^m, $m \in \mathbb{N}$, can be based on the Λ-lattice function for Δ by forming convolution integrals in the usual way as known from mathematical physics.

Definition 18.2. *The function* $G(\Delta^m; \cdot)$, $m = 1, 2, \ldots$, *defined by*

$$G\left(\Delta^1; x\right) = G^{(1)}\left(\Delta; x\right) = G\left(\Delta; x\right), \tag{18.9}$$

$$G\left(\Delta^m; x\right) = G^{(m)}\left(\Delta; x\right) = \int_{\mathcal{F}_\Lambda} G\left(\Delta^{m-1}; z\right) G(\Delta; x-z)\, dz,$$

$m = 2, 3, \ldots$, *is called the* Λ-lattice function for the operator Δ^m.

Obviously, for all $x \notin \Lambda$ and $g \in \Lambda$,

$$G\left(\Delta^m; x+g\right) = G\left(\Delta^m; x\right) \tag{18.10}$$

is satisfied, i.e., $G\left(\Delta^m; \cdot\right)$ is Λ-periodic. In analogy to well-known techniques of potential theory (see, e.g., N. Aronszaijn et al. [1983]) it can be deduced that

$$G\left(\Delta^m; x\right) = \begin{cases} O\left(|x|^{2m-q} \ln |x|\right) & , q = 2m, \\ O\left(|x|^{2m-q}\right) & , q > 2m. \end{cases} \tag{18.11}$$

The differential equation

$$\Delta_x G\left(\Delta^m; x\right) = G\left(\Delta^{m-1}; x\right), \quad x \notin \Lambda, \tag{18.12}$$

$m = 2, 3, \ldots$, represents a recursion relation relating the Λ-lattice function for the operator Δ^m to the Λ-lattice function for the operator Δ^{m-1}. The series expansion of $G\left(\Delta^m; \cdot\right)$ in terms of eigenfunctions, which is equivalent to the (formal) Fourier (orthogonal) expansion, is given for the iteration orders $m = 2, 3, \ldots$ by

$$\frac{1}{\sqrt{\|\mathcal{F}_\Lambda\|}} \sum_{\substack{h \in \Lambda^{-1} \\ \Delta^\wedge(h) \neq 0}} \frac{\Phi_h(x)}{-(\Delta^m)^\wedge(h)}, \tag{18.13}$$

where

$$(\Delta^m)^\wedge(h) = -(-\Delta^\wedge(h))^m = -(-4\pi^2 h^2)^m. \tag{18.14}$$

For $m > \frac{q}{2}$, therefore, it follows (cf. Theorem 17.7) that there is a constant $C > 0$ such that

$$\left| \sum_{\substack{h \neq 0 \\ h \in \Lambda^{-1}}} \frac{\Phi_h(x)}{-(\Delta^m)^\wedge(h)} \right| = \left| \frac{1}{\sqrt{\|\mathcal{F}_\Lambda\|}} \sum_{\substack{h \neq 0 \\ h \in \Lambda^{-1}}} \frac{e^{2\pi i h \cdot x}}{(-4\pi^2 h^2)^m} \right| \tag{18.15}$$
$$\leq C \sum_{h \in \Lambda^{-1}} \frac{1}{(1+h^2)^m} < \infty.$$

So it is clear that the Fourier series (18.13) converges absolutely and uniformly in $\mathbb{R}^q$.

Lemma 18.3. *For all $m > \frac{q}{2}$, the Λ-lattice function $G(\Delta^m; \cdot)$ is continuous in $\mathbb{R}^q$, and its bilinear series is given by*

$$G(\Delta^m; x - y) = \sum_{\substack{h \neq 0 \\ h \in \Lambda^{-1}}} \frac{\Phi_h(x)\overline{\Phi_h(y)}}{-(\Delta^m)^\wedge(h)}, \quad x, y \in \mathbb{R}^q. \tag{18.16}$$

Next, based on the properties of the multi-dimensional Λ-lattice functions for iterated Laplace operators the Euler summation formula will be developed by use of the Green's formula for the Laplacian.

Theorem 18.4 (Lattice Point-Generated Euler Summation Formula for the Laplace Operator Δ). *Let Λ be an arbitrary lattice in $\mathbb{R}^q$. Suppose that $\mathcal{G} \subset \mathbb{R}^q$ is a regular region. Let F be twice continuously differentiable on $\overline{\mathcal{G}}$, $\overline{\mathcal{G}} = \mathcal{G} \cup \partial\mathcal{G}$. Then*

$$\sum_{\substack{g \in \overline{\mathcal{G}} \\ g \in \Lambda}}{}' F(g) = \frac{1}{\|\mathcal{F}_\Lambda\|} \int_{\mathcal{G}} F(x)\, dx \tag{18.17}$$
$$+ \int_{\partial\mathcal{G}} F(x) \left(\frac{\partial}{\partial\nu} G(\Delta; x) \right) - G(\Delta; x) \left(\frac{\partial F}{\partial\nu}(x) \right)\, dS(x),$$
$$+ \int_{\mathcal{G}} G(\Delta; x) \Delta F(x)\, dx,$$

where

$$\sum_{\substack{g \in \overline{\mathcal{G}} \\ g \in \Lambda}}{}' F(g) = \sum_{\substack{g \in \overline{\mathcal{G}} \\ g \in \Lambda}} \alpha_{\mathcal{G}}(g) F(g) \tag{18.18}$$

and as always $\alpha_{\mathcal{G}}(g)$ is the solid angle subtended at $g \in \overline{\mathcal{G}}$ by the surface $\partial\mathcal{G}$.

The formula for the Laplacian Δ (Theorem 18.4) is a multi-dimensional generalization of the one-dimensional Euler summation formula, where $G(\Delta; \cdot)$ takes the role of the Bernoulli polynomial of degree 2. Analogously to the one-dimensional case, the formula provides a comparison between the integral over a regular region $\mathcal{G}$ and the sum over all functional values of the twice continuously differentiable function F in lattice points $g \in \overline{\mathcal{G}}$ under explicit knowledge of the remainder term in integral form.

Furthermore, in analogous way to the univariate theory, integration by parts enables us to extend the Euler summation to iterated Laplace operators (cf. W. Freeden [2011]).

Theorem 18.5 (Euler Summation Formula for the Operator Δ^m, $m \in \mathbb{N}$). *Let Λ be an arbitrary lattice in $\mathbb{R}^q$. Let $\mathcal{G} \subset \mathbb{R}^q$ be a regular region. Suppose that F is of class $\mathrm{C}^{(2m)}(\overline{\mathcal{G}})$, $\overline{\mathcal{G}} = \mathcal{G} \cup \partial\mathcal{G}$. Then,*

$$\begin{aligned} \sum_{\substack{g\in\overline{\mathcal{G}}\\ g\in\Lambda}}{}' F(g) = & \frac{1}{\|\mathcal{F}_\Lambda\|} \int_{\mathcal{G}} F(x)\, dx \\ & + \sum_{k=0}^{m-1} \int_{\partial\mathcal{G}} \left(\frac{\partial}{\partial\nu} G(\Delta^{k+1};x)\right) \Delta^k F(x)\, dS(x) \\ & - \sum_{k=0}^{m-1} \int_{\partial\mathcal{G}} G(\Delta^{k+1};x) \left(\frac{\partial}{\partial\nu}\Delta^k F(x)\right) dS(x) \\ & + \int_{\mathcal{G}} G(\Delta^m;x)\Delta^m F(x)\, dx. \end{aligned} \tag{18.19}$$

Replacing the lattice Λ by a shifted point lattice $\Lambda + \{x\}$ relative to $x \in \mathbb{R}^q$ we obtain as multi-dimensional counterpart to the extended Euler summation formula the following result.

Theorem 18.6. *Let Λ be an arbitrary lattice in $\mathbb{R}^q$. Let $\mathcal{G} \subset \mathbb{R}^q$ be a regular region, $\overline{\mathcal{G}} = \mathcal{G} \cup \partial\mathcal{G}$. Suppose that F is a member of class $\mathrm{C}^{(2m)}(\overline{\mathcal{G}})$, $m \in \mathbb{N}$. Then, for every $x \in \mathbb{R}^q$,*

$$\begin{aligned} \sum_{\substack{g+x\in\overline{\mathcal{G}}\\ g\in\Lambda}}{}' F(g+x) = & \frac{1}{\|\mathcal{F}_\Lambda\|} \int_{\mathcal{G}} F(y)\, dy \\ & + \sum_{k=0}^{m-1} \int_{\partial\mathcal{G}} \left(\frac{\partial}{\partial\nu} G\left(\Delta^{k+1};x-y\right)\right) \left(\Delta^k F(y)\right) dS(y) \\ & - \sum_{k=0}^{m-1} \int_{\partial\mathcal{G}} G\left(\Delta^{k+1};x-y\right) \left(\frac{\partial}{\partial\nu}\Delta^k F(y)\right) dS(y) \\ & + \int_{\mathcal{G}} G\left(\Delta^m;x-y\right)\Delta^m F(y)\, dy, \end{aligned} \tag{18.20}$$

where

$$\sum_{\substack{g+x\in\overline{\mathcal{G}}\\ g\in\Lambda}}{}' F(g+x) = \sum_{\substack{g+x\in\overline{\mathcal{G}}\\ g\in\Lambda}} \alpha_{\mathcal{G}}(g+x)F(g+x). \tag{18.21}$$

18.2 Zeta Function and Euler Summation

In Section 5.1 we discussed the one-dimensional relations between the Riemann zeta function and the $\mathbb{Z}$-lattice function for the operator Δ of the second derivative. Next we are interested in the close relationship between the zeta function and the Λ-lattice function for (iterated) Laplace operators. The essential tool is the lattice point generated Euler summation formula.

We start with some preparatory material (see W. Freeden [2011] and the literature therein) concerning the *zeta function* $\zeta^q(\cdot\,;\Lambda)$ *of dimension* $q \geq 2$ *and degree* n defined by

$$s \mapsto \zeta^q(s;\Lambda) = \sum_{\substack{|g|>0 \\ g\in\Lambda}} \frac{1}{|g|^s}, \tag{18.22}$$

where $s \in \mathbb{C}$ satisfies $\mathrm{Re}(s) > q$. For each positive value $\rho < \inf_{\mathcal{F}_\Lambda} |x|$, sufficiently large positive N, and for all $s \in \mathbb{C}$ with $\mathrm{Re}(s) > q$, the Euler summation formula yields in terms of the auxiliary function F defined by

$$x \mapsto F(x) = \frac{1}{|x|^s}, \quad x \neq 0, \tag{18.23}$$

the identity

$$\begin{aligned}
\sum_{\substack{\rho\leq|g|\leq N \\ g\in\Lambda}}{}' F(g) &= \frac{1}{\|\mathcal{F}_\Lambda\|} \int_{\substack{\rho\leq|x|\leq N \\ x\in\mathbb{R}^q}} F(x)\, dx \\
&+ \int_{\substack{\rho\leq|x|\leq N \\ x\in\mathbb{R}^q}} G(\Delta^m;x)\,(\Delta^m F(x))\, dx \\
&+ \sum_{k=0}^{m-1} \int_{\substack{|x|=N \\ x\in\mathbb{R}^q}} \left(\frac{\partial}{\partial\nu} G(\Delta^{k+1};x)\right) (\Delta^k F(x))\, dS(x) \\
&- \sum_{k=0}^{m-1} \int_{\substack{|x|=N \\ x\in\mathbb{R}^q}} G(\Delta^{k+1};x) \left(\frac{\partial}{\partial\nu} \Delta^k F(x)\right) dS(x) \\
&+ \sum_{k=0}^{m-1} \int_{\substack{|x|=\rho \\ x\in\mathbb{R}^q}} \left(\frac{\partial}{\partial\nu} G(\Delta^{k+1};x)\right) (\Delta^k F(x))\, dS(x) \\
&- \sum_{k=0}^{m-1} \int_{\substack{|x|=\rho \\ x\in\mathbb{R}^q}} G(\Delta^{k+1};x) \left(\frac{\partial}{\partial\nu} \Delta^k F(x)\right) dS(x),
\end{aligned} \tag{18.24}$$

where $m \in \mathbb{N}$ is chosen such that $m > q/2$ and ν is the outward unit normal to $\mathbb{B}^q_{\rho,N} = \{x \in \mathbb{R}^q : \rho < |y| < N\}$. First we want to calculate the second integral on the right side of (18.24). It is not difficult to see that

$$\Delta_x F(x) = s(s-q+2)\frac{F(x)}{|x|^2}, \quad x \neq 0, \tag{18.25}$$

such that

$$\Delta_x^m F(x) = A_{s,m}\frac{F(x)}{|x|^{2m}}, \quad x \neq 0, \tag{18.26}$$

where

$$A_{s,m} = \prod_{j=1}^{m}(s+2j-2)(s+2j-q). \tag{18.27}$$

For $m > q/2$ we find in connection with the absolutely and uniformly convergent bilinear expansion of the Λ-lattice function in $\mathbb{R}^q$ that

$$\begin{aligned} &\int_{\substack{\rho\leq|x|\leq N\\ x\in\mathbb{R}^q}} G(\Delta^m;x)\,\Delta^m F(x)\,dx \\ &\qquad = \frac{A_{s,m}}{\|\mathcal{F}_\Lambda\|}\sum_{\substack{|h|>0\\ h\in\Lambda^{-1}}}\frac{1}{(-4\pi^2h^2)^m}\int_{\substack{\rho\leq|x|\leq N\\ x\in\mathbb{R}^q}}\frac{e^{2\pi i x\cdot h}}{|x|^{s+2m}}\,dx. \end{aligned} \tag{18.28}$$

From the theory of Bessel functions (see, e.g., G.N. Watson [1944]) we obtain after some elementary calculations

$$\begin{aligned} &\int_{\substack{\rho\leq|x|\leq N\\ x\in\mathbb{R}^q}} G(\Delta^m;x)\,\Delta^m F(x)\,dx \\ &\qquad = A_{s,m}\frac{\|\mathbb{S}^{q-1}\|\Gamma(\frac{q}{2})}{2^{\frac{2-q}{2}}\|\mathcal{F}_\Lambda\|}\sum_{\substack{|h|>0\\ h\in\Lambda^{-1}}}\frac{1}{(2\pi|h|)^{q-s}}\int_{2\pi|h|\rho}^{2\pi|h|N}\frac{J_{\frac{q-2}{2}}(r)}{r^{s+2m-\frac{q}{2}}}\,dr. \end{aligned} \tag{18.29}$$

Remembering the asymptotic behavior of the Bessel function we see that the last series converges uniformly with respect to ρ and N for all $s \in \mathbb{C}$ with $-2m+\frac{q}{2}+\frac{1}{2} < \mathrm{Re}(s) < \min\{0,-2m+q\}$. Consequently, with $s \in \mathbb{C}$ indicated as before, it follows that

$$\begin{aligned} &\int_{\mathbb{R}^q} G(\Delta^m;x)\,\Delta^m F(x)\,dx \\ &\qquad = A_{s,m}\frac{\|\mathbb{S}^{q-1}\|\Gamma(\frac{q}{2})}{2^{\frac{2-q}{2}}\|\mathcal{F}_\Lambda\|}\sum_{\substack{|h|>0\\ h\in\Lambda^{-1}}}\frac{1}{(2\pi|h|)^{q-s}}\int_0^\infty\frac{J_{\frac{q-2}{2}}(r)}{r^{s+2m-\frac{q}{2}}}\,dr. \end{aligned} \tag{18.30}$$

In connection with the formula (see, e.g., W. Magnus et al. [1966])

$$\int_0^\infty \frac{J_\nu(r)}{r^{\nu+1-\mu}}\,dr = \frac{\Gamma(\frac{\mu}{2})2^{\mu-\nu-1}}{\Gamma(\nu-\frac{1}{2}\mu+1)}, \quad 0<\mu<\frac{3}{2}+\nu, \tag{18.31}$$

this allows to show the identity

$$\int_{\mathbb{R}^q} G(\Delta^m; x) \left(\Delta_x^m \frac{1}{|x|^s} \right) dx = \frac{1}{\|\mathcal{F}_\Lambda\|} \pi^{s-\frac{q}{2}} \frac{\Gamma(\frac{q-s}{2})}{\Gamma(\frac{s}{2})} \sum_{\substack{|h|>0 \\ h \in \Lambda^{-1}}} \frac{1}{|h|^{q-s}} \tag{18.32}$$

for all $s \in \mathbb{C}$ with $-2m + \frac{q+1}{2} < \mathrm{Re}(s) < \min(0, -2m+q)$. In connection with (18.24) and (18.32) we are able to show that the sum

$$\sum_{k=0}^{m-1} A_{s,k} \int_{\substack{|x|=N \\ x \in \mathbb{R}^q}} \left(\frac{\partial}{\partial \nu_x} G\big(\Delta^{k+1}; x\big) \right) \frac{F(x)}{|x|^{2k}} - G\big(\Delta^{k+1}; x\big) \left(\frac{\partial}{\partial \nu_x} \frac{F(x)}{|x|^{2k}} \right) dS(x) \tag{18.33}$$

is convergent as $N \to \infty$ provided that $\mathrm{Re}(s) > q$. Even more, for $\mathrm{Re}(s) > q$, the integrals (18.33) tend to zero as $N \to \infty$. Thus it follows for $\mathrm{Re}(s) > q$ that

$$\begin{aligned} \lim_{N\to\infty} \sum_{\substack{g \in \Lambda \\ \rho \le |g| \le N}}{}' \frac{1}{|g|^s} = {} & \frac{\|\mathbb{S}^{q-1}\|}{\|\mathcal{F}_\Lambda\|} \frac{\rho^{q-s}}{s-q} \\ & + \sum_{k=0}^{m-1} A_{s,k} \int_{\substack{|x|=\rho \\ x \in \mathbb{R}^q}} \left(\frac{\partial}{\partial \nu} G\left(\Delta^{k+1}; x\right) \right) \frac{F(x)}{|x|^{2k}} \, dS(x) \\ & - \sum_{k=0}^{m-1} A_{s,k} \int_{\substack{|x|=\rho \\ x \in \mathbb{R}^q}} G\left(\Delta^{k+1}; x\right) \left(\frac{\partial}{\partial \nu} \frac{F(x)}{|x|^{2k}} \right) dS(x). \end{aligned} \tag{18.34}$$

Consequently, the right side of (18.34) shows that $\zeta^q(\cdot;\Lambda)$ can be continued by the left side of (18.34) to the half plane $\mathrm{Re}(s) > -2m + \frac{q}{2} + \frac{1}{2}$. It remains to investigate the sums

$$\begin{aligned} & \sum_{k=0}^{m-1} A_{s,k} \int_{\substack{|x|=\rho \\ x \in \mathbb{R}^q}} \left(\frac{\partial}{\partial \nu} G(\Delta^{k+1}; x) \right) \frac{F(x)}{|x|^{2k}} \, dS(x) \\ & - \sum_{k=0}^{m-1} A_{s,k} \int_{\substack{|x|=\rho \\ x \in \mathbb{R}^q}} G\left(\Delta^{k+1}; x\right) \left(\frac{\partial}{\partial \nu} \frac{F(x)}{|x|^{2k}} \right) dS(x). \end{aligned} \tag{18.35}$$

Because of the singularity behavior of the Λ-lattice function at 0, the sums (18.35) tend to 0 as $\rho \to 0$, provided that $s \in \mathbb{C}$ with $\mathrm{Re}(s) < -2m+q$. Thus, for all values $s \in \mathbb{C}$ with $-2m + \frac{q}{2} + \frac{1}{2} < \mathrm{Re}(s) < \min(0, -2m+q)$, $\zeta^q(\cdot;\Lambda)$ admits the representation

$$\zeta^q(s;\Lambda) = \frac{\pi^{s-\frac{q}{2}} \Gamma(\frac{q-s}{2})}{\|\mathcal{F}_\Lambda\| \Gamma(\frac{s}{2})} \sum_{|h|>0} \frac{1}{|h|^{q-s}}. \tag{18.36}$$

Summarizing our results we therefore obtain

Theorem 18.7 (Functional Equation of the q-Dimensional Zeta Function). *$\zeta^q(\cdot\,;\Lambda)$ is a meromorphic function showing the single pole*

$$\frac{\|\mathbb{S}^{q-1}\|}{\|\mathcal{F}_\Lambda\|}\frac{1}{s-q}. \tag{18.37}$$

$\zeta^q(\cdot\,;\Lambda)$ satisfies the functional equation

$$\zeta^q(s;\Lambda) = \frac{1}{\|\mathcal{F}_\Lambda\|}\pi^{s-\frac{q}{2}}\frac{\Gamma(\frac{q-s}{2})}{\Gamma(\frac{s}{2})}\zeta^q(q-s;\Lambda^{-1}). \tag{18.38}$$

Proof. $\zeta^q(\cdot\,;\Lambda)$ is for $\mathrm{Re}(s) > -2m + \frac{q}{2} + \frac{1}{2}$ except for $s = q$ holomorphic. According to (18.36) the functional equation holds true for all $s \in \mathbb{C}$ with $-2m + \frac{q}{2} + \frac{1}{2} < \mathrm{Re}(s) < \min(0, -2m + q)$.

$\zeta^q(q-\cdot\,;\Lambda^{-1})$ is holomorphic (except for $s = q$) for all $s \in \mathbb{C}$ with $\mathrm{Re}(s) < 2m + \frac{q}{2} - \frac{1}{2}$. As the functional equation is valid in the aforementioned stripe, it is valid everywhere. □

Finally, by replacing the lattice Λ by its inverse lattice Λ^{-1} we obtain the functional equation

$$\zeta^q(s;\Lambda^{-1}) = \|\mathcal{F}_\Lambda\|\pi^{s-\frac{q}{2}}\frac{\Gamma(\frac{q-s}{2})}{\Gamma(\frac{s}{2})}\zeta^q(q-s;\Lambda). \tag{18.39}$$

18.3 Euler Lattice Point Cubature on Regular Regions

The Euler summation formulas can be exploited to approximate integrals over regular regions $\mathcal{G}$. In fact, this can be done in close orientation to the considerations known from the one-dimensional theory. The point of departure is the formula for the lattice $\tau\Lambda$ instead of Λ with $\tau > 0$ sufficiently small

$$\begin{aligned}\sum_{\substack{g\in\overline{\mathcal{G}}\\ g\in\tau\Lambda}}{}' F(g) = {} & \frac{1}{\|\mathcal{F}_{\tau\Lambda}\|}\int_{\mathcal{G}} F(y)\,dy \\ & - \sum_{k=0}^{m-1}\int_{\partial\mathcal{G}}\left(\frac{\partial}{\partial\nu}G\left(\tau\Lambda;\Delta^{k+1};y\right)\right)\left(\Delta^k F(y)\right)\,dS(y) \\ & + \sum_{k=0}^{m-1}\int_{\partial\mathcal{G}} G\left(\tau\Lambda;\Delta^{k+1};y\right)\left(\frac{\partial}{\partial\nu}\Delta^k F(y)\right)\,dS(y) + \mathcal{E}_m^\tau F,\end{aligned} \tag{18.40}$$

where the remainder term $\mathcal{E}_m^\tau F$ is given by the volume integral

$$\mathcal{E}_m^\tau F = \int_{\mathcal{G}} G\left(\tau\Lambda;\Delta^m;y\right)\left(\Delta^m F(y)\right)\,dy. \tag{18.41}$$

In particular, for integers m with $m > q/2$, we know that $G(\tau\Lambda; \Delta^m; y)$ is continuous in $\mathbb{R}^q$, and we have

$$G(\tau\Lambda; \Delta^m; x) = \frac{\tau^{2m-q}}{\|\mathcal{F}_\Lambda\|} \sum_{\substack{h \neq 0 \\ h \in \Lambda^{-1}}} \frac{e^{2\pi i \frac{x}{\tau} \cdot h}}{(-4\pi^2 h^2)^m} = \tau^{2m-q} G\left(\Lambda; \Delta^m; \frac{x}{\tau}\right). \quad (18.42)$$

Thus, for $m > q/2$, we are able to deduce that

$$|\mathcal{E}_m^\tau F| \leq \tau^{2m-q} \sup_{y \in \overline{\mathcal{G}}} \left| G\left(\Lambda; \Delta^m; \frac{x}{\tau}\right) \right| \int_{\mathcal{G}} |\Delta^m F(y)| \, dy. \quad (18.43)$$

Observing the Fourier expansion of $G(\Delta^m; \cdot)$ we obtain the estimate

$$|\mathcal{E}_m^\tau F| \leq \frac{\tau^{2m-q}}{\|\mathcal{F}_\Lambda\|} \sum_{\substack{h \neq 0 \\ h \in \Lambda^{-1}}} \frac{1}{(4\pi^2 h^2)^m} \int_{\mathcal{G}} |\Delta^m F(y)| \, dy, \quad (18.44)$$

i.e.,

$$|\mathcal{E}_m^\tau F| \leq \frac{\tau^{2m-q}}{\|\mathcal{F}_\Lambda\|} \frac{\zeta^q(2m; \Lambda^{-1})}{(4\pi^2)^m} \int_{\mathcal{G}} |\Delta^m F(y)| \, dy. \quad (18.45)$$

The kernel $G(\tau\Lambda; \Delta^m; \cdot)$ has a finite L^p-norm, i.e.,

$$\left(\int_{\mathcal{G}} |G(\tau\Lambda; \Delta^m; y)|^p \, dy \right)^{1/p} < \infty, \quad (18.46)$$

provided that p is chosen such that $p(q - 2m) < q$. Therefore, the Hölder inequality gives

Corollary 18.8. *For a given integer m, let $p > 1$, $p' > 1$ satisfy the conditions $q - 2m < \frac{p}{q}$, $\frac{1}{p} + \frac{1}{p'} = 1$. Then, we have*

$$|\mathcal{E}_m^\tau F| \leq \left(\int_{\mathcal{G}} |G(\tau\Lambda; \Delta^m; y)|^p \, dx \right)^{1/p} \left(\int_{\mathcal{G}} |\Delta^m F(y)|^{p'} \, dy \right)^{1/p'}. \quad (18.47)$$

Remark 18.9. *The identity* (18.40) *can be used to introduce discrepancies and variances associated to iterated Laplace operators (cf. W. Freeden [2011]) in close similarity to their spherical counterparts involving iterated Beltrami operators in* (14.68).

Going back to the Euler cubature formula (18.40) we notice that multi-dimensional numerical integration over a regular region, in general, does not mean to express the volume integral only by a sum of functional values of the integrand. Obviously, boundary surface integrals have to be taken into consideration. In other words, the numerical evaluation of a volume integral involves the problem of approximating surface integrals. This canonical difficulty of multi-dimensional numerical integration does not arise in the one-dimensional theory, where the boundary terms are functional values at end points.

Remark 18.10. *For $q = 3$ and $m > \frac{3}{2}$, the 3D-Fourier series*

$$G(\Delta^m; x) = \sum_{\substack{|h| \neq 0 \\ h \in \mathbb{Z}^3}} \frac{e(h \cdot x)}{(4\pi^2 h^2)^m}, \quad x \in \mathbb{R}^3, \tag{18.48}$$

should be discussed in more detail. The sum (18.48) may be rewritten formally as follows

$$G(\Delta^m; x) = \frac{1}{\Gamma(m)} \sum_{\substack{|h| \neq 0 \\ h \in \mathbb{Z}^3}} \frac{e(h \cdot x)}{(4\pi^2 h^2)^m} \Gamma(m, \pi h^2) + \frac{1}{\Gamma(m)} \sum_{\substack{|h| \neq 0 \\ h \in \mathbb{Z}^3}} \frac{e(h \cdot x)}{(4\pi^2 h^2)^m} \gamma(m, \pi h^2), \tag{18.49}$$

where $\Gamma(\cdot,\cdot)$, $\gamma(\cdot,\cdot)$ are the incomplete Gamma functions (see, e.g., M. Abramowitz, I.A. Stegun [1972], W. Freeden, M. Gutting [2013]). The first series on the right-hand side has a fast convergence, whereas the second series has the same rate of convergence as the original sum. Making use of the Fourier transform, however, we are able to convert this series into a rapidly convergent sum

$$G(\Delta^m; x) = \frac{1}{(4\pi^2)^m \Gamma(m)} \tag{18.50}$$

$$\times \left(\sum_{\substack{|h| \neq 0 \\ h \in \mathbb{Z}^3}} \frac{e(h \cdot x)}{|h|^{2m}} \Gamma(m, \pi h^2) - \frac{\pi^m}{m} + \pi^{2m - \frac{3}{2}} \sum_{g \in \mathbb{Z}^3} \frac{\Gamma\left(\frac{3}{2} - m, \pi |g - x|^2\right)}{|g - x|^{2m-3}} \right).$$

The expression in this form holds for $x \neq 0$. For the case $x = 0$, the second summation should be taken over all $g \in \mathbb{Z}^3$ with $|g| \neq 0$ and the term $\frac{\pi^m}{(m-(3/2))}$ should be added (cf. B.R.A. Nijboer, F.W. de Wette [1957]).

For $m > q/2$ and the special case $\Lambda = \tau \mathbb{Z}^q$ with $\tau > 0$ sufficiently small we obtain the following cubature rule, i.e., generalization of the trapezoidal rule,

$$\frac{1}{\|\mathcal{F}_\Lambda\|} \int_{\mathcal{G}} F(y)\, dy - \tau^q \sum_{\substack{\tau g \in \overline{\mathcal{G}} \\ g \in \mathbb{Z}^q}}{}' F(\tau g) \tag{18.51}$$

$$\begin{aligned} &= \sum_{k=0}^{m-1} \tau^{2k+2} \int_{\partial \mathcal{G}} \left(\frac{\partial}{\partial \nu} G\left(\mathbb{Z}^q; \Delta^{k+1}; \frac{y}{\tau}\right) \right) (\Delta^k F(y))\, dS(y) \\ &\quad - \sum_{k=0}^{m-1} \tau^{2k+2} \int_{\partial \mathcal{G}} G\left(\mathbb{Z}^q; \Delta^{k+1}; \frac{y}{\tau}\right) \left(\frac{\partial}{\partial \nu} \Delta^k F(y) \right) dS(y) \\ &\quad - \tau^{2m} \int_{\mathcal{G}} G\left(\mathbb{Z}^q; \Delta^m; \frac{y}{\tau}\right) (\Delta^m F(y))\, dy \end{aligned} \tag{18.52}$$

such that

$$\int_{\mathcal{G}} F(y)\, dy \approx \tau^q \sum_{\substack{\tau g \in \overline{\mathcal{G}} \\ g \in \mathbb{Z}^q}}{}' F(\tau g) \tag{18.53}$$

for sufficiently small $\tau > 0$.

18.4 Romberg Extrapolation

Applying the Euler summation formula to the dilated lattice leads to the following result.

Corollary 18.11. *Let $\mathcal{G} \subset \mathbb{R}^q$ be a regular region, $G(\Delta; \cdot)$ the Λ-lattice function for a lattice Λ in $\mathbb{R}^q$, $F \in C^{(2m)}(\overline{\mathcal{G}})$ for some $m \in \mathbb{N}$, and $\tau \in (0,1]$. Then*

$$\tau^q \sum_{\substack{g\in\overline{\mathcal{G}}\\ g\in\tau\Lambda}}{}' F(g) = \frac{1}{\|\mathcal{F}_\Lambda\|}\int_{\mathcal{G}} F(y)\, dy + \tau^{2m}\int_{\mathcal{G}} G\left(\Delta^m; \frac{y}{\tau}\right)\Delta^m F(y)\, dy \tag{18.54}$$
$$+ \sum_{k=0}^{m-1} \tau^{2k+2}\int_{\partial\mathcal{G}} (\Delta^k F(y)) \frac{\partial}{\partial\nu} G\left(\Delta^{k+1}; \frac{y}{\tau}\right)\, dS(y)$$
$$- \sum_{k=0}^{m-1} \tau^{2k+2}\int_{\partial\mathcal{G}} G\left(\Delta^{k+1}; \frac{y}{\tau}\right) \frac{\partial}{\partial\nu}\Delta^k F(y)\, dS(y).$$

Choosing the *cubature rule* $\mathcal{L}_\tau : \mathrm{C}^{(2m)}(\overline{\mathcal{G}}) \to \mathbb{R}$ with respect to Λ and τ as the sum

$$\mathcal{L}_\tau F = \tau^q \sum_{\substack{g\in\overline{\mathcal{G}}\\ g\in\tau\Lambda}}{}' F(g) - \tau^2 \int_{\partial\mathcal{G}} F(y)\frac{\partial}{\partial\nu} G\left(\Delta; \frac{y}{\tau}\right) dS(y), \tag{18.55}$$

Corollary 18.11 allows the following representation of the *cubature error*

$$\mathcal{E}_\tau F = \mathcal{L}_\tau F - \frac{1}{\|\mathcal{F}_\Lambda\|}\int_{\mathcal{G}} F(y)\, dy \tag{18.56}$$
$$= \tau^{2m}\int_{\mathcal{G}} G\left(\Delta^m; \frac{y}{\tau}\right)\Delta^m F(y)\, dy$$
$$+ \sum_{k=1}^{m-1} \tau^{2k+2}\int_{\partial\mathcal{G}} (\Delta^k F(y)) \frac{\partial}{\partial\nu} G\left(\Delta^{k+1}; \frac{y}{\tau}\right)\, dS(y)$$
$$- \sum_{k=0}^{m-1} \tau^{2k+2}\int_{\partial\mathcal{G}} G\left(\Delta^{k+1}; \frac{y}{\tau}\right) \frac{\partial}{\partial\nu}\Delta^k F(y)\, dS(y).$$

Since the integrals that occur on the right-hand side of (18.56) are bounded uniformly in $\tau \in (0,1]$, the cubature error behaves like

$$\mathcal{E}_\tau F = O(\tau^2) \tag{18.57}$$

if no further conditions other than $F \in \mathrm{C}^{(2m)}(\overline{\mathcal{G}})$, $m \geq 1$, are assumed. Under the assumption of periodicity (in the sense that the boundary integrals

$\int_{\partial\mathcal{G}} \Delta^k F(y) \frac{\partial}{\partial\nu} G(\Delta^{k+1}; \frac{y}{\tau})\, dS(y) = 0$ and $\int_{\partial\mathcal{G}} G(\Delta^{k+1}; \frac{y}{\tau}) \frac{\partial}{\partial\nu} \Delta^k F(y)\, dS(y) = 0$ for $k = 0, \ldots, m-1$), the cubature error is of order

$$\mathcal{E}_\tau F = O(\tau^{2m}). \tag{18.58}$$

The latter condition, however, is hardly ever satisfied in actual applications.

In the next section we, therefore, discuss how classical Romberg extrapolation (see, e.g., E.B. Anders [1966], H. Antes [1972], W. Romberg [1955], A. Schönhage [1970] for earlier work) transfers to Euler summation based cubature on regular regions (following the note due to W. Freeden, C. Gerhards [2017]). Accordingly, we continue with the first classical Romberg step, i.e., we substitute the cubature rule $\mathcal{L}_\tau$ from (18.55) by the linear combination

$$\bar{\mathcal{L}}_{c,\tau} = \frac{\mathcal{L}_{c\tau} - c^2 \mathcal{L}_\tau}{1-c^2}, \tag{18.59}$$

for some $c \in (0,1)$ and $\tau \in (0,1]$. According to (18.56), the remainder term for this cubature rule is given by

$$\begin{aligned}
\bar{\mathcal{E}}_{c,\tau} F &= \bar{\mathcal{L}}_{c,\tau} F - \frac{1}{\|\mathcal{F}_\Lambda\|} \int_{\mathcal{G}} F(y)\, dy \qquad\qquad (18.60)\\
&= \frac{1}{1-c^2} \mathcal{E}_{c\tau} F - \frac{c^2}{1-c^2} \mathcal{E}_\tau F \\
&= \frac{c^2}{1-c^2} \tau^{2m} \int_{\mathcal{G}} \left(c^{2m-2} G\left(\Delta^m; \frac{y}{c\tau}\right) - G\left(\Delta^m; \frac{y}{\tau}\right)\right) \Delta^m F(y)\, dy \\
&\quad - \sum_{k=1}^{m-1} \frac{c^2}{1-c^2} (1-c^{2k}) \tau^{2k+2} \int_{\partial\mathcal{G}} (\Delta^k F(y)) \frac{\partial}{\partial\nu} G\left(\Delta^{k+1}; \frac{y}{\tau}\right) dS(y) \\
&\quad + \sum_{k=0}^{m-1} \frac{c^2}{1-c^2} (1-c^{2k}) \tau^{2k+2} \int_{\partial\mathcal{G}} G\left(\Delta^{k+1}; \frac{y}{\tau}\right) \frac{\partial}{\partial\nu} \Delta^k F(y)\, dS(y) \\
&\quad + \bar{E}_{c,\tau} F,
\end{aligned}$$

where

$$\begin{aligned}
\bar{E}_{c,\tau} F &= \sum_{k=1}^{m-1} \frac{(c\tau)^{2k+2}}{1-c^2} \int_{\partial\mathcal{G}} (\Delta^k F(y)) \frac{\partial}{\partial\nu} D\left(\Delta^{k+1}; \frac{y}{\tau}\right) dS(y) \qquad (18.61)\\
&\quad - \sum_{k=0}^{m-1} \frac{(c\tau)^{2k+2}}{1-c^2} \int_{\partial\mathcal{G}} D\left(\Delta^{k+1}; \frac{y}{\tau}\right) \frac{\partial}{\partial\nu} \Delta^k F(y)\, dS(y),
\end{aligned}$$

and

$$D(\Delta^k; x) = G\left(\Delta^k; \frac{x}{c}\right) - G\left(\Delta^k; x\right). \tag{18.62}$$

Since $1 - c^{2k} = 0$ for $k = 0$, we can reformulate (18.60) as

$$\bar{\mathcal{E}}_{c,\tau}F = \frac{c^2}{1-c^2}\tau^{2m}\int_{\mathcal{G}}\left(c^{2m-2}G\left(\Delta^m;\frac{y}{c\tau}\right) - G\left(\Delta^m;\frac{y}{\tau}\right)\right)\Delta^m F(y)\,dy \tag{18.63}$$

$$-\sum_{k=1}^{m-1}\frac{c^2}{1-c^2}(1-c^{2k})\tau^{2k+2}\int_{\partial\mathcal{G}}\left(\Delta^k F(y)\right)\frac{\partial}{\partial\nu}G\left(\Delta^{k+1};\frac{y}{\tau}\right)\,dS(y)$$

$$+\sum_{k=1}^{m-1}\frac{c^2}{1-c^2}(1-c^{2k})\tau^{2k+2}\int_{\partial\mathcal{G}}G\left(\Delta^{k+1};\frac{y}{\tau}\right)\frac{\partial}{\partial\nu}\Delta^k F(y)\,dS(y)$$

$$+\bar{E}_{c,\tau}F.$$

Equation (18.63) shows that, for functions $F \in \mathrm{C}^{(2m)}(\overline{\mathcal{G}})$, $m \geq 2$, the convergence rate for of the modified cubature rule Q from (18.59) is given by

$$\bar{\mathcal{E}}_{c,\tau}F = \bar{E}_{c,\tau}F + O(\tau^4) \tag{18.64}$$

while the convergence rate for the original cubature rules $\mathcal{L}_\tau$ and $\mathcal{L}_{c\tau}$ is $\mathcal{E}_\tau F = O(\tau^2)$ and $\mathcal{E}_{c\tau}F = O(\tau^2)$, respectively. Therefore, it remains to investigate the term $\bar{E}_{c,\tau}F$ of (18.61) in more detail.

Remark 18.12. *In order to indicate the relation to the classical Romberg scheme (cf. Section 5.3), we return to the setup of one-dimensional theory with* $m = 2$, *and* $\mathcal{G} = [0, b]$ *(i.e., the boundary* $\partial\mathcal{G}$ *contains only the points* 0 *and* b*). We observe that the* Λ*-lattice function* $G(\Delta;\Lambda;\cdot)$ *and the iterated* Λ*-lattice function* $G(\Delta^2;\Lambda;\cdot)$ *for the lattice* $\Lambda = \mathbb{Z}$ *are given as in* (3.40) *and* (3.61)*, respectively. Furthermore, we choose* $\tau = \frac{b}{n}$ *and* $c = \frac{1}{p}$*, with fixed numbers* $n, p \in \mathbb{N}$*. Then* $\frac{b}{\tau}, \frac{b}{c\tau} \in \mathbb{Z}$*, which implies that* $b \in \tau\mathbb{Z}$*,* $b \in c\tau\mathbb{Z}$*. Since boundary integrals in the one-dimensional setting are just point evaluations in the end points* 0 *and* b*, we obtain from Equation* (18.61) *that*

$$\begin{aligned}\bar{E}_{c,\tau}F = &-\frac{(c\tau)^2}{1-c^2}\underbrace{\left(D(\Delta;\frac{y}{\tau})\frac{\partial}{\partial y}F(y)\right)_{y=0}^{y=b}}_{=0} \\ &+\frac{(c\tau)^4}{1-c^2}\underbrace{\left(\frac{\partial^2}{\partial y^2}F(y)\frac{\partial}{\partial y}D(\Delta^2;\frac{y}{\tau})\right)_{y=0}^{y=b}}_{=0} \\ &-\frac{(c\tau)^4}{1-c^2}\underbrace{\left(D(\Delta^2;\frac{y}{\tau})\frac{\partial^3}{\partial y^3}F(y)\right)_{y=0}^{y=b}}_{=0} = 0.\end{aligned} \tag{18.65}$$

In other words, the cubature rule $\bar{\mathcal{L}}_{c,\tau}F$ *satisfies the improved convergence rate* $\bar{\mathcal{E}}_{c,\tau}F = O(\tau^4)$.

Remark 18.12 tells us that the boundary term $\bar{E}_{c,\tau}$ vanishes in the one-dimensional setting. Unfortunately, this is not true for $\bar{E}_{c,\tau}$ in the general q-dimensional case and the boundary terms have an influence on the modified cubature rule $\mathcal{L}_{c,\tau} = \bar{\mathcal{L}}_{c,\tau} + \bar{E}_{c,\tau}$, which eventually satisfies

$$\mathcal{E}_{c,\tau}F = \mathcal{L}_{c,\tau}F - \frac{1}{\|\mathcal{F}_\Lambda\|}\int_{\mathcal{G}} F(y)\, dy = O(\tau^4). \tag{18.66}$$

A further improvement of the convergence rate to $O(\tau^{2\ell})$ requires an iteration of the Romberg step. We make the following more general definitions.

Definition 18.13. *Let $\bar{\mathcal{L}}_i^{(0)}$, $i \in \mathbb{N}_0$, denote the cubature rule $\mathcal{L}_{c^i\tau}$ from (18.55) with respect to the lattice Λ and $c^i\tau$, where $c \in (0,1)$ and $\tau \in (0,1]$ are fixed. Then, we iteratively introduce new cubature rules by*

$$\bar{\mathcal{L}}_i^{(\ell)} = \frac{\bar{\mathcal{L}}_i^{(\ell-1)} - c^{2\ell}\bar{\mathcal{L}}_{i-1}^{(\ell-1)}}{1-c^{2\ell}}, \quad \ell \in \mathbb{N}, i = \ell, \ell+1 \ldots . \tag{18.67}$$

Additionally, let $\bar{E}_i^{(0)} = 0$, $i \in \mathbb{N}_0$. Then, we iteratively define boundary terms for $\ell \in \mathbb{N}$, $i = \ell, \ell+1, \ldots$, by

$$\begin{aligned} \bar{E}_i^{(\ell)}F = {} & \frac{\bar{E}_i^{(\ell-1)}F - c^{2\ell}\bar{E}_{i-1}^{(\ell-1)}F}{1-c^{2\ell}} \\ & + (-1)^{\ell-1}\Bigg(\sum_{k=\max\{1,\ell-1\}}^{m-1} \frac{(c^{i-(\ell-1)}\tau)^{2k+2}\prod_{j=1}^{\ell-1} c^{2j}}{\prod_{j=1}^{\ell}(1-c^{2j})} \prod_{j=0}^{\ell-2}(1-c^{2(k-j)}) \\ & \times \int_{\partial\mathcal{G}} \Delta^k F(y) \frac{\partial}{\partial\nu} D\left(\Delta^{k+1}; \frac{y}{c^{i-\ell}\tau}\right) dS(y)\Bigg) \\ & - (-1)^{\ell-1}\Bigg(\sum_{k=\ell-1}^{m-1} \frac{(c^{i-(\ell-1)}\tau)^{2k+2}\prod_{j=1}^{\ell-1} c^{2j}}{\prod_{j=1}^{\ell}(1-c^{2j})} \prod_{j=0}^{\ell-2}(1-c^{2(k-j)}) \\ & \times \int_{\partial\mathcal{G}} D\left(\Delta^{k+1}; \frac{y}{c^{i-\ell}\tau}\right) \frac{\partial}{\partial\nu}\Delta^k F(y)\, dS(y)\Bigg), \end{aligned} \tag{18.68}$$

where $D(\Delta^k;\cdot)$ is given by (18.62).

In particular, $\bar{\mathcal{L}}_1^{(1)}$ denotes the first Romberg step $\bar{\mathcal{L}}_{c,\tau}$ as indicated in (18.59) and $\bar{E}_1^{(1)}$ denotes the correction term $\bar{E}_{c,\tau}$ as indicated in (18.61). For the iterates $\bar{\mathcal{L}}_i^{(\ell)}$, we obtain the following expression of the remainder term $\bar{\mathcal{E}}_i^{(\ell)}$.

Theorem 18.14. *Let $\mathcal{G} \subset \mathbb{R}^q$ be a regular region, $G(\Delta;\cdot)$ the Λ-lattice function for a lattice Λ in $\mathbb{R}^q$, and $F \in \mathrm{C}^{(2m)}(\overline{\mathcal{G}})$ for some $m \in \mathbb{N}$. Then, for*

$\ell = 1, \dots, m-1$, $i = \ell, \ell+1, \dots,$

$$\begin{aligned}
\bar{\mathcal{E}}_i^{(\ell)} F = & \bar{\mathcal{L}}_i^{(\ell)} F - \frac{1}{\|\mathcal{F}_\Lambda\|} \int_{\mathcal{G}} F(y)\, dy \qquad (18.69)\\
= & \frac{(c^{i-\ell})^{2m} \prod_{j=1}^{\ell} c^{2j}}{\prod_{j=1}^{\ell}(1-c^{2j})} \tau^{2m} \int_{\mathcal{G}} \left(\sum_{j=0}^{\ell} a_j^{(\ell)} G\left(\Delta^m; \frac{y}{c^{i-\ell+j}\tau}\right) \right) \Delta^m F(y)\, dy \\
& + (-1)^\ell \Bigg(\sum_{k=\ell}^{m-1} \frac{(c^{i-\ell})^{2k+2} \prod_{j=1}^{\ell} c^{2j}}{\prod_{j=1}^{\ell}(1-c^{2j})} \prod_{j=0}^{\ell-1} (1-c^{2(k-j)}) \\
& \times \tau^{2k+2} \int_{\partial\mathcal{G}} \Delta^k F(y) \frac{\partial}{\partial\nu} G\left(\Delta^{k+1}; \frac{y}{c^{i-\ell}\tau}\right)\, dS(y) \Bigg) \\
& - (-1)^\ell \Bigg(\sum_{k=\ell}^{m-1} \frac{(c^{i-\ell})^{2k+2} \prod_{j=1}^{\ell} c^{2j}}{\prod_{j=1}^{\ell}(1-c^{2j})} \prod_{j=0}^{\ell-1} (1-c^{2(k-j)}) \\
& \times \tau^{2k+2} \int_{\partial\mathcal{G}} G\left(\Delta^{k+1}; \frac{y}{c^{i-\ell}\tau}\right) \frac{\partial}{\partial\nu} \Delta^k F(y)\, dS(y) \Bigg) \\
& + \bar{E}_i^{(\ell)} F.
\end{aligned}$$

The coefficients $a_j^{(\ell)}$ *are iteratively defined via*

$$a_0^{(\ell)} = (-1)^\ell, \qquad (18.70)$$

$$a_j^{(\ell)} = \begin{cases} c^{2(m-\ell)} a_{j-1}^{(\ell-1)} - a_j^{(\ell-1)}, & j = 1, \dots, \ell, \\ 0, & j = \ell+1, \ell+2, \dots. \end{cases} \qquad (18.71)$$

Proof. For $\ell = 1$, the assertion can be checked by hand. Now, assuming that the representation (18.69) of $\bar{\mathcal{E}}_i^{(\ell)} F$ is true for ℓ and all $i = \ell, \ell+1, \dots,$ we show by induction that it also holds true for $\ell+1$ and all $i = \ell+1, \ell+2, \dots$. First, we observe

$$\bar{\mathcal{E}}_i^{(\ell+1)} F = \frac{1}{1-c^{2\ell+2}} \bar{\mathcal{E}}_i^{(\ell)} F - \frac{c^{2\ell+2}}{1-c^{2\ell+2}} \bar{\mathcal{E}}_{i-1}^{(\ell)} F. \qquad (18.72)$$

For the explicit computation of the right-hand side of Equation (18.72), we treat the different terms of $\bar{\mathcal{E}}_i^{(\ell)} F$ and $\bar{\mathcal{E}}_{i-1}^{(\ell)} F$, as appearing on the right-hand side of Equation (18.69), separately. For the first term, we obtain

$$\begin{aligned}
& \frac{(c^{i-\ell})^{2m} \prod_{j=1}^{\ell} c^{2j}}{(1-c^{2\ell+2}) \prod_{j=1}^{\ell}(1-c^{2j})} \tau^{2m} \int_{\mathcal{G}} \left(\sum_{j=0}^{\ell} a_j^{(\ell)} G\left(\Delta^m; \frac{y}{c^{i-\ell+j}\tau}\right) \right) \Delta^m F(y)\, dy \qquad (18.73)\\
& - \frac{c^{2\ell+2} (c^{i-1-\ell})^{2m} \prod_{j=1}^{\ell} c^{2j}}{(1-c^{2\ell+2}) \prod_{j=1}^{\ell}(1-c^{2j})} \tau^{2m} \int_{\mathcal{G}} \left(\sum_{j=0}^{\ell} a_j^{(\ell)} G\left(\Delta^m; \frac{y}{c^{i-1-\ell+j}\tau}\right) \right) \Delta^m F(y) dy
\end{aligned}$$

$$= \frac{(c^{i-(\ell+1)})^{2m}\prod_{j=1}^{\ell} c^{2j}}{\prod_{j=1}^{\ell+1}(1-c^{2j})}\tau^{2m}\int_{\mathcal{G}}\Bigg(c^{2m}\sum_{j=0}^{\ell} a_j^{(\ell)} G\left(\Delta^m; \frac{y}{c^{i-\ell+j}\tau}\right)$$

$$- c^{2\ell+2}\sum_{j=0}^{\ell} a_j^{(\ell)} G\left(\Delta^m; \frac{y}{c^{i-(\ell+1)+j}\tau}\right)\Bigg)\Delta^m F(y)\, dy$$

$$= \frac{(c^{i-(\ell+1)})^{2m}\prod_{j=1}^{\ell+1} c^{2j}}{\prod_{j=1}^{\ell+1}(1-c^{2j})}\tau^{2m}\int_{\mathcal{G}}\Bigg(c^{2m-2(\ell+1)}\sum_{j=0}^{\ell} a_j^{(\ell)} G\left(\Delta^m; \frac{y}{c^{i-\ell+j}\tau}\right)$$

$$- \sum_{j=0}^{\ell} a_j^{(\ell)} G\left(\Delta^m; \frac{y}{c^{i-(\ell+1)+j}\tau}\right)\Bigg)\Delta^m F(y)\, dy$$

$$= \frac{(c^{i-(\ell+1)})^{2m}\prod_{j=1}^{\ell+1} c^{2j}}{\prod_{j=1}^{\ell+1}(1-c^{2j})}\tau^{2m}\int_{\mathcal{G}}\Bigg(- a_0^{(\ell)} G\left(\Delta^m; \frac{y}{c^{i-(\ell+1)}\tau}\right)$$

$$+ \sum_{j=1}^{\ell+1}\left(c^{2(m-(\ell+1))} a_{j-1}^{(\ell)} - a_j^{(\ell)}\right) G\left(\Delta^m; \frac{y}{c^{i-(\ell+1)+j}\tau}\right)\Bigg)\Delta^m F(y)\, dy.$$

The representation of the coefficients $a_j^{(\ell+1)}$ can be obtained directly from the right-hand side of the equation above. For the second term of the right-hand side of Equation (18.69), we obtain

$$\frac{1}{1-c^{2(\ell+1)}}\sum_{k=\ell}^{m-1}\Bigg(\frac{(c^{i-\ell})^{2k+2}\prod_{j=1}^{\ell} c^{2j}}{\prod_{j=1}^{\ell}(1-c^{2j})}\prod_{j=0}^{\ell-1}(1-c^{2(k-j)}) \tag{18.74}$$

$$\times\, \tau^{2k+2}\int_{\partial\mathcal{G}}\Delta^k F(y)\frac{\partial}{\partial\nu}G\left(\Delta^{k+1}; \frac{y}{c^{i-\ell}\tau}\right) dS(y)\Bigg)$$

$$- \frac{c^{2(\ell+1)}}{1-c^{2(\ell+1)}}\sum_{k=\ell}^{m-1}\Bigg(\frac{(c^{i-1-\ell})^{2k+2}\prod_{j=1}^{\ell} c^{2j}}{\prod_{j=1}^{\ell}(1-c^{2j})}\prod_{j=0}^{\ell-1}(1-c^{2(k-j)})$$

$$\times\, \tau^{2k+2}\int_{\partial\mathcal{G}}\Delta^k F(y)\frac{\partial}{\partial\nu}G\left(\Delta^{k+1}; \frac{y}{c^{i-1-\ell}\tau}\right) dS(y)\Bigg)$$

$$= \sum_{k=\ell}^{m-1}\Bigg(\frac{c^{2k+2}-c^{2(\ell+1)}}{1-c^{2(\ell+1)}}\frac{(c^{i-1-\ell})^{2k+2}\prod_{j=1}^{\ell} c^{2j}}{\prod_{j=1}^{\ell}(1-c^{2j})}\prod_{j=0}^{\ell-1}(1-c^{2(k-j)})$$

$$\times\, \tau^{2k+2}\int_{\partial\mathcal{G}}\Delta^k F(y)\frac{\partial}{\partial\nu}G\left(\Delta^{k+1}; \frac{y}{c^{i-1-\ell}\tau}\right) dS(y)\Bigg)$$

$$+ \frac{1}{1-c^{2(\ell+1)}}\sum_{k=\ell}^{m-1}\Bigg(\frac{(c^{i-\ell})^{2k+2}\prod_{j=1}^{\ell} c^{2j}}{\prod_{j=1}^{\ell}(1-c^{2j})}\prod_{j=0}^{\ell-1}(1-c^{2(k-j)})$$

$$\times\, \tau^{2k+2}\int_{\partial\mathcal{G}}\Delta^k F(y)\frac{\partial}{\partial\nu}D\left(\Delta^{k+1}; \frac{y}{c^{i-1-\ell}\tau}\right) dS(y)\Bigg)$$

$$
= \underbrace{- \sum_{k=\ell+1}^{m-1} \left(\frac{(c^{i-(\ell+1)})^{2k+2} \prod_{j=1}^{\ell+1} c^{2j}}{\prod_{j=1}^{\ell+1}(1-c^{2j})} \prod_{j=0}^{\ell}(1-c^{2(k-j)}) \times \tau^{2k+2} \int_{\partial\mathcal{G}} \Delta^k F(y) \frac{\partial}{\partial\nu} G\left(\Delta^{k+1}; \frac{y}{c^{i-(\ell+1)}\tau}\right) dS(y) \right)}_{\text{part of } \bar{\mathcal{E}}_i^{(\ell+1)} F \text{ according to Equation (18.69)}}
$$

$$
\underbrace{+ \sum_{k=\ell}^{m-1} \left(\frac{(c^{i-\ell})^{2k+2} \prod_{j=1}^{\ell} c^{2j}}{\prod_{j=1}^{\ell+1}(1-c^{2j})} \prod_{j=0}^{\ell-1}(1-c^{2(k-j)}) \times \tau^{2k+2} \int_{\partial\mathcal{G}} \Delta^k F(y) \frac{\partial}{\partial\nu} D\left(\Delta^{k+1}; \frac{y}{c^{i-(\ell+1)}\tau}\right) dS(y) \right)}_{\text{part of } \bar{E}_i^{(\ell+1)} F \text{ according to Equation (18.69) and Definition 18.13}} .
$$

The contributions of the third term of the right-hand side of Equation (18.69) can be reformulated very analogously and the fourth term simply yields

$$
\frac{1}{1-c^{2\ell+2}} \bar{E}_i^{(\ell)} F - \frac{c^{2\ell+2}}{1-c^{2\ell+2}} \bar{E}_{i-1}^{(\ell)} F. \tag{18.75}
$$

Combining all computations above via (18.72) and observing Definition 18.13 (see also Figure 18.1) leads to the desired representation of $\bar{\mathcal{E}}_i^{(\ell+1)} F$. □

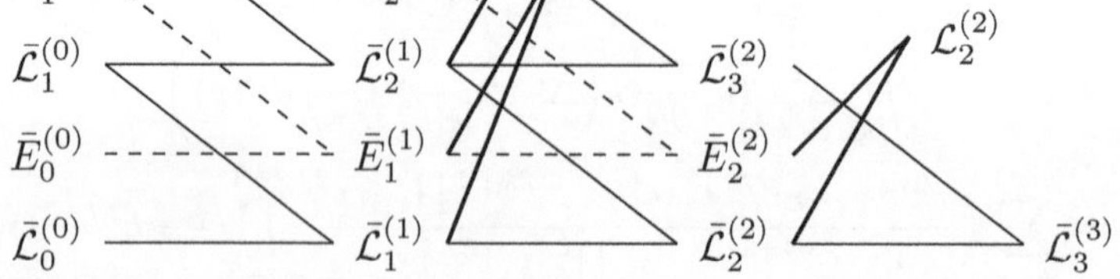

FIGURE 18.1: Illustration of the iterative construction of $\mathcal{L}_i^{(\ell)}$.

Corollary 18.15. *Let $\mathcal{G} \subset \mathbb{R}^q$ be a regular region, $G(\Delta; \cdot)$ the Λ-lattice function for a lattice Λ in $\mathbb{R}^q$, and $F \in \mathrm{C}^{(2m)}(\overline{\mathcal{G}})$ for some $m \in \mathbb{N}$. Then, for $\ell = 1, \ldots, m-1$, $i = \ell, \ell+1, \ldots$, the cubature rule $\mathcal{L}_i^{(\ell)} = \bar{\mathcal{L}}_i^{(\ell)} - \bar{E}_i^{(\ell)}$ satisfies*

$$
\mathcal{E}_i^{(\ell)} F = \mathcal{L}_i^{(\ell)} F - \frac{1}{\|\mathcal{F}_\Lambda\|} \int_{\mathcal{G}} F(y)\, dy = O(\tau^{2\ell+2}). \tag{18.76}
$$

Corollary 18.15 states the desired improved convergence rate for the Romberg extrapolated cubature rule $\mathcal{L}_i^{(\ell)}$ that holds true for general regular regions $\mathcal{G} \subset \mathbb{R}^q$.

Remark 18.16. *In the case of one-dimensional numerical integration, we proceed as in Remark 18.12 and choose the interval* $[0, b]$, $\tau = \frac{b}{n}$, *and* $c = \frac{1}{p}$, *with fixed numbers* $n, p \in \mathbb{N}$. *However, this time we do not investigate the first Romberg step but we look at its* ℓ*-th iterate for* $\ell \geq 2$. *Then, the term* $\bar{E}_\ell^{(\ell)}$ *from Definition 18.13 becomes the representation*

$$\bar{E}_\ell^{(\ell)} F = \frac{\bar{E}_{\ell-1}^{(\ell)} F - c^{2\ell} \bar{E}_{\ell-1}^{(\ell-1)} F}{1 - c^{2\ell}} \tag{18.77}$$

$$+ (-1)^{\ell-1} \sum_{k=\ell-1}^{m-1} \left(\frac{\tau^{2k+2} \prod_{j=1}^{\ell-1} c^{2j}}{\prod_{j=1}^{\ell} (1 - c^{2j})} \prod_{j=0}^{\ell-2} (1 - c^{2(k-j)}) \right.$$

$$\left. \times \underbrace{\left(\Delta^k F(y) \frac{\partial}{\partial \nu} D\left(\Delta^{k+1}; \frac{y}{\tau}\right) \right) \Bigg|_{y=0}^{y=b}}_{=0} \right)$$

$$- (-1)^{\ell-1} \sum_{k=\ell-1}^{m-1} \left(\frac{\tau^{2k+2} \prod_{j=1}^{\ell-1} c^{2j}}{\prod_{j=1}^{\ell} (1 - c^{2j})} \prod_{j=0}^{\ell-2} (1 - c^{2(k-j)}) \right.$$

$$\left. \times \underbrace{\left(D\left(\Delta^{k+1}; \frac{y}{\tau}\right) \frac{\partial}{\partial \nu} \Delta^k F(y) \right) \Bigg|_{y=0}^{y=b}}_{=0} \right)$$

$$= \frac{\bar{E}_\ell^{(\ell-1)} F - c^{2\ell} \bar{E}_{\ell-1}^{(\ell-1)} F}{1 - c^{2\ell}},$$

for $m \geq 3$ *and* $\ell \geq 2$. *Observing that* $\bar{E}_1^{(1)} F = \bar{E}_{c,\tau} F = 0$ *in Remark 18.12, we iteratively see that* $\bar{E}_\ell^{(\ell)} F$ *vanishes and obtain the classical one-dimensional Romberg method* $\mathcal{L}_\ell^{(\ell)} = \bar{\mathcal{L}}_\ell^{(\ell)}$ *with convergence rate* $O(\tau^{2\ell})$.

All in all, an explicit approach to a Romberg scheme on general regular regions in $\mathbb{R}^q$ was proposed that achieves the convergence rate known for the one-dimensional case and specific geometries like cuboids in higher dimensions. However, the cubature formulas $\mathcal{L}_i^{(\ell)}$ involve the Λ-lattice function $G(\Delta; \cdot)$ for which there are no explicit representations available in dimensions $q \geq 2$, to the best of our knowledge. This difficulty requires the use of approximations of $G(\Delta; \cdot)$ for the evaluation of the boundary integrals in $\mathcal{L}_i^{(\ell)}$. Therefore, the study of efficient evaluation methods of Λ-Green's functions is a crucial aspect in order to make use of the improved convergence rates for $\mathcal{L}_i^{(\ell)}$. One possibility is to use summability procedures such as Gauss–Weierstrass summability or lattice ball summability. This aspect will be analyzed in the next chapter. Nevertheless, we were able in this section to derive the framework for an improved Romberg-type cubature rule on general regular regions $\mathcal{G} \subset \mathbb{R}^q$ such as the ellipsoid, telluroid, geoid, the Earth's body, etc., that seems to be a new ingredient in modern cubature methodology.

Chapter 19

Integration by Averaged Euler Summation

Previously, we used the (pointwise) Euler summation formula corresponding to the Laplace operator to sum up values of a given function on a set of lattice points inside and on the boundary surface of a geoscientifically relevant regular region $\mathcal{G} \subset \mathbb{R}^q$. The essential difficulty is the handling of the convergence scenario of the Fourier expansion for the lattice function. This problem will be attacked now by application of averaging processes.

To be more concrete, we formulate a lattice ball generated Euler summation formula (in Section 19.1), thereby closely following W. Freeden [2011]. In doing so the summability effect is caused by jinc-functions, i.e., spherically understood multivariate analogs of sinc-functions. Then we go over to Gauss–Weierstrass integral means which show an exponential convergence acceleration. The chapter ends with averaged cubature over regular regions $\mathcal{G}$.

19.1 Lattice Ball Integral Means

The Euler summation formula corresponding to the Laplace operator is investigated for a sum of "mean integral values over (small) balls" with centers located at the lattice points of Λ. Our considerations are based on the so-called Λ-lattice τ-mean, which is nothing more than an average of the Λ-lattice function over a (small) ball around lattice points. Note that averaging over balls does not destroy the rotational symmetry that is inherent in a large number of geodata systems.

Definition 19.1. *Let Λ be an arbitrary lattice in $\mathbb{R}^q$. For sufficiently small $\tau > 0$ (i.e., $\tau \in \mathbb{R}$ with $0 < \tau < \inf_{x \in \partial \mathcal{F}_\Lambda} |x|$), let $G_\tau^{\mathfrak{I}}(\Delta; \cdot)$ be the function defined by*

$$G_\tau^{\mathfrak{I}}(\Delta; x) = \frac{1}{\|\mathbb{B}_\tau^q\|} \int_{\substack{|y| \leq \tau \\ y \in \mathbb{R}^q}} G(\Delta; x - y)\, dy, \quad x \in \mathbb{R}^q \setminus \Lambda, \tag{19.1}$$

where

$$\|\mathbb{B}_\tau^q\| = \int_{\substack{|y|\leq\tau \\ y\in\mathbb{R}^q}} dy = \frac{\pi^{\frac{q}{2}}}{\Gamma\left(\frac{q}{2}+1\right)}\tau^q. \tag{19.2}$$

Then, $G_\tau^{\mathfrak{I}}(\Delta;\cdot)$ *is called the* Λ-lattice τ-integral (ball) mean *in* $\mathbb{R}^q$ *with respect to* Δ.

From the definition of the Λ-lattice function (Definition 18.1) it is clear that the Λ-*lattice* τ-*mean* is Λ-periodic, i.e.,

$$G_\tau^{\mathfrak{I}}(\Delta;x) = G_\tau^{\mathfrak{I}}(\Delta;x+g), \quad x\in\mathbb{R}^q\setminus\Lambda, \quad g\in\Lambda. \tag{19.3}$$

Potential theoretic tools (see, e.g., W. Freeden, C. Gerhards [2013]), i.e., a variant of the Poisson differential equation (cf. Lemma 17.17), provide an interesting result for the Laplace derivative of the Λ-lattice τ-mean.

Theorem 19.2. *Let* $G_\tau^{\mathfrak{I}}(\Delta;\cdot)$ *be the* Λ-lattice τ-mean. *Then, for all* $x\in\mathcal{F}_\Lambda$, *we have*

$$\Delta_x G_\tau^{\mathfrak{I}}(\Delta;x) = \begin{cases} \frac{1}{\|\mathbb{B}_\tau^q\|} - \frac{1}{\|\mathcal{F}_\Lambda\|}, & |x|\leq\tau, \\ -\frac{1}{\|\mathcal{F}_\Lambda\|}, & |x|>\tau. \end{cases} \tag{19.4}$$

Theorem 19.2 can be rewritten by use of the so-called *characteristic lattice ball function* $\chi_{\overline{\mathbb{B}_\tau^q}+\Lambda}:\mathbb{R}^q\to\mathbb{R}$ given by

$$\chi_{\overline{\mathbb{B}_\tau^q}+\Lambda}(x) = \begin{cases} 1, & x\in\bigcup_{g\in\Lambda}\overline{\mathbb{B}_\tau^q(g)} \\ 0, & \text{otherwise.} \end{cases} \tag{19.5}$$

This notation yields the following reformulation of $\Delta G_\tau^{\mathfrak{I}}(\Delta;\cdot)$:

$$\Delta_x G_\tau^{\mathfrak{I}}(\Delta;x) = \frac{1}{\|\mathbb{B}_\tau^q\|}\left(\chi_{\overline{\mathbb{B}_\tau^q}+\Lambda}(x) - \frac{\|\mathbb{B}_\tau^q\|}{\|\mathcal{F}_\Lambda\|}\right), \quad x\in\mathbb{R}^q. \tag{19.6}$$

The representation (19.6) may be used to construct a variant of the Euler summation formula based on the Λ-*lattice* τ-*mean*: Let $\mathcal{G}$ be a regular region in $\mathbb{R}^q$. Furthermore, let F be a function of class $\mathrm{C}^{(2)}(\overline{\mathcal{G}})$, $\overline{\mathcal{G}} = \mathcal{G}\cup\partial\mathcal{G}$. Then integration by parts (i.e., the second Green's theorem) yields for every (sufficiently small) $\varepsilon>0$,

$$\begin{aligned} &\int_{\substack{x\in\overline{\mathcal{G}} \\ x\notin\mathbb{B}_\varepsilon^q+\Lambda}} F(x)\Delta G_\tau^{\mathfrak{I}}(\Delta;x) - G_\tau^{\mathfrak{I}}(\Delta;x)\,\Delta F(x)\,dx \qquad (19.7) \\ &= \int_{\substack{x\in\partial\mathcal{G} \\ x\notin\mathbb{B}_\varepsilon^q+\Lambda}} F(x)\frac{\partial}{\partial\nu}G_\tau^{\mathfrak{I}}(\Delta;x) - G_\tau^{\mathfrak{I}}(\Delta;x)\,\frac{\partial}{\partial\nu}F(x)\,dS(x) \\ &\quad+\sum_{\substack{g\in\overline{\mathcal{G}} \\ g\in\Lambda}}\int_{\substack{|x-g|=\varepsilon \\ x\in\overline{\mathcal{G}}}} F(x)\frac{\partial}{\partial\nu}G_\tau^{\mathfrak{I}}(\Delta;x) - G_\tau^{\mathfrak{I}}(\Delta;x)\,\frac{\partial}{\partial\nu}F(x)\,dS(x), \end{aligned}$$

where, as always, ν is the outer (unit) normal field on $\partial\mathcal{G}$. By observation of the differential equation (19.6) we obtain for $\varepsilon \to 0$

$$\int_{\mathcal{G}} F(x)\frac{1}{\|\mathbb{B}^q_\tau\|}\left(\chi_{\overline{\mathbb{B}^q_\tau}+\Lambda}(x) - \frac{\|\mathbb{B}^q_\tau\|}{\|\mathcal{F}_\Lambda\|}\right) - G^{\mathfrak{I}}_\tau(\Delta;x)\,\Delta F(x)\,dx \tag{19.8}$$
$$= \int_{\partial\mathcal{G}} F(x)\frac{\partial}{\partial\nu}G^{\mathfrak{I}}_\tau(\Delta;x) - G^{\mathfrak{I}}_\tau(\Delta;x)\frac{\partial}{\partial\nu}F(x)\,dS(x).$$

In terms of the characteristic lattice ball function (see (19.5)) this leads to the following statement (cf. W. Freeden [2011]).

Theorem 19.3 (Lattice Ball-Generated Euler Summation Formula for the Laplace Operator Δ). *Let Λ be an arbitrary lattice in $\mathbb{R}^q$. Let F be of class $\mathrm{C}^{(2)}(\overline{\mathcal{G}})$, $\overline{\mathcal{G}} = \mathcal{G}\cup\partial\mathcal{G}$, where $\mathcal{G}$ is a regular region in $\mathbb{R}^q$. Then, for sufficiently small $\tau > 0$, we have*

$$\frac{1}{\|\mathbb{B}^q_\tau\|}\int_{\mathcal{G}} \chi_{\overline{\mathbb{B}^q_\tau}+\Lambda}(x)F(x)\,dx \tag{19.9}$$
$$= \frac{1}{\|\mathcal{F}_\Lambda\|}\int_{\mathcal{G}} F(x)\,dx + \int_{\mathcal{G}} G^{\mathfrak{I}}_\tau(\Delta;x)\,\Delta F(x)\,dx$$
$$+ \int_{\partial\mathcal{G}} F(x)\frac{\partial}{\partial\nu}G^{\mathfrak{I}}_\tau(\Delta;x) - G^{\mathfrak{I}}_\tau(\Delta;x)\frac{\partial}{\partial\nu}F(x)\,dS(x).$$

In fact, Theorem 19.3 is the lattice ball counterpart of the Euler summation formula related to the Λ-*lattice* τ-*mean* and the Laplace operator Δ.

Next we go over to *iterated* Λ-*lattice* τ-*means*. Their definition is straightforward.

Definition 19.4. *For sufficiently small $\tau > 0$, let the function $G^{\mathfrak{I}}_\tau(\Delta^k;\cdot) : \mathbb{R}^q \to \mathbb{R}$, $k \in \mathbb{N}$, be defined by*

$$G^{\mathfrak{I}}_\tau(\Delta^k;x) = \frac{1}{\|\mathbb{B}^q_\tau\|}\int_{\substack{|y|\le\tau\\ y\in\mathbb{R}^q}} G(\Delta^k;x-y)\,dy, \quad x \in \mathbb{R}^q\setminus\Lambda, \tag{19.10}$$

Then, $G_\tau(\Delta^k;\cdot)$ is called the Λ-lattice τ-mean *in $\mathbb{R}^q$ with respect to Δ^k.*

From the definition of the Green's function it can be readily seen that the Λ-*lattice* τ-*mean* with respect to Δ^k is Λ-periodic, i.e.,

$$G^{\mathfrak{I}}_\tau(\Delta^k;x) = G^{\mathfrak{I}}_\tau(\Delta^k;x+g), \quad x\in\mathbb{R}^q\setminus\Lambda, \quad g\in\Lambda. \tag{19.11}$$

From the calculus of Bessel functions of dimension q (cf. W. Freeden, M. Gutting [2013]) we borrow

$$\int_{\substack{|x|\le\tau\\ x\in\mathbb{R}^q}} e^{-2\pi i h\cdot x}\,dx = \|\mathbb{S}^{q-1}\|\tau^q\frac{J_1(q;2\pi|h|\tau)}{2\pi|h|\tau}, \tag{19.12}$$

where

$$J_\nu(q;r) = \Gamma\left(\frac{q}{2}\right)\left(\frac{r}{2}\right)^{\frac{2-q}{2}} J_{\nu+\frac{q-2}{2}}(2;r) \tag{19.13}$$

and

$$J_\nu(2;r) = J_\nu(r) = \left(\frac{r}{2}\right)^\nu \sum_{k=0}^{\infty} \frac{(-1)^k}{k!} \frac{1}{\Gamma(\nu+k+1)} \left(\frac{r}{2}\right)^{2k}. \tag{19.14}$$

Using the bilinear expansion of $G\left(\Delta^k;\cdot\right)$ we are able to write down the (formal) Fourier series representation of $G_\tau^{\mathfrak{I}}\left(\Delta^k;\cdot\right)$ by means of the jinc-function $r \mapsto \frac{J_1(q;r)}{r}$, $r > 0$, namely

$$\frac{\|\mathbb{S}^{q-1}\|}{\|\mathbb{B}_\tau^q\|} \frac{\tau^q}{\|\mathcal{F}_\Lambda\|} \sum_{\substack{(\Delta^k)^\wedge(h)\neq 0 \\ h\in\Lambda^{-1}}} \frac{J_1(q;2\pi\tau|h|)}{2\pi\tau|h|} \frac{e^{2\pi i h\cdot x}}{-(\Delta^k)^\wedge(h)}. \tag{19.15}$$

For $k > \frac{q-1}{4}$, the well-known asymptotic behavior of Bessel functions (cf. W. Freeden, M. Gutting [2013], G.N. Watson [1944]) allows us to conclude that the series (19.15) converges absolutely and uniformly in $\mathbb{R}^q$, so that $G_\tau^{\mathfrak{I}}\left(\Delta^k;\cdot\right)$ is continuous in $\mathbb{R}^q$, and we have in jinc-function nomenclature

$$G_\tau^{\mathfrak{I}}\left(\Delta^k;x\right) = \frac{\|\mathbb{S}^{q-1}\|}{\|\mathbb{B}_\tau^q\|} \frac{\tau^q}{\|\mathcal{F}_\Lambda\|} \sum_{\substack{|h|>0 \\ h\in\Lambda^{-1}}} \frac{J_1(q;2\pi\tau|h|)}{2\pi\tau|h|} \frac{e^{2\pi i h\cdot x}}{(-4\pi^2 h^2)^k}. \tag{19.16}$$

Next we want to extend Theorem 19.3 to iterated operators Δ^m, $m \in \mathbb{N}$. Suppose that F is $2m$-times continuously differentiable in $\overline{\mathcal{G}}$. Then the second Green's theorem (see Theorem 17.9) yields for $k \in \{1,\ldots,m-1\}$ that

$$\begin{aligned} \int_{\mathcal{G}} G_\tau^{\mathfrak{I}}\left(\Delta^{k+1};x\right)\Delta^{k+1}F(x)\,dx &= \int_{\mathcal{G}} G_\tau^{\mathfrak{I}}\left(\Delta^k;x\right)\Delta^k F(x)\,dx \\ &\quad + \int_{\partial\mathcal{G}} G_\tau^{\mathfrak{I}}\left(\Delta^{k+1};x\right)\frac{\partial}{\partial\nu}\Delta^k F(x)\,dS(x) \\ &\quad - \int_{\partial\mathcal{G}} \left(\frac{\partial}{\partial\nu}G_\tau^{\mathfrak{I}}\left(\Delta^{k+1};x\right)\right)\Delta^k F(x)\,dS(x). \end{aligned} \tag{19.17}$$

Summing up over all k from 1 to $m-1$ we find in the well-known way known from the lattice point theory

$$\begin{aligned} \int_{\mathcal{G}} G_\tau^{\mathfrak{I}}\left(\Delta^m;x\right)\Delta^m F(x)\,dx &= \int_{\mathcal{G}} G_\tau^{\mathfrak{I}}\left(\Delta;x\right)\Delta F(x)\,dx \\ &\quad + \sum_{k=1}^{m-1}\int_{\partial\mathcal{G}} G_\tau^{\mathfrak{I}}\left(\Delta^{k+1};x\right)\left(\frac{\partial}{\partial\nu}\Delta^k F(x)\right)\,dS(x) \\ &\quad - \sum_{k=1}^{m-1}\int_{\partial\mathcal{G}} \left(\frac{\partial}{\partial\nu}G_\tau^{\mathfrak{I}}\left(\Delta^{k+1};x\right)\right)\Delta^k F(x)\,dS(x). \end{aligned} \tag{19.18}$$

Combining this result with Theorem 19.3 we obtain

Theorem 19.5 (Lattice Ball-Generated Euler Summation Formula for Iterated Laplace Operators Δ^m, $m \in \mathbb{N}$). *Let Λ be an arbitrary lattice in $\mathbb{R}^q$. Let F be of class $\mathrm{C}^{(2m)}\left(\overline{\mathcal{G}}\right)$, $m \in \mathbb{N}$, with $\mathcal{G}$ being a regular region. Then, for sufficiently small $\tau > 0$, we have*

$$\begin{aligned} \frac{1}{\|\mathbb{B}_\tau^q\|} \int_{\mathcal{G}} \chi_{\overline{\mathbb{B}_\tau^q}+\Lambda}(x) F(x)\, dx = {} & \frac{1}{\|\mathcal{F}_\Lambda\|} \int_{\mathcal{G}} F(x)\, dx \\ & + \int_{\mathcal{G}} G_\tau^{\mathfrak{I}}(\Delta^m; x)\, \Delta^m F(x)\, dx \\ & + \sum_{k=0}^{m-1} \int_{\partial\mathcal{G}} \left(\frac{\partial}{\partial\nu} G_\tau^{\mathfrak{I}}(\Delta^{k+1}; x)\right) \Delta^k F(x)\, dS(x) \\ & - \sum_{k=0}^{m-1} \int_{\partial\mathcal{G}} G_\tau^{\mathfrak{I}}(\Delta^{k+1}; x)\, \frac{\partial}{\partial\nu} \Delta^k F(x)\, dS(x). \end{aligned} \tag{19.19}$$

The last identity is the Euler summation formula related to the Λ-*lattice* τ-*mean* with respect to the iterated Laplacian Δ^m, $m \in \mathbb{N}$.

Remark 19.6. *The lattice ball-generated Euler summation formula for the iterated Laplace operators, i.e., Theorem 19.5, can be obtained by straightforward application of the τ-mean to the lattice point-generated Euler summation formula, too.*

19.2 Gauss–Weierstrass Integral Means

Remembering the Gauss–Weierstrass function W_τ, $\tau > 0$, (cf. C. Müller [1998]) given by

$$y \mapsto W_\tau(y) = (\tau\pi)^{-\frac{q}{2}} e^{-\frac{|y|^2}{t}}, \quad y \in \mathbb{R}^q,\ \tau > 0, \tag{19.20}$$

the Λ-*lattice* τ-*Gauss–Weierstrass mean with respect to the operator* Δ^m is understood to be the convolution of the Λ-lattice function with respect to Δ^m and the Weierstrass function.

Definition 19.7. *For $m \in \mathbb{N}$ and sufficiently small $\tau > 0$*

$$G_\tau^{\mathfrak{W}}(\Delta^m; x) = \int_{\mathbb{R}^q} G(\Delta^m; x - y) W_\tau(y)\, dy, \quad x \in \mathbb{R}^3 \tag{19.21}$$

is called the Λ-lattice (Green's) τ-Gauss–Weierstrass mean with respect to Δ^m (note that the superscript $\mathfrak{W}$ denotes the "Gauss–Weierstrass mean" (obtained via the Weierstrass convolution (19.21)).

The Λ-lattice τ-Gauss–Weierstrass mean $G_\tau^{\mathfrak{W}}(\Delta^m;\cdot)$ is continuous in $\mathbb{R}^q$ for all $m \in \mathbb{N}$. Furthermore, we have the absolutely and uniformly convergent series representation

$$G_\tau^{\mathfrak{W}}(\Delta^m;x) = \frac{1}{\|\mathcal{F}_\Lambda\|} \sum_{\substack{|h|\neq 0 \\ h\in\Lambda^{-1}}} \frac{e^{-\tau\pi^2|h|^2}}{(-4\pi^2|h|^2)^m} e^{2\pi i h\cdot x}, \quad x \in \mathbb{R}^q. \tag{19.22}$$

Note that the Gauss–Weierstrass mean does not only provide absolute and uniform convergence, but also an exponentially accelerated summation.

For $\tau \to 0$, the *Fourier inversion formula in the Gaussian nomenclature* shows (cf. Lemma 17.25) that

$$\lim_{\substack{\tau\to 0 \\ \tau>0}} (\tau\pi)^{-\frac{q}{2}} \int_{\mathcal{G}} e^{-\frac{|x-g|^2}{\tau}} F(x)\, dx = \alpha_{\mathcal{G}}(g) F(g), \tag{19.23}$$

where $\alpha_{\mathcal{G}}(g)$ is the solid angle subtended at $g \in \overline{\mathcal{G}}$ by the surface $\partial\mathcal{G}$. In connection with

$$\lim_{\substack{\tau\to 0 \\ \tau>0}} (\tau\pi)^{-\frac{q}{2}} \sum_{\substack{g\in\overline{\mathcal{G}} \\ g\in\Lambda}} \int_{\mathcal{G}} e^{-\frac{|x-g|^2}{\tau}} F(x)\, dx = \sum_{\substack{g\in\overline{\mathcal{G}} \\ g\in\Lambda}} \alpha_{\mathcal{G}}(g) F(g) = {\sum_{\substack{g\in\overline{\mathcal{G}} \\ g\in\Lambda}}}' F(g) \tag{19.24}$$

we therefore obtain

Theorem 19.8 (Gauss Averaged Euler Summation Formula for Δ^m). *Suppose $\mathcal{G} \subset \mathbb{R}^q$ is a regular region. For any $F \in \mathrm{C}^{(2m)}(\overline{\mathcal{G}})$, $\overline{\mathcal{G}} = \mathcal{G} \cup \partial\mathcal{G}$, and $m \in \mathbb{N}$, the equation*

$$\begin{aligned} &{\sum_{\substack{g\in\overline{\mathcal{G}} \\ g\in\Lambda}}}' F(g) - \frac{1}{\|\mathcal{F}_\Lambda\|} \int_{\mathcal{G}} F(x)\, dx \qquad\qquad (19.25) \\ &= \lim_{\substack{\tau\to 0 \\ \tau>0}} \left(\sum_{k=0}^{m-1} \int_{\partial\mathcal{G}} (\Delta^k F(x)) \frac{\partial}{\partial\nu} G_\tau^{\mathfrak{W}}(\Delta^{k+1};x)\, dS(x) \right. \\ &\qquad - \sum_{k=0}^{m-1} \int_{\partial\mathcal{G}} G_\tau^{\mathfrak{W}}(\Delta^{k+1};x) \frac{\partial}{\partial\nu} \Delta^k F(x)\, dS(x) \\ &\qquad \left. + \int_{\mathcal{G}} G_\tau^{\mathfrak{W}}(\Delta^m;x) \Delta^m F(x)\, dx \right) \end{aligned}$$

holds true.

19.3 Averaged Cubature over $3D$-Regular Regions

All averaged cubature formulas follow the approach presented by W. Freeden [2011]. The numerical calculations given later are due to W. Freeden, I. Ostermann [2012].

Ball Averaged Euler Summation. We remember the τ-integral-mean of the Λ-lattice function with respect to Δ^m (cf. Definition 19.1),

$$G_\tau^{\mathfrak{I}}(\Delta^m;x) = \frac{1}{\frac{4}{3}\pi\tau^3}\int_{\overline{\mathbb{B}_\tau(0)}} G(\Delta^m; x-y)\,dy, \quad x \in \mathbb{R}^3,\ \tau > 0. \tag{19.26}$$

The Λ-lattice τ-integral-mean is continuous in $\mathbb{R}^3$ for all $m \in \mathbb{N}$. Moreover, it has an absolutely and uniformly convergent series representation . Clearly, due to the definition of the characteristic lattice ball function and the mean value theorem of multivariate analysis, we obtain

$$\lim_{\substack{\tau\to 0\\ \tau>0}} \left(\frac{1}{\frac{4}{3}\pi\tau^3}\int_{\mathcal{G}} \mathcal{X}_{\overline{\mathbb{B}_\tau}+\Lambda}(x)F(x)\,dx\right) = \sum_{\substack{g\in\overline{\mathcal{G}}\\ g\in\Lambda}}{}' F(g). \tag{19.27}$$

In other words, the approximation error of the desired integral is given by volume and surface integrals depending on τ. In the limit $\tau \to 0$, however, it equals the lattice point sum (19.27). In extension to Theorem 19.3 we have

Theorem 19.9 (Ball Averaged Euler Cubature Formula for Δ^m in $\mathbb{R}^3$). *Let $\mathcal{G} \subset \mathbb{R}^3$ be a bounded, regular region (with Lipschitz boundary $\partial\mathcal{G}$). Suppose that $F \in \mathrm{C}^{(2m)}(\overline{\mathcal{G}})$, $\overline{\mathcal{G}} = \mathcal{G} \cup \partial\mathcal{G}$ and $m \in \mathbb{N}$. Then, for all sufficiently small $\tau > 0$, the equation*

$$\begin{aligned}
&\frac{1}{\frac{4}{3}\pi\tau^3}\int_{\mathcal{G}} \mathcal{X}_{\overline{\mathbb{B}_\tau}+\Lambda}(x)F(x)\,dx \\
&= \frac{1}{\|\mathcal{F}_\Lambda\|}\int_{\mathcal{G}} F(x)\,dx + \int_{\mathcal{G}} G_\tau^{\mathfrak{I}}(\Delta^m;x)\Delta^m F(x)\,dx \\
&\quad + \sum_{k=0}^{m-1}\int_{\partial\mathcal{G}} (\Delta^k F(x))\frac{\partial}{\partial\nu} G_\tau^{\mathfrak{I}}(\Delta^{k+1};x)\,dS(x) \\
&\quad - \sum_{k=0}^{m-1}\int_{\partial\mathcal{G}} G_\tau^{\mathfrak{I}}(\Delta^{k+1};x)\frac{\partial}{\partial\nu}\Delta^k F(x)\,dS(x)
\end{aligned} \tag{19.28}$$

is valid.

Considering a "compressed lattice" $\tau\Lambda$ with $\tau > 0$ instead of Λ, analogously to the lattice point generated case we observe that the approximation errors in (19.27) and (19.28) vanish for $\tau \to 0$. Thus, recalling the limit relation (19.27), the desired convergence result is obtained for $\tau \to 0$, i.e.,

$$\lim_{\substack{\tau\to 0\\ \tau>0}} \tau^3 \|\mathcal{F}_\Lambda\| \sum_{\substack{g\in\overline{\mathcal{G}}\\ g\in\tau\Lambda}}{}' F(g) = \int_{\mathcal{G}} F(x)\, dx. \tag{19.29}$$

Unfortunately, the convergence for τ tending to 0 is rather slow, as the introduced summation formulas reflect a lattice distribution of the points in $\mathbb{R}^3$ (i.e., they can be regarded as Riemann sums). Nevertheless, the cubature formulas are easy to handle and therefore practicable, e.g., in the solution of partial differential equations.

An algorithm such as the Romberg-type implementation in Section 18.4 which accelerates the convergence is a matter of particular interest. In this section we propose an alternative numerical realization (due to W. Freeden, I. Ostermann [2012]): For a function $F \in \mathrm{C}^{(2m)}(\overline{\mathcal{G}})$, $m \in \mathbb{N}$, satisfying

$$\Delta^k F|\partial\mathcal{G} \neq 0, \qquad \frac{\partial}{\partial\nu}\Delta^k F\Big|\partial\mathcal{G} \neq 0 \tag{19.30}$$

for $k = 0, \ldots, m-1$, the point of departure for our forthcoming considerations is the ball averaged Euler summation formula for the iterated Laplace operator as provided by Theorem 19.9. Choosing $1 \geq \tau_0 > \ldots > \tau_{2m} > 0$ and sufficiently small $t_0, \ldots, t_{2m} > 0$, the integral means $\mathcal{I}_i F$ given by

$$\mathcal{I}_i F = \frac{\tau_i^3 \|\mathcal{F}_\Lambda\|}{\frac{4}{3}\pi t_i^3} \int_{\mathcal{G}} \mathcal{X}_{\overline{\mathbb{B}_{t_i}}+\tau_i\Lambda}(x) F(x)\, dx. \tag{19.31}$$

satisfy the identity

$$\begin{aligned}\mathcal{I}_i F = {} & \int_{\mathcal{G}} F(x)\, dx + \tau_i^3 \|\mathcal{F}_\Lambda\| \int_{\mathcal{G}} G_{t_i}^{\mathfrak{I}}(\tau_i\Lambda; \Delta^m; x)\Delta^m F(x)\, dx \\ & + \tau_i^3 \|\mathcal{F}_\Lambda\| \sum_{k=0}^{m-1} \int_{\partial\mathcal{G}} (\Delta^k F(x)) \frac{\partial}{\partial\nu} G_{t_i}^{\mathfrak{I}}(\tau_i\Lambda; \Delta^{k+1}; x)\, dS(x) \\ & - \tau_i^3 \|\mathcal{F}_\Lambda\| \sum_{k=0}^{m-1} \int_{\partial\mathcal{G}} G_{t_i}^{\mathfrak{I}}(\tau_i\Lambda; \Delta^{k+1}; x) \frac{\partial}{\partial\nu}\Delta^k F(x)\, dS(x).\end{aligned} \tag{19.32}$$

For the sake of simplicity, we introduce the abbreviations

$$\mathcal{E}_{k,i}^{\mathfrak{I},+} F = \tau_i^3 \|\mathcal{F}_\Lambda\| \int_{\partial\mathcal{G}} (\Delta^k F(x)) \frac{\partial}{\partial\nu} G_{t_i}^{\mathfrak{I}}(\tau_i\Lambda; \Delta^{k+1}; x)\, dS(x) \tag{19.33}$$

and

$$\mathcal{E}_{k,i}^{\mathfrak{I},-} F = -\tau_i^3 \|\mathcal{F}_\Lambda\| \int_{\partial\mathcal{G}} G_{t_i}^{\mathfrak{I}}(\tau_i\Lambda; \Delta^{k+1}; x) \frac{\partial}{\partial\nu}\Delta^k F(x)\, dS(x) \tag{19.34}$$

for $k = 0, \ldots, m-1$ and $i = 0, \ldots, 2m$. Our objective is to construct a linear combination of the $2m+1$ integral means $\mathcal{I}_i(F)$ such that the resulting error of the approximate integral $\int_{\mathcal{G}} F(x)\, dx$ only depends on the weighted volume integral of $\Delta^m F$ with respect to the lattices $\tau_i\Lambda$. For this reason, coefficients $e_0, \ldots, e_{2m} \in \mathbb{R} \setminus \{0\}$ are sought which solve the following linear system

$$\begin{pmatrix} 1 & \cdots & 1 \\ \mathcal{E}^{\mathfrak{J},+}_{0,0}F & \cdots & \mathcal{E}^{\mathfrak{J},+}_{0,2m}F \\ \vdots & & \vdots \\ \mathcal{E}^{\mathfrak{J},+}_{m-1,0}F & \cdots & \mathcal{E}^{\mathfrak{J},+}_{m-1,2m}F \\ \mathcal{E}^{\mathfrak{J},-}_{0,0}F & \cdots & \mathcal{E}^{\mathfrak{J},-}_{0,2m}F \\ \vdots & & \vdots \\ \mathcal{E}^{\mathfrak{J},-}_{m-1,0}F & \cdots & \mathcal{E}^{\mathfrak{J},-}_{m-1,2m}F \end{pmatrix} \begin{pmatrix} e_0 \\ e_1 \\ \vdots \\ e_{2m} \end{pmatrix} = \begin{pmatrix} 1 \\ 0 \\ \vdots \\ 0 \end{pmatrix}. \tag{19.35}$$

The matrix occurring in (19.35) is denoted by $A_{\mathfrak{J}}$. If the linear system (19.35) is solvable, i.e., the matrix $A_{\mathfrak{J}}$ is non-singular, the coefficients are given by the first column of the inverse of $A_{\mathfrak{J}}$, namely $e_i = (A_{\mathfrak{J}}^{-1})_{i+1,1}$, $i = 0, \ldots, 2m$. Thus, the approximation of the desired integral is given as follows

$$\begin{aligned} \sum_{i=0}^{2m} (A_{\mathfrak{J}}^{-1})_{i+1,1} \mathcal{I}_i F &= \int_{\mathcal{G}} F(x)\, dx \\ &\quad + \sum_{i=0}^{2m} (A_{\mathfrak{J}}^{-1})_{i+1,1} \tau_i^3 \|\mathcal{F}_\Lambda\| \int_{\mathcal{G}} G^{\mathfrak{J}}_{t_i}(\tau_i\Lambda; \Delta^m; x) \Delta^m F(x)\, dx. \end{aligned} \tag{19.36}$$

In the special case of a function F satisfying $\Delta^m F|\mathcal{G} = 0$, i.e., F is *polyharmonic of degree m in $\mathcal{G}$*, the approximation error in (19.36) is zero. Hence, the integral under consideration can be determined exactly, provided that the integral means $\mathcal{I}_i F$ and the inverse matrix $A_{\mathfrak{J}}^{-1}$ are determined exactly. However, in most cases, both the integral means $\mathcal{I}_i F$ as well as the boundary integrals $\mathcal{E}^{\mathfrak{J},+}_{k,i}F$ and $\mathcal{E}^{\mathfrak{J},-}_{k,i}F$, which constitute the matrix $A_{\mathfrak{J}}$, can only be determined approximately. Due to the definition of the characteristic lattice ball function, the integral means $\mathcal{I}_i F$ correspond to integrals on the regular region $\mathcal{G}$ over small balls around the lattice points $g \in \tau_i\Lambda \cap \overline{\mathcal{G}}$. In order to determine the boundary integrals $\mathcal{E}^{\mathfrak{J},+}_{k,i}F$ and $\mathcal{E}^{\mathfrak{J},-}_{k,i}F$, a variety of integration methods can be applied. For example, a method presented in W. Freeden, H. Kersten [1980] (see also Chapter 16) is compatible to boundary surfaces of star-shaped regions in $\mathbb{R}^3$. It uses the transformation of surface elements of the surface of the star-shaped region onto surface elements of the unit sphere. In the special cases of $\mathcal{G}$ representing the three-dimensional open cube and open unit ball geometry-specific methods can be employed, e.g., the successive Gauss–Legendre quadrature (see, e.g., P.J. Davis, P. Rabinowitz [1975], P. Deuflhard, F. Bornemann [2002]) and a spherical equiangular (latitude-longitude) integration (see Section 11.3), respectively.

Obviously, for the "compressed lattice" $\tau\Lambda$ (with $\tau > 0$ sufficiently small), we find that

$$\lim_{\substack{\tau\to 0\\ \tau>0}} \tau^3 \|\mathcal{F}_\Lambda\| \sum_{\substack{g\in\overline{\mathcal{G}}\\ g\in\tau\Lambda}}{}' F(g) = \int_{\mathcal{G}} F(x)\, dx. \tag{19.37}$$

Gaussian Averaged Euler Summation. Again, an algorithm can be presented which results in an approximation error that only depends on the weighted volume integral of $\Delta^m F$ for a function $F \in \mathrm{C}^{(2m)}(\overline{\mathcal{G}})$, $m \in \mathbb{N}$. As in the consideration above, the function F is assumed to satisfy the boundary condition (19.30). Furthermore, we choose $1 \geq \tau_0 > \ldots > \tau_{2m} > 0$ and sufficiently small $t_0, \ldots, t_{2m} > 0$ such that the approximation error of the error term in summation formula (19.25) is negligible. As a consequence, the weighted sums $\mathcal{L}_i^{\mathfrak{W}}(F)$ given by

$$\mathcal{L}_i^{\mathfrak{W}}(F) = \tau_i^3 \|\mathcal{F}_\Lambda\| \sum_{\substack{g\in\overline{\mathcal{G}}\\ g\in\tau_i\Lambda}}{}' F(g), \quad i = 0, \ldots, 2m, \tag{19.38}$$

can be written in the form

$$\begin{aligned}
\mathcal{L}_i^{\mathfrak{W}}(F) = {} & \int_{\mathcal{G}} F(x)\, dx + \tau_i^3 \|\mathcal{F}_\Lambda\| \int_{\mathcal{G}} G_{t_i}^{\mathfrak{W}}(\tau_i\Lambda; \Delta^m; x) \Delta^m F(x)\, dx \\
& + \tau_i^3 \|\mathcal{F}_\Lambda\| \sum_{k=0}^{m-1} \int_{\partial\mathcal{G}} (\Delta^k F(x)) \frac{\partial}{\partial\nu} G_{t_i}^{\mathfrak{W}}(\tau_i\Lambda; \Delta^{k+1}; x)\, dS(x) \\
& - \tau_i^3 \|\mathcal{F}_\Lambda\| \sum_{k=0}^{m-1} \int_{\partial\mathcal{G}} G_{t_i}^{\mathfrak{W}}(\tau_i\Lambda; \Delta^{k+1}; x) \frac{\partial}{\partial\nu} \Delta^k F(x)\, dS(x)
\end{aligned} \tag{19.39}$$

for $i = 0, \ldots, 2m$. As before, for $k = 0, \ldots, m-1$ and $i = 0, \ldots, 2m$, the following abbreviations are useful:

$$\mathcal{E}_{k,i}^{\mathfrak{W},+} F = \tau_i^3 \|\mathcal{F}_\Lambda\| \int_{\partial\mathcal{G}} (\Delta^k F(x)) \frac{\partial}{\partial\nu} G_{t_i}^{\mathfrak{W}}(\tau_i\Lambda; \Delta^{k+1}; x)\, dS(x), \tag{19.40}$$

$$\mathcal{E}_{k,i}^{\mathfrak{W},-} F = -\tau_i^3 \|\mathcal{F}_\Lambda\| \int_{\partial\mathcal{G}} G_{t_i}^{\mathfrak{W}}(\tau_i\Lambda; \Delta^{k+1}; x) \frac{\partial}{\partial\nu} \Delta^k F(x)\, dS(x). \tag{19.41}$$

We analogously look for coefficients $e_0, \ldots, e_{2m} \in \mathbb{R} \setminus \{0\}$ which solve the linear system

$$\begin{pmatrix}
1 & \ldots & 1 \\
\mathcal{E}_{0,0}^{\mathfrak{W},+} F & \ldots & \mathcal{E}_{0,2m}^{\mathfrak{W},+} F \\
\vdots & & \vdots \\
\mathcal{E}_{m-1,0}^{\mathfrak{W},+} F & \ldots & \mathcal{E}_{m-1,2m}^{\mathfrak{W},+} F \\
\mathcal{E}_{0,0}^{\mathfrak{W},-} F & \ldots & \mathcal{E}_{0,2m}^{\mathfrak{W},-} F \\
\vdots & & \vdots \\
\mathcal{E}_{m-1,0}^{\mathfrak{W},-} F & \ldots & \mathcal{E}_{m-1,2m}^{\mathfrak{W},-} F
\end{pmatrix}
\begin{pmatrix} e_0 \\ e_1 \\ \vdots \\ e_{2m} \end{pmatrix}
= \begin{pmatrix} 1 \\ 0 \\ \vdots \\ 0 \end{pmatrix}. \tag{19.42}$$

The matrix occurring in (19.42) is denoted by $A_{\mathfrak{W}}$. If the matrix $A_{\mathfrak{W}}$ is non-singular, we obtain by the same procedure as already discussed earlier

$$\sum_{i=0}^{2m} (A_{\mathfrak{W}}^{-1})_{i+1,1} \mathcal{W}_i(F) = \int_{\mathcal{G}} F(x)\, dx \tag{19.43}$$
$$+ \sum_{i=0}^{2m} (A_{\mathfrak{W}}^{-1})_{i+1,1} \tau_i^3 \|\mathcal{F}_\Lambda\| \int_{\mathcal{G}} G_{t_i}^{\mathfrak{W}}(\Delta^m; \tau_i\Lambda; x) \Delta^m F(x)\, dx.$$

Similar to the case of the ball averaged Euler summation formula, the proposed approach to accelerate the convergence is interesting especially for polyharmonic functions of degree m, i.e., $\Delta^m F|\mathcal{G} = 0$.

All in all, the classical lattice function as well as ball and Gauss averaged lattice functions are seen to be adequate tools in volume integration under knowledge of boundary values. The corresponding summation formulas of Euler-type for regular regions $\mathcal{G}$ in $\mathbb{R}^3$ are useful with regard to their applicability for numerical integration on $\mathcal{G}$. Furthermore, accuracy enhancing algorithms can be realized. In particular, they are appropriate for polyharmonic functions F of degree $m \in \mathbb{N}$, as the remaining volume integral over $\Delta^m F$ vanishes.

Numerical examples and applications in W. Freeden, I. Ostermann [2012], I. Ostermann [2011] show the adequacy of this approach in geomathematically reflected applications. We omit the details here.

Chapter 20

Adaptive Integration by Euler and Poisson Summation

In one-dimensional theory we saw how the Helmholtz operator $\Delta + \lambda$ can be used to formulate quadrature rules adapted to certain oscillation properties of the integrand by a particular choice of the "wave parameter" λ (cf. Chapter 6). In this chapter we extend this approach to the multivariate case. To this end we introduce Λ-lattice functions $G(\Delta + \lambda; \cdot)$ with respect to the (Helmholtz) operators $\Delta + \lambda$, $\lambda \in \mathbb{R}$, and the "boundary condition of periodicity". We restrict our Helmholtz approach only to features which are different to the Laplace context.

It should be mentioned that, in similarity to the considerations of Chapter 19, all results of this chapter can be formulated in lattice ball or Gaussian averaged Euler summation as well. The details are omitted here (see W. Freeden [2011]).

20.1 Euler Summation Integration Involving Helmholtz Operators

As we already know, seen from mathematical physics, the Λ-lattice function is the Green's function to the operator $\Delta + \lambda$, $\lambda \in \mathbb{R}$, in $\mathbb{R}^q$ and the "boundary condition of Λ-periodicity".

Definition 20.1. $G(\Delta + \lambda; \cdot) : \mathbb{R}^q \setminus \Lambda \to \mathbb{R}$, $\lambda \in \mathbb{R}$ *fixed, is called* Λ-lattice function with respect to the operator $\Delta + \lambda$ *(briefly called Λ-lattice function for $\Delta + \lambda$), if it has the following properties:*

(i) *(*Periodicity*) For all $x \in \mathbb{R}^q \setminus \Lambda$ and $g \in \Lambda$*

$$G(\Delta + \lambda; x + g) = G(\Delta + \lambda; x) \tag{20.1}$$

holds true.

(ii) *(*Differential equation*) $G(\Delta + \lambda; \cdot)$ is twice continuously differentiable for all $x \notin \Lambda$ with*

$$(\Delta + \lambda) G(\Delta + \lambda; x) = 0 \tag{20.2}$$

if $\lambda \notin \mathrm{Spect}_\Delta(\Lambda)$, *and*

$$(\Delta+\lambda)G(\Delta+\lambda;x) = -\frac{1}{\sqrt{\|\mathcal{F}_\Lambda\|}} \sum_{\substack{(\Delta+\lambda)^\wedge(h)\neq 0 \\ h\in\Lambda^{-1}}} \Phi_h(x) \tag{20.3}$$

if $\lambda \in \mathrm{Spect}_\Delta(\Lambda)$ *(note that the summation on the right hand side is to be taken over all* $h \in \Lambda^{-1}$ *satisfying* $(\Delta+\lambda)^\wedge(h) = \Delta^\wedge(h) - \lambda \neq 0$, *i.e.,* $4\pi^2 h^2 \neq \lambda$*).*

*(iii) (*Characteristic singularity*) In the neighborhood of the origin*

$$G(\Delta+\lambda;x) = \begin{cases} -F_q(|x|) + O(1), & q=2, \\ -F_q(|x|)(1+O(|x|\ln|x|)), & q=4, \\ -F_q(|x|)(1+O(|x|^2)), & q\neq 2,4, \end{cases} \tag{20.4}$$

where F_q *is the fundamental solution* (17.45) *in* $\mathbb{R}^q$ *with respect to* Δ.

*(iv) (*Normalization*)*

$$\int_{\mathcal{F}_\Lambda} G(\Delta+\lambda;x)\overline{\Phi_h(x)}\, dx = 0 \tag{20.5}$$

for all $h \in \Lambda^{-1}$ *with* $(\Delta+\lambda)^\wedge(h) \neq 0$,

The main results, i.e., uniqueness and existence, of the theory of Λ-lattice functions with respect to the Helmholtz operator $\Delta+\lambda$, $\lambda \in \mathbb{R}$ are summarized in W. Freeden [2011]. They are omitted here.

Euler Summation Formulas Involving Helmholtz Operators. The Euler summation formulas for Helmholtz operators can be deduced in analogous way as for the Laplace case (cf. W. Freeden [2011]). However, as a key point of our approach, it should be noted that the differential equation (20.3) in Definition 20.1 leads to two different types of summation formulas dependent on the choice of the parameter λ, i.e., reflecting the cases $\lambda \notin \mathrm{Spect}_\Delta(\Lambda)$ or $\lambda \in \mathrm{Spect}_\Delta(\Lambda)$.

Theorem 20.2. *Let* $\mathcal{G} \subset \mathbb{R}^q$ *be a regular region. Let* F *be twice continuously differentiable on* $\overline{\mathcal{G}}$. *Then, for* $\lambda \notin \mathrm{Spect}_\Delta(\Lambda)$, *we have*

$$\sum_{\substack{g\in\overline{\mathcal{G}} \\ g\in\Lambda}}{}' F(g) = \int_{\mathcal{G}} G(\Delta+\lambda;x)\,(\Delta+\lambda)F(x)\, dx \tag{20.6}$$

$$+ \int_{\partial\mathcal{G}} F(x)\frac{\partial}{\partial\nu}G(\Delta+\lambda;x) - G(\Delta+\lambda;x)\,\frac{\partial F}{\partial\nu}(x)\; dS(x),$$

while, for $\lambda \in \mathrm{Spect}_\Delta(\Lambda)$, *we have*

$$\sum_{\substack{g\in\overline{\mathcal{G}}\\ g\in\Lambda}}{}' F(g) = \frac{1}{\sqrt{\|\mathcal{F}_\Lambda\|}} \sum_{\substack{(\Delta+\lambda)^\wedge(h)=0\\ h\in\Lambda^{-1}}} \int_{\mathcal{G}} F(x)\overline{\Phi_h(x)}\, dx \tag{20.7}$$
$$+ \int_{\mathcal{G}} G(\Delta+\lambda;x)\,(\Delta+\lambda)F(x)\, dx$$
$$+ \int_{\partial\mathcal{G}} F(x)\frac{\partial}{\partial\nu}G(\Delta+\lambda;x) - G(\Delta+\lambda;x)\frac{\partial F}{\partial\nu}(x)\, dS(x),$$

where $\frac{\partial}{\partial\nu}$ *denotes the derivative in the direction of the outer normal* ν *and*

$$\sum_{\substack{g\in\overline{\mathcal{G}}\\ g\in\Lambda}}{}' F(g) = \sum_{\substack{g\in\overline{\mathcal{G}}\\ g\in\Lambda}} \alpha_{\mathcal{G}}(g)\; F(g). \tag{20.8}$$

Analogously to the univariate settings, the difference between the two cases shows that we have to expect some kind of resonance phenomena in the summation procedures. Several examples will later bring out this structure more clearly. From an algorithmic point of view we see that the formulas express a sum in terms of integrals over $\mathcal{G}$ and its boundary $\partial\mathcal{G}$ involving derivatives of the function F up to the second order. Of particular importance is the case $\lambda = 0$, i.e., the Helmholtz operator reduces to the Laplace operator. As a matter of fact, the particular case $\lambda = 0$, i.e., the Euler summation formula corresponding to the Laplacian, turns out to be a "resonance case", i.e., $\lambda = 0$ is a member of $\mathrm{Spect}_\Delta(\Lambda)$.

Lattice Functions Involving Iterated Helmholtz Operators. The Λ-lattice functions for operators $(\Delta+\lambda)^m$, $\lambda \in \mathbb{R}$, $m \in \mathbb{N}$, can be based on the Λ-lattice functions for operators $\Delta+\lambda$, $\lambda \in \mathbb{R}$, by forming convolution integrals in the usual way

$$G\left((\Delta+\lambda)^1;x\right) = G(\Delta+\lambda;x)\,, \tag{20.9}$$
$$G\left((\Delta+\lambda)^m;x\right) = \int_{\mathcal{F}_\Lambda} G\left((\Delta+\lambda)^{m-1};z\right) G(\Delta+\lambda;x-z)\; dz,$$

for $m = 2, 3, \ldots$. We list some essential properties: For all $x \notin \Lambda$ and $g \in \Lambda$,

$$G\left((\Delta+\lambda)^m;x+g\right) = G\left((\Delta+\lambda)^m;x\right) \tag{20.10}$$

is satisfied, i.e., $G\left((\Delta+\lambda)^m;\cdot\right)$ is Λ-periodic. As in potential theory we have

$$G\left((\Delta+\lambda)^m;x\right) = \begin{cases} O\left(|x|^{2m-q}\ln|x|\right), & q = 2m\\ O\left(|x|^{2m-q}\right), & q > 2m. \end{cases} \tag{20.11}$$

As in (18.12) the differential equation

$$(\Delta+\lambda)G\left((\Delta+\lambda)^m;x\right) = G\left((\Delta+\lambda)^{m-1};x\right), \quad x \notin \Lambda, \tag{20.12}$$

$m = 2, 3, \ldots$, represents a recursion relation relating the Λ-lattice function for the operator $(\Delta + \lambda)^m$ to the Λ-lattice function for the operator $(\Delta + \lambda)^{m-1}$. The series expansion of $G((\Delta + \lambda)^m; \cdot)$ in terms of eigenfunctions, which is equivalent to the (formal) Fourier (orthogonal) expansion, is given for iteration orders $m = 2, 3, \ldots$

$$\frac{1}{\sqrt{\|\mathcal{F}_\Lambda\|}} \sum_{\substack{(\Delta+\lambda)^\wedge(h)\neq 0 \\ h\in\Lambda^{-1}}} \frac{\Phi_h(x)}{-((\Delta+\lambda)^m)^\wedge(h)}, \quad x \in \mathbb{R}^q, \tag{20.13}$$

where

$$((\Delta+\lambda)^m)^\wedge(h) = -(\lambda - \Delta^\wedge(h))^m = -(\lambda - 4\pi^2 h^2)^m. \tag{20.14}$$

For $m > \frac{q}{2}$, the Λ-lattice function $G((\Delta + \lambda)^m; \cdot)$ is continuous in $\mathbb{R}^q$, and its bilinear series admits the representation

$$G((\Delta+\lambda)^m; x-y) = \sum_{\substack{(\Delta+\lambda)^\wedge(h)\neq 0 \\ h\in\Lambda^{-1}}} \frac{\Phi_h(x)\overline{\Phi_h(y)}}{-((\Delta+\lambda)^m)^\wedge(h)}, \quad x, y \in \mathbb{R}^q. \tag{20.15}$$

Euler Summation for Iterated Helmholtz Operators. In connection with the summation formula (i.e., Theorem 20.2) we obtain the Euler summation formula with respect to the operator $(\Delta + \lambda)^m$, which at the time is a generalization of Theorem 18.5.

Theorem 20.3 (Euler Summation Formula for the Iterated Helmholtz Operator $(\Delta+\lambda)^m$, $\lambda \in \mathbb{R}$, $m \in \mathbb{N}$)**.** *Let Λ be an arbitrary lattice in $\mathbb{R}^q$. Let $\mathcal{G} \subset \mathbb{R}^q$ be a regular region. Suppose that F is of class $\mathrm{C}^{(2m)}(\overline{\mathcal{G}})$, $\overline{\mathcal{G}} = \mathcal{G} \cup \partial\mathcal{G}$. Then, for all $\lambda \in \mathbb{R}$,*

$$\begin{aligned}
\sum_{\substack{g\in\overline{\mathcal{G}} \\ g\in\Lambda}}{}' F(g) = {} & \frac{1}{\sqrt{\|\mathcal{F}_\Lambda\|}} \sum_{\substack{(\Delta+\lambda)^\wedge(h)=0 \\ h\in\Lambda^{-1}}} \int_{\mathcal{G}} F(x)\overline{\Phi_h(x)}\, dx \\
& + \sum_{k=0}^{m-1} \int_{\partial\mathcal{G}} \left(\frac{\partial}{\partial\nu} G\left((\Delta+\lambda)^{k+1}; x\right)\right) (\Delta+\lambda)^k F(x)\, dS(x) \\
& - \sum_{k=0}^{m-1} \int_{\partial\mathcal{G}} G\left((\Delta+\lambda)^{k+1}; x\right) \frac{\partial}{\partial\nu} (\Delta+\lambda)^k F(x)\, dS(x) \\
& + \int_{\mathcal{G}} G((\Delta+\lambda)^m; x)\, (\Delta+\lambda)^m F(x)\, dx,
\end{aligned} \tag{20.16}$$

where $\sum_{(\Delta+\lambda)^\wedge(h)=0} \cdots$ only occurs if $\lambda \in \mathrm{Spect}_\Delta(\Lambda)$.

The close relation of these identities to the one-dimensional Euler summation formulas is best seen by specializing the "wave number" $\lambda \in \mathbb{R}$. It turns out that the generalization of the Euler summation formula to Helmholtz operators $(\Delta + \lambda)^m$, $\lambda \in \mathbb{R}$, opens new perspectives to subtle questions of the convergence of multi-dimensional alternating series and the integration of multivariate alternating functions because the Helmholtz operator $(\Delta + \lambda)^m$, $\lambda \in \mathbb{R}$, can be closely adapted to oscillating properties of the summands or integrands, respectively. Note that, for the convergence of alternating series in one dimension, direct techniques are well-known. In multi-dimensional summation problems the situation is different because the one-dimensional concept of the alternating or oscillating series is usually not directly applicable.

Remark 20.4. *The Euler summation formulas can be formulated to more general differential operators (see V.K. Ivanow [1963], W. Freeden [1982a]). Throughout this work, we restrict ourselves to elliptic differential operators with constant coefficients. There are two ways of realization. On the one hand, we take Helmholtz operators and arbitrary lattices so that because of the choice of arbitrary lattices arbitrary second order elliptic operators are included. In this case, the* Λ*-lattice functions can be handled in the Helmholtz framework. On the other hand, we can also start with general elliptic differential operators. In this case, the generality of the approach is at the cost of the accessibility of the occurring settings.*

20.2 Adaptive Cubature over Regular Regions

Let $\mathcal{G}$ be a regular region in $\mathbb{R}^q$ (see Definition 16.1) and $\lambda \in \mathbb{R}$. From (20.16) in Theorem 20.3 we obtain for all $y \in \mathcal{F}_\Lambda$ that

$$
\begin{aligned}
&\sum_{\substack{g+y_k\in\overline{\mathcal{G}}\\ g\in\Lambda}}{}' F(g+y) \qquad (20.17)\\
&\quad = \sum_{\substack{4\pi^2h^2=\lambda\\ h\in\Lambda^{-1}}} \overline{\Phi_h(y)} \int_{\mathcal{G}} F(x)\Phi_h(x)\, dx\\
&\quad + \sum_{j=0}^{m-1} \int_{\partial\mathcal{G}} \left(\frac{\partial}{\partial\nu} G\left((\Delta+\lambda)^{j+1}; x-y\right)\right)(\Delta+\lambda)^j F(x)\, dS(x)\\
&\quad - \sum_{j=0}^{m-1} \int_{\partial\mathcal{G}} G\left((\Delta+\lambda)^{j+1}; x-y\right) \frac{\partial}{\partial\nu}(\Delta+\lambda)^j F(x)\, dS(x)\\
&\quad + \int_{\mathcal{G}} G\left((\Delta+\lambda)^m; x-y\right)(\Delta+\lambda)^m F(x)\, dx
\end{aligned}
$$

This identity enables us to conclude that Theorem 20.3 also allows the following extension

$$\sum_{k=1}^{L} a_k \sum_{\substack{g+y_k \in \overline{\mathcal{G}} \\ g \in \Lambda}}{}' F(g+y_k) \tag{20.18}$$

$$= \sum_{\substack{4\pi^2 h^2 = \lambda \\ h \in \Lambda^{-1}}} \sum_{k=1}^{L} a_k \overline{\Phi_h(y_k)} \int_{\mathcal{G}} F(x) \Phi_h(x) \, dx$$

$$+ \sum_{j=0}^{m-1} \int_{\partial\mathcal{G}} \left(\frac{\partial}{\partial\nu} \sum_{k=1}^{L} a_k G\left((\Delta+\lambda)^{j+1} ; x - y_k \right) \right) (\Delta+\lambda)^j F(x) \, dS(x)$$

$$- \sum_{j=0}^{m-1} \int_{\partial\mathcal{G}} \sum_{k=1}^{L} a_k G\left((\Delta+\lambda)^{j+1} ; x - y_k \right) \frac{\partial}{\partial\nu} (\Delta+\lambda)^j F(x) \, dS(x)$$

$$+ \int_{\mathcal{G}} \sum_{k=1}^{L} a_k G\left((\Delta+\lambda)^m ; x - y_k \right) (\Delta+\lambda)^m F(x) \, dx$$

holds for any choice of coefficients $a_1, \ldots, a_L \in \mathbb{R}$ and nodes $y_1, \ldots, y_L$ of the fundamental domain $\mathcal{F}_\Lambda$.

We now modify the Λ-lattice function to ensure a certain degree of exactness in our planned cubature rule.

Definition 20.5. *Let Ξ be a finite subset of Λ^{-1} containing all points $h \in \Lambda^{-1}$ with $4\pi^2 h^2 = \lambda$ as subset. Then the function $G\left((\Delta+\lambda)^m ; \Xi; \cdot\right) : \mathbb{R}^3 \to \mathbb{C}$ defined by*

$$G\left((\Delta+\lambda)^m ; \Xi; x\right) = G\left((\Delta+\lambda)^m ; x\right) - \frac{1}{\sqrt{\|\mathcal{F}_\Lambda\|}} \sum_{\substack{4\pi^2 h^2 \neq \lambda \\ h \in \Xi}} \frac{\Phi_h(x)}{(4\pi^2 h^2 - \lambda)^m} \tag{20.19}$$

is called Ξ-modified Λ-lattice function with respect to the operator $(\Delta+\lambda)^m$.

By repeated application of Green's theorem we obtain

$$0 = \sum_{\substack{4\pi^2 h^2 \neq \lambda \\ h \in \Xi}} \sum_{k=1}^{L} a_k \overline{\Phi_h(y_k)} \int_{\mathcal{G}} F(x) \Phi_h(x) \, dx \tag{20.20}$$

$$+ \sum_{j=0}^{m-1} \int_{\partial\mathcal{G}} \left(\frac{\partial}{\partial\nu} \sum_{\substack{4\pi^2 h^2 \neq \lambda \\ h \in \Xi}} \sum_{k=1}^{L} \frac{a_k}{(4\pi^2 h^2 - \lambda)^j} \overline{\Phi_h(y_k)} \Phi_h(x) \right) (\Delta+\lambda)^j F(x) \, dS(x)$$

$$- \sum_{j=0}^{m-1} \int_{\partial \mathcal{G}} \sum_{\substack{4\pi^2 h^2 \neq \lambda \\ h \in \Xi}} \sum_{k=1}^{L} \frac{a_k}{(4\pi^2 h^2 - \lambda)^j} \overline{\Phi_h(y_k)} \Phi_h(x) \frac{\partial}{\partial \nu} (\Delta + \lambda)^j F(x) \, dS(x)$$

$$- \int_{\mathcal{G}} \sum_{\substack{4\pi^2 h^2 \neq \lambda \\ h \in \Xi}} \sum_{k=1}^{L} \frac{a_k}{(4\pi^2 h^2 - \lambda)^m} \overline{\Phi_h(y_k)} \Phi_h(x) (\Delta + \lambda)^m F(x) \, dx.$$

Degree of Exactness. Combining (20.18) and (20.20) we therefore obtain

$$\sum_{k=1}^{L} a_k \sum_{g + y_k \in \overline{\mathcal{G}}}{}' F(g + y_k) \tag{20.21}$$

$$= \sum_{h \in \Xi} \sum_{k=1}^{L} a_k \overline{\Phi_h(y_k)} \int_{\mathcal{G}} F(x) \Phi_h(x) \, dx$$

$$+ \sum_{j=0}^{m-1} \int_{\partial \mathcal{G}} \left(\frac{\partial}{\partial \nu} \sum_{k=1}^{L} a_k G\left((\Delta + \lambda)^{j+1}; \Xi; x - y_k \right) \right) (\Delta + \lambda)^j F(x) \, dS(x)$$

$$- \sum_{j=0}^{m-1} \int_{\partial \mathcal{G}} \sum_{k=1}^{L} a_k G\left((\Delta + \lambda)^{j+1}; \Xi; x - y_k \right) \frac{\partial}{\partial \nu} (\Delta + \lambda)^j F(x) \, dS(x)$$

$$+ \int_{\mathcal{G}} \sum_{k=1}^{L} a_k G\left((\Delta + \lambda)^m; \Xi; x - y_k \right) (\Delta + \lambda)^m F(x) \, dx.$$

Definition 20.6. *A system $y_1, \ldots, y_L$ of L points of the fundamental cell $\mathcal{F}_\Lambda$ of the lattice Λ is called* Ξ-fundamental system *if $L \geq M = \sum_{h \in \Xi} 1$ and the rank of the $L \times M$-matrix of the linear system*

$$\sum_{k=1}^{L} a_k \overline{\Phi_h(y_k)} = d_h, \quad h \in \Xi, \tag{20.22}$$

is equal to M.

Let T be a periodic polynomial in $\mathbb{R}^q$ over Ξ, i.e., a function of the form

$$T = \sum_{h \in \Xi} d_h \Phi_h. \tag{20.23}$$

Given a function T of the form (20.23) and a Ξ-fundamental system in the sense of Definition 20.6. Then T allows the representation

$$T = \sum_{h \in \Xi} \sum_{k=1}^{L} a_k \overline{\Phi_h(y_k)} \, \Phi_h. \tag{20.24}$$

As a consequence, we are led to

Corollary 20.7. *Let $y_1, \ldots, y_L$ be a Ξ-fundamental system. Suppose that T is a periodic polynomial of the form (20.23). Then*

$$\sum_{k=1}^{L} a_k \sum_{\substack{g+y_k \in \overline{\mathcal{G}} \\ g \in \Lambda}}{}' F(g+y_k) \tag{20.25}$$

$$= \int_{\mathcal{G}} F(x)\, T(x)\, dx$$

$$+ \sum_{j=0}^{m-1} \int_{\partial \mathcal{G}} \left(\frac{\partial}{\partial \nu} \sum_{k=1}^{L} a_k G\left((\Delta+\lambda)^{j+1}; \Xi; x - y_k \right) \right) (\Delta+\lambda)^j F(x)\, dS(x)$$

$$- \sum_{j=0}^{m-1} \int_{\partial \mathcal{G}} \sum_{k=1}^{L} a_k G\left((\Delta+\lambda)^{j+1}; \Xi; x - y_k \right) \frac{\partial}{\partial \nu} (\Delta+\lambda)^j F(x)\, dS(x)$$

$$+ \int_{\mathcal{G}} \sum_{k=1}^{L} a_k G\left((\Delta+\lambda)^{m}; \Xi; x - y_k \right) (\Delta+\lambda)^m F(x)\, dx$$

holds for all coefficients $a_1, \ldots, a_L$ satisfying the linear equations (20.22).

The error term

$$\mathcal{E}_m^{\lambda,\Xi} F = \int_{\mathcal{G}} \sum_{k=1}^{L} G\left((\Delta+\lambda)^{m}; \Xi; x - y_k \right) (\Delta+\lambda)^m F(x)\, dx \tag{20.26}$$

can be estimated in different topologies. As an example we mention

Corollary 20.8. *For $m > q/2$,*

$$|\mathcal{E}_m^{\lambda,\Xi} F| \tag{20.27}$$

$$\leq \frac{1}{\sqrt{\|\mathcal{F}_\Lambda\|}} \sum_{h \notin \Xi} \frac{1}{|4\pi^2 h^2 - \lambda|^m} \left| \sum_{k=1}^{L} a_k \overline{\Phi_h(y_k)} \right| \int_{\mathcal{G}} |(\Delta+\lambda)^m F(x)|\, dx.$$

Again, we end up with a multiplicative splitting for the error estimate. The integral on the right hand side does not depend on the lattice Λ under consideration. Thus, the arbitrary choice of the parameter λ can be used to adapt the differential operator to specific features of the integrand.

Example 20.9. *Taking $F(x) = e(b \cdot x)$, $T(x) = 1$, $x \in \mathbb{R}^3$, $\Lambda = \tau \mathbb{Z}^3$, $\tau \in \mathbb{R}$, $b \in \mathbb{R}^3$, we are led to a simple, but instructive error estimate. The error bound*

$$\frac{\tau^{2m-3}}{(2\pi)^{2m}} \|\mathcal{G}\| \left|4\pi^2 |b|^2 - \lambda\right|^m \sum_{k=1}^{L} |a_k| \sum_{\substack{\tau^{-1} h \notin \Xi \\ h \in \mathbb{Z}^3}} \frac{1}{\left| |h|^2 - \frac{\lambda}{4\pi^2 \tau^2} \right|^m} \tag{20.28}$$

is zero for $\lambda = 4\pi^2 |b|^2$. In Table 20.1, the error estimate (20.28) is investigated for two parameters λ "close" to the value $4\pi^2|b|^2$ and different lattices widths τ and orders m. It shows that the choice of the parameter λ can be used to obtain adaptive cubatures.

TABLE 20.1: Error bound (20.28) for $b = (1, 1, 1)^T$, $\lambda = 118$, $\|\mathcal{G}\| = 1$, and $\Xi = \{0\}$

m	$\tau = 1.0$	$\tau = 0.5$	$\tau = 0.25$
2	2.501368e-02	3.537469e-02	5.002736e-02
3	1.956636e-04	6.917752e-05	2.445795e-05
4	1.925490e-06	1.701909e-05	1.504289e-08
5	2.028831e-08	4.483124e-10	9.906397e-12
10	3.187368e-18	6.878081e-23	1.484234e-27

20.3 Poisson Summation under Adaptive Criteria

Next multi-dimensional Euler summation formulas with respect to iterated Helmholtz operators are applied to lattice point sums inside spheres $\mathbb{S}_N^{q-1}$, $q \geq 2$, of radius N which provide interesting results particularly when N tends toward infinity, thereby adapting the iterated Helmholtz operator $(\Delta + \lambda)^m$, $\lambda \in \mathbb{R}$, to the specific (oscillating) properties of the weight function under consideration.

As a result, multi-dimensional analogs to the one-dimensional Poisson formula can be formulated in the lattice point context of Euclidean spaces $\mathbb{R}^q$, $q \geq 2$. Their special application to alternating integrals and series of Hardy–Landau type is of particular interest.

Convergence Criteria. It should be pointed out that we have to specify multi-dimensional convergence of integrals and series in a preassigned way, as far as the integrals and series are not absolutely convergent. We realize this requirement by applying the multi-dimensional Euler summation formula especially to sums of the spherical type

$$\sum_{g\in\Lambda} \cdots = \lim_{N\to\infty} \sum_{\substack{|g|\leq N \\ g\in\Lambda}} \cdots . \tag{20.29}$$

Now, let F be a $2m$-times ($m \in \mathbb{N}$) continuously differentiable function in $\overline{\mathbb{B}_N^q}$. Then, for each value $\lambda \in \mathbb{R}$, we have

$$\begin{aligned} \sum_{\substack{|g|\leq N \\ g\in\Lambda}} F(g) &= \frac{1}{\sqrt{\|\mathcal{F}_\Lambda\|}} \sum_{\substack{(\Delta+\lambda)^\wedge(h)=0 \\ h\in\Lambda^{-1}}} \int_{\substack{|x|\leq N \\ x\in\mathbb{R}^q}} F(x)\overline{\Phi_h(x)}\, dx \\ &+ \int_{\substack{|x|\leq N \\ x\in\mathbb{R}^q}} G\left((\Delta+\lambda)^m ; x\right) (\Delta+\lambda)^m F(x)\, dx + \mathcal{E}_N^{(m)} F \end{aligned} \tag{20.30}$$

where the sum $\Sigma_{(\Delta+\lambda)^\wedge(h)=0}$ on the right side of (20.30) occurs only if $\lambda \in \text{Spect}_\Delta(\Lambda)$. The term $\mathcal{E}_N^{(m)}F$ is explicitly given by

$$\mathcal{E}_N^{(m)}F = \frac{1}{2}\sum_{\substack{|g|=N\\ g\in\Lambda}} F(g) \tag{20.31}$$
$$+\sum_{k=0}^{m-1}\int_{\substack{|x|=N\\ x\in\mathbb{R}^q}} \left(\frac{\partial}{\partial\nu}G\left((\Delta+\lambda)^{k+1};x\right)\right)(\Delta+\lambda)^k F(x)\, dS(x)$$
$$-\sum_{k=0}^{m-1}\int_{\substack{|x|=N\\ x\in\mathbb{R}^q}} G\left((\Delta+\lambda)^{k+1};x\right)\frac{\partial}{\partial\nu}(\Delta+\lambda)^k F(x)\, dS(x).$$

Questions of the convergence require estimates for $N\to\infty$. In this respect, the following asymptotic integral relations borrowed from W. Freeden [2011] play an important part.

Theorem 20.10. *For all numbers $\lambda\in\mathbb{R}$, all lattices $\Lambda\subset\mathbb{R}^q$, and all positive integers k, the Λ-lattice functions for iterated Helmholtz operators satisfy the asymptotic estimates*

$$\int_{\substack{|x|=N\\ x\in\mathbb{R}^q}} \left|G\left((\Delta+\lambda)^k;x\right)\right| dS(x) = O\left(N^{q-1}\right), \quad N\to\infty, \tag{20.32}$$

and

$$\int_{\substack{|x|=N\\ x\in\mathbb{R}^q}} \left|\frac{\partial}{\partial\nu}G\left((\Delta+\lambda)^k;x\right)\right| dS(x) = O\left(N^{q-1}\right), \quad N\to\infty. \tag{20.33}$$

Since the "wave number" λ can be chosen arbitrarily in Theorem 20.10, the properties (20.32) and (20.33) can be used to formulate adaptive convergence theorems for multi-dimensional integrals and series, where the Euler summation formula is the key structure to be observed in our context.

We begin our discussion with the terms occurring in (20.31).

Lemma 20.11. *For given $m\in\mathbb{N}$ and $\lambda\in\mathbb{R}$, assume that the function $F\in C^{(2m)}(\mathbb{R}^q)$ satisfies the asymptotic relations*

$$\left|(\Delta_x+\lambda)^k F(x)\right| = o\left(|x|^{1-q}\right), \quad |x|\to\infty, \tag{20.34}$$
$$\left|\nabla_x(\Delta_x+\lambda)^k F(x)\right| = o\left(|x|^{1-q}\right), \quad |x|\to\infty, \tag{20.35}$$

for $k=0,\ldots,m-1$ (note that the case $m=1$ is independent of $\lambda\in\mathbb{R}$). Then, for $N\to\infty$, it follows that the remainder given by (20.31) satisfies

$$\mathcal{E}_N^{(m)}F = o(1). \tag{20.36}$$

Proof. Observing the classical Gaussian result that $\sum_{|g|=N, g\in\Lambda} 1 = O(N^{q-1})$ (cf. Corollary 17.6) we see that, due to the asymptotic relations of F, we obtain

$$\frac{1}{2}\sum_{\substack{|g|=N\\ g\in\Lambda}} F(g) = o\Bigg(N^{1-q}\sum_{\substack{|g|=N\\ g\in\Lambda}} 1\Bigg) = o(1), \quad N\to\infty. \tag{20.37}$$

Furthermore, we are allowed to conclude that

$$\begin{aligned}
&\sum_{k=0}^{m-1}\int_{\substack{|x|=N\\ x\in\mathbb{R}^q}}\left(\frac{\partial}{\partial\nu}G\left((\Delta+\lambda)^{k+1};x\right)\right)(\Delta+\lambda)^k F(x)\, dS(x)\\
&\qquad -\sum_{k=0}^{m-1}\int_{\substack{|x|=N\\ x\in\mathbb{R}^q}} G\left((\Delta+\lambda)^{k+1};x\right)\frac{\partial}{\partial\nu}(\Delta+\lambda)^k F(x)\, dS(x)\\
&\quad = o\Bigg(N^{1-q}\Bigg(\sum_{k=0}^{m-1}\int_{\substack{|x|=N\\ x\in\mathbb{R}^q}}\left|G\left((\Delta+\lambda)^{k+1};x\right)\right|\\
&\qquad\qquad +\left|\frac{\partial}{\partial\nu}G\left((\Delta+\lambda)^{k+1};x\right)\right|\, dS(x)\Bigg)\Bigg)
\end{aligned}\tag{20.38}$$

for $N\to\infty$. In connection with Theorem 20.10 and (20.34), (20.35) we therefore find that

$$\begin{aligned}
&\sum_{k=0}^{m-1}\int_{\substack{|x|=N\\ x\in\mathbb{R}^q}}\left(\frac{\partial}{\partial\nu}G\left((\Delta+\lambda)^{k+1};x\right)\right)(\Delta+\lambda)^k F(x)\, dS(x)\\
&\qquad -\sum_{k=0}^{m-1}\int_{\substack{|x|=N\\ x\in\mathbb{R}^q}} G\left((\Delta+\lambda)^{k+1};x\right)\frac{\partial}{\partial\nu}(\Delta+\lambda)^k F(x)\, dS(x)\\
&\quad = o(1)
\end{aligned}\tag{20.39}$$

for $N\to\infty$. Collecting all details we obtain the announced result of Lemma 20.11. □

Next we come to the discussion of the volume integral in (20.30) involving the iterated Helmholtz derivative $(\Delta+\lambda)^m F$.

Lemma 20.12. *For given $m\in\mathbb{N}$, $\varepsilon>0$, and $\lambda\in\mathbb{R}$, assume that the function $F\in \mathrm{C}^{(2m)}(\mathbb{R}^q)$ satisfies*

$$|(\Delta_x+\lambda)^m F(x)| = O\left(|x|^{-(q+\varepsilon)}\right), \quad |x|\to\infty. \tag{20.40}$$

Then the integral

$$\int_{\mathbb{R}^q} G\left((\Delta+\lambda)^m;x\right)(\Delta+\lambda)^m F(x)\, dx \tag{20.41}$$

is absolutely convergent. (Hence, it is of no significance how to understand the convergence of the integral over $\mathbb{R}^q$.)

Proof. From Theorem 20.10 we are immediately able to guarantee with suitable positive constants M, N that

$$\left| \int_{\substack{M\le|x|\le N\\ x\in\mathbb{R}^q}} G\left((\Delta+\lambda)^m;x\right)(\Delta+\lambda)^m F(x)\,dx \right| \tag{20.42}$$

$$= O\left(\int_M^N \frac{1}{(1+r)^{q+\varepsilon}} \int_{\substack{|x|=r\\ x\in\mathbb{R}^q}} \left|G\left((\Delta+\lambda)^m;x\right)\right|\,dS(x)\,dr \right)$$

$$= O\left(\int_M^N \frac{r^{q-1}}{(1+r)^{q+\varepsilon}}\,dr \right).$$

Consequently, if F satisfies (20.40), the absolute convergence of the integral (20.41) is guaranteed. □

Combining Lemma 20.11 and Lemma 20.12 we obtain as a first consequence.

Theorem 20.13. *Let Λ be an arbitrary lattice in $\mathbb{R}^q$. For given $m\in\mathbb{N}$, $\varepsilon>0$, and $\lambda\in\mathbb{R}$, assume that $F\in C^{(2m)}(\mathbb{R}^q)$ satisfies the properties (20.34), (20.35), and (20.40). Then, the limit*

$$\lim_{N\to\infty}\left(\sum_{\substack{|g|\le N\\ g\in\Lambda}} F(g) - \frac{1}{\sqrt{\|\mathcal{F}_\Lambda\|}} \sum_{\substack{(\Delta+\lambda)^\wedge(h)=0\\ h\in\Lambda^{-1}}} \int_{\substack{|x|\le N\\ x\in\mathbb{R}^q}} F(x)\overline{\Phi_h(x)}\,dx \right) \tag{20.43}$$

exists, and we have the limit relation

$$\lim_{N\to\infty}\left(\sum_{\substack{|g|\le N\\ g\in\Lambda}} F(g) - \frac{1}{\sqrt{\|\mathcal{F}_\Lambda\|}} \sum_{\substack{(\Delta+\lambda)^\wedge(h)=0\\ h\in\Lambda^{-1}}} \int_{\substack{|x|\le N\\ x\in\mathbb{R}^q}} F(x)\overline{\Phi_h(x)}\,dx \right)$$

$$= \int_{\mathbb{R}^q} G\Big((\Delta+\lambda)^m;x\Big)(\Delta+\lambda)^m F(x)\,dx, \tag{20.44}$$

where the sum $\sum_{(\Delta+\lambda)^\wedge(h)=0}\cdots$ has to be extended over all $h\in\Lambda^{-1}$ satisfying $(\Delta+\lambda)^\wedge(h)=0$.

Note that the convergence of the left-hand side of (20.44) has to be understood in spherical sense, while the convergence of the integral on the right-hand side of (20.44) is free of restrictions.

In the case that $\lambda\notin \mathrm{Spect}_\Delta(\Lambda)$, the sum $\sum_{(\Delta+\lambda)^\wedge(h)=0}\cdots$ is always understood to be zero, hence we have

$$\lim_{N\to\infty} \sum_{\substack{|g|\le N\\ g\in\Lambda}} F(g) = \int_{\mathbb{R}^q} G\left((\Delta+\lambda)^m;x\right)(\Delta+\lambda)^m F(x)\,dx. \tag{20.45}$$

In other words, Theorem 20.13 demonstrates that the convergence of multi-dimensional sums (understood in spherical summation)

$$\sum_{g\in\Lambda} F(g) = \lim_{N\to\infty} \sum_{|g|\le N} F(g) \tag{20.46}$$

is closely related to the spectrum $\mathrm{Spect}_\Delta(\Lambda)$. As a matter of fact, for the class of non-eigenvalues, i.e., $\lambda \notin \mathrm{Spect}_\Delta(\Lambda)$, Theorem 20.13 immediately guarantees the convergence of the infinite series on the left side of (20.45).

In order to ensure the convergence of the series (20.46) for an eigenvalue $\lambda \in \mathrm{Spect}_\Delta(\Lambda)$, however, an additional condition involving indefinite integrals has to come into play.

Theorem 20.14. *Let Λ be an arbitrary lattice in $\mathbb{R}^q$. For given $m \in \mathbb{N}$, $\varepsilon > 0$, and $\lambda \in \mathbb{R}$, assume that $F \in \mathrm{C}^{(2m)}(\mathbb{R}^q)$, $m \in \mathbb{N}$, satisfies the properties (20.34), (20.35), and (20.40). In addition, suppose that the integrals*

$$\int_{\mathbb{R}^q} F(x)\overline{\Phi_h(x)}\, dx \tag{20.47}$$

exist for all $h \in \Lambda^{-1}$ with $(\Delta + \lambda)^\wedge(h) = 0$ in the (spherical) sense

$$\int_{\mathbb{R}^q} F(x)\overline{\Phi_h(x)}\, dx = \lim_{N\to\infty} \int_{\substack{|x|\le N\\ x\in\mathbb{R}^q}} F(x)\overline{\Phi_h(x)}\, dx. \tag{20.48}$$

Then the series

$$\sum_{g\in\Lambda} F(g) = \lim_{N\to\infty} \sum_{\substack{|g|\le N\\ g\in\Lambda}} F(g) \tag{20.49}$$

is convergent.

Example 20.15. *We use the lattice $\Lambda = \tau\mathbb{Z}^3$ and the rule*

$$\mathcal{I}F = \int_{\mathbb{R}^3} F(y)dy \approx \mathcal{L}_\tau F = \tau^3 \sum_{g\in\mathbb{Z}^3} F(\tau g) \tag{20.50}$$

for evaluating the integral over $F(x) = e^{-|x|^2}$ in Table 20.2.

Summarizing our results we are confronted with the following situation: For a value $\lambda \notin \mathrm{Spect}_\Delta(\Lambda)$ and a function $F \in \mathrm{C}^{(2m)}(\mathbb{R}^q)$ satisfying (20.34), (20.35), and (20.40), we have

$$\sum_{g\in\Lambda} F(g) = \int_{\mathbb{R}^q} G\left((\Delta+\lambda)^m ; x\right)(\Delta+\lambda)^m F(x)\, dx, \tag{20.51}$$

TABLE 20.2: Approximation of $\mathcal{I}F = \pi^{3/2}$ with $F(x) = e^{-|x|^2}$.

τ	$\mathcal{L}_\tau F$	$\lvert\mathcal{E}_\tau F\rvert = \lvert\mathcal{I}F - \mathcal{L}_\tau F\rvert$
1.1	5.577915737665	9.5e-03
1.0	5.570056245595	1.7e-03
0.9	5.568498657686	1.7e-04
0.8	5.568334703541	6.7e-06
0.7	5.568328056576	6.0e-08
0.6	5.568327996873	4.1e-11
0.5	5.568327996832	6.8e-13
Exact value	5.568327996832	

whereas, for a value $\lambda \in \mathrm{Spect}_\Delta(\Lambda)$ and a function $F \in \mathrm{C}^{(2m)}(\mathbb{R}^q)$ satisfying (20.34), (20.35), (20.40), and (20.48), we have

$$\sum_{g\in\Lambda} F(g) = \frac{1}{\sqrt{\|\mathcal{F}_\Lambda\|}} \sum_{\substack{(\Delta+\lambda)^\wedge(h)=0\\ h\in\Lambda^{-1}}} \int_{\mathbb{R}^q} F(x)\overline{\Phi_h(x)}\, dx + \int_{\mathbb{R}^q} G\left((\Delta+\lambda)^m;x\right)(\Delta+\lambda)^m F(x)\, dx. \tag{20.52}$$

The convergence conditions (in Theorem 20.14) enable us to derive multi-dimensional analogs of the Poisson summation formula. As, for $m \in \mathbb{N}$ with $m > \frac{q}{2}$, the Λ-lattice function $G\left((\Delta+\lambda)^m;\cdot\right)$ permits an absolutely and uniformly convergent Fourier series in $\mathbb{R}^q$, Lebesgue's theorem allows to interchange summation and integration

$$\begin{aligned} &\int_{\mathbb{R}^q} G\left((\Delta+\lambda)^m;x\right)(\Delta+\lambda)^m F(x)\, dx \\ &= \frac{1}{\sqrt{\|\mathcal{F}_\Lambda\|}} \sum_{\substack{(\Delta+\lambda)^\wedge(h)\neq 0\\ h\in\Lambda^{-1}}} \frac{1}{-\left((\Delta+\lambda)^m\right)^\wedge(h)} \int_{\mathbb{R}^q} \overline{\Phi_h(x)}(\Delta+\lambda)^m F(x)\, dx \end{aligned} \tag{20.53}$$

thereby applying the conditions (20.34) and (20.35) for $F \in \mathrm{C}^{(2m)}(\mathbb{R}^q)$. Moreover, by observation of (20.34) and (20.35), repeated application of the second Green's theorem yields

$$\begin{aligned} \int_{\mathbb{R}^q} \overline{\Phi_h(x)}(\Delta+\lambda)^m F(x)\, dx &= \int_{\mathbb{R}^q} \left((\Delta+\lambda)^m \overline{\Phi_h(x)}\right) F(x)\, dx \\ &= -\left((\Delta+\lambda)^m\right)^\wedge(h) \int_{\mathbb{R}^q} \overline{\Phi_h(x)} F(x)\, dx. \end{aligned} \tag{20.54}$$

Inserting (20.54) into (20.53) we therefore find that

$$\int_{\mathbb{R}^q} G\left((\Delta+\lambda)^m; x\right)(\Delta+\lambda)^m F(x)\, dx \tag{20.55}$$
$$= \frac{1}{\sqrt{\|\mathcal{F}_\Lambda\|}} \sum_{\substack{(\Delta+\lambda)^\wedge(h)\neq 0\\ h\in\Lambda^{-1}}} \int_{\mathbb{R}^q} F(x)\overline{\Phi_h(x)}\, dx.$$

This finally leads to an adaptive realization of periodization, i.e., the *multi-dimensional Poisson summation formula over the Euclidean space* $\mathbb{R}^q$.

Theorem 20.16. *Let* Λ *be an arbitrary lattice in the Euclidean space* $\mathbb{R}^q$. *If, for* $\varepsilon > 0$ *and* $\lambda \in \mathbb{R}$, *the function* $F \in \mathrm{C}^{(2m)}(\mathbb{R}^q)$, $m > \frac{q}{2}$, *satisfies the properties (20.34), (20.35), and (20.40), then*

$$\lim_{N\to\infty}\left(\sum_{\substack{|g|\le N\\ g\in\Lambda}} F(g) - \frac{1}{\sqrt{\|\mathcal{F}_\Lambda\|}} \sum_{\substack{(\Delta+\lambda)^\wedge(h)=0\\ h\in\Lambda^{-1}}} \int_{\substack{|x|\le N\\ x\in\mathbb{R}^q}} F(x)\overline{\Phi_h(x)}\, dx \right)$$
$$= \frac{1}{\sqrt{\|\mathcal{F}_\Lambda\|}} \sum_{\substack{(\Delta+\lambda)^\wedge(h)\neq 0\\ h\in\Lambda^{-1}}} \int_{\mathbb{R}^q} F(x)\overline{\Phi_h(x)}\, dx. \tag{20.56}$$

More explicitly, if $\lambda \notin \mathrm{Spect}_\Delta(\Lambda)$ *and* $F \in \mathrm{C}^{(2m)}(\mathbb{R}^q)$, $m > \frac{q}{2}$, *satisfies (20.34), (20.35), and (20.40), then*

$$\sum_{g\in\Lambda} F(g) = \frac{1}{\sqrt{\|\mathcal{F}_\Lambda\|}} \sum_{h\in\Lambda^{-1}} \int_{\mathbb{R}^q} F(x)\overline{\Phi_h(x)}\, dx. \tag{20.57}$$

If $\lambda \in \mathrm{Spect}_\Delta(\Lambda)$ *and* $F \in \mathrm{C}^{(2m)}(\mathbb{R}^q)$, $m > \frac{q}{2}$, *satisfies (20.34), (20.35), (20.40), and (20.48), then*

$$\sum_{g\in\Lambda} F(g) = \frac{1}{\sqrt{\|\mathcal{F}_\Lambda\|}} \sum_{h\in\Lambda^{-1}} \int_{\mathbb{R}^q} F(x)\overline{\Phi_h(x)}\, dx. \tag{20.58}$$

The sum on the left side of the identities (20.57) and (20.58) is not necessarily absolutely convergent in $\mathbb{R}^q$, so the process of summation must be specified. Once more, following our approach the convergence of the series on the left side in Theorem 20.16 is understood in the spherical sense (20.29).

All in all, our approach implies a cubature rule of the form

$$\sum_{g\in\Lambda} F(g) = \frac{1}{\sqrt{\|\mathcal{F}_\Lambda\|}} \int_{\mathbb{R}^q} F(x)\, dx + \mathcal{E}F, \tag{20.59}$$

where the remainder term $\mathcal{E}F$ is given by

$$\mathcal{E}F = \frac{1}{\sqrt{\|\mathcal{F}_\Lambda\|}} \sum_{\substack{h\neq 0\\ h\in\Lambda^{-1}}} \int_{\mathbb{R}^q} F(x)\overline{\Phi_h(x)}\, dx \tag{20.60}$$

is free of derivatives.

Remark 20.17. *Once more, the "wave number" $\lambda \in \mathbb{R}$ reflects the specific character of the function under consideration. In fact, $\lambda \in \mathbb{R}$ can be adapted specifically to the oscillating properties which is of tremendous importance. In consequence, as already mentioned, our sufficient conditions to establish the validity of the Poisson summation formula are particularly suited for the discussion of multi-dimensional alternating integrals and series.*

Case-by-case studies within the number-theoretic framework of Hardy–Landau (alternating) identities are provided in W. Freeden [2011].

Two-Dimensional Hardy–Landau Type Alternating Integrals and Series. Next we present some important aspects reflecting cubature rules. As already known, the *one-dimensional Hardy–Landau identity* (see (6.57)) reads as follows:

$$\sideset{}{'}\sum_{\substack{|g|\leq R \\ g\in\mathbb{Z}}} 1 = 2R \sum_{h\in\mathbb{Z}} \frac{J_1(1;2\pi hR)}{2\pi hR}, \tag{20.61}$$

where (in accordance with the theory of Bessel functions as presented in (19.13))

$$\frac{J_1(1;2\pi hR)}{2\pi hR} = \sqrt{\frac{\pi}{2}}\,\frac{J_{\frac{1}{2}}(2;2\pi hR)}{\sqrt{2\pi hR}} = J_0(3;2\pi hR) = \operatorname{sinc}(2\pi hR). \tag{20.62}$$

In what follows we are interested in the classical *two-dimensional Hardy–Landau identity.* Our work starts with the proof of the two-dimensional identity for arbitrary lattices Λ, however, in strict verification of the sufficient convergence criteria as proposed above. Later on, we turn over to multi-dimensional extensions of Hardy–Landau identities on arbitrary lattices $\Lambda \subset \mathbb{R}^q$.

The point of departure is the infinitely often differentiable function F_R given by

$$F_R(x) = \frac{J_1(2\pi|x|R)}{|x|}, \quad x \in \mathbb{R}^2, \tag{20.63}$$

where R is a positive number and, in accordance with our theory of Bessel functions (cf. (19.14)). Our approach is essentially based on the differential equation (20.64) relating the Bessel function J_1 of order 1 to the Bessel function J_2 of order 2 (of course, both of dimension $q = 2$)

$$\Delta_x \frac{J_1(2\pi|x|R)}{|x|} + 4\pi^2 R^2 \frac{J_1(2\pi|x|R)}{|x|} = 4\pi R \frac{J_2(2\pi|x|R)}{|x|^2}\,. \tag{20.64}$$

This equation indicates to use the "wave number" $\lambda = 4\pi^2 R^2$ as an adaptive parameter to the oscillating properties of the function F_R as defined by (20.63). In fact, the well-known asymptotic estimates known from the theory of Bessel functions (see, e.g., W. Freeden, M. Gutting [2013]) tell us that

$$\frac{J_1(2\pi|x|R)}{|x|} = O\left(|x|^{-\frac{3}{2}}\right) \tag{20.65}$$

and

$$\left| \nabla_x \frac{J_1(2\pi|x|R)}{|x|} \right| = O\left(|x|^{-\frac{3}{2}} \right) \tag{20.66}$$

are valid for $|x| \to \infty$. Furthermore, we see from (20.64) that

$$(\Delta_x + 4\pi^2 R^2) \frac{J_1(2\pi|x|R)}{|x|} = O\left(|x|^{-\frac{5}{2}} \right) \tag{20.67}$$

for $|x| \to \infty$. Hence, for the particular choice

$$\varepsilon = \frac{1}{2}, \quad \lambda = 4\pi^2 R^2, \tag{20.68}$$

the function F_R satisfies the properties (20.34) and (20.35), where $m \in \mathbb{N}$ can be taken arbitrarily. Theorem 20.13, therefore, enables us to conclude that the limit

$$\lim_{N\to\infty} \left(\sum_{\substack{|g|\le N \\ g\in\Lambda}} \frac{J_1(2\pi|g|R)}{|g|} - \frac{2\pi}{\|\mathcal{F}_\Lambda\|} \sum_{\substack{|h|=R \\ h\in\Lambda^{-1}}} \int_0^N J_1(2\pi rR) J_0(2\pi|h|r)\, dr \right) \tag{20.69}$$

exists in spherical sense. Note that

$$\int_{\substack{|x|\le N \\ x\in\mathbb{R}^2}} \frac{J_1(2\pi|x|R)}{|x|} e(h\cdot x)\, dx = 2\pi \int_0^N J_1(2\pi rR) J_0(2\pi|h|r)\, dr. \tag{20.70}$$

Moreover, in connection with (20.64) we have for $m = 1$

$$\begin{aligned} &\lim_{N\to\infty} \left(\sum_{\substack{|g|\le N \\ g\in\Lambda}} \frac{J_1(2\pi|g|R)}{|g|} - \frac{2\pi}{\|\mathcal{F}_\Lambda\|} \sum_{\substack{|h|=R \\ h\in\Lambda^{-1}}} \int_0^N J_1(2\pi rR) J_0(2\pi|h|r)\, dr \right) \\ &\quad = 4\pi R \int_0^\infty \frac{J_2(2\pi rR)}{r^2} \left(\int_{\substack{|x|=r \\ x\in\mathbb{R}^2}} G\left(\Delta + 4\pi^2 R^2; x\right)\, dS(x) \right) dr. \end{aligned} \tag{20.71}$$

The last integral can be rewritten in the form

$$\begin{aligned} &4\pi R \int_0^\infty \frac{J_2(2\pi rR)}{r^2} \int_{\substack{|x|=r \\ x\in\mathbb{R}^2}} G\left(\Delta + 4\pi^2 R^2; x\right)\, dS(x)\, dr \\ &\quad = -\frac{2\pi}{\|\mathcal{F}_\Lambda\|} \sum_{\substack{|h|\ne R \\ h\in\Lambda^{-1}}} \frac{4\pi R}{-4\pi^2 h^2 + 4\pi^2 R^2} \int_0^\infty \frac{J_2(2\pi rR)}{r} J_0(2\pi|h|R)\, dr. \end{aligned} \tag{20.72}$$

Observing the following identity (see, e.g., G.N. Watson [1944])

$$\frac{4\pi R}{4\pi^2 h^2 - 4\pi^2 R^2} \int_0^\infty J_2(2\pi rR) J_0(2\pi|h|r) \frac{dr}{r} = \int_0^\infty J_1(2\pi rR) J_0(2\pi|h|r)\, dr \tag{20.73}$$

we obtain

$$\lim_{N\to\infty}\left(\sum_{\substack{|g|\le N\\ g\in\Lambda}}\frac{J_1(2\pi|g|R)}{|g|}-\frac{2\pi}{\|\mathcal{F}_\Lambda\|}\sum_{\substack{|h|=R\\ h\in\Lambda^{-1}}}\int_0^N J_1(2\pi rR)J_0(2\pi|h|r)\,dr\right)$$
$$=\frac{2\pi}{\|\mathcal{F}_\Lambda\|}\sum_{\substack{|h|\neq R\\ h\in\Lambda^{-1}}}\int_0^\infty J_1(2\pi rR)J_0(2\pi|h|r)\,dr. \tag{20.74}$$

Now, for arbitrary $R>0$, the Hankel transform of discontinuous integrals (see, e.g., G.N. Watson [1944]) guarantees the existence of all occurring integrals such that

$$\lim_{N\to\infty}\sum_{\substack{|g|\le N\\ g\in\Lambda}}\frac{J_1(2\pi|g|R)}{|g|}=\frac{2\pi}{\|\mathcal{F}_\Lambda\|}\sum_{\substack{|h|=R\\ h\in\Lambda^{-1}}}\int_0^\infty J_1(2\pi rR)J_0(2\pi|h|r)\,dr \tag{20.75}$$
$$+\frac{2\pi}{\|\mathcal{F}_\Lambda\|}\sum_{\substack{|h|\neq R\\ h\in\Lambda^{-1}}}\int_0^\infty J_1(2\pi rR)J_0(2\pi|h|r)\,dr.$$

Even more, the theory of discontinuous integrals (see, e.g., W. Freeden [2011], G.N. Watson [1944]) yields

$$2\pi R\int_0^\infty J_1(2\pi rR)J_0(2\pi|h|r)\,dr=\begin{cases}1, & |h|<R,\\ \frac{1}{2}, & |h|=R,\\ 0, & |h|>R.\end{cases} \tag{20.76}$$

Hence, we find

$$\lim_{N\to\infty}\sum_{\substack{|g|\le N\\ g\in\Lambda}}\frac{J_1(2\pi|g|R)}{|g|}=\frac{2\pi}{\|\mathcal{F}_\Lambda\|}\sum_{h\in\Lambda^{-1}}\int_0^\infty J_1(2\pi rR)J_0(2\pi|h|r)\,dr$$
$$=\frac{1}{\|\mathcal{F}_\Lambda\|R}\sum_{\substack{|h|\le R\\ h\in\Lambda^{-1}}}{}' \,1. \tag{20.77}$$

Summarizing our results we therefore obtain

Theorem 20.18. *For arbitrary, but fixed radius $R>0$, and arbitrary lattices $\Lambda\subset\mathbb{R}^2$*

$$\lim_{N\to\infty}\sum_{\substack{|g|\le N\\ g\in\Lambda}}\frac{J_1(2\pi|g|R)}{|g|}=\frac{1}{\|\mathcal{F}_\Lambda\|R}\sum_{\substack{|h|\le R\\ h\in\Lambda^{-1}}}{}' \,1. \tag{20.78}$$

Replacing $\Lambda\subset\mathbb{R}^2$ by its inverse lattice $\Lambda^{-1}\subset\mathbb{R}^2$ we finally get the *Hardy–Landau identity* in its canonical form

$$\sum_{\substack{|g|\le R\\ g\in\Lambda}}{}' \,1=\lim_{N\to\infty}\frac{R}{\|\mathcal{F}_\Lambda\|}\sum_{\substack{|h|\le N\\ h\in\Lambda^{-1}}}\frac{J_1(2\pi|h|R)}{|h|}. \tag{20.79}$$

Observing $J_1(r) = \frac{r}{2} + \ldots$ we are able to rewrite Equation (20.79) in its standard form comparing the number of lattice points inside a circle around the origin of radius R with the area of the circle under explicit specification of the remaining Hardy–Landau series.

Corollary 20.19 (Classical Two-Dimensional Hardy–Landau Identity). *For all positive numbers R and for each lattice $\Lambda \subset \mathbb{R}^2$*

$$\sum_{\substack{|g|\le R\\ g\in\Lambda}}{}' 1 = \frac{\pi}{\|\mathcal{F}_\Lambda\|}R^2 + \lim_{N\to\infty} \frac{R}{\|\mathcal{F}_\Lambda\|} \sum_{\substack{0<|h|\le N\\ h\in\Lambda^{-1}}} \frac{J_1(2\pi|h|R)}{|h|}. \tag{20.80}$$

As already mentioned, for $\Lambda = \mathbb{Z}^2$, the identity (20.80) is found in G.H. Hardy [1915], E. Landau [1915], G.H. Hardy, E. Landau [1924].

Remark 20.20. *Using the well-known asymptotic relation from the theory of Bessel functions (see, e.g., W. Magnus et al. [1966], C. Müller [1952])*

$$J_1(2\pi|h|R) = \frac{1}{\pi(|h|R)^{\frac{1}{2}}}\cos\left(2\pi|h|R - \frac{3\pi}{4}\right) + O\left((|h|R)^{-\frac{3}{2}}\right) \tag{20.81}$$

we are able to state that, apart from an additive remainder term, the Hardy–Landau series

$$\frac{R}{\|\mathcal{F}_\Lambda\|} \sum_{\substack{0<|h|\\ h\in\Lambda^{-1}}} \frac{J_1(2\pi|h|R)}{|h|} \tag{20.82}$$

is equal to

$$\frac{R^{\frac{1}{2}}}{\pi\|\mathcal{F}_\Lambda\|} \sum_{\substack{0<|h|\\ h\in\Lambda^{-1}}} \frac{\cos\left(2\pi|h|R - \frac{3\pi}{4}\right)}{|h|^{\frac{3}{2}}}. \tag{20.83}$$

In a certain sense, all investigations involving the classical two-dimensional Hardy–Landau identity, which have been done during the past century, are based on (partial sums of) the series (20.83).

Higher Dimensional Hardy–Landau Type Alternating Integrals and Series. Next we apply the sufficient convergence criteria (i.e., Theorem 20.16) to higher-dimensional extensions of the Hardy–Landau identity (Corollary 20.19). As the point of departure we choose a product function consisting of a radial and an angular weight function, i.e., the infinitely often differentiable function $F_R : \mathbb{R}^q \to \mathbb{R}$, $R > 0$ fixed, given by

$$F_R(x) = \frac{J_\nu(q; 2\pi|x|R)}{|x|^\nu}, \quad q \ge 2, \tag{20.84}$$

where $J_\nu(q;\cdot)$ is the *Bessel function of order* $\nu > 0$ *and dimension* q (see, e.g., W. Freeden, M. Gutting [2013], C. Müller [1998]). Expanding $J_\nu(q; 2\pi rR)$ in powers of its argument we have by (19.14)

$$\frac{J_\nu(q; 2\pi rR)}{r^\nu} = \frac{(\pi R)^\nu \Gamma\left(\frac{q}{2}\right)}{\Gamma\left(\nu + \frac{q}{2}\right)} \left(1 - \frac{(\pi r R)^2}{1!(\nu + \frac{q}{2})} + - \cdots\right). \tag{20.85}$$

For $n \in \mathbb{N}$ and $r \to \infty$ the standard estimates of the theory of Bessel functions yield the relation

$$\begin{aligned} J_\nu(q; 2\pi rR) = J_0(q; 2\pi rR) \cos\left(\nu\frac{\pi}{2}\right) \\ + J_1(q; 2\pi rR) \sin\left(\nu\frac{\pi}{2}\right) + O\left(r^{-\frac{q+1}{2}}\right). \end{aligned} \tag{20.86}$$

Under the assumption

$$q \geq 2, \quad \nu > \frac{q-1}{2}, \quad \nu > 0, \tag{20.87}$$

the theory of Bessel functions implies the asymptotic relations

$$\begin{aligned} |(\Delta + \lambda)^k F_R(x)| = o(|x|^{1-q}), \quad |x| \to \infty, \\ \left|\nabla(\Delta + \lambda)^k F_R(x)\right| = o(|x|^{1-q}), \quad |x| \to \infty \end{aligned} \tag{20.88}$$

for each non-negative integer k and all numbers $\lambda \in \mathbb{R}$. Moreover, from the differential equation of the Bessel function we are able to deduce that

$$\left|(\Delta + 4\pi^2 R^2) \frac{J_\nu(q; 2\pi|x|R)}{|x|^\nu}\right| = O\left(\frac{J_{\nu+1}(q; 2\pi|x|R)}{|x|^{\nu+1}}\right) \tag{20.89}$$

for $|x| \to \infty$. Thus, in view of the assumption (20.87), we obtain the estimate

$$(\Delta + \lambda) F_R(x) = O\left(|x|^{-(q+\varepsilon)}\right), \quad |x| \to \infty, \tag{20.90}$$

under the special choice

$$\lambda = 4\pi^2 R^2, \quad \varepsilon = \nu - \frac{q-1}{2} > 0. \tag{20.91}$$

Altogether, Theorem 20.16 leads to

Lemma 20.21. *Under the assumption (20.87),* $F_R \in \mathrm{C}^{(\infty)}(\mathbb{R}^q)$ *given by (20.84) satisfies the properties (20.34) and (20.35), where* $m \in \mathbb{N}$ *can be chosen arbitrarily, such that*

$$\begin{aligned} \lim_{N\to\infty} \left(\sum_{\substack{|g|\leq N \\ g\in\Lambda}} F_R(g) - \frac{1}{\sqrt{\|\mathcal{F}_\Lambda\|}} \sum_{\substack{|h|=R \\ h\in\Lambda^{-1}}} \int_{\substack{|x|\leq N \\ x\in\mathbb{R}^q}} F_R(x)\overline{\Phi_h(x)}\, dx \right) \\ = \frac{1}{\sqrt{\|\mathcal{F}_\Lambda\|}} \sum_{\substack{|h|\neq R \\ h\in\Lambda^{-1}}} \int_{\mathbb{R}^q} F_R(x)\overline{\Phi_h(x)}\, dx. \end{aligned} \tag{20.92}$$

For values R with $4\pi^2R^2 \notin \text{Spect}_\Delta(\Lambda)$, F_R already allows the identity

Lemma 20.22. *For all R with $4\pi^2R^2 \notin \text{Spect}_\Delta(\Lambda)$*

$$\lim_{N\to\infty} \sum_{\substack{|g|\le N\\ g\in\Lambda}} F_R(g) = \frac{1}{\sqrt{\|\mathcal{F}_\Lambda\|}} \sum_{h\in\Lambda^{-1}} \int_{\mathbb{R}^q} F_R(x)\overline{\Phi_h(x)}\, dx. \tag{20.93}$$

It remains to investigate the integrals on the right side of (20.93). Introducing standard polar coordinates we find

$$\begin{aligned} &\frac{1}{\sqrt{\|\mathcal{F}_\Lambda\|}} \int_{\mathbb{R}^q} F_R(x)\overline{\Phi_h(x)}\, dx \\ &= \frac{\|\mathbb{S}^{q-1}\|}{\|\mathcal{F}_\Lambda\|} \int_0^\infty J_\nu(q;2\pi rR) J_0(q;2\pi|h|r) r^{-\nu+q-1}\, dr. \end{aligned} \tag{20.94}$$

The theory of discontinuous integrals gives detailed information on the convergence of all terms

$$\int_0^\infty J_\nu(q;2\pi rR) J_0(q;2\pi|h|r) r^{-\nu+q-1}\, dr. \tag{20.95}$$

From (20.86) we deduce that the integrand is asymptotically equal to

$$\begin{aligned} &r^{-\nu+q-1} J_0(q;2\pi|h|r) J_0(q;2\pi rR) \cos\left(\nu\frac{\pi}{2}\right) \\ &= r^{-\nu+q-1} J_0(q;2\pi|h|r) J_1(q;2\pi rR) \sin\left(\nu\frac{\pi}{2}\right) + O\left(r^{-\nu-1}\right). \end{aligned} \tag{20.96}$$

Hence, under the following alternative conditions imposed on the values ν (see, e.g., C. Müller [1998], G.N. Watson [1944])

(i) $4\pi^2R^2 \notin \text{Spect}_\Delta(\Lambda)$, i.e., $R \neq |h|$ for all $h \in \Lambda^{-1}$:

$$\nu > 0. \tag{20.97}$$

(ii) $4\pi^2R^2 \in \text{Spect}_\Delta(\Lambda)$, i.e., $R = |h|$ for some $h \in \Lambda^{-1}$:

$$\nu > 1, \tag{20.98}$$

the integrals (20.95) are convergent. In accordance with our approach we therefore have to distinguish two cases, namely $4\pi^2R^2 \notin \text{Spect}_\Delta(\Lambda)$ and $4\pi^2R^2 \in \text{Spect}_\Delta(\Lambda)$:

First we deal with the case $4\pi^2R^2 \notin \text{Spect}_\Delta(\Lambda)$: Under the assumptions (20.87) and (20.97), i.e., $\nu > \frac{q-1}{2}$, the integrals (20.95) exist.

Even more, they are known from the Hankel transform of discontinuous integrals

$$\frac{1}{\sqrt{\|\mathcal{F}_\Lambda\|}} \int_{\mathbb{R}^q} F_R(x)\overline{\Phi_h(x)}\, dx \tag{20.99}$$

$$= \begin{cases} \frac{2}{\|\mathbb{S}^{q-1}\| \|\mathcal{F}_\Lambda\|} \frac{\pi^\nu}{R^{-\nu+q}\Gamma(\nu)} \left(1-\left(\frac{|h|}{R}\right)^2\right)^{\nu-1}, & |h| < R, \\ 0, & |h| > R. \end{cases}$$

Therefore, as a first remarkable result, we obtain from (20.92)

Theorem 20.23. *Let Λ be an arbitrary lattice in $\mathbb{R}^q$. Then, for all numbers R with $4\pi^2R^2 \notin \mathrm{Spect}_\Delta(\mathbb{R}^q)$ and for all values ν, n with $\nu > \frac{q-1}{2}$ (see (20.91)), we have*

$$\frac{2\pi^\nu}{\|\mathbb{S}^{q-1}\| \|\mathcal{F}_\Lambda\| R^{-\nu+q}\Gamma(\nu)} \sum_{\substack{|h|<R \\ h\in\Lambda^{-1}}} \left(1-\left(\frac{|h|}{R}\right)^2\right)^{\nu-1} = \sum_{g\in\Lambda} \frac{J_\nu(q; 2\pi|g|R)}{|g|^\nu}. \tag{20.100}$$

Next we turn over to the case $4\pi^2R^2 \in \mathrm{Spect}_\Delta(\Lambda)$: for all values ν with $\nu > \frac{q-1}{2}$ the identity (20.92) leads to the limit relation

$$\lim_{N\to\infty} \left(\sum_{\substack{|g|\le N \\ g\in\Lambda}} \frac{J_\nu(q; 2\pi|g|R)}{|g|^\nu} \right. \tag{20.101}$$

$$\left. - \frac{1}{\|\mathcal{F}_\Lambda\|} \sum_{\substack{|h|=R \\ h\in\Lambda^{-1}}} \int_{\substack{|x|\le N \\ x\in\mathbb{R}^q}} \frac{J_\nu(q; 2\pi|x|R)}{|x|^\nu} e^{-2\pi i h\cdot x}\, dx \right)$$

$$= \frac{2\pi^\nu}{\|\mathbb{S}^{q-1}\| \|\mathcal{F}_\Lambda\| R^{-\nu+q}\Gamma(\nu)} \sum_{\substack{|h|<R \\ h\in\Lambda^{-1}}} \left(1-\left(\frac{|h|}{R}\right)^2\right)^{\nu-1}.$$

For values R with $4\pi^2R^2 \in \mathrm{Spect}_\Delta(\Lambda)$, we obtain

Theorem 20.24. *Let Λ be an arbitrary lattice in $\mathbb{R}^q$. Let ν satisfy one of the following assumptions:*

$$\nu > \max\left\{1, \frac{q-1}{2}\right\}, \tag{20.102}$$

or

$$\nu > \frac{q-1}{2}, \quad \nu \text{ is an odd integer.} \tag{20.103}$$

Then the following q-dimensional analog to the Hardy–Landau identity *holds true:*

$$\frac{2\pi^\nu}{\|\mathbb{S}^{q-1}\| \|\mathcal{F}_\Lambda\| R^{-\nu+q}\Gamma(\nu)} \sum_{\substack{|h|\le R \\ h\in\Lambda^{-1}}}{}' \left(1-\left(\frac{|h|}{R}\right)^2\right)^{\nu-1} = \sum_{g\in\Lambda} \frac{J_\nu(q;2\pi|g|R)}{|g|^\nu}. \tag{20.104}$$

Replacing the inverse lattice $\Lambda^{-1}\subset\mathbb{R}^q$ by the lattice $\Lambda\subset\mathbb{R}^q$ we find

Corollary 20.25. *Under the assumptions of Theorem 20.24*

$$\frac{2\pi^\nu}{\|\mathbb{S}^{q-1}\| R^{-\nu+q}\Gamma(\nu)} \sum_{\substack{|g|\le R \\ g\in\Lambda}}{}' \left(1-\left(\frac{|g|}{R}\right)^2\right)^{\nu-1} = \frac{1}{\|\mathcal{F}_\Lambda\|} \sum_{h\in\Lambda^{-1}} \frac{J_\nu(q;2\pi|h|R)}{|h|^\nu}. \tag{20.105}$$

We conclude this section with some important two-dimensional examples.

Example 20.26. *For* $q=2$, *taking* $\Lambda\subset\mathbb{R}^2$ *simply as the unit lattice* $\mathbb{Z}^2$ *and setting* $\nu=1$, *we get the classical two-dimensional Hardy–Landau identity in its original form (cf. G.H. Hardy [1915], G.H. Hardy, E. Landau [1924])*

$$\sum_{(n_1,n_2)^T\in\mathbb{Z}^2} \frac{J_1(2\pi\sqrt{n_1^2+n_2^2}R)}{\sqrt{n_1^2+n_2^2}} = \frac{1}{R} \sum_{\substack{n_1^2+n_2^2\le R^2 \\ (n_1,n_2)^T\in\mathbb{Z}^2}}{}' 1. \tag{20.106}$$

20.4 Adaptive Cubature over Euclidean Spaces

We first mention an extension of the Poisson summation formula (cf. Theorem 20.16).

Theorem 20.27. *Suppose that* $F:\mathbb{R}^q\to\mathbb{R}$ *is a function of class* $\mathrm{C}^{(2m)}(\mathbb{R}^q)$ *satisfying the following properties:*

(C1) The asymptotic relations

$$|(\Delta+\lambda)^r F(x)| = o(|x|^{1-q}) \tag{20.107}$$

$$|\nabla(\Delta+\lambda)^r F(x)| = o(|x|^{1-q}) \tag{20.108}$$

hold for $r=0,\ldots,m-1$, $\lambda\in\mathbb{R}$, *and* $|x|\to\infty$.

(C2) There exists a positive number $\varepsilon>0$ *such that*

$$|(\Delta+\lambda)^m F(x)| = O(|x|^{-(q+\varepsilon)}) \tag{20.109}$$

Then the limit

$$\lim_{N\to\infty}\left(\sum_{\substack{|g+y|\le N\\ g\in\Lambda}} F(g+y) - \sum_{\substack{\lambda=4\pi^2h^2\\ h\in\Lambda^{-1}}} \Phi_h(y)\int_{|y|\le N} F(x)\overline{\Phi_h(x)}\,dx\right) \quad (20.110)$$

exists and is equal to

$$\int_{\mathbb{R}^q} G\left((\Delta+\lambda)^m; x-y\right)(\Delta+\lambda)^m F(x)\,dx. \quad (20.111)$$

In order to ensure the convergence of the series

$$\sum_{g\in\Lambda} F(g+y) \quad (20.112)$$

in the spherical sense we have to require an additional condition.

(C3) The integrals

$$\sum_{\substack{4\pi^2h^2=\lambda\\ h\in\Lambda^{-1}}} \Phi_h(y)\int_{\mathbb{R}^q} F(x)\overline{\Phi_h(x)}\,dx. \quad (20.113)$$

exist in the sense

$$\int_{\mathbb{R}^q} \ldots dx = \lim_{N\to\infty}\int_{|x|\le N} \ldots dx. \quad (20.114)$$

Theorem 20.28. *Let $F \in \mathrm{C}^{(2m)}(\mathbb{R}^q)$ satisfy conditions (C1), (C2), and (C3). Then, for each $y \in \mathbb{R}^q$*

$$\sum_{g\in\Lambda} F(g+y) = \sum_{\substack{4\pi^2h^2=\lambda\\ h\in\Lambda^{-1}}} \Phi_h(y)\int_{\mathbb{R}^q} F(x)\overline{\Phi_h(x)}\,dx \quad (20.115)$$
$$+\int_{\mathbb{R}^q} G\left((\Delta+\lambda)^m; x-y\right)(\Delta+\lambda)^m F(x)\,dx.$$

Moreover, if $m > q/2$, then for each $y \in \mathbb{R}^q$

$$\sum_{g\in\Lambda} F(g+y) = \sum_{h\in\Lambda^{-1}} \Phi_h(y)\int_{\mathbb{R}^q} F(x)\overline{\Phi_h(x)}\,dx. \quad (20.116)$$

It should be noted that the sufficient conditions (C1), (C2), (C3) do not necessarily imply the absolute convergence of the series on the left-hand side.

Degree of Exactness. Given a function T of the form (20.23) and a Ξ-fundamental system in the sense of Definition 20.6. Then we know that T allows the representation (20.24), i.e.,

$$T = \sum_{h\in\Xi}\sum_{k=1}^{L} a_k\overline{\Phi_h(y_k)}\,\Phi_h. \quad (20.117)$$

By observation of the characteristic singularity of iterated Green's functions (see Theorem 20.10) we are able to deduce that, for all integers r and all points $y \in \mathbb{R}^q$, the asymptotic relations

$$\int_{|x+y|=N} |G((\Delta+\lambda)^r; \Xi; x)| \, dS(x) = O(N^{q-1}), \ N \to \infty, \tag{20.118}$$

$$\int_{|x+y|=N} \left| \frac{\partial}{\partial \nu} G((\Delta+\lambda)^r; \Xi; x) \right| \, dS(x) = O(N^{q-1}), \ N \to \infty \tag{20.119}$$

are valid for any Ξ-modified Λ-lattice function, too.

The asymptotic expansions may be used to formulate appropriate conditions in problems of approximate integrals over $\mathbb{R}^q$. To this end we start from the following spherical specialization of Corollary 20.7.

Corollary 20.29. *Let $y_1, \ldots, y_L$ be a Ξ-fundamental system. Suppose that T is a periodic polynomial of the form (20.23). Then*

$$\sum_{k=1}^{L} a_k \sum_{\substack{|g+y_k| \le N \\ g \in \Lambda}}{}' F(g+y_k) \tag{20.120}$$

$$\begin{aligned} &= \int_{|x|\le N} F(x) T(x) \, dx \\ &+ \sum_{j=0}^{m-1} \int_{|x|=N} \left(\frac{\partial}{\partial \nu} \sum_{k=1}^{L} a_k G\left((\Delta+\lambda)^{j+1}; \Xi; x - y_k \right) \right) (\Delta+\lambda)^j F(x) \, dS(x) \\ &- \sum_{j=0}^{m-1} \int_{|x|=N} \sum_{k=1}^{L} a_k G\left((\Delta+\lambda)^{j+1}; \Xi; x - y_k \right) \frac{\partial}{\partial \nu} (\Delta+\lambda)^j F(x) \, dS(x) \\ &+ \int_{|x|\le N} \sum_{k=1}^{L} a_k G((\Delta+\lambda)^m; \Xi; x - y_k) (\Delta+\lambda)^m F(x) \, dx \end{aligned}$$

holds for all coefficients $a_1, \ldots, a_L$ satisfying the linear equations (20.22).

Hence, by virtue of the relations (20.118) and (20.119), we obtain

Corollary 20.30. *Let F be a function of class $\mathrm{C}^{(2m)}(\mathbb{R}^q)$ satisfying the following properties:*

(C1') There is a positive real number $\lambda \in \mathbb{R}$ such that

$$|(\Delta+\lambda)^r F(x)| = o(|x|^{1-q}), \tag{20.121}$$

$$|\nabla(\Delta+\lambda)^r F(x)| = o(|x|^{1-q}) \tag{20.122}$$

are valid for $r = 0, \ldots, m-1$ as $|x| \to \infty$.

(C2') There is a positive real number ε such that, for $\lambda \in \mathbb{R}$ specified by (C1')

$$|(\Delta+\lambda)^m F(x)| = O(|x|^{-(q+\varepsilon)}), \quad \text{for } |x| \to \infty. \tag{20.123}$$

Then, for any prescribed Ξ-fundamental system $y_1, \ldots, y_L$ and for all $a_1, \ldots, a_L$ fulfilling (20.22), *the limit*

$$\lim_{N\to\infty}\left(\sum_{k=1}^{L} a_k \sum_{\substack{|g+y_k|\le N\\ g\in\Lambda}} F(g+y_k) - \int_{|x|\le N} F(x)T(x)\,dx\right) \tag{20.124}$$

exists, and we have

$$\begin{aligned}\lim_{N\to\infty}&\left(\sum_{k=1}^{L} a_k \sum_{\substack{|g+y_k|\le N\\ g\in\Lambda}} F(g+y_k) - \int_{|x|\le N} F(x)T(x)\,dx\right)\\ &= \int_{\mathbb{R}^q} \sum_{k=1}^{L} a_k G\left((\Delta+\lambda)^m;\Xi;x-y_k\right)(\Delta+\lambda)^m F(x)\,dx,\end{aligned} \tag{20.125}$$

where T is a periodic polynomial of the form (20.23).

Proof. From condition (C1) and a result due to Gauss (Corollary 17.6) it follows that

$$\sum_{k=1}^{L} a_k \sum_{\substack{|g+y_k|=N\\ g\in\Lambda}} F(g+y_k) = o\left(\sum_{k=1}^{L} \sum_{\substack{|g+y_k|=N\\ g\in\Lambda}} 1\right) = o(1) \text{ for } N\to\infty. \tag{20.126}$$

For the surface integrals we find by virtue of (C1') that

$$\begin{aligned}&\sum_{j=0}^{m-1}\int_{|x|=N}\left(\frac{\partial}{\partial\nu}\sum_{k=1}^{L} a_k G\left((\Delta+\lambda)^{j+1};\Xi;x-y_k\right)\right)(\Delta+\lambda)^j F(x)\,dS(x)\\ &-\sum_{j=0}^{m-1}\int_{|x|=N}\sum_{k=1}^{L} a_k G\left((\Delta+\lambda)^{j+1};\Xi;x-y_k\right)\frac{\partial}{\partial\nu}(\Delta+\lambda)^j F(x)\,dS(x)\\ &= o\left(N^{1-q}\sum_{r=0}^{m-1}\sum_{k-1}^{L}\int_{|x+y_k|=R}\left|G\left((\Delta+\lambda)^r;\Xi;x-y_k\right)\right|\,dS(x)\right)\\ &= o\left(N^{1-q}\sum_{r=0}^{m-1}\sum_{k-1}^{L}\int_{|x+y_k|=R}\left|\frac{\partial}{\partial\nu}G\left((\Delta+\lambda)^r;\Xi;x-y_k\right)\right|\,dS(x)\right),\end{aligned} \tag{20.127}$$

for $N\to\infty$. Hence, by (20.118) and (20.119), it is clear that

$$\begin{aligned}&\sum_{j=0}^{m-1}\int_{|x|=N}\left(\frac{\partial}{\partial\nu}\sum_{k=1}^{L} a_k G\left((\Delta+\lambda)^{j+1};\Xi;x-y_k\right)\right)(\Delta+\lambda)^j F(x)\,dS(x)\\ &-\sum_{j=0}^{m-1}\int_{|x|=N}\sum_{k=1}^{L} a_k G\left((\Delta+\lambda)^{j+1};\Xi;x-y_k\right)\frac{\partial}{\partial\nu}(\Delta+\lambda)^j F(x)\,dS(x)\\ &= o(1)\end{aligned} \tag{20.128}$$

as N tends to infinity. Finally, in connection with (C2'), we obtain with a suitable constant C, and sufficiently large N, N'

$$\left|\sum_{k=1}^{L} a_k \int_{N' \le |x| \le N} G\left((\Delta+\lambda)^m; \Xi; x - y_k\right)(\Delta+\lambda)^m F(x)\, dx\right|$$
$$\le C \sum_{k=1}^{L} \int_{N'}^{N} \frac{1}{(1+r)^{q+\varepsilon}} \int_{|x|=r} \left|G\left((\Delta+\lambda)^m; \Xi; x - y_k\right)\right| dS(x)\, dr.$$

Thus, the integral

$$\int_{\mathbb{R}^q} \sum_{k=1}^{L} a_k G\left((\Delta+\lambda)^m; \Xi; x - y_k\right)(\Delta+\lambda)^m F(x)\, dx \tag{20.129}$$

is absolutely convergent. This proves Corollary 20.30. □

In order to ensure the convergence of the integral

$$\int_{\mathbb{R}^q} F(x)T(x)\, dx \tag{20.130}$$

in the spherical sense

$$\int_{\mathbb{R}^q} \dots\, dx = \lim_{N\to\infty} \int_{|x|\le N} \dots\, dx \tag{20.131}$$

we have to require, in addition,

(C3') the series

$$\sum_{k=1}^{L} a_k \sum_{g\in\Lambda} F(g+y_k) = \lim_{N\to\infty} \sum_{k=1}^{L} a_k \sum_{\substack{|g+y_k|\le N \\ g\in\Lambda}} F(g+y_k) \tag{20.132}$$

exists for some prescribed Ξ-fundamental system $y_1, \dots, y_L$ and coefficients $a_1, \dots, a_L$ determined by (20.22).

Corollary 20.31. *Let $F \in \mathrm{C}^{(2m)}(\mathbb{R}^q)$ satisfy conditions (C1'), (C2'), and (C3'). Then*

$$\sum_{k=1}^{L} a_k \sum_{g\in\Lambda} F(g+y_k) = \int_{\mathbb{R}^q} F(x)T(x)\, dx \tag{20.133}$$
$$+ \int_{\mathbb{R}^q} \sum_{k=1}^{L} a_k G\left((\Delta+\lambda)^m; \Xi; x - y_k\right)\left((\Delta+\lambda)^m F(x)\right) dx,$$

where a_k, y_k for $k = 1, \dots, L$ as well as T are given as in Corollary 20.30.

Corollary 20.31 may be used to derive adapted cubatures (as proposed earlier). We only give an estimate in analogy to Corollary 20.8: Allow $m > q/2$. Then $G\left((\Delta+\lambda)^m;\Xi;x\right)$ has an absolutely and uniformly convergent Fourier-expansion. Thus, Lebesgue's theorem yields

$$\int_{\mathbb{R}^q} \sum_{k=1}^{L} a_k G\left((\Delta+\lambda)^m;\Xi;x-y_k\right)(\Delta+\lambda)^m F(x)\,dx \qquad (20.134)$$
$$= \sum_{\substack{h\notin\Xi\\ h\in\Lambda^{-1}}} \sum_{k=1}^{L} a_k \frac{\overline{\Phi_h(y_k)}}{(\lambda-4\pi^2h^2)^m} \int_{\mathbb{R}^q} \Phi_h(x)(\Delta+\lambda)^m F(x)\,dx.$$

Hence, the remainder term

$$\mathcal{E}_m^{\lambda,\Xi}F = \int_{\mathbb{R}^q} \sum_{k=1}^{L} a_k G\left((\Delta+\lambda)^m;\Xi;x-y_k\right)(\Delta+\lambda)^m F(x)\,dx \qquad (20.135)$$

may be estimated, e.g., in the form

Corollary 20.32. *For $m > q/2$,*

$$\left|\mathcal{E}_m^{\lambda,\Xi}F\right| \qquad (20.136)$$
$$\leq \frac{1}{\sqrt{\|\mathcal{F}_\Lambda\|}} \sum_{\substack{h\notin\Xi\\ h\in\Lambda^{-1}}} \frac{1}{|\lambda-4\pi^2h^2|^m} \left|\sum_{k=1}^{L} a_k\overline{\Phi_h(y_k)}\right| \int_{\mathbb{R}^q} |(\Delta+\lambda)^m F(x)|\,dx.$$

As in one-dimensional computations of infinite integrals we are confronted with the problem that our approximate rule is an infinite series which must be truncated suitably.

Example 20.33. *Consider the integral*

$$I_{\nu,R}(q;b) = \int_{\mathbb{R}^q} \frac{J_{\nu+\frac{q-2}{2}}(q;2\pi|x|R)}{|x|^{\nu-\frac{q-2}{2}}} e(x\cdot b)\,dx, \qquad (20.137)$$

where R is a positive real number, b is an element of $\mathbb{R}^q$, and J_ν is the (regular) Bessel function of order ν and dimension q. For simplicity, we restrict ourselves to the case

$$\nu > \frac{q+1}{2}. \qquad (20.138)$$

$I_{\nu,R}(q;b)$ is a discontinuous integral of Weber–Schlafheitlin type. It has been evaluated exactly by (G.N. Watson [1944])

$$I_{\nu,R}(q;b) = \begin{cases} \frac{\pi^{\nu-1}}{\Gamma(\nu)R^{-\nu+\frac{q-2}{2}}}\left(1-\left(\frac{|b|}{R}\right)^2\right)^{\nu-1}, & |b|<R,\\ 0, & |b|\geq R.\end{cases} \qquad (20.139)$$

We are interested in calculating (20.137) numerically by use of adapted cubature formulas. From standard estimates of the theory of Bessel functions we deduce that, for each $\lambda \in \mathbb{R}$*, the function* $F_{\nu,R}$ *defined by*

$$F_{\nu,R}(x) = \frac{J_{\nu+\frac{q-2}{2}}(q; 2\pi|x|R)}{|x|^{\nu-\frac{q-2}{2}}}, \quad x \in \mathbb{R}^q, \tag{20.140}$$

satisfies property (C1') of Corollary 20.30. Employing the recursion relation of Bessel functions (see, e.g., G.N. Watson [1944]) we find after simple calculations

$$(\Delta + \lambda)F_{\nu,R}(x) = (\lambda - 4\pi^2 R^2)F_{\nu,R}(x) + 2\pi R\,(\nu + q - 2)\,F_{\nu+1,R}(x). \tag{20.141}$$

Consequently, property (C2') of Corollary 20.30 is valid for each $\lambda \in \mathbb{R}$.

Let Λ *be a lattice such that* $b \in \Lambda^{-1}$*. Then, for any given* λ *and all finite subsets* $\Xi \subset \Lambda$ *containing the elements* $b \in \Lambda^{-1}$ *and* $h \in \Lambda^{-1}$ *with* $\lambda = 4\pi^2 h^2$*, property (C3') is satisfied. Hence for any given* $\lambda \in \mathbb{R}$ *and all L-tuple* $(a_1, \ldots, a_L)$ *satisfying the linear equations*

$$\sum_{k=1}^{L} a_k \overline{\Phi_h(y_k)} = \delta_{h,b}, \quad h \in \Xi, \tag{20.142}$$

we find

$$\int_{\mathbb{R}^q} F_{\nu,R}(x)\Phi_b(x)\,dx = \sum_{k=1}^{L} a_k \sum_{g\in\Lambda} F_{\nu,R}(g + y_k) + \mathcal{E}_m^{\lambda,\Xi} F_{\nu,R}, \tag{20.143}$$

where

$$\mathcal{E}_m^{\lambda,\Xi} F_{\nu,R} \tag{20.144}$$
$$= \int_{\mathbb{R}^q} \sum_{k=1}^{L} a_k G\left((\Delta+\lambda)^m ; \Xi; x - y_k\right)(\Delta+\lambda)^m F_{\nu,R}(x)\,dx.$$

Chapter 21

Lattice Spline Interpolation and Monospline Integration

Next we are concerned with multivariate periodic spline interpolation guaranteeing some amount of polynomial exactness.

The layout of this chapter is as follows: First, in Section 21.1, we deal with the introduction of periodic splines in Sobolev spaces with respect to iterated Helmholtz operators and arbitrary lattices. Interpolating splines (as introduced in Section 21.2) are characterized by minimum norm conditions. Section 21.3 deals with periodic sampling. Finally (in Section 21.4), we describe Euler summation formulas corresponding to general second order elliptic differential operators with constant coefficients, which enable the characterization of monosplines in best approximate cubature.

21.1 Lattice Periodic Splines

Let $\lambda_1, \ldots, \lambda_m$ be not-necessarily distinct real numbers. We consider the iterated Helmholtz differential operator

$$\partial_{1,\ldots,m} = \partial_1 \ldots \partial_m = (\Delta + \lambda_1) \ldots (\Delta + \lambda_m). \tag{21.1}$$

For an arbitrary lattice Λ we let

$$\partial_{1,\ldots,m}^{\wedge}(h) = \partial_1^{\wedge}(h) \cdot \ldots \cdot \partial_m^{\wedge}(h), \quad h \in \Lambda^{-1}, \tag{21.2}$$

where we have used the abbreviation

$$\partial_j^{\wedge}(h) = (\Delta + \lambda_j)^{\wedge}(h) = \Delta^{\wedge}(h) - \lambda_j = 4\pi^2 h^2 - \lambda_j, \quad h \in \Lambda^{-1}, \tag{21.3}$$

for $j = 1, \ldots, m$. Moreover, we set

$$M = \sum_{j=1}^{m} \sum_{\substack{4\pi^2 h^2 = \lambda_j \\ h \in \Lambda^{-1}}} 1. \tag{21.4}$$

In other words, M is the total number of members in the set $\Xi \subset \Lambda^{-1}$ given by

$$\Xi = \left\{ h \in \Lambda^{-1} : 4\pi^2 h^2 = \lambda_j, \ j = 1, \ldots, m \right\} \tag{21.5}$$

such that $M = \sum_{h\in\Xi} 1$. $\mathrm{Pol}_{\Xi}^{\wedge}$ denotes the set of all linear combinations of the functions Φ_h, $h \in \Xi \subset \mathbb{R}^q$.

Sobolev Spaces. Our purpose is to introduce a Sobolev space associated to $\mathrm{C}_{\Lambda}^{(2m)}(\mathbb{R}^q)$. We start from the sequence $\{a_h\}_{h\in\Lambda^{-1}}$ defined by

$$a_h = \begin{cases} 1, & h \in \Xi, \\ -\partial_{1,\ldots,m}^{\wedge}(h), & h \notin \Xi. \end{cases} \tag{21.6}$$

Consider the set $\mathrm{E}(\{a_h\};\mathbb{R}^q)$ for all F of class $\mathrm{C}_{\Lambda}^{(\infty)}(\mathbb{R}^q)$ satisfying

$$\sum_{h\in\Lambda^{-1}} |a_h|^2 \left|F_{\mathcal{F}_\Lambda}^{\wedge}(h)\right|^2 < \infty, \tag{21.7}$$

where the Fourier coefficients $F_{\mathcal{F}_\Lambda}^{\wedge}(h)$, $h \in \Lambda^{-1}$, of a function $F \in \mathrm{E}(\{a_h\};\mathbb{R}^q)$ are defined as always by

$$F_{\mathcal{F}_\Lambda}^{\wedge}(h) = \int_{\mathcal{F}_\Lambda} F(x)\overline{\Phi_h(x)}\, dx. \tag{21.8}$$

From the Cauchy–Schwarz inequality it follows that

$$\left|\sum_{h\in\Lambda^{-1}} |a_h|^2 F_{\mathcal{F}_\Lambda}^{\wedge}(h)\overline{G_{\mathcal{F}_\Lambda}^{\wedge}(h)}\right|^2 \tag{21.9}$$
$$\leq \left(\sum_{h\in\Lambda^{-1}} |a_h|^2 |F_{\mathcal{F}_\Lambda}^{\wedge}(h)|^2\right)\left(\sum_{h\in\Lambda^{-1}} |a_h|^2 |G_{\mathcal{F}_\Lambda}^{\wedge}(h)|^2\right)$$

for all functions F, G of class $\mathrm{E}(\{a_h\};\mathbb{R}^q)$. As both sums on the right-hand side are finite for elements of class $\mathrm{E}(\{a_h\};\mathbb{R}^q)$, the left-hand side is also finite. We introduce an inner product $\langle\cdot,\cdot\rangle_{\mathrm{H}_{\Lambda}^{(2m)}(\mathbb{R}^q)}$ on $\mathrm{E}(\{a_h\};\mathbb{R}^q)$ by

$$\langle F, G\rangle_{\mathrm{H}_{\Lambda}^{(2m)}(\mathbb{R}^q)} = \sum_{h\in\Lambda^{-1}} |a_h|^2 F_{\mathcal{F}_\Lambda}^{\wedge}(h)\overline{G_{\mathcal{F}_\Lambda}^{\wedge}(h)}. \tag{21.10}$$

The corresponding norm is given by

$$\|F\|^2_{\mathrm{H}_{\Lambda}^{(2m)}(\mathbb{R}^q)} = \sum_{h\in\Lambda^{-1}} |a_h|^2 |F_{\mathcal{F}_\Lambda}^{\wedge}(h)|^2. \tag{21.11}$$

This enables us to introduce the Sobolev space $\mathrm{H}_{\Lambda}^{(2m)}(\mathbb{R}^q)$ as follows:

Definition 21.1. *The* Sobolev space $\mathrm{H}_{\Lambda}^{(2m)}(\mathbb{R}^q)$ *is the completion of* $\mathrm{E}(\{a_h\};\mathbb{R}^q)$ *under the* $\mathrm{H}_{\Lambda}^{(2m)}(\mathbb{R}^q)$*-norm, i.e.,*

$$\mathrm{H}_{\Lambda}^{(2m)}(\mathbb{R}^q) = \overline{\mathrm{E}(\{a_h\};\mathbb{R}^q)}^{\|\cdot\|_{\mathrm{H}_{\Lambda}^{(2m)}(\mathbb{R}^q)}}. \tag{21.12}$$

The space $\mathrm{H}_{\Lambda}^{(2m)}(\mathbb{R}^q)$ equipped with the inner product (21.10) is a Hilbert space. $\mathrm{Pol}_{\Xi}^{\Lambda}$ is the null space of the operator $\partial_{1,\ldots,m}$ in $\mathrm{H}_{\Lambda}^{(2m)}(\mathbb{R}^q)$.

Observing the special choice of the sequence $\{a_h\}_{h\in\Lambda^{-1}}$ in Equation (21.6) we can rewrite the norm (21.11) in integral form as follows:

$$\|F\|^2_{\mathrm{H}_{\Lambda}^{(2m)}(\mathbb{R}^q)} = \sum_{\substack{h\in\Xi \\ h\in\Lambda^{-1}}} \left| \int_{\mathcal{F}_\Lambda} F(x)\overline{\Phi_h(x)}\, dx \right|^2 + \int_{\mathcal{F}_\Lambda} |\partial_{1,\ldots,m}F(x)|^2\, dx. \tag{21.13}$$

Every $F \in \mathrm{H}_{\Lambda}^{(2m)}(\mathbb{R}^q)$ can be represented by its Fourier series

$$F(x) = \sum_{h\in\Lambda^{-1}} \int_{\mathcal{F}_\Lambda} F(z)\overline{\Phi_h(z)}\, dz\ \Phi_h(x). \tag{21.14}$$

Observing the fact that $\partial_{1,\ldots,m}\Phi_h(x) = -\partial^{\wedge}_{1,\ldots,m}(h)\Phi_h(x)$, $h \in \Lambda^{-1}$, $x \in \mathbb{R}^q$ we obtain

$$\begin{aligned} F(x) = & \sum_{\substack{h\in\Xi \\ h\in\Lambda^{-1}}} \int_{\mathcal{F}_\Lambda} F(x)\overline{\Phi_h(x)}\, dx\ \Phi_h(x) \\ & + \sum_{\substack{h\notin\Xi \\ h\in\Lambda^{-1}}} \frac{1}{-\partial^{\wedge}_{1,\ldots,m}(h)} \int_{\mathcal{F}_\Lambda} F(z)\partial_{1,\ldots,m}\overline{\Phi_h(z)}\, dz\ \Phi_h(x), \end{aligned} \tag{21.15}$$

such that we arrive at the following periodic *Euler summation-type formula in the framework of the Sobolev space* $\mathrm{H}_{\Lambda}^{(2m)}(\mathbb{R}^q)$.

Theorem 21.2. *For all F of class $\mathrm{H}_{\Lambda}^{(2m)}(\mathbb{R}^q)$ the integral formula*

$$\begin{aligned} F(x) = & \sum_{\substack{h\in\Xi \\ h\in\Lambda^{-1}}} \int_{\mathcal{F}_\Lambda} F(x)\overline{\Phi_h(x)}\, dx\Phi_h(x) \\ & + \int_{\mathcal{F}_\Lambda} \partial_{1,\ldots,m}G(\partial^2_{1,\ldots,m}, x-z)\partial_{1,\ldots,m}F(z)\, dz\ \Phi_h(x) \end{aligned} \tag{21.16}$$

holds true, where

$$G(\partial^2_{1,\ldots,m}, x-z) = \sum_{\substack{h\notin\Xi \\ h\in\Lambda^{-1}}} \frac{1}{(-\partial^{\wedge}_{1,\ldots,m}(h))^2}\Phi_h(x)\overline{\Phi_h(z)}, \tag{21.17}$$

and, under the assumption $m > \frac{q}{4}$, the bilinear series (21.17) is absolutely and uniformly convergent in $\mathbb{R}^q$.

We specify a $\mathrm{Pol}_{\Xi}^{\Lambda}$*-unisolvent set* $X_M = \{x_h \in \mathcal{F}_\Lambda : h \in \Xi\}$, i.e., a set of points $x_h \in \mathcal{F}_\Lambda$ such that the rank of the $M \times M$-matrix $(\Phi_h(x_{h'}))_{h,h'\in\Xi}$ is equal to M. Under the assumption of Pol_Ξ^Λ-unisolvence we are able to interpolate a given set $\{\alpha_h : h \in \Xi\}$ by a unique $P \in \mathrm{Pol}_\Xi^\Lambda$: $P(x_h) = \alpha_h$, $h \in \Xi$. However, for any set $X_N = \{x_h \in \mathcal{F}_\Lambda : h \in \Xi_N\}$ containing X_M as subset (so that $M \leq N$ and $\Xi \subset \Xi_N$), this interpolation property cannot be guaranteed canonically. In this case, we are led in straightforward way to a spline interpolation problem of discussing the (semi-)norm

$$\int_{\mathcal{F}_\Lambda} |\partial_{1,\ldots,m} F(x)|^2 \, dx \tag{21.18}$$

in the set $\mathcal{I}_N$ of all $\mathrm{H}_\Lambda^{(2m)}(\mathbb{R}^q)$-interpolants to the given data

$$\mathcal{I}_N = \left\{ F \in \mathrm{H}_\Lambda^{(2m)}(\mathbb{R}^q) : F(x_h) = \alpha_h, \ h \in \Xi_N \right\}. \tag{21.19}$$

This procedure of finding a minimum in the set $\mathcal{I}_N$ (cf. W. Freeden [1988]) is reasonable because it gives an interpolant which is as close to being a polynomial of class Pol_Ξ^Λ as can be achieved for the N data points. In addition, the spline interpolant is the "smoothest" in the sense of the (semi-)norm (21.18), thereby avoiding larger oscillations in the interpolant. Moreover, the previous case when $N = M$ also is included since the uniquely defined polynomial in Pol_Ξ^Λ is obviously the solution of the smoothest interpolation problem formulated above.

In what follows we are concerned with the problem of interpolating multi-dimensional periodic functions by periodic splines and developing an a priori error bound for the approximation. The periodic splines discussed here turn out to be multi-dimensional generalizations of the one-dimensional periodic splines (cf. Chapter 4) and to be natural analogs to the splines on the sphere (cf. Chapter 13). Our spline concept is based on the (iterated) operators given by (21.1) and on arbitrary lattices Λ. Therefore, in spite of restricting ourselves to the framework in terms of Helmholtz operators, all spline representations are included which are based on (iterations of) arbitrary second order elliptic differential operators with constant coefficients.

Consider a Pol_Ξ^Λ-unisolvent set $X_M = \{x_h \in \mathcal{F}_\Lambda : h \in \Xi\}$. There exists in Pol_Ξ^Λ a unique Lagrange basis $\{B_h : h \in \Xi\}$ given by

$$B_h(x) = \sum_{h' \in \Xi} C_{hh'} \Phi_{h'}(x), \quad x \in \mathbb{R}^q, \tag{21.20}$$

such that

$$B_h(x_{h'}) = \delta_{h,h'}, \quad h, h' \in \Xi. \tag{21.21}$$

For every $F \in \mathrm{H}_\Lambda^{(2m)}(\mathbb{R}^q)$, the unique Pol_Ξ^Λ-interpolant $\mathrm{Proj}_{\mathrm{Pol}_\Xi^\Lambda} F$ of F on the Pol_Ξ^Λ-unisolvent set X_M under consideration is given by the "Lagrange formula"

$$\mathrm{Proj}_{\mathrm{Pol}_\Xi^\Lambda} F = \sum_{h \in \Xi} F(x_h) B_h. \tag{21.22}$$

Note that $\mathrm{Proj}_{\mathrm{Pol}_{\Xi}^{\Lambda}}$ denotes the projection operator onto $\mathrm{Pol}_{\Xi}^{\Lambda} \subset \mathrm{H}_{\Lambda}^{(2m)}(\mathbb{R}^q)$. Any function $F \in \mathrm{H}_{\Lambda}^{(2m)}(\mathbb{R}^q)$ can be represented uniquely in the form

$$F = \mathrm{Proj}_{\mathrm{Pol}_{\Xi}^{\Lambda}} F + \mathring{F}, \quad \mathring{F} \in \mathring{\mathrm{H}}_{\Lambda}^{(2m)}(\mathbb{R}^q), \tag{21.23}$$

where

$$\mathring{\mathrm{H}}_{\Lambda}^{(2m)}(\mathbb{R}^q) = \left\{ F \in \mathrm{H}_{\Lambda}^{(2m)}(\mathbb{R}^q) : F(x_h) = 0, \quad h \in \Xi \right\}. \tag{21.24}$$

As a consequence, we obtain from Theorem 21.2 by straightforward considerations

$$\mathring{F}(x) = \int_{\mathcal{F}_\Lambda} \left(\partial_{1,\ldots,m} \mathring{G}(\partial^2_{1,\ldots,m}; x-y) \right) \partial_{1,\ldots,m} \mathring{F}(y)\, dy \tag{21.25}$$

for all $x \in \mathcal{F}_\Lambda$ and all $\mathring{F} \in \mathring{\mathrm{H}}_{\Lambda}^{(2m)}(\mathbb{R}^q)$, where the reproducing kernel $\mathring{G}(\partial^2_{1,\ldots,m}; x-y)$, $x, y \in \mathbb{R}^q$, is given by

$$\begin{aligned} \mathring{G}(\partial^2_{1,\ldots,m}; x-y) = {} & G(\partial^2_{1,\ldots,m}; x-y) \\ & - \sum_{h \in \Xi} G(\partial^2_{1,\ldots,m}; x_h - y) B_h(x) \\ & - \sum_{h \in \Xi} \overline{B_h(y)} G(\partial^2_{1,\ldots,m}; x - x_h) \\ & + \sum_{h \in \Xi} \sum_{h' \in \Xi} \overline{B_h(y)} G(\partial^2_{1,\ldots,m}; x_{h'} - x_h) B_h(x). \end{aligned} \tag{21.26}$$

Within this Lagrange approach the definition of splines can be given as follows:

Definition 21.3. *For a* $\mathrm{Pol}_{\Xi}^{\Lambda}$*-unisolvent system* $X_N = \{x_h \in \mathcal{F}_\Lambda : h \in \Xi_N\}$, *any function* S *of the form*

$$S(x) = \sum_{h \in \Xi} a_h B_h(x) + \sum_{h \in \Xi_N \setminus \Xi} a_h \mathring{G}(\partial^2_{1,\ldots,m}; x - x_h), \quad a_h \in \mathbb{C},\ x \in \mathbb{R}^q, \tag{21.27}$$

is called a Λ-periodic $\mathrm{H}_{\Lambda}^{(2m)}(\mathbb{R}^q)$-spline *relative to* X_N. *Note that, for* $N = M$, S *reduces to the first sum of the right hand side of (21.27). The space* $\mathrm{Spline}_{\mathrm{H}_{\Lambda}^{(2m)}(\mathbb{R}^q)}(\Xi_N)$ *of all* Λ*-periodic* $\mathrm{H}_{\Lambda}^{(2m)}(\mathbb{R}^q)$*-splines relative to* X_N *is an* N*-dimensional linear space containing the class* $\mathrm{Pol}_{\Xi}^{\Lambda}$.

An immediate consequence of Definition 21.3 is the following theorem.

Theorem 21.4. *Any* $S \in \mathrm{Spline}_{\mathrm{H}_{\Lambda}^{(2m)}(\mathbb{R}^q)}(\Xi_N)$ *can be represented in the form*

$$S(x) = P(x) + \sum_{h \in \Xi_N} b_h G(\partial^2_{1,\ldots,m}; x - x_h), \quad x \in \mathbb{R}^q,\ P \in \mathrm{Pol}_{\Xi}^{\Lambda}, \tag{21.28}$$

where the coefficients $b_h \in \mathbb{C}$ *have to satisfy the linear equations*

$$\sum_{h \in \Xi_N} b_h \overline{\Phi_{h'}(x_h)} = 0, \quad h' \in \Xi. \tag{21.29}$$

The proof follows easily from the definition of Λ-periodic $\mathrm{H}_{\Lambda}^{(2m)}(\mathbb{R}^q)$-splines by obvious calculation. Therefore, it is clear that the Λ-periodic splines discussed here form a multidimensional generalization of the already known one-dimensional periodic splines.

21.2 Minimum Norm Interpolation

Suppose now that there are given N prescribed data points (x_h, α_h), $h \in \Xi$, corresponding the a $\mathrm{Pol}_{\Xi_N}^{\Lambda}$-unisolvent system $X_N = \{x_h \in \mathcal{F}_\Lambda : h \in \Xi_N\}$. We consider the problem of finding the "smoothest" function in the set $\mathcal{I}_N$ of (21.19) containing all interpolants to the data, where by "smoothest" we mean that (21.18) becomes a minimum in $\mathrm{H}_{\Lambda}^{(2m)}(\mathbb{R}^q)$.

For that purpose we need the following preliminaries (cf. W. Freeden [1988]):

Lemma 21.5. *If $F \in \mathcal{I}_N$ and $S \in \mathrm{Spline}_{\mathrm{H}_{\Lambda}^{(2m)}(\mathbb{R}^q)}(\Xi_N)$, then*

$$\int_{\mathcal{F}_\Lambda} (\partial_{1,\ldots,m} S(y))(\partial_{1,\ldots,m} F(y))\, dy \tag{21.30}$$

$$= \sum_{h \in \Xi_N \setminus \Xi} a_h \left(\alpha_h - \sum_{h' \in \Xi} \alpha_{h'} B_{h'}(x_h) \right).$$

Lemma 21.6. *There exists a unique $S_N \in \mathcal{I}_N \cap \mathrm{Spline}_{\mathrm{H}_{\Lambda}^{(2m)}(\mathbb{R}^q)}(\Xi_N)$, denoted briefly by S_N.*

Proof. Any spline $S \in \mathrm{Spline}_{\mathrm{H}_{\Lambda}^{(2m)}(\mathbb{R}^q)}(\Xi_N)$ of the form (21.27) contains a total of N coefficients $a_{h'} \in \mathbb{C}$, $h' \in \Xi_N$. Thus, $S(x_h) = \alpha_h$ is equivalent to the linear equations

$$\sum_{h' \in \Xi_N \setminus \Xi} a_{h'} \mathring{G}(\partial^2_{1,\ldots,m}; x_h - x_{h'}) = \alpha_h - \sum_{h' \in \Xi} \alpha_{h'} B_{h'}(x_h), \quad h \in \Xi_N \setminus \Xi. \tag{21.31}$$

The coefficient matrix is (Hermitian) symmetric and positive definite as Gram matrix of a sequence of linearly independent elements. Hence, the linear system in (21.31) is uniquely solvable. □

We are now able to formulate the "smoothest spline interpolation property": If $F \in \mathcal{I}_N$, then

$$\int_{\mathcal{F}_\Lambda} |\partial_{1,\ldots,m} F(x)|^2\, dx = \int_{\mathcal{F}_\Lambda} |\partial_{1,\ldots,m} S_N(x)|^2\, dx \tag{21.32}$$

$$+ \int_{\mathcal{F}_\Lambda} |\partial_{1,\ldots,m} (S_N(x) - F(x))|^2\, dx.$$

This leads to the following result.

Theorem 21.7. *The* $\mathrm{H}_\Lambda^{(2m)}(\mathbb{R}^q)$*-spline interpolation problem*

$$\int_{\mathcal{F}_\Lambda} |\partial_{1,\ldots,m} S_N(x)|^2 \, dx = \min_{F \in \mathcal{I}_N} \int_{\mathcal{F}_\Lambda} |\partial_{1,\ldots,m} F(x)|^2 \, dx \tag{21.33}$$

is well-posed in the sense that its solution exists, is unique, and depends continuously on the data α_h, $h \in \Xi_N$.

21.3 Periodic Sampling

Next we come to periodic sampling based on spline interpolation. We assume that m is an integer satisfying $m > \frac{q}{4}$. From our spline interpolation procedure it is not difficult to verify the following result.

Theorem 21.8 (Shannon Sampling in $\mathrm{Spline}_{H_\Lambda^{(2m)}(\mathbb{R}^q)}(\Xi_N)$). *Let* F *be an element of* $\mathrm{Spline}_{H_\Lambda^{(2m)}(\mathbb{R}^q)}(\Xi_N)$. *Then* F *can be reconstructed from its samples at points* $x_h \in \mathcal{F}_\Lambda$, $h \in \Xi_N$, *by the formula*

$$F(x) = \sum_{h \in \Xi_N} F(x_h) L_h(x), \quad x \in \mathbb{R}^q, \tag{21.34}$$

where $L_h \in \mathrm{Spline}_{H_\Lambda^{(2m)}(\mathbb{R}^q)}(\Xi_N)$ *is given by*

$$L_h(x) = \sum_{h' \in \Xi} e_{h'}^h \Phi_h(x) + \sum_{l \in \Xi_N} a_l^h G(\partial_{1,\ldots,m}^2; x - x_l), \quad x \in \mathbb{R}^q \tag{21.35}$$

and the coefficients $c_{h'}^h$, a_l^h *satisfy the linear equations*

$$\sum_{h' \in \Xi} c_{h'}^h \Phi_{h'}(x_k) + \sum_{l \in \Xi_N} a_l^h G(\partial_{1,\ldots,m}^2; x_k - x_l) = \delta_{k,h}, \quad h, k \in \Xi_N, \tag{21.36}$$

$$\sum_{l \in \Xi} c_l^h \overline{\Phi_{h'}(x_l)} = 0, \quad h' \in \Xi. \tag{21.37}$$

Proof. The function $F \in \mathrm{Spline}_{H_\Lambda^{(2m)}(\mathbb{R}^q)}(\Xi_N)$ is uniquely characterized by

$$F(x) = \sum_{h' \in \Xi} c_{h'} \Phi_{h'}(x) + \sum_{l \in \Xi_N} a_l G(\partial_{1,\ldots,m}^2 (x - x_l)), \quad x \in \mathbb{R}^q, \tag{21.38}$$

where

$$\sum_{l \in \Xi_N} a_l G(\partial_{1,\ldots,m}^2; x_i - x_l) + \sum_{h' \in \Xi} c_{h'} \Phi_{h'}(x_i) = F(x_i), \tag{21.39}$$

$$\sum_{l \in \Xi_N} a_l \overline{\Phi_h(x_l)} = 0, \tag{21.40}$$

for $i \in \Xi_N$, $h \in \Xi$. It is not difficult to see that

$$\sum_{k\in\Xi_N} F(x_k)L_k(x) = \sum_{h'\in\sigma_M} c_{h'}\Phi_{h'}(x) + \sum_{l\in\Xi_N} a_l G(\partial^2_{1,\ldots,m}; x - x_l), \tag{21.41}$$

where

$$c_{h'} = \sum_{k\in\Xi_N} a^k_{h'} F(x_k) \tag{21.42}$$

and

$$a_l = \sum_{k\in\Xi_N} a^k_l F(x_k). \tag{21.43}$$

This yields

$$\sum_{h'\in\Xi} c_{h'}\Phi_{h'}(x_i) + \sum_{l\in\Xi_N} a_l G(\partial^2_{1,\ldots,m}; x_l - x_i) \tag{21.44}$$

$$= \sum_{k\in\Xi_N} F(x_k)\Bigg(\sum_{h'\in\Xi} c^k_{h'}\Phi_h(x_i) + \sum_{l\in\Xi_N} a^k_l G(\partial^2_{1,\ldots,m}, x_i - x_l)\Bigg) = F(x_i)$$

and

$$\sum_{l\in\Xi_N} a_l \Phi_{h'}(x_l) = \sum_{k\in\Xi_N} F(x_k) \sum_{l\in\Xi_N} a^k_l \Phi_{h'}(x_l) = 0. \tag{21.45}$$

This proves Theorem 21.8. □

In addition we mention the following sampling result.

Corollary 21.9 (Sampling Variant in $\mathrm{Spline}_{H^{(2m)}_\Lambda(\mathbb{R}^q)}(\Xi_N)$). *Any function $F \in \mathrm{Spline}_{H^{(2m)}_\Lambda(\mathbb{R}^q)}(\Xi_N)$ can be reconstructed from its samples at points $x_h \in \mathcal{F}_\Lambda$, $h \in \Xi_N$, by the combined polynomial and spline interpolation formula*

$$F(x) = \sum_{k\in\Xi} F(x_k)L_k(x) + \sum_{i\in\Xi_N\setminus\Xi} \Bigg(F(x_i) - \sum_{k\in\Xi} F(x_k)L_k(x_i)\Bigg) \mathring{L}_i(x) \tag{21.46}$$

with $x \in \mathbb{R}^q$, where the Lagrangians L_k, $k \in \Xi$, are given in the already described way and $\mathring{L}_i$, $i \in \Xi_N \setminus \Xi$, are characterized by

$$\mathring{L}_k(x) = \sum_{l\in\Xi_N\setminus\Xi} b^k_l \mathring{G}(\partial^2_{1,\ldots,m}; x - x_l) \tag{21.47}$$

and the coefficients b^k_l satisfy the linear equations

$$\sum_{l\in\Xi_N\setminus\Xi} a^k_l \mathring{G}(\partial^2_{1,\ldots,m}; x_i - x_l) = \delta_{i,k}. \tag{21.48}$$

In fact, it follows that

$$\sum_{k\in\Xi_N\setminus\Xi} F(x_k)L_k(x) = \sum_{i\in\Xi_N\setminus\Xi} \Big(F(x_i) - \sum_{k\in\Xi} F(x_k)L_k(x_i)\Big)\mathring{L}_i(x). \quad (21.49)$$

Hence, for $x \in \mathbb{R}^q$, we have

$$\sum_{k\in\Xi_N\setminus\Xi} F(x_k)\Big(L_k(x) + \sum_{i\in\Xi_N\setminus\Xi} L_k(x_i)\mathring{L}_i(x)\Big) = \sum_{i\in\Xi_N\setminus\Xi} F(x_i)\mathring{L}_i(x). \quad (21.50)$$

21.4 Elliptic Operators, Monospline Integration, and Remainder Terms

We denote by $L = L(\nabla)$ a linear differential operator of elliptic type and order p with constant (real) coefficients. $M = M(\nabla)$ denotes the adjoint operator of L. Assume that $\mathcal{G} \subset \mathbb{R}^q$ is a regular region. Then Green's formula tells us that

$$\int_{\mathcal{G}} \Phi(x)L\Psi(x) - \Psi(x)M\Phi(x)\, dx = \int_{\partial\mathcal{G}} P_L(\Phi(x), \Psi(x))\, dS(x) \quad (21.51)$$

is valid for its boundary $\partial\mathcal{G}$ provided that Φ, Ψ are sufficiently smooth. P_L is a bilinear form with respect to Φ and Ψ (and their partial derivatives) determined by the operator L.

Substituting ∇ by x, i.e., more explicitly $\frac{\partial}{\partial x_i}$ by x_i for each occurrence of $\frac{\partial}{\partial x_i}$ in $L(\nabla)$, respectively, $M(\nabla)$ we get polynomials $L(x)$, respectively, $M(x)$ of degree p in q variables. It is easy to see that

$$(L(\nabla) - L^\wedge(h))\,\Phi_h(x) = 0. \quad (21.52)$$

The functions Φ_h, $h \in \Lambda^{-1}$, are the only eigenfunctions, corresponding to the eigenvalues $L^\wedge(h) = -L(2\pi i h)$, of the differential equation $(L+L^\wedge(h))U(x) = 0$ satisfying the "boundary condition of periodicity". The eigenfunctions Φ_h with respect to the eigenvalues $L^\wedge(h)$ are eigenfunctions of Green's function $G(L;\cdot)$ for the lattice Λ with respect to the operator L in the sense of the integral equation

$$L^\wedge(h) \int_{\mathcal{F}_\Lambda} G(L;x,y)\overline{\Phi_h(x)}\, dx = \Phi_h(y), \quad L^\wedge(h) \neq 0. \quad (21.53)$$

Let L be an elliptic differential operator with order $2m > q$. Then the Green's function of the lattice Λ with respect to the operator L (briefly called: Λ-*lattice function for* L) is continuous in $\mathbb{R}^q$, and $G(L;x-y)$ admits the bilinear expansion

$$G(L; x-y) = \sum_{\substack{L^\wedge(h)\neq 0 \\ h\in\Lambda^{-1}}} \frac{\Phi_h(x)\overline{\Phi_h(y)}}{L^\wedge(h)}, \quad x, y \in \mathbb{R}^q, \tag{21.54}$$

which is absolutely and uniformly convergent both in x and y and uniformly in x and y together. As usual, the sum is to be taken over all lattice points $h \in \Lambda^{-1}$ for which $L^\wedge(h) \neq 0$. For an explicit discussion of the defining properties of the lattice functions with respect to the Laplacian Δ and the operators $(\Delta + \lambda)^m$ the reader is referred to Section 20.1.

Euler Summation Formulas. Standard considerations of the theory of elliptic differential operators enable us to deduce the following multi-dimensional analogs of the classical Euler summation formula.

Theorem 21.10. *Let $\mathcal{G} \subset \mathbb{R}^q$, $q \geq 2$, be a regular region. Suppose that L is an elliptic differential operator of order p with constant coefficients. Then, for $F \in \mathrm{C}^{(p)}(\overline{\mathcal{G}})$, $\overline{\mathcal{G}} = \mathcal{G} \cup \partial\mathcal{G}$,*

$$\sum_{g\in\overline{\mathcal{G}}}{}' F(g) = \frac{1}{\sqrt{\|\mathcal{F}_\Lambda\|}} \sum_{\substack{L^\wedge(h)=0 \\ h\in\Lambda^{-1}}} \int_{\mathcal{G}} F(x)\overline{\Phi_h(x)}\, dx \tag{21.55}$$

$$+ \int_{\partial\mathcal{G}} P_L(F(x), G(L;x))\, dS(x) + \int_{\mathcal{G}} G(L;x) MF(x)\, dx.$$

As usual, the symbol $\sum{}'$ means that we have to observe the solid angle $\alpha_{\mathcal{G}}(g)$ subtended at $g \in \overline{\mathcal{G}}$ by the surface $\partial\mathcal{G}$.

As special cases we mention two examples.

Example 21.11. *Let $L = M = \Delta$, $F \in \mathrm{C}^{(2)}(\overline{\mathcal{G}})$. Then we obtain*

$$\sum_{g\in\overline{\mathcal{G}}}{}' F(g) = \frac{1}{\|\mathcal{F}_\Lambda\|} \int_{\mathcal{G}} F(x)\, dx + \int_{\partial\mathcal{G}} P_\Delta(F(x), G(\Delta;x))\, dS(x)$$

$$+ \int_{\mathcal{G}} G(\Delta;x)\Delta F(x)\, dx, \tag{21.56}$$

where

$$\int_{\partial\mathcal{G}} P_\Delta(F(x), G(\Delta;x))\, dS(x) \tag{21.57}$$

$$= \int_{\partial\mathcal{G}} F(x)\, \frac{\partial}{\partial\nu} G(\Delta;x)\, dS(x) - \int_{\partial\mathcal{G}} G(\Delta;x) \frac{\partial}{\partial\nu} F(x)\, dS(x).$$

Example 21.12. *Consider now the following operators for $a \in \mathbb{R}^q$ and $\lambda \in \mathbb{R}$:*

$$L = \Delta + 4\pi i(a\cdot\nabla) + \lambda, \tag{21.58}$$

$$M = \Delta - 4\pi i(a\cdot\nabla) + \lambda = \overline{L}. \tag{21.59}$$

For all $y \in \mathbb{R}^q$ and $H \in \mathrm{C}^{(2)}(\overline{\mathcal{G}})$ we have

$$\sum_{\substack{g-y\in\overline{\mathcal{G}}\\ g\in\Lambda}}{}' H(g) = \frac{1}{\sqrt{\|\mathcal{F}_\Lambda\|}} \sum_{\substack{4\pi^2((h+a)^2-a^2)=\lambda\\ h\in\Lambda^{-1}}} \int_{x-y\in\mathcal{G}} H(x)\overline{\Phi_h(x)}\, dx$$

$$+ \int_{x-y\in\mathcal{G}} G(L;x)\, \overline{L}H(x)\, dx \tag{21.60}$$

$$+ \int_{x-y\in\partial\mathcal{G}} H(x) \left(\frac{\partial}{\partial\nu} + 4\pi i(a\cdot\nu)\right) G(L;x)\, dS(x)$$

$$- \int_{x-y\in\partial\mathcal{G}} G(L;x)\, \frac{\partial}{\partial\nu} H(x)\, dS(x).$$

$$\tag{21.61}$$

For $a = y = 0$, this case reduces to Example 21.11.

Choosing in Example 21.12 especially

$$H(x) = e^{2\pi i(a\cdot x)} F(x-y), \quad x-y\in\overline{\mathcal{G}}, \tag{21.62}$$

we are able to modify (21.60) by use of the elementary identities

$$\Delta_x e^{2\pi i(a\cdot x)} F(x-y) = e^{2\pi i(a\cdot x)}(\Delta_x + 4\pi i(a\cdot\nabla_x) - 4\pi^2 a^2) F(x-y), \tag{21.63}$$

as well as

$$(a\cdot\nabla_x) e^{2\pi i(a\cdot x)} F(x-y) = e^{2\pi i(a\cdot x)}((a\cdot\nabla_x) + 2\pi i a^2) F(x-y), \tag{21.64}$$

so that

$$(\Delta_x - 4\pi i(a\cdot\nabla_x)) e^{2\pi i(a\cdot x)} F(x-y) = e^{2\pi i(a\cdot x)} \left(\left(\Delta_x + 4\pi^2 a^2\right) F(x-y)\right). \tag{21.65}$$

As result we obtain

$$\sum_{\substack{g-y\in\overline{\mathcal{G}}\\ g\in\Lambda}}{}' e^{2\pi i(a\cdot g)} F(g-y) \tag{21.66}$$

$$= \frac{1}{\|\mathcal{F}_\Lambda\|} e^{2\pi i(a\cdot y)} \sum_{\substack{4\pi^2((h+a)^2-a^2)=\lambda\\ h\in\Lambda^{-1}}} e^{2\pi i(h\cdot y)} \int_{\mathcal{G}} F(x) e^{2\pi i((a+h)\cdot x)}\, dx$$

$$+ \int_{x-y\in\mathcal{G}} G(L;x) e^{2\pi i(a\cdot x)} \left(\left(\Delta_x + \lambda + 4\pi^2 a^2\right) F(x-y)\right)\, dx$$

$$- \int_{x-y\in\partial\mathcal{G}} e^{2\pi i(a\cdot x)} G(L;x) \frac{\partial}{\partial\nu_x} F(x-y)\, dS(x)$$

$$+ \int_{x-y\in\partial\mathcal{G}} e^{2\pi i(a\cdot x)} F(x-y) \left(\frac{\partial}{\partial\nu_x} + 4\pi i(a\cdot\nu)\right) G(L;x)\, dS(x).$$

We notice that both sides of (21.66) are Λ-periodic with respect to the variable $a \in \mathbb{R}^q$. This observation is the key point for multivariate Shannon-type sampling generalizations as proposed by W. Freeden, M.Z. Nashed [2015].

Monosplines and Truncation Error. Let W be a weight function defined and continuous on $\overline{\mathcal{G}}$, where $\mathcal{G}$ is a regular region. By V we denote a function of class $\mathrm{C}^{(p)}(\overline{\mathcal{G}})$ with $LV = W$, and it does not matter which we choose. Let y be a point of the fundamental cell $\mathcal{F}_\Lambda$ of the lattice Λ in $\mathbb{R}^q$. By combination of the Euler summation formula and Green's integral theorem we obtain

$$\begin{aligned}\sum_{\substack{g-y\in\overline{\mathcal{G}}\\ g\in\Lambda}}{}' F(g-y) = &\int_{\mathcal{G}} W(x)F(x)\,dx \\ &+ \sum_{\substack{L^\wedge(h)=0\\ h\in\Lambda^{-1}}} \overline{\Phi_h(y)} \int_{\mathcal{G}} F(x)\overline{\Phi_h(x)}\,dx \\ &+ \int_{\partial\mathcal{G}} P_L(F(x), G(L;x+y)-V(x))\,dS(x) \\ &+ \int_{\mathcal{G}} (G(L;x+y)-V(x))\,MF(x)\,dx.\end{aligned} \tag{21.67}$$

This establishes the following monospline formula.

Theorem 21.13. *Let L be an elliptic differential operator of order p. Suppose that W is a continuous weight function on $\overline{\mathcal{G}} = \mathcal{G}\cup\partial\mathcal{G}$. For given nodes $y_1,\ldots,y_n \in \mathcal{F}_\Lambda$ and weights $a_1,\ldots,a_n$, let K designate a function of the form*

$$K(x) = S(x) - V(x), \tag{21.68}$$

where

$$S(x) = \sum_{\substack{L^\wedge(h)=0\\ h\in\Lambda^{-1}}} c_h\overline{\Phi_h(x)} + \sum_{k=1}^{n} a_k G(L;x+y_k), \tag{21.69}$$

$V \in \mathrm{C}^{(p)}(\overline{\mathcal{G}})$ satisfies the equation $LV = W$, and $G(L;\cdot)$ is the Λ-lattice function with respect to L. Then, for each $F \in \mathrm{C}^{(p)}(\overline{\mathcal{G}})$, we have

$$\begin{aligned}\sum_{k=1}^{n} a_k \sum_{\substack{g-y_k\in\overline{\mathcal{G}}\\ h\in\Lambda^{-1}}}{}' F(g-y_k) = &\int_{\mathcal{G}} W(x)F(x)\,dx \\ &+ \sum_{\substack{L^\wedge(h)=0\\ h\in\Lambda^{-1}}} \sum_{k=1}^{n} a_k\overline{\Phi_h(y_k)} \int_{\mathcal{G}} F(x)\overline{\Phi_h(x)}\,dx \\ &+ \int_{\partial\mathcal{G}} P_L(F(x), K(x))\,dS(x) \\ &+ \int_{\mathcal{G}} K(x)MF(x)\,dx.\end{aligned} \tag{21.70}$$

The function K is called a *monospline relative to $y_1, \ldots, y_n$ and the operator L* (see Theorem 10.9 for the spherical case).

By specializing the summation formula (21.70) we are led to a large number of cubature formulas and numerical rules for a variety of regions $\mathcal{G}$.

Obviously, the truncation error $\mathcal{E}$ defined by

$$\mathcal{E}F = \int_{\mathcal{G}} K(x) MF(x)\, dx \tag{21.71}$$

annihilates all functions of the null space of M, i.e., $F \in \mathrm{C}^{(p)}(\overline{\mathcal{G}})$ with $MF = 0$.

Corollary 21.14. *Let L be of order p with $p \geq \lfloor q/2 \rfloor + 1$. Then*

$$|\mathcal{E}F| \leq \int_{\mathcal{G}} |K(x)|\, dx \quad \sup_{x \in \overline{\mathcal{G}}} |M\, F(x)| \tag{21.72}$$

Corollary 21.15. *Let L be of order p with $p \geq \frac{1}{2}(\lfloor \frac{q}{2} \rfloor + 1)$. Then*

$$|\mathcal{E}F| \leq \left(\int_{\mathcal{G}} |K(x)|^2\, dx \right)^{1/2} \left(\int_{\mathcal{G}} |M\, F(x)|^2\, dx \right)^{1/2}. \tag{21.73}$$

Best Approximation Cubature for Regular Region. Next we come to optimal rules for periodic functions. Consider an elliptic operator L of order p with $p \geq \frac{1}{2}(\lfloor \frac{q}{2} \rfloor + 1)$. As usual, let $\mathrm{H}_\Lambda^{(p)}(\mathbb{R}^q)$ designate the Sobolev space of all Λ-periodic functions of class $\mathrm{H}_\Lambda^{(p)}(\mathbb{R}^q)$ equipped with the semi-norm

$$\|F\|_{\mathrm{H}_\Lambda^{(p)}(\mathbb{R}^q)} = \left(\int_{\mathcal{F}_\Lambda} |M\, F(x)|^2\, dx \right) < \infty. \tag{21.74}$$

Then it is obvious that

$$\sup_{\substack{\|F\|_{\mathrm{H}^{(p)}(\mathbb{R}^q)} \leq 1 \\ F \in \mathrm{H}_\Lambda^{(p)}(\mathbb{R}^q)}} |\mathcal{E}F| \leq \left(\int_{F_\Lambda} |K(x)|^2\, dx \right)^{1/2}. \tag{21.75}$$

Moreover, for $V \in \mathrm{H}_\Lambda^{(p)}(\mathbb{R}^q)$, we have

$$\int_{\partial \mathcal{F}_\Lambda} P_L(F(x), K(x))\, dS(x) = 0. \tag{21.76}$$

For fixed nodes y_k, let the weights a_k satisfy

$$\sum_{k=1}^{n} a_k \overline{\Phi_h(y_k)} = 0 \tag{21.77}$$

for all $h \in \Lambda^{-1}$ with $L^\wedge(h) = 0$. Then, the cubature rule with error

$$\mathcal{E}'F = \int_{\mathcal{F}_\Lambda} W(x) F(x)\, dx - \sum_{k=1}^{N} \mathring{a}_k F(y_k) \tag{21.78}$$

represents a relatively minimum norm rule if

$$\int_{\mathcal{F}_\Lambda} |K(x)|^2 \, dx \longrightarrow \min . \tag{21.79}$$

The minimum is to be taken over all coefficients satisfying the linear equations stated in (21.77). The problem of determining such rules is equivalent to that of the best approximation by means of Λ-periodic splines.

Best Approximate Cubature in Euclidean Spaces. Under certain circumstances, Theorem 21.13 gives approximations to infinite integrals: Let L be an elliptic operator of order p. Suppose that, for fixed nodes $y_1, \ldots, y_n$, the weights $a_1, \ldots a_n$ satisfy

$$\sum_{k=1}^{n} a_k \overline{\Phi_h(y_k)} = 0, \tag{21.80}$$

for all $h \in \Lambda^{-1}$ with $L^\wedge(h) = 0$. Then, under the assumptions of Theorem 21.13 we obtain

$$\begin{aligned}\sum_{k=1}^{N} a_k \sum_{\substack{g \in \Lambda \\ |g-y_k| \leq N}}{}' F(g - y_k) &= \int_{|x| \leq N} W(x) F(x) \, dx \\ &+ \int_{|x|=N} P_L(F(x), K(x)) \, dS(x) + \int_{|x| \leq N} K(x) \, MF(x) \, dx. \end{aligned} \tag{21.81}$$

for all $N > 0$.

Theorem 21.16. *Suppose that $F \in \mathrm{C}^p(\mathbb{R}^q)$ satisfies the following properties:*

(i) $\int_{|x|=N} P_L(F(x), K(x)) \, dS(x) = o(1)$, $N \to \infty$,

(ii) $\int_{\mathbb{R}^q} |K(x) \, MF(x)| \, dx < \infty$,

(iii) the integral

$$\int_{\mathbb{R}^q} W(x) F(x) \, dx \tag{21.82}$$

exists in spherical sense

$$\int_{\mathbb{R}^q} \cdots = \lim_{N \to \infty} \int_{|x| \leq N} \cdots . \tag{21.83}$$

Then

$$\begin{aligned} \lim_{N \to \infty} &\left(\sum_{k=1}^{N} a_k \sum_{\substack{|g-y_k| \leq N \\ g \in \Lambda}} F(g - y_k) - \int_{|x| \leq N} W(x) F(x) \, dx \right) \\ &= \int_{\mathbb{R}^q} K(x) \, MF(x) \, dx. \end{aligned} \tag{21.84}$$

For $p \geq \frac{1}{2}(\lfloor \frac{q}{2} \rfloor + 1)$ and fixed nodes $y_1, \ldots, y_n$, best approximate integration can be achieved by letting

$$\int_{\mathbb{R}^q} |K(x)|^2 \, dx \longrightarrow \min . \tag{21.85}$$

in the coefficients $a_1, \ldots, a_N$.

Chapter 22

Shannon Sampling and Paley–Wiener Integration

In this chapter it is shown that the bridge between Gauss's and Shannon's work is constituted by certain extensions of the famous Hardy–Landau identities in geometric lattice point theory. Particular interest is laid on the matter dealing with bandlimited functions corresponding to, e.g., geoscientifically relevant regions (cf. W. Freeden, M.Z. Nashed [2015]). The routes to sampling expansions are exhibited in Paley-Wiener spaces, leading to multivariate sinc-type reproducing kernels and spline integration formulas over regular regions.

22.1 Gaussian Circle Problem and Hardy's Conjecture

Let us start our explanation of the bridge from Gaussian concepts of number theory to modern Shannon-type sampling with a recapitulation of some results on the number of lattice points inside circles $\mathbb{S}^1_N = \{x \in \mathbb{R}^2 : |x| = N\}$ of radii $N > \frac{\sqrt{2}}{2}$ around the origin 0 (for more background material and deeper number theoretical understanding the reader is referred, e.g., to the monographs W. Freeden [2011], F. Fricker [1982]).

We know from C.F. Gauss [1801, 1826] that the total number $\#_{\mathbb{Z}^2}\left(\overline{\mathbb{B}^2_N}\right)$ of lattice points in closed disks $\overline{\mathbb{B}^2_N} = \{x \in \mathbb{R}^2 : |x| \leq N\}$ is equal to the area of that circle plus a remainder of the order of the boundary, i.e.,

$$\#_{\mathbb{Z}^2}\left(\overline{\mathbb{B}^2_N}\right) = \pi N^2 + O(N). \tag{22.1}$$

Hence, as byproduct, a method of determining the irrational, transcendent number π becomes obvious (for alternative approaches to π within the history of analysis the reader is referred to T. Sonar [2011]):

$$\lim_{N\to\infty} \frac{\#_{\mathbb{Z}^2}\left(\overline{\mathbb{B}^2_N}\right)}{N^2} = \pi. \tag{22.2}$$

C.F. Gauss [1826] illustrated his result by taking $N^2 = 100{,}000$. In this case he calculated

$$\sum_{\substack{|g|^2 \leq 100{,}000 \\ g\in\mathbb{Z}^2}} 1 = 314{,}197. \tag{22.3}$$

This calculation determines the number π up to three digits after the comma.

The formula (22.1) due to C.F. Gauss [1826] allows the asymptotic expansion

$$\#_{\mathbb{Z}^2}\left(\overline{\mathbb{B}^2_{\sqrt{N}}}\right) = \pi N + O(N^{1/2}). \tag{22.4}$$

The so-called *circle problem* (see, e.g., W. Freeden [2011], F. Fricker [1982]) is concerned with the question of determining the bound

$$\alpha_2 = \inf\left\{\gamma : \#_{\mathbb{Z}^2}\left(\overline{\mathbb{B}^2_{\sqrt{N}}}\right) = \pi N + O(N^{\gamma})\right\}. \tag{22.5}$$

Until now, we know from (22.4) that $\alpha_2 \leq \frac{1}{2}$. An improvement of the Gaussian result, however, turns out to be very laborious, in fact, requiring a great effort. A first remarkable result is due to W. Sierpinski [1906], who proved by use of a method of his teacher G. Voronoi [1903] that

$$\#_{\mathbb{Z}^2}\left(\overline{\mathbb{B}^2_{\sqrt{N}}}\right) = \pi N + O\left(N^{\frac{1}{3}}\right), \tag{22.6}$$

i.e., $\alpha_2 \leq \frac{1}{3}$. The proof of Sierpinski is elementary (see, e.g., W. Freeden [2011], F. Fricker [1982] for more details), it is a link between geometry and number theory.

Hardy's conjecture claims

$$\#_{\mathbb{Z}^2}\left(\overline{\mathbb{B}^2_{\sqrt{N}}}\right) - \pi N = O\left(N^{\frac{1}{4}+\varepsilon}\right) \tag{22.7}$$

for *every* $\varepsilon > 0$. This conjecture seems to be still a challenge for future work.

Summarizing our results about lattice points inside circles (cf. W. Freeden [2011], F. Fricker [1982]) we are confronted with the following situation:

$$\frac{1}{4} \leq \alpha_2 \leq \frac{1}{4} + \varepsilon_2 \tag{22.8}$$

and

$$\#_{\mathbb{Z}^2}\left(\overline{\mathbb{B}^2_{\sqrt{N}}}\right) - \pi N \neq O\left(N^{\frac{1}{4}}\right), \tag{22.9}$$

$$\#_{\mathbb{Z}^2}\left(\overline{\mathbb{B}^2_{\sqrt{N}}}\right) - \pi N = O\left(N^{\frac{1}{4}+\varepsilon_2}\right), \tag{22.10}$$

where $0 < \varepsilon_2 \leq \frac{1}{4}$, for example, $\varepsilon_2 = 0.074009\ldots$ (G. Kolesnik [1985]) $\varepsilon_2 = 0.064903\ldots$ (M.N. Huxley [2003]).

22.2 Higher-Dimensional Variants of the Circle Problem

There are many perspectives to formulate variants of the Gaussian lattice point problem for the circle. It already was the merit of E. Landau [1924] to point out particularly interesting areas, such as

- *General two-dimensional lattices*

$$\Lambda = \{g = ng_1 + mg_2 \ : \ n, m \in \mathbb{Z}\} \tag{22.11}$$

with linearly independent $g_1, g_2 \in \mathbb{R}^2$ (see Figure 17.1) can be used instead of the lattice $\mathbb{Z}^2$,

- The *remainder term* can be represented as an alternating series, called the *Hardy–Landau series* in terms of the Bessel function J_1 of order 1 (for the different facets of the proof see, e.g., G.H. Hardy [1915], E. Landau [1915], C. Müller [1954a])

$$\sum_{\substack{|g|\le N\\ g\in\Lambda}}{}' 1 = \frac{\pi N^2}{\|\mathcal{F}_\Lambda\|} + \lim_{R\to\infty} \frac{\pi N^2}{\|\mathcal{F}_\Lambda\|} \sum_{\substack{0<|h|\le R\\ h\in\Lambda^{-1}}} \frac{J_1(2\pi|h|N)}{\pi|h|N}, \tag{22.12}$$

where Λ is an arbitrary lattice in $\mathbb{R}^2$. Moreover, the following convention

$$\sum_{\substack{|g|\le N\\ g\in\Lambda}}{}' 1 = \sum_{\substack{|g|< N\\ g\in\Lambda}} 1 + \frac{1}{2}\sum_{\substack{|g|= N\\ g\in\Lambda}} 1 \tag{22.13}$$

is used in lattice point theory (note that the last sum occurs only if there is a lattice point $g \in \Lambda$ with $|g| = N$).

- Lattice points can be affected by non-constant weights (see W. Freeden [1975, 1978a])

$$\sum_{\substack{|a+g|\le N\\ g\in\Lambda}}{}' e^{2\pi i y\cdot(a+g)} F(a+g) \tag{22.14}$$

$$= \lim_{R\to\infty} \frac{1}{\|\mathcal{F}_\Lambda\|} \sum_{\substack{|h-y|\le R\\ h\in\Lambda^{-1}}} e^{2\pi i a\cdot h} \int_{\substack{|x|\le N\\ x\in\mathbb{R}^2}} F(x)\, e^{-2\pi i x\cdot(h-y)}\, dy,$$

where $a, y \in \mathbb{R}^2$, F is twice continuously differentiable in $\overline{\mathbb{B}^2_N}$, $N > 0$, and the following convention has been used analogously to (22.13)

$$\sum_{\substack{|a+g|\le N\\ g\in\Lambda}}{}' \cdots = \sum_{\substack{|a+g|< N\\ g\in\Lambda}} \cdots + \frac{1}{2} \sum_{\substack{|a+g|= N\\ g\in\Lambda}} \cdots . \tag{22.15}$$

Note that, for $F = 1$, this formula leads back to

$$e^{2\pi i a\cdot y} \sum_{\substack{|g+a|\le N\\ g\in\Lambda}}{}' e^{2\pi i g\cdot y} = \lim_{R\to\infty} \frac{\pi N^2}{\|\mathcal{F}_\Lambda\|} \sum_{\substack{|h-y|\le R\\ h\in\Lambda^{-1}}} e^{2\pi i a\cdot h}\, \frac{J_1(2\pi|h-y|N)}{\pi|h-y|N}. \tag{22.16}$$

For $a = y = 0$, we obtain the *classical Hardy–Landau identity* (22.12) since J_1 satisfies the asymptotic relation $J_1(r) = \frac{r}{2} + \ldots$.

- Generalizations to lattices $\Lambda \subset \mathbb{R}^q$ and regular regions $\mathcal{G} \subset \mathbb{R}^q$, $q \geq 2$, and continuous functions on $\overline{\mathcal{G}} = \mathcal{G} \cup \partial\mathcal{G}$ can be formulated, for example, in Gaussian summability (see W. Freeden [2011] for the proof and a more detailed study)

$$\sum_{\substack{a+g\in\overline{\mathcal{G}}\\ g\in\Lambda}}{}' e^{2\pi i y\cdot(a+g)} F(a+g) = \lim_{\substack{\tau\to 0\\ \tau>0}} \frac{1}{\|\mathcal{F}_\Lambda\|} \sum_{h\in\Lambda^{-1}} e^{-\tau\pi^2 h^2} e^{2\pi i h\cdot a} \int_{\mathcal{G}} F(x) e^{-2\pi i x\cdot(h-y)}\, dx, \tag{22.17}$$

where $a, y \in \mathbb{R}^q$. Clearly, we have for $a, y \in \mathbb{R}^q$ that

$$\begin{aligned}\sum_{\substack{a+g\in\overline{\mathcal{G}}\\ g\in\Lambda}}{}' e^{2\pi i y\cdot(a+g)} F(a+g) &= \frac{1}{\|\mathcal{F}_\Lambda\|} \int_{\mathcal{G}} F(x) e^{2\pi i x\cdot y}\, dx \\ &\quad + \lim_{\substack{\tau\to 0\\ \tau>0}} \frac{1}{\|\mathcal{F}_\Lambda\|} \sum_{\substack{0<|h|\leq R\\ h\in\Lambda^{-1}}} e^{-\tau\pi^2 h^2} e^{2\pi i h\cdot a} \int_{\mathcal{G}} F(x) e^{-2\pi i x\cdot(h-y)}\, dx,\end{aligned} \tag{22.18}$$

where the following abbreviation has been used consistently

$$\sum_{\substack{a+g\in\overline{\mathcal{G}}\\ g\in\Lambda}}{}' \ldots = \sum_{\substack{a+g\in\mathcal{G}\\ g\in\Lambda}} \ldots + \sum_{\substack{a+g\in\partial\mathcal{G}\\ g\in\Lambda}} \alpha_{\mathcal{G}}(a+g) \ldots \tag{22.19}$$

with $\alpha_{\mathcal{G}}(a+g)$ denoting the *solid angle* subtended by $\partial\mathcal{G}$ at $a+g$.

Note that, as geoscientifically relevant regular regions, we may choose the interior of the (actual) Earth's body or parts of it, the interior of geoscientifically relevant surfaces such as the telluroid, geoid (see cover illustration), etc., but also ball, ellipsoid, cube, polyhedral bodies, etc., are included.

22.3 Multivariate Shannon Sampling

Let us continue with the observation that, for every $y \in \mathbb{R}^q$, the (for dimensions $q \geq 2$ formally understood) series

$$a \mapsto \lim_{N\to\infty} \frac{1}{\|\mathcal{F}_\Lambda\|} \sum_{\substack{|h-y|\leq N\\ h\in\Lambda^{-1}}} e^{2\pi i a\cdot h} \underbrace{\int_{\mathcal{G}} F(x) e^{-2\pi i x\cdot(h-y)}\, dx}_{F^\wedge_{\mathcal{G}}(h-y)}, \quad a \in \mathbb{R}^q \tag{22.20}$$

as well as the finite sum

$$a \mapsto \sum_{\substack{a+g\in\overline{\mathcal{G}}\\ g\in\Lambda}}{}' e^{2\pi i y\cdot(a+g)} F(a+g), \quad a \in \mathbb{R}^q \tag{22.21}$$

show Λ-periodicity, i.e., as functions of the variable $a \in \mathbb{R}^q$ they are periodic with respect to the lattice $\Lambda \subset \mathbb{R}^q$. As a consequence (see W. Freeden, M.Z. Nashed [2015] for the details), Shannon-type sampling procedures can be obtained by formal integration of the lattice point identity (22.17) over a regular region $\mathcal{H}$ that is not-necessarily equal to $\mathcal{G}$

$$\int_{\mathcal{H}} \sum_{\substack{a+g\in\overline{\mathcal{G}}\\ g\in\Lambda}}{}' e^{-2\pi i y\cdot(a+g)} F(a+g)\, da \tag{22.22}$$

$$= \sum_{\substack{(\mathcal{F}_\Lambda+\{g'\})\cap\overline{\mathcal{H}}\neq\emptyset\\ g'\in\Lambda}} \int_{\mathcal{G}\cap \bigcup\limits_{g\in\Lambda}(((\overline{\mathcal{H}}\cap(\mathcal{F}_\Lambda+\{g'\}))-\{g'\})+\{g\})} F(x) e^{-2\pi i y\cdot x}\, dx$$

$$= \sum_{\substack{(\mathcal{F}_\Lambda+\{g'\})\cap\overline{\mathcal{H}}\neq\emptyset\\ g'\in\Lambda}} F^{\wedge}_{\mathcal{G}\cap \bigcup\limits_{g\in\Lambda}(((\overline{\mathcal{H}}\cap(\mathcal{F}_\Lambda+\{g'\}))-\{g'\})+\{g\})}(y)$$

$$= \lim_{\substack{\tau\to 0\\ \tau>0}} \frac{1}{\|\mathcal{F}_\Lambda\|} \sum_{h\in\Lambda^{-1}} e^{-\tau\pi^2 h^2} \underbrace{\int_{\mathcal{G}} F(x) e^{-2\pi i h\cdot x}\, dx}_{=F^{\wedge}_{\mathcal{G}}(h)} \underbrace{\int_{\mathcal{H}} e^{2\pi i a\cdot(h-y)}\, da}_{=K_{\mathcal{H}}(h-y)}$$

The identity (22.22) has many interesting properties. For example, by virtue of the Gaussian summability, the convergence of the cardinal-type series on the right hand side of (22.22) is exponentially accelerated. All manifestations of over- and undersampling can be explicitly analyzed by the finite sum of Fourier transforms on the left side of the identity (see W. Freeden, M.Z. Nashed [2015]), dependent on the geometric configurations of the chosen regular regions $\mathcal{G}, \mathcal{H}$ (note that the identity (22.22) also seems to be unknown for the univariate case in this generality).

The Gaussian summability of the cardinal series on the right-hand side of (22.22) is of great importance from numerical point of view. It enables a fast computation of the series. Nonetheless, from W. Freeden, M.Z. Nashed [2015]) we know that the identity (22.22) additionally holds true in ordinary sense, i.e., we have

$$\int_{\mathcal{H}} \sum_{\substack{a+g\in\overline{\mathcal{G}}\\ g\in\Lambda}}{}' e^{-2\pi i y\cdot(a+g)} F(a+g)\, da \tag{22.23}$$

$$= \sum_{\substack{(\mathcal{F}_\Lambda+\{g'\})\cap\overline{\mathcal{H}}\neq\emptyset\\ g'\in\Lambda}} \int_{\mathcal{G}\cap \bigcup\limits_{g\in\Lambda}(((\overline{\mathcal{H}}\cap(\mathcal{F}_\Lambda+\{g'\}))-\{g'\})+\{g\})} F(x) e^{-2\pi i y\cdot x}\, dx$$

$$= \sum_{\substack{(\mathcal{F}_\Lambda + \{g'\}) \cap \overline{\mathcal{H}} \neq \emptyset \\ g' \in \Lambda}} F^\wedge_{\mathcal{G} \cap \bigcup_{g \in \Lambda} (((\overline{\mathcal{H}} \cap (\mathcal{F}_\Lambda + \{g'\})) - \{g'\}) + \{g\})}(y)$$

$$= \frac{1}{\|\mathcal{F}_\Lambda\|} \sum_{h \in \Lambda^{-1}} \underbrace{\int_{\mathcal{G}} F(x) e^{-2\pi i h \cdot x}\, dx}_{= F^\wedge_{\mathcal{G}}(h)} \underbrace{\int_{\mathcal{H}} e^{2\pi i a \cdot (h-y)}\, da}_{= K_{\mathcal{H}}(h-y)}$$

A simple, but significant case of (22.23) in Euclidean space $\mathbb{R}^q$ showing no phenomenon of aliasing is obtained under the special choice $\overline{\mathcal{G}} = \overline{\mathcal{H}} \subset \overline{\mathcal{F}_\Lambda}$ leading to the Shannon-type identity

$$F^\wedge_{\mathcal{G}}(y) = \frac{1}{\|\mathcal{F}_\Lambda\|} \sum_{h \in \Lambda^{-1}} F^\wedge_{\mathcal{G}}(h) K_{\mathcal{G}}(h-y), \quad y \in \mathbb{R}^q, \tag{22.24}$$

with

$$K_{\mathcal{G}}(x-y) = \int_{\mathcal{G}} e^{-2\pi i a \cdot (x-y)}\, da, \quad x, y \in \mathbb{R}^q. \tag{22.25}$$

In fact, the identity (22.24) is a multivariate variant of the Shannon sampling theorem (cf. C. E. Shannon [1949]), but now for (geoscientifically relevant) regions $\mathcal{G}$. The principal impact of Shannon sampling on information theory is that it allows the replacement of a bandlimited signal $F^\wedge_{\mathcal{G}}$ related to $\mathcal{G}$ by a discrete sequence of its samples without loss of any information. Also it specifies the lowest rate, i.e., the Nyquist rate (cf. W. Freeden, M.Z. Nashed [2015]), that it enables to reproduce the original signal. In other words, Shannon sampling provides the bridge between continuous and discrete versions of a bandlimited function.

The Shannon sampling theorem has many applications in engineering and physics, for example, in signal processing, data transmission, cryptography, constructive approximation. and inverse problems such as the multivariate discussion of a Fourier transform over a (geodetically relevant) regular region (see, e.g., M.Z. Nashed [1981] and the references therein for the univariate study of the antenna problem). For various perspectives and surveys on sampling expansions, see P.L. Butzer [1983], P. L. Butzer et al. [1988], P.L. Butzer, R.L. Stens [1992], J.R. Higgins [1985, 1996], J.R. Higgins, R.L. Stens [2000], J.A. Jerri [1969, 1976, 1977], A. Zayed [1993], and others. M.Z. Nashed, G.G. Walter [1991] introduced a new approach to general sampling theorems for functions in reproducing kernel Hilbert spaces and showed how many of the sampling results in the earlier literature are special cases of their approach. In another paper M.Z. Nashed, G.G. Walter [1995] showed how to construct a reproducing kernel Hilbert space from a function space that admits sampling expansions.

Finally, it should be remarked that sampling theory (in the sense of the representation of an analog signal in terms of its samples) has attracted considerable interest in the past three decades. Major advances in the theory and applications of sampling expansion in different function spaces have been made by mathematicians and engineers.

22.4 Paley–Wiener Integration

In modern systems, for example, in physical geosystems, we do not deal with continuous signals, but only with discrete versions. The classical theory of sampling aims at the following issue: Given discrete samples in lattice points, how can we capture the essential information about the underlying continuous phenomenon. The algebraic structure of a lattice relates Shannon sampling to lattice point theory. In the previous sections we made the attempt to work out their interrelations. New types of lattice point identities were proposed which, in turn, gave deeper insight in over- and undersampling processes in Euclidean spaces $\mathbb{R}^q$.

Many extensions of the sampling theorems as developed in our approach here can be studied in more detail. Some aspects of significance in cubature should be explained:

Restricting ourselves to regular regions $\mathcal{G} \subset \mathbb{R}^q$ with $\overline{\mathcal{G}} \subset \overline{\mathcal{F}_\Lambda}$, so that the continuous signal

$$F_{\mathcal{G}}^{\wedge}(y) = \int_{\mathcal{G}} F(a) e^{-2\pi i a \cdot y} \, da, \quad y \in \mathbb{R}^q \tag{22.26}$$

is recovered from the sampled signal over lattice points of the inverse lattice $F_{\mathcal{G}}^{\wedge}(h), h \in \Lambda^{-1}$ by (22.24). We are able to deduce some interesting results in the area of *approximate integration*: As a matter of fact, the bandlimited function $F_{\mathcal{G}}^{\wedge}$ allows to express its integral over the Euclidean space $\mathbb{R}^q$ by the product of the lattice density and the sum over all samples in points of the inverse lattice

$$\lim_{N \to \infty} \int_{\substack{|x| \leq N \\ x \in \mathbb{R}^q}} F_{\mathcal{G}}^{\wedge}(x) \, dx = \frac{1}{\|\mathcal{F}_\Lambda\|} \sum_{h \in \Lambda^{-1}} F_{\mathcal{G}}^{\wedge}(h). \tag{22.27}$$

Furthermore, the Parseval identity is valid:

$$\frac{1}{\|\mathcal{F}_\Lambda\|} \sum_{h \in \Lambda^{-1}} |F_{\mathcal{G}}^{\wedge}(h)|^2 = \int_{\mathcal{G}} |F(a)|^2 \, da. \tag{22.28}$$

From Fourier theory it follows that

$$\int_{\mathcal{G}} |F(a)|^2 \, da = \int_{\mathbb{R}^q} \left| F_{\mathcal{G}}^{\wedge}(y) \right|^2 \, dx. \tag{22.29}$$

In other words, if $F_{\mathcal{G}}^{\wedge}$ with $\overline{\mathcal{G}} \subset \overline{\mathcal{F}_\Lambda}$ belongs to the inner product space

$$C_{\overline{\mathcal{G}}}^{(0)} = \left\{ y \mapsto F_{\mathcal{G}}^{\wedge}(y) = \int_{\mathcal{G}} e^{-2\pi i a \cdot y} F(a) \, da : F \in \mathrm{C}^{(0)}\left(\overline{\mathcal{G}}\right), \; y \in \mathbb{R}^q \right\}, \tag{22.30}$$

then

$$\int_{\mathbb{R}^q} |F_{\mathcal{G}}^{\wedge}(y)|^2 \, dy = \frac{1}{\|\mathcal{F}_\Lambda\|} \sum_{h\in\Lambda^{-1}} |F_{\mathcal{G}}^{\wedge}(h)|^2. \tag{22.31}$$

Replacing Λ by its inverse lattice Λ^{-1} we find

$$\int_{\mathbb{R}^q} |F_{\mathcal{G}}^{\wedge}(y)|^2 \, dy = \|\mathcal{F}_\Lambda\| \sum_{g\in\Lambda} |F_{\mathcal{G}}^{\wedge}(g)|^2. \tag{22.32}$$

Looking at our approach critically we notice that Shannon sampling is formulated on the reference set $\mathrm{C}_{\overline{\mathcal{G}}}^{(0)}$, that is a strict subset of the associated Paley–Wiener space

$$\mathrm{B}_{\overline{\mathcal{G}}} = \left\{ y \mapsto \int_{\mathcal{G}} e^{-2\pi i a\cdot y} F(a) \, da, \ y \in \mathbb{R}^q : F \in \mathrm{L}^2(\mathcal{G}) \right\}. \tag{22.33}$$

This observation, however, does not bother us very much, since every $F \in \mathrm{L}^2(\mathcal{G})$ can be approximated (in $\mathrm{L}^2(\mathcal{G})$-sense) by a function $F_\varepsilon \in \mathrm{C}^{(0)}(\overline{\mathcal{G}})$ in ε-accuracy such that

$$\begin{aligned} &\sup_{y\in\mathbb{R}^q} \left| F_{\mathcal{G}}^{\wedge}(y) - (F_\varepsilon)_{\mathcal{G}}^{\wedge}(y) \right| \\ &\leq \sup_{y\in\mathbb{R}^q} \left| \int_{\mathcal{G}} e^{-2\pi i a\cdot y} (F(a) - F_\varepsilon(a)) \, da \right| \\ &\leq \left(\int_{\mathcal{G}} |F(a) - F_\varepsilon(a)|^2 \, da \right)^{1/2} \left(\int_{\mathcal{G}} |e^{-2\pi i a\cdot y}|^2 \, da \right)^{1/2} \\ &\leq \left(\int_{\mathbb{R}^q} \left| F_{\mathcal{G}}^{\wedge}(a) - (F_\varepsilon)_{\mathcal{G}}^{\wedge}(a) \right|^2 \, da \right)^{1/2} \left(\int_{\mathcal{G}} |e^{-2\pi i a\cdot y}|^2 \, da \right)^{1/2} \\ &= \sqrt{\|\mathcal{G}\|} \ \varepsilon. \end{aligned} \tag{22.34}$$

All in all, if $\mathcal{G} \subset \mathbb{R}^q$ is a regular region with $\overline{\mathcal{G}} \subset \overline{\mathcal{F}_\Lambda}$. Then the *Paley–Wiener space* $\mathrm{B}_{\overline{\mathcal{G}}}$ is the completion of the space $\mathrm{C}_{\overline{\mathcal{G}}}^{(0)}$ under the $\mathrm{L}^2(\mathbb{R}^q)$-topology:

$$\mathrm{B}_{\overline{\mathcal{G}}} = \overline{\mathrm{C}_{\overline{\mathcal{G}}}^{(0)}}^{\|\cdot\|_{\mathrm{L}^2(\mathbb{R}^q)}}. \tag{22.35}$$

Under the aforementioned assumption that $\mathcal{G}$ is a regular region with $\overline{\mathcal{G}} \subset \overline{\mathcal{F}_\Lambda}$, the set $\mathrm{B}_{\overline{\mathcal{G}}}$ forms a *reproducing kernel space* with kernel $K_{\mathcal{G}}$ of (22.25). In fact, by virtue of (22.29), we see that

$$|F_{\mathcal{G}}^{\wedge}(y)| \leq \sqrt{\|\mathcal{G}\|} \sqrt{\int_{\mathcal{G}} |F(x)|^2 \, dx} = \sqrt{\|\mathcal{G}\|} \sqrt{\int_{\mathbb{R}^q} |F_{\mathcal{G}}^{\wedge}(a)|^2 \, da}. \tag{22.36}$$

Moreover, standard Fourier inversion (see, e.g., M.A. Pinsky [2002], E.M. Stein, G. Weiss [1971]) yields

$$\begin{aligned} F_{\mathcal{G}}^{\wedge}(y) &= \int_{\mathbb{R}^q} F_{\mathcal{G}}^{\wedge}(x) \left(\int_{\mathcal{G}} e^{2\pi i a\cdot(x-y)} \, da \right) dx \\ &= \int_{\mathbb{R}^q} F_{\mathcal{G}}^{\wedge}(x) \, K_{\mathcal{G}}(x-y) \, dx \end{aligned} \tag{22.37}$$

for all $y \in \mathbb{R}^q$, where

$$\int_{\mathbb{R}^q} \cdots = \lim_{N\to\infty} \int_{\substack{|x|\le N \\ x\in\mathbb{R}^q}} \cdots . \tag{22.38}$$

Hence, $B_{\overline{\mathcal{G}}}$ is a reproducing kernel Hilbert space with kernel (22.25). Note that going over to the Paley–Wiener space $\mathrm{B}_{\overline{\mathcal{F}_\Lambda}}$ we are able to guarantee, in addition, that the reproducing kernels form an orthonormal system with discrete orthogonality property (see M.Z. Nashed, G.G. Walter [1995], M.Z. Nashed, Q. Sun [2010, 2013]).

Summarizing our number theoretical excursion starting from the Gaussian circle problem via the Hardy–Landau lattice point identities resulted in new geoscientifically relevant Shannon sampling procedures of high practical applicability enabling Paley–Wiener integration from discrete data, regularly located in lattice points. Moreover, we obtained the integration rules (22.31) and (22.32), which are remarkable results. These formulas tell us that band-limited functions of type (22.26) may be exactly integrated by lattice point sums.

22.5 Paley–Wiener Spline Interpolatory Integration

Under the aforementioned assumption that $\mathcal{G}$ is a regular region with $\overline{\mathcal{G}} \subset \overline{\mathcal{F}_\Lambda}$, standard Fourier inversion (see, e.g., C. Müller [1998]) guarantees that

$$\begin{aligned} F_{\mathcal{G}}^{\wedge}(y) &= \int_{\mathbb{R}^q} F_{\mathcal{G}}^{\wedge}(x) \left(\int_{\mathcal{G}} e^{2\pi i a\cdot(x-y)}\, da \right) dx \\ &= \int_{\mathbb{R}^q} F_{\mathcal{G}}^{\wedge}(x)\, K_{\mathcal{G}}(x-y)\, dx \end{aligned} \tag{22.39}$$

holds true for all $y \in \mathbb{R}^q$, where $\int_{\mathbb{R}^q} \ldots$ is understood in the following sense:

$$\int_{\mathbb{R}^q} \cdots = \lim_{N\to\infty} \int_{\substack{|x|\le N \\ x\in\mathbb{R}^q}} \cdots . \tag{22.40}$$

Hence, as we already know, the *Paley–Wiener space* $\mathrm{B}_{\overline{\mathcal{G}}}$

$$\mathrm{B}_{\overline{\mathcal{G}}} = \left\{ y \mapsto \int_{\mathcal{G}} e^{-2\pi i a\cdot y} F(a)\, da,\ y\in\mathbb{R}^q : F \in L^2(\mathcal{G}) \right\}. \tag{22.41}$$

is a reproducing kernel Hilbert space with the uniquely determined kernel

$$K_{\mathcal{G}}(x-y) = \int_{\mathcal{G}} e^{2\pi i a\cdot(x-y)}\, da. \tag{22.42}$$

The reproducing kernel framework enables us to realize minimum-norm-interpolation, i.e., *spline interpolation*, within the Paley–Wiener space $B_{\overline{\mathcal{G}}}$. More concretely, we are interested in finding a "smooth" spline interpolant to $F^{\wedge}_{\mathcal{G}} \in B_{\overline{\mathcal{G}}}$ from a given finite dataset

$$\{F^{\wedge}_{\mathcal{G}}(h) : h \in \Omega,\, \Omega \subset \Lambda^{-1}\}. \tag{22.43}$$

For that purpose, we consider the finite-dimensional space $\mathrm{Spline}^{\Omega}_{B_{\overline{\mathcal{G}}}}$ consisting of all *Paley–Wiener spline functions* $\hat{S}$ of the form

$$\hat{S}(y) = \sum_{h' \in \Omega} \hat{a}_{h'} \int_{\mathcal{G}} e^{2\pi i a\cdot(h'-y)}\, da, \quad y \in \mathbb{R}^q, \quad \hat{a}_{h'} \in \mathbb{C}. \tag{22.44}$$

It is easy to deduce from the theory of Gramian determinants (see, e.g., P.J. Davis, P. Rabinowitz [1975]) that there is one and only one spline in $\mathrm{Spline}^{\Omega}_{B_{\overline{\mathcal{G}}}}$, denoted by $\hat{S}_F$, with coefficients $\hat{a}^F_h$ satisfying the linear equations

$$\sum_{h' \in \Omega} \hat{a}^F_{h'} \int_{\mathcal{G}} e^{2\pi i a\cdot(h'-h)} da \;=\; F^{\wedge}_{\mathcal{G}}(h), \quad h \in \Omega. \tag{22.45}$$

$\hat{S}_F$ fulfills the *minimum-norm-property*

$$\left(\int_{\mathcal{G}} \left|\hat{S}_F(y)\right|^2 dy\right)^{1/2} = \min_{\substack{\hat{S}(h)=F^{\wedge}_{\mathcal{G}}(h) \\ \hat{S}\in B_{\overline{\mathcal{G}}},\, h\in\Omega}} \left(\int_{\mathcal{G}} |\hat{S}(y)|^2\, dy\right)^{1/2}.$$

As a consequence, suppose that the dataset

$$\left\{F^{\wedge}_{\mathcal{G}}(h) = \int_{\mathcal{G}} e^{-2\pi i a\cdot h} F(a)\, da : h \in \Omega, \Omega \subset \Lambda^{-1}\right\} \tag{22.46}$$

is given such that $\hat{S}_F$ is "close" to $F^{\wedge}_{\mathcal{G}}$, for example, if the nodal width of Ω is "small enough". Then, the Fourier inversion formula yields the identity

$$\begin{aligned} \alpha_{\mathcal{G}}(x)\, F(x) &\approx \int_{\mathbb{R}^q} e^{2\pi i x\cdot y}\, \hat{S}_F(y) dy \\ &= \sum_{h' \in \Omega} \hat{a}^F_{h'} \int_{\mathbb{R}^q} e^{2\pi i x\cdot y} \int_{\mathcal{G}} e^{2\pi i a\cdot(h'-y)}\, da\, dy \\ &= \sum_{h' \in \Omega} \hat{a}^F_{h'}\, e^{2\pi i x\cdot h'}, \quad x \in \overline{\mathcal{G}}, \end{aligned} \tag{22.47}$$

with coefficients $\hat{a}^F_h$ determined from the already known linear (spline) equations (22.45). In other words, our spline interpolation technique approximately solves the multivariate inversion problem of determining the function $\alpha_{\mathcal{G}} F \in \mathrm{L}^2(\mathcal{G})$ from a finite set of discrete values of $F^{\wedge}_{\mathcal{G}}$. Note that combined spline interpolation/smoothing (as proposed by W. Freeden, B. Witte [1982]) can be used instead of spline interpolation if the data are only (partially) noisy.

Integrating (22.47) over $\mathcal{G}$, we finally arrive at a Paley–Wiener (spline interpolation) cubature rule of the form

$$\int_{\mathcal{G}} F(x)\, dx \approx \sum_{h' \in \Omega} \hat{a}^F_{h'}\, K_{\mathcal{G}}(h'), \tag{22.48}$$

that is of high importance in geomathematically relevant practice.

Bibliography

Abeyratne, M.K., Freeden, W., Mayer, C., Multiscale Deformation Analysis by Cauchy-Navier Wavelets. *J. Appl. Math.*, **2003**:605–645, 2003.

Abramowitz, M., Stegun, I.A., *Handbook of Mathematical Functions.* Dover Publications, Inc., New York, 1972.

Ahrens, C., Beylkin, G., Rotationally Invariant Quadratures for the Sphere. *Proc. R. Soc. A.*, **465**:3103–3125, 2009.

Alberink, I.B., Bentkus, V., Berry-Esseen Bounds for von Mises and U-Statistics. *Liet. Mat. Rink.*, **41**:1–20, 2001.

Albrecht, J., Collatz, L., Zur numerischen Auswertung mehrdimensionaler Integrale. *ZAMM*, **38**:1–15, 1958.

Alfeld, P., Neamtu, M., Schumaker, L., Bernstein-Bézier Polynomials on Spheres and Sphere-Like Surfaces. *Comput. Aided Geom. Design (CAGD)*, **13**:333–349, 1996.

Anders, E.B., An Extension of Romberg Integration Procedures to N-Variables. *J. Assoc. Comput. Machinery (JACM)*, **13**:505–510, 1966.

Andrews, L.C., *Special Functions for Engineers and Applied Mathematicians.* Macmillan, New York, 1985.

Antes, H., Über die Romberg-Integration in n Dimensionen. *Computing*, **9**:203–209, 1972.

Antoine, J.-P., Demanet, L., Jaques, L., Vandergheynst, P., Wavelets on the Sphere: Implementations and Approximations. *Appl. Comput. Harm. Anal. (ACHA)*, **13**:177–200, 2002.

Antoine, J.-P., Vandergheynst, P., Wavelets on the 2-Sphere: A Group-Theoretic Approach. *Appl. Comput. Harmon. Anal.*, **7**:1–30, 1999.

Apostol, T.M., An Elementary View of Euler's Summation Formula. *The American Mathematical Monthly*, **106**:409–418, 1999.

Arenstorf, R., Johnson, D., Uniform Distribution of Integral Points on 3-Dimensional Spheres via Modular Forms. *J. Number Theory*, **11**:218–238, 1979.

Aronszajn, N., Theory of Reproducing Kernels. *Trans. Am. Math. Soc.*, **68**:337–404, 1950.

Aronszajn, N., Creese, T.M., Lipkin, L.J., *Polyharmonic Functions.* Clarendon Press, Oxford, 1983.

Atkinson, K., *An Introduction to Numerical Analysis.* John Wiley & Sons, New York, 1998.

Atkinson, K., Han, W., *Spherical Harmonics and Approximations on the Unit Sphere: An Introduction.* Springer, Berlin, Heidelberg, New York, 2012.

Atkinson, K., Sommariva, A., Quadrature Over the Sphere. *Electron. Trans. Numer. Anal. (ETNA)*, **20**:104–118, 2005.

Augustin, M., Freeden, W., Gerhards, C., Möhringer, S., Ostermann, I., Mathematische Methoden in der Geothermie. *Math. Semesterber.*, **59**:1–28, 2012.

Ayoub, R., Euler and the Zeta Function. *American Mathematical Monthly*, **81**:1067–1086, 1974.

Bajnok, B., Construction of Design on 2-Sphere. *European J. Combin.*, **12**:377–404, 1991.

Banday, A.J., Zaroubi, S., Bartelmann, M. (eds.), Mining the Sky. *ESO Astrophysics Symposia*, Springer, Berlin, Heidelberg, 2001.

Bannai, E., Damerell, R.M., Tight Spherical Designs I. *J. Math. Soc. Japan*, **31**:199–207, 1979.

Bannai, E., Damerell, R.M., Tight Spherical Designs II. *J. London Math. Soc.*, **21**(2):13–30, 1980.

Bannai, Ei., Bannai, Et., A Survey on Spherical Designs and Algebraic Combinatorics on Spheres. *European J. Combin.*, **30**:1392–1425, 2009.

Bellomo, N., Gatignol, R. (eds.), *Lecture Notes on the Discretization of the Boltzmann Equation.* World Scientific, 2003.

Benedetto, J.J., Ferreira, P.J.S.G. (eds.), *Modern Sampling Theory: Mathematics and Applications.* Birkhäuser, Boston, 2001.

Bennett, C.L., Banday, A., Gorski, K.M., Hinshaw, G., Jackson, P., Keegstra, P., Kogut, A., Smoot, G.F., Wilkinson, D.T., Wright, E.I., 4-Year COBE DMR Cosmic Microwave Background Observations: Maps and Basic Results. *Astrophys. J. Lett.*, **464**:1–4, 1996.

Berens, H., Butzer, P.L., Pawelke, S., Limitierungsverfahren mehrdimensionaler Kugelfunktionen und deren Saturationsverfahren. *Publ. Res. Inst. Math. Sci. Ser. A*, **4**:201–268, 1968.

Berz, E., *Verallgemeinerte Funktionen und Operatoren.* BI-Hochschultaschenbücher, 122/122a, Mannheim, 1967.

Biondi, B.L., *Three-Dimensional Seismic Imaging.* Society of Exploration Geophysicists, Tulsa, 2006.

Boal, N., Sayas, F.-J., Adaptive Numerical Integration on Spherical Triangles. *Monografias del Seminario Matemático Garcia de Galdeano*, **31**:61–69, 2004.

Bochner, S., Positive Zonale Functions on Spheres. *Proc. Nat. Acad. Sci. USA*, **40**:1141–1147, 1954.

Boyd, J.P., *Chebyshev and Fourier Spectral Methods.* Dover Publications, Mineola, NY, 2001.

Brauchart, J.S., Points on an Unit Sphere in $\mathbb{R}^{d+1}$, Riesz Energy, Discrepancy and Numerical Integration. PhD-thesis, Graz University of Technology (TU-Graz), 2005.

Buckley, M.J., Eagleson, G.K., An Approximation to the Distribution of Quadratic Forms in Normal Random Variables. *Austral. J. Statist., Spec. Issue*, **30A**:150–159, 1988.

Buhmann, M.D., *Radial Basis Functions on Grids and Beyond.* Intern. Workshop on Mesh Free Methods, 2003.

Burlisch, R., Bemerkungen zur Romberg-Integration. *Numer. Math.*, **6**:6–16, 1964.

Butzer, P.L., Nessel, R., *Fourier Analysis and Approximation Theory.* Birkhäuser, Basel, 1971.

Butzer, P.L., A Survey of the Whittaker Shannon Sampling Theorem and Some of its Extensions. *J. Math. Res. Exposition*, **3**:185–212, 1983.

Butzer, P.L., Splettstösser, W., Stens, R.L., The Sampling Theorem and Linear Prediction in Signal Analysis. *Jahresber. Dtsch. Math.-Ver.*, **90**:1–60, 1988.

Butzer, P.L., Stens, R.L., Sampling Theory for not Necessarily Band-Limited Functions: a Historical Overview. *SIAM Rev.*, **34**:40–53, 1992.

Butzer, P.L., Stens, R.L., The Euler-MacLaurin Summation Formula, the Sampling Theorem, and Approximate Integration over the Real Axis. *Linear Algebra Appl.*, **52/53**:141–155, 1983.

Caflisch, R.E., Monte Carlo and Quasi-Monte Carlo Methods. *Acta Numerica*, **7**:1–49, 1998.

Cassels, J.W.S., *An Introduction to the Geometry of Numbers.* Springer, Berlin, Heidelberg, New York, 1968.

Chen, D., Menegatto, V., Sun, X., A Necessary and Sufficient Condition for Strictly Positive Definite Functions on Spheres. *Proc. Amer. Math. Soc.*, **131**:2733–2740, 2003.

Chen, X., Frommer, A., Lang, B., *Computational Existence Proofs for Spherical t-Designs.* Department of Applied Mathematics, The Hong Kong Polytechnic University, 2009.

Chen, X., Womersley, R., Existence of Solutions to Systems of Underdetermined Equations and Spherical Designs. *SIAM J. Numer. Anal.*, **44**:2326–2341, 2006.

Choirat, C., Seri, R., The Asymptotic Distribution of Quadratic Discrepancies. In: Niederreiter, H., Talay, D. (eds.), *Monte Carlo and Quasi-Monte Carlo Methods.* Springer, 61–76, 2006.

Choirat, C., Seri, R., Computational Aspects of Cui-Freeden Statistics for Equidistribution on the Sphere. *Math. Comput.*, **82**:2137–2156, 2013.

Clenshaw, C.W., A Note on the Summation of Chebyshev Series. *Math. Comp.*, **9**:118–120, 1955.

Cohn, H., Kumar, A., Universally Optimal Distribution of Points on Spheres. *J. Amer. Math. Soc.*, **20**:99–148, 2007.

Cools, R., *Constructing Cubature Formulae: The Science Behind the Art.* Acta Numerica, 1–54, 1997.

Cotes, R., *Harmonia Mensurarum.* London, 1722.

Courant, R., Hilbert, D., *Methoden der Mathematischen Physik I, II.* Springer, Berlin, 2nd ed., 1968.

Coxeter, H.S.M., *Introduction to Geometry.* Wiley, 1961.

Cui, J., *Finite Pointset Methods on the Sphere and Their Application in Physical Geodesy.* PhD-thesis, University of Kaiserslautern, Geomathematics Group, 1995.

Cui, J., Freeden, W., Equidistribution on the Sphere. *SIAM J. Sci. Stat. Comput.*, **18**:595–609, 1997.

Dahlke, S., Maass, P., Continuous Wavelet Transforms with Applications to Analyzing Functions on Spheres. *J. Fourier Anal. Appl.*, **2**:379–396, 1996.

Dahlke, S., Dahmen, W., Schmitt, W., Weinreich, I., Multiresolution Analysis and Wavelets on S^2 and S^3. *Numer. Func. Anal. Opt.*, **16**:19–41, 1995.

Dahmen, W., *Wavelet and Multiscale Methods for Operator Equations.* Acta Numerica, 55–228, 1997.

Dahmen, W., Kurdila, A., Oswald, P., *Multiscale Wavelet Methods for Partial Differential Equations (Wavelet Analysis and Its Applications).* Academic Press, New York, 1997.

Damelin, S.B., A Walk through Energy, Discrepancy, Numerical Integration and Group Invariant Measures on Measurable Subsets of Euclidean Space. *Numerical Algorithms*, **48**:213–235, 2008.

Damelin, S.B., Grabner, P.J., Energy Functionals, Numerical Integration and Asymptotic Equidistribution on the Sphere. *J. Complexity*, **19**:231–246, 2003 (Corrigendum, J. Complexity, **20**:883–884, 2004).

Daubechies, I., Sweldens, W., Factoring Wavelet Transform into Lifting Steps. In: Klees, R., Haagmans, R. (eds.), Wavelets in Geosciences. *Lecture Notes in Earth Sciences*, 90, 131–157, 1999.

Davies, R.B., Numerical Inversion of a Characteristic Function. *Biometrica*, **60**:415–417, 1973.

Davis, P.J., *Interpolation and Approximation.* Blaisdell Publishing Company, Waltham, MA, 1963.

Davis, P.J., Rabinowitz, P., *Numerical Integration.* Blaisdell, Toronto, London, 1967.

Davis, P.J., Rabinowitz, P., Methods of Numerical Integration. *Computer Science and Applied Mathematics*, Academic Press, New York, 1975.

Davis, P.J., Rabinowitz, P., *Methods of Numerical Integration.* 2nd ed. Academic Press, Orlando, 1984.

Delsarte, P., Goethals, J., Seidel, J., Spherical Codes and Designs. *Geom. Dedicata*, **6**:363–388, 1977.

Deuflhard, P., Bornemann, F., *Numerische Mathematik II.* Walter de Gruyter, Berlin, New York, 1994.

Deuflhard, P., Bornemann, F., Scientific Computing With Ordinary Differential Equations. *Texts in Applied Mathematics*, Vol. 42. Springer, 2002.

Deuflhard, P., Hohmann, A., *Numerische Mathematik, eine algorithmisch orientierte Einführung.* Walter de Gruyter, Berlin, New York, 1991.

Driscoll, J.R., Healy, R.M., Computing Fourier Transforms and Convolutions on the 2-Sphere. *Advances in Applied Mathematics*, **15**:202–250, 1994.

Duchon, J., Spline Minimizing Rotation-Invariant Seminorms in Sobolev Spaces. In: Schempp, W., Zeller, K. (eds.), *Constructive Theory of Functions of Several Variables*. Springer, 85–100, 1977.

Dunham, W., Euler: The Master of Us All. *Mathematical Association of America*, Washington D.C., 1999.

Dyn, N., Interpolation and Approximation by Radial and Related Functions. In: Chui, C., Schumaker, L.L., Ward, J.D. (eds.), *Approximation Theory VI*. Vol. 1. Academic Press, 211–234, 1989.

Engels, H., *Numerical Quadrature and Cubature*. Academic Press, London, New York, Toronto, Sydney, San Francisco, 1980.

Englis, M., Peetre, J., Green's Functions for Powers of the Invariant Laplacian, Canad. *J. Math.*, **50**:40–73, 1998.

Erdelyi, A., Magnus, W., Oberhettingen, F., Tricomi, F.G., *Higher Transendental Functions. Vol. 2, Bateman Manuscript Project*, California Institute of Technology, MacGraw-Hill, New York, Toronto, London, 1953.

Eskin, G.I., *Boundary Value Problems for Elliptic Pseudodifferential Equations*. Translations of Math. Monographs, Vol. 52. American Mathematical Society, Providence, RI, 1981.

Euler, L., Methodus universalis serierum convergentium summas quam proxime inveniendi. *Commentarii academie scientiarum Petropolitanae*, **8**:3–9, Opera Omnia (XIV): 101–107, 1736a.

Euler, L., Methodus universalis series summandi ulterius promota, *Commentarii academie scientarium Petropolitana*, **8**:147–158; Opera Omnia (XIV), 124–137, 1736b.

Evans, G., *Practical Numerical Integration*. John Wiley & Sons, Chichester, New York, Brisbane, Toronto, Singapore, 1993.

Fasshauer, G.E., Schumaker, L.L., Scattered Data Fitting on the Sphere. In: Dahlen, M., Lyche, T., Schumaker, L.L. (eds.), *Mathematical Methods for Curves and Surfaces II*. Vanderbilt University, Nashville, 117–166, 1998.

Fehlinger, T., *Multiscale Formulations for the Disturbing Potential and the Deflections of the Vertical in Locally Reflected Physical Geodesy*. PhD-thesis, University of Kaiserslautern, Geomathematics Group, 2009.

Fehlinger, T., Freeden, W., Gramsch, S., Mayer, C., Michel, D., Schreiner, M., *Local Modelling of Sea Surface Topography From (Geostrophic) Ocean Flow. Z. Angew. Math. Mech. (ZAMM)*, **87**:775–791, 2007.

Felippa, C., A Compendium of FEM Integration Formulas for Symbolic Work. *Engineering Computation*, **21**:867–890, 2004.

Fengler, M.J., Freeden, W., A Nonlinear Galerkin Scheme Involving Vector and Tensor Spherical Harmonics for Solving the Incompressible Navier–Stokes Equation on the Sphere. *SIAM J. Sci. Comput.*, **27**:967–994, 2005.

Filbir, F., Themistaclakis, W., Polynomial Approximation on the Sphere Using Scattered Data. *Math. Nachr.*, **281**(5):650–668, 2008.

Fisher, N.I., Lewis, T., Embleton, B.J.J., *Stastical Analysis of Spherical Data.* Cambridge University Press, Cambridge, 1993.

Fliege, J., Maier, U., The Distribution of Points on the Sphere and Corresponding Cubature Formulae. *IMA J. Numer. Anal.* **19**(2):317–334, 1999.

Folland, G.B., How to Integrate a Polynomial Over a Sphere. *Am. Math. Mon.*, **108**:446–448, 2001.

Freeden, W., Eine Verallgemeinerung der Hardy-Landauschen Identität. PhD-thesis, RWTH Aachen University, 1975.

Freeden, W., Über eine Verallgemeinerung der Hardy-Landauschen Identität. *Manuscr. Math.*, **24**:205–216, 1978a.

Freeden, W., Über gewichtete Gitterpunktsummen in kreisförmigen Bereichen. *Mitt. Math. Seminar Giessen*, **132**:1–22, 1978b.

Freeden, W., An Application of a Summation Formula to Numerical Computation of Integrals Over the Sphere. *Bull. Géod.*, **52**:165–175, 1978c.

Freeden, W., *Über eine Klasse von Integralformeln der Mathematischen Geodäsie.* Veröff. Geod. Inst., RWTH Aachen University, Report No. 27, 1979.

Freeden, W., Über die Gaußsche Methode zur angenäherten Berechnung von Integralen. *Math. Method. Appl. Sci.*, **2**:397–409, 1980a.

Freeden, W., On Integral Formulas of the (Unit) Sphere and Their Application to Numerical Computation of Integrals. *Computing*, **25**:131–146, 1980b.

Freeden, W., On Spherical Spline Interpolation and Approximation. *Math. Meth. in the Appl. Sci.*, **3**:551–575, 1981a.

Freeden, W., On Approximation by Harmonic Splines. *Manuscr. Geod.*, **6**:193–244, 1981b.

Freeden, W., Eine Klasse von Kubaturformeln der Einheitssphäre. *Zeitschrift für Vermessungswesen (ZfV)*, **106**:200–210, 1981c.

Freeden, W., Multidimensional Euler Summation Formulas and Numerical Cubature. *ISNM*, **57**:77–88, 1982a.

Freeden, W., *On the Permanence Property in Spherical Spline Interpolation.* The Ohio State University, Department of Geodetic Science, Columbus, Ohio, OSU Report No. 341, 1982b.

Freeden, W., Spherical Spline Interpolation: Basic Theory and Computational Aspects. *J. Comput. Appl. Math.*, **11**:367–375, 1984a.

Freeden, W., Ein Konvergenzsatz in sphärischer Spline-Interpolation. *Zeitschrift für Vermessungswesen (ZfV)*, **109**:569–576, 1984b.

Freeden, W., A Spline Interpolation Method for Solving Boundary Value Problems of Potential Theory from Discretely Known Data. *Numer. Math. Partial Diff. Equations*, **3**:375–398, 1987.

Freeden, W., Interpolation by Multidimensional Periodic Splines. *J. Approx. Theory*, **55**:104–117, 1988.

Freeden, W., Spherical Spline Approximation and its Application in Physical Geodesy. In: Vogel, A., Ofeagbu, C.O., Gorenflo, R., Ursin, B. (eds.), *Geophysical Data Inversion Methods and Applications.* Vieweg Publication, 79–104, 1990.

Freeden, W., *Multiscale Modelling of Spaceborne Geodata.* B.G. Teubner, Leipzig, 1999.

Freeden, W., *Metaharmonic Lattice Point Theory.* CRC Press, Chapman & Hall, Boca Raton, FL, 2011.

Freeden, W., Geomathematics: Its Role, Its Aim, and Its Potential. In: Freeden, W., Nashed, M.Z., Sonar, T. (eds.), *Handbook of Geomathematics.* Vol. 1, 2nd ed. Springer, New York, Berlin, Heidelberg, 3–79, 2015.

Freeden, W., Fleck, J., Numerical Integration by Means of Adapted Euler Summation Formulas. *Numer. Math.*, **51**:37–64, 1987.

Freeden, W., Gerhards, C., *Geomathematically Oriented Potential Theory.* CRC Press, Chapman & Hall, Boca Raton, FL, 2013.

Freeden, W., Gerhards, C., Romberg Extrapolation for Euler Summation-Based Cubature on Regular Regions. *Int. J. Geomath.*, DOI 10.1007/s13137-017-0097-4, 2017.

Freeden, W., Gutting, M., *Special Functions of Mathematical (Geo-) Physics.* Birkhäuser, Basel, 2013.

Freeden, W., Hermann, P., Some Reflections on Multi-dimensional Euler and Poisson Summation Formulas. *ISNM*, **75**:166–179, 1985.

Freeden, W., Hermann, P., Uniform Approximation by Spherical Spline Interpolation. *Math. Z.*, **193**:265–275, 1986.

Freeden, W., Hesse, K., On the Multiscale Solution of Satellite Problems by Use of Locally Supported Kernel Functions Corresponding to Equidistributed Data on Spherical Orbits. *Studia Scientiarum Mathematicarum Hungarica*, **39**:37–74, 2002.

Freeden, W., Kersten, H., Best Approximate Integration Formulas for Boundary Surfaces of Star-Shaped Bodies. *Boll. Geod. e Scienze Affini*, **39**:37–48, 1980.

Freeden, W., Maier, T., On Multiscale Denoising of Spherical Functions: Basic Theory and Numerical Aspects. *Electron. Trans. Numer. Anal. (ETNA)*, **14**:40–62, 2002.

Freeden, W., Mayer, C., Wavelets Generated by Layer Potentials. *Appl. Comput. Harm. Analysis*, **14**:195–237, 2003.

Freeden, W., Michel, V., Constructive Approximation and Numerical Methods in Geodetic Research Today – An Attempt at a Categorization Based on an Uncertainty Principle. *J. Geodesy*, **73**:452–465, 1999.

Freeden, W., Michel, V., *Multiscale Potential Theory (With Applications to Geoscience)*. Birkhäuser Verlag, Boston, Basel, Berlin, 2004.

Freeden, W., Nashed, M.Z., Multi-Variate Hardy-Type Lattice Point Summation and Shannon-Type Sampling. *Int. J. Geomath.*, **6**:163–249, 2015.

Freeden, W., Nashed, M.Z., Operator-Theoretic and Regularization Approaches to Ill-Posed Problems. *Handbook of Mathematical Geodesy*, Geosystems Mathematics, Birkhäuser Basel, 2017 (in print).

Freeden,W., Nutz, H., Satellite Gravity Gradiometry as Tensorial Inverse Problem. *Int. J. Geomath.*, **2**:177–218, 2011a.

Freeden,W., Nutz, H., Mathematik als Schlüsseltechnologie zum Verständnis des Systems "Tiefe Geothermie". *Jahresber. Dtsch. Math.-Ver.*, **117**:45–84, 2011b.

Freeden, W., Ostermann, I., Integration on Three-Dimensional Regular Regions Based on (Modified) Euler Summation. *Numer. Func. Anal. Opt.*, **34**(6):613–634, 2013.

Freeden, W., Reuter, R., A Class of Multi-Dimensional Periodic Splines. *Manuscr. Math.*, **35**:371–386, 1981.

Freeden, W., Reuter, R., Remainder Terms in Numerial Integration Formulas of the Sphere. *ISNM*, **61**:151–170, 1982.

Freeden, W., Reuter, R., Spherical Harmonic Splines: Theoretical and Computational Aspects. In: Brosowski, B., Martensen, E. (eds.), *Methoden und Verfahren der mathematischen Physik*, Bd. 27, 79–103, 1983.

Freeden, W., Schneider, F., Regularization Wavelets and Multiresolution. *Inverse Probl.*, **14**:225–243, 1998.

Freeden, W., Schreiner, M., Non-Orthogonal Expansions on the Sphere. *Math. Meth. Appl. Sci.*, **18**:83–120, 1995.

Freeden, W., Schreiner, M., Orthogonal and Non-orthogonal Multiresolution Analysis, Scale Discrete and Exact Fully Discrete Wavelet Transform on the Sphere. *Constr. Approx.*, **14**:493–515, 1997.

Freeden, W., Schreiner, M., Local Multiscale Modelling of Geoid Undulations From Deflections of the Vertical. *J. Geodesy*, **79**:641–651, 2006.

Freeden, W., Schreiner, M., Biorthogonal Locally Supported Wavelets on the Sphere based on Zonal Kernel Functions. *J. Fourier Anal. Appl.*, **13**(6):693–709, 2007.

Freeden, W., Schreiner, M., *Spherical Functions of Mathematical Geosciences.* A Scalar, Vectorial, and Tensorial Setup. Springer, Berlin, Heidelberg, 2009.

Freeden, W., Windheuser, U., Spherical Wavelet Transform and Its Discretization. *Adv. Comput. Math.*, **5**:51–94, 1996.

Freeden, W., Windheuser, U., Combined Spherical Harmonic and Wavelet Expansion. *Appl. Comput. Harm. Anal. (ACHA)*, **4**:1–37, 1997.

Freeden, W., Witte, B., A Combined (Spline-) Interpolation and Smoothing Method for the Determination of the Gravitational Potential from Heterogeneous Data. *Bull. Géod.*, **56**:53–62, 1982.

Freeden, W., Fehlinger, T., Mayer, C., Schreiner, M., On the Local Multiscale Determination of the Earth's Disturbing Potential From Discrete Deflections of the Vertical. *Comput. Geosc.*, **4**:473–490, 2008.

Freeden, W., Gervens, T., Mason, J.C., A Minimum Norm Interpolation Method for Determining the Displacement Field of a Homogeneous Isotropic Elastic Body from Discrete Data. *IMA J. App. Math.*, **44**:55–76, 1990.

Freeden, W., Gervens, T., Schreiner, M., *Constructive Approximation on the Sphere (With Applications to Geomathematics).* Oxford Science Publications, Clarendon, Oxford, 1998.

Freeden, W., Mason, J.C., Uniform Piecewise Approximation on the Sphere. In: Mason, J.C., Cox, M.G. (eds.), *Algorithms for Approximation II.* Chapman & Hall, 320–333, 1990.

Freeden, W., Michel, V., Nutz, H., Satellite-to-Satellite Tracking and Satellite Gravity Gradiometry (Advanced Techniques for High–Resolution Geopotential Field Determination). *J. Engineering Math.*, **43**:19–56, 2002.

Freeden, W., Nashed, M.Z., Sonar, T. (eds.), *Handbook of Geomathematics*. Springer Verlag, Heidelberg, Vol. 1 and 2, 2010, 2nd edition, 2015.

Freeden, W., Schreiner, M., Franke, R., A Survey on Spherical Spline Approximation. *Surv. Math. Ind.*, **7**:29–85, 1996.

Freud, G., Error Estimates for Gauss-Jacobi Quadrature Formulae. In: Miller, J.J.H. (ed.), *Topics in Numerical Analysis*. Academic Press, London, 113–121, 1973.

Freund, R.W., Hoppe, R.W., *Stoer/Burlisch: Numerische Mathematik 1*. Springer, New York, Heidelberg, Berlin, 2007.

Fricker, F., Geschichte des Kreisproblems. *Mitt. Math. Sem. Giessen*, **111**: 1–34, 1975.

Fricker, F., *Einführung in die Gitterpunktlehre*. Basel, Birkhäuser, 1982.

Ganesh, M., Graham, I.G., Sivaloganathan, J., A New Spectral Boundary Integral Collocation Method for Three-Dimensional Potential Problems. *SIAM J. Numer. Anal.*, **35**:778–805, 1998.

Ganesh, M., Mhaskar, H.N., Matrix-Free Interpolation on the Sphere. *SIAM J. Numer. Anal.*, **44**:1314–1331, 2006.

Ganesh, M., Dominguez, V., Interpolation and Cubature Approximation and Analysis for a Class of Wideband Integrals on the Sphere, *Adv. Comput. Math.*, **39**:547–584, 2013.

Ganesh, M., Le Gia, Q.T., Sloan, I.H., A Pseudospectral Quadrature Method for Navier-Stokes Equations on Rotating Spheres. *Math. Comp.*, **80**:1397–1430, 2011.

Gauss, C.F., *Untersuchungen über Höhere Arithmetik (Disquisitiones Arithmeticae (Latin))*, Fleischer, G. jun., Leipzig, also: Gauss Werke, Band 1, 1801.

Gauss, C.F., *Disquisitiones generales circa seriem infinitam* $1 + ((\alpha \cdot \beta)/(1 \cdot \gamma)x + \ldots$ *etc.* Commentationes Societatis Regiae Scientarium Gottingensis Recentiores 2, Werke III, 123–162, 1812.

Gauss, C.F., *Methodus nova integralium valores per approximationem inveniendi*. Commentationes Societatis Regiae Scientarium Gottingensis Recentiores 3, Werke III, 163–196, 1814.

Gauss, C.F., De nexu inter multitudinem classicum, in quas formae binariae secondi gradus distribuuntur, earumque determinantem. *Werke*, **2**:269–291, 1826.

Gautschi, W., A Survey of Gauss-Christoffel Quadrature Formulas. In: Butzer, P.L., Feher, F. (eds.), E.B. Christoffel: *The Influence of His Work on Mathematics and the Physical Sciences.* Birkhäuser, Basel, 72–147, 1981.

Gautschi, W., *Orthogonal Polynomials: Computation and Approximation.* Oxford University Press, New York, 2004.

Gerhards, C., *Spherical Multiscale Methods in Terms of Locally Supported Wavelets: Theory and Application to Geomagnetic Modeling.* PhD-thesis, University of Kaiserslautern, Geomathematics Group, 2011.

Gerhards, C., Locally Supported Wavelets for the Separation of Vector Fields with Respect to Their Sources. *Int. J. Wavelets Multiresolut. Inf. Process.*, **10**, 1250034 (26 pages), 2012.

Gerhards, C., Multiscale Modeling of the Geomagnetic Field and Ionospheric Currents. In: Freeden, W., Nashed, M.Z., Sonar, T. (eds.), *Handbook of Geomathematics.* Springer, 2nd edition, 539–563, 2015.

Gine, E., Invariant Tests for Uniformity on a Compact Riemannian Manifolds Based on Sobolev Norms. *Ann. Statist.*, **3**(6):1243–1266, 1975.

Goldstine, H.H., *A History of Numerical Analysis from the 16th through the 19th Century.* Springer Verlag, New York, 1977.

Górski, K.M., Hivon, E., Banday, A.J., Wandelt, B.D., Hansen, F.K., Reinecke, M., Bartelmann, M., HEALPix: A Framework for High-Resolution Discretization and Fast Analysis of Data Distributed on the Sphere. *Astrophys. J.*, **622**:759–771, 2005.

Grabner, P.J., Erdös-Turan Type Discrepancy Bounds. *Monatsh. Math.*, **111**:127–135, 1991.

Grabner, P.J., Klinger, B., Tichy, R.F., Discrepancies of Point Sequences on the Sphere and Numerical Integration. In: Haußmann, W., Jetter, K., Reimer, M. (eds.), *Multivariate Approximation, Recent Trends and Results.* Proc. of the 2nd International Conference on Multivariate Approximation Theory, Witten-Bommerholz, Akademie Verlag, Berlin, 95–112, 1997.

Gradshteyn, I.S., Ryzhik, I.M., *Table of Integrals, Series, and Products.* Academic Press, New York, 1965.

Gräf, M., Kunis, S., Potts, D., On the Computation of Nonnegative Quadrature Weights on the Sphere. *Appl. Comput. Harm. Anal.*, **27**:124–132, 2009.

Gräf, M., Potts, D., Steidl, G., Quadrature Errors, Discrepancies, and Their Relations to Halftoning on the Torus and the Sphere. *SIAM J. Sci. Comput.*, **34**:2760–2791, 2012.

Grafarend, E.W., Three-Dimensional Deformation Analysis: Global Vector Spherical Harmonic and Local Finite Element Representation. *Tectonophysics*, **130**:337–359, 1986.

Grafarend, E.W., Ardalan, A.A., Sideris, M.G., The Spheroidal Fixed-Free Two-Boundary Value Problem for Geoid Determination (The Spheroidal Bruns Transform). *J. Geodesy*, **73**:173–195, 1999.

Gregory, G.G., Large Sample Theory for U-Statistics and Tests of Fit. *Ann. Statist.*, **5**:110–123, 1977.

Greville, T.N.E., Spline Functions, Interpolation and Numerical Quadrature. In: Ralston, A., Wilf, H.S. (eds.), *Mathematical Methods for Digital Computers*. Wiley, New York, 156–168, 1967.

Greville, T.N.E., *Theory and Applications of Spline Functions*. Academic Press, New York, London, 1969.

Gröbner, W., Hofreiter, N., *Integraltafeln, Erster Teil, Unbestimmte Integrale*. Springer, Wien, New York, 1975.

Gronwall, T., On the Degree of Convergence of Laplace's Series. *Trans. Amer. Math. Soc.*, **15**:1–30, 1914.

Gutting, M., *Fast Multipole Methods for Oblique Derivative Problems*. PhD-thesis, University of Kaiserslautern, Geomathematics Group, Shaker, Aachen, 2008.

Gutting, M., Fast Multipole Accelerated Solution of the Oblique Derivative Boundary Value Problem. *Int. J. Geomath.*, **3**:223–252, 2012.

Gutting, M., Fast Spherical/Harmonic Spline Modeling. In: Freeden, W., Nashed, M.Z., Sonar, T. (eds.), *Handbook of Geomathematics*. 2nd ed. Springer, Heidelberg, 2711–2746, 2014.

Gutting, M., Parameter Choices for Fast Harmonic Spline Approximation. *Handbook Math. Geod.*, Birkhäuser, Basel, 2017 (in print).

Haar, A., Zur Theorie der orthogonalen Funktionensysteme. *Math. Ann.*, **69**:331–371, 1910.

Hardin, R., Sloane, N., McLaren's Improved Snub Cube and Other New Spherical Designs in Three Dimensions. *Discrete Comput. Geom.*, **15**:429–441, 1996.

Hardy, G.H., On the Expression of a Number as the Sum of Two Squares. *Quart. J. Math. (Oxford)*, **46**:263–283, 1915.

Hardy, G.H., *Divergent Series*. Oxford University Press, Clarendon Press, Oxford, 1949.

Hardy, G.H., Landau, E., The Lattice Points of a Circle. *Proceed. of the Royal Society A*, **105**:244–258, 1924.

Hardy, R., Multiquadric Equations of Topography and Other Irregular Surfaces. *J. Geophysical Res.*, **76**:1905–1915, 1971.

Heiskanen, W.A., Moritz, H., *Physical Geodesy.* W.H. Freeman and Company, San Francisco, London, 1967.

Henrici, P., *Elements of Numerical Analysis.* Wiley, New York, 1964.

Hesse, K., Sloan, I.H., Optimal Lower Bounds for Cubature Error on the Sphere $\mathbb{S}^2$. *J. Complexity*, **21**:790–803, 2005.

Hesse, K., Sloan, I.H., Cubature Over the Sphere $\mathbb{S}^2$ in Sobolev Spaces of Arbitrary Order. *J. Approx Theory*, **141**:118–133, 2006.

Hesse, K., Sloan, I.H., Womersley, R.S., Numerical Integration on the Sphere. In: Freeden, W., Nashed, M.Z., Sonar, T. (eds.), *Handbook of Geomathematics.* Vol. 2. Springer, Berlin, Heidelberg, 1185–1219, 2010.

Heun, K., Neue Methode zur approximativen Integration der Differentialgleichungen einer unabhängigen Veränderlichen. *Z. Math. Phys.*, **45**:23–38, 1900.

Hielscher, R., Mainprice, D., Schaeben, H., Material Behavior: Texture and Anisotropy. In: Freeden, W., Nashed, M.Z., Sonar, T. (eds.) *Handbook of Geomathematics.* Vol. 2. Springer, Berlin, Heidelberg, 973–1003, 2010.

Higgins, J.R., Five Short Stories About the Cardinal Series. *Bull. Am. Math. Soc.*, **12**:45–89, 1985.

Higgins, J.R., *Sampling Theory in Fourier and Signal Analysis.* Volume 1: Foundations. Oxford University Press, Oxford, 1996.

Higgins, J.R., Stens, R.L., *Sampling Theory in Fourier and Signal Analysis.* Vol. 2: Advanced Topics. Oxford Science Publications, Oxford, 2000.

Hilbert, D., *Grundzüge einer allgemeinen Theorie der linearen Integralgleichungen.* Teubner, Leipzig, 1912.

Hinde R.J., Constructing Atom-Molecules Potential Surfaces from ab Initio Data: A Method Combining Quadrature and Interpolation. *Comput. Phys. Commun.*, **130**:1–11, 2000.

Hlawka, E., Zur Quantitativen Theorie der Gleichverteilung. *Österr. Akad. Wiss. math.-nat. Kl. s-B, II*, **184**:335–365, 1975.

Hlawka, E., *Theorie der Gleichverteilung.* B.I., Mannheim, 1979.

Hlawka, E., Gleichverteilung auf Produkten von Sphären. *J. Reine Angew. Math.*, **330**:1–45, 1981.

Hlawka, E., Schoißengeier, J., Taschner, R., *Geometric and Analytic Number Theory.* Springer, Berlin, Heidelberg, New York, 1991.

Hobson, E.W., *The Theory of Spherical and Ellipsoidal Harmonics.* Reprint Chelsea Publishing Company, New York, 1955.

Hofmann-Wellenhof, B., Moritz, H., *Physical Geodesy.* 2nd ed. Springer, Wien, New York, 2006.

Holschneider, M., Continuous Wavelet Transforms on the Sphere. *J. Math. Phys.*, **37**(8):4156–4165, 1996.

Hörmander, L., Pseudodifferential Operators. *Comm. Pure Appl. Math.*, **18**:501–517, 1965.

Hotine, M., *Mathematical Geodesy.* U.S. Environmental Science Services Administration, 1969.

Hua, L., Wang, Y., *Application of Number Theory to Numerical Analysis.* Springer Verlag and Science Press, Beijing, 1981.

Huxley, M.N., Exponential Sums and Lattice Points III. *Proc. London Math.*, **87**:591–609, 2003.

Ilk, K., Flury, J., Rummel, R., Schwintzer, P., Bosch, W., Haas, C., Schröter, J., Stammer, D., Zahel, W., Miller, H., Dietrich, R., Huybrechts, P., Schmeling, H.D., Wolf, H.G., Riegger, J., Bardossy, A., Güntner, A., Gruber, T., *Mass Transport and Mass Distribution in the Earth System, Contribution of the New Generation of Satellite Gravity and Altimetry Missions to Geosciences.* GOCE-Projektbüro TU Munich, GeoForschungsZentrum Potsdam 2005. `http://gfzpublic.gfz-potsdam.de/pubman/faces/viewItemOverviewPage.jsp?itemId=escidoc:231104:1`, cited 11 May 2017.

Ishtiaq, A., Michel, V., Pseudo-differential Operators, Cubature and Equidistribution on the 3D ball: An Approach Based on Orthonormal Basis Systems. *Numer. Funct. Anal. Optim.*, **38**(7):891–910, 2017.

Ivanow, V.K., A Generalization of the Voronoi-Hardy Identity. *Sibirsk. Math. Z.*, **3**:195–212, 1962.

Ivanow, V.K., Higher-Dimensional Generalization of the Euler Summation Formula (Russian). *Izv. Vyss. Ucebn. Zaved. Mathematika, 6,* **37**:72–80, 1963.

Jacobi, C.G.J., Über Gauß neue Methode, die Werthe der Integrale näherungsweise zu finden. *J. Reine Angew. Math.*, **1**:301–308, 1826.

Jerri, J.A., On the Application of Some Interpolating Functions in Physics. *J. Res. Nat. Bur. Standards, Sec. B*, **73B**:241–245, 1969.

Jerri, J.A., Sampling Expansion for Laguerre – L^2 Transforms. *J. Res. Nat. Bur. Standards, Sec. B*, **80B**:415–418, 1976.

Jerri, J.A., The Shannon Sampling Theorem – Its Various Extensions and Applications: A Tutorial Review. *Proc IEEE*, **65**:1565–1596, 1977.

Katz, V., *A History of Mathematics. 2nd ed.*, Addison-Wesley, Reading, MA, 1998.

Kellogg, O.D., *Foundations of Potential Theory*. Frederick Ungar Publishing Company, New York, 1929.

Kepler, J., Nova Stereomatrica Doliorum Vinarcortum. Linz, 1615. Latin translation into German by Klug, R., *Neue Stereometrie der Fässer*, Engelmann, Leipzig, 1908.

Kline, M., *Mathematical Thought from Ancient to Modern Times*. Oxford University Press, 1972.

Klinger, B., Tichy, R.F., Polynomial Discrepancy of Point Sequences, 1996.

Klug, M., *Integral Formulas and Discrepancy Estimates Using the Fundamental Solution to the Beltrami Operator on Regular Surfaces*. PhD-thesis, University of Kaiserslautern, Geomathematics Group, Verlag Dr. Hut, München, 2014.

Knopp, K., *Theory and Application of Infinite Series*, Dover Publications, New York, 1990.

Kolesnik, G., On the Method of Exponent Pairs. *Acta Arith.*, **45**:115–143, 1985.

Korevaar, J., Meyers, J., Spherical Faraday Cage for the Case of Equal Point Charges and Chebyshev-Type Quadrature on the Sphere. *Integral Transform Spec. Funct.*, **1**(2):105–117, 1993.

Kowalewski, A., *Newton, Cotes, Gauss, Jacobi: Vier grundlegende Abhandlungen über Interpolation und genäherte Quadratur*. von Veit, Leipzig, 1917.

Krylov, V.I., *Approximate Calculation of Integrals*. Russian edition: Izdat. Fiz.-Math. Lit., Moscow, 1959. Translation by A.H. Stroud, Collier McMillan, New York, 1962.

Kuijlaars, A., The Minimal Number of Nodes in Chebyshev Quadratur Formulas. *Indag. Math.*, **4**:339–362, 1993.

Kuipers, L., Niederreiter, H., *Uniform Distribution of Sequences*. John Wiley & Sons, 1974.

Kutta, W., Beitrag zur näherungsweisen Integration totaler Differentialgleichungen. *Z. Math. Phys.*, **46**:435–453, 1901.

Lagrange, J.-L., *Solution de différents problèmes de calcul intégral.* Miscellanea Taurinensia 3, Oeuvres 1, 1762–1765.

Lai, M.-J., Schumaker, L., *Spline Functions on Triangulations.* Encyclopedia of Mathematics and Its Applications. Vol. 110. Cambridge University Press, Cambridge, 2007.

Lain Fernández, N., Polynomial Bases on the Sphere. PhD-thesis, University of Lübeck, Logos, Berlin, 2003.

Landau, E., *Über die Gitterpunkte in einem Kreis (Erste Mitteilung).* Gött. Nachr., 148–160, 1915.

Landau, E., *Über die Gitterpunkte in einem Kreis IV.* Gött. Nachr., 58–65, 1924.

de Laplace, P.S., *Theorie des attractions des sphéroides et de la figure des planètes.* Mèm. de l'Acad., Paris, 1785.

Lebedev, N.N., *Spezielle Funktionen und ihre Anwendungen.* Bibliographisches Institut, Mannheim, 1973.

Lebedev, V.I., Values of the Nodes and Weights of Ninth to Seventeenth Order Gauss-Markov Quadrature Formulae Invariant under the Octahedron Group with Inversion. *Comput. Math. Math. Phys.*, **15**:44–51, 1975.

Lebedev, V.I., Laikov, D.N., A Quadrature Formula for the Sphere of the 131st Algebraic Order of Accuracy. *Dok. Math.*, **59**(3):477–481, 1999.

Leeb, H., Asymptotic Properties of the Spectral Test, Diaphony, and Related Quantities. *Math. Comp.*, **72**(237):297–309, 2002.

Legendre, A.M., Recherches sur l'attraction des sphéroides homogénes. *Mém. math. phys. prés. é l'Acad. Aci. Par. Divers Savantes*, **10**:411–434, 1785.

Le Gia, Q., Mhaskar, H.N., Localized Linear Polynomial Operators and Quadrature on the Sphere. *SIAM, J. Numer. Anal.*, **47**:440–466, 2008.

Le Gia, Q., Sloan, I., Wendland, H., *Multiscale analysis in Sobolev Spaces on the Sphere.* Applied Mathematics Report AMR09/20, University of New South Wales, 2009.

Lekkerkerker, C.G., *Geometry of Numbers.* North Holland, Amsterdam, London, 1969.

Lense, J., Kugelfunktionen. *Mathematik und ihre Anwendungen in Physik und Technik.* Reihe A, Bd. 23, Akad. Verlagsgesellschaft, Leipzig, 1954.

Leopardi, P., A Partition of the Unit Sphere into Regions of Equal Area ans Small Diameter. *Electron. Trans. Numer. Anal. (ETNA)*, **25**:309–327, 2006.

Leopardi, P., *Distributing Points on the Sphere: Partitions, Separation, Quadrature and Energy.* PhD-thesis, University of New South Wales, 2007.

Lighthill, M.J., *Fourier Transform and Generalized Functions.* Cambridge University Press, Cambridge, 1959.

Lubotzky, A., Phillips, R., Sarnak, P., Hecke Operators and Distributing Points on the Sphere. *Comm. Pure Appl. Math.*, **39**:148–186, 1986.

Lyche, T., Schumaker, L.L., A Multiresolution Tensor Spline Method for Fitting Functions on the Sphere. *SIAM J. Sci. Comput.*, **22**:724–74, 2000.

Magnus, W., Oberhettinger, F., Soni, R.P., *Formulas and Theorems for the Special Functions of Mathematical Physics.* Die Grundlehren der mathematischen Wissenschaften in Einzeldarstellungen, 52, 3rd ed. Springer, Berlin, 1966.

Malysev, A.V., The Distribution of Integer Points in a Four-Dimensional Sphere. *Doklady Akad. Nauk SSSR*, **114**:25–28, 1957.

Markov, A., Sur la méthode de Gauss pour le calcul approché des intégrales. *Math. Ann.*, **25**:427–432, 1885.

McEwen, J.D., Wiaux, Y., A Novel Sampling Theorem on the Sphere. *IEEE transactions of Signal Processing*, **59**:1–13, 2011.

McLaren, A.D., Optimal Numerical Integration on a Sphere. *Math. Comp.*, **17**(84):361–383, 1963.

McLaurin, C., A Treatise of Fluxions. Edinburgh, 1742.

Meissl, P., *On the Linearization of the Geodetic Boundary Value Problem.* Department of Geodetic Science, Ohio State University, Columbus, Ohio, No. 152, 1971.

Mhaskar, H.N., Local Quadrature Formulas on the Sphere. *J. Complexity*, **20**:753-772, 2004.

Mhaskar, H.N., Narcowich, F., Ward, J., Spherical Marcinkiewicz-Zygmund Inequalities and Positive Quadrature. *Math. Comp.*, **70**: 1113–1130 (Corrigendum (2002): *Math. Comp.*, **71**:453–454), 2001.

Mhaskar, H.N., Narcowich, F.J., Prestin, J., Ward, J.D. Polynomial Frames on the Sphere. *Adv. Comput. Math.*, **3**:378–403, 2003.

Micchelli, C., Interpolation of Scattered Data: Distance Matrices and Conditionally Positive Definite Functions. *Constr. Approx.*, **2**:11–22, 1986.

Michel, V., *A Multiscale Method for the Gravimetry Problem – Theoretical and Numerical Aspects of Harmonic and Anharmonic Modelling.* PhD-thesis, University of Kaiserslautern, Geomathematics Group, Shaker, Aachen, 1999.

Michel, V., Optimally Localized Approximate Identities on the 2-Sphere. *Numer. Func. Anal. Opt.*, **32**:877–903, 2011.

Michel, V., *Lectures on Constructive Approximation: Fourier, Spline, and Wavelet Methods on the Real Line, the Sphere, and the Ball.* Birkhäuser, Basel, 2013.

Michel, V., RFMP – An Iterative Best Basis Algorithm for Inverse Problems in the Geosciences In: Freeden, W., Nashed, M.Z., Sonar, T. (eds.), *Handbook of Geomathematics.* 2nd ed. Springer, Berlin, Heidelberg, 2121–2147, 2015.

Michlin, S.G., *Mathematical Physics, an Advanced Course.* North Holland, Amsterdam, London, 1970.

Mordell, L.J., Poisson's Summation Formula in Several Variables and Some Applications to the Theory of Numbers. *Cambr. Phil. Soc.*, **25**:412–420, 1928a.

Mordell, L.J., Some Applications of Fourier Series in the Analytic Theory of Numbers. *Cambr. Phil. Soc.*, **24**:585–595, 1928b.

Mordell, L.J., Poisson's Summation Formula and the Riemann Zeta Function. *J. London Math. Soc.*, **4**:285–296, 1929.

Morokoff, W., Caflisch, R.E, Quasi-Random Sequences and Their Discrepances. *SIAM J. Sci. Comput.*, **15**:1251–1279, 1994.

Müller, C., Über die ganzen Lösungen der Wellengleichung. (Nach einem Vortrag von G. Herglotz). *Math. Ann.*, **124**:235–264, 1952.

Müller, C., Eine Verallgemeinerung der Eulerschen Summenformel und ihre Anwendung auf Fragen der analytischen Zahlentheorie. *Abh. Math. Sem. Univ. Hamburg*, **19**:41–61, 1954.

Müller, C., Eine Formel der analytischen Zahlentheorie. *Abh. Math. Sem. Univ. Hamburg*, **19**:62–65, 1954.

Müller, C., Eine Erweiterung der Hardyschen Identität. *Abh. Math. Sem. Univ. Hamburg*, **19**:66–76, 1954.

Müller, C., Spherical Harmonics. *Lecture Notes in Mathematics*, 17. Springer, Berlin, 1966.

Müller, C., *Foundations of the Mathematical Theory of Electromagnetic Waves.* Springer, Berlin, 1969.

Müller, C., *Lectures on the Problem of Space and Time in Einstein's Theory of Gravitation.* University of Kaiserslautern, Technomathematics Group, 1995.

Müller, C., *Analysis of Spherical Symmetries in Euclidean Spaces.* Springer, New York, Berlin, Heidelberg, 1998.

Narcowich, F.J., Ward, J.D., Nonstationary Wavelets on the m-Sphere for Scattered Data. *Appl. Comput. Harmon. Anal.*, **3**:324–336, 1996.

Narcowich, F.J., Ward, J.D., Scattered Data Interpolation on spheres: Error Estimates and Locally Suppored Basis Functions. *SIAM J. Math. Anal.*, **33**(6):1393–1410, 2002.

Narcowich, F., Petrushev, P., Ward, J.D., Localized Tight Frames on Spheres. *SIAM J. Math. Anal.*, **38**(2):574–594, 2006.

Nashed, M.Z., Operator-Theoretic and Computational Approaches to Ill-Posed Problems with Applications to Antenna Theory. *IEEE Transactions on Antennas and Propagation*, **29**:220–231, 1981.

Nashed, M.Z., Scherzer, O., *Inverse Problems, Image Analysis and Medical Imaging (Contemporary Mathematics).* Vol. 313. American Mathematical Society, Providence, R.I., 2002.

Nashed, M.Z., Sun, Q., Sampling and Reconstruction of Signals in a Reproducing Kernel Subspace of $L^p(R^d)$. *J. Funct. Anal.*, **258**:2422–2452, 2010.

Nashed, M.Z., Sun, Q., Function Spaces for Sampling Expansions. In: Shen, X., Zayed, A.I. (eds.), *Multiscale Signal Analysis and Modeling.* Springer, New York, Heidelberg, 81–104, 2013.

Nashed, M.Z., Walter, G.G., General Sampling Theorems for Functions in Reproducing Kernel Hilbert Spaces. *Math. Control Signals Systems*, **4**:363–390, 1991.

Nashed, M.Z., Walter, G.G., Reproducing Kernel Hilbert Space from Sampling Expansions. *Contemporary Math.*, **190**:221–226, 1995.

Nerem, R.S., Tapley, B.D., Shum, C.K., A General Ocean Circulation Model Determined in a Simultaneous Solution With the Earth's Gravity Field. In: Sünkel, H., Baker, T. (eds.), *Sea Surface Topography and the Geoid.* International Association of Geodesy Symposia. Springer, 158–199, 1990.

Niederreiter, H., *Random Number Generation and Quasi-Monte Carlo Methods.* Vol. 63 of CBMS-NSF Regional Conference Series in Applied Math. SIAM, 1992.

Nijboer, B.R.A., de Wette, F.W., On the Calculation of Lattice Sums. *Physica*, **23**:309–321, 1957.

Nutz, H., *A Unified Setup of Gravitational Field Observables.* PhD-thesis, University of Kaiserslautern, Geomathematics Group, Shaker, Aachen, 2002.

Ostermann, I., *Modeling Heat Transport in Deep Geothermal Systems by Radial Basis Functions.* PhD-thesis, University of Kaiserslautern, Geomathematics Group, Dr. Hut Verlag, 2011.

Palczewski, A., Schneider, J., Bobylev, A., A Consistency Result for a Discrete Velocity Model of the Boltzmann Equation. *SIAM J. Numer. Anal.*, **34**:1865–1883, 1997.

Parzen, E., A Simple Proof and Some Extensions of the Sampling Theorem. Technical Report No. 7, Department of Statistics, Stanford University, 1–10, 1956.

Pavlis, N.K., Holmes, S.A., Kenyon, S.C., Factor, J.K., The Development and Evaluation of the Earth Gravitational Model 2008 (EGM2008). *J. Geophys. Res.*, **117**: B04406, 2012.

Peano, G., Resto nelle formule di quadratura espresso con un itegrale definito. *Atti R. Accad. Lincei Rend. Cl. Sci. Fis. Mat. Nat.*, **22**:562–569, 1913.

Peano, G., Residuo in formulas de quadratura. *Mathesis*, **34**:5–10, 1914.

Pedlowsky, J., *Geophysical Fluid Dynamics.* Springer, New York, Heidelberg, Berlin, 1979.

Pesenson, I., An Approach to Spectral Problems on Riemannian Manifolds. *Pacific J. Math.*, **215**:183–199, 2004.

Pesenson, I., *Splines and Wavelets on Geophysically Relevant Manifolds.* In: Freeden, W., Nashed, M.Z., Sonar, T. (eds.), Handbook of Geomathematics. Vol. 3, 2nd ed. Springer, New York, Berlin, Heidelberg, 2527–2562, 2015.

Pinsky, M.A., *Introduction to Fourier Analysis and Wavelets.* Brooks/Cole, Thomson Learning, Pacific Grove, CA, 2002.

Popov, A., Cubature Formulas on a Sphere Invariant Under the Icosahedral Rotation Group. *Num. Anal. Appl.*, **1**(4):355–361, 2008.

Potts, D., Tasche, M., Interpolatory Wavelets on the Sphere. In: Chui, C.K., Schumaker, L.L. (eds.), *Approximation Theory VIII.* Vol. 2. World Scientific, Singapore, 335–342, 1995.

Potts, D., Steidl, G., Tasche, M., Fast and Stable Algorithms for Discrete Spherical Fourier Transforms. *Linear Algebra Appl.*, **275/276**:433–450, 1998.

Prentice, M.J., On Invariant Tests of Uniformity for Directions and Orientations. *Ann. Statist.*, **6**:169–176 (Correction (1979): Ann. Statist. **7**:926), 1978.

Pycke, J.-R., A Decomposition for Invariant Tests of Uniformity on the Sphere. *Proc. Amer. Math. Soc.*, **135**: 2983–2993, 2007a.

Pycke, J.-R., U-Statistics Based on the Green's Function of the Laplacian on the Circle and the Sphere. *Statist. Probab. Lett.*, **77**:863–872, 2007b.

Quade, W., Collatz, L., *Zur Interpolationstheorie der reellen periodischen Funktionen.* Sitz. Berichte Preuss. Akad. Wiss., Berlin 49, 1938.

Radau, R., Étude sur les formules d'approximation qui servent à calculer la valeur numérique d'une intégrale définie. *J. Math. Pures Appl.*, **6**:283–336, 1880.

Rademacher, H., Topics in Analytic Number Theory. *Die Grundlehren der mathematischen Wissenschaften in Einzeldarstellungen*, 169, Springer, Berlin, Heidelberg, New York, 1973.

Radon, J., Restausdrücke bei Interpolations- und Quadraturformeln durch bestimmte Integrale. *Monatsh. Math.*, **42**:389–396, 1935.

Ragozin, D.L., Constructive Polynomial Approximation on Spheres and Projective *Spaces. Trans. Am. Math. Soc.*, **162**:157–170, 1971.

Rakhmanov, E., Saff, E., Zhou, Y., Minimal Discrete Energy on the Sphere. *Math. Res. Lett.*, **11**:647–662, 1994.

Reimer, M., On the Existence Problem for Gauss-Quadarature on the Sphere. In: Fuglede, F. (ed.), *Approximation by Solutions of Partial Differential Equations.* Kluwer Academic Publishing, Dordrecht, 169–184, 1992.

Reimer, M., Quadrature Rules for the Surface Integral of the Unit Sphere Based on Extremal Fundamental Systems. *Math. Nachr.*, **169**:235–241, 1994.

Reimer, M., *Multivariate Polynomial Approximation.* Birkhäuser Verlag, Basel, Boston, Berlin, 2003.

Reinsch, C., Smoothing by Spline Functions. *Numer. Math.*, **10**:177–183, 1967.

Reuter, R., *Über Integralformeln der Einheitssphäre und harmonische Splinefunktionen.* PhD-thesis, RWTH Aachen University, Veröff. Geod. Inst., Report No. 33, 1982.

Riemann, G.F.B., *Gesammelte Mathematische Werke.* 2nd ed. Dover, New York, 239–241, Dover, 1953.

Romberg, W., *Vereinfachte numerische Integration.* Norske, Vid. Selsk. Forh., Trondheim, **28**:30–36, 1955.

Rummel, R., Geodesy. In: Nierenberg, W.A. (ed.), *Encyclopedia of Earth System Science.* Vol. 2. Academic Press, 253–262, 1992.

Rummel, R., Spherical Spectral Properties of the Earth's Gravitational Potential and its First and Second Derivatives. In: Sansò, F., Rummel, R. (eds.), *Geodetic Boundary Value Problems in View of the one Centimeter Geoid.* Lecture Notes in Earth Sciences, 65. Springer, Berlin, 359–404, 1997.

Rummel, R., van Gelderen, M., Koop, R., Schrama, E., Sansò, F., Brovelli, M., Miggliaccio, F., Sacerdote, F., *Spherical Harmonic Analysis of Satellite Gradiometry.* Netherlands Geodetic Commission, New Series, No. 39, 1993.

Runge, C., Über empirische Funktionen und die Interpolation zwischen äquidistanten Ordinaten. *Zeitschrift für Mathematik und Physik,* **46**:224–243, 1901.

Runge, C., Willers, F.A., Numerische und graphische Integration. *Encyklopädie der Math. Wiss.*, **2**(3):47–176, Leipzig, 1915.

Saff, E.B., Kuijlaars, A.B.J., Distributing Many Points on a Sphere. *Math. Intell.*, **19**:5–11, 1997.

Sard, A., Integral Representations of Remainders. *Duke Math. J.*, **15**:333–345, 1948.

Sard, A., Best Approximate Integration Formulas. *Amer. J. Math.*, **71**:80–91, 1949.

Sard, A., *Linear Approximation.* American Mathematical Society, Providence, RI, 1963.

Schäferkotter, M.R., Kythe, P.K., *Handbook of Computational Methods for Integration.* CRC Press, Chapman & Hall, Boca Raton, FL, 2005.

Schoenberg, I.J., Metric Spaces and Completely Monotonic Functions. *Ann. Math.*, **39**:811–841, 1938.

Schoenberg, I.J., Positive Definite Functions on Spheres. *Duke Math. J.*, **9**:98–108, 1942.

Schoenberg, I.J., Spline Functions and the Problem of Graduation. *Proc. Nat. Acad. Sci. USA* **52**:947–950, 1964a.

Schoenberg, I.J., On Trigonometric Spline Interpolation. *J. Math. Mech.*, **13**:795–825, 1964b.

Schoenberg, I.J., Monosplines and Quadrature Formulae. In: Greville, T.N.E. (ed.), *Theory and Applications of Spline Functions.* Academic Press, New York, 157–207, 1969.

Schönhage, A., *Mehrdimensionale Romberg-Integration.* Numerische Mathematik, **14**:299–304, 1970.

Schreiner, M., On a New Condition for Strictly Positive Definite Functions on Spheres. *Proc. Am. Math. Soc.*, **125**:531–539, 1995.

Schreiner, M., Locally Supported Kernels for Spherical Spline Interpolation. *J. Approx. Theory*, **89**: 172–194, 1997.

Schröder, P., Sweldens, W., Spherical Wavelets: Efficiently Representing Functions on the Sphere. *Computer Graphics Proc. (SIG-Graph95)*, 161–175, 1995.

Schumaker, L., *Spline Functions: Basic Theory.* Cambridge University Press, 3rd edition, 2007.

Schwarz, L., *Théorie des distributions I.* Hermann, Paris, 1950.

Schwarz, L., *Théorie des distributions II.* Hermann, Paris, 1951.

Serfling, R.J., *Approximation Theorems of Mathematical Statistics.* John Wiley & Son, Inc., New York, 1980.

Seymour, P.D., Zaslavsky, T., Averaging Set: A Generalization of Mean Values and Spherical Designs. *Adv. Math.*, **52**:213–240, 1984.

Shannon, C.E., Communication in the Presence of Noise. *Proc. Institute of Radio Engineers*, **37**:10–21, 1949.

Shorack, G.R., *Probability for Statisticians.* Springer Texts in Statistics, Springer-Verlag, New York, 2000.

Shure, L., Parker, R.L., Backus, G.E., Harmonic Splines for Geomagnetic Modelling. *Phys. Earth Planet. Inter.*, **28**:215–229, 1982.

Sierpinski, W., O pewnem zagadnieniu z rachunku funckcyj asmptotycznych (Über ein Problem des Kalküls der asymptotischen Funktion (polnisch)), *Prace Math.-Fiz.*, **17**:77–118, 1906.

Simons, F.J., Plattner, A., Scalar and Vector Slepian Functions, Spherical Signal Estimation and Spectral Analysis. In: Freeden, W., Nashed, M.Z., Sonar, T. (eds.), *Handbook of Geomathematics.* 2nd ed. Springer, Heidelberg, 2563–2608, 2015.

Simpson, T., *Mathematical Dissertations on Physical and Analytic Subjects.* London, 1743.

Sloan, I.H., Joe, S., *Lattice Methods for Multiple Integration.* Clarendon Press, Oxford, 1994.

Sloan, I.H., Polynomial Interpolation and Hyperinterpolation over General Regions. *J. Approx. Theory*, **83**:238–254, 1995.

Sloan, I.H., Womersly, R.S., Polynomial Interpolation and Hyperinterpolation over General Region. *J. Approx. Theory*, **83**:238–254, 1995.

Sloan, I.H., Womersly, R.S., Constructive Polynomial Approximation on the Sphere. *J. Approx. Theory*, **103**:91–118, 2000.

Sloan, I.H., Womersly, R.S., How Good can Polynomial Interpolation on the Sphere be? *Adv. Comp. Math.,* **14**:195–226, 2001.

Sloan, I.H., Womersly, R.S., External Systems of Points and Numerical Integration on the Sphere. *Adv. Comput. Math.*, **21**:107–125, 2004.

Sloan, I.H., Womersly, R.S., A Variational Characterization of Spherical Design. *J. Approx. Theory*, **159**:308–318, 2009.

Sloane, N., Spherical Designs (2000) `http://neilsloane.com/sphdesigns/`, cited 11 May 2017.

Sobolev, S., Cubature Formulas on the Sphere Invariant with Respect to Any Finite Group of Rotations. *Dokl. Acad. Nauk. SSSR,* **146**:310–313, 1962.

Sobolev, S., Vaskevich, V., *The Theory of Cubature Formulas.* Kluwer Academic, Dordrecht, Boston, London, 1997.

Sommariva, A., Womersley, R.S., *Integration by RBF Over the Sphere.* Applied Mathematics Report AMR05/17, University of New South Wales, 2005.

Sonar, T., 3000 *Jahre Analysis.* Springer, Berlin, Heidelberg, New York, 2011.

Stein, E.M., Weiss, G., *Introduction to Fourier Analysis on Euclidean Spaces.* Princeton University Press, Princeton, NJ, 1971.

Stenger, F., Bounds on the Error of Gauss-Type Quadratures. *Numer. Math.*, **8**:150–160, 1966.

Stetter, H.J., Numerical Approximation of Fourier Transforms. *Numer. Math.*, **8**:235–249, 1966.

Stoer, J., *Numerische Mathematik 1.* Springer-Verlag, 5. Auflage, Berlin, Heidelberg, New York, 1989.

Stoer, J., Bulirsch, R., *Numerische Mathematik 2.* Springer, New York, Heidelberg, Berlin, 1980.

Stroud, A.H., *Approximate Calculation of Multiple Integrals.* Prentice-Hall, Englewood Cliffs, NJ, 1971.

Strubecker, K., *Differentialgeometrie II: Theorie der Flächenmetrik.* Walter de Gruyter & Co., Berlin, 1969.

Svensson, S.L., Pseudodifferential Operators – A New Approach to the Boundary Problems of Physical Geodesy. *Manuscr. Geod.*, **8**:1–40, 1983.

Svensson, S.L., Finite Elements on the Sphere. *J. Approx. Theory*, **40**:246–260, 1984.

Szegö, G., Orthogonal Polynomials. *American Mathematical Society Colloquium Publications*, 4th ed. Vol. 23. American Mathematical Society, Providence, Rhode Island, 1975.

Thomson, K., Generalized Spiral Points: Further Improvement (2007) `https://groups.google.com/d/topic/sci.math/CYMQX7HO1Cw/discussion`, cited 11 May 2017.

Tichy, R.F., Beiträge zur Polynomdiskrepanz. *Österr. Akad. Wiss., math. nat. Kl. Abt. II*, **193**(8–10):513–519, 1984.

Tichy, R.F., Ein Approximaionsverfahen zur Lösung spezieller partieller Differentialgleichunen. *Z. Angew. Math. Mech. (ZAMM)*, **68**:187–288, 1988.

Titchmarsh, E.C., *The Theory of the Riemann Zeta-Function.* Clarendon Press, Oxford, 1966.

Townsend, A., Wilber, H., Wright, G.B., Computing with Functions in Spherical and Polar Geometries I. The Sphere. *SIAM. J. Sci. Comput.*, **38**(4):C403–C425, 2016.

Turcotte, D.I., Willemann, R.J., Haxby, W.F., Nerberry, J., Role of Membrane Stresses in the Support of Planetary Topography. *J. Geophys. Res.*, **86**:3951–3959, 1981.

Vaaler, J.D., Some Extremal Functions in Fourier Analysis. *Bull. Amer. Math. Soc.*, **12**:183–216, 1985.

van der Corput, J.G., Zum Teilerproblem. *Math. Ann.*, **98**:697–716, 1928.

van der Corput, J.G., Verteilungsfunktionen I. *Proc. Kon. Ned. Akad. v. Wetensch.*, **38**: 813–821, 1935a.

van der Corput, J.G., Verteilungsfunktionen II. *Proc. Kon. Ned. Akad. v. Wetensch.*, **38**: 1058–1066, 1935b.

van der Vaart, A.W., Asymptotic statistics. *Cambridge Series in Statistical and Probabilistic Mathematics.* Vol. 3. Cambridge University Press, Cambridge, 1998.

von Mises, R., Zur mechanischen Quadratur. *Z. Angew. Math. Mech.*, **13**:53–56, 1933.

Voronoi, G., Sur un probleme du calcul des fonctions asymptotiques. *JRAM*, **126**:241–282, 1903.

Wahba, G., Spline Interpolation and Smoothing on the Sphere. *SIAM J. Sci. Stat. Comput.*, **2**:5–16 (Correction (1982): *SIAM J. Sci. Stat. Comput.*, **3**:385–386), 1981.

Wahba, G., *Spline Models for Observational Data.* CBMS-NSF Regional Conference Series in Applied Mathematics, 59, SIAM, Philadelphia, 1990.

Wahler, K.A., Downward Continuation of Magsat Lithosphere Anomalies to the Earth's surface. *Geophys. J.R. Astr. Soc.*, **116**:267–278, 1994.

Wangerin, A., *Theorie des Potentials und der Kugelfunktionen.* Walter de Gruyter & Co, Berlin, Leipzig, 1921.

Watson, G.N., *A Treatise on the Theory of Bessel Functions.* 2nd ed. Cambridge University Press, Cambridge, 1944.

Wendland, H., Piecewise, Positive Definite and Compactly Supported Radial Functions of Minimal Degree. *Adv. Comp. Math.*, **4**:389–396, 1995.

Wendland, H., *Scattered Data Approximation.* Cambridge Monographs on Applied and Computational Mathematics, Vol. 17. Cambridge University Press, Cambridge, 2005.

Werner, J., *Numerische Mathematik.* Vieweg Studium, Braunschweig, 1991.

Weyl, H., Über die Gleichverteilung von Zahlen mod. *Eins. Math. Ann.*, **77**:313–352, 1916.

Wieczorek, M.A., *Gravity and Topology of Terrestrial Planets.* Treatise on Geophysics, 2006.

Womersley, R.S., Interpolation and Cubature on the Sphere (2007) `http://web.maths.unsw.edu.au/~rsw/Sphere/`, cited 11 May 2017.

Womersley, R.S., *Spherical Designs with Close to the Minimal Number of Points.* Applied Mathematics Report AMR09, University of New South Wales, 2009.

Womersley, R.S., Sloan, I.H., How Good can Polynomial Interpolation on the Sphere be? *Adv. Comput. Math.*, **14**:195–226, 2001.

Wong, T.T., Luk, W.S., Heng, P.A., Sampling with Hammersley and Halton Points. *J Graph Tools*, **2**(2):9–24, 1997.

Xu, Y., Cheney, E.W., Strictly Positive Definite Functions on Spheres. *Proc. Am. Math. Soc.*, **116**:977–981, 1992.

Yamabe, H., On an Extension of Helly's Theorem. *Osaka Math. J.*, **2**:15–17, 1950.

Zaremba, S.K., Good Lattice Points in the Sense of Hlawka and Monte-Carlo Integration. *Monatsh. Math.*, **72**:264–269, 1968.

Zayed, A., *Advances in Shannon's Sampling Theory.* CRC Press, Chapman & Hall, Boca Raton, FL, 1993.

Zhan, J.-T., Approximate and Asymptotic Distributions of Chi-Squared-Type Mixtures with Applications. *J. Amer. Statist. Assoc.*, **100**(469):273–285, MR2156837(2006d:62050), 2005.

Index